FRACTURE PROCESSES IN POLYMERIC SOLIDS

FRACTURE PROCESSES IN POLYMERIC SOLIDS

Phenomena and Theory

Bernard Rosen, ***Editor***

Editorial Advisors:

R. W. Armstrong, J. J. Gilvarry, E. J. Mercado

Authors:

J. P. Berry
A. Charlesby
R. F. Fedors
P. H. Geil
J. J. Gilvarry
J. W. S. Hearle
C. C. Hsiao
H. H. G. Jellinek
S. Katz
A. Kobayashi
R. F. Landel
E. J. Mercado
S. B. Newman
J. R. M. Radok
M. Reiner
B. Rosen
K. Saito
A. V. Tobolsky
I. Wolock

INTERSCIENCE PUBLISHERS

a division of John Wiley & Sons
New York • London • Sydney

Library of Congress Catalog Card Number: 64-17053

PRINTED IN THE UNITED STATES OF AMERICA

PREFACE

This collective treatment of "Fracture Processes in Polymeric Solids" is being presented after a prolonged period of gestation. The aim of this book is twofold: (*1*) to bring the investigators of fracture processes in other materials into contact with the concepts, ideas, and results of the work of polymer scientists whose main interests are the relations between chemical structure and mechanical behavior, and (*2*) to familiarize the polymer scientists with the thinking, theories, and results of physicists and engineers who approach fracture as a feature of the solid state or as a limit to materials performance. In other words, this book attempts to serve the approach of two scientific worlds which unintentionally have been segregated by semantics, methods, and way of thinking, but have lately become aware of their common need to understand material behavior in stress-strain-time space. This aim could not have been accomplished without the understanding and cooperation of the contributors by being willing in many cases to accept editorial opinion in the interest of our common task.

The necessity to acquaint two different groups of readers or users of the book with the fundamental concepts and approaches of the other led to an inevitable and—I believe—welcome by-product. The volume became introductory in objective and tone without being too elementary or avoiding the treatment of special cases or special systems.

From whatever point of departure the reader follows the course of this treatise, therefore, he should meet with relevant material either by seeing well-known facts and phenomena in a new context, or by becoming familiar with related concepts and theories which have been developed in areas of science and engineering other than his own.

Apart from general considerations regarding the intent of the book, it is desirable to give the reader some idea of the character and scope of the work. The first chapter ("Polymeric Solids") sets the stage by introducing the special features of macromolecular materials. The two contributions by the Editor ("Physical Status" and "Mechanical Aspects") thus are prefatory in nature; essentially, they presuppose no prior close contact of the reader with the physics or chemistry of polymeric solids. The picture is rounded by Geil's article, "Morphology," Hearle's contribution "Mechanics of Fibers," and Tobolsky's "Aspects of Viscoelasticity." None of these articles makes special demands on a reader's mathematical sophistication; they try to present the basic physics and mechanics (possibly at the expense of the chemistry) at the outset.

For many readers, Chapter II will represent a prime point of interest of the book. In a review ("General Theory of Brittle Fracture"), Berry summarizes the now-classical theory of Griffith for the actual strength of brittle solids based on the picture of microscopic flaws as points of origin of propagating fracture surfaces and compares its conclusions with the high values of cohesive energies implied by the findings on real solids. Following his theoretical discussion, Berry presents a treatment ("Brittle Behavior of Polymeric Solids") which emphasizes the experimental aspects; it considers the effect of such variables as temperature, molecular size, and orientation on the fracture process which can not be readily treated by theory and thus must be considered experimentally or semi-empirically. The chapter is concluded by an unusually thorough discussion ("Fracture Topography") by Wolock and Newman of the properties of fracture surfaces in solids and the variables entering as causal or modifying agents.

From the standpoint of investigators concerned directly with properties of polymeric solids, Chapter III ("Time-Dependent Fracture Processes") should be of immediate interest. The chapter consists of two sections, "Phenomenological Aspects of Progressive Weakening in the Highly Viscoelastic Glasses," by the Editor, and "Rupture of Amorphous, Unfilled Polymers," by Landel and Fedors. The former is concerned primarily with precursory processes rather than actual macroscopic rupture itself and is marked by a treatment of crazing phenomena emphasizing the ancillary physical effects (such as those on the reflectivity of light) which accompany them. The latter emphasizes viscoelasticity and time-temperature superposition in an integrated discussion of the phenomena observed in fracture of rubberlike solids. The treatment of this collective chapter emphasizes both the theoretical and experimental aspects of the fatigue problem; at every salient point, confrontation of theoretical conclusions with experimental results is made in graphical form or in textual discussion.

With Chapter IV, the book becomes further specialized. Thus, Kobayashi and Saito discuss fracture phenomena in cutting, Reiner presents a thermodynamic theory of strength, Hsiao analyes fracture strength in its dependence on molecular orientation, Geil treats morphological aspects of the deformation and fracture of crystalline polymers, Jellinek considers the general role of chemical changes in fracture, while Charlesby describes the results of irradiation.

The concluding Chapter V discusses special topics pertaining directly or indirectly to fracture. Gilvarry shows theoretically and experimentally how the distribution function for fragment size in fracture follows directly from Griffiith's model on the basis of the theory of probability; the physical discussion should be of interest to physicists and physical chemists, while the results are of direct concern to mining engineers, ceramic engineers, and metallurgists (who are concerned with grinding processes). Radok represents an interesting contribution, "Effects of Stress Raisers in Linear Viscoelastic Bodies," and Mercado and Katz treat an important question, the effect of nonlinearities, in "Plane Waves in Nonlinear Viscoelastic Materials." The latter section should be of obvious interest to applied mathematicians, rheologists, physicists, and chemists, as well as to geophysicists and astrophysicists.

Although the underlying ideas and physical implications of the Griffith theory have been known for nearly half a century, actual use of its concepts seems to have been limited to a relatively few investigators with a close interest in fracture of amorphous, crystalline, or metallic solids (in fact, the first application was to fracture of glass, by Griffith himself). This book aims to define the applicability of Griffith's model and in particular to estimate its validity for polymeric solids. It is hoped that the present volume will go far toward demonstrating to investigators the limitations and applications of Griffith's and subsequent theories. Thus, in some small measure, this volume may contribute to the accelerating pace of interdisciplinary contact between the different sciences and engineering fields.

The Editor, confronted with the formidable problem of presenting a promising, yet clearly embryonic, branch of materials science, is aware that this book contains many shortcomings.

B.R.

Dallas, Texas
Fall, 1964

ACKNOWLEDGMENTS

The cooperation needed in order to generate a volume of this particular kind extended well beyond that implied in the Preface regarding author-author interchanges. I wish to take this opportunity to express our personal gratitudes to a great many individuals and organizations that have been of indispensable help to us.

Because of the considerable difficulty in assembling a collective volume in the particular manner deemed necessary, this book's publishers had to double their anticipated efforts in helping all of us to bring it forth.

Cooperating through the authors, several industrial, academic, and governmental organizations have played key roles. I am personally indebted to the Research Laboratories of the Westinghouse Electric Corporation.

The Editorial Advisors—one a metallurgical engineer and plasticity rheologist; one a versatile physicist, engineer, and theorist; and one a geophysicist and theoretical viscoelastician—have been my personal consultants. Through their constructive advice, they have had a profound influence on this highly interdisciplinary undertaking.

Many individuals have also been of direct assistance to me in my multifarious editorial functions, a particularly significant fraction of whom are listed below:

A. M. Freudenthal	E. Orowan	F. R. Eirich
F. Bueche	J. A. Sauer	A. G. Thomas
H. F. Mark	P. I. Vincent	J. C. Halpin
H. Kolsky	E. A. Davis	L. Mandelcorn
T. L. Smith	J. B. Howard	R. G. Black
W. N. Findley	H. F. Minter	J. Mercado
G. Salomon	B. Campbell	M. T. Rosen

Thanks, with apologies, are extended to those who were exploited as sounding boards or who were otherwise instrumental in generating any portion of this volume; whether physicist, chemist, biologist, or engineer; whether Japanese, Dutch, Israeli, French, Russian, Scandinavian, Canadian, Chinese, German, Australian, English, American, or whatever.

Permission and valuable assistance in reproducing figures, tables, or text was generously extended by a score or more publishing firms and journals, and by many scores of authors. The readers, publishers, authors, and I are all fully appreciative of that scholarly cooperation.

Finally I wish to acknowledge in advance, all creative readers who will report important blunders or make constructive suggestions to the publishers, the authors, or to me, so as to help us to improve any subsequent printings, editions, or extensions of this work.

B.R.

Authors and Editorial Staff

R. W. Armstrong, Westinghouse Research Laboratories, Pittsburgh, Pennsylvania

J. P. Berry, Rubber & Plastics Research Association of Great Britain, Shropshire, England

A. Charlesby, Physics Department, Royal Military College of Science, Wilts., England

R. F. Fedors, Jet Propulsion Laboratory, California Institute of Technology, Pasadena, California

P. H. Geil, Engineering Division, Case Institute of Technology, Cleveland, Ohio

J. J. Gilvarry, Astronautics Division, General Dynamics Corp., San Diego, California

J. W. S. Hearle, Manchester College of Science & Technology, University of Manchester, Manchester, England

C. C. Hsiao, Institute of Technology, University of Minnesota, Minneapolis, Minnesota

H. H. G. Jellinek, Department of Chemistry, University of Windsor, Windsor, Ontario, Canada

S. Katz, Department of Geology, Rensselaer Polytechnic Institute, Troy, New York

A. Kobayashi, Technology Section, Electrotechnical Laboratory, Tokyo, Japan

R. F. Landel, Jet Propulsion Laboratory, California Institute of Technology, Pasadena, California

E. J. Mercado, Exploration Division, Gulf Research & Development Co., Pittsburgh, Pennsylvania

S. B. Newman, Polymer Characterization Section, National Bureau of Standards, Washington, D. C.

J. R. M. Radok, Department of Mathematics, University of Adelaide, Adelaide, South Australia

M. Reiner, Israel Institute of Technology, Haifa, Israel

B. Rosen, Senior Scientist in Polymer Physics. Westinghouse Research Laboratories, Pittsburgh, Pennsylvania (More recently: Consultant, Research Division, The Western Company, Dallas, Texas)

K. Saito, Hokkaido University, Sapporo, Japan

A. V. Tobolsky, Frick Chemical Laboratory, Princeton University, Princeton, New Jersey

I. Wolock, Mechanics Division, Naval Research Laboratory, Washington, D. C.

CONTENTS

CHAPTER I

Polymeric Solids

Sections

IA. PHYSICAL STATUS

B. Rosen

The concepts of "giant molecules" evolved (3) from careful studies of the thermodynamics and hydrodynamics of the colloidal solutions of certain interesting, yet uncategorized, substances and from the keen ability of the chemical synthesist to prepare many original varieties by employing a consistent approach to "molecular architecture." The findings suggested that a single molecule might actually resemble a chain; whether "linear," "branched," or in a "crosslinked" network (3,5). The repeatedly successful exploitation of this unorthodox notion in the interpretation of several unconventional aspects of the behavior of these bodies, contributed much to confirm this molecular chain approach (3,5). It has now become clear that a wide variety of natural and synthetic substances (3,5,6), such as those considered in this volume, are composed of giant molecules.

Each of the *monomer*ic "building blocks" or potential segments for these *polymer*ic molecular chains possess the facility to undergo chemical changes which join it with two or more monomers by covalent, *primary* bonding (3,5,6). Depending on the segmental constitution and environment, the energies of its "linkages" are ca. 50–125 kcal./mole, while the maximum energies of *secondary* interaction between any two segments of different molecular chains, usually are smaller by one or two orders of magnitude (5).

IA1. Molecular Chain Integrity

The chemical or detailed molecular structural implications of the term "giant molecule," however, do not of themselves bring an immediate focus on its *physical* implications unless this complex molecular entity is also to be responsible for macroscopic phenomena which characteristically reflect its integrity or cohesion. It is the large difference between the primary (chain) and secondary (interchain) bonding energies which frequently permits even this physical connotation to be strongly felt.

Hence it is often possible to separate giant molecules by extensive dilutions with solvent so as to create true solutions in the form of *macro*molecular colloids (3,5). Their thermal energy fluctuation-induced, segmental motions not only afford intermolecular diffusions, but frequently can generate a "chain *flexibility*" in the sense of a sympathetic or coherent motion of the adjacent segments of a given macromolecule. (See Figure IC-6, p. 53.)

However, our task in this volume is related to "polymeric *solids*," not colloids, and it would not be immediately obvious to a student of the solid state who has been dealing with crystalline or amorphous nonpolymeric matter, that the mechanical or ultimate mechanical properties of polymeric substances deserved separate attention. The major justification for the existence of this book is simply that even as solids, polymeric substances frequently can reflect characteristic physical contributions of this molecular chain integrity, either spontaneously or in response to a force field.

The clearest example of spontaneous molecular chain action is found with the amorphous, rubberlike polymeric substances (1,8). (See Subsection IC3, p. 59.) Despite their general appearance as "solids," the polymeric molecules have sufficiently low secondary bonding energies so as to be periodically exceeded by the localized changes in thermal energy at ordinary temperatures. By this facilitation of *interdependent* chain-segmental rotations (3,5), these complex molecules can exhibit a chain flexibility even without the addition of solvents. A vulcanized (i.e., crosslinked) rubber behaves somewhat like a fluid that is equipped with a prodigious, molecular structural "memory" which is derived from its permanent, flexible network of primary-bonded chains. Its "rubber elasticity" is not that conventional variety of elasticity which is derived from extremely short-range resilient interatomic displacements, but one that is generated by long-range entropy changes on the strain-imposed restriction of the number of possible chain configurations.

The probability for spontaneous chain flexibility in the amorphous regions of polymeric bodies, however, necessarily decreases with decreasing temperatures, increasing secondary bonding energies, or with the existence of a flexibility-hindering macromolecular architecture (1,3,7,8). Yet even under these circumstances, it is frequently possible to supplement the local fluctuations in thermal energy by applied force-couples to achieve "chain translations" in the sense of a shear stress-biased, coherent self-diffusion of adjacent segments without altering the primary-bonded integrity of the chain itself. (See Subsection IC2, p. 40.) This directionally biased transport of molecular chains can be forced to a degree which is sufficient to generate an axially aligned chain texture; the limiting case being an anisotropic molecular crystallization. (See Section IVD.) A homogeneous strain-induced anisotropy, which generates a *crystallization-induced* hardening, is quite common with polymeric solids. In the interesting case of a crosslinked rubber, crystallizations obtained on elongations of several hundred per cent can actually be reversible on unloading (8).

Whether accomplished thermally or by the removal of shearing forces, the melting of polymeric crystals does not necessarily entail a destruction (i.e., "degradation") of the primary-bonded chain since, like chain flexibility or biased chain translations, melting is a matter of surmounting the potential energy barriers of *secondary* bonding (5). However, if one attempts to generate the vapor state of these giant molecules, their primary bonds become taxed. The macromolecule as a whole cannot attain a kinetic energy that is equal to the sum of all of the secondary bonding energies interacting with its very many segments without first exceeding the activation energy required to disengage one or more of its linkages (3,5). Therefore, the thermally enhanced "vapor pressures" of (previously out-gassed) macromolecular substances are due to the partial pressures of the gaseous products of those *chemical* changes involved in the chain degradation.

From what has been stated above, it can be sensed that the cohesion of polymeric solids frequently can be a dual consideration of the *phenomenologically* distinguishable primary and secondary bonding energies and the complexities thereof. In general, it is the profound role of this "two-headed" cohesion and the frequently long-range molecular memory of these kinds of substances which will predominate in the delivery of this volume, since these macroscopic reflections of molecular chain integrities are the origins of many of the mechanical and ultimate mechanical properties of polymeric bodies.

IA2. "States" of Polymeric Solids

Since chainlike molecules necessarily have large hydrodynamic frictions, their solutions and melts have low mobilities (1). When this factor is coupled with the low probability of efficiently *aligning* these complex molecules solely by a lowering of temperature, it is not surprising to find that they have a definite affinity for supercooling. This becomes particularly true when their secondary bonding energies are low or they are irregular or abnormally complex in chain structure or in segmental constitution (5). Despite this, a large variety of "highly" crystalline and even single crystalline polymeric solids have been prepared or found in nature. However, by and large, and particularly with the synthetic substances, this profound supercooling tendency generates amorphous solids or those which contain generous mixtures of properly and poorly crystallized regions which are frequently called "semicrystalline" bodies. (See Section IB.)

Whether wholly or partly amorphous, an extensive supercooling which generates a viscosity greater than 10^{13} poises (producing a maxwellian relaxation time greater than 10^2 sec.) in the uncrystallized regions, initiates the transition to a glasslike and sensibly "brittle" solid (4). At temperatures between the melting point T_m and this glass transition T_g, these supercooled liquids exhibit an interesting combination of extreme ductility, rubberlike elasticity and tensile strain-induced hardening. Thus T_g takes on much of the character of a brittle–ductile, or more properly, a "glassy–leathery" transition. (See Figure IC-10, p. 71.)

At temperatures much lower than T_g, it can be stated with sufficient generality that *isotropic* polymeric solids lose virtually all physical properties which can be readily associated with a "molecular chain integrity." Whether or not polymeric, the salient mechanical character of all glasses approaches that of a hard elastic body. However, at temperatures greater than T_g (and even above T_m) the longer-range roles of these coherently chained entities are strongly felt in their mechanical behaviors. Hence, whenever a polymeric substance is fluid, rubbery, or leathery, it is possible to appreciate the salient modes of motion of the polymer molecules themselves in their contributions to the macroscopic properties. The sharp contrast in the physical behaviors below and immediately above T_g has prompted the use of the loose terms, glassy and leathery "states." The rubbery "state" is achieved by a thermal softening of the leather, etc. (See Subsection IC4b, p. 71.)

By virtue of the apparent viscoelastic character of T_g in wholly or partly amorphous solids however, the temperature for this *time-dependent* transformation in mechanical behavior can be shifted by anything which affects the mobility of the chain segments (2). Hence, the "state" of an amorphous or semicrystalline polymeric solid (in regard to a given period for observations at a given temperature) is sharply influenced by the specific nature of its *interchain* bonding and macromolecular architecture. For example, a polymeric solid can be "plasticized" (i.e., softened) by swelling with a fluid solute so that its viscoelastic contribution, and hence its "state," becomes *hypersensitively* dependent on solute concentration as well as on temperature and time. Moreover, this "state" will shift if the mobility of the amorphous regions reflect changes in *primary* bonding in the sense of chain degradation, further chain growth or in an enhanced chain (branching) complexity. (See Sections IC and IE.)

It also has evolved that the initial synthetic preparation (polymerization) of the substance, depending on the imposed circumstances of synthesis or *choice* of chain segments, can be made to generate entirely new materials in a *predetermined* "state" at a given temperature (3). Therefore, now we find ourselves within the peculiar domain of "tailor-made" substances; whether fibers, foams, fluids, rubbers, films, leathers, laminates, or glasses, which are prepared with an order of facility and versatility that is not achieved elsewhere in materials technology today. This technological revolution has evolved from the extreme *molecular structure-sensitive* character of the mechanical properties of wholly or partly amorphous polymeric substances, and from the molecular architect's amazing ability to modify this structure virtually at will.

IA3. Real Solids

Just as the solution behavior of *macro*molecules is necessarily "abnormal" relative to that of *micro*molecules (or even associated micromolecular colloids) (3), it is also true that polymeric *solids*, whether amorphous, semicrystalline, or single crystalline have their "abnormalities" relative to nonpolymeric bodies.

The rubberlike elastic behaviors of certain amorphous or sparingly semicrystalline polymeric bodies are reflections of long-range, molecular structural contributions and are not merely automatic consequences of a pronounced viscosity. To the contrary, "ideal *rubbery* behavior" requires extremely low secondary bonding energies, hence extremely small resistances to stress-biased, segmental self-diffusions. (See Subsection IC2b, p. 51.) The heat which is to be evolved on stretching an ideal rubber is also to be totally regained on the subsequent *instantaneous* contraction in reflection of the chains' configurational entropy changes alone, and without a dependency on either the internal friction or the internal energy.

Furthermore, even when in the glassy "state" polymeric solids (in special cases) can exhibit properties which are atypical of nonpolymeric glasses. If the chainlike molecules are first directionally aligned in shear (yet, short of an anisotropic molecular crystallization) at a temperature $>T_g$ and subsequently cooled to a temperature $\ll T_g$, the molecules are brought under an effective residual strain of an order which is unobtainable with nonpolymeric glasses. This is reflected in their unusually strong optical and mechanical anisotropies (5,7). Now, on bringing this highly anisotropic solid above T_g once again, although it thermally expands in bulk, it *contracts* directionally (sometimes, several *hundred* per cent) to reach a state of maximum chain configurational probability; hence, it reattains its isotropy.

One also can observe "abnormalities" with single crystalline polymeric solids, since they do not always experience the same varieties of crystalline plasticity that are characteristic of single or polycrystalline, nonpolymeric solids. (See Section IVD.) With polymeric crystals which are equipped with "long-range memories" for their molecular packing (mainly simple chain folding or helically twisted chains), there is a tendency for "chain fibrillation" as well as for simple interplanar slip. If "crystalline plasticity" in polymeric bodies is also to imply a stress-biased melting along a plane of packing, this primarily falls within a consideration of *secondary* bonding and not primarily within one of chain integrity. How, for example, should one regard the "plasticity" in molecular crystals of the crystalloidal proteins (such as tobacco mosaic virus) which are frequently regarded as *single* molecules (6)? Here again, although with the "ideally" single crystalline forms of polymeric solids, we must still be concerned with a "two-headed" cohesion.

It seems rather clear that the real problem is that we now are to deal with different *kinds* of solids than those usually implied by the term " normal." However, these "abnormalities" *become* quite normal when viewed in the light of their molecular architecture, segmental constitution, and inter- and intrachain bonding. Thus, the physics of polymeric solids does not, nor can it ever effectively, divorce itself from the dominant role played by the specific character of the chainlike molecule itself. It will be appreciated throughout this volume that even the highly formal topic of fracture must, with polymeric bodies, place an unusually strong emphasis on the chemical physics of the *molecular processes* involved. We make no apology for this firm trend since this same general path is now being travelled in the fracture studies of all *real* solids of which polymeric varieties are quite interesting and exceptionally challenging examples.

General References

1. Bueche, F., *Physical Properties of Polymers*, Interscience (Wiley), New York–London, 1962.
2. Ferry, J. D., *Viscoelastic Properties of Polymers*, Wiley, New York–London, 1961.
3. Flory, P. J., *Principles of Polymer Chemistry*, Cornell Univ. Press, Ithaca, N. Y., 1953.
4. Jones, G. O., *Glass*, Meuthuen, London; Wiley, New York, 1956.
5. Mark, H. F., and A. V. Tobolsky, *Physical Chemistry of Polymeric Systems*, 2nd ed., Interscience, New York–London, 1950.
6. Tanford, C., *Physical Chemistry of Macromolecules*, Wiley, New York–London, 1961.
7. Tobolsky, A. V., *Properties and Structure of Polymers*, Wiley, New York–London, 1960.
8. Treloar, L. R. G., *The Physics of Rubber Elasticity*, 2nd ed., Clarendon Press, Oxford, 1958.

Editor's Note: As stated in the Preface, throughout this entire introductory chapter, i.e., Sections IA through IE, considerable efforts have been expended to avoid terminology or approaches which would be unfamiliar to physicists or engineers. In view of the broad interdisciplinary scope encompassed by these readers, however, these efforts will not be fruitful in all cases, and in those instances of insufficient clarity the reader is directed to the General References above.

IB. MORPHOLOGY

P. H. GEIL

IB1. Introduction

As mentioned in the previous section, the interatomic forces within the macromolecule are considerably greater than those between the atoms of different chains. This *molecular* anisotropy in the cohesive forces strongly affects the physical properties of macromolecular substances in their solid and liquid states. For the purposes of this and the subsequent sections, it is also important to keep in mind the large anisotropy in the dimensions of these macromolecules. Typical linear polymer molecules are on the order of 1,000 to 10,000 A. in length and 4 to 20 A. in diameter. Although they are often treated as being simply "threadlike," this diameter is of considerable significance in determining their packing, whether as liquids or solids. It is this particular problem; how the macromolecules pack in the liquid and solid states, i.e., their morphology, which we will consider in this section.

In general, those polymer molecules which are sterically able to crystallize, i.e., those which consist of regularly repeating, identical chemical groups, will do so when the melt or solution is cooled. Although, in comparison with many low molecular weight materials, the crystals that form are relatively small and imperfect, nevertheless they are perfect enough and large enough to permit determination of the unit cell. This unit cell can be determined by straightforward x-ray techniques and its dimensions are known for numerous polymers (26,61). However, this will not be our concern here. The size, shape, and perfection of the polymer crystals, on the other hand, have been under intensive study, and the bulk of this section will consist of a summary of recent results in this area. Several modern reviews of this subject have been published (26,45,50,62), and the reader is directed to them for details which cannot be included here in view of the objectives of this necessarily abbreviated treatment.

There are a number of polymers which, for various reasons, are *unable* to crystallize, or which can do so only partially (see below) and/or with difficulty. Details of the morphology of noncrystalline polymer solids, usually termed "amorphous," are not satisfactorily understood at present. Direct means of observation, such as electron microscopy (which has had the major role in the investigation of the morphology of crystalline polymers) and x-ray diffraction, have not yet proven suitable for "amorphous" polymers. Although there is believed to be structure present on a scale ca. 100 A., as will be discussed in Subsections IB2 and IB3, it is insufficiently well-developed for observation by present techniques. The ideas to be discussed in this section concerning the morphology of "amorphous" polymers are thus the result of indirect observation combined with recognition of the relative dimensions of polymer molecules.

We shall now consider, in order: contemporary concepts of the structure of melts and solutions, and the mobility of polymer molecules therein; a probable structure of wholly noncrystalline polymers; and experimental details concerning the structure of "partly" (40–80%) and "highly" (>80%) crystalline polymeric solids.

In general, this discussion will follow that which has been previously employed by the present author (26). In brief, it will be suggested that: (*1*) in solutions and melts, polymer molecules have a high degree of mobility along the approximate axis of their backbones, with a considerably lower degree of mobility in the transverse directions; (*2*) in polymer melts, density considerations require that portions of neighboring molecules form non-cohesive bundles in which they are more or less aligned; and (*3*) as the temperature is increased (or, in solution, concentration decreased), fewer bundles exist and the molecules assume a more isotropically random, coiled and intercoiled conformation. Noncrystalline polymeric solids are tentatively suggested to have a structure similar to that of molten polymer but with sufficient restraint on the molecular motion so that at $T \geq T_g$ the bundles have a considerable degree of permanence unless directionally strained, while at $T \ll T_g$ pronounced translational motion of the molecular chain is sensibly impossible on an ordinary time-scale for observations. In crystalline polymers it will be shown that the crystallizing molecules fold into thin platelets or lamellae on the order of 100 A. thick; that a considerable number of defects can be incorporated within these lamellae; and that uncrystallizable molecules which may be present are, in many cases, segregated as isolated "amorphous" areas. These uncrystallizable or poorly crystallizable molecules which are present in many polymers [as short molecules and "atactic" (36a) or "stereoblock" molecules] are, in a sense, "impurities" even though they are made from the same monomers as the crystallizable molecules.

IB2. Structure of Polymeric Solutions and Melts

It is well known that as the temperature of any material (polymeric or otherwise, crystalline or amorphous) is raised from 0°K., the vibration and libration of its constituent atoms or groups of atoms increase with the increase in thermal energy. Because the vibrational motions are anharmonic, the macroscopic body expands with a reduction in the effective energy of interatomic cohesion.

The molecular motions which occur in polymer crystals are closely related to those in molecular crystals of simpler substances. For instance, large-scale free rotation of entire molecules occurs in crystals of a number of low molecular weight substances consisting of nearly spherical molecules; oftentimes the lattice is sustained even though the molecules are rotating. In addition, short chain molecules, such as those of the paraffins, ketones and alcohols, start to twist or rotate in the crystal well below the melting point [see, for instance, Daniel (10)]. In the case of polymer crystals, the motion should differ only in that it would involve limited portions or segments of the molecule. Furthermore, this motion is expected to occur most readily in the vicinity of defects in the crystal. Evidence for molecular motion in polymers at temperatures well below their melting points T_m can be found with NMR studies (13,59). In polytetrafluoroethylene (cf. Teflon), for instance, molecular libration within the crystal has a sufficient magnitude to significantly affect x-ray diffraction at temperatures 300°C below T_m (8).

In general, rotation of the segments of one molecule about its axis within a lattice will require some cooperative motion by the neighboring molecules, resulting in a restraint on that motion.

The long-chain nature of polymers causes thermal motions involving rotation about, and screwlike motions along, the molecular backbone to be considerably more important than translations. Even in the molten state, translation normal to the molecular backbone is highly unfavored.

As the temperature increases, the restraint of interaction decreases as a result of thermal expansion until the molecules are sufficiently separated so that the restraining energy of the lattice is comparable to the thermal energy, there is then a catastrophic increase in the number and size of rotating groups. As in the case of low molecular weight substances, long-range order is lost, and the sample melts. In the vicinity of defects, the restraints on the motion are expected to differ from those in the perfect crystal and, in many cases, will be lower; thus permitting greater motion of the neighboring molecules at a given temperature.

In a polymer melt, as in melts of low molecular weight materials, there is little or no restriction on the rotation of molecular segments due to the proximity of neighboring molecules (unless these molecules have a two-dimensional character). However, as mentioned above, large-scale translation of the segments of the polymer molecule normal to their chain axes is highly restricted; requiring the cooperative motion of a large number of segments of many molecules and giving rise to the high viscosity of these melts. Translation parallel to the axis can probably occur by the motion of "point defects" that will be discussed in more detail later in connection with the molecular motion which occurs when polymeric crystals are annealed.

A polymer melt has often been treated as consisting of more or less randomly coiled and intercoiled molecules (18a). There is some evidence from crystallization kinetics studies (30) and retention of birefringence in fibrils above the melting point (27), that the general molecular orientation in the crystalline solid is *maintained* for considerable periods of time at temperatures above T_m. More significantly, the density of a polymer melt is sufficiently close to that of its crystal that, in this author's opinion, there must be many localized regions in which molecular segments are more or less parallel, approximating the structure of a crystal but without a crystal's relatively rigidly defined lateral and longitudinal order. These clusters would not greatly affect the physical properties of a melt because of their lack of cohesion. However, they would be retained as a static representation of the originally fluctuating situation if the sample were rapidly cooled below T_g. The absence of any but vibratory or oscillatory motion sufficiently below T_g prevents the rotation and translation of the segments required to form a more perfect lattice.

A polymer solution is also believed to consist of more or less coiled and intercoiled molecules (18a). Here, molecular diffusion is less restricted than in polymer melts because fewer cooperative motions are required. However, depending on concentration and solvent–chain interactions, the present author expects that clusters of aligned molecular segments exist similar to those postulated in the melt.

IB3. Morphology of Noncrystalline Polymeric Solids

At least four different types of "amorphous" or "partially amorphous," noncrystalline polymer solids should be considered: (*a*) linear or slightly branched polymer molecules which cannot crystallize because of steric considerations [most atactic (36a) polymer chains belong in this group]; (*b*) crystallizable polymer molecules which crystallize so slowly that they can be cooled to below T_g before crystallization occurs (the resulting glass will usually crystallize upon a subsequent warming to temperatures above T_g, i.e., to a temperature at which segmental rotation can take place rather rapidly upon treatment with a swelling agent, or on uniaxial straining in tension); (*c*) crosslinked polymer molecules which are prevented from crystallizing because of their three-dimensional networks (the uncrosslinked parent molecules, i.e., linear molecules, may or may not be crystallizable); and (*d*) noncrystalline regions within "partly" or "highly" crystalline polymer solids. (These regions may consist of segregated, uncrystallizable molecules whose morphology will resemble that to be discussed for case *a;* regions which have not yet crystallized when glasslike polymer solids as in case *b* are warmed; or regions in the vicinity of crystal defects.) The last situation will be considered in more detail in Subsection IB4, while the first three are considered in items *a–c*.

a. As an uncrystallizable polymer is gradually cooled, its density and viscosity gradually increase. When the viscosity becomes $\sim 10^{12}$–10^{13} poises, this body may be considered to be a tough, leathery solid. No large spontaneous change in morphology occurs, and this "amorphous" polymeric solid should essentially retain whatever structure it had at higher temperatures. The segments' rotational motion is reduced in amount and frequency, but "persists" until the sample is cooled to below T_g. It would thus be expected that the bundle structure postulated previously for the polymer melt would be retained; the bundles probably becoming somewhat larger and more permanent as the sample is cooled.

The concept that "amorphous" polymeric solids contain packets or bundles of nearly parallel chains has been most highly developed by Kargin and co-workers (40). Probably the best model for the proposed packet structure would be a slight modification of the outdated "fringed micelle" model for polymer "crystallites" (32) (the molecules within the micelles or packets would be aligned but would not have the three-dimensional order found in a crystal lattice, see Fig. IB-1). Although the randomly coiled and intercoiled model for

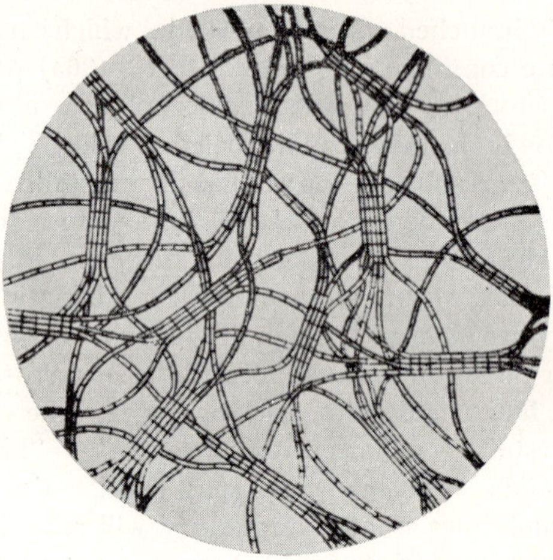

Fig. IB-1. Proposed static model of packet structure in noncrystalline, linear polymers. However, the bundles shown here would be dynamically dissociating and reforming at $T > T_g$. The drawing shown here was actually employed to depict the out-dated "fringed micelle" model for crystalline polymers (32).

amorphous polymers (18a) cannot be completely ruled out, simple considerations of the lateral dimensions of polymer molecules, the relative denisties of noncrystalline and "highly" crystalline polymeric solids, and the presence of packets in molten samples of low molecular weight linear molecules such as paraffins (10), all suggest the presence of bundles in the longer chained polymer melts as well.

b. Crystallizable polymer molecules when quenched to their glassy state are also expected to retain their packet structure. With a slight amount of segmental rotation, obtained by subsequent heating just above T_g, the packets would presumably become small crystals. However, experiments (see Subsection IB4) show that only small further increases in temperature may result in a recrystallization of the material in what is apparently a more stable structure than the crystalline packet; i.e., the molecules *fold* and form lamellar crystals.

c. The morphology of substances composed of crosslinked polymer molecules will depend on a number of factors such as density of crosslinks, morphology (crystal, amorphous solid, liquid) when crosslinked, and their subsequent thermal and mechanical treatments. Rubber elasticity theory is largely based on the assumption that any interactions between neighboring chains do not greatly affect the statistical origin of the elastic properties. (See Subsection IC3, p. 59.) However, as Gee pointed out (20), some inconsistencies between theory and experiment are probably due to precisely this interaction between aligned molecules. If this explanation were actually proved to be correct, it could be assumed that polymers which are crosslinked within the "amorphous" solid or liquid states retain the aforementioned packet structure; at least to some extent. In the case of polymeric solids which are crosslinked in the crystalline state, recent results of Salovey and Keller (75,76) are interpreted in terms of primary bonds formed between the folds in adjacent lamellar crystals. With only small degrees of crosslinking, the molecular orientation can be sustained well above the melting point of the uncrosslinked polymer (31).

IB4. Morphology of Crystalline Synthetic Polymers

In the last few years it has been shown rather conclusively that many species of polymer molecules can crystallize from solution or melt in the form of thin platelets or lamellae on the order of 100 A. thick, and in which the molecules are folded (26,45,50,62). Beyond simply this, it tentatively appears that for most synthetic polymers lamella formation is the *typical* form of crystallization. In the remainder of this section we will briefly outline recent results concerning the effect of various factors on the structure of these lamellae as observed in single crystals, hedrites and spherulites which are grown from both solution and melt. Detailed discussions of the various factors and their effects as well as the historical development of the subject are found elsewhere (26). Very little is known concerning the morphology of synthetic fibers; however, see Section IVD.

IB4a. Single Crystals

The growth of single crystals from dilute solution has been reported for polymer molecules which are representative of most classes of those which are crystallizable (Table IB-I), and there have even been

TABLE IB-I
Polymers Crystallized from Solution in the Form of Lamellae

	References
Polyolefins	
Linear polyethylene	16, 37, 49, 82
Branched polyethylene	14, 24
Polypropylene	41, 71
Polybutene	26
Poly-4-methylpentene-1	19
Polyamides	
Nylon 6	23
Nylon 7	35
Nylon 8	35
Nylon 66	23
Nylon 610	23
Polyesters	
Poly(hexamethylene terephthalate)	77
Poly(pentaglycol terephthalate)	77
Linear polyesters	6, 64
Polyacrylonitrile	36
Poly(acrylic acid)	43, 44, 61
Polydioxolane	26
Poly(ethylene oxide)	34, 69
Poly(propylene oxide)	27
Polyoxymethylene	28
Poly(3,3-bis chloromethyl oxacyclobutane)	26a
Polychlorotrifluoroethylene	42
Polytetrafluoroethylene	26, 67
Polystyrene	41
Cellulose triacetate	58
Amylose	57
Xylan	29

reports of similar single crystals of polypropylene (25), polytetrafluoroethylene (81), and nylons (55) grown from the melt. In general however, they all have the same basic appearance, consisting of lamellae on the order of 100 A. thick (Fig. IB-2). Thickening of the crystals during isothermal crystallization occurs by spiral growth of additional lamellae by means of a type of screw dislocation.

Electron diffraction patterns (Fig. IB-3) indicate that the molec-

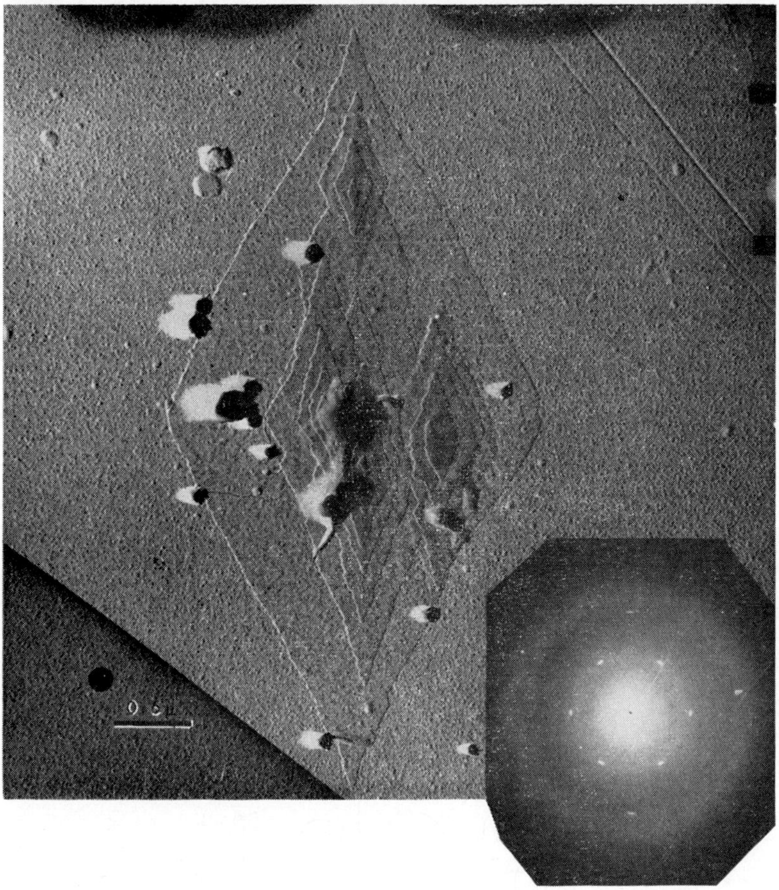

Fig. IB-2. Single crystal of nylon 6 grown from a dilute glycerin solution. The lamellae are approximately 60 A. thick (23).

Fig. IB-3. Inset. Electron diffraction pattern for a nylon 6 single crystal. The **b** axis (hydrogen bond direction) is parallel to the short axis of the crystal (23).

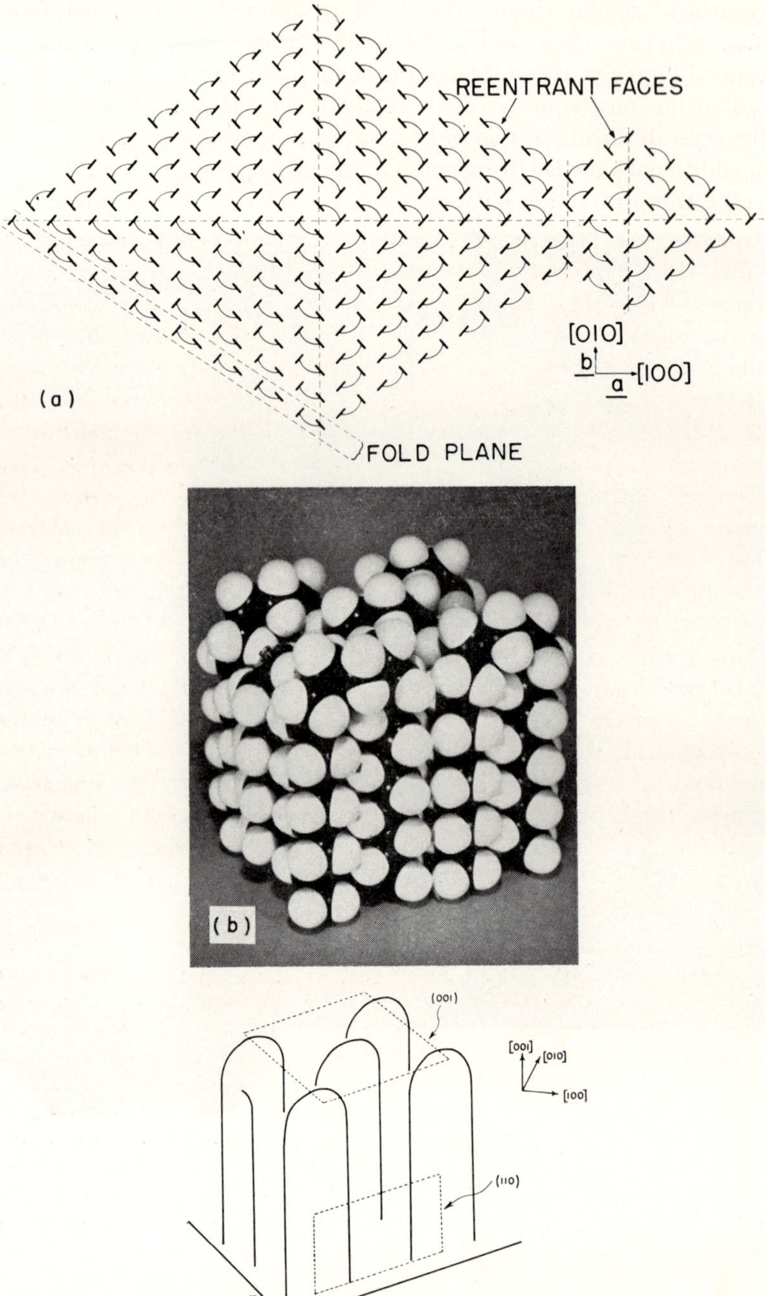
REENTRANT FACES
[010]
b
a
[100]
(a)
FOLD PLANE
(b)
(001)
[001]
[010]
[100]
(110)
(001)
(c)

ular axes, i.e., chain backbones, in the portions of the crystal contributing to the pattern (the entire crystal in the case of nylon 6, but only portions in crystals of some polymers, including polyethylene) are, to within a few degrees, *normal* to the lamellae. Since, in the crystals of the various polymers for which diffraction patterns have been reported, the molecules are a few to some tens of times longer than the thickness of the lamellae, they must be folded within these lamellae (49). The folded arrangement is known to be sterically possible. In crystals of polyethylene it probably resembles the fold formed when cyclic paraffins crystallize (64). In both of these cases it is found that the fold can occur between adjacent molecules in the direction of closest packing, i.e., in the $\{110\}$ planes of the orthorhombic lattice. The atoms in the fold are in a higher energy state than those in the straight segments of the molecule. In addition, on employing atomic models, the fold appears to be somewhat larger than the corresponding cross-sectional areas of the two neighboring chain segments that it joins. In those cases in which the molecules have a helical conformation (e.g., polypropylene, polyoxymethylene) the straight segments within the lamella are helical but the portions of the molecule within the folds presumably are not.

Although no direct confirmation of the orientation of the folds has been obtained, it is believed that they are parallel to the observed growth faces of the crystals; i.e., along $\{110\}$ planes in the nylon and polyethylene crystals shown in Figures IB-2 and IB-5. The *hk*0 planes containing the folds have been termed "fold planes" (29) (Fig. IB-4). The cohesion between adjacent fold planes, as expected, is low since only van der Waals forces are active. It has been found (17,29) that tensions applied normal to the fold planes may result in a cleavage

Fig. IB-4. Diagrams and model of folded chains in single crystal of polyethylene. (*a*) Projection of the crystal onto a (001) plane. The straight bar (-) corresponds to the projection of the planar zigzag of the molecule's "backbone," while the arcs (⌒) infer the fold on the upper fold surface. The dotted lines (--) are fold domain boundaries. The reentrant faces, shown, develop during dendritic crystal growth. (*b*) Photograph of a Stuart-Briegleb model showing the folds and their orientation in several adjacent fold planes. (Only a portion of a fold-period is shown for each segment.) (*c*) Sketch showing the crystallographic planes and directions for (*b*). The diagrams shown here correspond to a flat crystal. A certain type of hollow pyramid is formed if each successive fold plane is displaced vertically by one repeat distance. Another type of fold packing results if the folds in every other fold plane are shifted along the fold plane by one lattice spacing (73a). (See also ref. 5.)

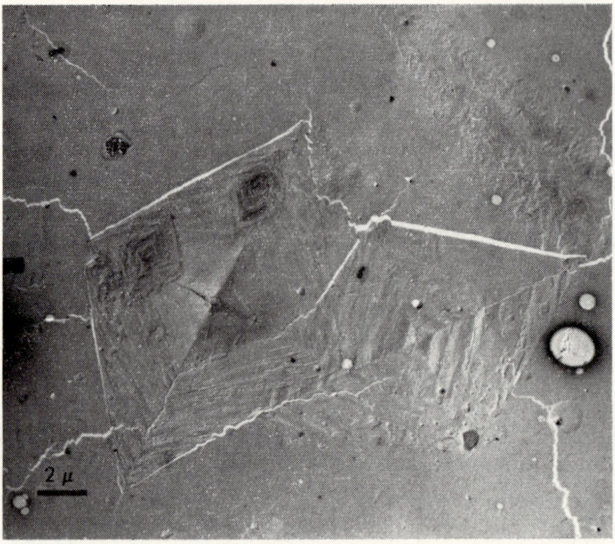

Fig. IB-5. Hollow pyramidal crystals of polyethylene partially supported to prevent their total collapse. The pyramidal nature is particularly clear near the center of the crystals. The **a** and **b** axes are parallel to the long and short diagonals, respectively (26).

between these planes (Fig. IVD-13, p. 567), whereas tensions directed elsewhere may result in the formation of fibers; presumably arising from an unfolding of the molecules.

The fold planes in the different sectors of a crystal (defined by diagonals from the center to the apices) must have different directions. As a result, a crystal will be composed of four or more "fold domains," i.e., regions with a common fold plane orientation (Fig. IB-4). A diamond-shaped polyethylene crystal consists of four $\{110\}$ fold domains, all of which have essentially the same properties (the domains, however, may deform differently in response to the direction of applied tensile forces). From solutions employing certain solvents, six-sided polyethylene crystals are formed; having two $\{100\}$ fold domains as well as four $\{110\}$ fold domains (4,26,50). In the resulting truncated, diamond-shaped crystals, the $\{100\}$ fold domains are found to have different physical properties such as thermal stability (2) and response to tensile stresses (26,27), than the $\{110\}$ fold domains.

By means of dark-field electron (29,50), and dark-ground optical (5,50), microscopy it has been possible to show that some monolamellar polyethylene crystals are not flat but actually are hollow pyramids (Fig. IB-5). Bassett, Frank, and Keller (5) have shown that in some cases the sides of the pyramids have a monotonic slope whereas in other cases, as in Figure IB-5, the slope in each quadrant alternates between given positive and negative values. They determined the crystallographic indices of the "fold surfaces" (i.e., the surfaces containing the folds) and found, in the crystals examined, that the hollow pyramids have a non-flat-based topology; i.e., if the apices of the short diagonals rest on a plane which is perpendicular to the axis of the pyramid, the long diagonals will extend below it.

The formation of the hollow pyramids is presumed (5,26,50, 73a) to be a result of the aforementioned bulkiness of the folds. Flat-based pyramidal polyethylene crystals with two different slopes, depending on the details of the packing of the folds on the fold surfaces, are formed if adjacent folds in successive fold planes are displaced in the same direction by a constant amount (73a) (Fig. IB-4). Non-flat-based pyramids result if adjacent folds within a fold plane are also displaced in a similar manner (5,50). In addition to this behavior in polyethylene, there are also suggestions that some crystals of polyoxymethylene, poly-4-methylpentene-1 (26) and poly(3,3-bis chloromethyl oxacyclobutane) (26a) are pyramidal in nature.

The study of the structure and properties of polyethylene crystals, including their deformation under directionally applied forces, is considerably complicated by their pyramidal nature. In general, the crystal must be placed on a substrate for study. The resulting surface tension is sufficient to cause it to collapse as the solvent evaporates, even if it had not done so from its weight alone. This collapse can occur in various ways: by slip between fold planes, resulting in a flat crystal in which the molecular axes are normal to the substrate; by tilting of the various fold domains toward the center, resulting in the formation of a fold or pleat of material at the center with the molecular axes at an angle of inclination to the substrate; or by various combinations of these methods. Even in a single fold domain, following collapse, the molecular axes may make angles between 0 and 35° or more with respect to a normal to the substrate, and considerable distortion of the lattice may occur. Electron diffraction patterns from such crystals contain a larger number of reflections than ex-

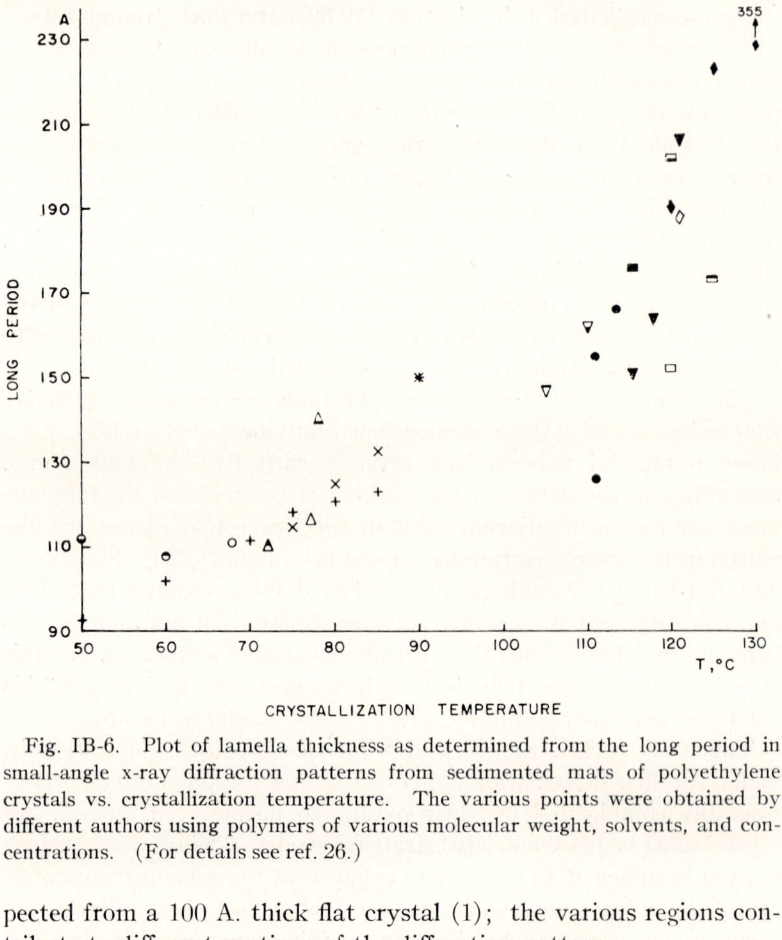

Fig. IB-6. Plot of lamella thickness as determined from the long period in small-angle x-ray diffraction patterns from sedimented mats of polyethylene crystals vs. crystallization temperature. The various points were obtained by different authors using polymers of various molecular weight, solvents, and concentrations. (For details see ref. 26.)

pected from a 100 A. thick flat crystal (1); the various regions contribute to different portions of the diffraction pattern.

For most polymers, the thickness of individual lamellae depends on the temperature of crystallization; increasing with increasing temperature (Fig. IB-6). The effect in the case of polyethylene has been shown both by quenching the hot solution to different temperatures (4,52,69,70), and by crystallizing from different solvents (71,72,74). Solvent–polymer interaction as well as variations in the number of molecular branches, may be responsible for the scatter in the results shown in Figure IB-6. In addition, some of the differences observed among the sets of data from the various authors may be

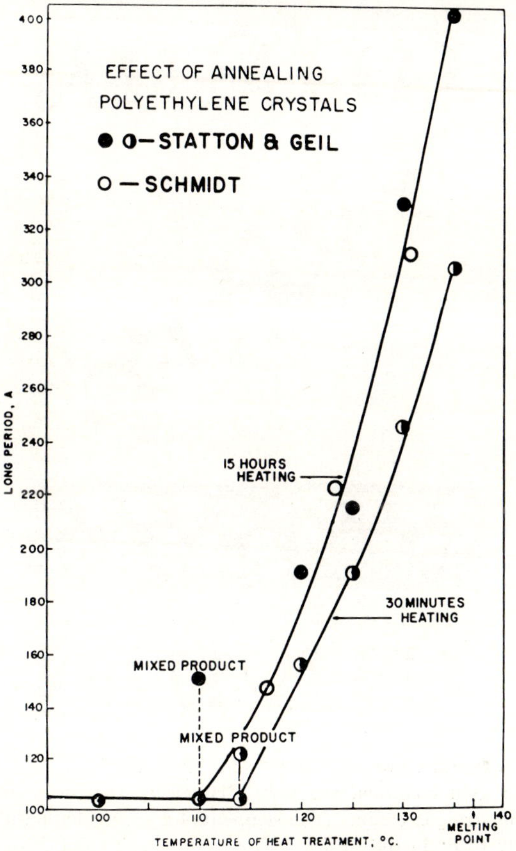

Fig. IB-7. Plot of long period vs. annealing temperature of polyethylene crystal mats. The data of Statton and Geil (79) were obtained after the samples had cooled to room temperature, whereas those of Schmidt (17) were obtained at the annealing temperature (26).

due to differences in morphology (as a result of concentration effects, for instance) and in the preparation of mats of sedimented crystals used for small-angle x-ray diffraction determinations of their thickness. Variations in molecular weight appear to have little, if any, effect (53).

For a given molecular weight, concentration and solvent, the morphology of polyethylene crystals depends on the temperature of

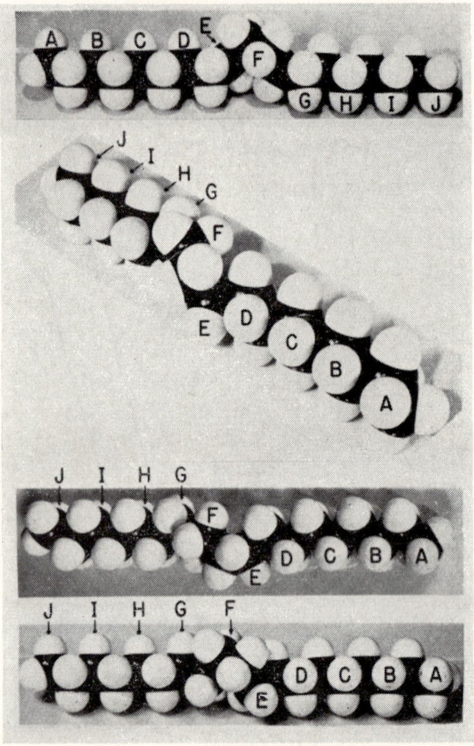

Fig. IB-8. Various views of a model of a crowdion in a polyethylene chain. The distorted region contains one more CH_2 group than a corresponding distance along the straight chain. The segments on each side of the crowdion are spaced properly to fit in the lattice. The crowdion, involving an excess energy of about 0.2 e.v., is relatively free to move along the molecule and does not greatly distort the surrounding lattice (73).

crystallization. Bassett and Keller (4) using 0.1% xylene solutions, indicated that at 90°C. (the highest temperature at which crystals grew from this particular solution) truncated, diamond-shaped crystals formed. As the crystallization temperature was reduced, the relative size of the {100} (truncated) growth faces decreased with respect to the {110} faces in an apparently continuous manner. At 80°C., true diamond-shaped crystals were found, whereas at lower temperatures (larger cooling rates) dendritic crystals were formed. With increased concentration and/or molecular weight, the tempera-

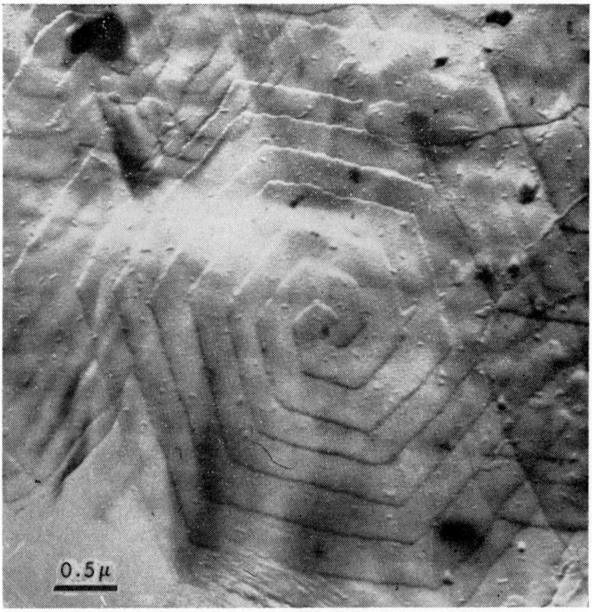

Fig. IB-9. Spiral growth on a single crystal of polypropylene grown from a thin molten film by a slow cooling (25).

ture increases for the transition between simple monolayer crystal growth and dendritic crystallization. If the temperature is changed during growth, abrupt steps form on the crystal which correspond to a change in "fold period," and the growth habit may also change.

Annealing of the crystals following their growth from solution, at temperatures between that of crystal formation and T_m, results in an increase in fold period which can be observed with both small-angle x-ray scattering (Fig. IB-7) and electron microscopy (79). At a given annealing temperature, the long period increases linearly with logarithmic time; the rate increasing with increasing temperature (18). During annealing, the crystals develop holes but the chain segments often retain their orientation and mutual alignment; suggesting that the segmental transport (that must occur as the fold period increases) takes place along the backbone of these chainlike molecules. Reneker has suggested (73) a possible mechanism for this "*intra*molecular" self-diffusion, involving the motion of a point dislocation or crowdion along the chain (Fig. IB-8). Similar molecular motion has been suggested to occur in both polymer melts and solutions, and possibly even during the deformation of polymer crystals (26).

Isolated single crystals grown from the melt appear to be basically similar to those grown from solution (25,81). Those of polypropylene (Fig. IB-9), grown by the slow cooling of molten films, have a perfection that appears to exceed those of polypropylene grown from solution. To date, comparatively little research has been done in characterizing these melt-grown crystals but presumably they will be affected by crystallization temperature, molecular weight, etc., in a manner similar to solution-grown varieties.

IB4b. Hedrites

Hedrites, which are crystallized from the melt (Fig. IB-10) and solution, are significant primarily because they appear to be intermediate in size and complexity between the single crystal and the often considerably larger spherulites (discussed below) that normally form during crystallization from the melt (21). Hedrites have been

Fig. IB-10. Hedrites and oval structures of polyoxymethylene crystallized from a molten film (incident, steep oblique illumination) (21).

grown from molten thin films of polyoxymethylene (21), isotactic polystyrene (11), poly-4-methylpentene-1 (54), poly(ethylene oxide) (21), and gutta-percha (51,78), and from solutions of polyoxymethylene (21) and concentrated polyethylene (3,65). Where their morphology has been examined with the electron microscope, the melt-grown hedrites have been shown to be composed of a number of non-branching lamellae having a common orientation and appearing to develop from a central spiral growth. The solution-grown hedrites or "axialites," as termed by Bassett et al. (3), also appear to consist of a number of lamellae with a common orientation and source, which may, however, diverge while in suspension. For instance, polyethylene hedrites may take on the approximate appearance of two books placed spine to spine and with the pages spread open loosely.

IB4c. Spherulites

The normal mode of polymer crystallization from the melt is through the growth of complex, nearly macroscopic bodies termed spherulites (Fig. IB-11). Crystallization starts at various points in the sample; the nuclei generally being of heterogeneous origin, and spreads out in a grossly spherical fashion from each point. Because of the change in specific volume, the body contracts toward each

Fig. IB-11. Spherulites of polypropylene formed by air-quenching a film. The polymer in the brighter spherulites has crystallized in a hexagonal unit cell, whereas that in the darker spherulites is in a monoclinic unit cell (crossed polarizers) (26).

Fig. IB-12. Portion of a spherulite of polyoxymethylene. The nucleus is to the upper right, the lamellae tending to twist and turn irregularly as they grow out from the central sheaf. Several voids can be seen between the lamellae (Cr replica) (26).

nucleus during this crystallization. Of significance with respect to a discussion of spherulite growth are the observations that gutta-percha (78) and polypropylene (47,66) (both of which can crystallize with at least two distinct unit cells) can crystallize at a given temperature with two different spherulitic structures. Individual spherulites are composed entirely of one or the other crystal modification (Fig. IB-9). The two types differ, for instance, in their birefringence, melting point, mechanical properties and morphology. As suggested by Schuur (78), the crystal structure (unit cell) of each nucleus is propagated by a continuous crystallization mechanism throughout the entire spherulite. In many, or possibly all, polymers it can be shown that this mechanism involves the growth and branching of lamellae or ribbons in which the molecules are folded (26).

When spherulites of polyoxymethylene (which is representative of a "highly" crystalline polymer) are observed in the electron microscope, lamellae on the order of 100 A. thick can be observed on both exterior (Fig. IB-12) and fracture surfaces (22). The spherulite nuclei appear to consist of sheaves of lamellae. It has been suggested (22,26) that they develop in a manner similar to the hedrites, i.e., the lamellae in the sheaf originate by spiral growth from a primary, single crystal nucleus. Regardless of the details of their origin, however, the polyoxymethylene spherulite develops through the irregular twisting and branching of these lamellae. Regular, hexagonal spiral growths develop if the branching takes place near the exterior surface. The lamellae from adjacent spherulites interleave at the boundaries indicating that the growth-front of the spherulites consists of individual separated "secondary" lamellae growing out into the melt. Large volumes of as yet uncrystallized polymer in the intervening regions are included in "tertiary" lamellae which result from the branching of the primary and secondary lamellae. Smaller pockets of uncrystallized polymer may either nucleate and crystallize at lower temperatures, or be incorporated into neighboring lamellae during the secondary stage of crystallization (26,33). Voids are found to form not only at the spherulite boundaries but between the lamellae as a result of the crystallization process. These voids may weaken the polymeric solid.

The branching of the lamellae during growth may result from a "cellulation" type of growth, as described by Keith and Padden (45, 48), which results in the segregation of less crystallizable molecules in pockets along the growth-front of the lamella. These less crystallizable molecules consist of ones with short chains, atacticity, branches, stereoblocks, crosslinks, "permanent" entanglements, etc. As a result, a single broad lamella would break up into a number of narrower lamellae which need not remain coplanar. The number of potentially crystallizable polymer molecules which remain excluded from the lamellae after a sample has crystallized, and the number of "tie" molecules (i.e., molecules included in more than one lamella) should both depend on such factors as the length of time at the crystallization temperature, its magnitude and rate of lowering, the room temperature mobility of the molecules (as reflected by the relative position of T_g), as well as molecular weight, number and type of branches, and the stereoregularity of the segments. In "highly" crystalline polymeric solids, such as linear polyethylene and polyoxymethylene, the number of excluded molecules is believed to be

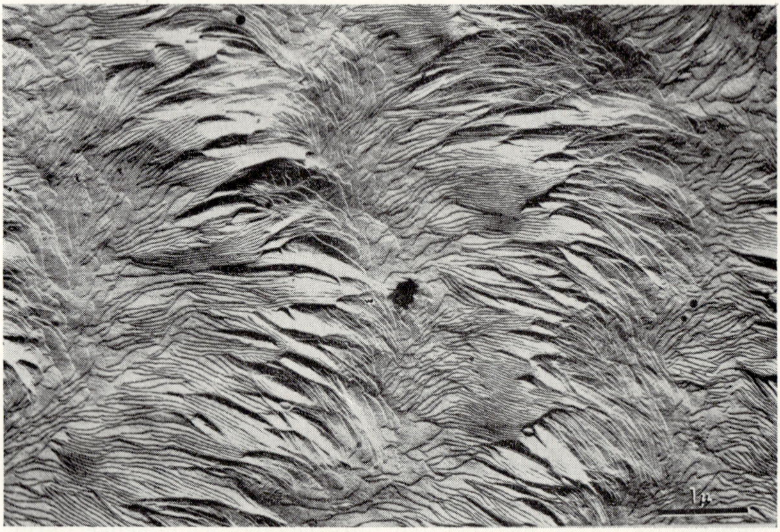

Fig. IB-13. Surface replica of a portion of a linear polyethylene spherulite. The lamellae are seen to twist by 180° within each ring. Portions of two rings are shown, the nucleus being to the left of the micrograph (15).

negligible, while in those polymeric solids with lower degrees of crystallinity, such molecules may have a significant role. It is thus suggested (22,26,80) that the "amorphous" properties (low density, diffuse x-ray diffraction, etc.) of "highly" crystalline polymeric bodies should be attributed primarily to defects in the lamellae and the fold surfaces. In less crystalline polymeric solids, these "amorphous" properties are probably in part due to defects in the lamellae and in part to segregated molecules in *truly* amorphous regions. (See Subsection IB3h.)

In spherulites of a number of polymers, including polyethylene, there is a regular, cooperative twist to the lamellae, resulting in a "ring structure" (16) (Fig. IB-13). In most polymers the periodicity of the rings depends on the temperature, increasing with increasing temperature (24,46). The thicknesses of the lamellae also increase with increasing temperature and, as in the case of single crystals, with subsequent annealing treatments (26).

A number of polymers can be quenched from the melt to their glassy state and then crystallized by annealing at temperatures between T_g and T_m. Kampf has published (38,39) electron micrographs showing that polycarbonate treated in this manner crystallizes in the form of well developed lamellae. Related small-angle diffraction evidence has been obtained for polyurethane (14) and poly(ethylene terephthalate) (15). In Table IB-II a list is given of polymers which

TABLE IB-II
Polymers Crystallized from the Melt or Glassy State in the Form of Lamellae

	References
Polyolefins	
Linear polyethylene	9, 16
Branched polyethylene	24, 68
Polypropylene	25
Poly-4-methylpentene-1	26
Polybutene	26
Polyoxymethylene	22
Poly(ethylene oxide)	15, 26
Poly(propylene oxide)	27
Polydioxalene	26
Poly(ethylene adipate)	12
Polycarbonate	38, 39
Polytetrafluoroethylene	7, 26
Polychlorotrifluoroethylene	14, 26
Poly(3,3-bis chloromethyl oxacyclobutane)	26a
Polyurethane	14

have been shown by electron microscopy to crystallize from the melt or glassy state in the form of lamellae or ribbons in which the molecules are presumably folded.

The mechanical properties of a crystalline polymeric solid will depend largely on the properties and interaction of the lamellae.

IB5. Morphology of Crystalline Natural Polymers

A great deal of effort has been devoted to determining the molecular conformation of such natural polymers as proteins, nucleic acids, and polysaccharides. In some cases, as for instance the determination of the conformation of the globular proteins hemoglobin and myoglobin, this work has been done using x-ray diffraction patterns from single crystals. (See, for instance, ref. 53a.) These crystals bear no relation to the crystals of synthetic polymers discussed above; the molecules pack as *individual* ellipsoidal entities.

In other cases the conformation studies have been based on diffraction patterns obtained from fibrillar material; the polymer molecules either being found in nature in this form (as in cellulose, wool, silk, collagen, etc.) or being drawn from concentrated solutions for the purpose of study (as has been done for nucleic acids, for instance).

In Section ID, models for the morphology of cellulose and wool fibers are shown in Figures ID-2 (p. 111) and ID-8 (p. 119). As Hearle discusses in that section, these models can be interpreted in terms of a "fringed micelle" or "fringed fibril" morphology. As in the case of fibers of synthetic polymers (Section IVD), these models, although presently accepted, are beginning to be questioned in view of the known, folded molecular conformation found in unoriented synthetic polymers. In particular, for instance, Dolmetsch (12a) has suggested that the cellulose molecules in the microfibrils making up a cellulose fiber (see Fig. ID-2, p. 111) are folded. The fold surface, it is suggested, is normal to the fiber axis; the molecular segments between the folds being parallel to the fiber axis. The growth of lamellar single crystals of cellulose esters (57) shows that folding is sterically possible. The sensitivity of the cellulose microfibrils to chemical attack at periodic positions along the fibril can be interpreted in terms of the fold period. Dolmetsch indicates (12a) that the observed tensile strength of the cellulose requires that only 2% or less of the molecules pass from one "lamella" to the next along the fibril axis. Whether or not a similar model might apply to other fibers, such as wool, silk, and collagen in which there are also periodic structural features, is not known at present.

IB6. Summary

We have been concerned here with the structure of polymers in the size range between the unit cell and the macroscopic body. Although this morphology of "partly" and "highly" crystalline polymers has recently been under intensive study, comparatively little is known of the microstructure of the noncrystalline species. It is suggested that neighboring molecules in these latter polymers form non-cohesive bundles in which they are more or less aligned. These bundles are essentially permanent below T_g because of insufficient segmental mobility, semipermanent somewhat above T_g (unless strained), and impermanent at higher temperatures.

Synthetic polymer molecules which are sterically *able* to crystallize, appear to do so preferentially in the form of thin platelets or lamellae from both melt and solution. The molecules are folded within these lamellae. For many polymers it is possible to obtain isolated and relatively simple, single crystals. The larger and more complex spherulites, which normally form during crystallization from the melt or concentrated solution, are found individually to consist of an aggregate of these lamellae growing from a single nucleus. Properties often attributed to amorphous regions in "semicrystalline" polymers are tentatively suggested as being due to defects in the lamellae, and to segregated, "uncrystallizable" molecules.

References

1. Agar, A. W., F. C. Frank, and A. Keller, *Phil. Mag.*, **4**, 32 (1959).
1a. Bassett, D. C., personal communication.
2. Bassett, D. C., F. C. Frank, and A. Keller, *Nature*, **184**, 810 (1959).
3. Bassett, D. C., A. Keller, and S. Mitsuhashi, *J. Polymer Sci.*, **A1**, 763 (1963).
4. Bassett, D. C., and A. Keller, *Phil. Mag.*, **7**, 1553 (1962).
5. Bassett, D. C., F. C. Frank, and A. Keller, *Phil. Mag.*, **8**, 1739 (1963).
6. Belavtseva, E. M., and Yu. V. Mnyukh, *Vysokomol. Soedin.*, **3**, 213 (1961).
7. Bunn, C. W., A. J. Cobbold, and R. P. Palmer, *J. Polymer Sci.*, **28**, 365 (1958).
8. Clark, E. S., and L. T. Muus, *Z. Krist.*, **117**, 119 (1962).
9. Claver, G. C., Jr., R. Buchdahl, and R. L. Miller, *J. Polymer Sci.*, **20**, 202 (1956).
10. Daniel, V., *Advan. Phys.*, **2**, 450 (1953).
11. Danusso, F., and F. Sabbioni, *Rend. Ist. Lombardo Sci. Lettere*, **A92**, 435 (1958).
12. Dauscher, R., E. W. Fischer, and H. A. Stuart, *Z. Naturforsch.*, **15a**, 116 (1960).
12a. Dolmetsch, H., *Kolloid-Z.*, **176**, 63 (1961).
13. Eby, R. K., and K. M. Sinnott, *J. Appl. Phys.*, **32**, 1765 (1961).
14. Eppe, R., E. W. Fischer, and H. A. Stuart, *J. Polymer Sci.*, **34**, 721 (1959).
15. Fischer, E. W., personal communication.
16. Fischer, E. W., *Z. Naturforsch.*, **12a**, 753 (1957).
17. Fischer, E. W., *Ann. N. Y. Acad. Sci.*, **83**, 1 (1959).
18. Fischer, E. W., and G. F. Schmidt, *Angew. Chem.*, **74**, 551 (1962).
18a. Flory, P. J., *Principles of Polymer Chemistry*, Cornell Univ. Press, Ithaca, N. Y., 1953.
19. Frank, F. C., A. Keller, and A. O'Connor, *Phil. Mag.*, **4**, 200 (1959).
20. Gee, G., *Trans. Faraday Soc.*, **42**, 585 (1946).

21. Geil, P. H., in *Growth and Perfection of Crystals,* R. H. Doremus and B. W. Roberts, Eds., Wiley, New York, 1958.
22. Geil, P. H., *J. Polymer Sci.,* **47,** 65 (1960).
23. Geil, P. H., *J. Polymer Sci.,* **44,** 449 (1960).
24. Geil, P. H., *J. Polymer Sci.,* **51,** S10 (1961).
25. Geil, P. H., *J. Appl. Phys.,* **33,** 642 (1962).
26. Geil, P. H., *Polymer Single Crystals,* Interscience (Wiley), New York, 1963.
26a. Geil, P. H., *Polymer,* **4,** 404 (1963).
27. Geil, P. H., unpublished data.
28. Geil, P. H., N. K. J. Symons, and R. G. Scott, *J. Appl. Phys.,* **30,** 1516 (1959).
29. Geil, P. H., and D. H. Reneker, *J. Polymer Sci.,* **51,** 569 (1961).
30. Gubler, M., J. Rabiesiaki, and A. J. Kovacs, to be published.
31. Hammer, C. F., W. W. Brandt, and W. L. Peticolas, *J. Polymer Sci.,* **24,** 291 (1957).
32. Hermann, K., O. Gerngross, and W. Abitz, *Z. Physik. Chem.,* **B10,** 371 (1930).
33. Hoffman, J. D., and J. J. Weeks, *J. Res. Natl. Bur. Std.,* **66A,** 13 (1962).
34. Holland, V. F., personal communication.
35. Holland, V. F. ,paper presented at Electron Microscopy Society of America Meeting, Milwaukee, Wisc., August 1960.
36. Holland, V. F., S. B. Mitchell, W. L. Hunter, and P. H. Lindenmeyer, *J. Polymer Sci.,* **62,** 145 (1962).
36a. Huggins, M. L., G. Natta, V. Desreux, and H. Mark, *J. Polymer Sci.,* **56,** 153 (1962).
37. Jaccodine, R., *Nature,* **176,** 305 (1955).
38. Kampf, G., *Kolloid-Z.,* **172,** 50 (1960).
39. Kampf, G., *Kolloid-Z.,* **185,** 6 (1962).
40. Kargin, V. A., A. I. Kitaigorodskii, and G. L. Slonimsky, *Colloid J.* (*USSR*) (Engl. transl.), **19,** 141 (1957).
41. Kargin, V. A., N. F. Bakev, L. Li-Shen, and T. S. Ochapovskaya, *Vysokomol. Soedin.,* **2,** 1280 (1960).
42. Kargin, V. A., N. F. Bakev, and L. Li-Shen, *Vysokomol. Soedin.,* **3,** 1100 (1961).
43. Kargin, V. A., V. A. Kabanov, S. Ya. Mirlina, and A. V. Vlasov, *Vysokomol. Soedin.,* **3,** 134 (1961).
44. Kargin, V. A., S. Ya. Mirlina, V. A. Kabanov, and G. A. Mikheleva, *Vysokomol. Soedin.,* **3,** 139 (1961).
45. Keith, H. D., in *Physics and Chemistry of the Organic Solid State,* D. Fox, M. M. Labes, and A. Weissberger, Eds., Interscience (Wiley), New York, 1963.
46. Keith, H. D., and F. J. Padden, Jr., *J. Polymer Sci.,* **2,** 547 (1947).
47. Keith, H. D., F. J. Padden, Jr., N. M. Walter, and H. W. Wyckoff, *J. Appl. Phys.,* **30,** 1485 (1959).
48. Keith, H. D., and F. J. Padden, Jr., *J. Appl. Phys.,* **34,** 2409 (1963).
49. Keller, A., *Phil. Mag.,* **2,** 1171 (1957).
50. Keller, A., *Polymer,* **3,** 393 (1962).
51. Keller, A., and J. R. S. Waring, *J. Polymer Sci.,* **17,** 447 (1955).
52. Keller, A., and A. O'Connor, *Discussions Faraday Soc.,* No. **25,** 114 (1958).

53. Keller, A., and A. O'Connor, *Polymer,* **1,** 163 (1960).
53a. Kendrew, J. C., *Sci. Am.,* **205,** 96 (1961).
54. Leugering, H. J., *Kolloid-Z.,* **172,** 184 (1960).
55. MaGill, J. H., and P. H. Harris, *Polymer,* **3,** 252 (1962).
56. Manley, R. St. J., paper presented at 142nd American Chemical Society Meeting, Atlantic City, September 1962.
57. Manley, R. St. J., *J. Polymer Sci.,* **A1,** 1875 (1963).
58. Marchessault, R. H., F. F. Morehead, N. M. Walter, C. P. J. Glaudemans, and T. E. Timell, *J. Polymer Sci.,* **51,** S66 (1961).
59. McCall, D. W., and W. P. Slichter, *J. Polymer Sci.,* **26,** 171 (1957).
60. Miller, M. L., M. C. Betty, and C. E. Rauhut, *J. Colloid Sci.,* **15,** 83 (1960).
61. Miller, R. L., and L. E. Nielsen, *J. Polymer Sci.,* **55,** 643 (1961).
62. R. L. Miller in *Crystalline Olefin Polymers,* R. Raff and R. W. Doak, Eds., Interscience (Wiley), New York, in press.
63. Mnyukh, Yu. Y., E. M. Belavtseva, and A. I. Kitaigoradskiĭ, *Proc. Acad. Sci.* (*USSR*), (Engl. Transl.) **133,** 739 (1960).
64. Muller, A., *Helv. Chim. Acta,* **16,** 155 (1933).
65. Niegisch, W. D., *J. Polymer Sci.,* **40,** 263 (1959).
66. Padden, F. J., Jr. and H. D. Keith, *J. Appl. Phys.,* **30,** 1479 (1959).
67. Palmer, R. P., and A. J. Cobbold, to be published.
68. Peck, V., and W. Kaye, *J. Appl. Phys.,* **25,** 1465 (1954).
69. Price, F. P., personal communication.
70. Price, F. P., *J. Chem. Phys.,* **35,** 1884 (1961).
71. Rånby, B. G., F. F. Morehead, and N. M. Walter, *J. Polymer Sci.,* **44,** 349 (1960).
72. Rånby, B. G., and H. Brumberger, *Polymer,* **1,** 399 (1960).
73. Reneker, D. H., *J. Polymer Sci.,* **59,** S39 (1962).
73a. Reneker, D. H., and P. H. Geil, *J. Appl. Phys.,* **31,** 1916 (1960).
74. Ryan, J. personal communication.
75. Salovey, R., *J. Polymer Sci.,* **61,** 463 (1962).
76. Salovey, R., and A. Keller, *Bell System Tech. J.,* **40,** 1397, 1409 (1961).
77. Schick, M. J., personal communication.
78. Schuur, G., *Rubber Stichting Comm.* (*Delft*), No. **276,** 1955.
79. Statton, W. O., and P. H. Geil, *J. Appl. Polymer Sci.,* **3,** 357 (1960).
80. Stuart, H. A., *Ann. N. Y. Acad. Sci.,* **83,** 3 (1959).
81. Symons, N. K. J., *J. Polymer Sci.,* **A1,** 2843 (1963).
82. Till, P. H., *J. Polymer Sci.,* **24,** 301 (1957).

IC. MECHANICAL ASPECTS

B. Rosen

IC1. Introduction

As emphasized in the previous sections, polymeric substances: (a) usually possess a pronounced noncrystalline contribution, (b) are composed of molecules which are basically "chainlike" in the sense of having remarkably high orders of molecular size and asymmetry, and (c) frequently can exhibit properties that readily reflect the macroscopic contributions of this novel molecular constituency.

When considering the salient deformational responses of bodies of this generalized type, one repeatedly discovers the need for focusing his attentions on the distortions experienced by the noncrystalline zones in shear. This point becomes particularly clear when these changes in the shape of the body are abnormally large, whether generated by a pronounced inelasticity, or a "high" (i.e., "configurational" or "rubberlike") elasticity.

However, the shear strains of a polymeric body may be solidlike (i.e., with a high rigidity), liquidlike (i.e., inelastic), or gaslike (i.e., a shear which is sensibly independent of internal friction and changes in internal energy). While little more need be said here regarding solidlike behaviors in shear (33), considerable clarification is required concerning their liquidlike and gaslike properties. Moreover, these last two contributions permit the roles of molecular size and shape to be appreciated within that deformation.

The following presentation of considerable breadth and of limited length, in a clear reflection of the inherent failings of all introductory perspectives, is necessarily overgeneralized or oversimplified. While adequate general references are provided for further study, the conceptual continuity is salvaged at all costs in view of the primary objectives of the average reader.

IC2. Thermally Activated, Self-Transport Processes

IC2a. Random and Biased Self-Diffusions

The atoms of a "solid" usually are regarded as vibrating about equilibrium positions while being bounded "elastically" by force fields derived from their interacting environments. This model is to be nonconservative since the atoms would absorb or dissipate thermal energy depending on the amplitude and mode of these vibrations (42). The concept of "quanta" places limitations on the energy levels available for such motions.

The equilibrium relative location of two interacting atoms at a given temperature is achieved at that separation which is suitable for maxima in stability and bond energy, i.e., for a minimum in the total energy. At sufficiently low levels of kinetic energy, this positioning of atoms results from the minimized energy of interaction of the attractive and repulsive contributions to the variation of the potential energy $\phi(s)$ with the interatomic separation. If both of these contri-

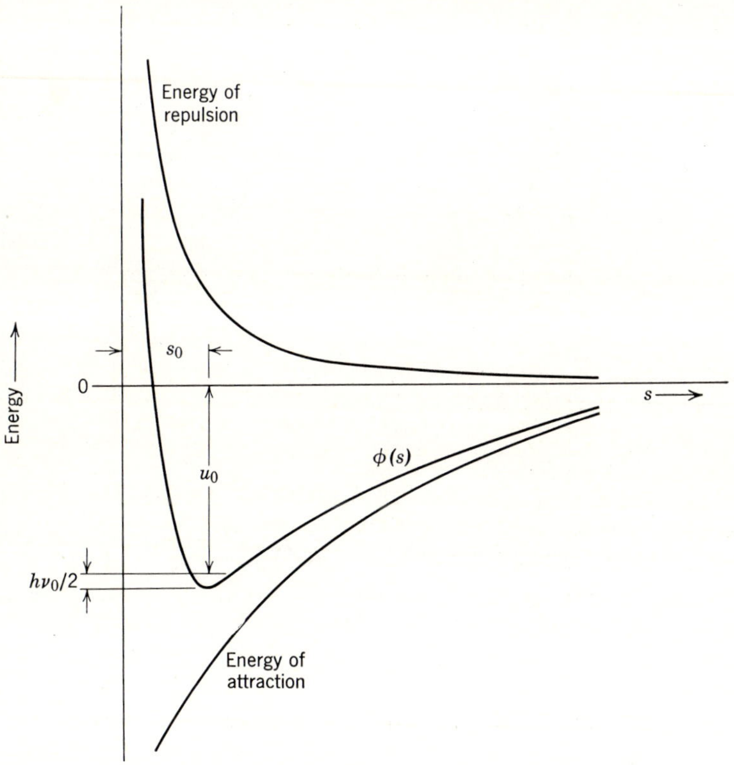

Fig. IC-1. Energies of interatomic interaction. A schematic illustration of eq. (1).

butions were inversely dependent on the distance of separation, we could employ (18)

$$\phi(s) = -(A/s^a) + (R/s^r) \qquad \text{IC2-(1)}$$

where A, R, a and r are constants, and a net attraction, i.e., a negative $\phi(s)$, must attain. Figure IC-1 provides a schematic illustration of this rather versatile function.

As the temperature is elevated, however, the increased amplitude of the anharmonic vibrations shifts this equilibrium positioning toward greater interatomic distances as macroscopically realized in a thermal expansion. When the kinetic energy of the atom momentarily attains a level which is equivalent to its interatomic dissocia-

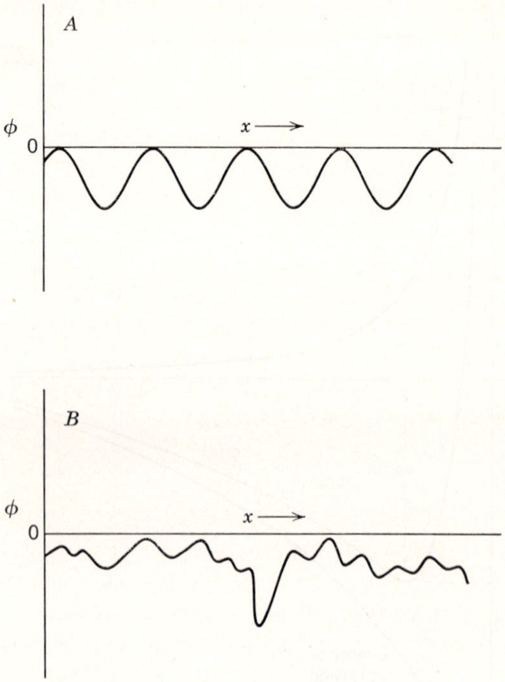

Fig. IC-2. The variations of the interatomic potential energy along an arbitrary direction: (*A*) along a plane in an ordered crystalline lattice; (*B*) in an amorphous body. Highly schematic.

tion energy u_0, the atom is said to be "activated" (16), hence freer to seek a lower total energy either by a relocation so as to lower its potential energy, or by a transmission of its kinetic energy by impact with, and displacement of, neighboring atoms. The mechanism for energy-lowering by a favorable relocation or transport of the atom may entail a subsequent dissipation of some of its dissociation energy by heat evolution, while the transmission of its kinetic energy may entail a heat redistribution or thermal conduction.

If the potential energy "troughs" in question refer to those of a perfectly ordered crystalline lattice, simple melting occurs at a critically balanced level of thermal energy. If these troughs are to control the relative positioning of the dynamic atomic centers of an already violently vibrating and rotating polyatomic molecule, an activated state here implies a possible molecular "degradation" in the sense of primary bond dissociation or chemical change.

The macroscopic phenomena associated with thermal activation in noncrystalline bodies, however, usually are far less dramatic than the above events. These particles continuously self-diffuse to and from energy troughs of unequal depths (Fig. IC-2). This is a highly dynamic process, yet one with an unprogressive future, for the isothermal and thermodynamically stable liquid or gas of fixed density. For a supercooled liquid, however, the average interparticle bond energy may spontaneously drift to a higher level (i.e., to a lower total energy) with a resulting decrease in the volume of the progressively devitrifying system (26).

Consider $N = \sum N_i$ particles with a total energy $\sum N_i e_i$, where e_1 is one of the several possible energy states. The probability for finding a particle in any particular state is (15)

$$p_i = N_i/N = \exp\{(u - e_i)/\bar{e}\}$$

where u is a constant that is characteristic of the potential energy of the distribution of particles, while $\bar{e}$ is the mean thermal energy kT of the vibrating particles for each degree of freedom; hence, $p_i =$ const. $\exp[-e_i/kT]$. If there are dN/N particles of this incremental energy level, we may employ

$$p = dN/N = \text{const. } \exp\{-e/kT\} \qquad \text{IC2-(2)}$$

which is in the form of the Maxwell-Boltzmann distribution.

Where $e = u_0 =$ the interparticle dissociation energy, e may be re6arded as the potential energy barrier against thermal activation. Let $Q = Nu_0 =$ the activation energy for a mole of particles, the probability for which is

$$p = \text{const. } \exp\{-Q/RT\} \qquad \text{IC2-(3)}$$

The order of the facility for spontaneous particle transport can be represented by a coefficient of self-diffusion (16)

$$\mathfrak{D} = \text{const. } \delta^2\bar{\nu}p \qquad \text{IC2-(4)}$$

where δ is the minimum distance of particle transport and $\bar{\nu} =$ const. kT/h is the mean frequency of the vibrations. Hence, $\bar{\nu}p$ implies the frequency of successful transports and,

$$\mathfrak{D} = \text{const. } \delta^2(kT/h) \exp\{-Q/RT\} \qquad \text{IC2-(5)}$$

The superimposure of externally applied forces affects this dynamic situation somewhat. A *classical solid* can be regarded as responding in infinitesimal strain by a mechanism of fully reversible, linear interatomic displacements within their energy troughs; this trivial repositioning being caused by a balanced set of imposed forces acting on a "continuous," differential volume element within the macroscopic body. (An *un*balanced system of forces or torques does not generate a strain alone, but also acts to accelerate the whole body in translation or rotation.)

The net *cohesive* force of atomic interaction may be approximated from eq. (1) by

$$f(s) = d\phi(s)/ds = (aA/s^{a+1}) - (rR/s^{r+1}) \qquad \text{IC2-(6)}$$

which is the variety of function schematically illustrated in Figure IC-3. The term ξ/s_0 is the slope of the force–displacement plot at the equilibrium interatomic separation. Hence ξ is an intensity factor for interatomic resiliency in one dimension (cf. Young's modulus for uniaxial stress in compression or tension upon a differential, *continuous* volume element). A slightly increased temperature here, while increasing the probability for anharmonic thermal activity within the potential energy trough, decreases ξ only to a degree on the order of the subtly positive coefficient of linear thermal expansion of the "elastic" body. Since the periods for atomic vibrations are on the approximate order of 10^{-13} sec., these "hookean" responses are treated as sensibly independent of experimental time.

Thus a classical solid would be a perfect *storer* of the energy required for the deformation, i.e., of the "strain energy." The interatomic contribution to the energy stored in this body is the differential area under the force-interatomic displacement plot in the immediate region of s_0 where $f = \pm s\xi/s_0$, and

$$W = \pm\int f\,ds = \pm(\xi/s_0)\int s\,ds = \pm\xi s^2/2s_0 = \pm s_0 f^2/2\xi \qquad \text{IC2-(7)}$$

The "storage" of strain energy implies a contribution to the *potential* energy of the atoms hence to the *internal* energy of the macroscopic body.

The application of a hydrostatic system of forces on a continuous, isotropic body (whether liquid or solid) adjusts the mean potential energy of the particles in a directionally *unbiased* manner through the strain energy, $W = \pm\int P\,dV$, thus imposed. Hence, an isothermal

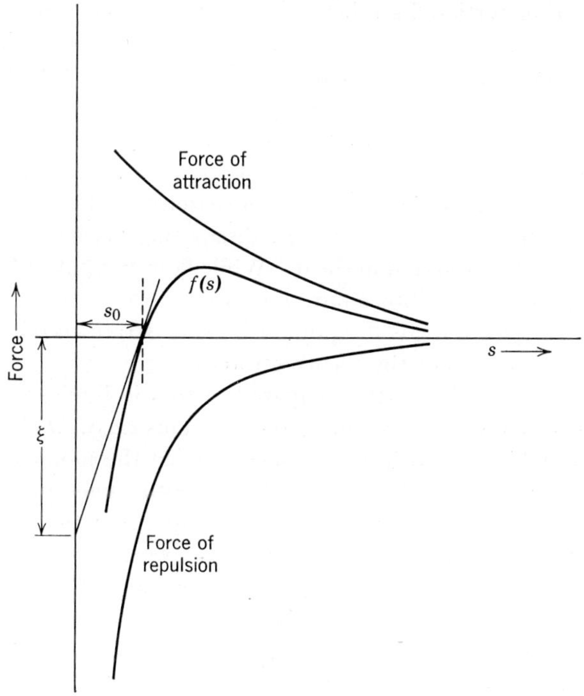

Fig. IC-3. Forces of interatomic interaction. A schematic illustration of eq. (6). In modification after Houwink (23).

hydrostatic compression, by doing work on the body, tends to compact it molecularly, to supplement its interparticle bond energy, and to make its thermal energy-induced activated state less probable. (This kind of compression would tend to raise the melting point of a crystalline body.) Conversely, an isothermal hydrostatic tension tends to increase the probability for a net, fully random, particle self-diffusion. Short of catastrophic consequences reflected on the macroscopic level, however, a continuous, isotropic body under hydrostatic stressing merely dilates elastically in bulk while exhibiting no directionally biased deformation, i.e., no distortion or *warping* of the body's shape.

Shearing forces, or the shearing components of axially applied tensions or compressions, tend to distort the shape of a solid body in shear. This *biases* the potential energy distribution along the directions of distortional strain; precisely where the strain energy in shear is stored. If the body which is subjected to this influence should exhibit either a permanently irrecoverable and/or a *time-dependently* recoverable distortion on the subsequent release of the shearing forces, then the term "inelasticity" applies to those *kinds* of shear. The irrecoverable shear is called "flow," whereas the retarded recovery of shear has been termed "anelasticity" (51).

Inelastic deformations act toward dissipating, as heat, the strain energy momentarily stored in shear. While flow dissipates this energy completely and irreversibly, a pure anelastic deformation evolves some heat in shear, yet absorbs it again on its slow strain recovery subsequent to the removal of the shearing forces.

Flow in solids is frequently compared to that in liquids; the idealization for which is a "newtonian body." This *classical liquid*, however, is formally regarded as incapable of even the momentary storage of strain energy in shear. This hypothetical entity cannot sustain a finite shearing force without flowing at a rate which is proportional to it. While it is indeed true that work is done in moving the liquid under the thus generated velocity gradient, this energy is fully dissipated by heat losses during that same flow. Its frictional "drag" are forces acting counter to the applied force-couple in reaction to the transfer of momentum via the dynamic particle motion from one velocity level to the next during the liquid's laminar slip (29). Hence, a newtonian liquid would be a perfect *dissipator* of work under flow. A frictional force is generated on flow in the sense: $f = \eta/\dot{\gamma}_f$, where $\dot{\gamma}_f$ is the velocity gradient. The rate of energy dissipation on shearing a unit cube of this liquid becomes (39)

$$\dot{W} = f\dot{\gamma}_f = \eta\dot{\gamma}_f^{\,2} \qquad \text{IC2-(8)}$$

Real liquids, however, tend to store some strain energy in shear; at least momentarily. Moreover, real solids invariably exhibit some order of inelastic deformation; hence they can be dissipators of shear strain energy. Thus real *bodies* are partly solidlike and partly liquidlike, differing mainly in the order of their contributions in the one sense or the other.

When subjected to a *fixed* shear, real bodies tend towards a gradual reduction or "relaxation" of the applied forces which are required to maintain that strain level. Thus a real body can be assigned a "relaxation time" during a portion of which it is sensibly solidlike.

Bodies which tend towards classical solid behavior appear to have virtually infinite relaxation times, while those which tend towards classical liquid behavior appear to have virtually no period for solid-like properties. The vast domain of real materials between these hypothetical extremes, i.e., to some degree a feature of all bodies, may be designated by terms such as "viscoelasticity" or by others which imply both elastic and inelastic contributions (39). Some of these will be further defined within the following.

In the very general sense employed above, flow in real bodies is a shearing stress-*biased*, thermal energy-activated, particle self-diffusion. The potential energy contribution of the applied shearing forces acts toward favoring a net *relocation* of particles rather than maintaining a dynamic, yet macroscopically balanced self-diffusion. Hence, after Eyring (12), we will say that there exists an E_s which is to represent this energy contribution of the imposed shear. The probability for finding particles in the thermally activated state is increased by $Q - E_s$ and decreased by $Q + E_s$, so that eq. (3) becomes

$$p' = \text{const. } [\exp \{-(Q - E_s)/RT\} - \exp \{-(Q + E_s)/RT\}]$$

or

$$p' \propto \exp \{-Q/RT\} \sinh (E_s/RT)$$

and eq. (5) becomes

$$\dot{\gamma}_f = \mathfrak{D}' \propto \delta^2(kT/h) \exp \{-Q/RT\} \sinh (E_s/RT) \qquad \text{IC2-(9)}$$

where $\dot{\gamma}_f$ is the shearing stress-biased coefficient of self-diffusion or velocity gradient of *flow* of the macroscopic body. Equation (9) is perfectly general (if not overly general) and presupposes no knowledge of the body's structure. Entropy considerations are relegated to the (rather potent) proportionality factor.

Under the condition $E_s/RT \ll 1$, $\sinh (E_s/RT) \cong E_s/RT$, and eq. (9) can be expressed

$$E_s \propto \exp \{Q/RT\}\dot{\gamma}_f; \quad E_s\, RT \ll 1$$

If, as has been assumed (12), $E_s \propto \mathcal{S}$ = the shearing stress, this equation becomes

$$\mathcal{S} = \eta\dot{\gamma}_f; \quad E_s/RT \ll 1,\ E_s \propto \mathcal{S} \qquad \text{IC2-(10)}$$

where η is the coefficient of viscosity for newtonian flow in the sense,

$$\eta \propto \exp \{Q/RT\} \qquad \text{IC2-(11)}$$

However, for the case where the simplification $E_s/RT \ll 1$ cannot be employed, within the repeated assumption $E_s \sim \mathcal{S}$ eq. (9) becomes

$$\sinh\,(\text{const. } \mathcal{S}) \sim \dot{\gamma}_f; \quad E_s \propto \mathcal{S} \qquad \text{IC2-(12)}$$

where this is the general relation for the flow rates of "non-newtonian" liquids (38). Note that a higher level of shearing stress appears to have made our body into a different "kind" of liquid. Actually, the liquid is precisely as before, but its energetic environment has changed.

The assumption $E_s \propto \mathcal{S}$, however, has a questionable validity for a newtonian liquid since, as mentioned before, this hypothetical body is not *permitted* to store strain energy in shear. The use of this assumption is justified only by the agreement of eq. (11) with salient fact and the apparent validity of eq. (9). Yet we know that if $E_s \propto \mathcal{S}$ is to be "adequate" for liquids, the condition $E_s \propto \mathcal{S}^2$ is appropriate for classical solids in the general sense of eq. (7). However, it is equally useless to employ $E_s \propto \mathcal{S}^2$ in eq. (9) since this hypothetical solid is not *permitted* to experience a flow.

An idealized "plastic" solid, i.e., a "Bingham body" (22), is to behave as a classical solid below a certain shear stress, $\mathcal{S}_y$, yet is to be a classical liquid at higher stress levels in the general sense of a modified eq. (10):

$$\mathcal{S} - \mathcal{S}_y = \eta\dot{\gamma}_f \qquad \text{IC2-(13)}$$

Below the yield stress, $\mathcal{S}_y$, we can presume $E_s \propto \mathcal{S}^2$, yet above $\mathcal{S}_y$ we possibly should presume $(E_s - E_y)/RT \ll 1$, and $E_s \propto \mathcal{S}$. If this were true, then when $(E_s - E_y)/RT \ll 1$ cannot be employed, perhaps

$$\sinh\,[\text{const. } (\mathcal{S} - \mathcal{S}_y)] \sim \dot{\gamma}_f \qquad \text{IC2-(14)}$$

in a modification of eq. (12). Equation (14) represents the overgeneralized flow rate of a "pseudoplastic solid" (1,15).

The more likely situation, however, is represented in Figure IC-4, where we see the general form of eq. (9) schematically reproduced for hypothetical orders of $\dot{\gamma}_f$, E_s, Q, and RT. (Actually, these curves are the *same* hyperbolic sine curve rotated about the origin.) Here, the *idealized* bodies whether liquid, viscoelastic or plastic are visualized as occurring under trivial flow rates, while the unideal, i.e., *real*, behaviors are pictured as occurring at larger rates of flow. Within this particular perspective, plastic yielding may be regarded as spontane-

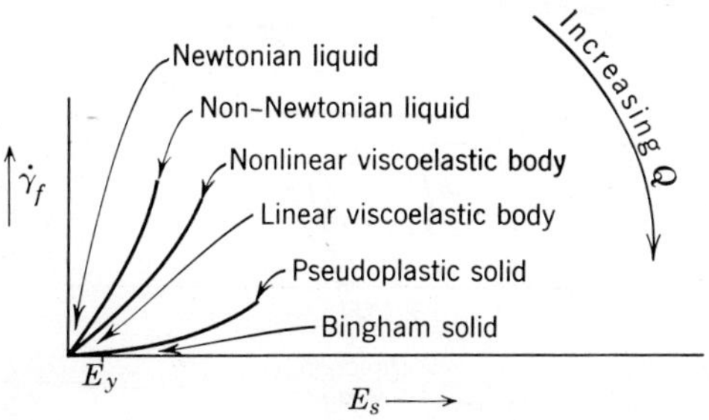

Fig. IC-4. The flow rates of several kinds of idealized and real bodies in a highly schematic illustration of eq. (9) and its subsequent discussion.

ously initiating after an enormously prolonged relaxation time, one which is well beyond human patience.

Regardless of one's particular perspective of real bodies, all these perspectives must have a wide ground in common. Viscoelastic bodies can also be made to "yield" when forced to achieve strain rates which are too great for their stress relaxation processes, see Subsection IC5c. The softer of the crystalline bodies, with sufficiently low Q values, can appear to be viscoelastic. Common glass, while an orthodox *liquid*, is still regarded by many to be our most perfectly brittle *solid*. Hence, whether amorphous or crystalline, real bodies exhibit some of the features of both hypothetical liquids and hypothetical solids.

It is frequently true, however, that some solids can appear differently during the macroscopic event of yielding. A single crystalline body containing small molecules, with $(E_s - E_y)/RT \ll 1$, in which Q is very formidable relative to RT, tends to slip discontinuously on experiencing a "critically resolved shearing stress" and, in general, closely approaches the properties of the hypothetical Bingham body (6) (Fig. IC-5).

On the other hand, the initiation of "necking" usually reflects the enhanced flow along the planes of *maximized* shearing stress (i.e., the flow processes are less directionally selective; proceeding best where "hardest pushed"). In either "plastic" or commonly "ductile" bodies, the location along the sample's principal axis at which the pure slip or neck initiates is a reflection of the locality of an efficient inhomogeneity; a consideration which is entirely unacademic in real solids, whether or not supposedly "ideal." Hence a *perfect* Bingham body

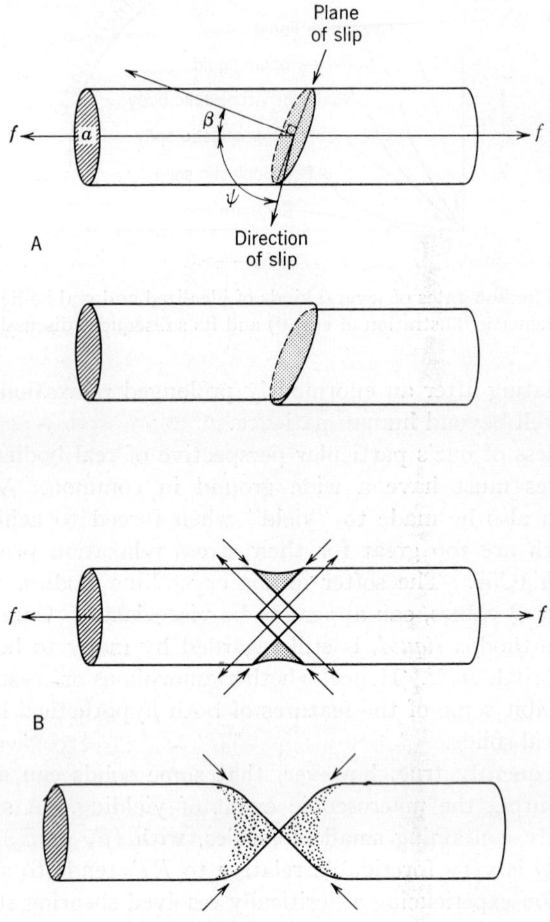

Fig. IC-5. General appearance of yielded solids: (A) a single crystalline body with a critically resolved shearing stress, $\mathcal{S}_y = (f/a) \cos \beta \cos \psi$; (B) a "ductile" body. Highly schematic.

would not be expected to appear in yield as illustrated in Figure IC-5A.

Solids which tend to approach the properties of formal plasticity have yield stresses which are sensibly independent of ordinary temperature levels. This is frequently termed (15) "athermal inelasticity;" the implication being that the order of E_s or Q usually is far greater than that of RT in defining that stress level sufficient for the initiation of a macroscopically obvious flow. Conversely, "thermal inelasticity" connotes a strongly temperature-dependent shearing stress for flow initiation. It is clear, however, that an athermally inelastic body *becomes* more thermally inelastic as its temperature is raised sufficiently so that RT approaches Q.

All real bodies have finite *thicknesses*. This factor plus a stagnant ambient atmosphere can generate significant thermal transport lags due to slow moving temperature gradients. Polymeric bodies, generally being good thermal insulators (35), suffer greatly in this particular regard. Aside from heating on internal or surface frictions (Section IVA), even classical solids tend to absorb or evolve heat *reversibly* on altering the internal energy of the body under the dilative components of the strain energy. These solids would absorb heat in expansion and evolve heat in contraction (39). An *ideal* rubber, however, even though incompressible and free of internal friction, *evolves* heat in distortion, yet absorbs it again on the recovery of this shear in much the thermodynamic manner of an anelastic deformational response. (See Subsection IC3.)

With all of these possible heating and cooling considerations associated with finite coefficients of linear and volumetric thermal expansion or contraction, athermally inelastic bodies as well as thermally inelastic bodies tend to deform differently depending on whether the process is performed "isothermally" (i.e., sufficiently slow) or "adiabatically" (i.e., sufficiently fast). The thermal lags in the faster experiments, and those performed with the better thermally insulating or thicker bodies, frequently generate the *appearances* of experiencing an inelastic deformation at a rate which is controlled by the thermal diffusion (15,51).

IC2b. Self-Transport in Polymeric Bodies

With *poly*atomic molecules, and especially with those varieties that are sufficiently complex to be termed "chainlike," the mechanisms of transport are relatively complicated (10,52). Here, we will briefly outline only a few of the salient factors. Some of the other considerations will be encountered later in the present section as well as elsewhere in this volume.

The primary point to bear in mind with the transport properties of amorphous, polymeric molecular chains is that it is quite useless to regard the entire molecule as residing in a single potential energy well. Attentions are necessarily focused on the mobility of the chain *segments* rather than that of the molecule as a whole. The behavior of these segments then controls the chain's motions and its consequent transport. This indirect mechanism demands a completely fresh perspective of transport processes in these kinds of substances.

As emphasized by Rosen in Section IA, there is a "two-headed cohesion" to be met with those polymeric bodies which phenomenologically exhibit the physical integrity of their molecular chains. The "scission" of primary bonds should be viewed as a formal chemical change, while the surmounting of the smaller potential energy barriers of intersegmental secondary bonding are less catastrophic in consequence. However, a given segment of a chain can accomplish three kinds of intersegmental transport. It can disengage itself from its own chain and thus generate the chemical change, it can transport itself over one of the many segments of its own chain, or it can transport itself over one of the segments of a neighboring chain. Hence, two of the three possible varieties of segmental self-diffusion can be accomplished without losses in the chain's integrity.

A given segment of a given chain momentarily finds itself in a given free energy trough which is defined by its steric and energetic circumstances. When "in" such a trough, the segment is fully subject to (at least) all of the considerations outlined in Subsection IC2a, above. It is only when it attempts to leave its trough that this polyatomic aggregate appreciates the complexities of being only a segment of a molecular chain, and not a whole molecule of itself.

If it is to remain an interdependent part of that chain, the motions of that segment must fully respect the primary-bonded integrity of the adjacent segments of its own chain. However, this same consideration is met by each and every segment of that complex chain so that a given segment, in effect, has rather long-ranged commitments in its motions; at least ones that cannot always be fully reconciled by its immediate neighbors. Needless to overstate, such complications increase the probability for that segment to remain in its own trough, hence the mobilities of chainlike molecules are quite inhibited.

One perspective of chain transport can be relegated to a given chain of itself: free of neighboring molecules! This consideration is peculiar to this kind of complex particle since it entails a reflection of

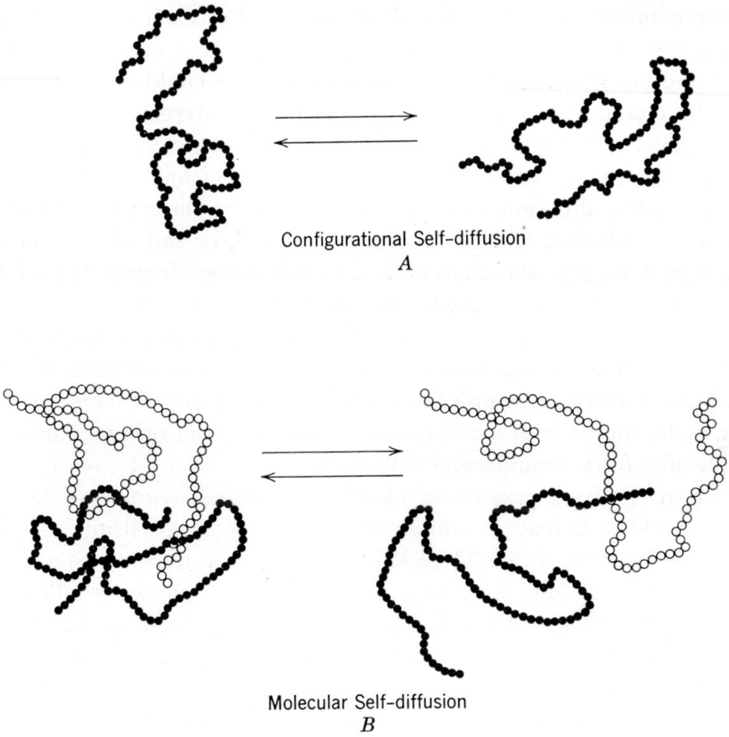

Fig. IC-6. Self-transport of polymeric chains: (*A*) interconfigurational diffusion; (*B*) intermolecular diffusion.

its *configurational uncertainty*. Within time and fluctuations in thermal energy, a molecular chain will spontaneously seek a maximized "flexibility" in the sense of an optimized degree of "inter*configurational* diffusion," the limiting case of which is a *micro*-brownian (i.e., a sub-molecular) randomized motion (Fig. IC-6*A*).

Another perspective of chain transport is the more classical variety, inter*molecular* diffusion, the limiting case of which is *macro*-brownian (i.e., a full-molecular) randomized motion (Fig. IC-6*B*).

We have employed the term "perspectives of transport" since these two varieties of motion result from a single event, namely inter*segmental* diffusion within each coherent chain, yet are viewed within two different frames of reference. For example, consider a unit volume-element containing several million chainlike molecules in their equilibrium, fully amorphous state. Let each chain segment experience small randomly directed translations at certain randomly selected intervals. Let us now suppose that we could watch the ends of the chains dance randomly about under interconfigurational or intermolecular diffusion (the reader may have his choice). If these randomized transports occur rapidly enough, at any one *instant* the chain-ends will measure to be, *on the average*, at the same separations as before whether head-to-head, head-to-tail, or tail-to-tail; hence, as stated above, the system is in *equilibrium* despite the highly dynamic processes involved. We imply here, that while it is quite true that the molecules are continuously changing their configurations and interchain separations, one cannot distinguish between the two randomized situations unless his attentions are focused on one particular molecule for interconfigurational diffusion, and on two particular molecules for intermolecular diffusion.

In the limiting case of an ideally randomized segmental motion, one would anticipate a simple problem in gaussian statistics. Let us imagine that we could pick up one of these dancing chains and spread its ends slightly apart. From time to time we would feel a tugging together or a pushing apart, depending upon the end-to-end chain separation imposed. The further the ends are separated, the greater is the frequency of tugging over that of pushing. If we superimpose the two ends, the frequencies of tugging and pushing, granting sufficient patience, become quite equal.

If we do not stretch the chain too much, the *potential energies* of all possible configurations are identical and the time-averaged work done in separating the ends of the chain is only that required for lowering its *configurational entropy*. Hence, we have restricted the number of possible configurations to be made by that chain; at least, its *segments* dance freer if the chain-ends are superimposed rather than separated.

Now if we release one end of our playful chain and observe the periodically changing distances between the free and fixed ends, their frequency and order of separations, as projected on any straight line passing through the fixed end, will generate the gaussian error function about that fixed point. In other words, the fixed end becomes a target at which the free end haphazardly aims. If we hold the heads of *two* chains a fixed distance apart so as to permit their tails to dance randomly about, however, these tails also will eventually average-out to be no farther separated than their heads.

In this specific sense, the instantaneous state of inter*molecular* diffusion is an automatic consequence of the instantaneous state of inter*configurational* diffusion; the cause of the whole affair being the interdependently randomized inter*segmental* transports within the continuous chains themselves.

If the segments of a given chain were actually limited to three-dimensional vibrations alone, the chains would have negligible flexibilities relative to that known to be experienced in rubberlike bodies or in polymeric solutions and melts. It has become clear that the *rotation* of segments about their primary-bonded "backbones" is that which generates "chain flexibility" as normally implied. Hence the segment's problem of leaving its trough is greatly simplified by its rotation relative to its neighbor's. These rotations would depend on the highly specific character of the polyatomic segments in the specific chain under consideration, and these have been treated and summarized elsewhere (1,14,48,49) in quite elaborate detail.

It has frequently been proposed that the glassy transition T_g demarks the *time-dependent* temperature for the empirically sufficient onset of the segmental rotations in the molecular chains of supercooled polymeric bodies (8,14). Then, T_g would also imply the minimum temperature for "significant" intersegmental, interconfigurational, or intermolecular diffusion; again, the reader may have his choice.

While the *time*-averaged, unstressed, end-to-end, chain length projection is zero for an ideally flexible, linear polymeric molecule, the *instantaneous* distribution of projected chain-end separations, i.e., the *spread* of the gaussian curve, is obviously dependent on the diameters and number of chain segments. Hence, for a random *instant* the chain's "most probable," "average," and "standard," projected end-to-end deviations are necessarily finite. When viewing an infinite number of flexible amorphous chains, the macroscopic reality of "sample-length" evolves as a composite reflection of the instantaneous, end-to-end chain separation as averaged throughout the entire sample in its dynamic state of equilibrium. One commonly expresses the "*size*" of a chain-like molecule by its mean-square, end-to-end chain length $\langle \overline{R^2} \rangle$ (its square root, $\langle \overline{R^2} \rangle^{1/2} \neq \langle \overline{R} \rangle$, is the *standard* deviation of chain-ends) where the instantaneous squaring, summing and dividing is done over an infinite population of molecules. For the fully flexible case, it has been shown (10) that,

$$\overline{\langle R^2 \rangle} = na^2 \qquad \text{IC2-(15)}$$

where n is the number of segments, each of an effective diameter a.

Again, the whole concept of the transport of a molecular chain hinges on intersegmental transport. Employing a as the minimum distance for this segmental self-diffusion, we have from eqs. (5) and (15),

$$\mathfrak{D} \propto [kT\overline{\langle R^2 \rangle}/n] \exp \{-Q/RT\} \qquad \text{IC2-(16)}$$

where Q is largely a reflection of the activation energy barriers for the *rotation* of one segment relative to that of the adjacent segments of its *own* chain; this usually being the simplest avenue to either interconfigurational or intermolecular diffusion. However, it is not always clear as to *how many* of such rotations are required to "transport" an Avogadro Number of segments. This depends on what frame of reference one need employ for the singular "transport" under consideration. Equation (16) also requires a different frame of reference for the case of diffusion of an Avogadro Number of segments subsequent to chain degradation by primary bond scission. Here Q would reflect the *dissociation* energy of a particular kind of primary bond, etc.

It should be quite clear, however, that eq. (16) has a critically important entropy contribution in the transport of these extremely complex molecules (10).

The application of a shearing force directionally *biases* all three varieties of intersegmental diffusion. Hence, primary bond scission, as well as intersegmental slip over its own or neighboring chain, becomes *directionally* favored.

The scission of primary bonds in polymeric bodies, if not accompanied by a comparable order of chemical *repolymerization* at the newly formed chain-ends, is a contributor to the flow of the macroscopic body. Frequently, this inelasticity on primary bond scission is enhanced in the presence of reacting gases, and then it is termed "chemorheology" (see Section IVE). Subordinately, the chemical reactivity of polymeric bodies is accelerated by stress components in shear. (However, components in hydrostatic tension, just as they tend to lower the melting points of crystalline bodies, also play an important role here.)

Now, a shear stress-biased, inter*molecular* diffusion may also be a contributor to flow, but *not* when that self-diffusion is permanently *reversible!*

The high order of conceptual unorthodoxy displayed here, however, is quite clear from the foregoing. When dealing with molecules which cannot be relegated to a single potential energy trough, a given segment may leave its trough and then prefer to *return* to it on release of the shearing forces. This is due to the long-ranged commitments suffered by any segment in the amazingly cooperative, chainlike molecule. We do not imply that each and every segment of a given chain will return to its unaltered trough, yet some indeed must at, say, a point of "chain entanglement" (10).

Thus one is forced into the realization that, insofar as a given chain segment can *slowly* return to its original trough within the cooperation of all other segments in that chain, this stress-biased intermolecular diffusion also can contribute to an *anelastic*, i.e. a retarded elastic, response. While this important consideration is easiest to appreciate with soft, amorphous varieties of polymeric bodies, it may possibly be a contributor to the anelasticity of all *real* bodies whether or not polymeric, and perhaps one should begin to term this interesting event "*structural anelasticity.*"

Presumably, structural anelasticity also would be a contributor to the *non* newtonian flow of polymeric bodies when, say, the chain entanglements gradually loosen in shear (10). However, let us briefly consider the *newtonian* flow of a polymeric fluid. Employing eqs. (11) and (16) we have

$$\mathcal{S}/\dot{\gamma}_f = \eta \propto kT\overline{\langle R^2 \rangle}/n\mathfrak{D} \qquad \text{IC2-(17)}$$

Treating the coefficient of intersegmental diffusion in the "fully random" sense such as that implied when employing $\mathfrak{D} = kT/f_0$ where f_0 is a coefficient of intersegmental friction, on employing eq. (17) we have

$$\eta \propto f_0\overline{\langle R^2 \rangle}/n \qquad \text{IC2-(18)}$$

which characterizes the salient flow behavior of many polymeric solutions and melts (10); the more generalized case being obtained by the use of $\overline{\langle R^2 \rangle}/n$ for δ^2 in eq. (9).

Perhaps most important is the point that a shear stress-biased, inter*configurational* diffusion *cannot*, of itself, contribute to the "flow" of polymeric bodies, i.e., not in the normally employed sense of this term. While it is quite true that this *distortion of a chain's flexibility* can be made to be residual, aside from the conflicting, dual perspective of intermolecular flow, a given chain will wish to reattain its entropy-maximized state and thus contribute to a residual stress in that body, at that same temperature, if its configurational probability is not regained on the release of the shearing forces. Expressed another way: if one lowers the configurational entropy of a chain under shear and simultaneously facilitates its flow to another locality, and if, when at that new locality, the release of the shearing forces does not permit that chain to regain an equally high level of configurational entropy, there is an enhanced residual force acting upon that particular chain.

If we are given an ideally flexible chain system with no *irreversible* stress-biased, inter*molecular* diffusion, i.e., with no flow, we automatically have a system which is under pure, stress-biased, inter*configurational* diffusion. This micro-brownian system would have *no* internal friction; it *cannot* experience classical solid behavior under small strains, yet it can distort instantaneously and reversibly. This structural anelastic body which is *without* a significant relaxation time is called "an ideal rubber." More descriptively, it experiences a pure "*unretarded* structural anelasticity," or less ambiguously, a pure "*configurational elasticity*." This remarkable concept will be treated further in the next subsection, IC3.

Short of the inelastic contributions of chain scission, flow in the amorphous regions of polymeric bodies is aggravated by a small $\langle\overline{R^2}\rangle$ in approximate accord with eq. (18). Hence, short chains are strong contributors to the flow character of polymeric bodies. Aside from their more obvious hydrodynamic influences on an enhanced η, extremely long chains tend to entangle, and $\langle\overline{R^2}\rangle$ generates a higher order of influence on η, beyond a *critical* value of $\langle\overline{R^2}\rangle$. The effects of chain entanglements on η and $\mathfrak{D}$ have been summarized (10).

While chain entanglements frequently serve to *remind* a partly distorted molecule from whence it came, a far more reliable "structural memory" is imparted by the introduction of primary-bonded "crosslinks." Here, the chains are held together in a long-range, three-dimensional molecular network with the same kind of continuity as exhibited by the chains themselves. Under sufficiently high orders of crosslinking, such as with a *vulcanized* rubber, short of primary bond scission, true molecular flow is sensibly impossible. In this sense, crosslinks can be viewed as "permanent entanglements." The synthesist of chainlike molecules can introduce crosslinks, to a large extent, at will. Hence, he can chemically control the order of flow in the body virtually *as he chooses* (14).

Chains are also "linked" together as they meet in crystallized domains in semicrystalline bodies. Thus the flow contribution of partly crystallized bodies is greatly limited (see Section IE). With such "long-ranged molecules," a small amount of crystallinity can have pronounced effects in that regard since even small portions of a single molecule may enter a crystallized region (see Section IB).

Chain branching, which also can be controlled by the molecular architect, tends to decrease the geometric asymmetry of the chainlike molecule and this frequently decreases both its density and probability for crystallization.

The secondary bonding energies are greatly affected by the *segmental constitution*. This also can be controlled in a given synthesis, along with other factors including molecular size and size-distribution (14). Beyond this, the secondary bonding energies may be decreased by anything which increases the volume of the body; whether this is accomplished by chain branching, swelling the sample with a fluid solute, or by hydrostatic components in tension. These all tend to lower the T_g of a supercooled polymeric body and enhance its configurational elastic, structural anelastic and flow contributions to deformational responses in shear.

Now if the reader superimposes over all of this a finite contribution from the more classical, or harder, solids in shear, together with *their* inelastic overtones, he will then become at least conceptually oriented to the salient complexities in the mechanics of *real* polymeric bodies. However, let us now retrace our steps to consider several of these same problems from a modified perspective; one which is somewhat closer to empirical observations.

IC3. Configurational Elasticity

IC3a. Elastic Character of Rubberlike Solids

The term "rubberlike" has evolved to be equally as general and equally as descriptive as the term "glasslike." As we look about us, we find many varieties of substances which have an elastic character which is unlike that idealized by the hard, classical solid. Their shear and Young's moduli are remarkably small and their ultimate and recoverable elongations are remarkably large. Whether they are proteins of the fibrous muscles, skins, or gelatins; whether they are natural or synthetic "elastomers;" or whether we use them in the body, the home, the playground, or within our industrious and industrial undertakings, rubberlike solids are rather important to us. If macroscopically continuous, all known varieties of rubberlike bodies are polymeric in molecular structure.

Table IC-I gives a very approximate comparison of Poisson's ratios μ and Young's moduli E for a spectrum of real solids.

TABLE IC-I

Solid	μ	$E \times 10^{-6}$, dynes/cm.^{2}
Vitreous silica	0.14	700,000
Cast iron	0.27	900,000
Granite	0.30	300,000
Polystyrene (glasslike)	0.33	30,000
Polyethylene (leatherlike)	0.38	2,000
Rubber (vulcanized, natural)	0.49	20

In view of the formal relations for isotropic bodies (33),

$$E = 3K(1 - 2\mu) = 2G(1 + \mu) \qquad \text{IC3-(19)}$$

where K and G are the bulk and shear moduli respectively, rubberlike solids characteristically approach the conditions: $E = 3G$, $K/E = \infty$, and $K/G = \infty$, as μ approaches the liquidlike value, $^1/_2$. This implies that the solid is sensibly free of a dilative strain contribution in response to a directional stress. Thus, on either elongations or shear, we need be concerned primarily with changes in shape rather than in volume. In place of the (symmetrical) maximum of 21 elastic constants for anisotropic bodies (24), or for the two for formal isotropic solids, the deformation of an isotropic, differential volume element of a rubberlike solid under small strains can, rather effectively, be represented by a single constant which characterizes its facility for distortion. Thus for our case, $K \gg G$, the strains as functions of stresses can be represented simply (1) by

$$\begin{array}{cccccc} J/3 & -J/6 & -J/6 & 0 & 0 & 0 \\ & J/3 & -J/6 & 0 & 0 & 0 \\ & & J/3 & 0 & 0 & 0 \\ & & & J & 0 & 0 \\ & & & & J & 0 \\ & & & & & J \end{array}$$

where $J = 1/G$ = shear compliance. This also may be written as,

$$3G\epsilon_1 = \sigma_1 - {}^1/_2(\sigma_2 + \sigma_3)$$
$$3G\epsilon_2 = \sigma_2 - {}^1/_2(\sigma_1 + \sigma_3)$$
$$3G\epsilon_3 = \sigma_3 - {}^1/_2(\sigma_1 + \sigma_2)$$
$$G\epsilon_4 = \sigma_4; \quad G\epsilon_5 = \sigma_5; \quad G\epsilon_6 = \sigma_6$$

where ϵ is strain and σ is stress, while the subscripts 1, 2, and 3 imply normal components, and 4, 5, and 6 are the components in shear, *e.g.*, $\epsilon_5 = (\partial u/\partial z) + (\partial w/\partial y)$ for the new positions $x + u$, $y + v$, $z + w$. (See refs. 1, 40, and 49 for a fuller development in this regard, whether for infinitesimal or finite strains.)

We are dealing with a *kind* of solid, however, that also is overly easy to distort, and with one that can recover its shear after strains which are several hundred per cent beyond its hookean limit.

The thermodynamic features of the elasticity of rubberlike solids also are unusual. Figure IC-7 shows the equilibrium, residual tensile stress under 350% elongation of a typical vulcanized rubber as a function of temperature. This elongation was imposed at the highest temperature; time-dependent stresses being relaxed prior to the cooling. The body was noncrystalline at all temperatures shown (36). Here, the approximate temperature of the glassy transition T_g also demarks the transition from a positive to a negative coefficient of linear thermal expansion in the direction of the principal strain.

The general situation above T_g is a reflection of the extreme predominance of the entropy term dS in the Helmholtz Free Energy changes dA in a reversible and isothermal extension; $dA = dW = fdL = d\mathcal{E} - TdS$, where dW is the tensile work. If we let $\mu \equiv {}^1/_2$, the body is like an incompressible or unexpandable liquid which cannot store strain energy by resilient interatomic displacements in shear, and the changes in internal energy $d\mathcal{E} = 0$. Thus,

$$f = -T(\partial S/\partial L)_{V,T} \qquad \text{IC3-(20)}$$

which is reasonably well followed for common vulcanized rubbers at extensions even well above 100%. The deviations are discussed in Subsection IC3d.

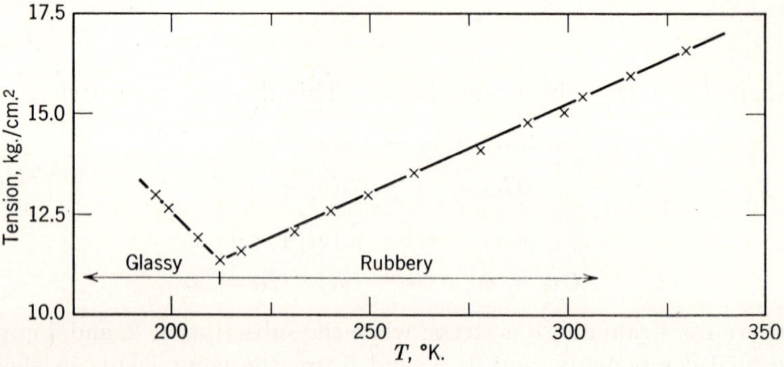

Fig. IC-7. The residual tensile stress on cooling a vulcanized (8% sulfur) natural rubber which was maintained at a 350% elongation. After Meyer and Ferri (36).

It is interesting to compare the internal energy situations in the two systems,

For a perfect gas:

$$(\partial \mathcal{E}/\partial V)_T = 0$$

For an ideal rubber:

$$(\partial \mathcal{E}/\partial L)_{T,V} = 0$$

Just as the pressure of a perfect gas in isothermal compression, $P = T(\partial S/\partial V)_T$, is derived from the decrease in entropy with decreasing volume, the tensile or shearing resistance of an ideal rubber in isothermal deformation arises from a decrease in entropy at constant volume. Thus, in strong contrast with classical solids, the elasticity of rubberlike solids is not predominantly dependent on changes in the internal energy.

A further appreciation of the general situation met here requires an insight into the origins of the entropy contributions in these novel kinds of solids. However, first we will consider a highly simplified case in which the salient factors still can be appreciated.

IC3b. Configurational Probabilities of an Idealized Chainlike Molecule (50)

Consider a single chainlike polymeric molecule with several hundred segments which experiences a thermal energy fluctuation-induced, intersegmental diffusion-executed, ideal, or micro-brownian chain flexibility. This chain is isolated in space and although interconnected, each segment of the chain is free to take single steps forward or backward with equal facility.

Each of the n segments of length a undergo a single-stepped random walk. There are n_+ steps made forward and n_- steps made backward, and $n = n_+ + n_-$. The net forward progress is obtained by, $m = n_+ - n_-$. The probability for this net forward progress in the n single steps is (14)

$$W(m) = (1/2)^n n!/\{[(n+m)/2]!\,[(n-m)/2]!\}$$

Using Stirling's approximation for the factorials, where both n and m are large while $n \gg m$, this eventually becomes

$$W(m) \cong (2/\pi n)^{1/2} \exp\{-m^2/2n\}$$

The probability that the vector connecting the ends of the chain has a component in the x direction is, $W(x) = W(m)/\Delta x$. If the length of each step in this particular direction is its root-mean-square projection, $\pm a/(3)^{1/2}$, at the end of the n single steps the total forward progress is simply, $ma/(3)^{1/2}$. However, $\Delta x = 2ma/(3)^{1/2}$, since the smallest value for Δm is 2. This leads to,

$$W(x) \cong [3/(2\pi na^2)]^{1/2} \exp\{-3x^2/(2na^2)\}$$

The probability that the component of the distance between the chain ends will be x, y, and z is given by a product of three such probability densities (10):

$$W(x\,y,z) \cong [3/(2\pi na^2)]^{3/2} \exp\{-3(x^2+y^2+z^2)/(2na^2)\} \quad \text{IC3-(21)}$$

If the end-to-end chain distance is the radius of the sphere, $r^2 = x^2 + y^2 + z^2$, this probability density may be expressed:

$$W(r) \cong [3/(2\pi na^2)]^{3/2} \exp\{-3r^2/(2na^2)\}$$

The entropy of the chain is determined by the *number* of its possible configurations, and this is dependent on $W(r)dV$, where dV

is the volume element (49). If V is held constant, however, this number of configurations is dependent on r, and is proportional to $\exp\{-3r^2/(2na^2)\}$. From statistical thermodynamics, the chain entropy for a given value of r is, $S \cong \text{const.} - \{3kr^2/(2na^2)\}$, where k is Boltzmann's constant. When the time-averaged value of r equals 0, the entropy is maximized. As the time-averaged positioning of the chain-ends deviates from this coincidence, the number of possible chain configurations, hence the entropy, decreases. The tensile force necessary to move one end of the chain from a distance r to $r + dr$ is obtained from eq. (20)

$$f = -T(\partial S/\partial r)_{T,V} \cong 3kTr/na^2 \qquad \text{IC3-(22)}$$

This tension is proportional to the absolute temperature and the chain is *hookean* at sufficiently low elongations. This idealized chain behaves like a dynamic spring with a spring constant, $3kT/na^2$, and is one whose time-averaged unstressed, end-to-end length is zero.

IC3c. Elasticity of an Idealized Network of Molecular Chains

In order to achieve a true configurational elasticity in a given *sample* of a solid, however, the many chains must be crosslinked, i.e., they must have certain permanent points of juncture where the flexibility is cooperative. Moreover, flow must be eliminated completely. The crosslinking of linear, *cis*-polyisoprene, for example, gives the common vulcanized natural rubber. This becomes a three-dimensional network which, in the simplest case, can be treated in a similar manner as for a single idealized chain, except that this treatment has to be more generalized to include *cooperative* behavior in three dimensions. However, the crosslinks must be sufficiently rare so that the "chainlets" between the juncture points contain a sufficiently large number of segments to exhibit flexibilities which are reminiscent of our former case. For example, not too much chemically reactive sulfur should be incorporated into the vulcanizate or else a glassy ebonite, rather than a soft rubber, will be obtained. (See Section IE.)

A sphere of isotropic rubber is defined by the equation, $r_0^2 = x_0^2 + y_0^2 + z_0^2$. The sphere is now stretched along its x-axis to give the ellipsoid, $r_1^2 = x_1^2 + y_1^2 + z_1^2$, so that $x_1 = \lambda x_0$, $y_1 = y_0/(\lambda)^{1/2}$, and $z_1 = z_0/(\lambda)^{1/2}$, with a consequent conservation of volume. On treating the rubber sample as a hypothetical aggregate of freely orienting chains, each of which displays the properties of eq. (22) in three dimensions, the work done on each chain is

$$W_c = \int_{x_0}^{x_1} f_x\,dx + \int_{y_0}^{y_1} f_y\,dy + \int_{z_0}^{z_1} f_z\,dz$$

where $f_x = 3kTx/na^2$, etc. This becomes (10)

$$W_c = [3kT/(2na^2)][(\lambda^2 - \lambda^{-1})x_0^2 + (\lambda^{-1} - 1)r_0^2]$$

On multiplying by eq. (21) and integrating over all values of x_0, y_0, and z_0, the averaged work on the total number of chains, N, becomes

$$W_N = (NkT/2)[\lambda^2 - \lambda^{-1} + 3(\lambda^{-1} - 1)]$$

The differentiation, $\mathcal{S}_x = dW_N/Vd\lambda$, gives the principal tensile stress (48):

$$\mathcal{S}_x = NkT(\lambda - \lambda^{-2})/V \qquad \text{IC3-(23)}$$

which is somewhat like an equation of state for a gas. The shearing stress in the limit of an infinitesimal shear is

$$\mathcal{S}_{xy} = NkT\gamma_{xy}/V \qquad \text{IC3-(24)}$$

where NkT/V is the shear modulus G, and Young's modulus E becomes

$$E = 3G = 3NkT/V \qquad \text{IC3-(25)}$$

IC3d. Behavior of Real Rubberlike Solids

The deviations of real rubberlike solids from the behavior described by the highly oversimplified approach employed in the above treatment are plentiful indeed, and they have been discussed elsewhere (10,14,48,49) in sufficiently comprehensive detail so as to be too lengthy for inclusion here in view of the purposes of our present introductory perspective. Some of these will be considered by Tobolsky, Section IE, and Landel and Fedors, Section IIIB. There are two major areas in regard to real rubberlike solids, however, which need to be touched upon here in view of their direct pertinence in this volume. These relate to the phenomena met at extremely high elongations and the influence of time in the overly idealized picture which has been presented thus far.

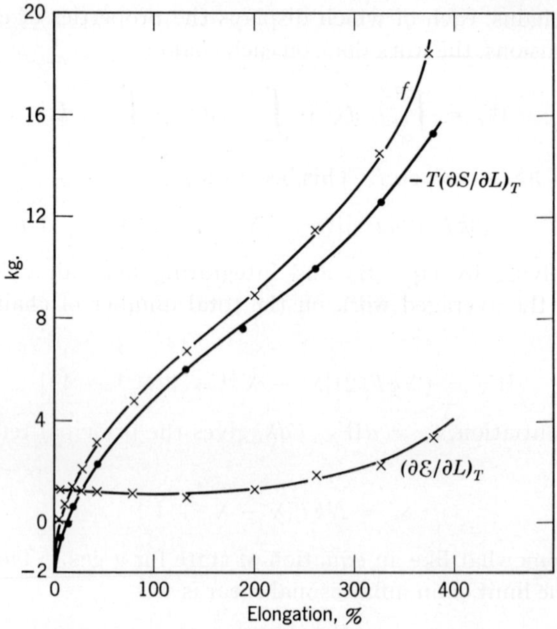

Fig. IC-8. Changes in internal energy $\mathcal{E}$ and entropy S with the uniaxial extension of rubber at 20°C. After Anthony et al. (5).

Figure IC-8 shows a tension–elongation curve for a typical vulcanized rubber when dissected into its entropy and internal energy contributions. From this, it will be appreciated that our previous discussions of the entropy alone must be weighted with a finite internal energy factor which is particularly important at the lowest elongations. Furthermore, the entropy consideration goes beyond the hookean limit of the rubber without necessitating an irreversible deformation. The "strain-hardening" of the rubber is lost on strain recovery. Crystallization on this pronounced entropy-lowering may occur at elongations approximating 600%, although this is not an absolute necessity for all rubberlike solids since this would depend on the molecular *favorability* for crystallization.

The advent of a finite internal energy contribution implies that the volume is not wholly conserved and $\mu < {}^1/_2$. The heat evolved on the configurational entropy-lowering, as well as that evolved on the highly anisotropic, molecular crystallization, short of flow contributions or fatigue, are regained on the release of the applied tension, and a full strain recovery is achieved. The internal energy contribution also suggests a small role for classical solid behavior here, with its consequent heat absorption on reversible dilatations. This reflects some primary bond stretching or bending which *detract* from the randomized "flexibility" previously employed in the intentional overgeneralization.

In real rubberlike solids, the stress-strain curves also have a finite dependency on *rate* of strain which will be discussed in detail by Landel and Fedors in Section IIIB. This dependency is due to the

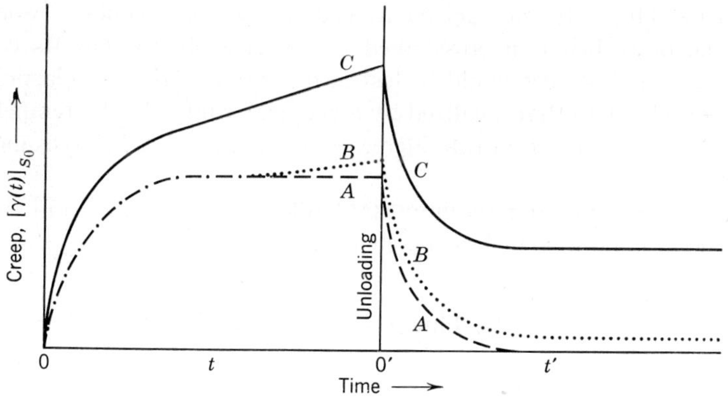

Fig. IC-9. Schematic, comparative tensile creep and creep-recovery plots for three rubbers of different molecular constitution: (*A*) lightly crosslinked chains; (*B*) long linear chains experiencing temporary entanglements; (*C*) short linear chains with a pronounced flow contribution.

time required for the stress-biased, thermally activated self-diffusion over the remarkably small, although existing, energy barriers of secondary or interchain bonding, especially at the network junctions. (See Subsection IC4c.) This is also reflected in the creep behavior in tension or shear schematically illustrated in Figure IC-9. As shown in Curve *A*, however, a crosslinked rubber attains an equilibrium or residual creep strain unless there is evidence of simultaneous chain degradation aggravated by the stress, concurrent chemical reaction with ambient gases, or both. Generally, however, the release of load also generates a time-dependent recovery. In other words, the crosslinks of the three-dimensional network can act as a fully reliable aspect of the network's structural memory, hence as the origin of its structural anelasticity.

Curves *B* and *C* of Figure IC-9 represent the generalized creep character of uncrosslinked, i.e., linear-chained, rubbers such as unvulcanized varieties. The chains can flow because of the absence of crosslinks. However, as these chains increase in length there is increasing evidence of temporary chain entanglements which greatly retard the onset of flow process in the sense of "temporary crosslinks." On rapid stressing, these three rubbers would behave similarly, however their observed elastic moduli would depend strongly on the frequency or "density" of the crosslinking (14,44) or entanglements (10). In fact, as mentioned before the "rubber" would become quite brittle if crosslinked too extensively so that its configurational elasticity would be lost in favor of a variety of elasticity which is closer to that idealized by a classical solid. As the temperature decreases, or strain-rate increases, the real rubber behaves more brittly.

Let us now consider the important roles of temperature and time in general polymeric bodies somewhat more carefully.

IC4. Linear Viscoelasticity

IC4a. Preliminary Notions

While the onset of inelasticity in a classical plastic body requires a critical shearing stress, the inelasticity of a classical liquid is experienced as long as a finite shearing stress persists. As the viscosity of a *real* liquid increases (or as its shear becomes increasingly dependent on a lowering of the configurational entropy), a distortion which involves flow alone can be experienced only at decreasing rates of strain. At shear-rates which are too high to dissipate efficiently the potential energy contribution of the imposed shearing forces, short of inertial effects (27), a real liquid exhibits a character of deformation which is *concurrently* elastic and inelastic, or simply "viscoelastic."

This viscoelastic body has not now evolved into a formal plastic solid since it still can deform time-dependently at any given level of shearing stress. Although now possessing a time- or temperature-dependent rigidity, this body is not necessarily nonnewtonian since, aside from its concurrent elastic character, its inelastic contribution might still be viewed within that classical perspective of liquidlike behavior.

When a body of finite rigidity experiences a concurrent newtonian contribution or, more generally, when its moduli or compliances are functions of time alone, then the characterization of its deformation falls within the *mechanical framework* of "linear viscoelasticity." This also implies that its strains in response to stresses can be quantitatively characterized by a given assembly of simple hookean and newtonian contributions. However, if these elastic and inelastic deformations must, at least in small part, be likened to the properties of a formal plastic body or nonnewtonian body, etc., then its viscoelasticity is said to be "nonlinear." (See Subsection IC5a.)

Aside from the modifications in notations and approaches which are required to depict the dependencies on time, the relationships between stresses and infinitesimal strains of a linear viscoelastic body can be treated in a manner which is quite analogous to that employed for classical solids (2,19,27). It is true, however, that new formalisms must appear.

There are several systems of notation for representing linear viscoelasticity (Subsection IC4h), each having certain advantages and disadvantages in the application to given mechanical problems. Since we are attempting to outline the conceptual trends in a given area of study, it seems highly proper that we employ the system of representation which is most frequently used by investigators of that area.

Primarily, however, we do not wish to relegate the reader's attentions to highly formalistic approaches which are treated far more properly in standard texts (9,19), but to bring an immediate focus on phenomenological situations, the *interrelations* between which can be achieved by following well-trodden, hence formalistic, pathways. Here, too, our approach remains overgeneralized within the need for brevity and a sufficient conceptual continuity; adequate references being provided for the fully required, higher levels of clarification.

The infinitesimal deformation of a macroscopically isotropic and continuous, linear viscoelastic body can be characterized by the *time-dependent* tensile and shear moduli, $E(t)$ and $G(t)$, and by their compliances, $D(t)$ and $J(t)$, respectively. The results of stress-relaxation experiments at constant strain are frequently expressed in terms of the moduli, while those for creep under constant stress are denoted by the compliances. For example, the *relaxation modulus* in uniaxial tension can be represented by (13)

$$E(t) = [\mathcal{S}(t)]_{\gamma_0}/\gamma_0 \qquad \text{IC4-(26)}$$

while the *creep compliance* in simple shear becomes

$$J(t) = [\gamma(t)]_{\mathcal{S}_0}/\mathcal{S}_0 \qquad \text{IC4-(27)}$$

Hence these relaxation moduli or creep compliances are the *entire time plots* of the appropriate stress(t)/(fixed strain), or strain(t)/(fixed stress) ratios. With these kinds of designations it is generally true that $G(t) \neq 1/J(t)$ for relaxation and creep in shear, since the loading patterns differ in these different types of experiments. (The moduli and compliances become negative in compression.)

The reader will take note that this particular variety of notation places an emphasis on *how* the moduli or compliances are obtained. This is a natural consequence of the time, hence of the mechanism, for that deformation; a consideration which is entirely unnecessary in time-independent, classical elastic notation.

Both relaxation and creep usually involve irrecoverable strains. In a *linear* viscoelastic body, this flow contribution *must* be characterized solely by newtonian processes. For the case of uniaxial tensile creep, for example, the compliance can be expressed as

$$D(t) = [\gamma(t)]_{\mathcal{S}_0}/\mathcal{S}_0 = D_R(t) + D_I(t) = D_R(t) + (t/\eta^*) \qquad \text{IC4-(28)}$$

where $D_I(t)$ is the irrecoverable *portion* of the creep compliance due to the newtonian flow contribution, $\gamma/\mathcal{S}_0 = t/\eta^*$; cf. eq. (10), where η^* is the coefficient of viscosity in uniaxial tension (i.e., in the shear components of this applied tension).

The remarkably larger order of viscoelastic contribution exhibited by the "leathery" and "rubbery" polymeric solids necessitates the use of log plots of the moduli or compliances. Thus, a steep slope in the *arithmetic* creep strain may become an inflection with a *logarithmic* compliance, etc. (This is illustrated in the following discussion.) It should be made quite clear, however, that empirical representations in the regions of *finite* strains, while helpfully descriptive, are far less useful within the presently employed framework of their classical mechanical treatments. (See Subsection IC5a.)

IC4b. Examples of Behavior

Figure IC-10 is a plot of the stress relaxation modulus in uniaxial tension for an amorphous, linear polystyrene as a function of temperature. $E(t)$ was evaluated here at $E(5)$, $E(10)$, $E(100)$, and $E(1000)$, where t is in seconds. This implies that four readings of the

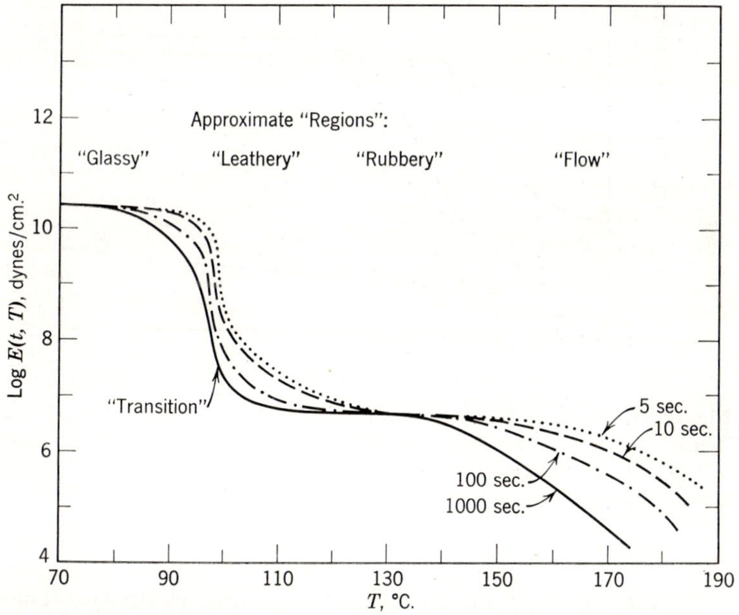

Fig. IC-10. The relaxation moduli of amorphous polystyrene in uniaxial tension at four different times of observation as a function of temperature. In modification after Tobolsky (48).

residual stress were periodically recorded for a given sample of polystyrene (subjected to a known and fixed, small strain within its macroscopically "linear" viscoelastic response) at each temperature; a new sample being required for each temperature (due to residual strains).

This body exhibits two temperature ranges over which this relaxation modulus is sensibly independent of these *times* of observation. Hence, below 80°C. the amorphous body is glasslike and "elastic" in the "classical solid" sense, while at approximately 130°C., the poly-

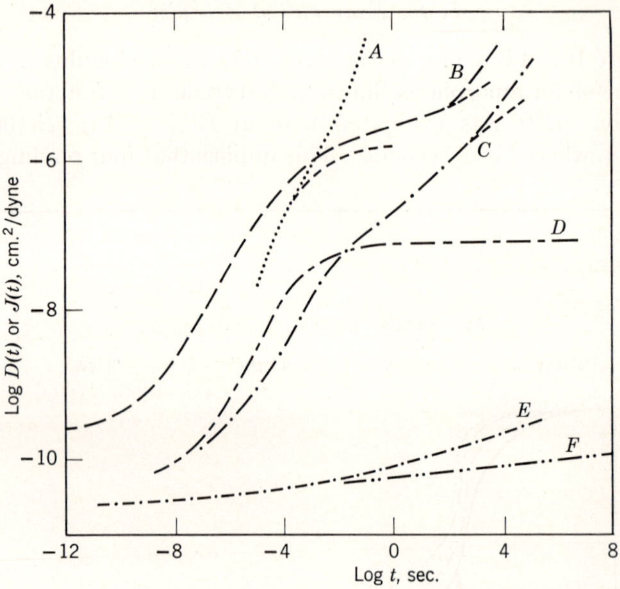

Fig. IC-11. The time-dependent creep compliances of various polymeric bodies in tension or shear. The short dashed lines show the effect of subtracting the flow. *A:* Linear polymer (short-chained). *B:* Branched polymer. *C:* Linear polymer (long-chained). *D:* Crosslinked rubber. *E:* "Highly" crystalline polymer. *F:* Glassy polymer. In modification after Ferry (13).

styrene exhibits its configurational or rubberlike elasticity. The relaxation moduli in these two temperature-defined, elastic zones differ by well over three orders of magnitude.

All other portions of these curves are strongly time-dependent, hence "highly viscoelastic." The "temperature" (i.e. the time-dependent temperature range) for the fairly abrupt loss of glasslike properties is the T_g. Under these particular times for observation, $T_g \cong 97 \pm 2°C$. This T_g implies the thermal energy level for an empirically sufficient onset of intersegmental diffusion; hence that temperature for a "significant" intermolecular or interconfigurational diffusion.

Between 97 and 120°C., the polystyrene exhibits an amazingly versatile combination of classical solidlike and configurational elasticities coupled with a time-dependent toughness derived from a pronounced structural anelastic contribution. In this temperature range the sample is "leatherlike."

In the rubbery temperature (region in the "plateau zone") its anelastic contribution is (within these experimental limits of time) time-independent, while at higher temperatures, those chain entanglements which are responsible for its three-dimensional structural memory are gradually lost in favor of a time-dependent stress relaxation. (See Section IE for a further clarification and analysis of these relaxation moduli.)

Figure IC-11 shows an approximate plot of the creep compliances in uniaxial tension or shear (within a relatively minor factor of 2–3, these are the same on this rough log compliance plot) for several kinds of polymeric bodies as a function of log time. The dashed lines indicate the results of subtracting the contributions to newtonian flow, see eq. (28).

The glassy and semicrystalline polymeric bodies, Curves *F* and *E*, exhibit a much lower order of creep than experienced by the other polymeric bodies shown. The crosslinked rubber, Curve *D*, appears to have a residual compliance (i.e., it would tend to support a load indefinitely; cf. Fig. IC-9) while the linear and branched chains (with no structural memory other than that derived from their temporary chain entanglements) eventually flow towards infinite strains under the fixed loads imposed.

The log time values for the most pronounced enhancement of the creep of Curves *A–D*, may be viewed as a *time-defined* T_g or "transition zone" for those samples. Thus there is a certain order of interplay between time and temperature in viscoelastic bodies which is discussed further in Subsection IC4e. The time-*spread* for this marked inelastic contribution is a reflection of the distribution of times required for the many modes of creep retardation such as those discussed in Subsection IC4c.

Aside from variations of the viscoelastic contributions of polymeric bodies within the factors of time, temperature, molecular architecture, and "state," such as briefly alluded to above in connection with Figures IC-10 and IC-11, another factor to consider is the concentration of fluid solute (i.e., diluent) in the amorphous zones. As mentioned in Subsection IC2b, the dilatation of the bulk (in this particular sense, a swelling) enhances the softness of the amorphous body. One perspective of this is provided in Figure IC-12, where the T_g of poly-

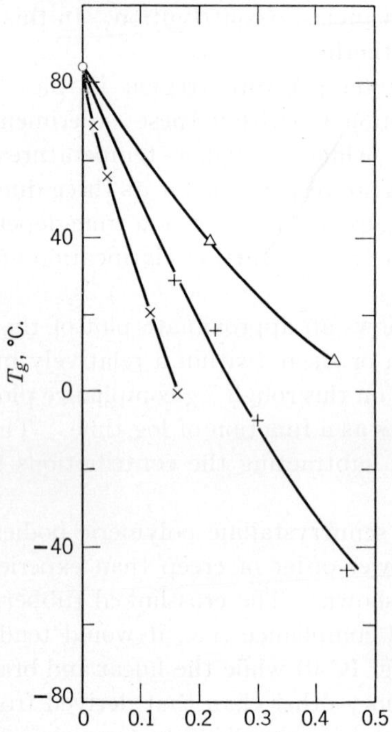

Fig. IC-12. The lowering of the glass temperature of polystyrene by absorption of various liquids as a function of weight fraction of diluent. (△) β-Naphthyl salicylate. (+) Nitrobenzene. (×) Carbon disulfide. In modification after Jenckel and Heusch (25).

styrene is markedly lowered by added solute. This should be regarded as a true *ab*sorption in the sense of an admixture of the liquids with the polymeric molecules in the amorphous zones. This is a method for lowering the effective secondary bonding energy and enhancing the intersegmental diffusion (1).

For the special cases of hydrophilic polymeric bodies such as the proteins, poly(vinyl acetate), nylon, and cellulose and its chemical derivatives, even an increased *humidity* can play an important role in this softening action.

Thus the viscoelastic contribution of those polymeric solids having an amorphous composition are hypersensitively dependent on their vapor environment as well as on their thermal environment. In fact, the effective brittle–ductile transitions of polymeric bodies are vapor pressure- as well as temperature- and time-dependent.

IC4c. Configurational Anelasticity

After outlining the topic of configurational elasticity of an idealized network of flexible molecules, we mentioned in Subsection IC3d that time, quite *unideally*, played a significant role in the distortion of real rubbers. This generated a *retarded* elastic contribution in the sense of a "configurational anelasticity." This would immediately suggest that time and shear strain energy aid in enhancing the probability for successful fluctuations in thermal energy toward achieving a stress-biased interconfigurational diffusion of the chainlets between the network junctions or crosslinks.

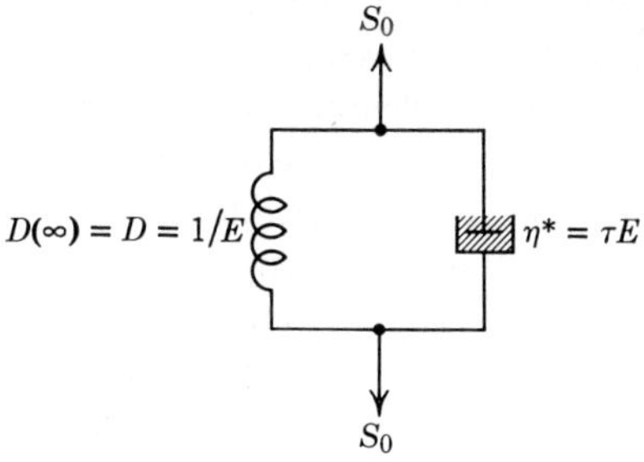

Fig. IC-13. A Kelvin-Voigt mechanical model in uniaxial tension.

The concept of a tangible frictional drag added to an otherwise time-independent, configurational elastic process has encouraged the introduction of the Kelvin-Voigt mechanical model (Fig. IC-13) by way of descriptive analogy to a rubber's behavior in tensile creep. The spring is to represent the *configurational*, yet *hookean* (see eqs. 22 and 25) elasticity of the network under small strains in uniaxial tension, while the dashpot is to depict a newtonian contribution to the internal frictional processes involved.

Since these two elements are to be permanently coupled in parallel, this model cannot be used to represent an irrecoverable strain. Moreover, each element must bear a portion of the tensile stress:

$$\mathcal{S} = (\gamma/D) + \eta^*\dot{\gamma} \qquad \text{IC4-(29)}$$

where D is the spring's tensile compliance and η^* is the coefficient of viscosity in tension, cf. eq. (28). A convenient solution of this equation at constant stress $\mathcal{S}_0$ is (9)

$$\gamma(t) = \mathcal{S}_0 D + \text{const. } \exp\{-t/\eta^* D\}$$

The constant here is equal to $-\mathcal{S}_0 D$ as appreciated on setting $t = 0$. The time-dependent compliance of the composite model is

$$D(t) = [\gamma(t)]_{\mathcal{S}_0}/\mathcal{S}_0 = D(\infty)\,(1 - \exp\{-t/\tau\}) \qquad \text{IC4-(30)}$$

where $\tau = \eta^* D$, is the "retardation time" of the tensile creep, and $D(\infty) = D$ is the equilibrium tensile compliance, since $(1 - \exp\{-t/\tau\})$ becomes unity at $t = \infty$. When $t = \tau$, $D(t) = 0.63D(\infty)$, and the creep strain of the model is just short of two-thirds completed. Since this is a *linear* viscoelastic model (i.e., a combination of hypothetical hookean and newtonian elements) these conditions are to be independent of the stress level experienced.

Unlike the behavior of a real rubber as illustrated in Figure IC-11D, however, a plot of eq. (30) would reveal that most of the creep of this simple model occurs over a single decade of time (in the region of τ). For example, from eq. (30) it can be appreciated that when t is altered from 0.10τ to 1.00τ, $D(t)$ changes from $0.10D(\infty)$ to $0.63D(\infty)$, which encompasses 53% of the total creep compliance. Since several decades of time are actually required to accomplish this percentage of creep in a real rubber, adequate attention should be given to the many possible *modes* of motion to be experienced by the chainlet and its associated internal frictions. Obviously, the shorter-ranged or less complex configurational modes are not expected to be as hindered as those of longer-range, so that we actually need an entire series of τ values to account for all the frictional drags on the interconfigurational diffusion of the average chainlet.

This kind of reasoning leads to a complex Kelvin-Voigt model representation as illustrated in Figure IC-14. This also can be expressed as a series of relations of the type given by eq. (30) in the sense:

$$D(t) = \sum^{i} D(\infty)_i\,(1 - \exp\{-t/\tau_i\}) \qquad \text{IC4-(31)}$$

Moreover, from eq. (25) we know something of the nature of $D(\infty)_i$ since, in the most ideal case, $D(\infty) = D = 1/E = V/(3NkT)$. In

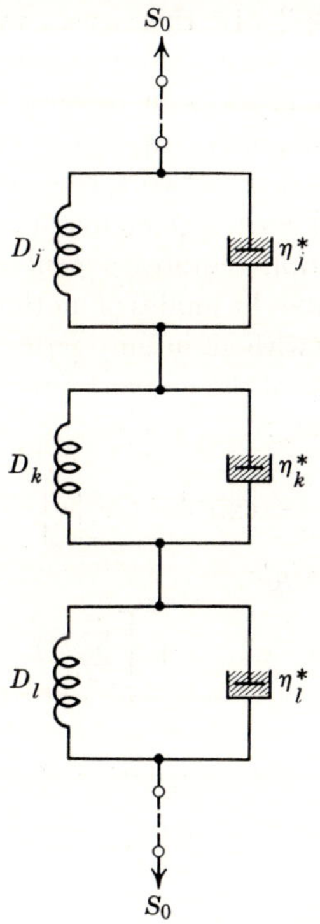

Fig. IC-14. A series of Kelvin-Voigt mechanical models in uniaxial tension.

addition, a further insight into τ_i can be obtained from eqs. (15), (17), and (25), since $\tau = \eta^*D \propto a^2f_0V/(3NkT)$.

However, while complex models referring to the modes of motion of the chainlets broaden the $D(t)$ versus ln t curve, they do this mainly toward the range of the *shorter* times, and thus they still do not include the most time-dependent frictional processes (10).

It was finally appreciated that the motions of the *network junctions*, not those of the chainlets, were the most retarded events. While this would appear to be entirely reasonable in view of the chain complexity in the regions of these junctions, their roles were initially ignored since they were not critical in the calculation of congfiurational elasticity itself, and only become critical in these *time-dependent* phenomena.

These junctions cannot move with the same agility as do the chainlets, and yet their motions condition the movements of the chainlets in a given "subnetwork." In this sense, we need to consider the longest of the long-range contributions, namely, the fullest participation of each of the subnetworks in bearing its share of the applied forces or, expressed differently: we need to consider the gradual migration of the macroscopic network towards its equilibrium distribution of network junctions in a given distortion.

This fuller consideration requires a generalized series of Kelvin-Voigt models that describe the modes of motion of the chainlets which, themselves, are in series with an infinite series of Kelvin-Voigt models that describe the motions of the network-junctions. This complexity leads to the expression (11):

$$D(t) = \underbrace{\left[\sum_{\text{odd } i}^{n} D(\infty)_i(1 - \exp\{-t/\tau_i\})\right]}_{\textit{Chainlets}} + \underbrace{\left[\sum_{i=1}^{\infty} D(\infty)_i(1 - \exp\{-t/\tau_i\})\right]}_{\textit{Junctures}}$$

where the compliances and retardation times for the chainlets are

$$D(\infty)_i = 8V/(3NkT\pi^2 i^2) \qquad \text{IC4-(32)}$$

and

$$\tau_i = a^2 f_0 n^2/(3kT\pi^2 i^2)$$

Those for the network junctions are:

$$D(\infty)_i = V/(3NkT\lambda i)$$

and

$$\tau_i = \tau_1 \lambda^{2i}$$

where λ is related to the specific geometry of the network junctures and τ_1 is the retardation time for the first mode of the chainlets' motions. These kinds of relations can adequately describe the creep of lightly crosslinked rubbers where $\lambda \cong 2$–3.

IC4d. Spectra and Interrelation of Time Constants

Time constants which characterize the internal frictions, or reflect the modes of delayed motions, usually are spaced sufficiently close in real solids to constitute continuous spectra. Thus, for example, the tensile creep compliance for pure anelastic deformation is more generally written:

$$D(t) = \int_0^\infty D(\infty,\tau)(1 - \exp\{-t/\tau\})d\tau \qquad \text{IC4-(33)}$$

where $D(\infty,\tau)$ is a continuous function of τ. However, since ln τ plots are commonly employed, eq. (33) is frequently expressed:

$$D(t) = \int_{-\infty}^\infty \tau D(\infty,\tau)(1 - \exp\{-t/\tau\})d \ln \tau \qquad \text{IC4-(34)}$$

where $\tau\, d \ln \tau = d\tau$. This also can be written (13)

$$D(t) = \int_{-\infty}^\infty L^*(\tau)(1 - \exp\{-t/\tau\})d \ln \tau \qquad \text{IC4-(35)}$$

where $L^*(\tau)$ can be regarded as the "strength" toward $D(\infty)$, which is contributed by the increment of retardation times $d \ln \tau$, or the relative importance of $d \ln \tau$ in the total distribution of τ values. $L^*(\tau)$ is also called the "spectrum of retardation times" (in tensile creep).

For the case of anelastic creep in simple shear one employs

$$J(t) = \int_{-\infty}^\infty L(\tau)(1 - \exp\{-t/\tau\})d \ln \tau \qquad \text{IC4-(36)}$$

where $\tau = \eta^* D(\infty) = \eta J(\infty)$. If the infinite Kelvin-Voigt model were modified to include contributions to flow and *un*retarded elasticity, a modification of eq. (36) can be employed:

$$J(t) = J(0) + (t/\eta) + \int_{-\infty}^\infty L(\tau)(1 - \exp\{-t/\tau\})\, d \ln \tau$$

The first and third terms of this equation contribute to the residual, yet fully recoverable, portion of the equilibrium compliance. Therefore, at $t = \infty$,

$$[J(\infty)]_R = J(0) + \int_{-\infty}^\infty L(\tau)\, d \ln \tau$$

where the τ-dependent contribution to this residual compliance is the total area under an *arithmetic* $L(\tau)$ versus ln τ plot.

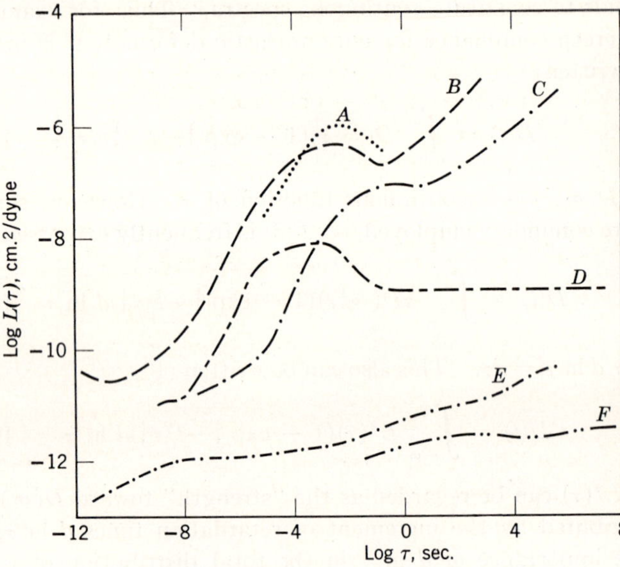

Fig. IC-15. The spectra of retardation times for the same polymeric bodies shown in Fig. IC-11. In modification after Ferry (13).

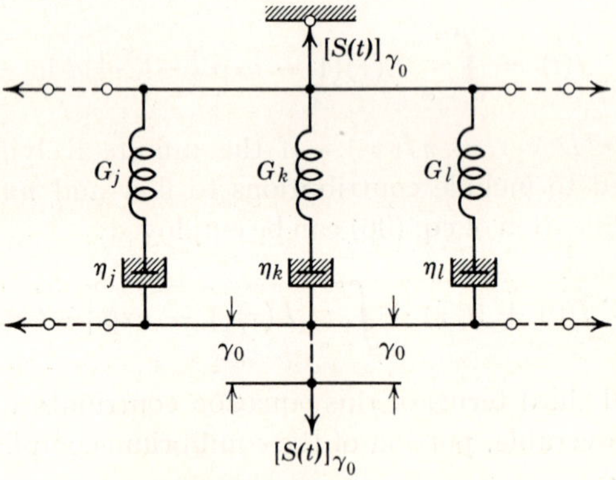

Fig. IC-16. A parallel arrangement of Maxwell mechanical models in shear.

Figure IC-15 shows the *log* $L(\tau)$ versus log τ plot for the same bodies characterized by log $J(t)$ versus log t in Figure IC-11. Note that the two figures are similar, but the flows are missing and the former plateau (rubbery) zones now initiate with maxima.

While Kelvin-Voigt models and the *retardation times* for their compliances are employed for characterizations of behavior in creep experiments, Maxwell models and the *relaxation times* for their moduli are used in stress relaxations. The Maxwell model also consists of hookean and newtonian elements, but in *series;* the complex model being a *parallel* arrangement of such models (Fig. IC-16).

On the application of a shear stress to a *single* Maxwell model, the rate of strain of the model is the sum of the strain-rates of each element so that,

$$\dot{\gamma} = (\dot{\mathcal{S}}/G) + (\mathcal{S}/\eta) \qquad \text{IC4-(37)}$$

This is equated to zero for relaxation at constant strain, and a solution is (9)

$$G(t) = [\mathcal{S}(t)]_{\gamma_0}/\gamma_0 = G(0) \exp\{-t/\tau'\} \qquad \text{IC4-(38)}$$

where $\tau' = \eta/G(0)$ is the relaxation time. At $t = \infty$, the residual stress $G(\infty) = 0$; while at $t = \tau'$, $G(t) = G(0)e^{-1}$, so that the stress has decayed to slightly more than a third of its original value, $G(0)$, by this time.

The complex model of Figure IC-16 can be represented by

$$G(t) = \sum^{i} G(0)_i \exp\{-t/\tau'_i\} \qquad \text{IC4-(39)}$$

which, for a continuous distribution of τ'_i gives:

$$G(t) = \int_{-\infty}^{\infty} H(\tau) \exp\{-t/\tau'\} d \ln \tau' \qquad \text{IC4-(40)}$$

where $H(\tau) = \tau' G(0,\tau') =$ the "spectrum of relaxation times." Further characterizations of $G(t)$ are provided in Section IE.

Equation (38) will, in general, give the same kind of $G(t)/G(0)$ function as eq. (39) except that the major portion of the stress decay for the more realistic model usually requires several decades of time. Since $G(t)$ is known for only an experimental time interval, the integral of eq. (40) is not readily evaluated. Thus several kinds of approximations of this equation are employed, the simplest of which assigns a linear value of $H(\tau)$ to approximate the experimental observation. For example, let the term $\exp\{-t/\tau'\}$ be a step-function so that it is unity for $t < \tau'$ and zero for $t > \tau'$. Equation (40) becomes

$$G(t) = \int_{t}^{\infty} H(\tau)\, d \ln \tau'$$

which on differentiation becomes (3)

$$H(\tau) \cong -[dG(t)/d \ln t]_{t=\tau'}$$

Thus, $H(\tau)$ is evaluated from the observed slope of the $G(t)$ versus $\ln t$ plot in the region of τ'. By an analogous procedure employing eq. (36) one obtains

$$J(\tau) \cong [dJ(t)/d \ln t]_{t=\tau}$$

Higher order approximations are frequently desirable and the following have been employed with greater suitability (43):

$$H(\tau) = -\{[dG(t)/d \ln t] - [d^2G(t)/d (\ln t)^2]\}_{t=2\tau'}$$

$$L(\tau) = \{[dJ(t)/d \ln t] - [d^2J(t)/d (\ln t)^2]\}_{t=2\tau}$$

Obviously, there is a great advantage to be gained if one can interrelate the spectra of time constants from *different* experimental approaches. Any attempt in this direction, however, must be conditioned by assumptions related to the different loading patterns. Boltzmann's superposition principle, to be discussed later (Subsection IC4f), is generally employed. The exact relationships between $H(\tau)$ and $J(\tau)$ have been found to be (19):

$$L(\tau) = \frac{H(\tau)}{\left[\int_{-\infty}^{\infty} H(\theta)\, d \ln \theta/(1 - \theta/\tau')\right]^2 + \pi^2 H^2(\tau)}$$

$$H(\tau) = \frac{L(\tau)}{\left[\int_{-\infty}^{\infty} L(\theta)\, d \ln \theta/(1 - \theta/\tau)\right]^2 + \pi^2 L^2(\tau)}$$

from which approximate methods for these transformations have been derived under special circumstances. For example, in the vicinity of T_g, $H(\tau) \propto (\tau')^{-m}$ and $L(\tau) \propto \tau^m$, while $m \cong 1/2$. Under these conditions we can employ (13)

$$H(\tau)\, L(\tau) \cong 1/\pi^2$$

When $m \ll 1$, as in the prolonged rubbery plateau zone, $H(\tau)L(\tau) \cong m^2$.

Of more fundamental importance than this, however, is the need to define these spectra in terms of *molecular* parameters. With simple chainlike, amorphous solids this frequently can be done as implied in the previous subsection.

IC4e. Time–Temperature Superposition

Since the time constants for mechanical response are primarily derived from internal friction processes, there necessarily are direct connections between these delays and the probability-level for successful fluctuations in thermal energy. Furthermore, it is clear that these interrelationships can be stated in terms of "time-temperature factors," hence they can be employed to predict the order of inelastic behavior at other times or temperatures. With *linear* viscoelastic responses, this becomes a particularly useful consideration.

Since η = (modulus)τ, for two different temperatures T and T_0 we may write (48):

$$\tau(T)/\tau(T_0) = \eta(T)/\eta(T_0) = \kappa \qquad \text{IC4-(41)}$$

where $\kappa < 1$ for $T > T_0$. Since the moduli also are strongly temperature-dependent, $\kappa = f(T,T_0)$.

If a creep experiment were performed with a single Kelvin-Voigt model which had a retardation time $\tau(T)$ at a lower temperature T_0, the retardation time would be larger, $\tau(T)/\kappa$, since $\tau(T) = \kappa\tau(T_0)$. This is further illustrated by the following relations:

$$[J(t)]_T = [J(t/\kappa)]_{T_0} = [J(\infty)(1 - \exp\{-t/\kappa\tau\})]_{T_0}$$

In other words, an experimental observation at time t with temperature T is equivalent to an observation at time t/κ with temperature T_0. For the converse experimental situation we can write

$$[J(t)]_{T_0} = [J(\kappa t)]_T$$

A slightly more refined version of this same approach can be expressed (13)

$$\rho_0 T_0 \tau(T)/\rho T \tau(T_0) = \kappa' = a_T \qquad \text{IC4-(42)}$$

where ρ is the solid's density. T_0 is frequently taken as T_g. The use of a_T is further illustrated in Figure IC-17. Landel and Fedors, Section IIIB, will later attempt to treat its empirical use in connection with the prediction of the rupture times of amorphous solids at various temperatures and strain rates despite the clear "nonlinear" situation

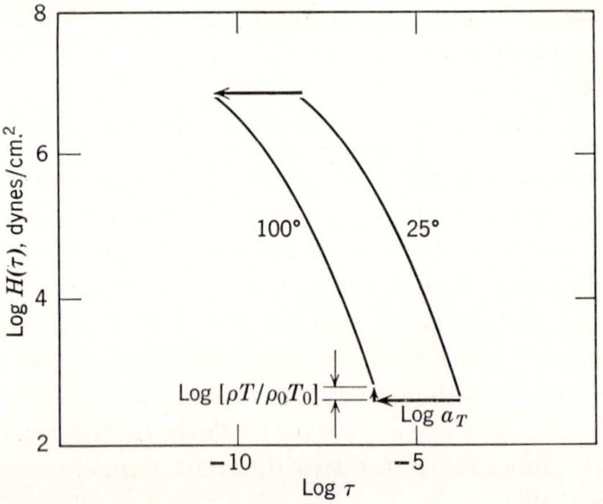

Fig. IC-17. Effect of change in temperature from T_0 to T on the relaxation spectrum of a low molecular weight polyisobutylene. An illustration of eq. (42). After Ferry (13).

there. A further discussion of time–temperature superposition is also given in Section IE.

IC4f. Transient Stressing or Straining Patterns

Up to this point, we have considered only patterns of stressing or straining which were "statically" imposed, i.e., at fixed stresses or fixed strains at $t = 0$. Since hookean springs and newtonian dashpots strain in a reproducible, hence in a predictable manner in response to a given stressing pattern, howsoever complex, within the idealized mechanical framework of linear viscoelastic behavior we are able to employ the *principle of superposition* (19), after Boltzmann, which is briefly illustrated among the following remarks.

Consider a relaxation experiment in shear which is quantitatively characterized by a *single* Maxwell model of Figure IC-14. However, let this model be strained *progressively* at times t_1, t_2, t_3, . . ., t_n, so that the increasing strains, γ_1, γ_2, γ_3, . . ., γ_n, are experienced for times $(t - t_1)$, $(t - t_2)$, $(t - t_3)$, . . ., $(t - t_n)$, respectively. In view of eq. (38), the stress history that is required to accomplish this is expressed

$$\mathcal{S}(t) = G(0) \sum_{i=1}^{i=n} \gamma_i \exp\left\{-(t - t_i)/\tau'\right\}$$

If the straining pattern is far more complex than this, yet is continuous in the sense that the slopes $d\gamma_i/dt_i$ describe a smooth curve, then we can employ

$$\mathcal{S}(t) = G(0) \int_0^t [d\gamma(\theta)/d\theta] \exp \{(\theta - t)/\tau'\} d\theta \quad \text{IC4-(43)}$$

Let us now perform a stress–strain experiment on the Maxwell model at a *constant rate* of strain. Since this may be regarded as a *relaxation* experiment which is performed under a known, transient pattern of straining, eq. (43) can be employed directly. Setting $d\gamma(\theta)/d\theta = \dot{\gamma}_{\text{const}}$, and multiplying by τ'/τ' we have

$$\mathcal{S}(t) = \dot{\gamma}_{\text{const}}\, \tau' G(0) \int_0^t \exp \{(\theta - t)/\tau'\} d(\theta/\tau')$$

By adjusting the limits we can write

$$\mathcal{S}(t) = \dot{\gamma}_{\text{const}}\, \tau' G(0) \int_{-t/\tau'}^0 \exp \{(\theta - t)/\tau'\} d[(\theta - t)/\tau']$$

which gives

$$\mathcal{S}(t) = \dot{\gamma}_{\text{const}}\, \tau' G(0)(1 - \exp \{-t/\tau'\}) \quad \text{IC4-(44)}$$

Equation (44) is also the solution of eq. (37) where $\dot{\gamma} = \text{const.}$ By straining the *complex* Maxwell model with a continuous distribution of τ_i' values one obtains (45)

$$\mathcal{S}(t) = \dot{\gamma}_{\text{const}} \int_0^\infty H(\tau)(1 - \exp \{-t/\tau'\}) d\tau'$$

For the continuous tensile *stressing* of a single Kelvin-Voigt model we have by comparison of eqs. (30) and (43),

$$\gamma(t) = D(\infty) \int_0^t [d\mathcal{S}(\theta)/d\theta][1 - \exp \{(\theta - t)/\tau\}] d\theta \quad \text{IC4-(45)}$$

which, by the same procedure as above, becomes

$$\gamma(t) = \dot{\mathcal{S}}_{\text{const}}\, D(\infty)[t - \tau(1 - \exp \{-t/\tau\})] \quad \text{IC4-(46)}$$

for a stress–strain experiment at a *fixed* rate of stressing. The continuous general model then experiences (13)

$$\gamma(t) = \dot{\mathcal{S}}_{\text{const}} \int_{-\infty}^{\infty} L^*(\tau)[t - \tau(1 - \exp \{-t/\tau\})] d \ln \tau$$

We have attempted to illustrate that the Boltzmann superposition principle, by providing relations in the form of eqs. (43) and (45), can be useful in obtaining solutions such as in eqs. (44) and (46), when the transient patterns of stressing or straining are known.

IC4g. Dynamic Patterns

It also is quite common to employ oscillating stresses as a loading pattern. In this case the viscoelastic body experiences an *acceleration* as well as being strained elastically and inelastically.

If $\mathcal{S}(\omega)$ and $\gamma(\omega)$ are the oscillating stresses and strains, and η' and $G'(\omega)$ are the dynamic coefficient of viscosity and dynamic shear modulus, then (48)

$$\mathcal{S}(\omega) = M\ddot{\gamma}(\omega) + \eta'\dot{\gamma}(\omega) + G'(\omega)\gamma(\omega) \qquad \text{IC4-(47)}$$

where M is the mass of a unit area upon which the oscillating shear stress acts. Let $\mathcal{S}(\omega) = \mathcal{S}_0 \exp\{i\omega t\}$ and $\gamma(\omega) = \gamma_0 \exp\{i\omega t\}$, where $\exp\{i\omega t\} = \cos \omega t + i \sin \omega t$, and the stress in eq. (47) is the real part of $\mathcal{S}(\omega)$. The strain rate, $\dot{\gamma}(\omega) = i\omega\gamma_0 \exp\{i\omega t\} = i\omega\gamma(\omega)$, is 90° out of phase with respect to both $\gamma(\omega)$ and $\ddot{\gamma}(\omega) = -\omega^2\gamma(\omega)$, and eq. (47) can be written as

$$\mathcal{S}(\omega) = [-M\omega^2 + iG''(\omega) + G'(\omega)]\gamma(\omega) \qquad \text{IC4-(48)}$$

where

$$G''(\omega) = \omega\eta'$$

Thus, $G''(\omega)$ is the 90° out-of-phase shear modulus ("loss modulus") and the dynamic shear modulus, $G'(\omega)$, is frequently termed the "storage modulus" in shear.

Because of losses in strain energy on heat dissipation, $\gamma(\omega)$ will generally not be in phase with $\mathcal{S}(\omega)$ and the phase angle of the lag of the strain with respect to the stress is called the "loss angle" δ. A solution of eq. (48) can be expressed

$$\gamma(\omega)G^*(\omega) = \mathcal{S}_0 \cos(\omega t - \delta)$$

where $G^*(\omega)$ is the complex shear modulus which has the magnitude

$$|G^*(\omega)| = \mathcal{S}(\omega)/\gamma(\omega) = \mathcal{S}_0/\gamma_0 = \{[G'(\omega) - M\omega^2]^2 + [G''(\omega)]^2\}^{1/2} \qquad \text{IC4-(49)}$$

and its direction in phase space is defined by the "loss tangent," $\tan \delta$:

$$\tan \delta = G''(\omega)/[G'(\omega) - M\omega^2] \qquad \text{IC4-(50)}$$

Let us now consider the behavior of a *weightless*, single Maxwell model in response to an oscillating shear stress of the same type as above. In view of eq. (37) we can write

$$\dot{\gamma}(\omega) = [\dot{\mathcal{S}}(\omega)/G(\omega)] + [\mathcal{S}(\omega)/\tau' G(\omega)] \qquad \text{IC4-(51)}$$

This also can be written

$$i\omega\gamma(\omega) = [i\omega\mathcal{S}(\omega)/G(\omega)] + [\mathcal{S}(\omega)/\tau' G(\omega)] \qquad \text{IC4-(52)}$$

From eq. (52) we can define a complex shear modulus by

$$G^*(\omega) = \mathcal{S}(\omega)/\gamma(\omega) = iG(\omega)\omega\tau'/(1 + i\omega\tau') \qquad \text{IC4-(53)}$$

By multiplying eq. (53) by $(1 - i\omega\tau')/(1 - i\omega\tau')$ we obtain

$$G^*(\omega) = [G(\omega)\omega^2\tau'^2/(1 + \omega^2\tau'^2)] + i[G(\omega)\omega\tau'/(1 + \omega^2\tau'^2)] \qquad \text{IC4-(54)}$$

where the complex shear modulus is separated into its real and imaginary parts. Equation (54) can also be written as,

$$G^*(\omega) = G'(\omega) + iG''(\omega)$$

where

$$G'(\omega) = G(\omega)\omega^2\tau'^2/(1 + \omega^2\tau'^2)$$

and

$$G''(\omega) = G(\omega)\omega\tau'/(1 + \omega^2\tau'^2)$$

so that

$$G' = \omega\tau' G''(\omega)$$

which gives

$$\tan \delta = G''(\omega)/G'(\omega) = 1/\omega\tau'$$

For a number of Maxwell models in parallel we can, for example, write for $G'(\omega)$

$$G'(\omega) = \sum^{j} G(\omega)_j\ \omega^2\tau_j'^2/(1 + \omega^2\tau'^2_j)$$

while for a continuous distribution of τ'_j values we have

$$G'(\omega) = \int_{-\infty}^{\infty} [H(\tau)\omega^2\tau'^2/(1 + \omega^2\tau'^2)]d \ln \tau'$$

and

$$G''(\omega) = \int_{-\infty}^{\infty} [H(\tau)\omega\tau'/(1 + \omega^2\tau'^2)]d \ln \tau' \qquad \text{IC4-(55)}$$

where $H(\tau) = \tau' G(\omega,\tau')$. It is clear that losses by heat dissipation will be maximized when $\tau' = 1/\omega$. Here, $\omega\tau'/(1 + \omega^2\tau'^2) = {}^1/_2$, and eq. (55) becomes (13)

$$G''(\omega) \cong (\pi/2)[H(\tau)]_{\tau'=1/\omega}$$

For a weightless, single Kelvin-Voigt model, we have from equation (29),

$$\mathcal{S}(\omega) = [\gamma(\omega)/J(\omega)] + [\tau\dot{\gamma}(\omega)/J(\omega)] \qquad \text{IC4-(56)}$$

for the case of oscillations in shear. Since $\dot{\gamma}(\omega) = i\omega\gamma(\omega)$, it is clear that

$$J^*(\omega) = \gamma(\omega)/\mathcal{S}(\omega) = J(\omega)/(1 + i\omega\tau) \qquad \text{IC4-(57)}$$

Multiplying eq. (57) by $(1 - i\omega\tau)/(1 - i\omega\tau)$ as before, we have

$$J^*(\omega) = [J(\omega)/(1 + \omega^2\tau^2)] - i[J(\omega)\omega\tau/(1 + \omega^2\tau^2)] \qquad \text{IC4-(58)}$$

and

$$\tan \delta = J''(\omega)/J'(\omega) = \omega\tau$$

For a continuous distribution of τ_j values we have from eq. (58)

$$J'(\omega) = \int_{-\infty}^{\infty} [L(\tau)/(1 + \omega^2\tau^2)]d \ln \tau$$

and

$$J''(\omega) = \int_{-\infty}^{\infty} [L(\tau)\omega\tau/(1 + \omega^2\tau^2)]d \ln \tau$$

Plots of log $J'(\omega)$ versus log $(1/\omega)$ bear a good resemblance to plots of log $J(t)$ versus log t as shown in Figure IC-11, providing that they are compared at the same stress amplitude. However, plots of log

$J''(\omega)$ versus log $(1/\omega)$ differ from these, especially in the range of $\tau = 1/\omega$ where the glassy-leathery transitions appear as maxima. When the logarithm of the ratio, $J''(\omega)/J'(\omega) = \tan \delta$, is plotted against log ω, as shown in Figure IC-18 for the same solids shown in Figure IC-11, the relative behaviors contrast considerably.

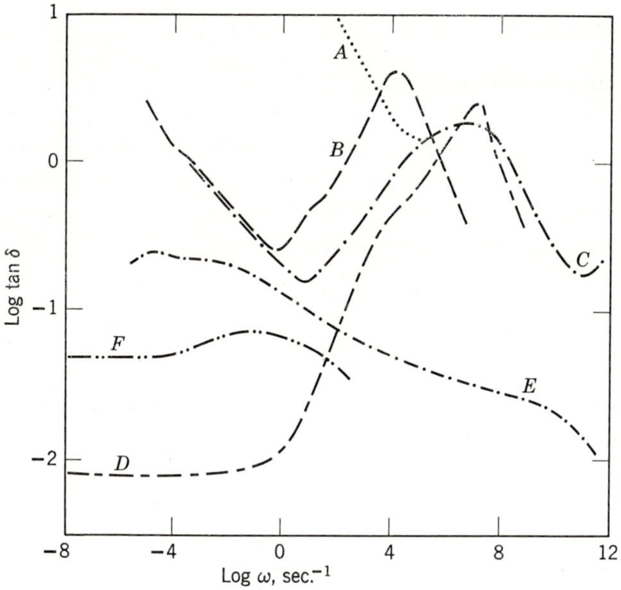

Fig. IC-18. The loss tangents vs. frequency for the same polymeric bodies shown in Figs. IC-11 and IC-15. In modification after Ferry (13).

If a viscoelastic body is first strained and then the force is removed so that a free vibration takes place, the oscillating strain will be damped by internal friction. Setting eq. (48) equal to zero, we have

$$0 = M\ddot{\gamma}(\omega) + \eta'\dot{\gamma}(\omega) + G'(\omega)\gamma(\omega) \qquad \text{IC4-(59)}$$

A solution can be expressed (48)

$$\gamma(\omega) = \gamma_0(\exp\{-\eta'/2M\}t) \cos \omega t \qquad \text{IC4-(60)}$$

If the damping is not too rapid, $[G'(\omega)/M]^{1/2} = \omega$, and the term in parentheses in eq. (60) takes the form, $\exp\{[\text{const. } G''(\omega)/G'(\omega)]t\}$.

The logarithm of the relative amplitudes of any two successive vibrations follows the relation

$$\Delta = \ln (A_i/A_j) = \ln (A_j/A_k) \propto G''(\omega)/G'(\omega) \qquad \text{IC4-(61)}$$

where Δ is the "logarithmic decrement." It is clear that Δ has many of the same virtues as $\tan \delta$.

IC4h. Various Representations of Linear Viscoelasticity

As mentioned in Subsection IC4a, there are several approaches available in the specification of linear viscoelastic functions. Some overly abbreviated remarks will be made here in that regard; fuller treatments of this same domain being readily available (1,2,4.19,44).

The stress tensor σ_{ij} can be resolved into "mean normal" σ and "deviatoric" s_{ij} tensors in the sense,

$$\sigma = \sum \sigma_{ij}/3; \quad i = j$$

and

$$s_{ij} = \sigma_{ij} - g\sigma; \; (g = 1, i = j; \; g = 0, i \neq j)$$

where σ is a dilative contribution and s_{ij} is a distortional contribution. Similarly, the strain tensor ϵ_{ij}, defined by

$$\epsilon_{ij} = [(\partial u_i/\partial x_j) + (\partial u_j/\partial x_i)]/2$$

where u is the displacement vector, can be resolved into a mean normal strain ϵ and a deviatoric strain e_{ij}:

$$\epsilon = \sum \epsilon_{ij}/3; \quad i = j$$

and,

$$e_{ij} = \epsilon_{ij} - g\epsilon$$

This leads to the designations of the bulk and shear moduli:

$$K = \sigma/3\epsilon$$

and,

$$G = s_{ij}/2e_{ij} \qquad \text{IC4-(62)}$$

In view of eqs. (27), (29), and (62), the single Kelvin-Voigt model can be represented as

$$s_{ij}(t) = 2\{[e_{ij}(t)/J] + \eta\dot{e}_{ij}(t)\}$$

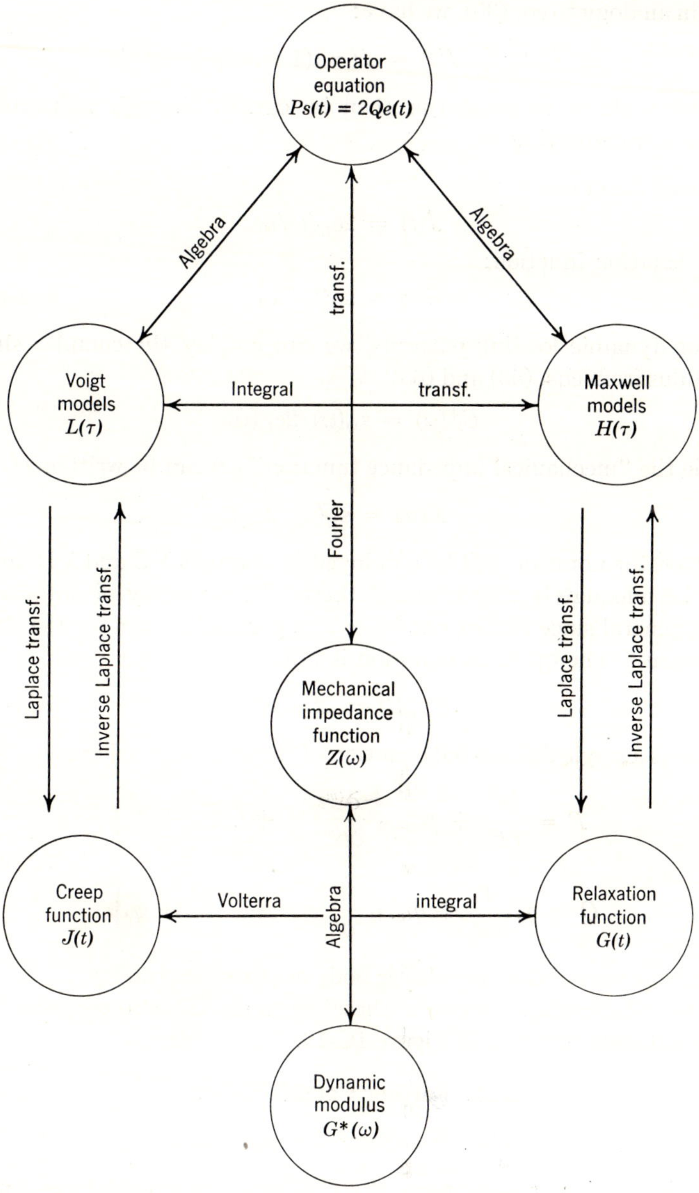

Fig. IC-19. Relationships between various methods for representing linear viscoelastic behavior. In modification after Alfrey and Gurnee (2).

and in analogy to eq. (30), we have:

$$2e_{ij}(t)/s_0 = J(t) = J(\infty)(1 - \exp\{-t/\tau\})$$

This style of notation can then be carried through Subsections IC4a-g by employing

Creep function:

$$J(t) = 2e_{ij}(t)/s_0$$

Relaxation function:

$$G(t) = s_{ij}(t)/2e_0 \qquad \text{IC4-(63)}$$

For dynamic loading patterns, we can employ the complex shear modulus from eqs. (53) and (63)

$$G^*(\omega) = s_{ij}(\omega)/2e_{ij}(\omega)$$

while the "mechanical impedance function" (2) can be written as

$$Z(\omega) = s_{ij}(\omega)/2\dot{e}_{ij}(\omega)$$

Operator notation will be employed in Sections VC and VD and it will be adequately characterized there. This variety of notation is very general since it does not favor any particular form for the stress or strain. The operator equation is (19)

$$Ps_{ij}(t) = 2Qe_{ij}(t) \qquad \text{IC4-(64)}$$

which is a single differential equation of the nth order, where

$$\left.\begin{aligned} P &= \frac{\partial^m}{\partial t^m} + p_{(m-1)}\frac{\partial^{(m-1)}}{\partial t^{(m-1)}} + \cdots + p_0 \\ Q &= q_n\frac{\partial^n}{\partial t^n} + q_{(n-1)}\frac{\partial^{(n-1)}}{\partial t^{(n-1)}} + \cdots + q_0 \end{aligned}\right\} \qquad \text{IC4-(65)}$$

The properties of the viscoelastic body are fixed by p and q.

The relationships between these methods of representation are schematically outlined in Figure IC-19.

IC5. Further Considerations

IC5a. Nonlinear Viscoelasticity

Finite strains or finite rates of strain generate irregularities within the *mechanical framework* for the formal representation of viscoelastic behavior. These complications give rise to "nonlinear" viscoelasticity.

While a solid may *appear* to be "hookean" under a finite strain, the geometric definitions of stresses and strains, hence those of elasticity itself, begin to fail as soon as the strain is greater than a true differential quantity (33). This also is true of a "newtonian" liquid which experiences a finite velocity gradient of flow, e.g., finite time and displacement derivatives of a "differential" strain. Being synthesized from an assembly of truly hookean and truly newtonian contributions, a "linear viscoelastic body" is an *abstraction* which can be no less idealized than its component parts.

As the strain or strain rate increases even further, the deviations from the phenomenological aspects of "hookean" and "newtonian" behaviors can also be *macroscopically* obvious. Here a mechanical framework for the concise representation of this behavior, if at all possible, would be extremely complicated. For example, consider a "liquid" which, under small strains and strain-rates, behaves as a sensibly obedient, "linear viscoelastic body." Now, we increase the strain and rate of strain, and the sample necks in ductile failure. Even if we enormously simplify the formal problem by stating that there is to be no macroscopic discontinuity in the necked sample, we still have at least three sources of nonlinearities: Firstly and secondly, the elaborate formalisms for representing both the inhomogeneous (33) hookean and the newtonian responses must be radically amended to include those strains actually experienced at the neck. Thirdly, any strain or time dependency in the relative order of hookean or newtonian contributions (whether derived from primary bond scission, crystallization or other structural changes) must now be introduced into this clumsy mechanical framework. (Of course, we refer here to the difficulty in the explicit mechanical representation of the process; a "prediction" of necking time, necking stress, or necking strain from prior experiments always being quite possible.)

Figure IC-20 shows a series of plots for the rates of stress relaxation in uniaxial tension for glasslike polystyrene under different levels of fixed strain (34). The lack of coincidence of these curves implies a *macroscopic* violation of eq. (26), since this plot requires a further dependency on the order of γ_0. However, the samples employed here exhibited signs of fatigue during these stress-relaxation experiments (this kind of progressive failure under relaxing stress levels is quite

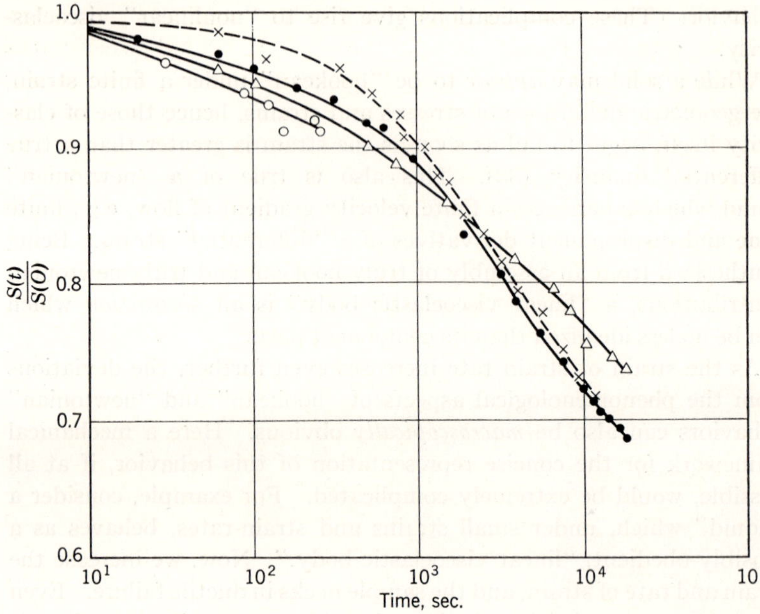

Fig. IC-20. The decay of tensile stresses in polystyrene at 50°C. at different strain levels. Elongation, %: (O) 0.90; (△) 0.75; (●) 0.50; (×) 0.25. After Maxwell and Rahm (34).

common with polymeric bodies, see Section IIIA), since the intensity of light reflections from the surfaces of the microfracture-like "crazing" marks in this otherwise transparent solid, vary with both time and strain level.

The type of macroscopic nonlinearity one obtains, however, depends on which combination of a gamut of possible complications is affecting the momentary observation. Functions of the type (7,20, 30,31,46,47,48)

$$E(t) = [\mathcal{S}(t, \gamma_0)]_{\gamma_0} / F(\gamma_0) \qquad \text{IC5-(66)}$$

have been reported and discussed in which the time and strain dependency of the stress are separable.

Section VD is dedicated to the extremely complex problem of reconciling the notational complications of nonlinear viscoelasticity. We place great emphasis on this topic in a book of this kind since we are primarily dealing with viscoelastic bodies here, and it is rather difficult to appreciate how a real viscoelastic solid can fail either brittly or ductilly without first becoming "nonlinear" in the most formalistic sense of this term. Moreover, the sooner one is equipped with the proper approaches to the concise mechanical representation of these nonlinearities, the closer will he be to the understanding of the true causes for them in view of the succinctly defined, experimental variables imposed.

IC5b. Bulk Viscoelasticity and T_g

A real body may experience a time-dependent hydrostatic compressibility (32,39). While the reasons for this would become rather obscure if the body were isotropic and continuous, no real body has yet been discovered which is either truly isotropic or continuous.

An anisotropic or real body may alter its shape under hydrostatic pressure, such as readily observed with samples of wood (21). A discontinuous or real body under hydrostatic pressure can have an inhomogeneous distribution of stress, not excluding components in shear.

The shear experienced during "hydrostatic stressing" can be either elastic or inelastic. When these distortions are inelastic and this same shear is a prerequisite for the completion of the compression, then the compression itself would appear to be inelastic.

A "hydrostatic" compression which appears to be inelastic in part, may be viewed as one which possesses a "bulk viscoelasticity," whether or not macroscopically "linear." The terms "bulk anelasticity" and "bulk flow" are natural consequences in terminology. If one employs a Kelvin-Voigt model to represent the isotropic stress for bulk creep, it is done in the sense

$$-\mathcal{S}_v = (\gamma_v/B) + \eta_v\dot{\gamma}_v \qquad \text{IC5-(67)}$$

(cf. eq. 29), where the spring's $B = B(\infty)$ is the bulk compliance and η_v is the coefficient of "volume viscosity." This approximate kind of bulk anelasticity has been observed (39), while other varieties of viscoelasticity in bulk are frequently encountered (13,32,39).

Far more mechanically unorthodox than the mere notion of bulk viscoelasticity, yet unavoidably related to it, is the point that glass-like bodies can exhibit anomalous time lags in bulk dilatations and contractions when these bulk deformations are induced by heat or solute absorptions and desorptions, even when they are externally unrestrained (28,41). These considerations are related to the basic peculiarities of the nonequilibrium status of these glasses, not excluding the very nature of T_g itself. Such profound complications lead to the use of mechanical models in bulk response to a mass or heat flux despite the frequent *absence* of applied forces, transient stresses or externally imposed restraints. Needless to overstate, this causes unreconcilable complexities in the consideration of both transient and residual thermal or solute stresses, as well as in the kinetics of heat and solute transport in glasslike bodies.

IC5c. Behavior on Large Extensions

Noncrystalline or partly crystalline polymeric bodies become of particular interest under large strains. Indeed, a very significant portion of their applicability stems from their properties under extremely large extensions. In order to facilitate the reader's appreciation of many of the interesting phenomena and situations to be treated throughout this volume, we will briefly introduce some of the more salient factors here. Others will be treated elsewhere, as they arise.

If one were to *abuse* the inherent meanings of linear viscoelastic models by extending them to large strains, he still could not represent an unambiguous "yield point" by them (37). The introduction of static friction elements into the model, which thus generates a gross nonlinearity would, of course, help somewhat. But even there, the strain required for the activation of a given static friction element would be time-sensitive. Hence our brief and overgeneralized discussion here should not be misconstrued as a fully logical extension of the treatment presented in Subsection IC4 concerning *linear* viscoelasticity. Nevertheless, the physical *intuition* gained from time-temperature substitution procedures becomes conceptually or qualitatively applicable here.

Figure IC-21 displays a highly schematic caricature of the nominal stress-elongation curves obtainable with polymeric bodies, together with the generalized appearances of the specimens at the highest-strain portions of these curves. Let us say that we have a fully amorphous, linear chained, polymeric body which is capable of crystallizing on a pronounced inelastic shear. Conceivably, all of the curves shown here could be experienced by this same body depending

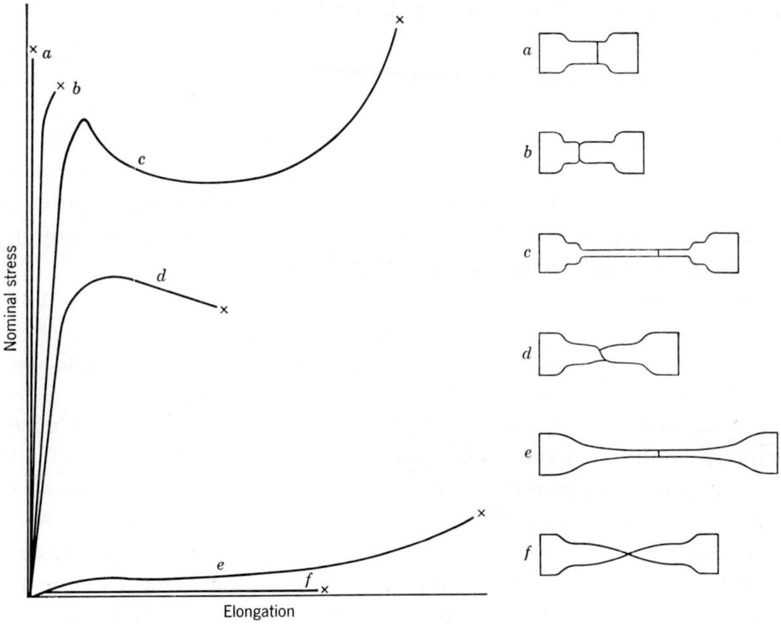

Fig. IC-21. Stress-extension plots and sketches characterizing the behavior and appearance of polymeric bodies at large strains. (*a*) A glass, (*b*) a soft glass, (*c*) the cold-drawing of a fiber, (*d*) a leather, (*e*) a rubber, and (*f*) a semi-solid. Highly schematic.

on the rate of strain (i.e., the time involved), the imposed ambient temperature, or the body's concentration of fluid solute or "plasticizer."

For example, let us suppose further that the rate of strain and the concentration of plasticizer were fixed so that Curves $a \rightarrow f$ were to be generated at different experimental temperatures. (We will neglect considerations of adiabatic heating, etc.) At the temperature and conditions for Curve *a*, we might say that the sample behaved "glasslike" and subsequently ruptured brittly. For Curve *b;* like a "soft glass" which yielded with the sample's necking, and later failed with an abrupt "necking rupture." For Curve *c;* like a "hard leather" which yielded on a necking in which the "shoulders" of the neck progressively collapsed so as to generate a "cold-drawn" *fiber* which was considerably "strain-hardened" such that it later failed rather "brittly." For Curve *d;* like a "leather" which yielded less abruptly on a less abrupt necking than that for Curve *c,* and later failed by a necking rupture having a less "brittle-appearing" fracture surface than for Curve *b.* (See Fig. IIC-15, p. 251.) For Curve *e;* like a "rubber" which did not "yield" in the usual sense of this term, yet subtly "narrowed" during its more pronounced elongation prior to its sensibly "brittle" rupture. For Curve *f;* like a highly viscous liquid or "semi-solid" which also did not "yield" *per se,* but simply flowed-apart while generating its ductile rupture.

If the specimens had been sparingly *semicrystalline* at the outset, all curves would likely have shifted somewhat to the left, at those same temperatures for that same variety of polymeric molecular composition.

If the specimens had been "highly" crystalline at the outset, Curves *e* and *f* would not have been experienced unless the imposed ambient temperature for those experiments made those specimens somewhat less crystalline by a *partial* melting.

If the polymer specimens had been fully amorphous but "uncrystallizable" in shear, Curve *c* would have been missing; true fiber-formation on a pronounced inelastic extension not being experienced. (See Fig. IIC-18, p. 253.)

An efficiently crosslinked polymeric body, whether or not semicrystalline, would not have experienced a Curve *f* unless the temperature were sufficiently elevated to induce chain rearrangements or chain degradation.

At a *constant* temperature for all experiments, the strain rate would have needed to be decreased in very generous logarithmic steps in going from Curve $a \rightarrow f$; or the concentration of plasticizer, systematically increased (see Fig. IC-12).

IC5d. Some New States of Matter

The advent of large, asymmetric molecules necessitates several conceptual modifications on the part of the student of the physical sciences. One might even say that these peculiar entities have forced a conceptual revolution. The explicit implications of the general notions: "macromolecular colloids," "micro-brownian motion," "configurational elasticity," "structural anelasticity," and "molecular structure sensitivity" are merely a few of the gamut of repercussions of that conceptually intriguing state of affairs. Depending on one's perspective of physical phenomena, it may be worthwhile to emphasize here that large, asymmetric molecules also can generate completely different states of matter than those available to the more "conservative," small, or sensibly symmetrical molecules.

In Section IB, it was mentioned that some polymeric molecules were "uncrystallizable." The implication there was that even within infinite patience during a given attempted crystallization, the molecular architecture of certain giant molecules simply did not *permit* a crystallization. This can be most readily appreciated if the molecule is to be part or all of a three-dimensional network, yet a highly branched or segmentally atactic chain also would satisfy that unorthodox condition.

Now, the question might be raised as to how one should proceed in an attempt to *supercool* an uncrystallizable body. This question leads us precisely nowhere since in order to supercool something one must go below its known melting point; yet none is available to it. Hence, certain polymeric bodies are, let us say, "permanent liquids." No; this is in equally poor conceptual taste. For example, let us cool this "permanent liquid" until it becomes "glasslike." Here we have a situation where a body *becomes glassy without being supercooled;* no supercooling being allowed for this "permanent liquid." Here, T_g is the only "melting point" available to that remarkably rigid body.

Now, let us aggravate this situation by taking this same body above its T_g again and elongating it uniaxially. If we extend it sufficiently, our uncrystallizable body (i.e., "uncrystallizable" on a cooling which is free of uniaxial tension) may or may not, as it structurally chooses, become "partly crystallized" in the sense of a uniaxially textured, semicrystalline solid. However, at elongations which are insufficient to generate a crystallization of any kind, our "permanent liquid" is anisotropic, i.e., it possesses a *chain-textured, noncrystalline status.*

Note here, as also implied in the previous section, that we cannot employ the term "amorphous," since this connotes "formless." The problem is simply that we must come to the appreciation of two major varieties of liquids: those which are isotropic and those which are *anisotropic.*

A liquid becomes "isotropic" if it can experience the fully randomized, macro-brownian or intermolecular motion. If a body is melted (i.e., if it is fully noncrystalline) it still can possess certain senses of molecular packing short of this idealized isotropic status *providing*

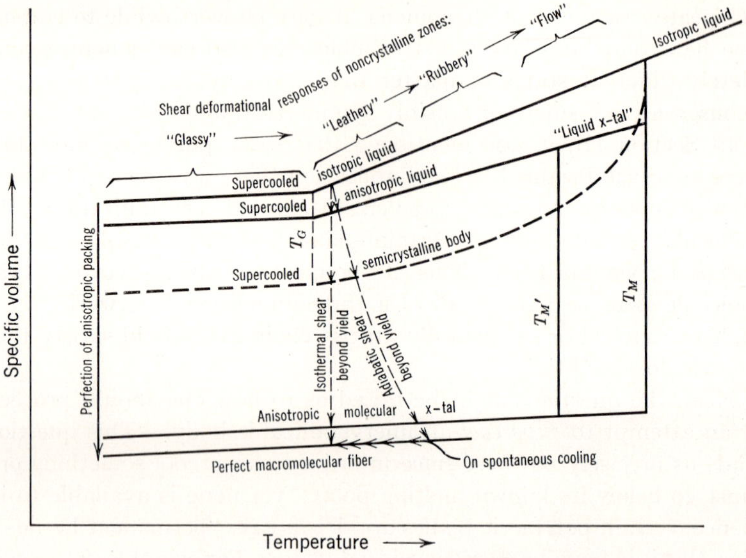

Fig. IC-22. A caricature of merely a portion of the gross complexity frequently encountered with volume-temperature plots dealing with substances composed of asymmetric molecules. Among a variety of other things, this schematically illustrates: (*1*) the supercooling of an isotropic polymer liquid; (*2*) the supercooling of an anisotropic liquid such as a low molecular weight "liquid crystal" having two or more melting points; (*3*) the supercooling of a semicrystalline polymeric body (cf. the supercooling of a low molecular weight semicrystallized, liquid crystalline body); (*4*) the melting, double-melting, or triple-melting, etc., of an anisotropic molecular crystal composed of asymmetric molecules, whether or not polymeric; (*5*) the generation of a supercooled liquid crystalline polymer, sometimes called an "oriented, noncrystalline polymer," by inelastic shearing; (*6*) the inelastic generation of a cold-drawn polymeric fiber towards a hypothetically perfect, anisotropic crystalline packing. Highly schematic.

that the molecule itself is sufficiently asymmetric to experience a formidable barrier against a free molecular rotation. Hence, the asymmetric soap molecules may be molecularly aligned near the surface of a soap film; experience a fluidity in certain directional senses, yet an elasticity in others. Moreover, on heating certain varieties of anisotropic molecular crystals which are composed of *asymmetric* organic molecules, they can exhibit two or three well-defined melting points, the anisotropic *viscoelastic* states between which are called "liquid crystalline" (17).

Figure IC-22, together with its legend, tends to caricature *some* of the very many states available to various varieties of molecularly asymmetric bodies.

Thus, when dealing with asymmetric molecules of appreciable size, the classical perspectives of "liquid" and "solid" states are even more inadequate than usual.

IC6. Summary

Several points with firm mechanical bearings on polymeric solids have been outlined in an introductory style. These have included the segmental mobilities of the molecular chains as reflected in their intermolecular transport and uncertainties in chain configuration. These liquidlike and gaslike behaviors in the diffusion of the amorphous chains were employed in the clarification of the characteristics of rubberlike, leatherlike, and glasslike polymeric bodies in shear. The idealities of "hookean" and "newtonian" behaviors were weighed against actual events in real substances, whether or not polymeric. Polymeric bodies are viewed as prime examples of the present inadequacies in the designation of the states of matter; merely being somewhat more "real" than the commonplace "real liquid" or "real solid."

Emphasis has been placed on thermally activated transport properties, configurational elasticity and the linear viscoelastic representations of polymeric bodies. This presentation has been designed to supplement the other sections of this first chapter; hence to provide a highly abbreviated, cross-sectional perspective of those salient concepts and polymeric behaviors which are deemed most pertinent to an adequate delivery of this volume.

IC7. List of Key Symbols

a	Segmental diameters
$a_T = \kappa'$	Time-temperature factors
A	Helmholtz free energy
B	Bulk compliances
$\mathfrak{D}$	Diffusion coefficients
D	Tensile compliances
e	Energy levels
e	Deviatoric strain tensor
E	Young's moduli

$\mathcal{E}$	Internal energies
E_s	Shear strain energies
E_y	Shear strain energies for yielding
f	Forces
f_0	Coefficients of segmental friction
G	Shear moduli
h	Planck's constant
$H(\tau)$	Spectra of relaxation times
J	Shear compliances
k	Boltzmann's constant
K	Bulk moduli
L	Extensions
$L(\tau)$	Spectra of retardation times
m	Forward progress of chain segments
M	Masses
n	Number of chain segments
N	Large numbers
p	Probabilities
P	Pressures; operator symbol
Q	Activation energy; operator symbol
r	Radii of spheres
R	Gas constant
$\overline{\langle R^2 \rangle}$	Mean-square, chain-ends separations
s	Interatomic distances
s	Deviatoric stress tensor
$\mathcal{S}$	Principal stresses
S	Entropies
t	Times
T	Temperatures
T_g	Glass temperatures
u	Potential energy levels
V	Volumes
$W; W(\ )$	Work; probability functions
$Z(\omega)$	Mechanical impedance functions
$u,v,w,$ x,y,z	Directional terms
γ	Principal strains
$\dot{\gamma}_f$	Velocity gradients of flow
δ	Transport distances; loss angles

Δ	Logarithmic decrement; linear changes
ϵ	Strain tensors (other than deviatoric)
η	Coefficients of viscosity
κ	Time–temperature factors
λ	Extension ratios
μ	Poisson's ratio
ν	Particle vibration frequencies
ξ	Interatomic resiliencies
ρ	Densities
σ	Stress tensors (other than deviatoric)
τ	Retardation times
τ'	Relaxation times
$\phi(s)$	Interatomic potential energy functions
ω	Dynamic mechanical frequencies

*References**

1. Alfrey, T., *Mechanical Behavior of High Polymers*, Interscience, New York, 1948.
2. Alfrey, T., and E. F. Gurnee in *Rheology*, Vol. I, F. R. Eirich, Ed., Academic Press, New York, 1956.
3. Alfrey, T., and P. Doty, *J. Appl. Phys.*, **16**, 700 (1945).
4. Alfrey, T., *Quart. Appl. Math.*, **3**, 143 (1945).
5. Anthony, R. L., R. H. Caston, and E. Guth, *J. Phys. Chem.*, **46**, 826 (1942).
6. Azaroff, L. V., *Introduction to Solids*, McGraw-Hill, New York, 1960.
7. Bergen, J. T., D. C. Messersmith, and R. S. Rivlin, *J. Appl. Polymer Sci.*, **3**, 153 (1960).
8. Billmeyer, F. W., Jr., *Textbook of Polymer Chemistry*, Interscience, New York, 1957.
9. Bland, D. R., *The Theory of Viscoelasticity*, Pergamon Press, New York–London, 1960.
10. Bueche, F., *Physical Properties of Polymers*, Interscience (Wiley), New York, 1962.
11. Bueche, F., *J. Appl. Polymer Sci.*, **1**, 240 (1959).
12. Eyring, H., *J. Chem. Phys.*, **4**, 283 (1936); T. Ree and H. Eyring, in *Rheology*, Vol. II, F. R. Eirich, Ed., Academic Press, New York, 1958; S. J. Hahn, T. Ree, and H. Eyring, in *Noncrystalline Solids*, V. D. Frechette, Ed., Wiley, New York, 1960.

* For reader and editorial conveniences, these are mainly general references; i.e., where a given book is referred to repeatedly, the page numbers are not specified and these are not necessarily the original sources, but *convenient* sources for information. In all cases the reader will find the general topic under consideration in that book's subject index or table of contents.

13. Ferry, J. D., *Viscoelastic Properties of Polymers*, Wiley, New York, 1961.
14. Flory, P. J., *Principles of Polymer Chemistry*, Cornell Univ. Press, Ithaca, N. Y., 1953.
15. Freudenthal, A. M., *The Inelastic Behavior of Engineering Materials and Structures*, Wiley, New York, 1950.
16. Glasstone, S., K. J. Laidler, and H. Eyring, *The Theory of Rate Processes*, McGraw-Hill, New York, 1941.
17. Gray, G. W., *Molecular Structure and Properties of Liquid Crystals*, Academic Press, New York, 1962; H. Sobotka, *Monomolecular Layers*, American Association for the Advancement of Science, Washington, 1954.
18. Grimm, H., and H. Wolff, *Handbuch der Physik*, Vol. 24, Part 2, Geiger-Scheel, 1933, p. 958.
19. Gross, B., *Mathematical Structures of the Theories of Viscoelasticity*, Hermann, Paris, 1953.
20. Guth, E., P. E. Wack, and R. L. Anthony, *J. Appl. Phys.*, **17,** 347 (1946).
21. Hearmon, R. F. S., in *Mechanical Properties of Wood and Paper*, R. Meredith, Ed., North-Holland, Amsterdam (Interscience, New York), 1953.
22. Hill, R., *Plasticity*, Oxford, 1950.
23. Houwink, R., *Elasticity, Plasticity and Structure of Matter*, Cambridge Univ. Press, 1937.
24. Jaeger, J. C., *Elasticity, Fracture and Flow*, 2nd Ed., Methuen, London (Wiley, New York), 1962.
25. Jenckel, E., and R. Heusch, *Kolloid-Z.*, **130,** 89 (1953).
26. Kingery, W. D., *Introduction to Ceramics*, Wiley, New York, 1960.
27. Kolsky, H., *Stress Waves in Solids*, Clarendon Press, Oxford, 1953.
28. Kovacs, A. J., *Trans. Soc. Rheol.*, **5,** 285 (1961).
29. Lamb, H., *Hydrodynamics*, Cambridge, 1932.
30. Landel, R. F., and P. J. Stedry, *J. Appl. Phys.*, **17,** 1885 (1960).
31. Leaderman, H., *Trans. Soc. Rheol.*, **6,** 361 (1962).
32. Litovitz, T. A., in *Noncrystalline Solids*, V. D. Frechette, Ed., Wiley, New York, 1960.
33. Love, A. E. H., *A Treatise on the Mathematical Theory of Elasticity*, Dover, New York, 1944.
34. Maxwell, B., and L. F. Rahm, *Ind. Eng. Chem.*, **41,** 1988 (1949).
35. McKelvey, J. M., *Polymer Processing*, Wiley, New York, 1962.
36. Meyer, K. H., and C. Ferri, *Helv. Chim. Acta*, **18,** 570 (1935).
37. Nielsen, L. E , *Mechanical Properties of Polymers*, Reinhold, New York, 1962.
38. Oldroyd, J. G., in *Rheology*, Vol. I, F. R. Eirich, Ed., Academic Press, New York, 1956.
39. Reiner, M., *Deformation, Strain and Flow*, 2nd Ed., Lewis, London, 1960.
40. Rivlin, R. S., in *Rheology*, Vol. I, F. R. Eirich, Ed., Academic Press, New York, 1956.
41. Rosen, B., *J. Polymer Sci.*, **58,** 821 (1962); *ibid.*, **A1,** 1472 (1963).
42. Schaefer, C., and F. Matossi, *Das ultrarote Spectrum*, Springer, Berlin, 1930; C. Kittel, *Elementary Solid State Physics*, Wiley, New York, 1962.
43. Schwarzl, F., and A. J. Staverman, *Appl. Sci. Res.*, **A4,** 127 (1953).
44. Simha, R., *J. Appl. Phys.*, **13,** 201 (1942).

45. Sips, R., *J. Polymer Sci.*, **5**, 69 (1950).
46. Smith, T. L., *Trans. Soc. Rheol.*, **6**, 61 (1962).
47. Staverman, A. J., and F. Schwarzl, in *Die Physik der Hochpolymeren*, Vol. IV, H. A. Stuart, Ed., Springer, Berlin, 1956.
48. Tobolsky, A. V., *Properties and Structure of Polymers*, Wiley, New York, 1960.
49. Treloar, L. R. G., *The Physics of Rubber Elasticity*, 2nd Ed., Clarendon Press, Oxford, 1958.
50. Volkenstein, M. V., *Configurational Statistics of Polymer Chains*, Interscience (Wiley), New York, 1963.
51. Zener, C., *Elasticity and Anelasticity of Metals*, Univ. of Chicago Press, Chicago, 1948.
52. Zimm, B. H., in *Rheology*, Vol. III, F. R. Eirich. Ed., Academic Press, New York, 1960.

ID. MECHANICS OF FIBERS

J. W. S. HEARLE

ID1. Fiber Structure

Most organic fibers are axially oriented, partly crystalline polymers, but their structure depends on how the fundamental processes of crystallization, discussed by Geil in Section IB, occur in practice in the living growth of plant and animal fibers or in the formation of man-made fibers in the laboratory or the factory. The crystallization is far from complete: on one model of structure, the proportions of perfectly crystalline material are estimated to be about $^1/_5$ in wool, $^1/_3$ in regenerated cellulose fibers, $^1/_2$ in polyamide and polyester fibers, and $^2/_3$ in cotton and other natural cellulose fibers. However, the exact nature of the deviations from perfect crystalline order is still a controversial subject.

Some workers, such as Stuart (37), suggest that the crystalline lattice is essentially continuous throughout the fiber but contains some degree of disorder; others, such as Kargin (26), view the structure as noncrystalline or "amorphous," but with some degree of ordering similar to the local ordering that occurs in a liquid. In both of these points of view, the structure is continuous except for locally random variations in the relative position of neighboring portions of the molecules, and is not differentiated into distinct regions. Although the idea of a continuous structure was favored by Staudinger (35a) in the 1920's and 1930's, it has not been the generally accepted view since then, and little work has been done on the interpretation of mechanical properties of fibers in terms of a continuous structure, whether of the imperfectly crystalline form or the partially ordered, noncrystalline form.

Over the last three decades, the conservative view of fiber structure includes a clear division between crystalline and noncrystalline regions. The crystalline regions are regarded as blocks or micelles of the order of 50 A. wide and 500 A. long; fringing at their ends as the polymer molecules pass into the noncrystalline regions. Each long-chain molecule will pass through a number of crystalline regions. One view of the fringed micelle structure is shown in Figure ID-1*a*.

The fringed micelle structure fitted in with many observations on fibers and other crystalline polymers, but it failed to explain satisfactorily the occurrence of the microfibrils (long, fiberlike structures with widths on the order of 100 A.) which were observed in electron microscopic studies of most natural fibers and some man-made fibers. In order to reconcile this difficulty, Hearle (14,17) proposed a fringed fibril structure. In this form of structure, the crystalline regions are regarded as long, imperfect crystals with about 100 molecules in their cross sections, and with the molecules branching off at intervals to pass into noncrystalline regions and then back again into crystalline fibrils. A typical view of the fringed fibril structure is shown in Figure ID-1*b*.

As the only major difference between the fringed micelle structure and the fringed fibril structure is in the lengths of the crystalline regions many observations explained by one theory can also be explained by the other. Both theories differ from the "continuous theories" in that the regions of order and disorder are localized and not spread uniformly through the material. In a review of fiber structure and mechanical properties, Hearle (16) has pointed out that there are continuous gradations between all four forms of structure. More detailed reviews of fiber structure (23) and physical properties (31) have recently been published.

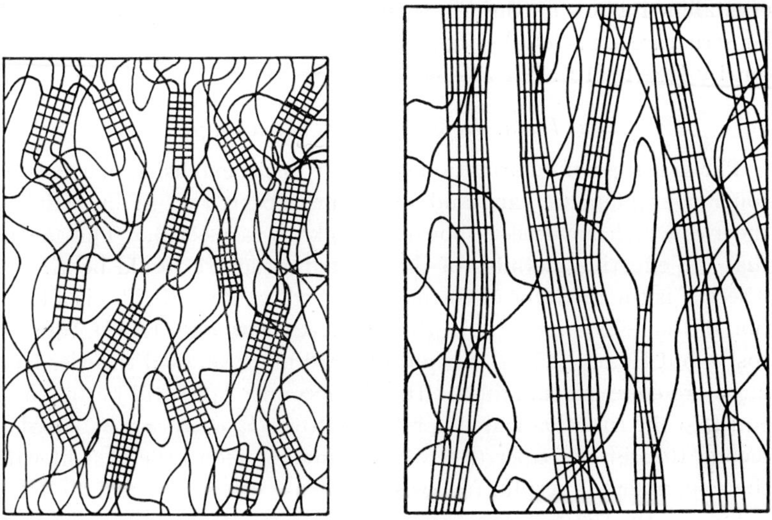

(*a*) (*b*)

Fig. ID-1*a*. Fringed micelle structure.
Fig. ID-1*b*. Fringed fibril structure.

ID2. Plant Fibers

In discussing the mechanism of fiber deformation, it is convenient to deal first with the natural cellulosic fibers which have a spiral structure with mean helix angles ranging from about 6° in flax to about 30° in cotton and to more than 40° in coir and some selected leaf fibers. In the fringed fibril theory the structure may be regarded schematically as in Figure ID-2, the crystalline fibrils embedded in (and molecularly continuous with) a matrix of noncrystalline regions. The orientation of the molecules is along the fibrils. On extension of a fiber with this structure, the deformation may occur in the following ways: (*1*) by an increase in length of the fibrils and of the noncrystalline regions in between, (*2*) (*a*) by extension like a spiral spring, with bending and twisting of the fibrils accompanied by (*b*) a reduction in volume of the fibrils, of the interfibrillar matrix of noncrystalline material, and of any voids present, and (*3*) by shearing of the noncrystalline regions to conform with the new configuration of the fibrillar structure. A detailed theory along these lines has been worked out by Hearle (19).

ID2a. Deformation by Stretching

The first type of deformation is similar to the extension of a twisted continuous filament yarn and would be expected to fit in with the theories which have been shown to apply to yarns (22). This means that the effective modulus of the twisted structure would be $EF_1(\theta)$, where E is the modulus of the untwisted structure and $F_1(\theta)$ is a function of the helix angle θ*. In the simplest treatment (12,32), $F_1(\theta) = \cos^2 \theta$, but Hearle (15,22) has put forward more detailed theories having more complex functions. In this account, $F_1(\theta)$ will be taken as equal to the function which is referred to as $F_2(\alpha, \sigma_1, \sigma_y)$ in Hearle's second theoretical paper (22). This function shows the best fit with the experimental results for twisted yarns. It is assumed that the volume of the twisted structure remains constant during the deformation, as illustrated in Figure ID-3.

The combined effects of deformation due to stretching in both crystalline and noncrystalline regions can be encompassed by setting

$$E = \gamma E_c + (1 - \gamma)E_n$$

where γ = the relative volume of crystalline material in the total volume of fiber, E_c = Young's modulus of the crystalline regions, parallel to the fibril axis, and E_n = Young's modulus of the noncrystalline regions. It is possible that E_n will be much smaller than E_c, so that the second term will be almost negligible and the effective modulus would be γE_c, i.e., approximately $^2/_3$ E_c in natural cellulose fibers.

ID2b. Deformation in the Manner of a Spring

It can be shown (19) that, for the extension of each fibril like that of an independent spiral spring, the effective modulus would be:

$$\text{Effective modulus} = \gamma(r/R)^2 \cos \theta[n_c E_c/(E_c \sin^2 \theta + 2n_c \cos^2 \theta)]$$

* In this rather general analysis, the problem of the exact definition of θ is not examined. In the theories on twisted yarn the angle used is α: the surface-angle of twist in a structure in which the length of one turn of twist is constant throughout the cross section, so that the helix angle decreases from α at the outside to zero at the center of the yarn. In natural plant fibers it may usually be more correct to take θ as a constant throughout the cross section; this will cause $F_1(\theta)$ to decrease somewhat more rapidly with θ.

where r = fibril radius, R = helix radius, and n_c = shear modulus. The last term, in brackets, must be less than either n_c or E_c (whichever is greater) and if, as is probable, $n_c \ll E_c$, then it tends towards $n_c/\sin^2\theta$. The factors γ and $\cos\theta$ will have values slightly less than unity. However the second term $(r/R)^2$ will equal 10^{-6} for a fibril of radius 100 A. at the outside of a fiber of radius 10 μ. This means that the

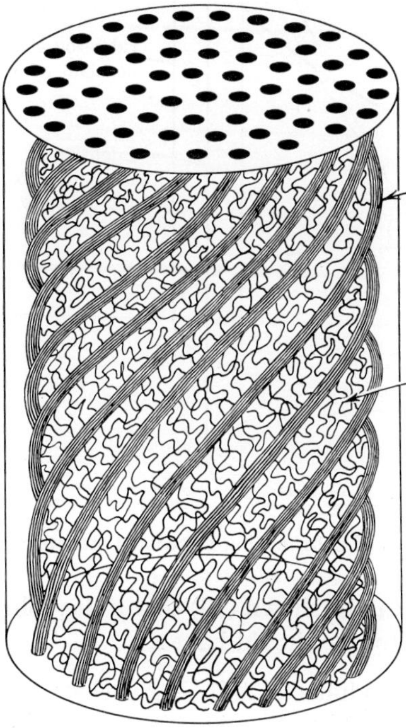

Fig. ID-2. Helical arrangement of fibrils in a natural cellulose fiber.

modulus will be very small except for an infinitesimal area at the center of the fiber; thus the resistance to deformation due to the stretching of fibrils like springs will be negligible.

The springlike extension with a constant fibrillar length, however, involves a reduction in volume. In an ordinary spring this can occur because the center of the spring is open, but in a closely packed array of springs (as in the spiral fibrillar structure) there will be a resistance

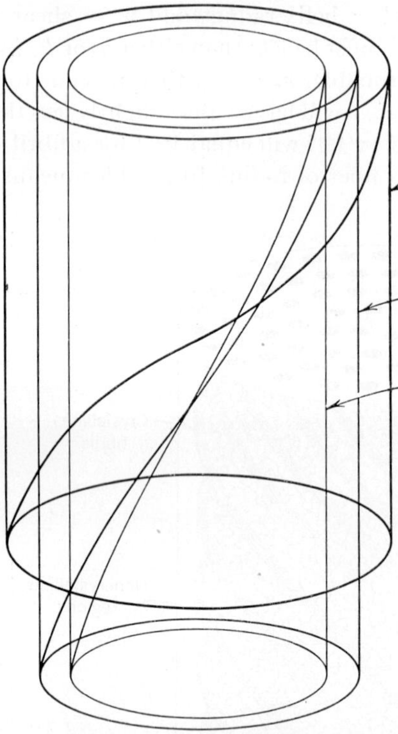

Fig. ID-3. Types of deformation in twisted structure.

to volume reduction. It can be shown (19) that if the volume changes were the determining factor we should have:

$$\text{Effective modulus} = K(1 - 2\cot^2\theta)^2$$

where,

$$K = [(1 - \alpha)^2/(1 - \gamma)]k_n + (\alpha^2/\gamma)k_c$$

k_n being the bulk modulus of the noncrystalline regions, k_c the bulk modulus of the crystalline regions, and α the fraction of total volume change occurring in the noncrystalline regions, equal to $\gamma k_n/[(1 - \gamma)k_c + \gamma k_n]$.

If, as would be expected, $k_c \gg k_n$, i.e., if the fibrils can be regarded as incompressible, then $\alpha \ll 1$ and $K = k_n/(1 - \gamma) \simeq 3k_n$ for a natural cellulose fiber.

The above argument is based on the assumptions that (*a*) the fiber cannot untwist when it is extended, and (*b*) the fiber is a solid cylinder. If either of these conditions is not satisfied, then the forces developed will be less. If untwisting is possible, as it may well be locally in cotton where the direction of the spiral alternates frequently, then the volume reduction will be less and, in some circumstances, could be zero.

If there is a cylindrical void at the center of the fiber, the volume reduction could occur by a contraction of this void, and the compression of the fiber *material* would be greatly decreased. However, such a deformation is likely to be resisted by other smaller forces, and a factor proportional to $(1 - 2 \cot^2 \theta)^2$ will still represent the magnitude of the deformation which has to be accommodated. The effect of voids can therefore be regarded as causing a reduction in the effective value of K; the parameter including the bulk moduli.

ID2c. Shear Deformation in Noncrystalline Regions

In considering the extension of a twisted yarn or of a system of spiral springs, it is tacitly assumed that the individual filaments are free to slide over one another in adjusting themselves to the new conformation. This will not be possible in the fringed fibril structure, because of the continuity of the molecular network through crystalline and noncrystalline regions. Some shear deformation of the noncrystalline matrix must take place, and this will add a further resistance to the extension of the fiber. The contribution will not be easy to calculate, but it will probably not be large.

ID2d. Combination of Mechanisms

Neglecting shear effects, we see that there are two possible mechanisms of deformation of the spiral structure: (*1*) due to an increase in length of fibrils at constant volume, with a modulus of $EF_1(\theta)$, and (*2*) due to springlike extension at constant fibril length, accompanied by volume compression, with a modulus $K(1 - 2 \cot^2 \theta)^2$. These two functions are plotted for various values of E and K in Figure ID-4. It will be seen that for low values of θ, $K(1 - 2 \cot^2 \theta)^2 \gg EF_1(\theta)$, showing that the springlike deformation with volume reduction can be ignored, and the fibril modulus will equal $EF_1(\theta)$. But the term $K(1 - 2 \cot^2 \theta)^2$ is much more sensitive to θ than is $EF_1(\theta)$, so that at high values of θ, where $K(1 - 2\cot^2\theta)^2 \ll EF_1(\theta)$, the springlike mechanism will predominate and the fiber modulus will be $K(1 - 2\cot^2\theta)^2$.

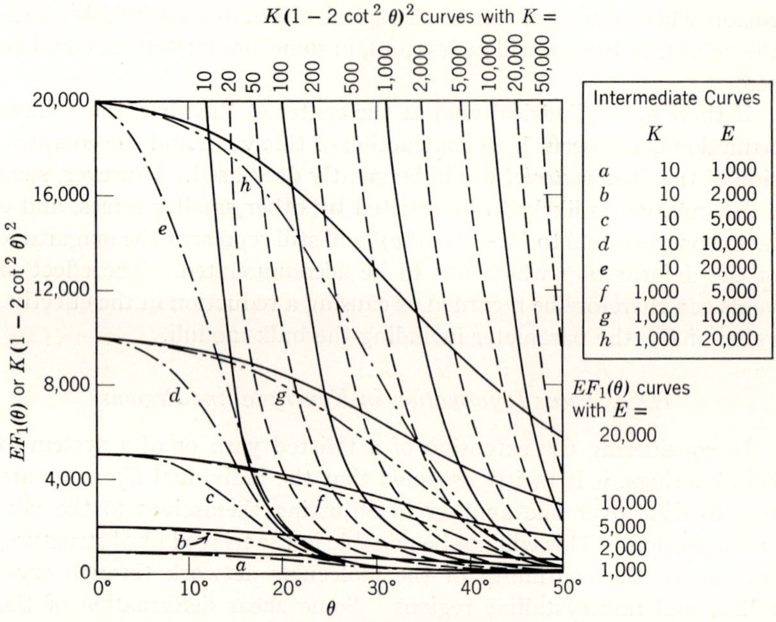

Fig. ID-4. Graphs of $EF_1(\theta)$ and $K(1 - 2\cot^2\theta)^2$ versus θ.

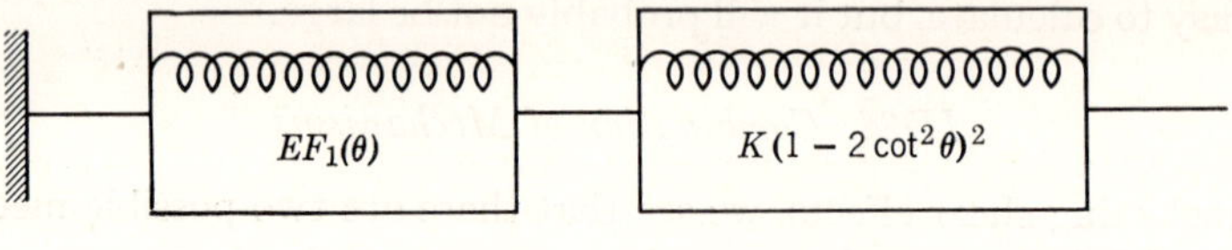

Fig. ID-5. Combination of two mechanisms in series.

The two mechanisms will act like two springs in series, as in Figure ID-5, each tending to relieve the stresses in the other by an interchange of strain, and so for intermediate values of θ we shall have:

$$\begin{array}{l}\text{Effective modulus due}\\ \text{to combined factors}\end{array} = \frac{EF_1(\theta)\,[K(1 - 2\cot^2\theta)^2]}{EF_1(\theta) + K(1 - 2\cot^2\theta)^2}$$

The form of this equation will not necessarily be proper, since the interchange of strain between the two mechanisms is not quite as simple as between two springs in series because of the geometry of the system, but it should be a reasonable approximation for intermediate values. A selection of intermediate curves is included in Figure ID-4.

ID2e. Comparison with Experiment

In Fig. ID-6, this equation has been fitted to values found by Spark et al. (35) for the modulus of two sisal fiber cells having spiral angles of 10 and 42°, respectively. For the fitted curve, $E = 2180$ and $K = 229$ kg./mm.^{2}. Neglecting the contribution of the noncrystalline regions, we should have $E = \gamma E_c \simeq {}^2/_3 E_c$. Treloar (39) has calculated a value of 5770 kg./mm.^{2} for the modulus of cellulose crystals, giving ${}^2/_3 E_c = 3900$ kg./mm.^{2}. The fitted value of E is of the correct order, although somewhat lower than the theoretical value (which was itself considered by Treloar to be too low an estimate). The difference may, in part, be explained by the occurrence of creep; dynamic values of modulus of sisal are appreciably higher.

If the fibrils are assumed to be incompressible, K should equal $k_n/(1 - \gamma) \simeq 3k_n$. This fitted value of K would thus correspond to a bulk modulus of 76 kg./mm.^{2} in the noncrystalline regions. It is rather low in comparison, for example, with a bulk modulus approximating 200 kg./mm.^{2} for a soft rubber, perhaps owing partly to the fact that the full volume change in the material does not occur because of untwisting or a compression of the void at the center of the fiber. Barkas (2) has pointed out that the equilibrium moisture absorption of wood, which is reduced from about 22%, when swelling is allowed, to 13% at constant gross volume, is limited to about 1.5% when the volume of the wood cell wall is held constant; this illustrates the great potentialities for expansion into the central void.

It may also be noted that increasing the value of k_n to a more reasonable value, i.e., 200 kg./mm.^{2}, so that $K = 600$, would give a modulus of 520 kg./mm.^{2} at 42° as against the experimental value of 315 kg./mm.^{2}. Considering the crudity of the analysis this is still a reasonable fit.

Included in Figure ID-6 are values of the modulus of a variety of cotton fibers measured by Meredith (29) and of some bast, leaf, and coir fibers measured by Stout and Jenkins (36). They are not very far

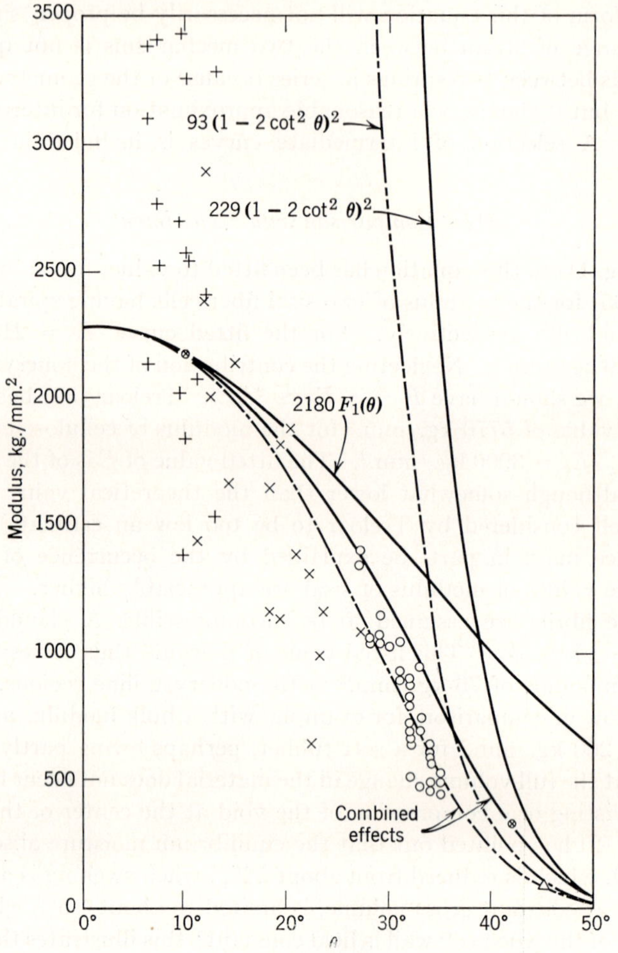

Fig. ID-6. Comparison of theory and experiment for various fibers: (×) leaf; (⊗) sisal; (○) cotton; (+) bast; (△) coir.

from the theoretical curve combining the two mechanisms, except that there is a very wide spread of values for the bast fibers. Some of these results will be influenced by the fact that the commercial bast and leaf fibers are made up of an aggregate of plant cells and contain varying amounts of lignin, pectin, and other materials. The values for cotton are better fitted by a curve with $K = 93$.

ID2f. Influence of Time and Moisture

The mechanical behaviors of fibers are time-dependent. Provided the stresses are not too great, the best agreement between experiment and theory has been the three-element model given by Tobolsky and Eyring (38) (Fig. ID-7) which, on the basis of reaction rate theory, incorporates a dashpot with a hyperbolic sine law of flow rate. Such a model could be very crudely interpreted according to the ideas given above in the following way. The arm containing the spring would represent the elastic extension of the crystalline fibrils, given by the term, $\gamma E_c F_1(\theta)$, and to a lesser extent by the compression of fibrils,

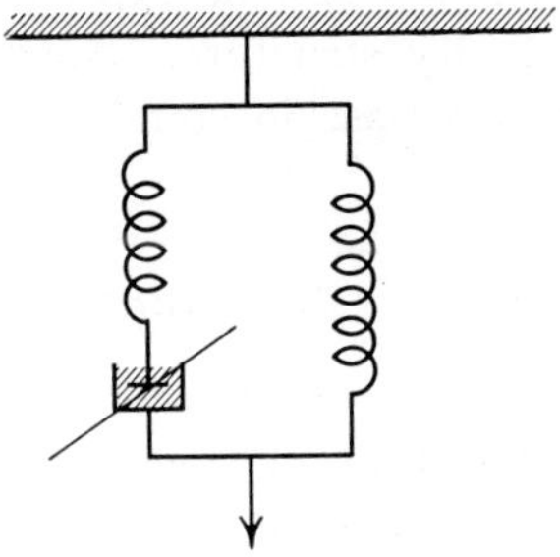

Fig. ID-7. Eyring's three-element mechanical model with two hookean springs and a nonlinear, i.e., nonnewtonian, dashpot.

that is, the term $(\alpha^2/\gamma)k_c(1 - 2\cot^2\theta)^2$. The other arm would crudely represent the viscoelastic compression (Subsection IC5*b*, p. 95) of the noncrystalline material, given by the term $[(1 - \alpha)^2/(1 - \gamma)]k_n$ $(1 - 2\cot^2\theta)^2$ and also by the contributions of other deformations in the noncrystalline regions, namely $(1 - \gamma)E_n F_1(\theta)$ and any shear effect. There may also be some time-dependent element in the deformation of the crystalline fibrils due to rearrangement of the molecules in the imperfect parts of the crystal.

When the stresses become larger, secondary (irrecoverable) creep also occurs, so that Eyring's model ceases to be applicable. This secondary creep would be a further feature of the deformation of the noncrystalline regions.

The moduli of plant fibers decrease as water is absorbed. Since the water does not readily penetrate the crystalline regions, this can effectively be due only to changes in the interfibrillar material. In cotton and other fibers with high spiral angle, it can be explained by a change in k_n, the bulk modulus of the noncrystalline regions. However, the change is as marked in a highly oriented fiber, such as ramie, as it is in cotton (30), showing that the noncrystalline regions are playing an important part even where the extension of the fibrils appears to be the dominant mechanism. This must be due partly to the influence of the second term $(1 - \gamma)E_n$ in the expression for E (in other words, in the dry fibers there is appreciable resistance to extension of the noncrystalline regions, with E_n comparable to E_c) and partly to the influence of the neglected factor, namely the shearing forces arising from the fitting of the noncrystalline regions into the new configuration, which will also be expected to be much greater in a heavily crosslinked, dry structure. Another factor which will have some effect is the decrease in spiral angle resulting from the swelling of the fiber.

ID2g. Fiber Breakage

One may expect breakage to occur when the stress concentrations rise above a critical level, so that, in general, the fiber strength should decrease with spiral angle in the same way that the modulus decreases. This, in fact, is found, although there is a good deal of scatter which reflects the varying incidence of flaws and points of weakness in the structures of different fibers.

If the fibers are dry, the comparative rigidity of the noncrystalline material will restrict the freedom of movement of the fibrils and so promote high stress concentrations whereas, when the fiber is wet, there will be greater freedom and lower stress concentrations and the fiber strength will increase, as is found experimentally in the natural cellulosic fibers. A similar effect was found in model yarns when the strength fell on glueing together of the filaments (21).

ID3. Protein Fibers

Wool fibers have an extremely complex morphology at all levels of structure. The finest observable units are microfibrils with widths of the order of 100 A. This is entirely compatible with a fringed fibril structure. However, the detailed interpretation of the wool structure is still a matter of controversy. Feughelman (6) describes a model of the wool fiber in the following terms:

"These mechanical properties of the set fibre can be understood in terms of a molecular model for wool, which has already been proposed to explain other mechanical properties. The structure of wool as displayed by its mechanical and allied properties was considered to consist of two phases in parallel [*see Fig. ID-8*], one phase being highly water-absorbing and also mechanically very much weakened by the water, and the other phase water-impenetrable and hence unaffected mechanically by water [*5*]. From the work on electron

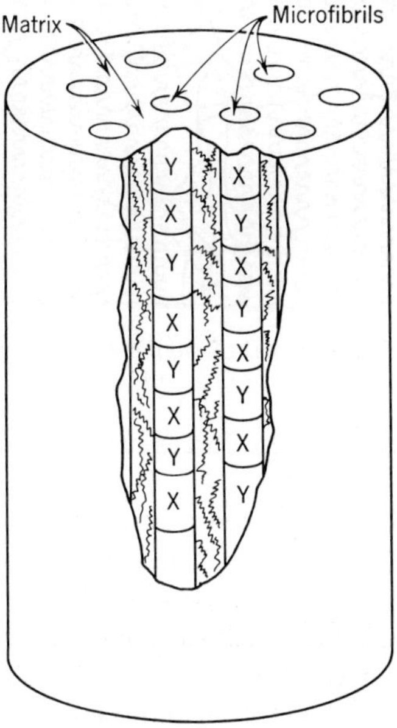

Fig. ID-8. Diagrammatic representation of model of wool. [After Feughelman (6).]

microscopy of Mercer, Rogers, Sikorski, and others [*1,3,25,29,33*] together with the infrared and x-ray results of Fraser and MacRae [*10,11*] the water-impenetrable phase was identified with the microfibrils in wool and the water-penetrable with the matrix. The microfibrils are considered to consist of well-organized keratin α-helices, whereas the matrix consists of opened-up or distorted coils of polypeptide chains held together by covalent links, inter- and intrachain hydrogen bonds, salt links and van der Waals forces. Except

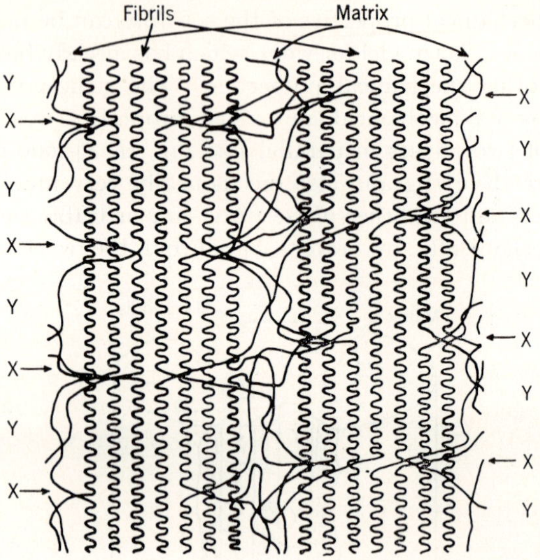

Fig. ID-9. Interpretation of model of wool structure on fringed fibril theory.

for the covalent links, all these forces are drastically affected by the presence of water.

"In the microfibrils, it has been postulated [7,9] that two types of zones exist in series (X and Y). These zones are present alternately along the microfibril. Both zones, it was proposed, consist of α-helices in an organized packed state with the zones X capable of opening out prior to zones Y when the fibre is stressed. These latter zones are possibly cross-linked to themselves and their surrounding matrix, and may have more bulky side chains, which introduce steric hindrance. In this model the yield region of the load–extension curve of a wool fibre in water at room temperature (~20°C.) corresponds to the opening out of the X zones, and the post-yield region to the opening up of the Y zones. If the temperature (T) of the water in which the fibre is being tested is raised above the transition temperature (T_c) of the fibre [8], cross-linking and the effect of steric hindrance associated with zones of type Y, disappear progressively with increase in temperature above this critical temperature."

This model could be modified and interpreted on the fringed fibril theory in the following way. The microfibrils are the crystalline regions, with the protein chain molecules fringing off to pass through the interfibrillar matrix. The zones X, which open out first, might represent regions in the fibril in which the branching of the molecules from the fibril occurs; this would give the structure illustrated in Figure ID-9. Alternatively, the alternation in zones X and Y may be caused by differences in the chemical constitution of the keratin.

It may be noted that Woods (40) considers that Feughelman's model is too simple and suggests that the interfibrillar matrix may be an active elastic component in parallel with the microfibrils. This would be the case if the portions of molecules in the matrix are normally coiled into an approximation to the α-helix, but can be stretched out into the β-chain. A consequence of the extension of the protein molecules, whether in the crystalline or noncrystalline regions, from the coiled α-helix to the fully extended β-chain is the great recoverable extensibility shown by wool and hair fibers.

In silk, the protein chains are fully extended, and the structure is likely to be a more or less oriented fringed fibril structure. There is evidence that silk is a natural "block copolymer" in which only certain segments will fit into the crystalline regions, while others will always be in the noncrystalline regions. Mechanically, silk is similar to synthetic fibers.

Regenerated protein fibers are, in general, very poorly crystalline and weak.

ID4. Man-Made Fibers

Although their production is more or less under human control, and thus accessible to experiment, apparently less is known about the structure of man-made fibers than is known about natural fibers. The details of the fine structure will depend on how the molecules come together from the comparatively disordered configurations in the solution or melt to the more ordered solid form, and on the rearrangement which takes place during the stretching or drawing of the fibers.

Regenerated cellulose fibers offer an interesting variety of structures and properties. Ordinary viscose rayon is made by a process in which the viscose (cellulose xanthate) solution is extruded into an acid bath where the cellulose is precipitated. The fibers are subsequently stretched to give them adequate stiffness. Both the mode of formation and many structural investigations suggest that the crys-

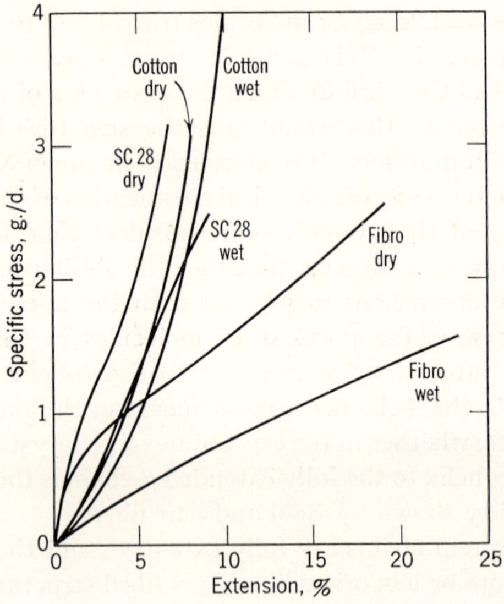

Fig. ID-10. Load-extension curves for rayon. [After Griffiths (13).] "Fibro" is an ordinary viscose rayon staple fiber; "SC28" is a polynosic rayon fiber.

talline regions are small, and the fringed micelle theory is a reasonable view of the structure. The stress–strain curves of rayon are shown in Figure ID-10. When wet, ordinary rayon shows a low Young's modulus since the deformation is that of a noncrystalline network of cellulose molecules surrounded and effectively lubricated by absorbed water molecules. The crystalline regions rigidify the structure only to a limited extent. When dry, the molecules in the noncrystalline regions are "linked" by hydrogen bonds, and so the material is initially much stiffer; when the stresses rise to a level at which the crosslinks break, however, there is a "yield point," and deformation becomes much easier. Cumberbirch and Mack (4) have developed a theory of breakage of ordinary wet rayon based on the fringed micelle theory, and this shows good agreement with experimental results.

By contrast, in the newer polynosic rayons (and in some older regenerated cellulose fibers such as Fortisan and Lilienfeld rayon) the solid fiber is formed as a cellulose derivative, which is then stretched to orient the molecules before the regeneration of the cellulose occurs. The crystallization accompanying the regeneration thus takes place in a solid oriented fiber, and it is not surprising that it yields a struc-

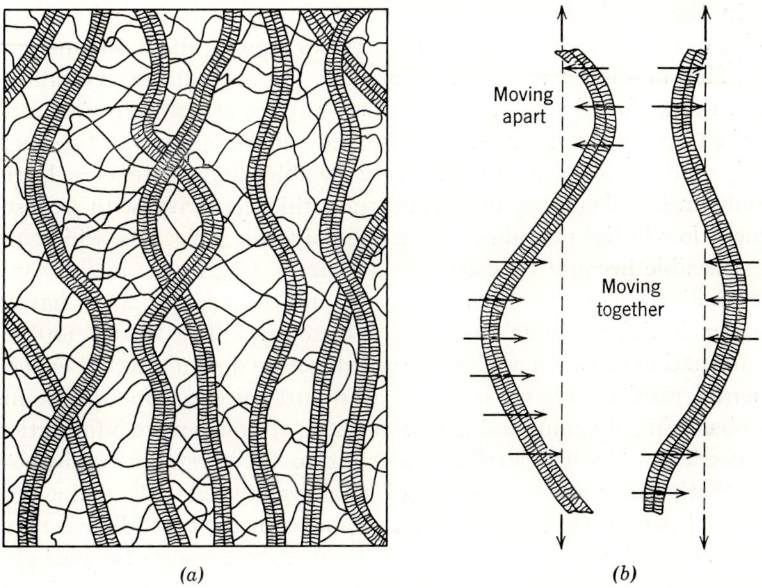

Fig. ID-11. (*a*) Irregular orientation of fibrils. (*b*) Movements of fibrils apart and together in different sections as the fibrils are straightened.

ture different from that resulting from the precipitation of cellulose from solution. The structure of the polynosic fibers, and of the others made by a similar sequence of processes, is markedly fibrillar; the fringed fibril structure appears the most likely. A schematic view, with somewhat irregularly oriented fibrils, is shown in Figure ID-11*a*. As with the plant fibers, two mechanisms of deformation are possible; either the fibrils will extend in length, or they will straighten with a resistance to deformation coming from the movement of some parts of fibrils towards one another and some parts away from one another, as illustrated in Figure ID-11*b*. A detailed theory (20) worked out on these lines shows reasonable agreement with experimental results for polynosic rayons. Mann and Roldan-Gonzalez (27) have shown that some extension of the crystalline lattice does take place. It is also significant that the stress–strain curves of polynosic rayon, shown in Figure ID-10, are much less affected by water, since the crystalline fibrils provide an unaltered reinforcement of the structure, whether wet or dry.

In the synthetic fibers, the fine structure will depend on the relative rates of crystal nucleation, crystal growth, and adjustment of the fiber in a temperature gradient (18). In polyamides, crystallization takes place in the undrawn fiber, but the structure is then disturbed by the three- or four-fold drawing which is given to the fiber. Polyamide molecules are relatively flexible, and there is a large, configurational entropy-dependent rubberlike elasticity, with a consequent low initial modulus. With the polyester molecules, which are less flexible because they contain a benzene ring in the main chain, crystallization occurs only on drawing, and the resulting fibers have a higher initial modulus than the polyamides. The theory (20) of deformation based on a fringed fibril structure does not fit the experimental results and, on balance, a fringed micelle structure seems probable in polyamide and polyester fibers, although more information is needed. The deformation occurring in the crystalline regions of synthetic fibers has been studied by Sakurada et al. (34), and a substantial part of the total extension appears in the crystalline lattice. For the acrylic fibers, Hosemann (24) strongly supports the idea of a paracrystalline structure. All the synthetic fibers are strong and relatively extensible; a combination giving a very high work of rupture.

ID5. Summary

Four views of the fine structure of fibers are described: continuous, imperfectly crystalline; continuous, noncrystalline with some ordering; fringed micelle; and fringed fibril. A mechanical analysis of the uniaxial extension of plant fibers is presented as based on a model consisting of a helical array of crystalline fibrils embedded in a noncrystalline matrix. Deformation of this structure can take place in two ways: by an extension of the fibrils at constant volume (analogous to the deformation of a twisted yarn); or by a springlike deformation of the fibril helices at constant length, with the major resistance to deformation coming from the reduction in volume. The theoretical relation can be fitted to experimental values of sisal, cotton, and coir, using values of the elastic constants which are of the correct order. The behavior of the system can be related to Eyring's three-element model of nonlinear viscoelasticity. The application of similar ideas to other fibers is discussed—in particular to a model of the wool fiber and to polynosic rayons. For many synthetic fibers and for rayons, the fringed micelle structure is more likely to be correct, while polyacrylonitrile appears to be paracrystalline.

ID6. List of Symbols

E	Modulus with zero twist
E_c	Young's modulus (crystalline)
E_n	Young's modulus (noncrystalline)
$F_1(\theta)$	Function of helix angle (twisted yarns)
k_c	Bulk modulus (crystalline)
k_n	Bulk modulus (noncrystalline)
K	Weighted-mean bulk modulus
n_c	Shear modulus (crystalline)
r	Fibril radius
R	Helix radius
α	Fraction of volume change occurring in crystalline region
γ	Relative volume of crystalline material
θ	Helix angle

References

1. Andrews, M. W., and J. Sikorski, *Proc. Conf. Electron Microscopy*, Stockholm 1956, p. 300.
2. Barkas, W. W., *The Swelling of Wood under Stress*, Her Majesty's Stationery Office, London, 1949.
3. Birbeck, M. S. C., and E. H. Mercer, *J. Biophys. Biochem. Cytol.*, **3**, 203 (1957).
4. Cumberbirch, R. J. E., and C. Mack, *J. Textile Inst.*, **51**, T458 (1960); *ibid.*, **52**, T382 (1961).
5. Feughelman, M., *Textile Res. J.*, **29**, 223 (1959).
6. Feughelman, M., *J. Textile Inst.*, **51**, T589 (1960).
7. Feughelman, M., and A. R. Haly, *Biochim. Biophys. Acta*, **32**, 596 (1959).
8. Feughelman, M., A. R. Haly, and B. J. Rigby, *Textile Res. J.*, **29**, 311 (1959).
9. Feughelman, M., and A. R. Haly, *Kolloid Z.*, **168**, 107 (1960).
10. Fraser, R. D. B., and T. P. MacRae, *Nature*, **179**, 732 (1957).
11. Fraser, R. D. B., and T. P. MacRae, *J. Chem. Phys.*, **28**, 1120 (1958).
12. Gégauff, C., *Bull. Soc. Ind. Mulhouse*, **77**, 153 (1907).
13. Griffiths, J. D., *J. Textile Inst.*, **52**, P575 (1961).
14. Hearle, J. W. S., *J. Polymer Sci.*, **28**, 432 (1958).
15. Hearle, J. W. S., *J. Textile Inst.*, **49**, T389 (1958).
16. Hearle, J. W. S., *J. Textile Inst.*, **53**, P449 (1962).
17. Hearle, J. W. S., *J. Appl. Polymer Sci.*, **7**, 1175 (1963).
18. Hearle, J. W. S., *J. Appl. Polymer Sci.*, **7**, 1193 (1963).
19. Hearle, J. W. S., *J. Appl. Polymer Sci.*, **7**, 1207 (1963).

20. Hearle, J. W. S., *J. Appl. Polymer Sci.*, **7**, 1635 (1963).
21. Hearle, J. W. S., H. M. A. E. El-Behery, and V. M. Thakur, *J. Textile Inst.*, **51**, T299 (1960).
22. Hearle, J. W. S., H. M. A. E. El-Behery, and V. M. Thakur, *J. Textile Inst.*, **52**, T197 (1961).
23. Hearle, J. W. S., and R. H. Peters, Eds., *Fibre Structure*, Textile Institute, Manchester; Butterworths, London, 1963.
24. Hosemann, R., *Polymer*, **3**, 349 (1962).
25. Jeffrey, G. M., J. Sikorski, and H. J. Woods, *Textile Res. J.*, **25**, 714 (1955).
26. Kargin, V. A., *J. Polymer Sci.*, **30**, 247 (1958).
27. Mann, J., and L. Roldan-Gonzalez, *Polymer*, **3**, 549 (1962).
28. Mercer, E. H., *Textile Res. J.*, **27**, 860 (1957).
29. Meredith, R., *J. Textile Inst.*, **37**, T205 (1946).
30. Meredith, R., *The Mechanical Properties of Textile Fibres*, North Holland, Amsterdam, 1956, p. 79.
31. Morton, W. E., and J. W. S. Hearle, *Physical Properties of Textile Fibres*, Textile Institute, Manchester; Butterworths, London, 1962.
32. Platt, M., *Textile Res. J.*, **20**, 1 (1950).
33. Rogers, G. E., *Summarized Proc. of Conf. Electron Microscopy of Fibres* by C. E. Challice and J. Sikorski, *Brit. J. Appl. Phys.*, **8**, 1 (1957).
34. Sakurada, I., Y. Nukushina, and T. Ito, *J. Polymer Sci.*, **57**, 651 (1962).
35. Spark, L. C., G. Darnbrough, and R. D. Preston, *J. Textile Inst.*, **49**, T309 (1958).
35a. Staudinger, H., *Die Hochmolekularen Organischen Verbindungen*, Springer–Verlag, Berlin, 1932.
36. Stout, H. P., and A. Jenkins, *Ann. Sci. Textiles Belges*, **1955**, 231 (Dec. 1955).
37. Stuart, H. A., *Kolloid Z.*, **165**, 3 (1959).
38. Tobolsky, A. V., and H. Eyring, *J. Chem. Phys.*, **11**, 125 (1943); Halsey, G., H. J. White, and J. Eyring, *Textile Research J.*, **15**, 295 (1945); see also Meredith, R., *The Mechanical Properties of Textile Fibres*, North Holland, Amsterdam, 1956, pp. 87 et seq.
39. Treloar, L. R. G., *Polymer*, **1**, 290 (1960).
40. Woods, H. J., *J. Textile Inst.*, **51**, T601 (1960).

IE. ASPECTS OF VISCOELASTICITY*

ARTHUR V. TOBOLSKY

IE1. Modulus–Temperature Curves for Linear Amorphous Polymers

Aside from single crystalline species, polymers divide into two categories: the semicrystalline and the wholly amorphous. (The term "amorphous" simply implies noncrystalline here, as in most other places in this book. See Subsections IB3, IC5; pp. 15, 98). Clearly, the semicrystalline polymers are amorphous above their melting points and sometimes remain amorphous if quickly quenched from their molten condition. In this subsection, however, we shall discuss linear polymeric solids which are wholly amorphous under *all* conditions. These bodies contain polymer molecules which generally are stereochemically or structurally irregular, so that they cannot be made to crystallize.

A plot of specific volume–temperature for an amorphous polymer shows a time-dependent change of slope at an experimentally sensitive temperature called the glass transition, T_g. As emphasized in the previous sections, this is one of the most important of those parameters

* Cf. ONR Technical Report RLT-58; Contr. #Nonr-1858(07), NR356–377.

which define the properties of wholly or partly amorphous polymers. Sufficiently below T_g, the polymer is hard and glassy. Above T_g, an amorphous linear polymeric body of sufficiently great chain length successively behaves as a leather, a rubber, and finally as a liquid, with increasing temperatures.

In some respects the molecular aggregation of a linear amorphous polymer crudely resembles the contents of a bowl of cooked spaghetti. Above the T_g of the polymer, however, our spaghetti must be thought of as being in constant wriggling motion more like an entangled mass of live earthworms. A dynamic, quasi-network is forming and re-forming by the periodic entanglements between the molecules.

The simplest way to characterize the mechanical properties of a polymer is to measure one of its directional elastic moduli as a function of temperature. Since polymers are viscoelastic, however, these moduli will depend on both the time and the method of measurement. For the moment we shall discuss the tensile *relaxation* modulus, $E(t)$. As mentioned in Section IC, this is obtained by measuring the stress, as a function of time, acting upon a sample maintained at a constant uniaxial extension and under a constant ambient temperature.

For an immediate appraisal of the temperature dependency of the viscoelasticity, we will standardize the time of measurement at 10 sec. Therefore, the quantity being examined as a function of temperature is $E(10)$ or, more precisely, we will consider $[E(t,T)]_{t=10}$.

Other methods for measuring the moduli will give equally useful and sensibly similar results if the time of measurement is the same. For example, determination of the torsional creep modulus at 10 sec. can be utilized equally well. The *shear* relaxation modulus, $G(10) \cong E(10)/3$; this relation being nearly exact for values of $E(10) < 10^9$ dynes/cm^2. The reciprocal of the compliance measured in tensile *creep*, $1/D(10) \cong E(10)$, in most cases. Mathematical transformations exist to convert relaxation to creep and vice versa (see Subsection IC4d, p. 78).

The results obtained for $E(10)$ with a typical linear amorphous polymer, atactic polystyrene, are displayed in Figure IE-1. One of the samples shown in this figure is denoted as Sample C and has a weight-average molecular weight, $\bar{M}_n = 325{,}000$, and a number-average molecular weight, $\bar{M}_n = 217{,}000$. (For a definition of these averages, see ref. 4.) The other sample shown in this figure A, has $\bar{M}_w \doteq 210{,}000$; $\bar{M}_n = 140{,}000$. The larger the value of $\bar{M}_w/\bar{M}_n$

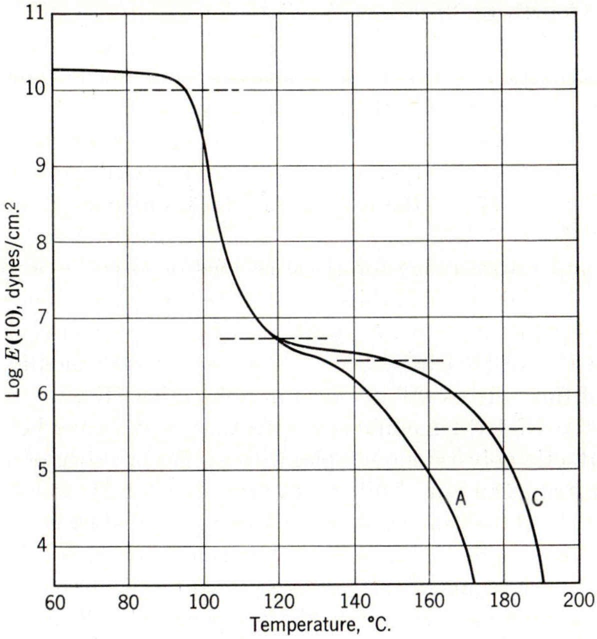

Fig. IE-1. Tensile relaxation modulus–temperature curves of amorphous polystyrene samples A and C.

("heterogeneity index") the broader is the distribution of molecular sizes. Where $\bar{M}_w/\bar{M}_n \equiv 1$, all molecules have the same size (4).

Let us first consider sample C. The modulus–temperature curve shows five regions of viscoelastic behavior (6). The first region, below 97°C., is the "glassy" region in which $E(10) = 10^{10.5}$ to $10^{10.0}$ dynes/cm.^{2}. In the second or "transition region," $E(10) = 10^{10.0}$ to $10^{6.7}$ in the 97–120°C. range. In this temperature-sensitive region of $E(10)$, the mechanical properties are best described as "leathery." The onset of this region starts at $\sim T_g$ which is 100°C. for atactic polystyrene (under the usual methods of evaluating T_g).

In the third or "rubbery plateau" region, $E(10)$ remains fairly independent of temperature at a value of $10^{6.7}$ to $10^{6.4}$. The spread of temperatures for this region depends on the molecular weight of the polymer; for polymer C, this spread is 120–150°C. In this temperature range, the polymer behaves in a quite rubbery manner with long-range elastic behavior if it is not deformed for long periods.

In the fourth or "rubbery flow" region, $E(10) = 10^{6.4}$ to $10^{5.5}$. Here, the polymer still is somewhat rubbery, but also has a marked flow contribution when under shearing components for periods $\cong$ 10 sec. For Sample C, this region persists between 150 and 177°C.

Finally, in the fifth region where $E(10) < 10^{5.5}$, the polymer exhibits very little elasticity and manifests an apparent state of "liquid flow." Above 177°C., the maximum relaxation time of Sample C is less than 10 sec.

The temperature range for these regions of viscoelastic behavior, however, depends on our arbitrarily selected reference time of 10 sec. This is why the division into the various regions of viscoelastic behavior in Figure IE-1 have been defined in terms of moduli. A discussion of time effects will be given in Subsection IE2.

In Figure IE-1 the modulus–temperature curves have been shown for two atactic polystyrene samples with similar breadths of molecular weight distribution but of different average molecular weights. It is clear that the glassy transition regions are unaffected by the molecular weight of the sample (for sufficiently large chain lengths). The modulus value in the region of the rubbery plateau is likewise independent of molecular weight. However, the temperature regions of rubbery flow and liquid flow are markedly dependent on molecular weight or chain length.

In the glassy region, the chain segments are in fixed positions on the sites of a sensibly disordered quasi-lattice. They vibrate about these fixed positions just as the molecules of a molecular crystal. However, they undergo little if any shear stress-biased, segmental diffusion in periods <10 sec. This translational motion sets in efficiently at $T > T_g$.

In the transition region, the segments of the chain are undergoing short-range translations, the time for segmental self-diffusion being ~10 sec. and independent of chain length. In the transition region, the modulus changes markedly with both time and temperature.

In the quasi-static rubbery region, the short-range transport of the segments is very rapid. However, the stress-biased, thermally activated transport of the molecules as a whole, i.e., a true fluidity is retarded (particularly by entanglements between the chains which act as "temporary crosslinks"). In fact, one can compute the average

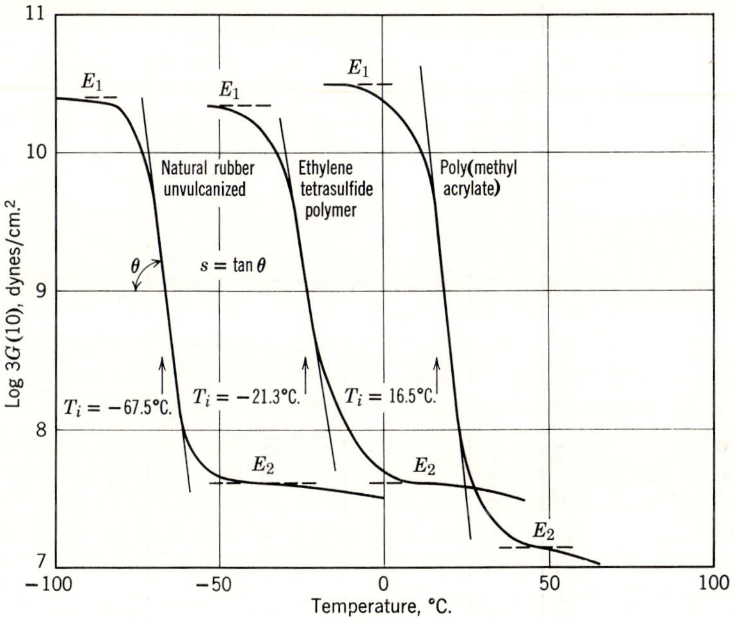

Fig. IE-2. Modulus–temperature curves of some linear amorphous polymers.

length of the subnetwork chains between points of entanglement from the value of the rubbery plateau modulus.

In the region of rubbery flow, the shear-biased translation of whole molecules becomes important. Major long-range configurational changes, including the slippage of entanglements, are taking place in $\sim$10 sec.

Finally, in the region of liquid flow the long-range configurational changes and consequent translations of the molecules are occurring in $<$10 sec. The rubber elasticity is virtually negligible here for shearing components which are maintained for more than 10 sec.

The regions of rubbery flow and liquid flow can be suppressed completely if chemical crosslinks are introduced to serve as permanent network junctures.

For other linear amorphous polymers, the main features of the above discussion based on polystyrene are still applicable; changes in T_g being the major difference from one wholly amorphous, linear

TABLE IE-I
Characteristic Parameters for Linear Amorphous Polymers

	From stress relaxation master curves in tension					From modulus–temperature curves in torsion			
Polymer	$E_1 \times 10^{-10}$, dynes/cm.²	$E_2 \times 10^{-7}$, dynes/cm.²	T_i, °C.	q	P	$3G_1 \times 10^{-10}$, dynes/cm.²	$3G_2 \times 10^{-7}$, dynes/cm.²	T_i, °C.	s
Polyisobutylene	3.0	0.76	−64	0.75	0.21	3.4	0.89	−62	0.13
Poly(methyl methacrylate)	2.8	2.5	113	0.54	0.33	1.4	1.9	..	0.14
Natural rubber, unvulcanized	2.6	1.0	−67	1.23	0.31	2.5	4.0	−68	0.22
Polystyrene	..	..	..	..	..	1.8	0.53	101	0.21
Poly(butyl acrylate)	..	..	..	..	..	1.5	0.56	−52	0.22
cis-Polybutadiene	..	..	..	..	..	2.0	1.1	−106	0.19
Polypropylene, atactic	..	..	..	..	..	1.7	2.5	−16	0.17

polymer to the next. In Figure IE-2 we show the log $3G(10)$–temperature curves for three other linear amorphous polymers with emphases on the glassy, transition and rubbery plateau regions. On the graph we also indicate how one may obtain four characteristic parameters for each polymer (7).

The most important parameter is T_i, the temperature at which $E(10) \cong 3G(10) = 10^9$ dynes/cm.². The value 10^9 is arbitrarily chosen, but it lies more or less in the center of the transition region. Furthermore, T_i is generally close to, perhaps 2–5°C. greater than, T_g as obtained from customary volume–temperature plots.

Two other characteristic parameters shown in Figure IE-2 are E_1 and E_2, the "glassy- and rubbery-plateau moduli," respectively. Finally we define s as the slope, at T_i, of log $E(10)$ versus T.

The values of these characteristic parameters for some amorphous polymers are shown in Table IE-I. It is interesting to note that the variations of E_1, E_2, and s for amorphous polymers of very different molecular structure are not very great.

Many of the measurements reported here were made in torsion rather than in tension. This type of measurement gives a shear modulus. The data in shear are reported as $3G$ in order to approximate the magnitude of E. For $E > 10^9$, i.e., approaching the glassy region, the relation between E and G depends on Poisson's ratio and there, $E = 2$–$3G$. The precision of the measurements of the glassy modulus reported in Table IE-I is not very great, i.e., not great enough to attach significance to the differences between E and $3G$.

IE2. Master Curves of Amorphous Polymers: Transition Region

The times of direct measurement of $E(t)$ which are experimentally feasible are 1–10^5 sec. Values of $E(t)$ for shorter and longer *times*, however, are obtained by utilizing data obtained at different *temperatures*. (See Subsection IC4e, p. 83.)

Stress relaxation master curves for amorphous polymers are obtained by horizontal displacement of log $E(t)$ data obtained at different temperatures along the log t axis. A typical example is shown in Figure IE-3 for polyisobutylene in the transition region. The solid lines represent the master curve while the white and black circles represent the experimental points. See Section IMB for a comprehensive treatment of this time–temperature shifting procedure.

The general appearance of stress relaxation master curves for various amorphous polymers in the transition region is quite similar.

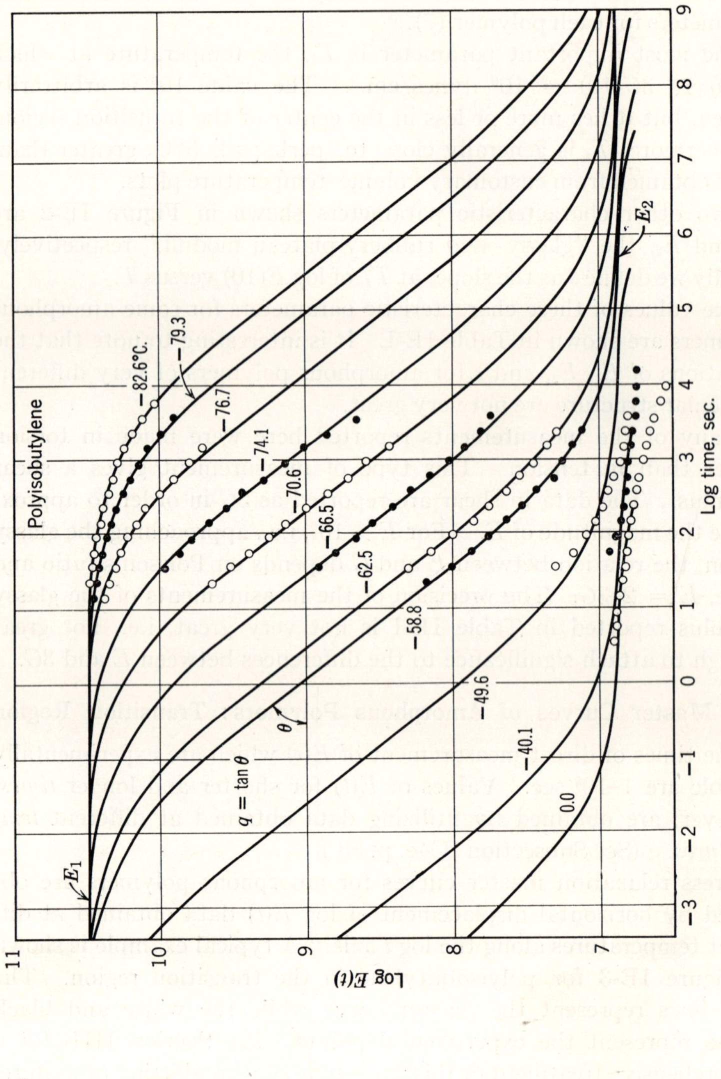

Fig. IE-3. Stress relaxation master curves for polyisobutylene in the transition region.

Linear amorphous polymers or very slightly crosslinked amorphous polymers generally show a relatively linear region in the log $E(t)$–log t plot when $E(t) \cong 10^9$, as seen in Figure IE-3.

The characteristic relaxation time, $\tau(T)$, at any temperature is the time required for $E(t)$ to attain a value of 10^9. By definition, $\tau(T_i) = 10$ sec.

According to the semi-empirical Williams-Landel-Ferry (WLF) equation (16),

$$\log[\tau(T)/\tau(T_g)] = -17.44(T - T_g)(51.6 + T - T_g)^{-1} \quad \text{IE2-(1)}$$

The constants in this equation vary somewhat from polymer to polymer. We therefore rewrite the equation as follows:

$$\log[\tau(T)/\tau(T_g)] = -P(T - T_g)[1 + P'(T - T_g)]^{-1} \quad \text{IE2-(2)}$$

Inasmuch as $P' \ll P$, where $T \cong T_g$, eq. (2) becomes,

$$\log[\tau(T)/\tau(T_g)] = -P(T - T_g) \quad \text{IE2-(3)}$$

Inasmuch as $T_i \cong T_g$, eqs. (1), (2), and (3) are not greatly altered if one employs $T_i \equiv T_g$.

Plots of log $[\tau(T)/\tau(T_i)]$ versus $(T - T_i)$ are, in fact, nearly linear in the region where $T \cong T_i$. A plot of this type is shown for polyisobutylene in Figure IE-4. The negative slope of this plot at $T = T_i$ is defined as P, which we take to be another characteristic parameter.

$$\log[\tau(T)/\tau(T_i)] = -P(T - T_i) \quad \text{IE2-(3A)}$$

The WLF relation, eq. (1), predicts $P \cong 0.34$. The actual values of P for three polymers are shown in Table IE-I.

We define another characteristic parameter, q; the negative slope of the log $E(t)$ versus log t master curve at $E(t) = 10^9$. (Cf. s for log $E(10)$ versus T plots.) According to the Rouse-Bueche theory (1,5), $q \cong 0.5$. Values of q for some amorphous polymers are also given in Table IE-I.

Two simple empirical equations are presented for the relaxation master curve and modulus–temperature curve in the transition region (7):

$$E(t) \cong 10^9\{[t/\tau(T)]^q + (10^9/E_1)\}^{-1} + E_2 \quad \text{IE2-(4)}$$

$$E(10) \cong 10^9[10^{(sT - Ti)} + (10^9/E_1)]^{-1} + E_2 \quad \text{IE2-(5)}$$

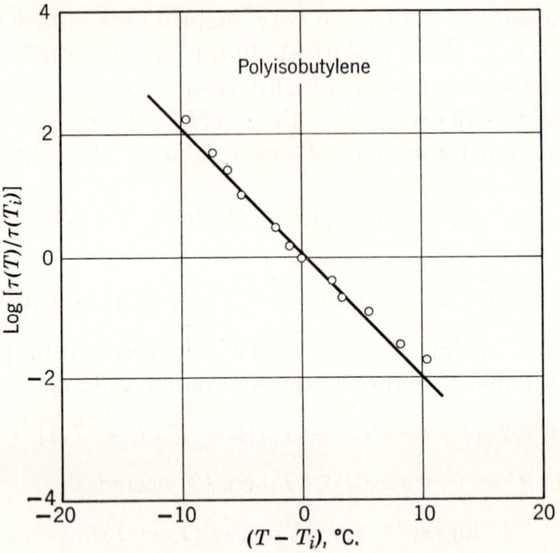

Fig. IE-4. Log $[\tau(T)/\tau(T_i)]$ for polyisobutylene in the temperature region near T_i.

Equations (4) and (5) are far from exact. However, they have the virtue of being simple, employ only the characteristic parameters, and present the essential behaviors of the stress relaxation master curve and the modulus–temperature plot.

If eq. (3) is taken as valid, one can easily show that the following holds:

$$s = qP \qquad \text{IE2-(6)}$$

The use of the well-defined characteristic parameters presented here (7,8) permits a rapid and convenient survey of the viscoelastic properties of linear amorphous polymers (or very lightly crosslinked polymers) in the transition region. This is especially true if only approximate results are required.

IE3. Stress Relaxation Master Curves: Flow Region

Stress relaxation master curves in the "rubbery plateau–rubbery flow–liquid flow region" at 115°C. for 15 samples of polystyrene of differing molecular weights are shown in Figure IE-5. All samples

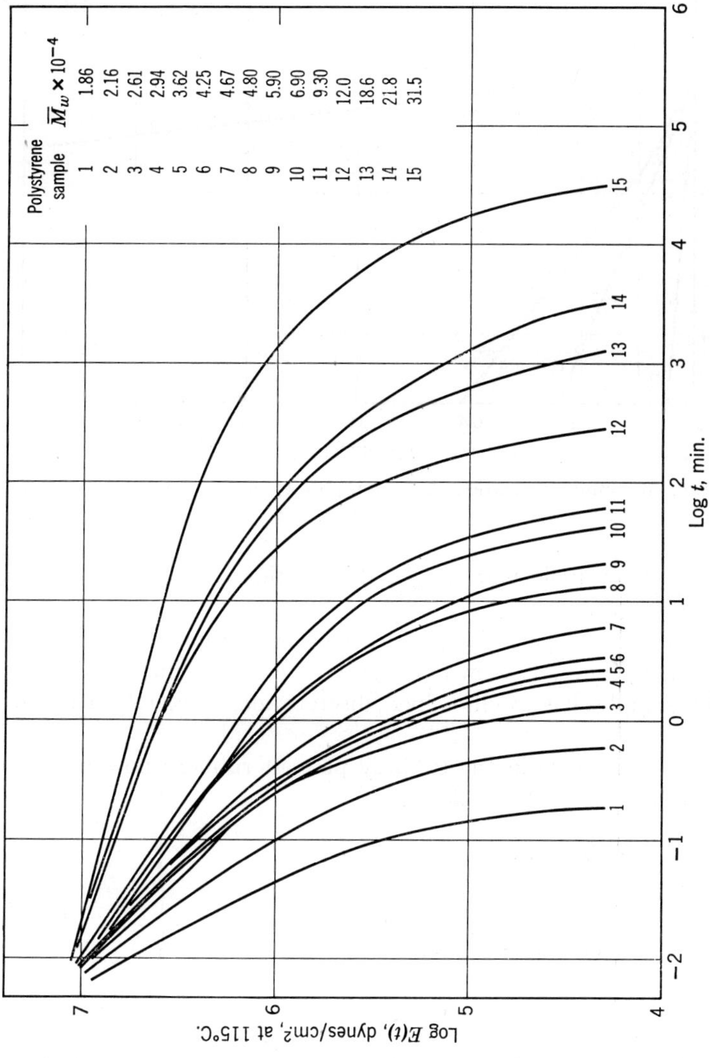

Fig. IE 5. Stress relaxation master curves for polystyrene in the rubbery → flow regions.

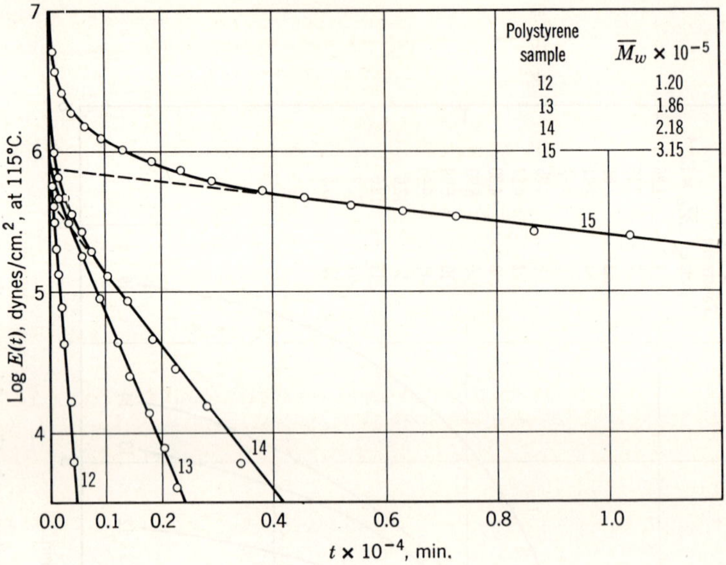

Fig. IE-6. Maximum relaxation times for polystyrene in "rubbery flow-liquid flow" region.

have a narrow molecular weight distribution, i.e., a heterogeneity index of 1.5 (9).

In this region, a plot of log $E(t)$ versus *linear* time approaches linearity at long times, as shown in Figure IE-6. This implies that each of these samples has a well-defined, discrete, maximum relaxation time (9).

The master curve in the rubbery plateau–rubbery flow–liquid region can be expressed as follows:

$$\left.\begin{aligned} E(t) &= \phi(t) \\ \phi(t) &= (E_2)_{t \cong \tau_c} \\ \phi(t) &\cong -E_0 \ln (t/\tau_m); \quad \tau_c < t < \tau_m \\ \phi(t) &\cong E_m \exp \{-t/\tau_m\}; \quad t > \tau_m \end{aligned}\right\} \qquad \text{IE3-(7)}$$

$\phi(t)$ is a decay or relaxation function which has the properties shown in eq. (7). τ_c is the minimum relaxation time of the rubbery flow region, and E_2 is the rubbery plateau modulus. The quantities τ_m and E_m are the maximum relaxation time and its associated partial modulus. The quantity E_0 correlates inversely with the heterogeneity index of the polymer; being larger for a smaller index.

There are theoretical grounds for expecting a monodisperse (a uniform chain-length) polymer to have a sharply defined maximum relaxation time, associated with the motion of entire molecules. Experimentally, even polydisperse polystyrene shows a sharply defined maximum relaxation time, i.e., a plot of log $E(t)$ versus linear t approaches a straight line for sufficiently long times.

One can express $\tau_m \cong f(T, \bar{M}_w)$ as follows:

$$\log \tau_m = \log A - 17.44(T - T_g)(51.6 + T - T_g)^{-1} + 3.4 \log \bar{Z}_w \qquad \text{IE3-(8)}$$

The second term on the right is from the WLF relation, eq. (1). $\bar{Z}_w$ is the weight-average number of links in the polymer chain. It was found that $\log A = -0.70$ for polystyrene, -0.52 for polyisobutylene, and 0.86 for poly(vinyl acetate).

The following relation has been obtained among τ_m, $\langle \overline{R^2} \rangle$, T, and T_g.

$$\log \tau_m = \log B - 17.44(T - T_g)(51.6 + T - T_g)^{-1} + 3.4 \log (\langle \overline{R^2} \rangle / l_0) \qquad \text{IE3-(9)}$$

In eq. (9) $\langle \overline{R^2} \rangle$ is the mean-square end-to-end distance of the polymer chains in a "Θ-solvent" (a solvent composition for which the polymer is just at the point of precipitation) (8) and l_0 is the length of the C—C bond, or 1.54 A. (See Subsection IC2b, p. 51, regarding $\langle \overline{R^2} \rangle$.)

Whereas the value of τ_m reflects the $\bar{M}_w$ of the polymer (or the end-to-end dimensions of the polymer molecule), the *shape* of $E(t)$ reflects the spread of the molecular weight distribution. (Further work is required to ascertain whether τ_m is perhaps best represented as varying with $\bar{M}_w$ or with some other type of average molecular weight.)

IE4. Rouse-Bueche Theory

An analysis of the dynamic mechanical properties of an idealized, linear amorphous polymer chain in colloidal solution leads to the following relation for the relaxation modulus (1,5):

$$E(t) = 3NkT \sum_{p=1}^{n} \exp \{-tp^2/n^2\tau_{\min}\} \qquad \text{IE4-(10)}$$

$$\tau_{\min} = f_0 a^2 / 6\pi^2 kT \qquad \text{IE4-(11)}$$

In eq. (10), N is the number of polymer molecules per cubic centimeter, n is the number of freely rotating segments per molecule, k is Boltzmann's constant, T is the absolute temperature, and $\tau_{\min}$ as defined by eq. (11) is the minimum relaxation time of the system. In eq. (11), f_0 is a frictional coefficient which is proportional to the "segmental viscosity" in a given solvent and is presumed to have a similar temperature dependency to that specified by the WLF equation. The quantity a^2 is the mean-square length of the freely rotating segments of the polymer chain. (See Subsection IC2b, p. 51.)

The system obviously has n discrete relaxation times defined as follows:

$$\tau_p = \tau_{\min}\, n^2/p^2; \quad p = 1, 2, 3, \ldots, n \qquad \text{IE4-(12)}$$

Equations (10)–(12) have been applied to interpret the $E(t)$ of undiluted polymer. One must consider f_0 as pertaining to a particular segment moving through a hypothetical fluid consisting of the neighboring polymer segments.

In its unmodified form, eq. (10) does give a reasonable explanation of the behavior of $E(t)$ in the lower modulus portion of the transition region. In fact, for $E(t) = 10^9$ to 10^7, it predicts a relation for $E(t)$ similar to eq. (4) with $q \cong 0.5$.

In its *unmodified* form, however, the Rouse-Bueche equation does not explain the high modulus portion of the transition, the rubbery plateau, the rubbery flow or liquid flow regions. Modifications of the theory have been proposed to explain the experimental facts, but these are not yet nearly as fundamentally based or as convincing as the original theory.

IE5. Crosslinked Amorphous Polymers

A very wide range of "crosslink densities" in an amorphous polymer system can, for example, be achieved by *copolymerizing* ethyl acrylate with ethyleneglycol dimethacrylate (EGDM) with the use of varying mole fractions of EGDM.

In Figure IE-7 are shown the modulus–temperature curves for a series of these polymers (2). In this case the *shear* relaxation modulus measured at 10 sec. is plotted against temperature.

As with linear polystyrene, the modulus–temperature curve for the "homopolymer" of ethyl acrylate, i.e., pure poly(ethyl acrylate)

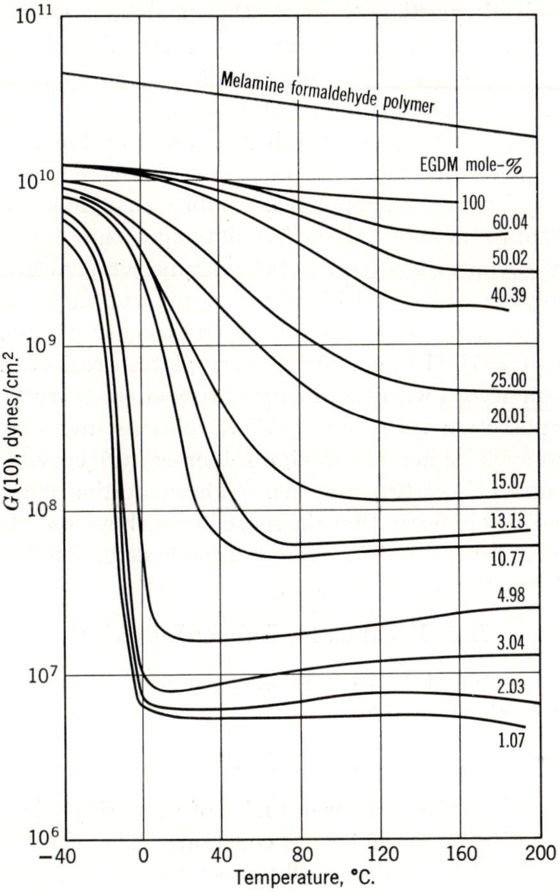

Fig. IE-7. Modulus–temperature curves of copolymers of ethyl acrylate–ethyleneglycol dimethacrylate and of melamine formaldehyde polymer.

shows the usual five regions of viscoelastic behavior. Introduction of small amounts of "crosslinking agent" (EGDM) suppresses the regions of rubbery flow and liquid flow. What remains is a glassy, a transition, and a rubbery plateau region.

Increasing the amount of the EGDM increases the temperature of onset of the transition region, as can be seen in a qualitative way from Figure IE-7. Increasing the amount of EGDM also increases the rubbery plateau modulus, as also is clear from Figure IE-7.

An unresolved question concerns the maximum extent of crosslinking consistent with maintaining a rubbery plateau region. At present, this can be determined only qualitatively by an examination of the shape of the modulus-temperature curve.

For comparative purposes we have shown in Figure IE-7 the modulus–temperature curve of melamine formaldehyde polymer. This very highly crosslinked polymer exhibits a modulus–temperature curve which appears to be "glassy" at all temperatures until the chain *degrades* by chemical changes. The modulus–temperature curve of the homopolymer of EGDM likewise does not appear to have a rubbery plateau region. For EGDM contents below 15 mole-%, the copolymers of EGDM have true rubbery plateau regions which are, indeed, in full accord with the theory of rubber elasticity as next discussed in Subsection IE6. For EGDM contents up to 25 mole-%, and perhaps even higher, the modulus–temperature curves appear to indicate "rubbery" plateau regions. (The quotation marks around the word rubbery indicate that the sense in which we use the term in the above discussion is only very qualitatively defined from the shape of the modulus–temperature curve.)

IE6. Equation for Rubber Elasticity

The equation of state for rubber elasticity gives the following formula for the shear modulus:

$$G = \mathcal{E}NRT \qquad \text{IE6-(13)}$$

In eq. (13), N is the number of moles of network chains per cubic centimeter, and R is the molar gas constant. The quantity $\mathcal{E}$ is the "front factor" which is unity at all temperatures for "ideal" rubbers. (See Subsection IC3, p. 59.) For nonideal rubbers, $\mathcal{E} \neq 1$, and may or may not be temperature-dependent (10).

For the ethyl acrylate–EGDM copolymers, the molar concentration of EGDM crosslinks can be readily computed from the copolymer composition and density. N in eq. (13) is simply twice the number of moles of EGDM crosslinks per cubic centimeter; the latter being obtained from the stoichiometric quantity corrected by the crosslinking efficiency.

All the quantities in eq. (13) except $\mathcal{E}$ are directly measurable. For the copolymers shown in Figure IE-7, $\mathcal{E}$ was computed at 426°K. from the experimentally determined G and N. It was found that $\mathcal{E} = 1$

for <13 mole-% EGDM, or $N = 2 \times 10^{-3}$ moles/cm.^{3}. This perhaps is an unexpectedly high concentration of crosslinks for a polymer exhibiting ideal rubber elasticity.

For other polymers, $\mathcal{E}$ also appears to be relatively independent of crosslink concentration up to 10^{-3} moles/cm.^{3}, but definitely is a function of chain structure. For a series of acrylates and methacrylates, it was found that $\mathcal{E}$ decreased with increasing length of the side chain. Also, $\mathcal{E}$ was temperature-independent for acrylates, but increased with decreasing temperature for methacrylates. This indicates that the end-to-end dimensions of polyacrylate chains are independent of temperature, but that the end-to-end dimensions of methacrylates decrease with decreasing temperatures (10).

IE7. Interchanging Networks

The introduction of crosslinks into an amorphous polymer completely suppresses the regions of rubbery flow and liquid flow until such temperatures are reached at which the network degrades chemically, or at which it will undergo reversible interchange reactions. An excellent example of the latter are the polysulfides.

A series of copolymers was prepared by reacting various amounts of ethylene dichloride (CH_2ClCH_2Cl), 1,2,3-trichloropropane (CH_2Cl-$CHClCH_2Cl$), and sodium tetrasulfide (Na_2S_4). Networks are produced whose repeating segments are

—$SSCH_2CH_2SS$— (A)

—$SSCHCH_2SS$— with —S—S— attached to CH (B)

The modulus–temperature curves of a series of these tetrasulfide polymers with varying mole fractions of the crosslinking unit B are shown in Figure IE-8 (11). There is a downturn of the rubbery plateau modulus which is due to interchange between neighboring tetrasulfide linkages in the network.

All of these polymers obey the following equation:

$$E(t) = E(0) \exp \{-k\beta t\} \qquad \text{IE7-(14)}$$

In eq. (14), $E(0)$ is the modulus at $t = 0$, calculable from the statistical theory of rubberlike elasticity and the density of crosslinking

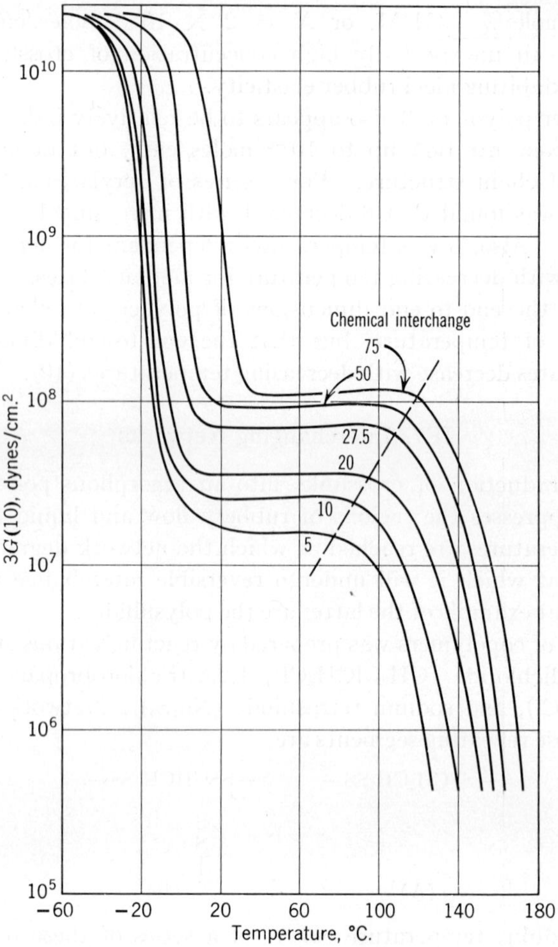

Fig. IE-8. Modulus–temperature curves for crosslinked ethylene tetrasulfide polymers showing the region of chemical interchange.

(related to the mole fraction of B). The quantity β is the number of tetrasulfide links for each network chain, and also can be calculated from the mole fraction of B. The quantity k is the specific rate constant for the tetrasulfide interchange reaction and was shown to be given (11) by

$$k = 5.86 \times 10^{12} \exp \{-25{,}900/RT\} \text{ sec.}^{-1} \qquad \text{IE7-(15)}$$

IE8. Modulus–Temperature Curves: Crystalline Polymers

In Figure IE-9, log $G(10)$ versus temperature is shown for a few homopolymers and copolymers of ethylene and propylene.

The ethylene-propylene copolymer (mole ratio, 2:1) and atactic polypropylene are both linear amorphous polymers and show the typical amorphous, modulus–temperature curves.

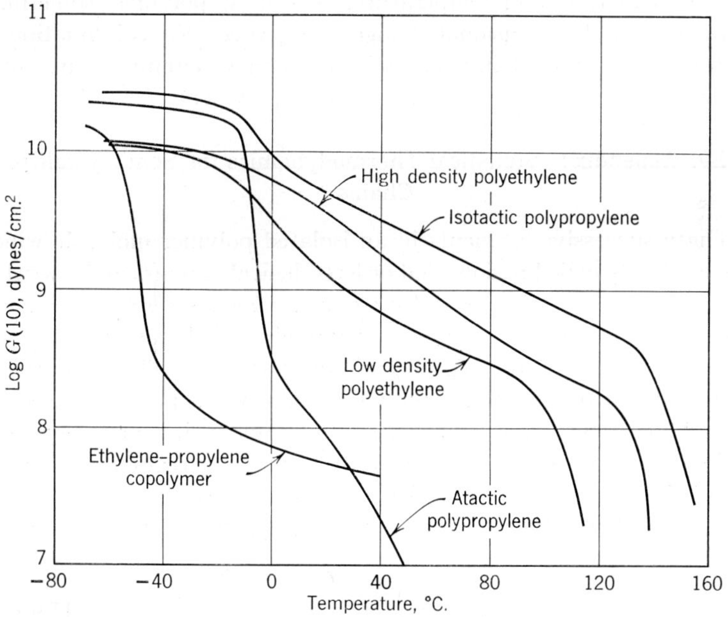

Fig. IE-9. Modulus–temperature curves for some semicrystalline and some amorphous varieties of the homopolymers and copolymers of ethylene and propylene.

On the other hand, low density polyethylene, high density polyethylene, and isotactic polypropylene are semicrystalline polymers with melting points of 110, 137, and 175°C., respectively.

For polypropylene, $T_g = -20$°C., while for polyethylene it has been estimated (from the T_g for polypropylene and that for ethylene–propylene copolymer) to be −81°C.

Between T_g and T_m, semicrystalline polymers have a relatively high modulus, as seen in Figure IE-9, and also a high impact strength due to the resiliency contributions of the rubbery amorphous regions. The crystalline regions act, in part, like crosslinks and, in part, like rigid filler particles (14).

The exact shape of the modulus–temperature curve of a semicrystalline polymer depends on the thermal history of the polymer, i.e., whether quickly quenched from the melt or carefully annealed.

Because of the possibility of melting or growth of crystalline regions with time and temperature, crystalline polymers sometimes show undesired dimensional changes. However, the combination of high modulus and high impact strength over a wide temperature range is a very useful property of semicrystalline polymers.

IE9. Appendix: Statistical Thermodynamics for Semicrystallized Chains*

The n successive segments in an isolated polymer molecule which has randomly coiled regions and ordered helical regions can be written as a sequence $r\,r\,r\,h\,h\,h\,h\,r\,r\,h\,h\,h\,r\,h\,r\,r\,r\,r\,r\,h\,h\,h\,h\,r\,h\,r\,h\,h\,h\,h\ldots$, where r represents a segment in a randomly coiled state and h represents a segment in an ordered crystalline region. The same representation can be used for a polymer chain which is part of a semicrystalline polymer; r representing a segment in an amorphous region and h representing a segment in a crystalline region.

Tobolsky and Gupta (12) have treated the sequence of r and h states as a Markoff chain whose normalized probability matrix is:

$$\mathbf{p}^{(1)} = \begin{array}{c|cc} & r & h \\ \hline r & (1-p) & p \\ h & (1-\alpha) & \alpha \end{array} \qquad \text{IE9-(16)}$$

The quantities p and α are probability parameters: for an r segment converting to an h segment along the sequence, and for an h segment continuing as an h segment along the sequence. These definitions are inherent in the above matrix.

By this treatment we obtained expressions for: the expected value n_h of segments in the crystalline state; the expected value n_r of segments in the amorphous state; the average sequence length $\bar{h}$ of segments in crystalline regions; the average sequence length $\bar{r}$ of segments in amorphous regions; and expressions for the expected

* Reprinted in part from *J. Polymer Sci.*, **A1**, 301 (1963).

number of crystalline sequences $n_{\text{seq},h}$ and expected number of amorphous sequences $n_{\text{seq},r}$.

$$n_h = np/(1 - \alpha + p) \qquad \text{IE9-(17)}$$

$$n_r = n(1 - \alpha)/(1 - \alpha + p) \qquad \text{IE9-(18)}$$

$$\bar{h} = 1/(1 - \alpha) \qquad \text{IE9-(19)}$$

$$\bar{r} = 1/p \qquad \text{IE9-(20)}$$

$$n_{\text{seq},h} = np(1 - \alpha)/(1 - \alpha - p) \qquad \text{IE9-(21)}$$

$$n_{\text{seq},r} = np(1 - \alpha)/(1 - \alpha + p) \qquad \text{IE9-(22)}$$

We also derived an expression for the mean-square length $\langle \overline{R^2} \rangle$ of the polymer chain (12,13):

$$\langle \overline{R^2} \rangle = \left\{ \left[\frac{3 - 3\alpha + p}{3(1 - \alpha + p)} \right] \frac{3 + p}{3 - p} + \frac{2p(2 - \alpha)}{(1 - \alpha)(3 - p)(1 - \alpha + p)} \right\} na^2 \qquad \text{IE9-(23)}$$

where a is the segment length. In addition, we derived a more complex formula which included as a parameter the pitch of the helix in the crystalline regions: in eq. (23) the helices of the crystalline region are assumed to have the minimum possible coiling. The problem of Keller-Till type folding has also been treated by the present author (14). A simplified form of eq. (23) has also been presented (14,15).

(The term segment, as used here, has the same multiple significance as it frequently does in polymer science. It may mean the link of the equivalent freely rotating chain; the chemically repeating unit of the polymer chain; or the atoms associated with an individual bond along the main polymer chain. One or the other uses of the term may be more useful or appropriate for a particular problem.)

The values of α and p that actually obtain for the r and h sequence may result from kinetic conditions (as in rapidly quenched specimens) or the values may, in some cases, be equilibrium values that can be determined by statistical thermodynamic considerations. It is of interest in the latter case to relate the quantities α and p to appropriate segment partition functions.

I now consider the sequence of segments $r\ r\ r\ h\ h\ h\ h\ r\ r\ .\ .\ .$ from the point of view of a linear Ising problem. This would appear to be valid for an isolated chain; for a chain which is a part of a semicrystalline polymer we *assume* that the partition functions of each chain can be computed independently of the others.

We may first rewrite the sequence of segments given at the beginning of the treatment as *r r r k h h h r r k h h r k r r r r r k h h h r k k h h h* . . ., where *k* now represents the first of a sequence of *h*-type segments. This notation is used because we shall assign a special partition function to these segments.

By the matrix method (3), the basic matrix **M** for this problem is

$$\mathbf{M} = \begin{array}{c|ccc} & r & k & h \\ \hline r & f_r^{1/2} f_r^{1/2} & f_r^{1/2} f_k^{1/2} & 0 \\ k & f_k^{1/2} f_r^{1/2} & 0 & f_k^{1/2} f_h^{1/2} \\ h & f_h^{1/2} f_r^{1/2} & 0 & f_h^{1/2} f_h^{1/2} \end{array} \qquad \text{IE9-(24)}$$

The quantities f_r, f_h, and f_k which appear above are the partition functions for the *r*, *h*, and *k* segments.

The partition function *Q* for the chain is:

$$Q = \lambda_1^n \qquad \text{IE9-(25)}$$

where λ_1 is the largest root of the matrix **M**. This root is

$$\lambda_1 = \left[f_r + f_h + \sqrt{(f_r - f_h)^2 + 4f_k f_r}\right]/2 \qquad \text{IE9-(26)}$$

When f_k/f_r and f_k/f_h are both smaller than unity it is easy to see that λ_1 is approximately f_r or f_h, depending on which is the larger. There is a region of transition between these two values whose sharpness depends on how small f_k/f_r and f_k/f_h are. If these quantities approach zero, the transition becomes perfectly sharp.

The number of *r* segments, *k* segments, and *h* segments exclusive of *k* segments will be denoted as n_r, n_k, and n_h^*, respectively. These values are given by:

$$\begin{aligned} n_r/n &= d \ln \lambda_1/d \ln f_r \\ &= (f_r/\mathrm{R})(f_r - f_h + 2f_k + \mathrm{R})(f_r + f_h + \mathrm{R}) \end{aligned} \qquad \text{IE9-(27)}$$

$$\begin{aligned} n_k/n &= d \ln \lambda_1/d \ln f_k \\ &= (f_k/\mathrm{R})\ 2f_r/(f_r + f_h + \mathrm{R}) \end{aligned} \qquad \text{IE9-(28)}$$

$$\begin{aligned} n_h^*/n &= d \ln \lambda_1/d \ln f_h \\ &= (f_h/\mathrm{R})(f_h - f_r + \mathrm{R})/(f_r + f_h + \mathrm{R}) \end{aligned} \qquad \text{IE9-(29)}$$

where R is defined as

$$R = +\sqrt{(f_r - f_h)^2 + 4f_kf_r} \qquad \text{IE9-(30)}$$

The above statistical thermodynamic treatment of a chain with "amorphous segments" and "crystalline segments" is conceptually and mathematically very similar to the solution of the helix–coil transition in synthetic polypeptides proposed by Zimm and Bragg (17). The matrix proposed by these authors is somewht more complex than the matrix presented here because in the polypeptide problem it was necessary to exclude *h* states preceded by only one or two *r* states, i.e., the configurations *h r h* or *h r r h*. Nevertheless, the mathematical results are very similar whether or not one includes this restriction. Moreover, if the Ising model can be applied to a polymer chain which is a part of a semicrystalline polymer, there is no physical reason to include these restrictions inasmuch as a chain can touch a crystallite boundary and immediately return to the amorphous region.

I now turn to the problem of relating the partition functions f_r, f_h, and f_k to the quantities α and p which appear in the markoffian matrix. Inasmuch as the number of amorphous sequences is equal to the number of crystalline sequences,

$$\bar{r} = n_r/n_k = (f_r - f_h + 2f_k + R)/2f_k \qquad \text{IE9-(31)}$$

Similarly,

$$\bar{h} = (n_h{}^* + n_k)/n_k = [f_h(f_h - f_r + R) + 2f_kf_r]/2f_kf_r \qquad \text{IE9-(32)}$$

However, simple expressions for $\bar{r}$ and $\bar{h}$ in terms of α and p have previously been given in eqs. (19) and (20). Combining eq. (20) with eq. (31), and eq. (19) with eq. (32) leads to

$$p = 2f_k/(f_r - f_h + 2f_k + R) \qquad \text{IE9-(33)}$$

$$\alpha = [f_h(f_h - f_r + R)]/[f_h(f_h - f_r + R) + 2f_kf_r] \qquad \text{IE9-(34)}$$

These relations are valid only if statistical thermodynamic equilibrium is reached. Under these conditions eq. (23) for the mean-square length of the polymer chain can in principle be determined from the segment partition functions f_r, f_h, and f_k.

I shall now specialize to consider the segments of our chain to be the individual bonds along the main polymer chain. For many

TABLE IE-II

x	n_h^*/n			n_k/n			n_r/n			α			p		
	$f_rf_k =$			$f_rf_k =$			$f_rf_k =$			$f_rf_k =$			$f_rf_k =$		
	0.1	0.01	0.0001	0.1	0.01	0.0001	0.1	0.01	0.0001	0.1	0.01	0.0001	0.1	0.01	0.0001
0.10	1.000	1.000	1.000	0.000	0.000	0.000	0.000	0.000	0.000	1.000	1.000	1.000	1.000	1.000	1.000
0.50	0.999	1.000	1.000	0.000	0.000	0.000	0.000	0.000	0.000	1.000	1.000	1.000	0.78	0.78	0.78
0.75	0.989	1.000	1.000	0.002	0.000	0.000	0.008	0.000	0.000	0.998	1.000	1.000	0.290	0.45	0.45
0.90	0.915	0.999	1.000	0.009	0.000	0.000	0.077	0.001	0.000	0.991	1.000	1.000	1.114	0.196	0.196
0.95	0.781	0.995	1.000	0.014	0.000	0.000	0.205	0.005	0.000	0.983	1.000	1.000	0.068	0.050	0.100
1.00	0.482	0.498	0.503	0.018	0.002	0.000	0.500	0.501	0.540	0.965	0.996	1.000	0.036	0.004	0.000
1.05	0.217	0.059	0.000	0.015	0.000	0.000	0.768	0.994	1.000	0.934	0.954	0.95	0.020	0.000	0.000
1.10	0.106	0.002	0.000	0.012	0.000	0.000	0.882	0.998	1.000	0.901	0.918	0.92	0.013	0.000	0.000
1.50	0.011	0.000	0.000	0.005	0.000	0.000	0.989	1.000	1.000	0.711	0.75	0.71	0.005	0.000	0.000
2.00	0.001	0.000	0.000	0.003	0.000	0.000	0.996	1.000	1.000	0.61	0.61	0.61	0.003	0.000	0.000

crystalline polymers the entropy of fusion ΔS_f per mole of chain bonds is R; the heat of fusion ΔH_f per mole of chain bonds ranges from 350 to 1500 cal. For explicit calculation I shall take $f_h = e^{\Delta H_f/RT}$ and $f_r = e$. This corresponds to a value $\Delta S_f = R$ for each mole of bonds. For $f_k f_r$ we shall take values of 0.1, 0.01, and 0.0001. In Table IE-II values of n_h^*/n, n_k/n, n_r/n, α, and p are presented as functions of the variable $x = RT/\Delta H_f$ for these three cases.

In eq. (23) is presented the result for the value of $\langle \overline{R^2} \rangle$ for the molecules in an unoriented crystalline polymer. The value of $\langle \overline{R_x^2} \rangle$ for the molecules of a semicrystalline polymer whose crystallites are completely oriented in the x direction has also been calculated (12). $\langle \overline{R_x^2} \rangle$ is the value of $\langle \overline{R^2} \rangle$ projected in the x direction. $\langle \overline{R_{x,0}^2} \rangle$ is the value of $\langle \overline{R_x^2} \rangle$ for the same polymer chain when the polymer is brought into a completely amorphous condition. For this we gave

$$\langle \overline{R_x^2} \rangle / \langle \overline{R_{x,0}^2} \rangle = [2p(2-\alpha)/(1-\alpha)(1-p)(1-\alpha+p)] + [(1+p)/(1-p)] \quad \text{IE9-(35)}$$

In Table IE-III values of $\langle \overline{R^2} \rangle$ are given from eq. (23) and $\langle \overline{R_x^2} \rangle$ $\langle \overline{R_{x,0}^2} \rangle$ from eq. (35) for various values of $x = RT/\Delta H_f$, computed from the values of α and p listed in Table IE-II. These values of α and p are, of course, obtained from our statistical thermodynamic equilibrium theory.

TABLE IE-III

	$\langle \overline{R^2} \rangle / na^2$ from eq. (23)			$\langle \overline{R_x^2} \rangle / \langle \overline{R_{x,0}^2} \rangle$ from eq. (35)		
		$f_r f_k =$			$f_r f_k =$	
x	0.1	0.01	0.0001	0.1	0.01	0.0001
0.75	215.	—	—	832.	—	—
0.90	72.	—	—	238.	—	—
0.95	34.	—	—	107.	—	—
1.00	10.8	—	—	32.1	—	—
1.05	3.38	1.02	1.00	8.64	1.00	1.00
1.10	1.80	1.02	1.00	3.63	1.00	1.00
1.15	1.50	1.00	1.00	1.16	1.00	1.00
2.00	1.18	1.00	1.00	1.06	1.00	1.00

The values of $\langle \overline{R^2} \rangle / na^2$ and $\langle \overline{R_x^2} \rangle / \langle \overline{R_{x,0}^2} \rangle$ approach infinity as α approaches unity. Inasmuch as α was calculated only to three places we do not tabulate $\langle\overline{R^2}\rangle/na^2$ and $\langle\overline{R_x^2}\rangle/\langle\overline{R_{x,0}^2}\rangle$ for α = 1.000.

It should again be emphasized that eqs. (23) and (35) are the simplest forms of a more general equation which (12) accounts for short-range and long-range folding of the molecules in the crystalline regions. Equations (23) and (35) are valid if there is no folding and if the unit cell of the crystallites is fully extended. The application of these results to more complex crystalline structures presents no difficulties in principle.

Using the theory for molecular dimensions presented here, a theory of fiber shrinkage was developed (14,15).

Dr. M. Takahashi is to be acknowledged for his aid in the computations for Tables IE-II and IE-III, while the parent paper from which this Appendix has been taken in part was made possible through a grant from the California Research Institute.

IE10. Summary

Illustrations of the viscoelastic behaviors of several synthetic polymeric solids are provided for stress relaxations in tension, shear, and torsion. Five mechanical regions are characterized in temperature and time with linear amorphous, polymeric bodies while crosslinked and semicrystalline polymers also are considered. Time-temperature master curves, interchanging networks, and the Rouse-Bueche theory are outlined. Some statistical aspects of semicrystalline chains are treated in the Appendix.

This is an abbreviated, broad perspective of several of the key experimental and theoretical factors met in the mechanics of polymeric solids. Emphasis is placed on the structure and properties of chain-like molecules together with their segmental compositions and physical contributions.

References

1. Bueche, F., *J. Chem. Phys.*, **22**, 603 (1954).
2. Katz, D., and A. V. Tobolsky, *J. Polymer Sci.*, **A2**, 1595 (1964).
3. Kramers, H. A., and G. Wannier, *Phys. Rev.*, **60**, 252 (1941); E. W. Montroll, *J. Chem. Phys.*, **9**, 706 (1941); E. N. Lassettre and J. P. Howe, *J. Chem. Phys.*, **9**, 747 (1941).
4. Mark, H. F., and A. V. Tobolsky, *Physical Chemistry of High Polymeric Systems*, 2nd Ed., Interscience, New York, 1950, p. 274.
5. Rouse, P. E., *J. Chem. Phys.*, **21**, 1272 (1953).

6. Tobolsky, A. V., and J. R. McLoughlin, *J. Polymer Sci.*, **8**, 543 (1952).
7. Tobolsky, A. V., and M. Takahashi, *J. Appl. Polymer Sci.*, **7**, 1341 (1963).
8. Tobolsky, A. V., *Structure and Properties of Polymers*, Wiley, New York, 1960, pp. 160–166.
9. Tobolsky, A. V., and K. Murakami, *J. Polymer Sci.*, **40**, 443 (1959); *ibid.*, **47**, 55 (1960).
10. Tobolsky, A. V., D. Carlson, and N. Indicator, *J. Polymer Sci.*, **54**, 175 (1962).
11. Tobolsky, A. V., R. B. Beevers, and G. D. T. Owen, *J. Colloid Sci.*, **18**, 353, 359 (1963).
12. Tobolsky, A. V., and V. D. Gupta, *J. Chem. Phys.*, **36**, 430 (1962).
13. Tobolsky, A. V., and V. D. Gupta, *Textile Res. J.*, **32**, 85 (1962).
14. Tobolsky, A. V., *J. Chem. Phys.*, **37**, 1139 (1962).
15. Tobolsky, A. V., *Textile Res. J.*, **32**, 252 (1962).
16. Williams, M. L., R. F. Landel, and J. D. Ferry, *J. Am. Chem. Soc.*, **77**, 3701 (1955).
17. Zimm, B. H., and J. K. Bragg, *J. Chem. Phys.*, **28**, 1246 (1958); *ibid.*, **31**, 526 (1959).

CHAPTER II

Brittle-Like Fracture

Sections

IIA. General Theory of Brittle Fracture, J. P. Berry

IIB. Brittle Behavior of Polymeric Solids, J. P. Berry

IIC. Fracture Topography, I. Wolock and S. B. Newman

IIA. GENERAL THEORY OF BRITTLE FRACTURE

J. P. BERRY

IIA1. General Considerations of Fracture and Strength

One of the principal aims of theories on behavior of materials is to relate the observable effect of the imposed conditions to the response of the material on a molecular or atomic level or, conversely, to predict the macroscopic material properties from the known structure. In the case of the ultimate (macroscopic) mechanical strength it would appear that this property should be related to the cohesive forces acting between the elementary constituents and their arrangement in the body. The fracture of the material under stress involves the rupture of those bonds which intersect the plane defined by the growing fracture surfaces. Then, the theoretical strength should presumably be the sum of the forces required to break the individual bonds. Hence the computation of the strength involves the estimation of the number and nature of the bonds which are to be broken, and the forces required to do so, from the known or presumed molecular structure of the material and the fracture locus. The various treatments of this problem, while they differ somewhat in complexity, give values for the strength which are much greater than those usually observed (17,37,41). However, in experiments on specially prepared samples under carefully controlled laboratory conditions, it is possible to approach the theoretical strength (19,23,32).

The discrepancy between the calculated and normally observed values of the strength is due to the assumption, implicit in the theoretical analyses, that the stress arising from the applied load is distributed uniformly through the sample. In general, however, the sample is not physically perfect; structural imperfections arise either as a result of the nature of the material, as in crystal defects, or as a result of mechanical handling or fabrication procedures, when they take the form of adventitious scratches, cracks, etc.

In a great many cases the imperfections introduced during sample preparation are large relative to molecular dimensions. Under these conditions, the body's responses to applied mechanical loads frequently can be represented adequately as those of an elastic continuum; hence these cases can be treated within formal elasticity. Of course, this procedure immediately precludes interpretation of observed behavior directly in molecular terms, although attempts have been made to bridge this conceptual chasm (15,39).

The relatively low value of the empirical strength can then be attributed to the presence of defects which locally affect the stress distribution in the body. In an isotropic medium, the failure point is reached when the largest defect increases in size, and it is necessary to define the condition of stress at which this occurs. Two approaches are possible:

From the stress distribution equations, it is known that the local stress in the vicinity of a flaw is much higher than the average value throughout the body. Hence a limiting value, corresponding to the "true" strength, will be attained at this point when the average stress is still comparatively low. Then local failure will occur, causing an increase in the size of the defect which, in turn, results in an increase in the local stress. Consequently the system becomes progressively less stable and rupture ensues.

Alternatively, the energetics of the process may be considered, the criterion for failure being defined in terms of the conditions under which the total energy of the (conservative) system decreases with increasing flaw size. The condition that the system be conservative demands that when it is subjected to constant imposed stress, allowance must be made for the work done on the system by the stress as the external boundaries are displaced. This is essentially a thermodynamic argument, and the conditions so defined are expected to be necessary for failure. In this energetic case as well, flaw extension becomes increasingly favorable with increasing flaw size.

It is clear that the viewpoints adopted in the two types of theory are quite different. In the first, attention is directed specifically to a particular point in the system where the stress attains a maximum value. In the continuum models normally considered, the flaw is assumed to have the form of a line crack of zero thickness, and consequently the radius of curvature of the tip of the crack is zero. This geometry gives rise to an infinite singularity in the stress field. Of course, in real systems the limiting radius of curvature at the crack tip will be related to, and limited by, the atomic nature of the material. By the same token, the "true" strength of the material, and therefore the limiting stress, will be related to interatomic bond strengths. Consequently, in applying this criterion for fracture, it is necessary at some stage to depart from the concepts embodied in the continuum model.

In the energetic treatment, the system is considered as a whole; no conceptual difficulties arise when the limiting case of a line crack is considered. Furthermore, slight departures from the limiting condition have a negligible effect on the predictions of the theory; a result which makes possible the application of the theory to real materials (19).

The two criteria for failure and the relation between them were discussed by Griffith (19), who pointed out that "the application of the mathematical theory of elasticity on the basis that the crack is assumed to be a traction-free surface must give the stresses correctly at all points of the body, with the exception of those near the ends of the crack," i.e., in the region which is of greatest interest in the limiting stress theory. Griffith went on to say "in a sufficiently large crack, the error in the strain energy so calculated must be negligible" (19).

The relation between these two approaches will be considered in more detail below. The second has received more attention than the first, most flaw theories being developed from this point of view. In spite of the shortcomings inherent in the continuum mechanics treatment, the theories which attempt to describe strength and fracture phenomena in such terms have provided valuable insight into the factors which determine the ultimate mechanical properties of materials.

IIA2. The Griffith Theory of Brittle Fracture

The influence of flaws on the stress field in a two-dimensional body was first treated by Inglis (24). The body considered was an infinite sheet containing a central elliptical hole, the imposed stresses being applied at the external boundaries of the sheet. An elliptical coordinate system was employed, the boundary of the defect (in the undeformed condition) being defined by the constant ξ_0. This is related to the dimensions of the corresponding ellipse by the relations

$$\begin{aligned} a &= c \cosh \xi_0 \\ b &= c \sinh \xi_0 \end{aligned} \qquad \text{IIA2-(1)}$$

where a and b are the lengths of the major and minor semi-axes respectively, and $2c$ is the interfocal distance.

The classical treatment of the fracture problem using the energetic approach is that by Griffith (19,20), who considered the model of Inglis, above, assuming that the defect was a line crack of length $2c$ ($\xi_0 = 0$). Two contributions to the energy of the system were considered: the elastic strain energy U, and the surface energy T. The former arises from the work done on the system by the imposed stresses, while the latter arises from the energy required for the formation of the fracture surfaces as the original defect increases in size. (This is analogous to the surface tension of a liquid and has the same dimensions.) Both contributions to the energy are functions of the defect's size, and it is necessary to establish the nature of this dependence. The form of the strain energy function was obtained by Griffith from the equations of the stress and displacement fields given by Inglis.

The elastic strain energy of an infinite sheet of unit thickness and containing an elliptical hole ξ_0, is greater than that of a continuous sheet by an amount

$$\Delta U = (\pi S^2 c^2 / 2G)\, q \cosh 2\xi_0 \qquad \text{IIA2-(2)}$$

when both are subjected to a uniform stress S at the infinite boundary (19). G is the shear modulus, $q = (1 - \mu)$ for a condition of plane strain, and $q = 1/(1 + \mu)$ for a condition of plane stress, where μ is Poisson's ratio. These conditions are determined by the ratio of the size of the hole to the thickness of the sheet. If this ratio is large, the system is under plane stress, and if it is small, the system is under plane strain. In the limit where the hole becomes a line crack ($\xi_0 = 0$),

$$\Delta U = (\pi S^2 c^2/2G)q \qquad \text{IIA2-(3)}$$

Although the strain energy of the sheet is increased because of the presence of the crack, its deformability (stiffness) is reduced. If a crack is introduced into an infinite sheet subjected to a constant uniform stress S, the points of application of the stress will be displaced, and the work done by the stress can be shown (39) to be $2\Delta U$; thus the net energy of the system is reduced by an amount ΔU because of the crack. Similarly, if the sheet boundaries are considered fixed, the insertion of the crack will reduce the strain energy of the system by ΔU.

The increase in the surface area of the system is $4c$, since the sheet is of unit thickness. If the specific surface energy γ is defined as that required for the formation of unit area of fracture surface, by analogy to the surface tension, the corresponding increase in the surface energy of the system is $\Delta T = 4\gamma c$. Hence the change in the total energy is

$$\Delta H = \Delta T - \Delta U = 4\gamma c - (\pi S^2 c^2/2G)q \qquad \text{IIA2-(4)}$$

Griffith then made the assumption that the system would become unstable and the crack would increase in size if $(\partial \Delta H/\partial c) = 0$. Applying this criterion to the expression given above, the critical stress for failure is found to be

$$S_c = 2(G\gamma/\pi c q)^{1/2} \qquad \text{IIA2-(5)}$$

which reduces to

$$S_c = \{2E\gamma/[\pi c(1 - \mu^2)]\}^{1/2} \qquad \text{IIA2-(6)}$$

for plane strain, and

$$S_c = (2E\gamma/\pi c)^{1/2} \qquad \text{IIA2-(7)}$$

for plane stress, where E is Young's modulus. It was noted by Griffith that the stress field associated with the crack was unaffected by a transverse stress, and consequently it was concluded that the failure point was determined solely by the imposed stress acting normal to the axis of the crack.

In his first paper on this topic, the form of the strain energy function derived by Griffith was not, in fact, that given above but was (19); $\Delta U = \pi S^2 c^2 (1 - q)/2G$. Consequently the expressions for the critical stress reduced to

$$S_c = 2[G\gamma/\pi c\mu]^{1/2}; \quad \text{for plane strain} \qquad \text{IIA2-(8)}$$

$$S_c = [2E\gamma/\pi c\mu]^{1/2}; \quad \text{for plane stress} \qquad \text{IIA2-(9)}$$

These expressions were subsequently modified to those quoted above since, according to Griffith, "the expressions used for the stresses gave values at infinity differing from the postulated uniform stress at infinity by an amount which, though infinitesimal, yet made a finite contribution to the energy when integrated round the infinite boundary. This difficulty has been overcome by slightly modifying the expressions for the stresses so as to make this contribution to the energy vanish" (20). However, the point is not further elaborated and no indication is given of the way in which the stress relations were changed. The modified expressions for the critical imposed stress have persisted in the literature and apparently without question until recently, and have now come to be generally accepted as correct.

The derivation of the strain energy function for a system subjected to a uniform biaxial stress and to a uniaxial stress has been given by Wolf (55). In the first case the result agrees with Griffith's original expression, and in the second it is

$$\Delta U = \pi S^2 c^2 (1 + \mu)/4E = \pi S^2 c^2/8G; \quad \text{plane stress}$$

Consequently, it is concluded that the critical stress for fracture is not the same for the two stress systems considered. This point has been further explored by Swedlow, who showed that the generally accepted critical stress equation corresponded to a model subjected to a two-dimensional hydrostatic pressure within the crack rather than a uniform tensile stress at the external boundary (50). It was also shown that the form of the final equation depended on the details of the loading arrangement. The interaction between longitu-

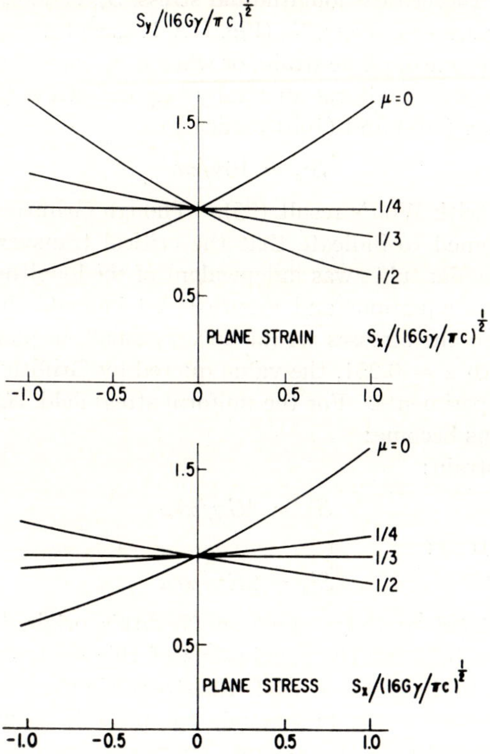

Fig. IIA-1. Loci of brittle fracture criteria for various degrees of biaxial loading (Swedlow, 1962).

dinal and transverse stresses was shown by considering the effect of applying an arbitrary biaxial stress field to the Griffith-Inglis infinite sheet model. [Griffith himself considered this case but adopted a limiting stress rather than an energy criterion for failure (20).] The stress S_x is imposed parallel and S_y is imposed normal to the direction of the central crack of length $2c$. The derived equations for instability are (49):

For plane strain:

$$S^2_y - (1 - 4\mu)S_xS_y = 16G\gamma/\pi c \qquad \text{IIA2-(10a)}$$

For plane stress:

$$S^2_y - [(1 - 3\mu)/(1 + \mu)]\, S_xS_y = 16G\gamma/\pi c \quad \text{IIA2-(10b)}$$

In general, the critical longitudinal stress, S_y, is dependent on the value of the transverse stress, S_x, (Fig. IIA-1) except when $\mu = {}^1/_4$, and the condition is one of plane strain, or when $\mu = {}^1/_3$, and the condition is one of plane stress. For a uniaxially imposed stress ($S_y = S_c$; $S_x = 0$), both eqs. (10a) and (10b) reduce to

$$S^2_c = 16\gamma/\pi c \qquad \text{IIA2-(11)}$$

in agreement with Wolf's result (55). Though Griffith's experimental results seemed to indicate that the critical transverse stress for fracture of circular tubes was independent of the longitudinal loading (19), the above equations and Figure IIA-1 indicate that the interaction between the stresses would be very small, in plane stress, for a material with $\mu = 0.251$, the value quoted by Griffith for the glass used in his experiments. For the uniform stress field, $S_y = S_x = S_c$, these equations become:

For plane strain:

$$S^2_c = 4G\gamma/\pi c\mu \qquad \text{IIA2-(12a)}$$

For plane stress:

$$S^2_c = 2E\gamma/\pi c\mu \qquad \text{IIA2-(12b)}$$

which correspond to those given in Griffith's original paper (19). The difference between the conclusions of this analysis and that of Griffith is due to the fact that while the stresses in the complex system can be superposed, it is, in general, not possible to superpose the energies associated with the separate stress systems (50).

The three-dimensional analog of the sheet model discussed above also has been treated (43,47). In this case the defect in the infinite body is a flat disk crack lying in the x-z plane, and its boundary is defined by $x^2 + z^2 = c^2$; $y = 0$. The stress field associated with such a defect has been obtained from the Westergaard (53) stress functions, hence the dependence of the strain energy of the body on the size of the defect (47). Application of the Griffith criterion for mechanical instability then gave the values of the imposed stresses which would lead to failure. It was found that, in contrast to the two-dimensional model discussed above (49,50,55), the stresses acting parallel to the plane of the crack had no influence on the critical value of the stress acting normal to the crack plane (43). The expression for this stress is

$$S_c = \{\pi E\gamma/[2c(1 - \mu^2)]\}^{1/2} \qquad \text{IIA2-(13)}$$

Comparison of this equation with eq. (11) given above for the two-dimensional model indicates that the ratio of the critical stresses for the two cases is,

$$S_2/S_3 = 4\ (1 - \mu)^{1/2}/\pi \cong 1 \qquad \text{IIA2-(14)}$$

where S_2 is the critical stress for the two-dimensional model and S_3 that for the three-dimensional model.

IIA3. The Irwin Theory of Brittle Fracture

A related, but somewhat different approach to the flaw hypothesis has been adopted by Irwin (25,26). In this case attention is confined to the stress field in the immediate vicinity of the tip of the flaw in the systems, which, as in the models discussed above, is considered to be a line crack or disk crack of zero thickness. The appropriate stress functions for such a model are those of Westergaard (53) which have conveniently been given for a variety of crack and stress systems. In discussing the stress field associated with the flaw, Irwin adopts a two-dimensional polar coordinate system with the origin at the end of the crack. It can then be shown that for values of the radius vector $r << c$, the half-length of the crack, the stresses normal σ_y and parallel σ_x to the crack plane are given by

$$\sigma_y = \frac{K}{(2r)^{1/2}} \cos\frac{\theta}{2}\left(1 + \sin\frac{\theta}{2}\sin\frac{3\theta}{2}\right) \qquad \text{IIA3-(15a)}$$

$$\sigma_x = \frac{K}{(2r)^{1/2}} \cos\frac{\theta}{2}\left(1 - \sin\frac{\theta}{2}\sin\frac{3\theta}{2}\right) \qquad \text{IIA3-(15b)}$$

where θ is the angle between the axis of the crack and the radius vector. The corresponding displacements normal to the crack plane are given by

$$v = (1 - \mu)\ K\ (2r)^{1/2}/G$$

Under the influence of the imposed stress, the boundary of the crack takes the shape of an ellipse. The form of eqs. (15a) and (15b) is independent of the nature of the stress functions used; thus they are applicable to a wide range of crack configurations and stress systems. The parameter K is designated the "stress intensity factor" and is determined by the details of the model considered (26). It is mathematically possible to close up a short segment of length α from the tip of the opened crack, by applying forces in the y direction. These forces are given by

$$f_y(p) = pK/(2x)^{1/2} \qquad \text{IIA3-(16)}$$

along the segment from $x = 0$ to $x = \alpha$, where p is a loading parameter which increases from 0 to 1 as the crack is closed. Simultaneously, the resulting displacements of the boundary in the y direction are given by

$$v(p) = 2(1-p)K[2(\alpha - x)]^{1/2}/E \qquad \text{IIA3-(17)}$$

The work done in closing up the crack segment is then

$$\int_0^\alpha v(0)f_y(1)\,dx = (2K^2/E)\int_0^\alpha [(\alpha - x)/x]^{1/2}\,dx = \alpha K^2\pi/E \qquad \text{IIA3-(18)}$$

Consequently, the strain energy of the system is increased by this amount as the crack length is decreased by α. Conversely, the decrease in the strain energy of the system for a unit increase in crack length, $(\partial\Delta U/\partial c)$, is clearly $K^2\pi/E$. This corresponds to the strain energy derivative in the Griffith theory. In the present case, it should be noted that the change in strain energy is obtained by considering only the stress field in the immediate vicinity of the crack rather than that which obtains over the entire model. It is clear that in performing the integration above, the stress intensity factor, and therefore the imposed stress, is considered constant. Since the effective stiffness of the sheet is increased as the length of the crack is reduced, the external boundaries of the sheet will be displaced and work will be done by the system. This contribution to the total energy is ignored in the theory, though it may be significant in magnitude. In the Irwin theory, the quantity $\mathcal{G} = \pi K^2/E$ is designated the "strain energy release rate" or crack extension force. Its magnitude is controlled by the stress intensity factor K, the form of which is determined by the details of the model and the stress system. In general, for dimensional reasons, $K \propto S$, where S is the imposed stress. In the Griffith theory, the strain energy derivative is equated to the surface energy derivative in the instability criterion to give the value of the critical imposed stress for failure. In the Irwin theory, while the strain energy release rate corresponds to the strain energy derivative of the Griffith theory, it is not equated to a surface energy derivative since Irwin has emphasized that the term includes all contributions to the energy dissipated as the crack advances, including those which are not related directly to that required for the formation of new surface area (26). Another generalization of the Griffith theory was proposed independently by Orowan (38). Because of the versatility of the Westergaard stress functions, the Irwin treatment yields expressions for a critical stress in terms of the strain energy release rate $\mathcal{G}$ for a number of systems of interest. The equivalent Griffith equations for these systems can be obtained by equating $\mathcal{G}$ to the surface energy derivative, thereby introducing into the equations the specific surface energy γ. If the infinite sheet model is considered as containing the central crack of length $2c$ and subjected to a uniform stress S, the stress intensity factor K is then $Sc^{1/2}$, while $\mathcal{G} = (\partial \Delta T/\partial c) = 2\gamma$. Therefore, the value of the critical stress in a form equivalent to that discussed above is

$$S_c = (\mathcal{G}E/\pi c)^{1/2} = (2E\gamma/\pi c)^{1/2} \qquad \text{IIA3-(19)}$$

This agrees with the revised expression given by Griffith, but in view of the fact that it is based on a method of deriving the strain energy derivative which appears open to question, it casts further doubt on the validity of the revised Griffith equation. Furthermore, it is a corollary of the Irwin approach that transverse stresses acting parallel to the plane of the crack do not contribute to the strain energy release rate; hence have no effect on the critical longitudinal stress. This is at variance with the result of Swedlow's analysis (50).

Because of the uncertain validity of the method of deriving the strain energy derivative, it should be noted that the expressions which result from its application may be somewhat in error. However, the discrepancy will be only a relatively small factor associated with the strain energy release rate (or equivalently, the surface energy) and will not influence the general form of the expressions given.

An alternative mode of loading which has been considered in terms of Irwin's theory is that in which wedge forces, P, are applied to the crack boundaries at the midpoint, acting normally to the crack's axis (Fig. IIA-2*a*) (25). Then $K = P/\pi c^{1/2}$ and the critical condition becomes

$$P_c = (\pi c E \mathcal{G})^{1/2} = (2\pi c E\gamma)^{1/2} \qquad \text{IIA3-(20)}$$

A number of other models of interest are summarized below.

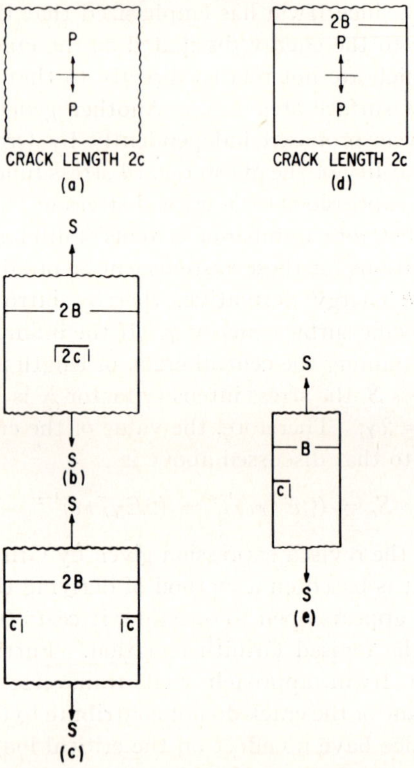

Fig. IIA-2. Elastic models treated by the Irwin theory of brittle fracture.

(*a*) For a sheet of width $2B$ containing a central crack of length $2c$ loaded by a stress S which acts normal to the crack's axis (Fig. IIA-2*b*) (25):

$$K = S\,[(2B/\pi)\tan(\pi c/2B)]^{1/2} \qquad \text{IIA3-(21)}$$

$$S_c = \left[\frac{E\mathcal{G}}{2B\tan(\pi c/2B)}\right]^{1/2} = \left[\frac{E\gamma}{B\tan(\pi c/2B)}\right]^{1/2} \qquad \text{IIA3-(22)}$$

The same relations apply for the case of a sheet of width $2B$ containing two colinear cracks, each of length c, penetrating from opposite edges of the sheet to which a stress S is applied normal to the crack lengths (Fig. IIA-2*c*) (25).

(*b*) For a sheet of finite width $2B$ containing a central crack of length $2c$ with P applied to the midpoint of the crack, acting normally to its axis (Fig. IIA-2*d*) (25):

$$K = P[\pi B \sin(\pi c/B)]^{-1/2} \qquad \text{IIA3-(23)}$$

$$P_c = [E\mathcal{G}B \sin(\pi c/B)]^{1/2} = [2E\gamma B \sin(\pi c/B)]^{1/2} \qquad \text{IIA3-(24)}$$

The stress field equations obtained from the Westergaard functions for a sheet of finite width are somewhat approximate; however this disadvantage is offset by the convenience of the compact form of the equations which results in relatively simple expressions for the critical stress. As B becomes large, the equations approach those which have been given for the infinite sheet models.

Most of the systems discussed above have been selected for theoretical treatment because the mathematical problems encountered have proved tractable. The models have indicated the kind of experimental procedure required to confirm the deductions and to obtain the characteristic parameters which govern the process; the specific surface energy γ (in the Griffith theory) or the strain energy release rate $\mathcal{G}$ (in the Irwin theory). The results of these experimental investigations also help to establish confidence in the theories as applied to conventional test samples and to failure phenomena occurring in service. In some cases, however, it may be difficult to establish the relation between the details of the particular test or service failure and the model system which has been treated theoretically. In particular, the geometry of the naturally occurring flaw may not correspond at all with a line crack penetrating the center of a thin sheet or a disk crack at the center of a more third dimensionally-dependent body. It is more likely to take the form of a crack at one edge of a sheet or a crack of generally semi-elliptical shape penetrating from, but not necessarily normal to, the face of a three-dimensional body. Unfortunately the elastic problems involved in these types of configurations are much more severe, and, in general, a rigorous solution of the stress field equation is not available. The two-dimensional model has received greater attention and perhaps the simplest, although inexact, solution of the problem has been obtained from a Westergaard stress function. For a sheet of width B containing a crack of length c penetrating from one edge of a sheet subjected to a stress S imposed normal to the crack axis, the stress intensity factor is (Fig. IIA-2e) (51)

$$K = Sc^{1/2}\left\{1 - (c/B)^2\left[6 \cosh^{-1} (B/c) - 1/2\right]\right\}^{1/2} \qquad \text{IIA3-(25)}$$

Hence, according to the Irwin theory, the critical stress for failure is

$$S_c = (E\mathcal{G}/\pi c)^{1/2} \{1 - (c/B)^2 [6 \cosh^{-1} (B/c) - 1/2]\}^{-1/2}$$
$$= (2E\gamma/\pi c)^{1/2}\{1 - (c/B)^2 [6 \cosh^{-1} (B/c) - 1/2]\}^{-1/2} \qquad \text{IIA3-(26)}$$

While more exact (and more complex) expressions for the stress field have been obtained, the determination of the corresponding strain energy expressions has not been carried out. Consequently, a rigorous analysis of the problem is not available in terms of the Griffith-Irwin theory. The relation of real systems to theoretical models has been the subject of a recent comprehensive review (23).

IIA4. Fracture By Cleavage

According to the flaw theories discussed above, the fracture surface energy γ or the strain energy release rate $\mathcal{G}$ are important parameters of the material which characterize its resistance to the extension of a flaw or crack. The model systems which have been treated have formed the basis of experiments to determine the value of the material parameters (7,8,16). Another, which has been used extensively for this purpose, is that in which a crack is propagated along the median plane of a strip sample by forces applied at its open ends (Fig. IIA-3). This is usually termed "cleavage" (4,9,18, 36,48). A method somewhat different from those discussed above is used to calculate the strain energy of the system. The sample is

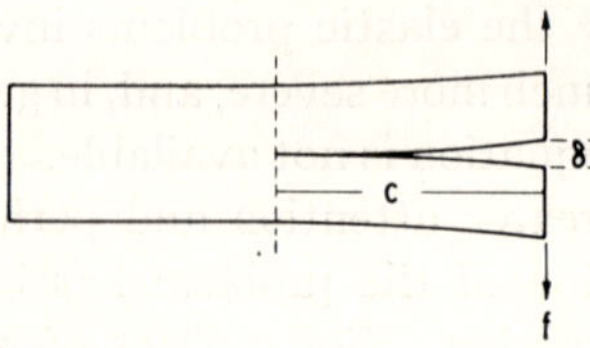

Fig. IIA-3. Cleavage sample.

symmetrical about the median plane and may be regarded as a pair of identical opposed cantilever beams. Therefore, the strain energy has usually been calculated from simple beam theory, although there are objections to this approach arising from the fact that the equations used assume that the beams under consideration are built into a perfectly rigid support; a boundary condition which is not appropriate to the cleavage sample (9). Neglecting the shearing components of the stress, the relation between the applied force f, the length of the beams, i.e., the length of the crack in the cleavage sample c, and the total deflection which results, δ, is, according to the simple beam theory,

$$f = (3EI/2c^3)\delta \qquad \text{IIA4-(27)}$$

where E is Young's modulus and I is the moment of inertia of cross section of half of the sample. Since the validity of this relation as applied to the cleavage system is open to question, it is convenient and appropriate to represent the behavior by an empirical relation of the form $f = (\phi c^{-\nu})\delta$, where ϕ and ν are arbitrary constants (9). For a given crack length, both equations imply that the deflection is proportional to the force and that the proportionality constant m $(=f/\delta)$ is inversely proportional to some power of the crack length, given by the slope of the linear relation, $\log m = \log \phi - \nu \log c$. In order for beam theory to apply, $\nu = 3$, while ϕ is identified with $3EI/2$. Experimental application of this relation usually necessitates a $\nu \cong 2.7$, while ϕ frequently differs from its beam theory-predicted value by a factor approximating two (9).

If the empirical relation given above is assumed, the strain energy of the system is given by

$$U = f\delta/2 = \phi\delta^2/2c^\nu \qquad \text{IIA4-(28)}$$

hence

$$(\partial U/\partial c) = -\nu\phi\delta^2/2c^{\nu+1} = -\nu f\delta/2c \qquad \text{IIA4-(29)}$$

The surface energy is given by $T = T_0 + 2wc\gamma$, where w is the thickness of the sample, i.e., the width of the fracture plane, and T_0 is the surface energy of the system in the absence of the crack.

Then $\partial T/\partial c = 2w\gamma$, and applying the Griffith criterion yields

$$f_c = 4\gamma wc/\nu\delta_c \qquad \text{IIA4-(30)}$$

where f_c and δ_c are values of the applied force and deflection when the crack becomes unstable. Although this expression is perhaps aesthetically less satisfying than those derived from the analytical functions, it has proved most useful for determining values of the fracture surface energy from cleavage experiments (9). Possibly the most important advantage of the technique arises from the fact that the crack in a cleavage sample at fixed deflection is inherently stable (6). If a constant deflection of sufficient magnitude to cause instability is imposed on the sample, the crack will grow until a new equilibrium condition, as given by eq. (30), is attained. Thus the crack can be made to propagate across the sample in a controlled manner and much data can be obtained from that one sample. Subsection, IIB6, p. 217, and Figs. IIB-19 and -20. When failure occurs under the influence of tensile forces applied to a conventional sheet or strip sample, the system becomes progressively less stable as the crack advances. Since no control is possible, only one data point is obtained for each sample.

IIA5. The Limiting-Local-Stress Criterion for Fracture

It was noted above that the condition of instability in a model system involving a flaw can be treated in two ways, and the energy viewpoint has been explored at length. The second criterion which may be adopted is that local failure occurs when the local stress attains a particular limiting value, corresponding to the "true" strength of the material. Since the local stress attains its highest value in the vicinity of the tip of a flaw, it is obvious that this determines the immediate locus of fracture. Application of the limiting stress principle therefore involves consideration of the stress field, particularly in the region of the flaw.

A considerable simplification in the theory is achieved by utilizing Inglis' observation that the stress concentration due to a flaw is essentially independent of its location, at the edge or center of a sheet, and of the details of its shape (24). In the latter case it need not be elliptical; the stress concentration factor being given, to a good approximation, by

$$\sigma_m = 2S(c/\rho)^{1/2} \qquad \text{IIA5-(31)}$$

where S is the imposed stress, c is the length of an edge crack, or the half-length of a central crack, and ρ is the radius of curvature at its tip. It is necessary, however, that the crack length be small compared with the width of the sheet; otherwise consideration must be given to the distortion of the stress field by the further edge of the sheet. It is this factor that gives rise to the various modifications of the simple Griffith-Irwin equation discussed previously. The effect of complex stress systems can also be treated without difficulty since the stresses can be superposed, while the strain energies associated with those stresses cannot, except under certain restricted conditions (50).

In its simplest form, the critical stress for fracture, S_c, can be written in terms of the limiting stress criterion:

$$S_c = \sigma_c(\rho/4c)^{1/2}$$

where σ_c is the true strength, i.e., molecular cohesion of the material, and ρ is a length parameter which defines the effective radius of curvature of the tip of the flaw of length c. This theory then predicts that the critical stress for fracture is inversely proportional to the square root of the flaw size, assuming that the radius of curvature is independent of that size. Since the equations derived from the energetic criterion predict the same relation between the critical imposed stress and flaw size, it is clearly impossible to distinguish between the validity of the criteria by the experimental determination of the dependence of the critical stress on flaw size.

It has been suggested, and originally by Griffith (20), that these two criteria for failure are, in fact, equivalent. It was pointed out that on substituting into eq. (31) the value of the imposed stress for instability, i.e., eq. (7)

$$S_c = (2E\gamma/\pi c)^{1/2}$$

the length of the crack is eliminated and,

$$\sigma_m = (8E\gamma/\pi\rho)^{1/2} \qquad \text{IIA5-(32)}$$

Therefore, it was concluded that σ_m is constant at failure, provided that the ρ is constant also. As noted by Griffith, however, the equations for the stresses would be expected to be invalid for the regions centered around the tip of the crack, hence the application of the stress concentration equation is open to question when, as in a real material, the radius of curvature of the crack tip approaches molecular dimensions. The concept of a point stress is only valid when the gradient of the stress is such that no significant change occurs over a distance which corresponds to interatomic dimensions. Whenever an infinite stress is predicted by the application of classical elasticity considerations, this condition is not met, and it is then necessary to consider the average stress over an arbitrarily small area which can be made to approximate that defined by the interatomic spacing (21).

The limiting-local-stress criterion predicts that the critical imposed stress to engender instability in a continuum model containing a line crack is infinitesimal, whereas the energy criterion yields the expression for a finite stress given in eq. (7), above. Because of the thermodynamic nature of the energy criterion, it must be concluded that the prediction of the limiting stress criterion is incorrect. In this particular limiting case, the two criteria do not correspond, and it is therefore of interest to establish the relation between them under less restrictive conditions, i.e., where the radius of curvature of the tip of the flaw is sufficiently large for the equations of continuum mechanics to be strictly applicable over all parts of the field. To do this it is necessary to consider a more general model in which it is assumed that the defect in the infinite sheet is an elliptical hole, the shape of which is characterized by the ratio R of the lengths of the minor, b, to major, a, semi-axis. Then, $R = 0$ corresponds to a line crack, and $R = 1$ to a circular hole. It will also be assumed that the sheet is loaded uniformly in its plane by an imposed stress S.

Then the local stress is:

$$\sigma_m = 2S(c/\rho)^{1/2} = 2Sa/b = 2S/R \qquad \text{IIA5-(33)}$$

Assuming a limiting-local-stress σ_c for failure

$$S_c = \sigma_c R/2 \qquad \text{IIA5-(34)}$$

It is now necessary to apply the energy criterion for failure to this model. It was shown by Griffith (19) that the increase in the elastic strain energy due to the cavity is, for plane stress:

$$\Delta U = (\pi S^2 c^2/E) \cosh 2\xi_0 \qquad \text{IIA5-(35)}$$

or,

$$\Delta U = (\pi S^2/E)\,(a^2 + b^2) \qquad \text{IIA-5(36)}$$

Then

$$\partial \Delta U/\partial a = (2\pi S^2 c/E)(1 + Rb')/(1 - R^2)^{1/2} \quad \text{IIA-5(37)}$$

where $R = b/a$ and $b' = db/da$. It will be assumed that the radius of curvature of the cavity at the ends of the major axis remains constant as the length of the axis increases; hence $b' = R/2$. Then,

$$\partial \Delta U/\partial a = (\pi S^2 c/E)\,(2 + R^2)/(1 - R^2)^{1/2} \quad \text{IIA5-(38)}$$

If the sheet is of unit thickness, the surface area s bounding the cavity is

$$s = 4\int_0^{\pi/2} (a^2 \cos^2\theta + b^2 \sin^2\theta)^{1/2}\, d\theta = 4aJ \quad \text{IIA5-(39)}$$

where J is the complete elliptical integral of the second kind, of modulus $(1 - R^2)^{1/2}$.

The increase in the surface energy due to the cavity is $\Delta T = 4\gamma a J$. Hence,

$$(\partial \Delta T/\partial a) = [4\gamma/(1 - R^2)]\,[R(b' - R)F + (1 - Rb')J] \quad \text{IIA5-(40)}$$

where F is the complete elliptical integral of the first kind, of modulus $(1 - R^2)^{1/2}$. Since $b' = R/2$,

$$(\partial \Delta T/\partial a) = [2\gamma/(1 - R^2)]\,[(2 - R^2)J - R^2 F] \quad \text{IIA5-(41)}$$

Application of the instability criterion $(\partial \Delta U/\partial a) = (\partial \Delta T/\partial a)$ yields the relation

$$S_c = (2\,E\gamma/\pi c)^{1/2}\,\mathfrak{F} = S_g \mathfrak{F} \quad \text{IIA5-(42)}$$

where

$$\mathfrak{F} = \{[(2 - R^2)J - R^2 F]/[(2 + R^2)\,(1 - R^2)^{1/2}]^{1/2}$$

and S_g is the critical imposed stress for the sheet containing a crack of length $2c$, $(R = 0)$. The dependence of the critical stress for failure S_c on the shape of the cavity R according to this equation, is illustrated in Figure IIA-4. The result is quite different from the direct proportionality predicted by the limiting stress criterion, and one must conclude that for a continuum model to which the principles of classical elasticity apply the two criteria for instability are not equivalent. In terms of the energy criterion, it is interesting to note that the value of the critical applied stress changes by only 1% over the

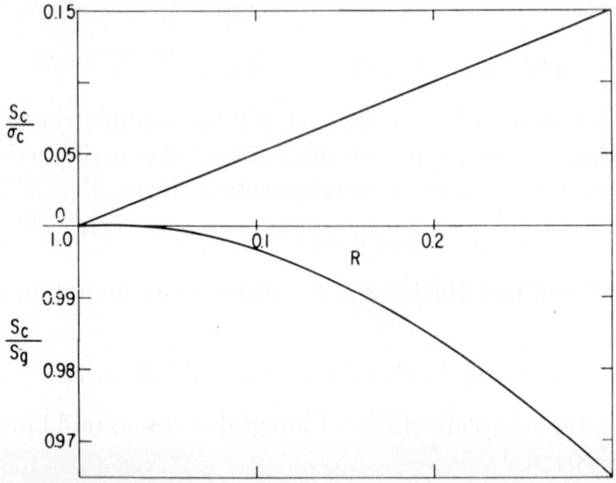

Fig. IIA-4. Loci of instability conditions for a limiting-local-stress criterion (upper curve) and an energetic criterion (lower curve)

range of R from 0 to 0.16. In the limiting case of a line crack ($R = 0$), the relations reduce to the simple case already considered [eq. (7)].

IIA6. Limitations of the Line Crack Elastic Model

In the above analyses it is assumed that the laws of continuum mechanics apply strictly to the system under consideration. In real systems containing sharp cracks, this assumption cannot be justified for the region at the crack tip at distances comparable to molecular dimensions. It is therefore necessary to consider the relation between the two criteria for failure when the atomistic nature of the material is taken into account. The thermodynamic nature of the energy criterion implies that it is a necessary condition for failure to occur. From a consideration of the conditions at the tip of a crack in an atomistic model, Orowan (39) has attempted to show that a sufficient condition for failure also obtains since the local stress is then equal to the theoretical strength (molecular cohesion). The value of the molecular cohesion is calculated by considering the rupture of a rod of unit cross-sectional area about a smooth plane perpendicular to the rod axis. The work required to create the new surface is then 2γ and is that necessary to overcome the intermolecular attractive forces. The force displacement relation for the planes of atoms bounding the

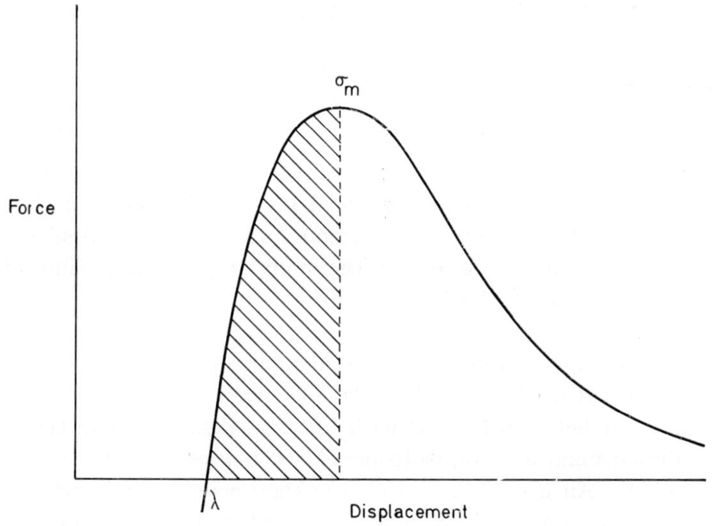

Fig. IIA-5. Force-displacement relation for the planes of atoms bounding a potential fracture surface (diagrammatic).

crack surface is illustrated in Figure IIA-5, where λ is the interatomic spacing in the equilibrium unstressed (and unbroken) state of the body. The total area beneath the curve represents the surface energy, and the amount of energy corresponding to the shaded portion must be available between the planes defining the location of the fracture surfaces immediately prior to rupture. If the assumption is made that the force-deflection relation is approximately sinusoidal, then it can be shown that:

$$\gamma \cong \lambda \sigma_m{}^2/E \qquad \text{IIA6-(43)}$$

hence the molecular cohesion is given approximately by

$$\sigma_m = (E\gamma/\lambda)^{1/2} \qquad \text{IIA6-(44)}$$

This equation has been used to calculate the theoretical strength of a medium (37), but in the present case it is more pertinent to consider the relation between the imposed stress S_c and σ_m. It is assumed that the stress concentration factor of Inglis can be applied to the system [eq. (31)], the effective radius of curvature of the crack tip being approximated by the equilibrium interatomic distance λ. The stress concentration factor was, of course, derived from a continuum model; as justification for its application to an atomistic model, Neuber's theory of notch stresses is invoked (35). The final equation is

$$S_c = {}^1/_2(E\gamma/c)^{1/2} \quad \text{IIA6-(45)}$$

This equation is in the same form as that given by the energy criterion and differs only by a small numerical factor. Therefore, it is concluded that when the energy criterion for fracture is satisfied, the local stress at the crack tip in the body approximates molecular cohesion, hence that the limiting stress criterion is satisfied simultaneously.

A more complete and sophisticated analysis of this problem has been published by Elliott who derived equations for the stress and displacement fields associated with line and disk cracks in two- and three-dimensional infinite, isotropic elastic bodies subjected to tensile stress (15). An atomistic model was then considered in which the crack lay between lattice planes separated by an equilibrium spacing a in the unstressed state. The stresses at and displacements of the planes when the body was subjected to tensile stress were then ob-

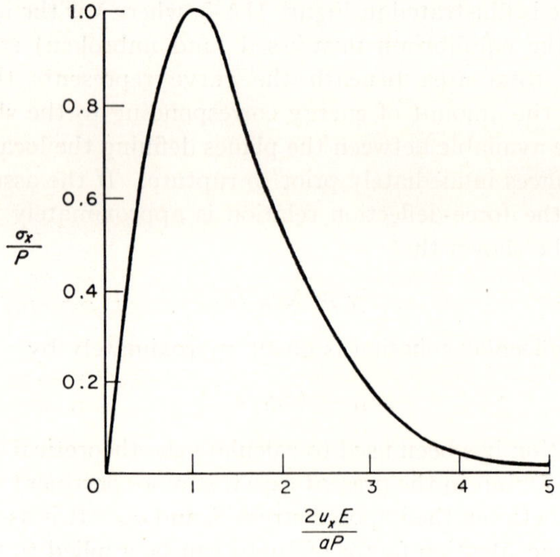

Fig. IIA-6. Theoretical force-displacement relation for the planes of atoms bounding a fracture surface (Elliott, 1947).

tained from the relations already derived. The form of the stress displacement relation corresponded quite closely to that expected for interatomic forces and is illustrated in Figure IIA-6.

The rate of increase of force with displacement was equal to E at zero displacement, but decreased with increasing displacement until a maximum in force was observed. On further displacement, the force decreased and asymptotically approached zero. It then was assumed that the maximum in the curve corresponded to the maximum interatomic force; hence, the surface energy was calculated by integration. When this was included in a form analogous to the Griffith equation, it gave

$$S_c = 0.840\ (E\gamma/c)^{1/2} \qquad \text{IIA6-(46)}$$

In comparison, the numerical factor in the Griffith equation is $(2/\pi)^{1/2} = 0.796$. Since a critical limiting stress condition for failure is assumed by Elliott, it appears that this criterion for his model is equivalent to the energy criterion for the corresponding Griffith model.

It is clear that the approaches of Orowan and Elliot are quite analogous. In both cases the maximum stress in the system, obtained by the application of theories assuming an elastic continuum model, is equated to the molecular cohesion. The differences between the derived relations (and the conventional Griffith expression) is small and oniy involves a numerical factor. The agreement is perhaps surprising in view of the considerable difference in the complexity of the derivations.

IIA7. The Influence of Inelastic Responses

In deducing the conditions for instability, above, it was assumed that the material behaved elastically and that there was no mechanism for energy dissipation other than that of creating new surfaces even though the local stress may have approached that corresponding to molecular cohesion. With fracture in such materials as the very highly supercooled glasses, it is probable that this assumption is justified. However, in metals and in glassy polymers there is evidence that local plastic and viscoelastic processes, respectively, accompany the formation of the new surface (7,16) and, under these conditions, the apparent surface energy may be much greater than the energy of interatomic bonding. It is clear that such materials can respond inelastically to the high local stresses at the apex of a defect; hence the maximum effective stress concentration is reduced to that for inelastic deformation. Under these conditions, the application of classical fracture theories which assume truly elastic behavior is inappropriate. However, if the region of inelastic response is small compared with the dimensions of the crack, the nature of the stress field, hence the energy associated with it, will be affected negligibly. Under these conditions it is presumably still satisfactory to apply the energy criterion for instability, but the apparent surface energy will include all that which is dissipated in the net inelastic process. However, such apparent surface energies give limited information on the details of the actual phenomena involved. On the other hand, since the stress field is most affected in the region which is important in the critical stress criterion, some revision of that theory is obviously necessary. For perfectly plastic, single crystalline bodies, it may be assumed that the maximum point shear stress which can be sustained is the critically resolved yield stress in shear. Hence, the relation between this localized stress and the macroscopically imposed stress would be determined by the frequently unknown details of the geometry of the crack and the yielded region, and the precise plastic properties of that and of neighboring regions.

Some attempts have been made to describe the behavior in terms of more realistic models. Perhaps the simplest approach is to assume that there is a region centered about the tip of the crack over which the stress is constant and equal to that at the boundary of the elastic region. It has been shown that the stress concentration factor, i.e., the ratio of the local constant stress to that imposed on the boundary, is the same as for a crack of the same length, but with a radius of curvature equal to that of the inelastically deformed region, in an otherwise purely elastic sheet (35). In assessing the strain energy of the system in a rigorous analysis, however, it is necessary to subtract the elastic strain energy in the inelastically deformed region from the value for an otherwise elastic sheet containing a crack, and to add the energy dissipated in that abnormal region. The usual energy balance techniques are then applied to give the expression for the critical stress under these conditions (49). For a plastic body, equations have been derived for the critical tensile stress applied normal to the crack axis as well as for that in uniform tension.

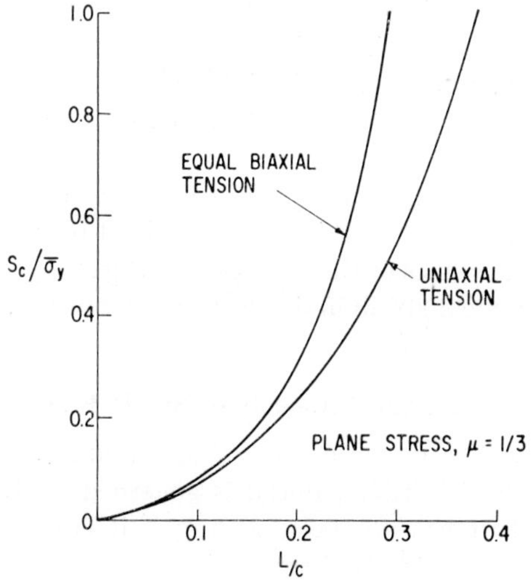

Fig. IIA-7. Loci of inelastic failure criteria for various amounts of crack-localized yielding (Swedlow, 1962).

For uniaxial tension:

$$S_c = \omega/[1 - (3L/2c)] \qquad \text{IIA7-(47a)}$$

For uniform tension:

$$S_c = \omega/[1 - (3L/2c) - 4.8\ (L/2c)^{3/2}] \qquad \text{IIA7-(47b)}$$

where

$$\omega = (16\ \bar{\sigma}_y/\pi)\ (L/2c)^{3/2}$$

L is the radius of an assumed circular yielded region, $\bar{\sigma}_y$ is the yield stress of the material, and c is the half-length of the crack. The dependence of the critical stress on the size of the yielded region is illustrated in Figure IIA-7.

A more refined treatment which also has been considered suggests a detailed mechanism of the failure process (31). In this case as well, the plastic body is assumed to have yielded within a constant stress boundary as the first stage in the failure process. However, the yielded material would not be expected to rupture until some later stage. It is assumed that the rupture of the material is governed by a limiting strain condition; the stress over the yielded region being assumed constant. Thus crack extension does not occur until the second criterion is satisfied. This mechanism has the appeal of conceptual simplicity, but the analysis of the system presents considerable mathematical difficulty. The resulting expression for the critical stress has the same form as the Griffith equation, but the parameter which corresponds to the specific surface energy is a rather complicated function, the form of which is determined by the details of the mechanism. The values of the parameters which appear in this function, such as the yield strain and the strain at rupture, are somewhat difficult to assess, *a priori*.

IIA8. Statistical Aspects of Brittle Fracture

All of the above discussion hinges on the premise that the strength of a brittle or quasi-brittle material is governed by the presence of flaws. Although it has not been so-stated, it is normally assumed that the flaws are randomly disposed in size and position in a particular sample and also in any population of samples. For this reason the ultimate strength is not strictly a material constant, but should be treated as a statistical quantity, since the strength distribution of a population of samples will be determined by the numbers and severity of the flaws. The problems with which statistical theories of fracture are concerned are associated with the nature of the probability distribution function for the severity of the flaws which best represents the strength values (12). Whether the distribution functions have any fundamental significance depends on the origin of the flaws in the samples. Numerous forms of distribution functions have been proposed, each of which is capable of representing the observed behavior over some part of the range of variables studied. The form of all of the functions is such that the strength increases as the size of the sample is reduced (12), a result consistent with the findings of many experimental investigations on the strength of glass fibers (1,34). However, recent careful work has shown that this effect is due to the conditions under which the fibers are produced, and that intrinsically the strength is independent of fiber diameter (40). In this case it appears that the flaws are produced entirely fortuitously during fabrication of the samples, and under these conditions the statistical theories provide no more than a convenient and succinct way of representing the experimental data. How generally true is this conclusion has yet to be ascertained.

The situation is more complicated in the case of glassy polymers. There is some evidence that the most effective flaws are not entirely of adventitious origin but are determined, in part, by properties of the material (8). Furthermore, it appears that the size of the effective flaw at the failure point is determined by the stress level and the time for which it is imposed (8). The influence of mechanical handling and surface abrasion is slight, but fabrication conditions which lead to "orientation" and frozen-in stress affect the properties significantly. However, there is much less variability in the strength data of glassy polymers than in that of the inorganic glasses, for instance (22). The statistical aspects of the ultimate properties of glassy polymers have received very little attention, and it is not possible, at present, to assess the relative contributions of the fabrication conditions and the intrinsic properties of the materials to the resultant distribution of the strength data. The statistical theories have been the subject of recent reviews and will not be further discussed here (12).

IIA9. Fracture Dynamics

In all of the foregoing considerations, attention has been confined to the system up to the point of mechanical instability and, in particular, to defining the conditions which govern the onset of that instability. Historically, there has been less concern with the behavior of the system subsequent to this point, when the crack is propagating, though it is of considerable theoretical interest. In general, two different aspects of the problem have been considered.

In the first, attempts have been made to deduce the expressions for the stress and strain fields by the solution of the dynamic elasticity equations. In these treatments the assumption is usually made that the crack is propagating with a constant velocity, and the influence of this velocity on the stress and strain fields is deduced.

The second aspect that has been treated is the determination of the equations of motion of the crack by an extension and generalization of Griffith's treatment of the energetics of the system. These two approaches are not entirely independent since any change in the stress field, because of the motion of the crack, will result in a change in the strain energy function, though this effect has usually been ignored (5,6,42).

In deducing the influence of the (constant) crack velocity on the stress field at its tip, Yoffe (56) made the assumption that double-

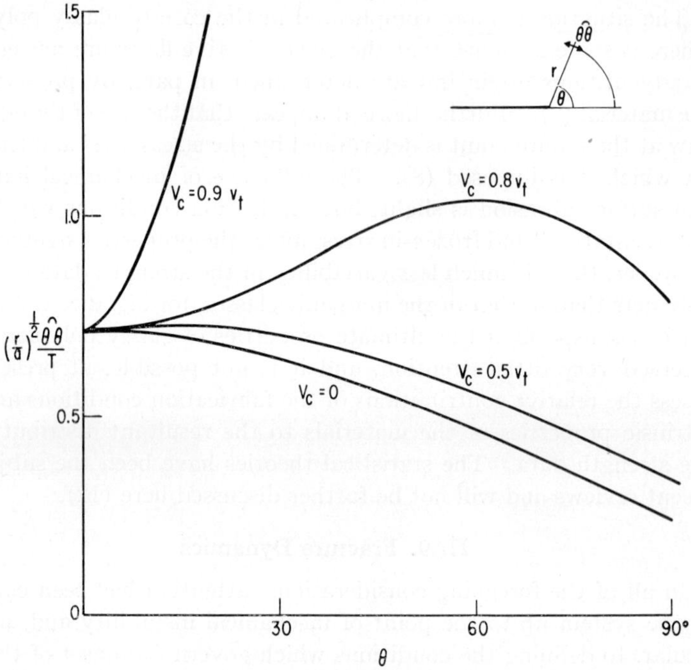

Fig. IIA-8. The dependence of the stress field at the tip of a moving crack on its velocity (after Yoffe, 1951).

ended crack remained of constant length as it moved across the sheet model, subjected to a uniform tensile stress normal to the crack axis. This simplification is admissible provided that the crack is sufficiently long so that there is no interference between the stress fields at the two ends. The stress field was then mapped on a polar coordinate system centered on the tip of the crack; the tangential stress $\widehat{\theta\,\theta}$ being obtained as a function of the crack velocity. When the crack was stationary, the maximum value of $\widehat{\theta\,\theta}$ coincided with the crack axis ($\theta = 0$). As the velocity increased, this maximum broadened, and then divided into two separate maxima symmetrically disposed about the crack axis (Figure IIA-8). The velocity at which this occurred was $\sim 0.6 v_t$. On the assumption that the crack propagates normal to the maximum tensile stress, this analysis provides an explanation for the observed bifurcation of cracks moving with a high velocity in an isotropic medium. More general treatments of the problem that have been given include the above result as a special case (11,13). It has also been shown that the applied force required to maintain a given crack velocity decreases as the crack velocity increases (13); a conclusion consistent with the equations of motion derived from simple energy considerations (5). The dynamic stress field equations at the tip of a crack subjected to a longitudinal shear have also been obtained, and it has again been shown that as the velocity of the crack increases, the applied force necessary to maintain that velocity decreases (30).

The application of the Griffith condition for instability to a stressed sheet containing a crack, described above, establishes the condition of imposed stress at which the total energy of the system will decrease as the crack begins to increase in size. However, it gives no information of the behavior of the crack as it propagates across the sample. By an extension of the theoretical treatment to include the contribution of the kinetic energy associated with the moving crack, it is possible to describe the behavior of the dynamic system in terms of an equation for the motion of the crack. While detailed discussion of this phase of the problem lies outside the scope of this article, an elementary consideration of some aspects of the dynamics of crack growth provides information on the nature of the instability and the conditions which obtain at the point defined by the Griffith criterion.

It was shown by Mott (33) on dimensional grounds, that the form of the kinetic energy term for the Griffith model was

$$\kappa = kdS_c^2c^2v_c^2/2E^2 \qquad \text{IIA9-(48)}$$

where k is an (undetermined) numerical constant, d is the density of the material and v_c is the crack velocity. On the assumption that energy was conserved in the system, an expression for the velocity of the crack was derived. The derivation was repeated by Roberts and Wells (42), who obtained the same result although it was, in fact, incorrect. However, both this result and the correct expression, deduced later, predict that the velocity should tend to a limiting value at crack sizes large compared with the initial value. The terminal velocity is

$$v_m = (2m/k)^{1/2}(E/d)^{1/2} = (2\pi/k)^{1/2}v_s \qquad \text{IIA9-(49)}$$

where $(E/d)^{1/2}$ is the velocity of the longitudinal elastic wave v_s in the material. This result implies that the crack velocity is limited by the rate at which elastic energy can be supplied to the region at the tip of the travelling crack.

From a consideration of the stress field, Roberts and Wells deduced that the factor $(2\pi/k)^{1/2}$ had a value of 0.38, which corresponds reasonably well with the experimentally determined values of terminal crack velocities in a number of materials (44). It was also shown that the result is independent of the boundary conditions (constant stress or constant strain) of the model (42). Some of the shortcomings of this treatment have been discussed by Anderson (2). Perhaps the most important is the assumption that the expression for the strain energy of the system is independent of the crack velocity. Clearly, the strain energy is a function of the stress distribution associated with the crack and, as discussed above, this changes as the crack velocity increases, particularly as the velocity approaches the terminal value predicted by the treatment of Roberts and Wells. The effect of the change in stress distribution on the strain energy function has not been estimated quantitatively, and this factor has been discounted (42). In support of this contention, it has been pointed out that photoelastic studies indicate that the static and dynamic stress fields associated with the crack are closely similar (52).

It is also possible to establish the velocity characteristics of the moving crack from a consideration of the stress field at the tip without reference to its detailed geometry or the stress conditions at the external boundary of the sample (3). This treatment bears the same relation to that of Mott, discussed above, as the limiting local stress criterion for fracture does to the energy criterion. In this treatment it is implied that the predominant component of the stress field at the crack boundary is a shear τ_{xy}. Consequently the kinetic energy is derived as a function of the transverse (shear) wave velocity v_t, i.e.,

$$\kappa = \left(\frac{v_c}{v_t}\right)^2 \frac{\tau_{xy}^2}{2G} \qquad \text{IIA9-(50)}$$

where G is the shear modulus and v_c is the crack velocity. The expression for the crack velocity is then shown to be

$$v_c = \frac{v_l}{2}\left[\frac{2}{1+\mu}\right]^{1/2}\left[1 - \frac{\sigma_c}{\sigma_m}\right]^{1/2} = \frac{v_s}{2(1+\mu)}\left[1 - \frac{\sigma_c}{\sigma_m}\right]^{1/2}$$

IIA9-(51)

where v_s is the longitudinal wave velocity, μ is Poisson's ratio, σ_c is the critical local stress (molecular cohesion), and σ_m is the maximum stress at the crack tip at any time. The equation is similar in form to that of Roberts and Wells discussed above, but attention here is directed to the local stress as governing the propagation of the crack. Thus the initial flaw only begins to increase in size when the local stress becomes equal to the molecular cohesion. It appears that the limiting value of σ_m depends on the composition of the material, hence the ratio of the limiting crack velocity to the longitudinal wave velocity is likewise dependent, and is not a universal constant. There is some experimental evidence in support of this conclusion (44,45, 46).

While there may be some question as to the validity of the static strain energy function for the Griffith model when the crack is moving with a high velocity, the discrepancy in that function will decrease as the crack velocity decreases, vanishing at the point where crack propagation is initiated. In considering the application of the energy criterion for instability, the conventional approach, as outlined above, is to consider the static system and deduce the conditions under which it becomes dynamic. From the equations of motion which represent the dynamic condition, it is possible to deduce the conditions under which the velocity tends to zero, hence to approach the point of instability from the opposite direction. For the reasons stated above, the static form of the strain energy function will then be valid. It can be shown that under a constant imposed stress, $S_g = (2E\gamma/\pi c_0)^{1/2}$, the correct expression for the crack velocity is (5,14)

$$v_c = (2\pi/k)^{1/2}(E/d)^{1/2}[1 - (c_0/c)] \qquad \text{IIA9-(52)}$$

When this equation is integrated and the appropriate initial conditions inserted ($t = 0$, when $c = c_0$), an infinite integration constant is obtained. This implies that a crack of initial length $2c_0$ in a sheet subjected to a constant uniform stress S_g will only begin to increase in size after infinite time (5,10). As pointed out by a number of authors (15), this is a consequence of the fact that the condition defined by the Griffith energy criterion is one of unstable equilibrium. To cause the crack to grow, the stress must be increased (infinitesimally) to some value, S_c. The discrepancy between S_c and S_g then determines the equations of motion of the crack, and particularly its initial acceleration. Defining a parameter $n = 2S_g^2/S_c^2$, it can be shown that the motion of the crack is represented (5) by the equations:

$$v_m t/c_0 = (\beta - 1)^{1/2}[\beta - (n - 1)]^{1/2} + n \ln \{(\beta - 1)^{1/2} + [(\beta - (n - 1)]^{1/2}\} - (n/2) \ln (2 - n) \quad \text{IIA9-(53)}$$

where $\beta = c/c_0$

$$v_c^2 = v_m^2[1 - (c_0/c)][1 - (n - 1)(c_0/c)] \quad \text{IIA9-(54)}$$

$$a_c = (v_m^2 c_0/2c^2)[n - 2(n - 1)(c_0/c)] \quad \text{IIA9-(55)}$$

where a_c is the acceleration of the crack. Its initial value is clearly

$$a_c^0 = (v_m^2/2c_0)(2 - n)$$

Similar considerations to those described above have been shown to apply to the cleavage system (5,27).

By analogy with the problem of instability and crack growth initiation in viscoelastic systems, as already discussed, it is of interest to consider the corresponding dynamic situations. The analysis of the stress field associated with the crack tip is simplified by reason of the fact that for given stress boundary conditions, the elastic solution is applicable to the viscoelastic case (28). For the sheet model, subjected to a tensile stress normal to the crack axis, it is possible to deduce the crack propagation equation on the assumption that the behavior of the material through which the crack moves can be represented by a series of discrete Voigt elements (54). The fracture criterion in this case is defined in terms of a characteristic limiting strain which is sustained, at rupture, by the Voigt elements of the model. The analysis indicates that the velocity increases exponentially with time, without limit, the latter result arising, presumably, because inertia effects are neglected. A similar conclusion is reached by thermodynamic reasoning analogous to the generalization of the Griffith theory to include the kinetic energy, as described above. In the present case the contribution of the kinetic energy is neglected and replaced by a viscous dissipative term (54). It is concluded that crack propagation is only possible if the imposed stress exceeds that given by the Griffith criterion; the analogy with the purely elastic case is clear. A remarkably similar but independent analysis of the behavior of a viscoelastic medium subjected to longitudinal shear leads to very similar conclusions (29).

IIA10. Summary

The aim of theories of brittle fracture is to account for the discrepancy between the strength as measured in conventional tests and that as calculated from the assumed molecular structure. The most widely accepted explanation for this discrepancy is that experimental samples are imperfect, and that the flaws they contain reduce the imposed stress for fracture from the theoretical to the observed value. It is convenient to treat representative model systems by the methods of classical elasticity, and two criteria are available for defining the condition of mechanical instability. The first assumes that this will occur when the stress at some point in the body attains a characteristic limiting value. The second postulates that the onset of instability is governed by energetic considerations. Of these, the latter has received more attention and equations have been derived for the critical imposed stress for a number of different systems.

The application of the limiting stress theory to complex systems is simplified by the fact that the stress concentration at a flaw is largely independent of the details of its geometry and position. Furthermore, complex stress systems can be treated without difficulty because of the superposability of stresses.

It is not possible to distinguish between the two fracture criteria by experiments which determine the dependence of the critical stress for failure on flaw (crack) size. Although it has been stated that the two criteria are equivalent, it appears that for a continuum model this is not so. However, in the more realistic cases where consideration is given to the atomistic structure of matter, it has been shown that the two criteria are satisfied simultaneously, hence can be considered equivalent. The equations which support this conclusion have been derived on the assumption that the equations of continuum mechanics are valid at the molecular level.

The local stress in a loaded, perfectly elastic body containing a crack attains a high value, approaching the molecular cohesion. It is possible that an inelastic response will be induced in some systems under these conditions and the criterion for failure has then to take this effect into account. Some success has been achieved in the derivation of appropriate theories.

In general, the flaws which are believed to be responsible for the low observed strength values may be fortuitous and/or an intrinsic property of the material. Because they are of random origin, the strength must be considered a statistical property, and there are many theories available that provide distribution functions for the number and severity of flaws in the samples tested which would give rise to the observed distribution of strengths. Only if the flaws are related to the material structure can such theories be of any fundamental significance. In glassy polymers there is some evidence that the flaws are not entirely adventitious, but are due to the properties of the materials. However, the statistical aspects of the ultimate properties have received little attention in this case.

In most practical situations it is the fracture stress which commands the greatest interest, and the behavior of the system after the critical condition is attained is of secondary importance. For this reason, rather less attention has been paid to the kinetic aspects of fracture and the dynamics of crack propagation. However, it is possible to obtain information on the condition of the system at the point of instability from such considerations, since this condition can be approached, theoretically, from a dynamic as well as the conventional static point of view. It has been shown that the stress field of a loaded sheet containing a crack is influenced by the velocity of the crack. This result offers an explanation for the observed bifurcation of rapidly moving cracks and it also indicates that the strain energy of the system is likewise a function of the crack velocity. The equations of motion of the crack have been obtained from energetic considerations, though the arguments used assume that the static strain energy equation is applicable to the dynamic case. Some success has been achieved in deducing the equations of motion of a crack in a viscoelastic medium subjected to tension or shear. In this case a limiting strain criterion for failure leads to essentially the same result as obtained on the assumption of an energy criterion. In both the elastic and viscoelastic systems, it is concluded that the instability (energy) criterion as normally expressed does not lead to crack propagation, hence sample failure. Rather, it defines a condition of unstable equilibrium from which the system must be displaced, by the application of an infinitesimally greater stress, for fracture to occur.

The dynamics of the system are then governed by the amount by which the unstable equilibrium stress is exceeded.

IIA11. List of Key Symbols

a, b	Lengths of major and minor semiaxes, respectively, of planar elliptical hole.
A	Surface area of (infinite) sheet model
B	Width of finite sheet model
$c,\ c_0$	Half of interfocal distance of ellipse; also crack length.
d	Density of material of body
E	Young's modulus
f, f_c, f_y, P, P_c	External (applied) forces
F	Complete elliptical integral of the first kind
G	Shear modulus
$\mathcal{G}$	Strain energy release rate
H	Total energy of system
I	Moment of inertia of cross section of beam (cleavage)
J	Complete elliptical integral of the second kind
k	Proportionality constant
K	Stress intensity factor
L	Radius of plastically deformed region at crack tip
m	Slope of force deflection relation (cleavage)
n	Ratio $2S_g^2/S_c^2$
p	Loading factor
$q = (1 - \mu)$	Condition for plane strain
$q = 1/(1 + \mu)$	Condition for plane stress
r, θ	Polar coordinates
R	Ratio b/a
s	Surface area of elliptical hole
$S, S_c, S_x, S_y, S_2, S_3$	External (applied) stresses
t	Time
T, T_0	Surface energy
U	Elastic strain energy
$v,\ u_z$	Displacement in deformed body
v_c	Crack velocity

v_m	Maximum crack velocity
v_s	Velocity of longitudinal (sound) wave
v_t	Velocity of transverse (shear) wave
w	Width of fracture plane (cleavage)
x, y, z	Cartesian coordinates
α	Increment of crack length
β	Ratio c/c_0
γ	Specific surface energy
δ, δ_c	Total deflection of cleavage sample
κ	Kinetic energy
λ	Characteristic molecular (interatomic) distance
μ	Poisson's ratio
ξ_0	Elliptical coordinate
ρ	Radius of curvature of crack tip
$x, \sigma_y, \sigma_z, \sigma_c, \sigma_m, \bar{\sigma}_y, \tau_{xy}, \widehat{\theta\theta}$	Point stresses in deformed body
ϕ, ν	Arbitrary constants (cleavage)

References

1. Anderegg, F. O., *Ind. Eng. Chem.*, **31,** 290 (1939).
2. Anderson, O. L., in *Fracture,* B. L. Averbach, D. K. Felbeck, G. T. Hahn, and D. A. Thomas, Eds., Wiley, New York, 1959, p. 331.
3. Bateson, S., *Phys. Chem. Glasses,* **1,** 139 (1960).
4. Benbow, J. J., and F. C. Roesler, *Proc. Phys. Soc.* (*London*), **B70,** 201 (1957).
5. Berry, J. P., *J. Mech. Phys. Solids,* **8,** 194 (1960).
6. Berry, J. P., *J. Mech. Phys. Solids,* **8,** 207 (1960).
7. Berry, J. P., *J. Polymer Sci* , **50,** 107 (1961).
8. Berry, J. P., *J. Polymer Sci.,* **50,** 313 (1961).
9. Berry, J. P., *J. Appl. Phys.,* **34,** 62 (1963).
10. Berry, J. P., *J. Appl. Phys.,* **33,** 226 (1962).
11. Bromberg, K. B., *Arkiv. Fysik,* **18,** 159 (1960).
12. Charles, R. J., and J. C. Fisher, in *Non-Crystalline Solids,* V. D. Frechette, Ed., Wiley, New York, 1960, p. 491.
13. Craggs, J. W., *J. Mech. Phys. Solids,* **8,** 66 (1960).
14. Dulaney, E. N., and W. F. Brace, *J. Appl. Phys.,* **31,** 2233 (1960).
15. Elliott, H. A., *Proc. Phys. Soc.* (*London*), **59,** 208 (1947).
16. Felbeck, D. K., and E. Orowan, *Welding J. Res. Suppl.,* **34,** 510 (1955).
17. Frenkel, J., *Z. Physik,* **37,** 572 (1926).
18. Gilman, J. J., *J. Appl. Phys.,* **31,** 2208 (1960).
19. Griffith, A. A., *Phil. Trans. Roy. Soc.,* **A221,** 163 (1921).
20. Griffith, A. A., *Proc. Intern. Congr. Appl. Mech.* (*Delft*), 1924, p. 55.
21. Gurney, C., *Phil. Mag.,* **39,** 71 (1948).

22. Gurney, C., *Properties of Metallic Surfaces*, Institute of Metals, London, 1953, p. 145.
23. Hillig, W. B., *Modern Aspects of the Vitreous State*, Vol. II, Butterworths, London, 1962, p. 152.
24. Inglis, C. E., *Trans. Inst. Nav. Arch. London*, **55**, 219 (1913).
25. Irwin, G. R., *J. Appl. Mech.*, **61**, A49 (1939).
26. Irwin, G. R., *Handbuch der Physik*, Vol. 6, Springer, Berlin, 1958, p. 551.
27. Krause, I., private communication.
28. Lee, E. H., *Structural Mechanics*, Pergamon, New York, 1960, p. 456.
29. McClintock, F. A., *J. Mech. Phys. Solids*, **8**, 187 (1960).
30. McClintock, F. A., Research Memorandum 34, Fatigue and Plasticity Laboratory, Mass. Inst. Technol., February, 1962.
31. McClintock, F., private communication.
32. Morley, J. J., P. Andrews, and I. Whitney, Union Scientifique Continentale du Verre, Symposium, 1961.
33. Mott, N. F., *Engineering*, **165**, 16 (1948).
34. Murgatroyd, J. B., *J. Soc. Glass Technol.*, **28**, 368T (1944).
35. Neuber, H., *Kerbspannungslehre*, Springer, Berlin, 1937.
36. Obreimoff, J. W., *Proc. Roy. Soc. (London)*, **A127**, 290 (1930).
37. Orowan, E., *Trans. Inst. Eng. Shipb. Scot.*, **89**, 165 (1946).
38. Orowan, E., *Proceedings of the Symposium on Fatigue and Fracture of Metals*. Wiley, New York, 1950, p. 139.
39. Orowan, E., *Welding J. Res. Suppl.*, **34**, 157-S (1955).
40. Otto, W. H., *J. Am. Ceram. Soc.*, **38**, 122 (1955).
41. Polanyi, M., *Z. Physik*, **7**, 323 (1921).
42. Roberts, D. K., and A. A. Wells, *Engineering*, **178**, 820 (1954).
43. Sack, R. A., *Proc. Phys. Soc. (London)*, **58**, 729 (1946).
44. Schardin, H., in *Fracture*, B. L. Averbach, D. K. Felbeck, G. T. Hahn, and D. A. Thomas, Eds., Wiley, New York, 1959, p. 297.
45. Shand, E. B., *J. Am. Ceram. Soc.*, **37**, 52 (1954).
46. Shand, E. B., *J. Am. Ceram. Soc.*, **37**, 559 (1954).
47. Sneddon, I. N., *Proc. Roy. Soc. (London)*, **187A**, 229 (1946).
48. Svennson, N. L., *Proc. Phys. Soc. (London)*, **77**, 876 (1961).
49. Swedlow, J. L., GALCIT Report No. SM62-16, Calif. Inst. Technol., 1962.
50. Swedlow, J. L., GALCIT Report No. SM63-8, Firestone Flight Sciences Laboratory, Cal. Inst. Technol., March 1963.
51. Wells, A. A., N. R. L. Report No. 4705, March 1956.
52. Wells, A. A., and D. Post, *Proc. Soc. Exptl. Stress Anal.*, **16**, 69 (1958).
53. Westergaard, H. M., *J. Appl. Mech.*, **61**, A49 (1939).
54. Williams, M. L., GALCIT Report No. SM 62-21, Firestone Flight Sciences Laboratory, Cal. Inst. Technol., August 1962.
55. Wolf, K., *Z. Angew. Math. Mech.*, **3**, 107 (1923).
56. Yoffe, E. H., *Phil. Mag.*, **42**, 739 (1951).

IIB. BRITTLE BEHAVIOR OF POLYMERIC SOLIDS

J. P. BERRY

IIB1. Characteristics of Brittle Failure

Under sufficiently large applied tensile forces, glassy polymers may fracture in a brittle manner, depending on the experimental conditions, including time and temperature. A number of different criteria are available for defining the failure process, and the ones most frequently employed are those which take account of the changes in the sample and the form of the force–deflection relation (Fig. IC-21, p. 97). In terms of this essentially engineering definition, brittle failure is characterized by uniform deformation until the rupture which occurs by the rapid propagation of a fracturing crack across the sample, i.e., in a plane normal to an applied uniaxial tensile force. Examination of the fractured sample reveals that there is essentially no reduction in cross-sectional area at the fracture plane, nor is there any residual longitudinal strain of the broken halves of the sample. The force–deflection curve is still linear at the failure strain, which normally is below 5%. In common with most mechanical properties, the nature of the failure process depends on the conditions under which it occurs. Brittle fracture in viscoelastic bodies is observed at sufficiently low temperatures and short times (or high strain rates).

IIB2. Application of Brittle Fracture Theories to Glassy Polymers

So far as their macroscopic mechanical properties are concerned, glasslike polymeric bodies can be regarded as sensibly hookean under those conditions where they display a brittle mode of failure. On a macroscopic scale they frequently are isotropic. It is reasonable and justifiable therefore, to apply to these materials the theories of brittle fracture which have been derived from consideration of classically elastic models. Naturally this approach does not permit experimental results to be interpreted in molecular terms directly, but in terms of continuum properties. Further theoretical considerations and supplementary experimental evidence are necessary to elucidate the relation between the observable macroscopic, and the resultant molecular, responses.

As has been discussed in Section IIA, the "flaw theories" assume that the strength of brittle materials is limited by the presence of flaws in the sample, and these discontinuities so distort the stress or energy field as to cause the initiation of the failure process by extension of the largest, properly oriented flaw, when a critical tensile stress is attained. The relevant equation which describes this behavior is that due to Griffith (21,22), i.e.,

$$S_c = K'(E\gamma/c)^{1/2} \qquad \text{IIB2-(1)}$$

where S_c is the critical tensile stress, E is Young's modulus, γ is the energy required for the production of a unit area of fracture surface (i.e., the surface energy of fracture) and c is the length of the flaw.

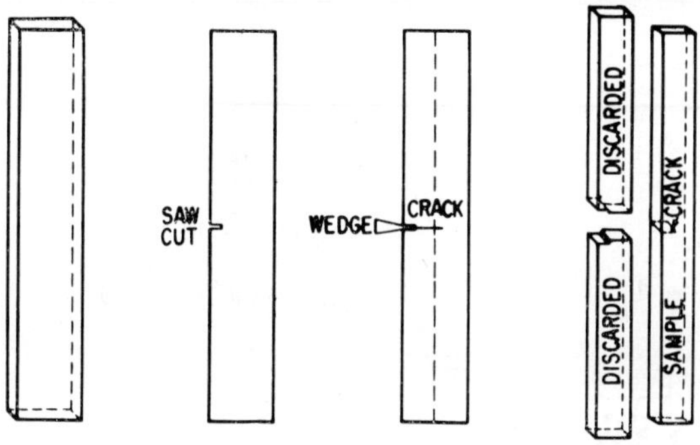

Fig. IIB-1. Method of preparation of tensile samples.

K' is a dimensionless geometrical factor determined by the details of the model treated and the applied force system.

E and γ both would be expected to be properties of the material; hence this theory predicts that the critical tensile stress for brittle failure should be inversely proportional to the square root of the length of the flaw, where the constant of proportionality is dependent on $E\gamma$. To establish whether the theory applies to a particular material, it is necessary to determine the critical failure stress for a series of samples containing defects of known size. This procedure has been adopted for metal (17,29) and silica-glass (21) specimens and more recently (7,8) for samples of the glassy polymers, poly(methyl methacrylate)* (PMMA) and polystyrene (PS). The nature of the flaw that is introduced into the sample is important, particularly if the conclusions are to be applied to the behavior of conventional, untreated samples under similar loading conditions. Attempts have been made to produce flaws which, in their morphology, are as close as possible to the type of flaw which would be expected to arise naturally. The common technique has been to produce the flaw as a "natural crack" by wedging open a relatively blunt notch which typically takes the form of a saw-cut (7,8,17,29). This technique may be used to give an "internal flaw" (29), or one which penetrates the sample from one of the external boundaries (7,17). The latter permits somewhat better control of the final length (i.e., the depth) of the flaw, and a representative method which gives a flaw of this type is illustrated in Figure IIB-1. One advantage of most glassy materials for this type of study arises from their transparency which

* Plexiglass, Lucite, Prespex, Diakon, etc.

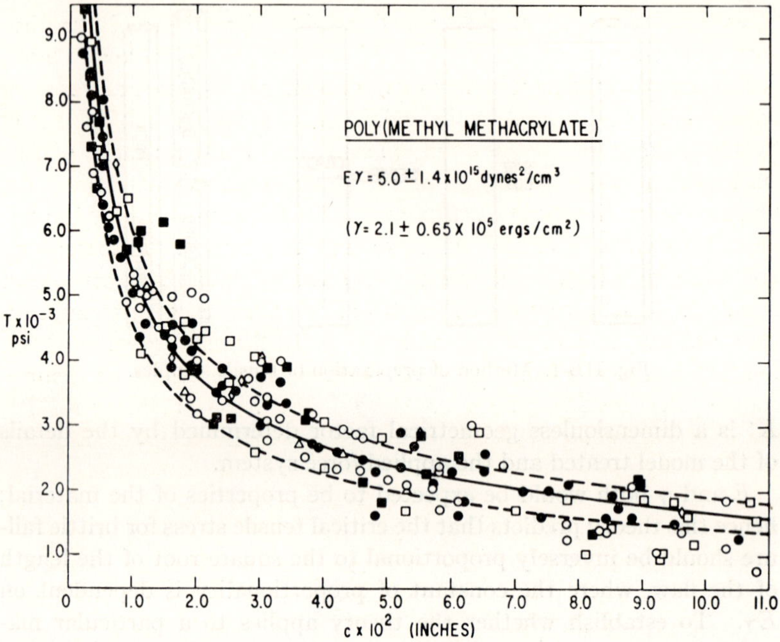

Fig. IIB-2. The dependence of tensile strength on crack-size in PMMA samples for various sample cross sections and extension rates.

enables the moving crack to be seen and, to a large extent, controlled by judicious manipulation of the entering wedge. Furthermore, it is easy to machine the polymers to remove the extraneous material and to confer on the sample the appropriate shape for testing.

For the purpose of establishing the validity of the theory, the distance of penetration of the crack from the edge of the sample is measured with a travelling microscope. Ideally, this crack should lie entirely in the plane normal to the long axis of the sample, along which the uniaxial tensile force is applied. The whole leading edge of the crack should be parallel to the edge of the sample from which it entered. Such an ideal morphology is never encountered and even in the most nearly "perfect" of cracks, the leading edge is an arc rather than a straight line, due to the retardation of propagation at the faces of the sample. In rupturing the samples in tension, it is necessary to avoid imposing any bending or torsional strains. The results of such tensile tests on samples of PMMA at room temperatures, are illustrated in Figure IIB-2. The agreement of the data with the theoretical $S_c \propto c^{-1/2}$ relation of eq. (1) (solid curve) is reasonably satisfactory, although the scatter of the experimental points does not permit a precision greater than is approximated by the separation of the two dashed lines. It appears that at least part of this scatter is due to the deviation of the crack morphology from the theoretical form. Minor changes in the experimental variables such as rate of testing and sample dimensions do not strongly influence the general agreement with the theory, and it may be concluded that the dependence of the tensile strength on flaw size can be represented adequately by the inverse square root relation implied in eq. (1). The position of the theoretical curve enables the proportionality constant to be determined, hence the material parameter $E\gamma$. By employing the macroscopic value for Young's modulus, a fracture surface energy is readily calculated. The value of this parameter can also be obtained by cleavage (2,3,10,42), which involves an entirely different mechanical system. Quite good agreement is found between the two methods. This latter technique offers several advantages over the tension method, and cleavage will be discussed in Subsection IIB6.

The fact that the relation between the tensile strength and flaw size can be represented by the Griffith theory equation does not, by itself, prove that the point of instability of the sample is governed by energy considerations, since a critical stress criterion for instability predicts precisely the same dependence (Subsec. IIA5, p. 172). However, the equations used to interpret the data obtained in tension and cleavage experiments were derived on the assumption of an energy criterion. Since agreement is found between the values of γ obtained by the two methods, even though the mechanical systems are distinctly different, it may be assumed that the energy criterion is, in fact, valid for these systems.

There remains the problem of assessing the significance of the value of γ obtained by application of the theory. Here it becomes necessary to introduce considerations which lie outside the scope of the continuum theory in order to attempt to relate the observed phenomena to molecular parameters. The quantity γ has the dimensions of a surface energy and, in terms of the theory, is that which is required for the formation of new surfaces by the extension of the flaw which is present in the sample. However, in interpreting the experimental data in terms of the theory, it is merely an arbitrary adjustable parameter, its value being obtained by a curve-fitting procedure. It is necessary, therefore, to consider this value in relation to the molecular constitution of the material. In view of the complexity of the molecular structure in polymeric bodies, the elucidation of the theoretical surface energy is a matter of considerable difficulty, and simplifying assumptions are necessary. Although these assumptions are such as to provide a theoretical upper limit for the estimated surface energy, on the premise that only *covalent, primary bonds* are ruptured, the value calculated is much less than that deduced from the experimental results (7). Similar calculations have been performed for other materials that have been studied in relation to the Griffith theory (16,17), and the experimental and theoretical values of the fracture surface energies are summarized in Table IIB-I.

TABLE IIB-I
Fracture Surface Energies at 25°C.

Material	$\gamma_{calc} \times 10^{-3}$, ergs/cm.^{2}	$\gamma_{obs} \times 10^{-3}$, ergs/cm.^{2}
Steel[a]	1.0	1,000
Glass[b]	1.7	0.55
PMMA	0.5	200

[a] Ref. 17.
[b] Ref. 21.

It is clear that the large discrepancy for the organic polymer is comparable to that for the steel, but distinctly different from that for the inorganic glass. This large discrepancy in PMMA and steel is attributed to the energy dissipated in inelastic processes which occur under the influence of the high stresses at the tip of the flaw (7,17). The mechanical properties and molecular structure of the inorganic glass are such as to preclude a pronounced, local ductile response. A large discrepancy in γ values is not observed; indeed, the experimental value is lower than that calculated. In the steel, the inelastic processes include the plastic distortion of the crystalline lattice at the fracture plane; an effect which is detected by back-reflection x-ray diffraction techniques (15,33,35). It is possible from such measurements to obtain an independent estimate of the energy dissipated by this mechanism, and the discrepancy between the observed and calculated values of γ can be accounted for in this way (17,25). Unfortunately, such techniques are of little avail for the examination of a noncrystalline material. However, in the glassy polymer the local stresses near the crack tip produce a modification of the microstructure which gives rise to optical phenomena, as considered below.

Examination of the fracture surfaces of PMMA reveals brilliant color effects, particularly in those cases where the fracture is initiated at an inserted, and preferably large, flaw (4,26,44). The characteristics of the colors indicate that they are due to optical interference in a thin layer at the fracture plane. It is not yet possible to define precisely the nature of the structure of this layer, but since it probably resembles that of a "craze" mark, its consideration will be deferred until the structure of these craze marks is later discussed. (See Subsection IIB9.) It appears that the surface layer is a consequence of the stress conditions which obtain at the crack tip, and that the material adjacent to the fracture plane is affected to a depth which approximates the wavelengths of visible light. Strain energy is dissipated in forming the structure since it persists when the stress is removed from the sample, and it may be assumed that the amount of energy is directly related to the amount of material so affected. In most cases the color observed changes only slowly in moving along the surface from the crack origin; hence the film is approximatelv of uniform thickness. Therefore, the amount of energy dissipated should be proportional to the area of the fracture surface, and in the analysis of the experimental data in terms of the Griffith theory, this energy will be included in the surface energy term while, in the strictest sense, it is not directly associated with the formation of surface

It may be objected that it is inappropriate to apply to such systems as steel and PMMA a relation which has been derived on the assumption of elastic behavior, particularly since the values of the parameter "γ" obtained by this application indicate that the systems are, in fact, not behaving in a perfectly elastic manner. On the other hand, if the volume of material which responds inelastically is small compared with the total volume of the body, the corresponding error in the computed *strain* energy is negligible. Provided that the total energy consumed is proportional to the area of surface produced, it is legitimate to apply the elastic analysis (36). However, a more detailed analysis of the conditions under which the surface and the underlying structure are produced is necessary for a complete understanding of the fundamental processes that occur in fracture. At present, it is not possible to obtain an independent estimate of the energy required to form the structure at the fracture plane of PMMA

Fig. IIB-3. Fracture surfaces of a tensile sample of PMMA (blue filter) (9).

and to account for the discrepancy in the experimental and theoretical surface energies. However, some of the more important phenomenological factors here are considered by Wolock and Newman, Section IIC.

IIB3. Relation Between Surface Colors and Topography

Close examination of the fracture surfaces of PMMA ruptured in tension reveals that a specific relation exists between the surface features and the colors which are observed. The color areas have distinct boundaries and, within these boundaries, the colors are essen-

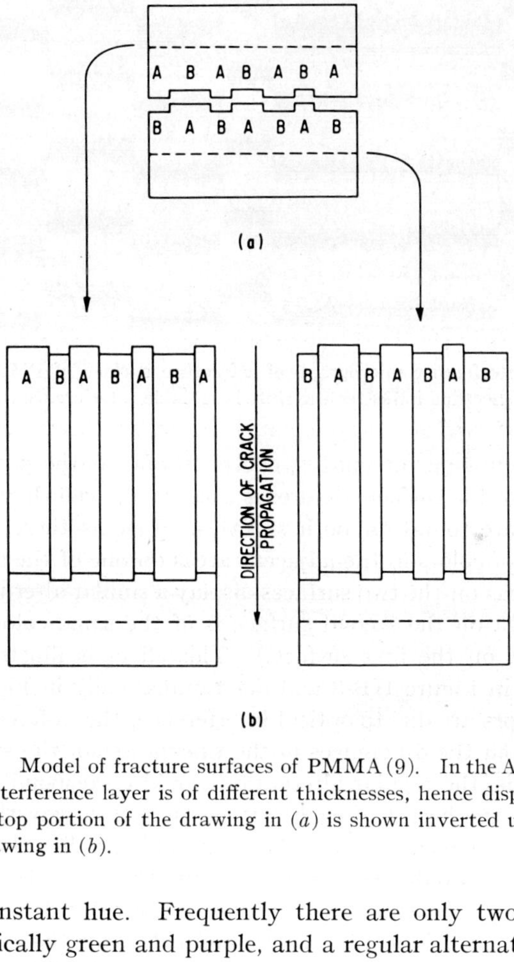

Fig. IIB-4. Model of fracture surfaces of PMMA (9). In the A and B regions the surface interference layer is of different thicknesses, hence displays different colors. The top portion of the drawing in (*a*) is shown inverted upward at the left in the drawing in (*b*).

tially of constant hue. Frequently there are only two colors displayed, typically green and purple, and a regular alternation is found in the plane of the fracture surface, but along a line which is normal to the direction of crack propagation. However, the two fracture

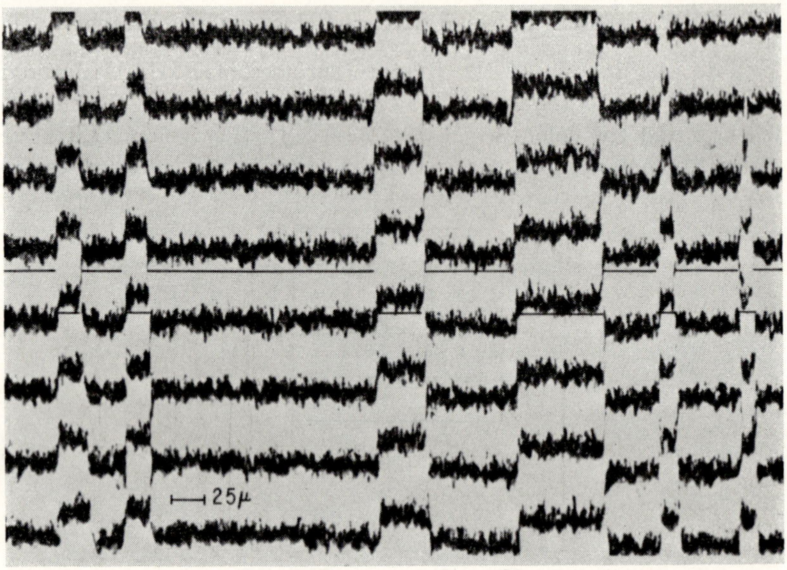

Fig. IIB-5. Interference micrograph of a fracture surface of PMMA (9). The reference line indicates the shift in a single interference fringe.

surfaces are not mirror images, except insofar as the position of the boundaries of the surface features is concerned, even though the same two colors are found on both surfaces. Just as there is a regular alternation of colors in the adjacent areas on one of the surfaces, the opposite areas on the two surfaces display a similar alternation. (An opposite area on the second surface is of the same color as the adjacent areas on the first surface.) This effect is illustrated photographically in Figure IIB-3 and diagrammatically in Figure IIB-4.

If the colors are due to optical interference, the differences in color can arise from the differences in the thicknesses of the surface films. On this basis, the regular alternation would correspond to a regular variation in the height of each surface, in such a way that they are spatially complementary (Fig. IIB-4). This implies that the crack propagates on two distinct levels only, and that the mean level of the main crack lies centrally within this inelastically deformed region. Since the stress field would be symmetrical about the mean fracture plane, the latter condition is quite reasonable. The topography of the fracture surface has been examined by interference microscopy

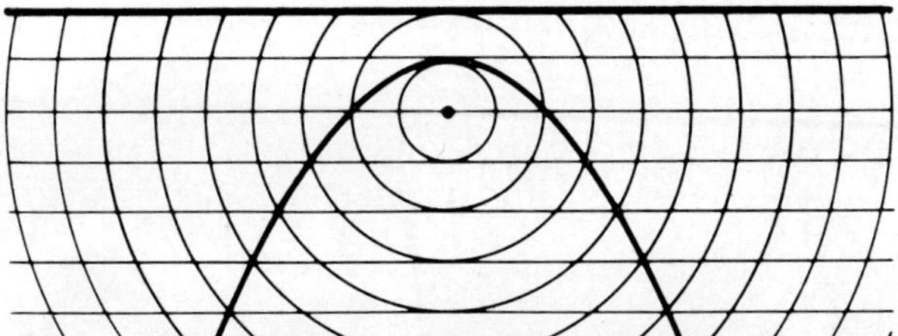

Fig. IIB-6. Generation of parabolic figures by the intersection of primary and secondary crack-fronts.

(Fig. IIB-5) and found to correspond very closely to that suggested by the disposition of the color areas on the surfaces (9). Even so, no satisfactory explanation has been given for this somewhat peculiar mode of crack propagation, which is hardly to be expected, *a priori*, from the amorphous nature of the material. In view of the magnitude of the areas over which the color is constant, it seems most likely that the effect arises from the mechanical rather than molecular considerations.

The fracture surfaces of PMMA and other polymers frequently display parabolic figures which are believed to arise when a secondary fracture is initiated from a point flaw ahead of the primary fracture front. The secondary fracture propagates uniformly as a system of concentric circles which, as they intersect the primary crack-front,

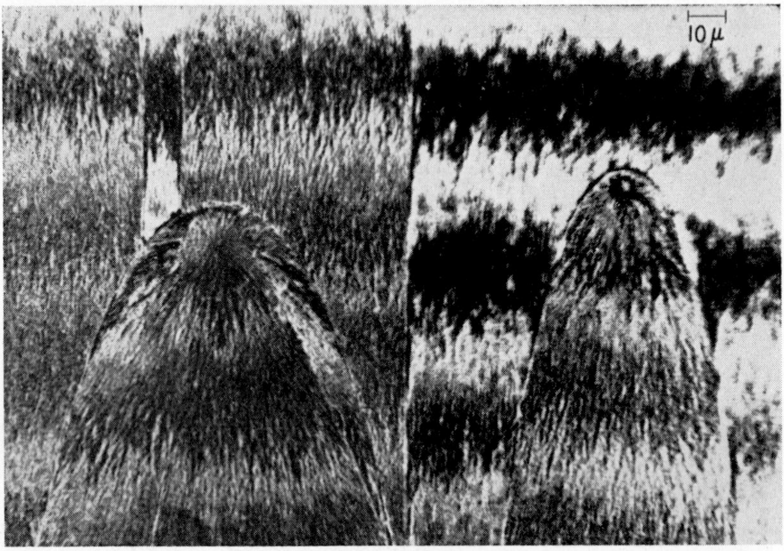

Fig. IIB-7. Interference micrograph of parabolic surface features, showing secondary crack-front (9).

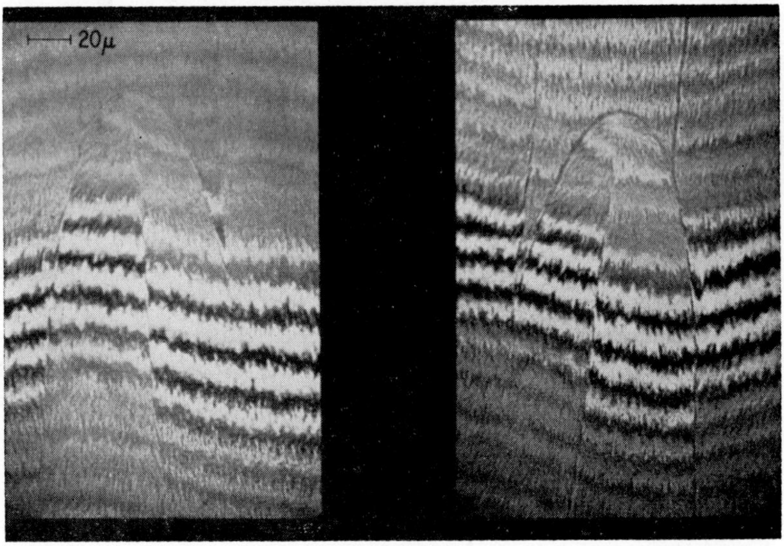

Fig. IIB-8. Interference micrographs of a parabolic surface feature, taken at corresponding areas on the matching fracture surfaces (9).

produce the parabolic figures (Fig. IIB-6). The initiating flaw of the secondary fracture then appears as the focus of the parabola. This hypothesis is supported by the observation of circular features within the parabola, and centered about the focus, which represent a hesitation in the secondary fracture while it was still isolated from the primary fracture (Fig. IIB-7).

The parabolic figures have also been examined by interference microscopy (Fig. IIB-8), and a model of the fracture surface at this point constructed from the information in the micrographs (Fig. IIB-9). It is evident from this model that the primary fracture was

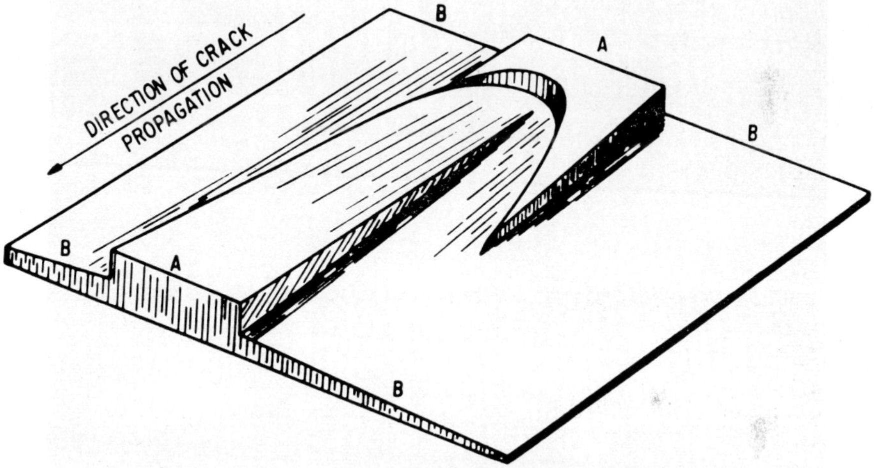

Fig. IIB-9. Model of the parabolic surface feature, based on the data relative to Fig. IIB-8 (9).

propagating with a biplanar configuration as discussed above, and that the plane of the secondary fracture was tilted with respect to that of the first. However, as soon as the secondary fracture attained the levels (upper and lower) of the primary fracture, these levels persisted so that, within a very short distance, the biplanar surface structure was re-established. It is evident, therefore, that this mode of crack propagation is inherently stable, and that the controlling mechanism is sufficiently powerful to rapidly overcome the disturbing influence of the secondary fracture which gives rise to the parabolic features.

A somewhat different situation may arise when the crack is propagating slowly, as in a cleavage experiment. In this case, the fracture surfaces frequently display areas which apparently are of the *same* color on the opposite surfaces. It is found in this case that the relative heights of the gross surface features are much greater ($\sim 50\ \mu$) than those of the color areas on the surfaces of tensile samples ($\sim 0.2\ \mu$). A closer examination of the opposite areas of the "same" color

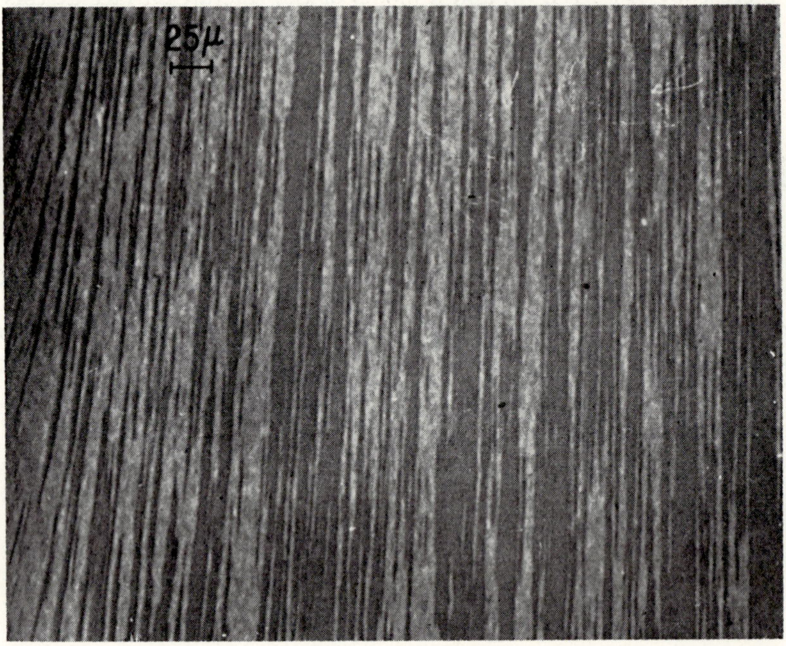

Fig. IIB-10. Detail of fracture surface in "same color" area of a cleavage sample of PMMA.

reveals a fine structure, which closely resembles that of the tensile fracture surfaces, as discussed above, but on a much smaller scale. Thus there is essentially a two-color system (Fig. IIB-10), and interference microscopy again reveals a relation between surface height and color (Figure IIB-11). It appears, therefore, that the difference in the surface is merely one of degree and, in the cleavage case, the matching surfaces appear to be of the same color because the eye is not capable of resolving the individual color areas.

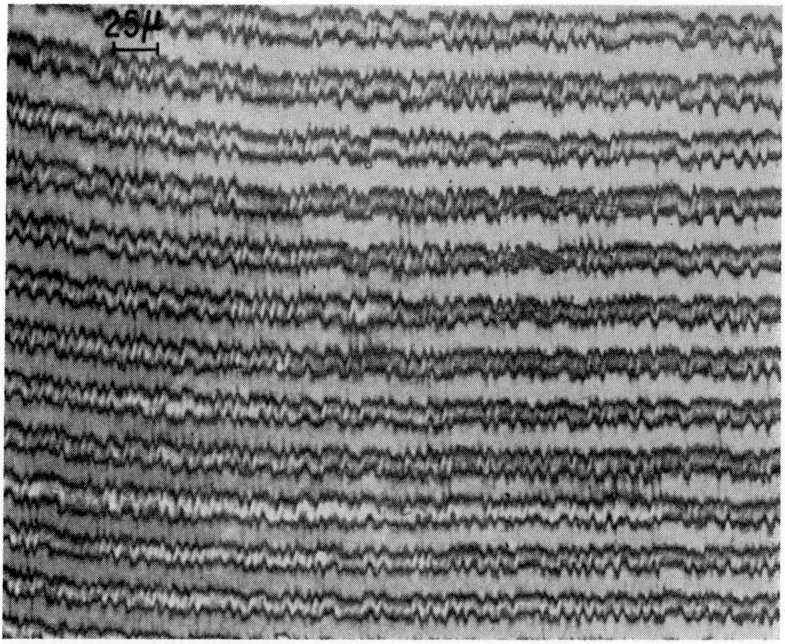

Fig. IIB-11. Interference micrograph of surface displayed in Fig. IIB-10.

IIB4. Influence of Other Variables on the Apparent Surface Energy of Fracture

The discrepancy between the theoretical and experimental fracture surface energies and the appearance of the fracture surfaces lead to the conclusion that *the material responds to the local stress at the tip of the flaw in a ductile rather than in a brittle manner,* despite the point that, macroscopically, the behavior would be characterized as brittle; at least according to the criteria discussed above. Such a ductile response involves a considerable amount of chain segmental transport, and it is the energy dissipated in this process that makes the largest contribution to the apparent γ. This response would involve the van der Waal's secondary forces, rather than the covalent primary bonds. Consequently, it would be expected that the experimental conditions and changes in the molecular structure would have a significant influence on the behavior. The effects of some of these variables have been determined.

IIB4a. Temperature

The experiments to determine the dependence of the critical tensile stress for failure on the size of the flaw have been performed at temperatures over the range −196°C. (77°K.) to +50°C. (323°K.) (11). The upper temperature limit chosen avoids significant macroscopic inelastic (i.e., viscoelastic) response of the sample, which would invalidate the theory used to interpret the data. At the lowest temperature studied, the samples were wholly immersed in liquid nitrogen to ensure good thermal contact. The parameter $E\gamma$ was obtained from the position of the curve of best fit of eq. (1) to the experimental data at any temperature, and E was determined independently over the same temperature range; hence $\gamma(T)$ was established over a wide range. These data are plotted in Figure IIB-12. The results of cleavage experiments performed over a somewhat smaller range of temperatures agree with those obtained in tension (13).

In view of the point that γ is believed to represent, almost entirely, the energy dissipiated with segmental transport, it is perhaps surprising that it increases as the temperature is *reduced* and, indeed, at the lowest temperature studied, there does not appear to be any leveling off in its value. Over this temperature range, it would be expected that the possibility of segmental transport would be reduced to a negligible degree. However, examination of the fracture surfaces produced by cleavage at the lowest temperature (77°K.) reveals color effects; hence there is evidence that a structure can be formed at the fracture plane at this temperature which has the same optical properties as that formed at the higher temperatures. In view of the sharply decreased segmental mobility at the lower temperatures, the energy that needs to be dissipated in forming this inelastic structure should increase and, in these terms, it is perhaps surprising that the observed increase in γ is so relatively *modest* (11).

It is of interest to compare this behavior with that of metals. It was noted above that at room temperature (25°C.) there was a resemblance in the trend of γ_{calc} and γ_{obs} with PMMA and a steel. It is found that this analogy does not persist over the wide temperature range described above. The experimental surface energies of metals decrease as the temperature is reduced, and approach their theoretical values (19). At room temperatures, it is believed that an analogous, local inelastic response is responsible for the similar discrepancy in the experimental and theoretical surface energy values. The mechanisms for energy dissipation involve a shear stress-biased, self-diffusion process in both instances, but the nature of the diffusing entity and the characteristics of the energy field in which it diffuses are distinctly different in the two types of materials. In the case of the steel, the diffusion process is described in terms of the motion of dislocations in the metal crystal lattice, whereas polymer segmental diffusion occurs by translations in a van der Waal's force field imposed by neighboring molecules. In view of these differences in the details of the processes concerned, the difference in the effects of temperature on the fracture surface energy is perhaps not surprising, though, unfortunately, it is not yet possible to predict, *a priori*, what the effect of temperature should be on any one system.*

IIB4b. Molecular Structure

It was noted above that it was expected that at low temperatures the internal energy of the molecules would have been reduced to the point where the energy barriers to segmental transport, as represented by the energy of secondary bonding, could not be surmounted efficiently. The experimental values of γ indicated that this was not the case. A similar effect could be produced, in principle, by increasing the strength of the intermolecular interaction by the introduction

* **Editor's Note:** The dilemma actuated by the unanticipated trend of the experimental values of γ in Figure IIB-12 would be considerably lessened if the fracture surfaces of the glassy PMMA were effectively "rubberlike," in part, at the *instant* of fracture under a 50°C. ambient, yet "leatherlike," in part, at that same instant under a -196°C. ambient. While perhaps seemingly extraordinary, this still appears to be quite possible in view of the extraordinary inelastic phenomena reported here, and those reported by Wolock and Newman in Section IIC and by Rosen in Section IIIA. Overstress and dissipation of strain energy under fracture instability in a rather good thermal insulator with a viscoelastic mechanical response can generate truly remarkable localized effects; cf. Figure IIIA-3, p. 298, etc.

A rubberlike fracture locus would be a *more efficient* storer of shear strain energy than one which is leatherlike; see Figure IC-10, p. 71, etc. While both a glassy polymer and a metal tend to become "softer" at elevated temperatures, the polymer may be approaching its "elastomeric state"; the metal merely becoming more inelastic or ductile. Despite its pronounced softness, a true rubber cannot be properly regarded as "ductile"; see Figure IC-21, p. 97, etc.

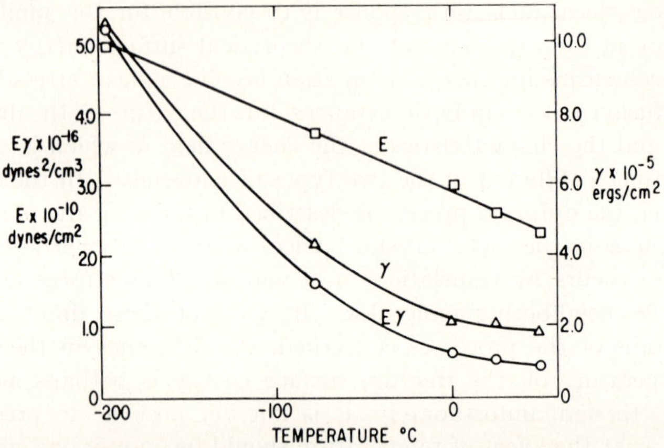

Fig. IIB-12. The temperature dependence of the tensile modulus E, the fracture surface energy γ, and the product $E\gamma$ for PMMA.

of primary bonded crosslinks. This approach would then impose much higher energy barriers to molecular translational motion, and the effect should be observable at a correspondingly higher temperature. The study of crosslinked systems is therefore of interest, particularly those systems corresponding to the linear polymers which have been studied for γ. Network polymers have been produced (11) by copolymerizing methyl methacrylate (I) and ethylene glycol dimethacrylate (II).

$$\underset{\text{I}}{CH_2 = C(CH_3)COOCH_3} \qquad \underset{\text{II}}{CH_2 = C(CH_3)COOCH_2{-}CH_2OOC(CH_3)C = CH_2}$$

The dependence of the critical tensile stress for failure on the size of the crack inserted into the samples of the copolymer can again be represented by eq. (1). The experimental value of γ (4.5×10^4 ergs/cm.^{2}) is about one-fourth of that determined for the samples of linear polymer and, in the present case, no colors could be seen on the fracture surfaces. The reduction in γ is comparatively modest in view of the relatively high degree of crosslinking of the system. However, although the concentration of crosslinks is such as to give an average of one crosslink for every ten linear chain segments, assuming complete reaction, the crosslinks will be distributed at random in the system, and the nature of this (number-average) distribution is such as to have a maximum at the shortest chain lengths. Correspondingly, there will be a significant fraction of relatively long chains and hence a capacity for energy dissipation by pronounced intersegmental motion. In this case, however, the resultant change in structure at the fracture surfaces is not sufficient to give rise to readily detectable optical effects.

IIB4c. Molecular Weight

The cleavage technique has been used to determine γ in a series of samples of PMMA with molecular weights varying from 0.98×10^5 to 60×10^5 (Table IIB-II) (12). In common with many other mechanical properties, the dependence of γ on (viscosity average) molecular weight M_v can be represented by the equation, $\gamma = A - (B/M_v)$, where A and B are constants. Using this relation for extrapolation purposes, the limiting value of γ for a material of infinite molecular weight is $\sim 1.55 \times 10^5$ ergs/cm.^{2}, while γ becomes zero for a polymer of 25,000 molecular weight. In view of the theoretical dependence of tensile strength on γ, the observed strength should likewise vanish for such a polymer. Vincent (43) has determined the variation in tensile strength in flexure of PMMA with molecular weight in this range and his results are in systematic accord with the conclusions derived from the extrapolated surface energy data. Most of the samples used in this study had been specially prepared. The sole exception was the commercially avail-

TABLE IIB-II
Dependence of Brittle Fracture Properties of PMMA on Molecular Weight

$M_v \times 10^{-5}$	$\gamma \times 10^{-5}$ ergs/cm.^{2}	$E \times 10^{-5}$ psi	T_0, psi	c_0, mm.
0.98	1.14	3.33	9,880	0.036
1.1	1.24	3.56	9,260	0.047
1.8	1.33	3.58	10,490	0.040
4.2	1.45	3.67	10,810	0.042
13	1.50	4.25	10,160	0.058
30	1.35	4.44	10,120	0.056
60	1.56	4.24	10,010	0.061

able Plexiglas II ($M_v = 3 \times 10^6$), which displays somewhat anomalous behavior, for reasons that are unknown at present.

IIB4d. Molecular Orientation

It is well known that the tensile strength of polymeric materials is markedly influenced by orientation of the molecules within the sample. See Section IVC. Such molecular orientation may be a result of the processing technique, or it may be introduced deliberately by stretching the polymer near its glass temperature T_g. The strength is increased when the tension is applied parallel to the orientation, but is little influenced when the tension is applied normal to it (30). In view of these points, it is of interest to determine the influence of orientation on γ.

The cleavage technique has been used to determine γ in sheets of PMMA which were *biaxially* oriented in its plane, and which were of sufficient thickness for samples to be cut out of the cross sections as well as out of the plane of the sheets (13). It was possible, therefore, for γ to be obtained for three mutually perpendicular planes in

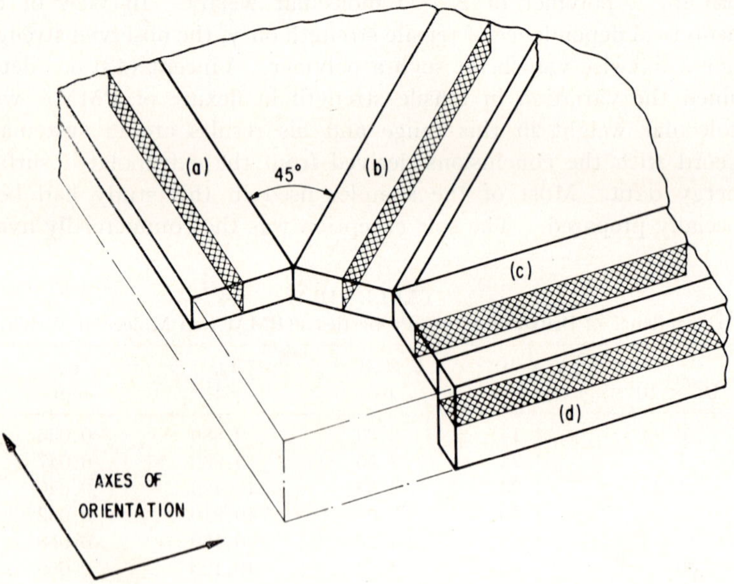

Fig. IIB-13. Disposition of cleavage samples in the sheet of oriented PMMA.

a sample (Fig. IIB-13). The degree of biaxial extension during orientation was 65% as determined by the change in the dimensions of the sheets. The results are summarized in Table IIB-III which also includes the result of a similar experiment on an unoriented sample. The variation in γ with the rotation of the samples (a, b, c in Fig. IIB-13) with respect to the direction of molecular orientation in the plane of the film indicates that this orientation was, in fact, not uniform. These values of γ are significantly higher than that for the unoriented sample, which is, itself, much larger than that obtained in the oriented sample (d in Fig. IIB-13) where the fracture plane is parallel to the face of the oriented sheet. These findings are considered below in relation to the structure of the surface layer, the formation of which is believed to be responsible for the dissipation of most of the energy which appears as γ.

TABLE IIB-III
Dependence of Fracture Properties of Poly(methyl methacrylate) on Molecular Orientation

Sample[a]	$E \times 10^{-5}$, psi	$\gamma \times 10^{-5}$, ergs/cm.^{2}	$T_0 \times 10^{-3}$, psi	$c_0 \times 10^2$, cm.
a	4.65	6.5	12.7	9.2
b	4.30	5.8	12.5	9.2
c	4.70	7.8	12.5	10.0
d	—	0.17	—	—
Unoriented	4.45	1.4	12.2	3.6

[a] See Figure IIB-13.

IIB5. Surface Energy of Fracture and Polymer Tensile Strength

According to the Griffith theory, the tensile strength is a property which is a function of the properties of the material, as reflected in Young's modulus and the surface energy, and also a function of the size of the adventitious flaw in the sample tested (21,22). In terms of its definition, the surface energy is also a measure of the resistance of the material to the initiation of the fracture process. In this respect, it is a more satisfactory index of the limiting tensile properties of a material than the tensile strength since it is independent of the local characteristics of the sample. By the same token, it is independent of the nature of the mechanical system used to determine its value and, indeed, the use of a tension technique for the

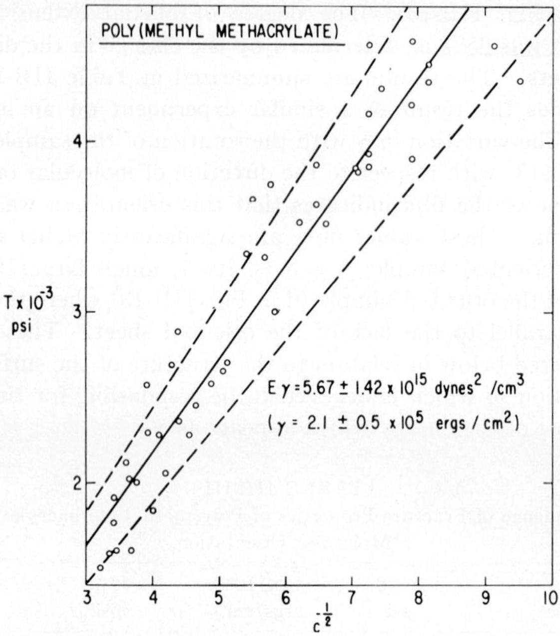

Fig. IIB-14. Dependence of the tensile strength of PMMA samples on crack size (10).

purpose leaves a great deal to be desired. Part of the data of Figure IIB-2 is plotted in Figure IIB-14 to give a linear relation. The severe amount of scatter of the results is clearly apparent and, because of the inherent low precision of this type of experiment, a large amount of data is required to find the position of the theoretical line with any degree of assurance. The nature of the stress field of samples in tension is such that the system is inherently unstable (6), at least for the small flaw sizes considered here. Consequently, each sample provides only a single data point; hence a large number of samples must be studied to obtain the position of the theoretical line. Accordingly, a large amount of material and time is required for this purpose. Application of eq. (1) would enable the material constant $E\gamma$ to be calculated. The uncertainty in its value, as approximated by the separation of the dashed lines in Figure IIB-14, is about $\pm 25\%$ in this instance. Furthermore, to estimate the value of γ, it is necessary to assume or determine a value for E. Any uncertainty in this latter value will further increase that of γ. Although the precision of the tension technique is sufficiently good when the discrepancy between the experimental and theoretical values of γ is under consideration, it is quite inadequate for distinguishing between the tensile properties of a series of polymers.

IIB6. Surface Energies of Fracture by the Cleavage Technique

As discussed in Section IIA, the Griffith criterion for mechanical instability can be applied to any system provided that the strain energy and surface energy can be expressed as a function of crack length. A system which has received considerable attention in the

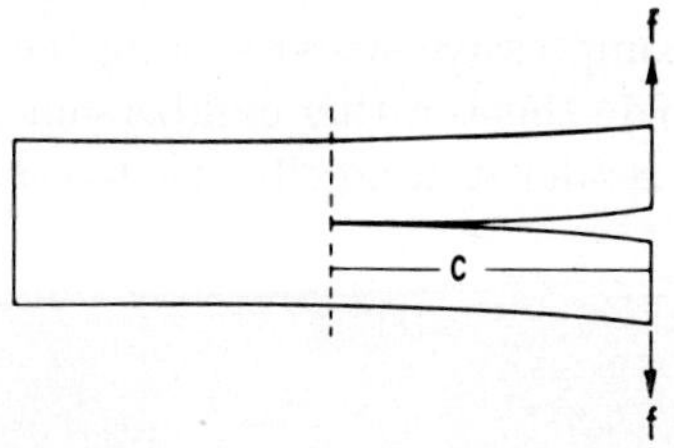

Fig. IIB-15. Conventional cleavage sample (10).

past is that in which a crack is propagated along the median plane of a strip sample by forces applied to the free ends (Fig. IIB-15) (1,2,3,10, 18,34,42). Much of the work with this system has suffered from both theoretical and experimental shortcomings. The theoretical shortcomings are associated with the assumption that the system could be represented as a pair of identical opposed simple cantilever beams, as has been discussed in some detail in Section IIA. The elastic behavior of the system can be represented adequately by the empirical relation

$$f = \phi c^{-\nu} \delta$$

where f is the force applied to the sample, δ is the resulting deflection when the crack length is c, and ϕ and ν are arbitrary constants. This relation may be expressed as $\log m = \log (f/\delta) = \log \phi - \nu \log c$, and the condition for instability is $(f\delta/w) = (4\gamma/\nu)c$, where w is the width of the fracture plane (10).

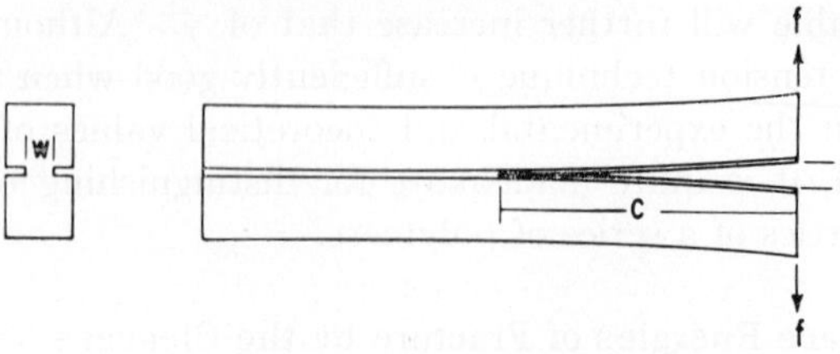

Fig. IIB-16. Modified cleavage sample (10).

The experimental shortcoming arises from the fact that the stress field in the vicinity of the tip of the crack in an isotropic sample is such as to cause the crack to turn out of the plane of symmetry towards one edge of the sample (23). It is possible to counteract this effect by imposing compressive stresses along the sample, but the apparatus necessary to do this is rather cumbersome and introduces frictional factors which render it impossible to determine the value of the

Fig. IIB-17. Experimental arrangement for studying fracture by cleavage (10).

force corresponding to the condition of instability (2,3,42). Consequently, the inexact beam theory equations must be used to evaluate the results. The use of such constraints can be avoided by modifying the configuration of the sample slightly, so that the cross section is reduced along the median plane (Fig. IIB-16) (10). The simplified theoretical relation can then be used for analyzing the experimental results since all the variables which appear therein are directly observable.

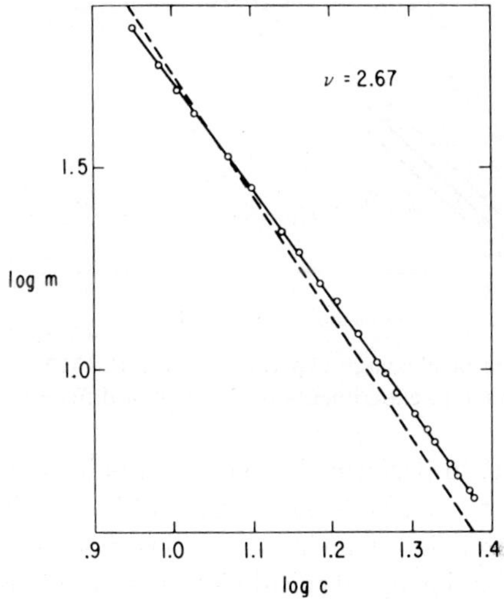

Fig. IIB-18. Dependence of the slope of the force-deflection relation in cleavage on crack length (10).

The principal advantage of the cleavage technique is that the system is inherently stable at a constant deflection (6). If the deflection imposed on the sample is so great as to cause the central crack to increase its length, and this deflection is maintained constant, the crack will grow only until the experimental variables satisfy the theoretical equation and there it will stop, and the corresponding equilibrium values of the variables can be recorded. A slight increase in the deflection will result in a new equilibrium position with a new set of values for the variables. Thus a large number of experimental points can be obtained on a single sample in a reasonable period of time.

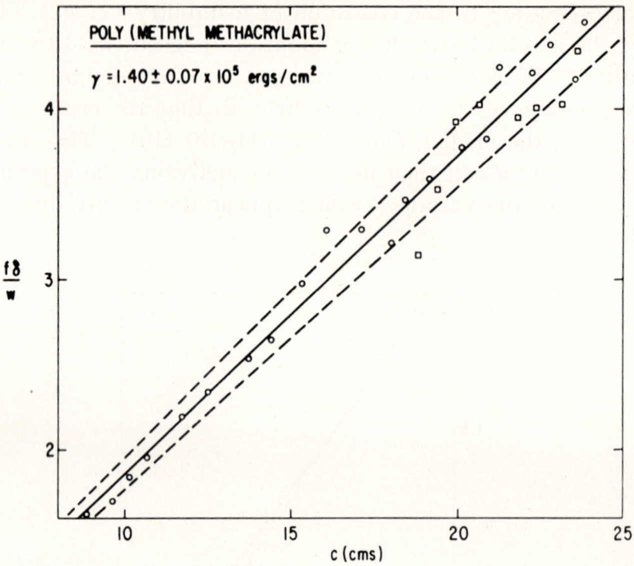

Fig. IIB-19. Results of cleavage experiment with PMMA. The points of different shape refer to experiments performed on different samples (10).

It is convenient to perform the experiments in conventional tensile testing equipment (Fig. IIB-17) which readily permits the necessary corrections to be made for inelastic responses of the material (10).

The elastic behavior of a typical cleavage sample is represented in Figure IIB-18, which displays the excellent linearity of the data. The slope of the line is somewhat less than that required by the beam theory (dashed line), and although this may appear to be a somewhat minor correction, it should be noted that the discrepancy between the values of the theoretical and experimental intercepts is quite considerable, frequently a factor of two. To evaluate γ, the value of the derived quantity $f\delta/w$ is plotted against c (Figs. IIB-19 and IIB-20). γ is obtained from the slope of the best line through the points and the value of ν from the log m versus log c plot. In contrast to the tensile technique, γ is obtained directly rather than as a product with another material constant; hence one source of error in its value is eliminated. It will also be observed from Figures IIB-19 and IIB-20 that γ can be obtained easily from experiments on a single sample, and the scatter of the data is much less than was found in the

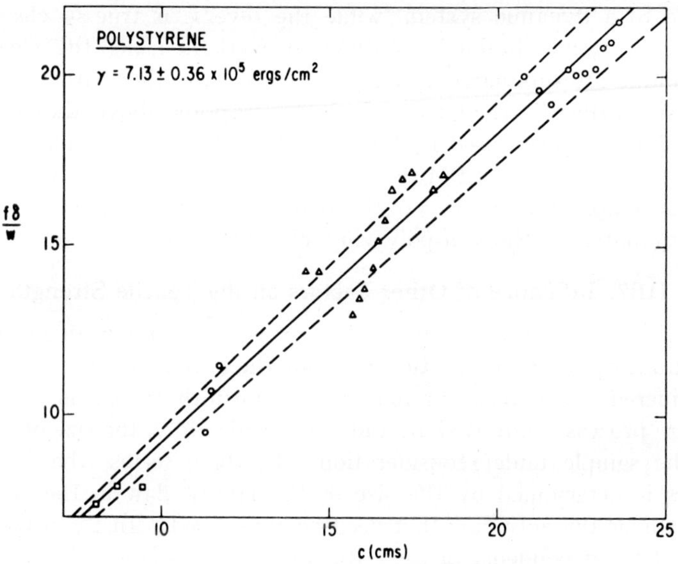

Fig. IIB-20. Results of cleavage experiments on PS. The points of different shape refer to experiments performed on different samples.

tension experiments. In these figures the dashed lines define the approximate limits of uncertainty in the value of γ, e.g., $\pm 5\%$.

The results obtained by this technique are comparable to those found in the tension experiments but tend to be somewhat lower, as might be expected from theoretical considerations (6). The difference here arises from the difference in the condition of the system at the point at which the experimental data are taken. In tension, this condition is defined, in terms of the crack, by the transition from a

TABLE IIB-IV
Fracture Surface Energies γ of Glassy Polymers

Method	$\gamma \times 10^{-5}$, ergs/cm.^{2} PMMA	PS	Reference
Cleavage	4.9 ± 0.5	25.5 ± 3	Benbow and Roesler (3)
Cleavage	4.5	9.0	Svensson (42)
Cleavage	1.40 ± 0.07	7.13 ± 0.36	Berry (10)
Tension	2.1 ± 0.5	17 ± 6	Berry (7)

static to a dynamic system, while the reverse is true in cleavage. Typical results obtained by different workers using the cleavage technique are summarized in Table IIB-IV, together with the comparative values for the tension method. As noted above, the cleavage technique has been used to determine the effect of molecular weight and orientation on γ of PMMA. It has also been used to study the effect of temperature on the γ of this polymer (13,40,42), and also for the determination of the γ of poly(vinyl chloride) (40).

IIB7. Influence of Other Factors on the Tensile Strength

Although the fracture surface energy is important in determining the limiting tensile properties, it is not the only factor that must be considered. The basic premise of the Griffith theory is that the failure process is initiated at, and is determined by, the size of a flaw in the sample under consideration. In these terms, the fracture stress is determined by the size of the largest flaw in the sample, and this is the only flaw that need be considered. In the investigation of the dependence of the critical tensile stress for failure on the size of artificial flaws, a deviation from the theoretical behavior must, therefore, be expected when the size of the artificial flaw becomes less than that of the flaw which arises naturally (8). This effect is quite obvious in PS (Fig. IIB-21), but much less so in PMMA (Fig. IIB-2). It is possible to define the size of the "inherent flaw" from the tensile strength of a conventional sample since, in this case, the failure point is determined by the natural flaw. It is found that the variation in the tensile strength evaluations is much less in polymers than in most other materials, which indicates that there is a much smaller variation in the size of the inherent flaw from sample to sample (24). The tensile strength determines the point of deviation from the theoretical line, and the value of the flaw size corresponding to this is that of the inherent flaw. From the data on PS the inherent flaw size is about 1 mm., while in PMMA the value is ca. 0.5 mm. Indeed, in the latter case it is difficult to detect any deviation from theoretical behavior (Fig. IIB-2), since the size of the smallest flaw which can conveniently be introduced artificially is about the size of the inherent flaw. The large difference in the value for these two polymers indicates that the inherent flaw size may be a material property, and it is therefore of interest to determine on what it depends.

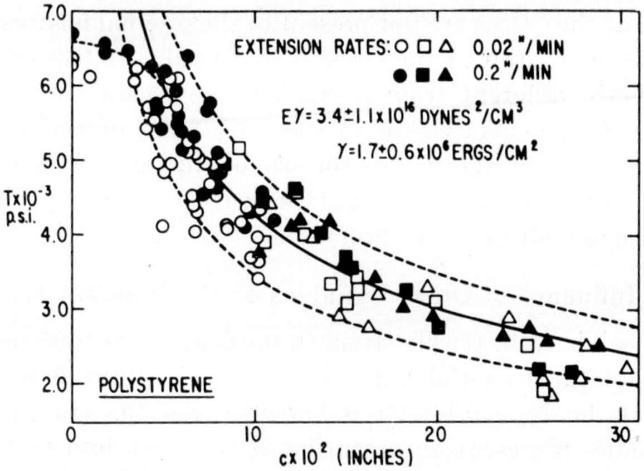

Fig. IIB-21. The dependence of tensile strength on crack-size in PS samples for various cross sections and extension rates.

The results obtained with PS give a clear indication that the inherent flaw size is related to the tendency of the material to display crazing behavior. See Sections IIC and IIIA. By "crazing" is meant the development of microscopic planar defects in the sample at relatively low tensile stress. The defects are highly reflective and are oriented in a plane normal to the applied tension, frequently penetrating the sample from the surface. In the tensile experiments on PS (Fig. IIB-21), agreement with the theory is found for large inserted cracks, crazing is absent, and the fracture plane always coincides with the location of the inserted crack. At small inserted crack sizes, when deviation from the theory is observed, crazing is extensive, and frequently the location of the fracture plane and the inserted defect do not coincide. Moreover, PMMA is much less susceptible to crazing than PS, and the size of the inherent flaw is much less in the former material.

In the elastic models discussed in Section IIA, the crack which is present remains at its initial length until a sufficiently large stress is imposed to render it unstable. On the other hand, the statement that the inherent flaw size in PS is 1 mm. does not imply that a physical crack of this size is present in the original unstressed sample. Such a crack would be readily visible to the unaided eye, and yet none of this size is detected. Since the inherent flaw is attributed to the effect of crazing, and perhaps to an individual craze mark, it is important to note that the crazes are not present in the original unstressed state, but develop and grow as the stress is increased. In this respect they are distinctly different from a crack in an elastic model. Consequently, it is the effective size of the craze mark immediately before rupture occurs that determines the size of the inherent flaw. In the unstressed sample, the flaw is not present as such; it requires the application of a stress to develop it.

IIB8. Influence of Other Variables on the Inherent Flaw Size

It is possible from tensile strength measurements to determine the influence of experimental variables on the inherent flaw size, provided that the $E\gamma$ parameter is known under the same conditions. These studies represent an extension of the work undertaken to determine the influence of these variables on the fracture surface energy, as discussed above.

IIB8a. Temperature

The tensile strength of PMMA has been determined over the temperature range for which γ was obtained, and hence the corresponding variation in the inherent flaw size has been calculated (11). This is illustrated in Figure IIB-22. It appears that the inherent flaw size passes through a minimum at about $-20°$C., and increases markedly as the temperature is further reduced so that, at the lowest temperature, it has a value of 0.25 mm., which is comparable to that of PS at room temperatures. Unfortunately there is no information on the crazing behavior of the polymer over this temperature range. However, examination of samples fractured in liquid nitrogen reveals a considerable degree of crazing, comparable with that of a PS sample fractured at room temperatures. The form of the force–deflection relation at the failure point is also similar in that it displays a tendency to level off, due to the stress–relieving effect of rapid crazing under the high stress conditions. In liquid nitrogen, crazing is enhanced to such an extent that it is impracticable to establish the dependence of the tensile strength of PS on flaw size under these conditions.

IIB8b. Molecular Structure

The tensile strength of a crosslinked copolymer of methyl methacrylate and 10 mole-% of ethylene glycol dimethacrylate is 8100 psi,

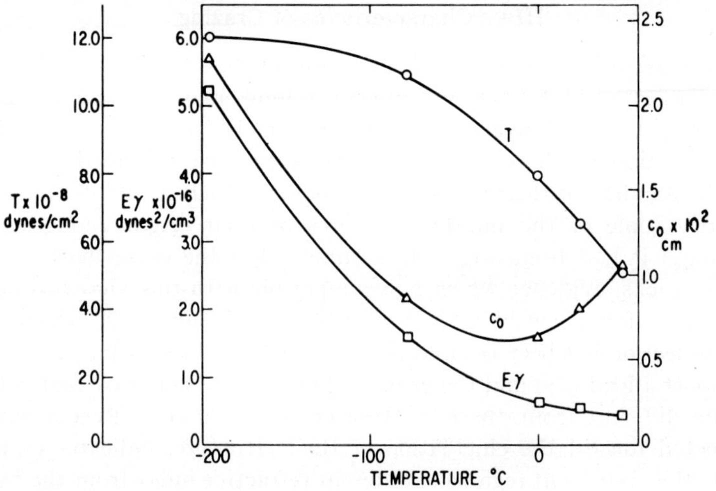

Fig. IIB-22. The temperature dependence of the tensile strength T, the inherent flaw size c_0, and the product $E\gamma$ of PMMA.

which, with the experimental value of $E\gamma$ of 8.85×10^{14} dynes^2/cm.^{3}, gives a value (11) of the inherent flaw size of 1.81×10^{-2} mm. This is considerably less than the value for the linear polymer (5×10^{-2} mm.), but it is not known what is the relative tendency to crazing of the two materials.

IIB8c. Molecular Weight

The inherent flaw size c_0 shows only a slight dependence on molecular weight (Table IIB-II); increasing by only 70% for a 60-fold change in molecular weight.

IIB8d. Molecular Orientation

The tensile strengths of molecularly oriented samples of PMMA corresponding to those used for the determination of γ have also been obtained, and it has been possible to establish the dependence of the inherent flaw size on this variable. The results given in Table IIB-III indicate that in these samples the tensile strength and E are little affected by the orientation, and hence that the inherent flaw size increases in about the same ratio as γ on going from the unoriented to the oriented polymer.

IIB9. Characteristics of Crazing

In view of the above, it is clearly desirable to establish what factors are important in determining and controlling crazing behavior. This aspect of the problem is discussed in Sections IIC and IIIA. Detailed examination of the craze marks themselves has provided a better understanding of their nature and structure which should do much to clarify some of the puzzling features of their origin and growth. Though it had been originally assumed that the craze marks were true cracks, evidence which is incompatible with this view has been available for a considerable time (28,39). A sample in which the cross section has been completely traversed by crazes will continue to support a load (28) and the growth characteristics of craze marks are quite different from those of true cracks (5,39,41). Recent work, directed toward the elucidation of their structure, indicates clearly that they represent regions of different refractive index from the bulk of the sample (hence their reflectivity), but that they are mechanically continuous with the rest of the sample and can sustain a stress (31,41). It appears that they represent isolated local regions where yielding has occurred on a microscopic scale. Examination of craze marks induced by alcohol (and stress) in a polycarbonate resin reveals that in this case the material of the craze is not homogeneous but contains about 50% of voids (31). These voids have been penetrated with silver and their distribution determined by examination of ultramicrotomed sections in the electron microscope (32). This technique reveals a wide range of void sizes, down to the limit of resolution of the microscope, and it appears that these voids are distributed randomly throughout the craze mark. How far the details of this structure apply to craze marks in other materials has yet to be determined, but theoretical considerations indicate that they should be quite general (38). It appears that the craze mark is the result of two types of response to external stress: (*1*) the formation of voids and (*2*) extension and orientation of the polymer molecules in the direction of the local stress. The second type of response is analogous to the macroscopic response of a sample as it undergoes cold-drawing.

If this view of the structure of a craze mark is correct, the relation of crazing to the size of the inherent flaw can be readily understood. As the stress on the sample is increased, and the craze grows in size, it would be expected that a point would be reached when the voids begin to coalesce until they constitute a true crack which is then activated by the imposed stress. Since a rate process is involved (see Section IIIA), it would be expected that the observed fracture stress would depend on the rate of stressing and, in fact, such a dependence can be detected in the region where the behavior of the sample is determined by the crazing tendency (Fig. IIB-21). Once the system becomes unstable, it is unlikely that there will be a discontinuous change in the stress field at the tip of the (true) crack and thus the structure produced as the crack passes through the sample would be expected to resemble that of its logical precursor, the craze mark.

It has been suggested that the structure at the *fracture plane* arises from orientation of the molecules under the influence of the local shear stress at the crack tip, in a manner analogous to that which occurs in macroscopic cold-drawing (7). There are a number of facts which, however, are difficult to reconcile with this simple hypothesis:

(*1*) The brilliance of the interference colors demands a pronounced difference in refractive index on passing from the surface layer to the substrate. This, in turn, demands a high degree of molecular orientation in the surface layer, but no evidence of any significant degree of ordering can be detected by x-ray or electron diffraction techniques. On the other hand, the inclusion of only a small concentration of voids in the structure markedly influences the refractive index (31).

(*2*) A high degree of orientation of a polymer results from an effectively large elongation in the longitudinal direction, which demands a correspondingly large contraction in the transverse direction. However, the surface layer is so constrained by the unoriented substrate that it cannot undergo a significant reduction in cross-sectional area. The material is therefore in a state of hydrostatic tension; hence void formation is encouraged by this stress system.

(*3*) The surface layer is produced at ambient temperatures at which molecular translational motion is greatly retarded. However, it is energetically much easier to create a void by separation of molecules than to produce any significant relative lateral motion of the molecules; hence void formation is possible under conditions where molecular orientation is prohibited. It has been suggested that the local temperature at the tip of the moving crack may be much higher than the ambient, due to strain energy being dissipated as heat and that, for this reason, molecular motion may be possible under apparently unfavorable conditions (27). However, in terms of the orientation hypothesis, it is necessary for the molecules to remain in their oriented condition at these relatively high local temperatures. The velocity of crack propagation is so high that it is difficult to appreciate how the heat can be removed sufficiently quickly behind the crack-front to prevent disorientation of the molecules.

(*4*) It has been reported that exposing the fracture surface of PMMA to high vacuum influences the colors which are observed (2). This effect cannot be understood in terms of the orientation hypothesis alone, whereas if the presence of voids is permitted it is not surprising that high vacuum influences the system. It is possible that the effect of such inert liquids as water and silicone oil on the surface colors can be explained in a similar way (2).

(*5*) If the local stress near the crack tip merely produces orientation of the molecules and it is the energy dissipated in this way that makes the largest contribution to the observed γ, then it would be expected that prior orientation of the sample would reduce the apparent surface energy, since the amount of molecular motion required to produce the oriented surface layer is thereby decreased, and the energy dissipated in this way should show a corresponding reduction. As discussed above, γ is, in fact, increased by orientation when the fracture is propagated in a plane normal to the direction of prior orientation of the molecules. Unfortunately little information is available on the fracture surface energy for a crack propagated in a plane parallel to the direction of orientation of the sample. (It would be of interest to study sheets with uniaxial orientation in this regard.)

In considering the energy required to produce the effects which are presumed to be important in forming the structure at the fracture plane, it is probable that orientation would influence the two factors in opposite ways. As discussed above, the energy dissipated in molecular orientation should decrease. It may be assumed that the voids are produced by separating layers of molecules which lie in a plane approximately normal to the imposed tensile stress. Consequently, if the molecules tend to be oriented in that direction, the probability of occurrence of a group of favorably disposed molecular segments will be reduced. Hence, it will be more difficult to create the voids, since more energy will be required for this purpose. However, in view of the differences in the energy required for relative translation and separation of the molecules, the primary processes in orientation and void formation, the experimental γ will probably represent a rather delicate balance of the two effects, so that the relative contributions are difficult to estimate, *a priori*. Furthermore, the two effects will make different contributions to the optical properties of the structure which is formed, different from one another and probably different from their relative contributions to γ. For this reason the effects of, for instance, crosslinking on γ and on the optical properties of the fracture surface are difficult to interpret unequivocally (11).

IIB10. Factors Governing the Fracture of Glassy Polymers

It was concluded in the discussion above that there is a close relation between the structure of a craze mark and that of the interference layer which is found at the fracture plane. Indeed, the first is essentially the precursor of the second. This relationship is important in regard to the Griffith brittle fracture theory which indicates the importance of two factors in determining the tensile strength: (*1*) the fracture surface energy and (*2*) the inherent flaw size. It has been shown above that the first is related to the energy dissipated in forming the structure at the fracture plane, while the second is related to the crazing behavior of the sample. Hence these two factors cannot be considered completely independent, since they both depend on essentially the same type of response of the material to the stress. It is somewhat paradoxical that this response is essentially ductile in character, in spite of the point that the macroscopic sample response would be considered as brittle.

The effect of the two factors operating together is illustrated in Figure IIB-23, which depicts the relative position of the theoretical curves of Figures IIB-2 and IIB-21. It is clear, from this figure, that PS has a higher intrinsic strength due to its higher γ. However, because of its larger inherent flaw size, deviation from the theoretical curve occurs at a relatively low stress; thus the tensile strength of a conventional sample is, in fact, about half that of PMMA. This difference is still comparatively small and, indeed, since the factors governing the tensile strength are both related to the same microscopic or molecular response, it is probable that a degree of compensation will occur.

According to the Griffith theory, since the tensile strength is related directly to the surface energy but inversely to the inherent flaw

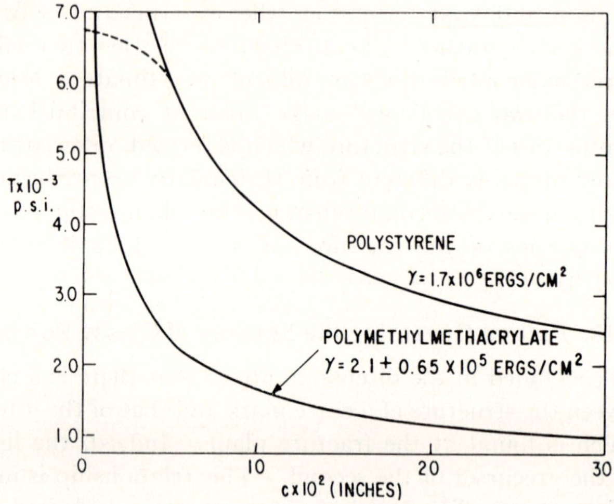

Fig. IIB-23. Comparison of the dependence of tensile strength on crack-size in PS and in PMMA samples.

size, the two factors influence the strength in opposite directions. However, it is expected that a high γ would be associated with a tendency to form a structure at the fracture plane and to dissipate energy in so doing. In view of the relation of this structure to that of a craze mark, it tentatively seems probable that there would be a correspondingly strong tendency to craze; hence the inherent flaw size would be large. It is the ratio of these two factors which determines the fracture properties, and it is probable that this ratio is relatively insensitive to changes in the material structure or the experimental conditions. This conclusion is borne out, in large part, by the results quoted above.

IIB11. Application of the Griffith Theory to Other Polymeric Systems

The treatment of failure in terms of the energetics of the system is capable of wide application, and the development of the theoretical equations discussed above for tension and cleavage experiments are only two of many possible examples. However, in applying a theoretical equation care must be taken to ensure that the assumptions made in deriving that equation are not violated in the particular system under consideration. For this reason the interpretation of *the tensile strength of rubbers by the Griffith equation (1) is not justified* since the dependence of the strain energy on flaw size in a material which undergoes large deformations is not the same as that for a classically elastic material where the strains are assumed to be small, i.e., infinitesimal (14). The stress concentration factor in such a system is not the same as in a classically elastic body; hence the interpretation of the data in this way is likewise invalid (14).

On the other hand, *provided that the nature of the elastic properties is taken into consideration, the Griffith criterion for instability can be applied to rubbery materials.* The treatment of tearing from this point of view has been particularly successful (37). Special sample configurations greatly simplify the determination of the strain energy derivative, and it is possible to obtain expressions for the tearing energy in terms of the conditions at the point of instability. The tearing energy is analogous to the fracture surface energy of a rigid material, in that it is the energy required for the formation of unit area of surface as the initial defect increases in size. The values obtained for natural rubber (1.1×10^7 ergs/cm.^{2}) indicate that in this case also, energy is being dissipated in inelastic processes, which are probably associated with crystallization at the tip of the tear because of the high strain at that point. This work has been extended to include the study of tearing under dynamic conditions, and at different temperatures, the tearing energy being taken as a parameter characteristic of the behavior (20).

IIB12. Summary

The physical condition of polymers in the glassy state is such that classical flaw theories of brittle strength are applicable, and experimental data establish the validity of this approach. In terms of the theory, a surface energy of fracture γ can be defined as that required for the formation of unit area of surface by extension of an existing flaw. Its value can be obtained from experiments in either tension or cleavage, and good agreement is found between the results of the two techniques. The magnitude of γ indicates that inelastic processes occur at the tip of the flaw, under the influence of the local high stress concentration. In this respect, it appears that the behavior of a glassy polymer is more analogous to that of a metal, rather than to that of a conventional inorganic glass.

The local inelastic process in PMMA primarily results in the formation of a structure at the fracture plane which gives rise to optical interference, with the production of brilliant surface colors. The arrangement of these colors is consistent with a biplanar surface topography, the existence of which has been established by interferometry. However, the reason for the peculiar surface configuration remains an enigma. It is believed that the structure of the interference layer is similar to that of the "craze" marks produced at lower stress levels.

The energy dissipated in producing this surface structure, and which appears in γ, is one involving molecular motion in the van der Waal's secondary force field and is influenced by the experimental variables. Over a temperature range, 77 to 323°K., γ for PMMA decreases by a factor of five and appears to approach a limiting value at the higher temperature. Increase of temperature above that quoted results in a significant macroscopic inelastic response of the samples, which invalidates the theory used to determine γ. Crosslinking the polymer also results in a reduction in γ, while it varies linearly with the reciprocal of the molecular weight of the linear polymer; a dependence common to several other of its mechanical properties. Using this relation, it would be expected that γ, hence the tensile strength, should vanish for a material of 25,000 molecular weight. This result is in agreement with independent experimental observations of the dependence of tensile strength on the molecular weight of PMMA. Prior molecular orientation of the sample results in an increase in γ when the fracture plane is normal to the plane of orientation, and a large reduction in γ when it is in the direction of that plane.

γ appears to be a more satisfactory index of strength properties than the observed tensile strength itself since it is not dependent on the vagaries of the samples tested. For its evaluation, the cleavage technique is preferable to the tensile technique.

According to the flaw theories, the fracture stress is limited by the size of the largest flaw present immediately prior to rupture. In glassy polymers, it appears that the "inherent flaw size" is related to the crazing behavior, i.e., the development of planar defects in the sample at stresses below the critical value. Unfortunately crazing behavior is little understood and difficult to express quantitatively. The influence of temperature, molecular structure, molecular weight, and prior molecular orientation on the inherent flaw size have been determined, but the effects are relatively small, and somewhat difficult to interpret.

Recent work on the structure of craze marks indicates that they represent regions of local yielding where molecular orientation and void formation have occurred. It is believed that the layer of material at the fracture surface, which gives rise to optical interference, possesses a similar structure. In view of the underlying mechanistic and morphological relation between the surface energy of fracture and the inherent flaw size, it seems probable that these two factors, which determine the strength properties, cannot vary independently. The relatively small range of tensile strengths of glassy polymers lends support to this conclusion.

The *principles* embodied in the classical fracture theories can be applied to other polymer systems. Particular success has been achieved in the study of the tearing of rubber by this approach.

References

1. Bailey, A. I., *Proc. 2nd Intern. Congr. Surface Activity*, **3**, 406 (1957).
2. Benbow, J. J., *Proc. Phys. Soc. (London)*, **78**, 970 (1961).
3. Benbow, J. J., and F. C. Roesler, *Proc. Phys. Soc. (London)*, **B70**, 201 (1957).
4. Berry, J. P., *Nature*, **185**, 91 (1960).
5. Berry, J. P., *J. Mech. Phys. Solids*, **8**, 194 (1960).
6. Berry, J. P., *J. Mech. Phys. Solids*, **8**, 207 (1960).
7. Berry, J. P., *J. Polymer Sci.*, **50**, 107 (1961).
8. Berry, J. P., *J. Polymer Sci.*, **50**, 313 (1961).
9. Berry, J. P., *J. Appl. Phys.*, **33**, 1741 (1962).
10. Berry, J. P., *J. Appl. Phys.*, **34**, 62 (1963).
11. Berry, J. P., *J. Polymer Sci.*, **A1**, 993 (1963).
12. Berry, J. P., *J. Polymer Sci.*, in press.
13. Broutman, L. J., private communication.
14. Bueche, A. M., and J. P. Berry, in *Fracture*, B. L. Averbach, D. K. Felbeck, G. T. Hahn, and D. A. Thomas, Eds., Wiley, New York, 1959, p. 265.
15. Chang, L., *J. Mech. Phys. Solids*, **3**, 212 (1955).
16. Charles, R. J., *Progress in Ceramic Science*, Vol. I, Pergamon Press, London, 1961.
17. Felbeck, D. K., and E. Orowan, *Welding J. Res. Suppl.*, **34**, 510 (1955).
18. Gilman, J. J., in *Fracture*, B. L. Averbach, D. K. Felbeck, G. T. Hahn, and D. A. Thomas, Eds., Wiley, New York, 1959, p. 193.
19. Gilman, J. J., *J. Appl. Phys.*, **31**, 2208 (1960).
20. Greensmith, H. W., and A. G. Thomas, *J. Polymer Sci.*, **18**, 189 (1955).
21. Griffith, A. A., *Phil. Trans. Roy. Soc.*, **A221**, 163 (1921).

22. Griffith, A. A., *Proc. Intern. Congr. Appl. Mech. (Delft)*, 1924, p. 55.
23. Guernsey, R., and J. J. Gilman, *Proc. Soc. Exptl. Stress Anal.*, **18**, No. 2, 50 (1961).
24. Gurney, C., *Properties of Metallic Surfaces*, The Institute of Metals, London, 1953, p. 145.
25. Hall, E. O., *J. Mech. Phys. Solids*, **1**, 227 (1953).
26. Higuchi, M., *Rept. Res. Inst. Appl. Mech. (Japan)*, **6**, 173 (1958).
27. Higuchi, M., *J. Japan. Soc. Test. Mat.*, **9**, 79 (1960).
28. Hsiao, C. C., and J A. Sauer, *J. Appl. Phys.*, **21**, 1071 (1950).
29. Irwin, G. R., *Handbuch der Physik*, Vol. 6, Springer, Berlin, 1958, p. 551.
30. Jackson, G. B., and R. L. Ballman, paper presented at 16th Antec, Society of Plastics Engineers, Chicago, 1960.
31. Kambour, R. P., *Nature*, **195**, 1299 (1962).
32. Kambour, R. P., private communication.
33. Klier, E. P., *Trans. Am. Soc. Metals*, **43**, 935 (1951).
34. Obreimoff, J. W., *Proc. Roy. Soc. (London)*, **A127**, 290 (1930).
35. Orowan, E., *Trans. Inst. Engrs. Shipb. Scot.*, **89**, 165 (1946).
36. Orowan, E., *Proceedings of Symposium on Fatigue and Fracture of Metals*, Wiley, New York, 1950, p. 139.
37. Rivlin, R. S., and A. G. Thomas, *J. Polymer Sci.*, **18**, 177 (1955).
38. Robertson, R. E., private communication.
39. Sauer, J. A., and C. C. Hsiao, *Trans. A.S.M.E.*, **75**, 895 (1953).
40. Shoulberg, R. H., private communication.
41. Spurr, O. K., Jr., and W. D. Niegisch, *J. Appl. Polymer Sci.*, **6**, 585 (1962).
42. Svensson, N. L., *Proc. Phys. Soc. (London)*, **77**, 876 (1961).
43. Vincent, P. I., *Polymer*, **1**, 425 (1960).
44. Wolock, I., J. A. Kies, and S. B. Newman, in *Fracture*, B. L. Averbach, D. K. Felbeck, G. T. Hahn, and D. A. Thomas, Eds., Wiley, New York, 1959, p. 250.

IIC. FRACTURE TOPOGRAPHY*

IRVIN WOLOCK AND SANFORD B. NEWMAN

* **Editor's Note:** A sizable portion of the material treated in this section concerns topics also considered by Berry, Section IIB, and Rosen, Section IIIA. This is intentional since the approaches employed in the present section are oftentimes considerably different; hence refreshing as regards firm contrasts in viewpoint as well as being of supplementary information value on those particular topics.

IIC1. General Morphology of the Fracture Surface

The processes involved in the fracture of solids are reflected in the appearance of the fracture surface. By studying this surface, one can gain valuable information to supplement theoretical studies and provide additional insight into the failure mechanisms.

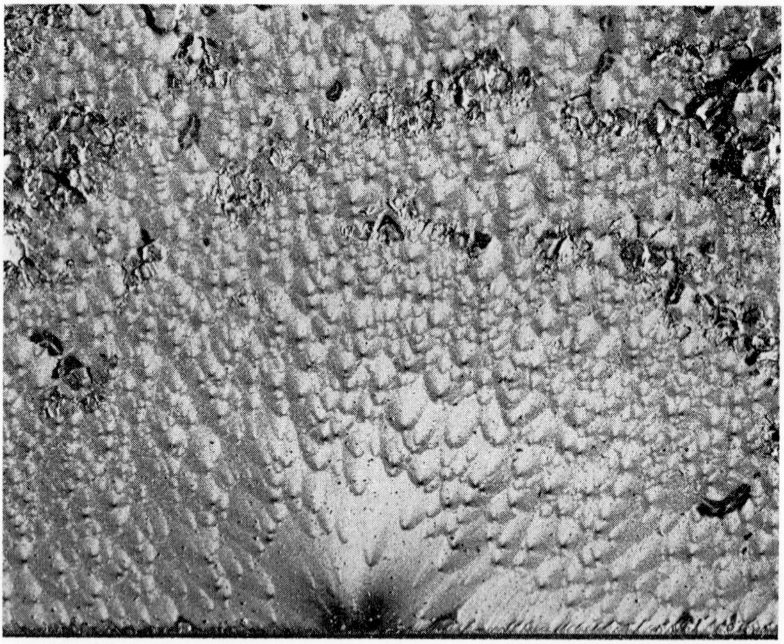

Fig. IIC-1. Fracture surface of a tensile specimen of PMMA (×8).

Fracturing begins at "flaws." The initial stages are usually accompanied by local plastic deformation. The progressive extension of fracture may be locally discontinuous. New separations of fracture elements can occur in a scattered, independent fashion on different levels in regions of high tensile stress in advance of the main crack-front. These separately initiated elements can join up to form the crack. The lines which mark the boundaries between the fracture elements spreading forward on different levels have been called tear lines or lag lines (because the fracture lags behind). The tear lines and the fracture origins represent the major deformation processes

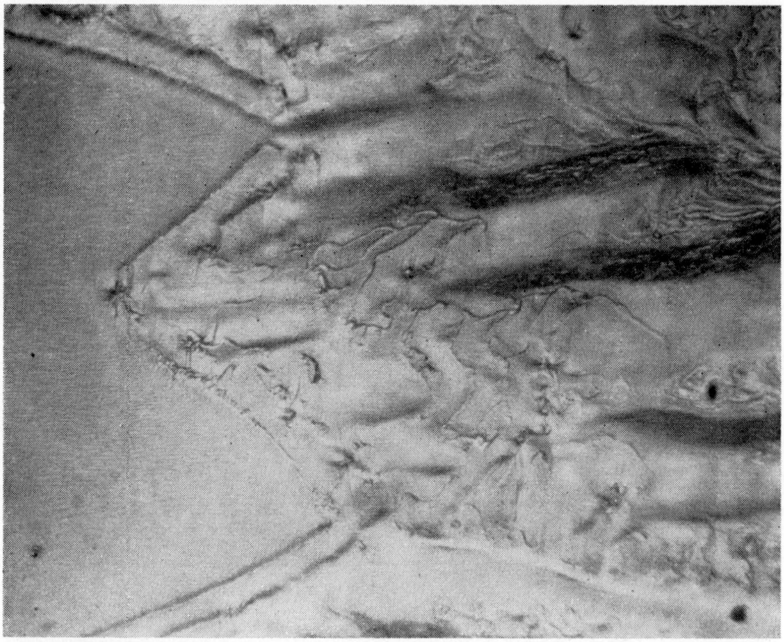

Fig. IIC-2. Smooth, slow-growth area on fracture surface of centrally notched, plate specimen of PMMA (×50).

for which energy must be supplied to maintain or increase the velocity of fracturing (25).

The fracture surfaces of polymers can exhibit certain characteristic features which record the fracture process as described above. A "typical" fracture surface of a glassy polymer, in this case poly(methyl methacrylate), is shown in Figure IIC-1. In the fracture surface shown, from an ASTM tensile specimen (3), the different areas comprising the surface are compressed somewhat. These areas can be extended and more readily differentiated if the tensile specimen is notched or if a centrally notched or edge-notched plate specimen is used.

The fracture progresses slowly from its origin until it reaches a size which is ctitical for the applied stress, at which point it becomes unstable and propagates rapidly. The area of slow growth that contains the origin of failure in a tension specimen is smooth and of high specular reflectivity. This is known as the "mirror area."

This area is usually located at an edge, but in some cases—when there is a rather large internal flaw—in the interior. The mirror area appears to be featureless under low magnifications, but usually exhibits some markings under higher magnifications. The slow-growth area in a centrally notched specimen (which also is rather smooth) exhibits two types of markings, as shown in Figure IIC-2. The portion at the left consists of a fine ground, having a single color when

Fig. IIC-3. Second area, of irregular lines, on fracture surface of centrally notched, plate specimen of PMMA. Lines probably represent level-differences between fracture planes ($\times 100$).

viewed macroscopically, oriented in the direction of fracture propagation. At the right, the fine structure is complicated somewhat by body corrugations, as well as by some strongly defined linear structures, and is covered by bright multicolored areas. Because of the rather low stress levels at which the crack-front propagates initially, secondary fractures are initiated close to the advancing front and the resulting surface is relatively featureless. The markings ob-

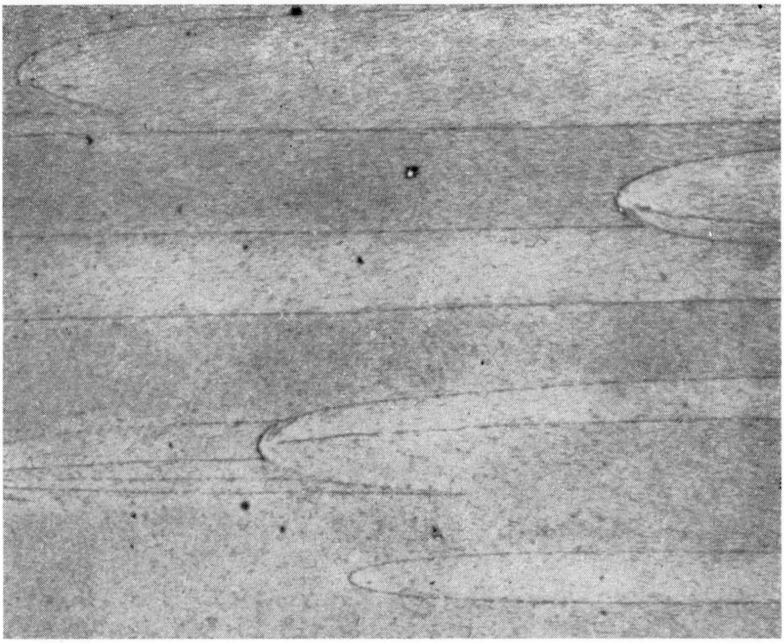

Fig. IIC-4. Third area, showing well-defined elongated parabolas, on fracture surface of centrally notched, plate specimen of PMMA. Boundaries of parabolas represent level-differences (×100).

served are probably level-differences at which fractures progressing on different planes meet.

Beyond the mirror or slow-growth area, there is a transition region which is usually merely an area of increasing roughness in a tension specimen. In a notched specimen, however, this area is expanded and is seen as a region of irregular lines, as shown in Figure IIC-3, oriented in the direction of crack propagation. These lines also appear to represent level-differences but are quite numerous and more pronounced than in the mirror or slow-growth area. In this area, the stress level and crack velocity are probably higher than in the slow-growth area and may cause some slight advance nucleation of flaws and the formation of the lines or figures observed.

The next area contains a number of geometric figures, resembling primarily parabolas and hyperbolas. These are the classical fracture markings. In this region, the stress level and crack velocity

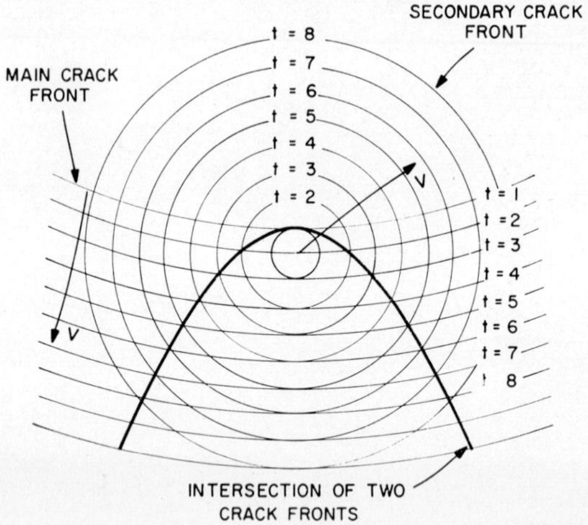

Fig. IIC-5. Diagram showing a possible mechanism for formation of parabolas on fracture surface. Secondary fracture starts at $t = 0$ at focus of parabola and intersects main crack-front. Both fractures are traveling at the same velocity V. [After Feltner (18).]

are higher, and the stress is sufficient to actuate flaws well in advance of the main crack-front. These geometric markings are produced by the intersection of the main crack-front with secondary cracks propagating radially from the numerous local flaws which were activated by the advance stress wave, as shown in Figures IIC-4 and IIC-5. The two crack-fronts form a conic section at their juncture. If the velocities are equal, a parabola is formed. If the secondary front is slower, the primary front will completely enclose it and an ellipse will result. Slight level differences for the two advancing fractures make the figures easily visible and are the source of much energy absorption, as stated previously.

Each parabola or geometric marking contains a flaw or pit at the focus, at which secondary fracture is initiated. An electron micrograph of the focus of a parabola from the fracture surface of a specimen of cast poly(methyl methacrylate) is shown in Figure IIC-6. In this case, the flaw or source of failure is probably beyond the limit of resolution of the electron microscope and may well be at the *molecular* level. There is a semblance of a mirror area surrounding the focus; this area often contains streamers or tear lines radiating from

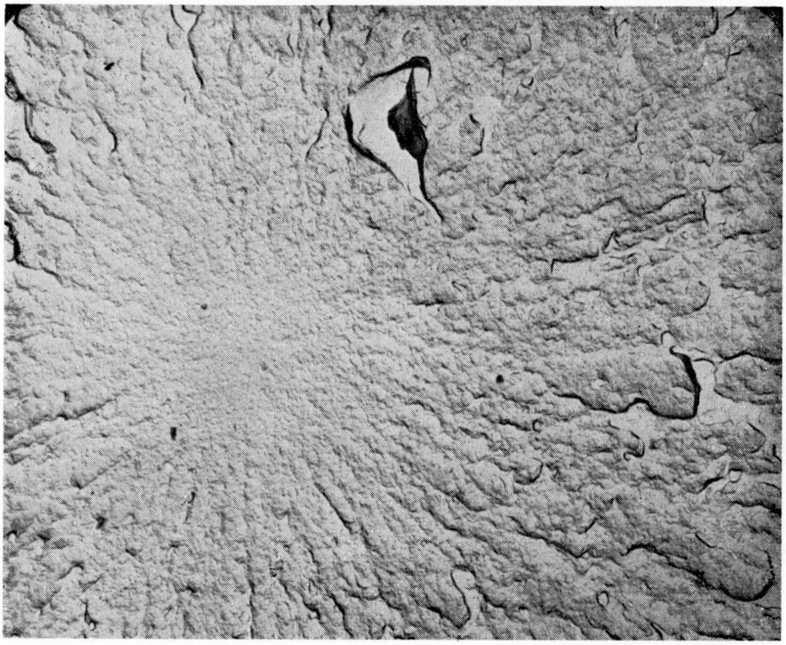

Fig. IIC-6. Electron micrograph of the focus of a parabola on the fracture surface of PMMA. Note the relatively featureless nature of the center (×7500).

the focus and extending through the entire parabola orthogonal to the secondary fracture-front. These tear lines mark the boundaries between fracture elements spreading forward on different levels within the parabolas. The tear lines and the fracture origins represent the major deformation processes for which energy must be supplied for the fracture process.

Finally, beyond this area, the stress level, and presumably the crack velocity, increases still further and a shattering type of failure is observed. Only the foci of the parabolas are very evident in this region, as shown in Figure IIC-7.

This description of fracture markings actually presents only some of the basic features observed in polymers and is therefore somewhat oversimplified. Other more complex markings have been observed, such as "scallops," "chevrons" or "herringbones," and "river systems" (30). Many factors may operate to modify the mechanics of the formation of fracture markings. Defects of many kinds in both

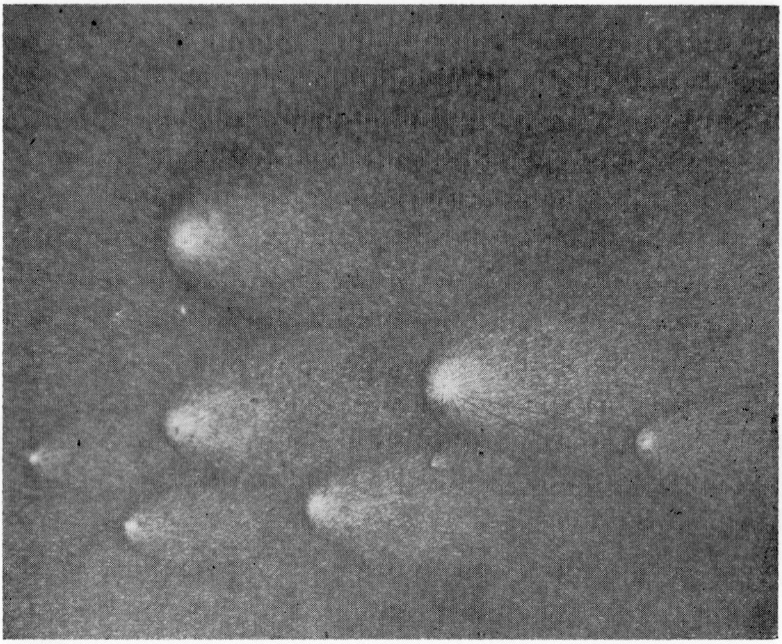

Fig. IIC-7. Fourth area on fracture surface of centrally notched, plate specimen of PMMA. Failure is probably the shattering-type with only the foci of the parabolas readily evident (×70).

amorphous and crystalline bodies, gross inhomogeneities in the matrix, notches, specimen geometry, and even various interactions between elements on the developing fracture all complicate the analysis of topographical data. Nevertheless, the mechanism of formation of the markings is similar and the markings observed are merely special cases of the general features described above.

Fracture surfaces have often been used to elucidate the structure of polymeric solids (4,42,43), in conjunction with other so-called "resinographic" methods. Useful data can be obtained from many polyphase systems but increased emphasis on the contributions of fracture dynamics to the plane of failure seems desirable in such work.

Another type of marking observed on fracture surfaces is "Wallner lines" (53). They are observed mainly in the mirror area. They are most prominent on the fracture surfaces of glass, but have also been detected in less prominent fashion in poly(methyl methacrylate). They are interpreted as being the loci of the intersection of the primary fracture-front with transverse ultrasonic waves that are released during the fracture process. Kerkhof (29) was able to produce a set of Wallner lines on a glass fracture surface by application of a high frequency transverse ultrasonic wave to a fracturing specimen, thus tending to verify this interpretation.

The method of formation of Wallner lines is shown in Figure IIC-8. The fracture-front progresses radially from the fracture origin P at a velocity c. At the point U on the periphery, an ultrasonic wave is released, having a velocity v. While the material is being deformed by the ultrasonic wave, the fracture-front progresses through this

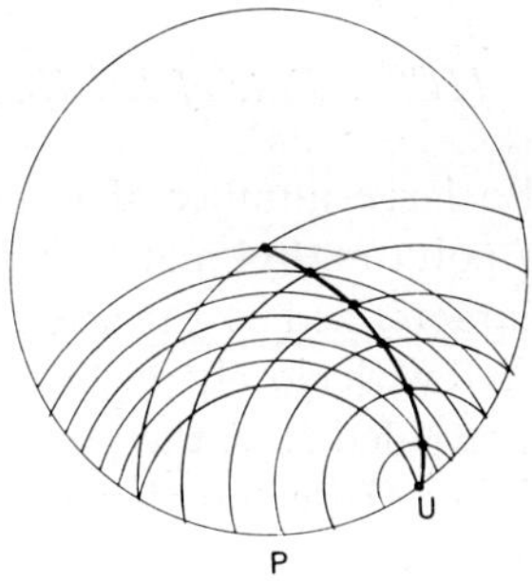

Fig. IIC-8. Diagram showing mechanism of formation of Wallner lines. The fracture-front progresses radially from P, releasing ultrasonic wave at U. [After Leeuwerik and Schwarzl (33).]

oscillating material and a wavy field is formed. The shape of the resulting curve obviously depends on the c/v ratio. As this ratio decreases, the Wallner line tends to indicate the instantaneous fracture-front. Using this principle, Smekal (49) showed that it is possible to calculate the velocity of propagation of the primary fracture-front.

Leeuwerik and Schwarzl (33) observed Wallner lines on the fracture surface of poly(methyl methacrylate). The configuration of the Wallner lines was almost identical with the instantaneous fracture fronts, indicating that the ratio c/v was locally small, i.e., approximately 0.1.

Andrews (5) observed Wallner waves on the fracture surface of glassified rubber containing 30% by weight of colloidal carbon black, broken in flexure. He found that the c/v ratio was approximately 0.5.

IIC2. Variables Affecting the Appearance of Fracture Surfaces

The most detailed observations on variables affecting topography appear to be those of Zandman (59), who has analyzed fractures in a variety of specimens of poly(methyl methacrylate) broken by different methods of loading. Effects of temperature on the fracture topography of a given material have been demonstrated by Wolock et al. (57), and the changes in the fracture surface produced by varying the molecular weight have also been documented (39).

IIC2a. Rate of Loading

Zandman examined a large number of fracture surfaces produced in tensile specimens of poly(methyl methacrylate) by simple tension. These surfaces were characterized by the usual mirror area surrounded by concentric ribs covered with closely packed hyperbolic figures.

Zandman's analysis revealed that the portion of the surface of the broken section containing these circular ribs can be reduced considerably if the tension specimen is loaded at a slow rate. It is even possible to eliminate the ribs completely by using a very slow rate of loading (requiring about 2 months for failure). Under these conditions the center of fracture has very large dimensions, is poorly delimited, and is often situated at the interior of the section. His results are shown graphically in Figure IIC-9.

Kies, Sullivan, and Irwin (30) have indicated the modifications found in fractures produced at high rates of loading. In high-speed shattering-fracture in poly(methyl methacrylate) they found a multiplicity of starlike components on the fracture surface. These have spread more or less concurrently, so that a system of honeycomb boundaries separates the individual elements. There may be no evidence of a main crack-front. It is emphasized that operative flaws may be created anywhere in a solid, provided that the appropriate stresses operate for sufficient time. Under conditions of high overstress, many flaws become operative simultaneously, and the material fractures in a brittle fashion, regardless of its behavior under static loading.

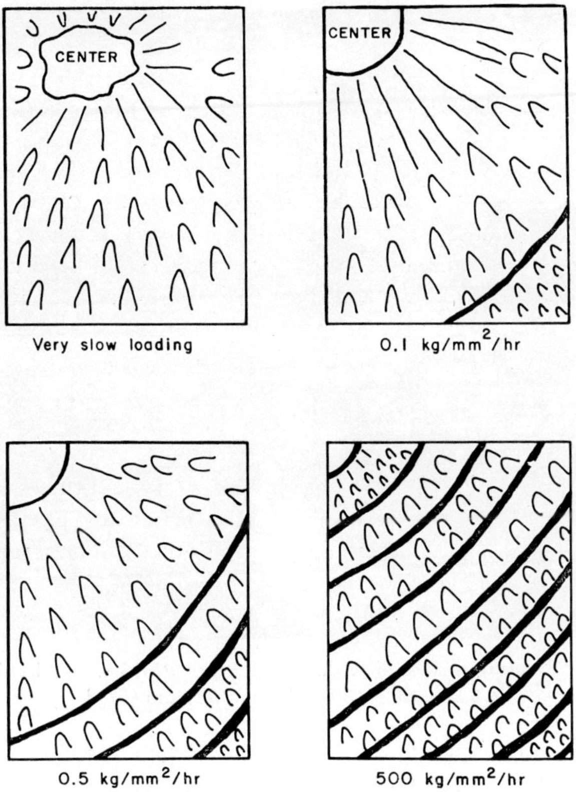

Fig. IIC-9. Effect of loading rate on fracture surface of PMMA. [After Zandman (59).] [From (38) and (57).]

IIC2b. Molecular Weight

When the fracture surfaces of tensile specimens of poly(methyl methacrylate) of varying molecular weight are loaded to failure at a given rate of loading and are examined at low magnifications ($\cong 30\times$), a developmental series, similar to that shown for rate of loading in the preceding subsection, can be easily distinguished. Figures IIC-10 to IIC-14 were all photographed at the same magnification, and the features are dimensionally comparable. The fields covered in the photomicrographs are areas immediately surrounding the center of fracture and represent only about $^1/_{10}$ of the entire

Fig. IIC-10. Fracture surface of tensile specimen of PMMA. Molecular weight $\cong$ 3,200,000 ($\times$30). [From (38) and (57).]

fracture surface. The highest molecular weight polymer, shown in Figure IIC-10 (molecular weight $\cong$ 3,200,000), is similar to the specimen shown for the slowest rate of loading in Figure IIC-9. Here the hyperbolic-figured area occupies the entire field. When the molecular weight drops to ~490,000, several well-defined ribs of pulled-out material, referred to by Kies et al. as "hesitation lines," are present (Fig. IIC-11). The ribs increase in number and resolution in Figure IIC-12 (molecular weight $\cong$ 200,000) and Figure IIC-13 (molecular weight $\cong$ 120,000). At a molecular weight of ~90,000 (Fig. IIC-14) the ribs have become so compressed that they can barely be resolved around the glassy area. A series of concentric bands appears even to have invaded the glassy area itself. These latter appear identical with the Wallner lines which have been observed in poly(methyl methacrylate).

Coincident with the development of ribs is the compression of the hyperbola-figured area immediately surrounding the fracture origin.

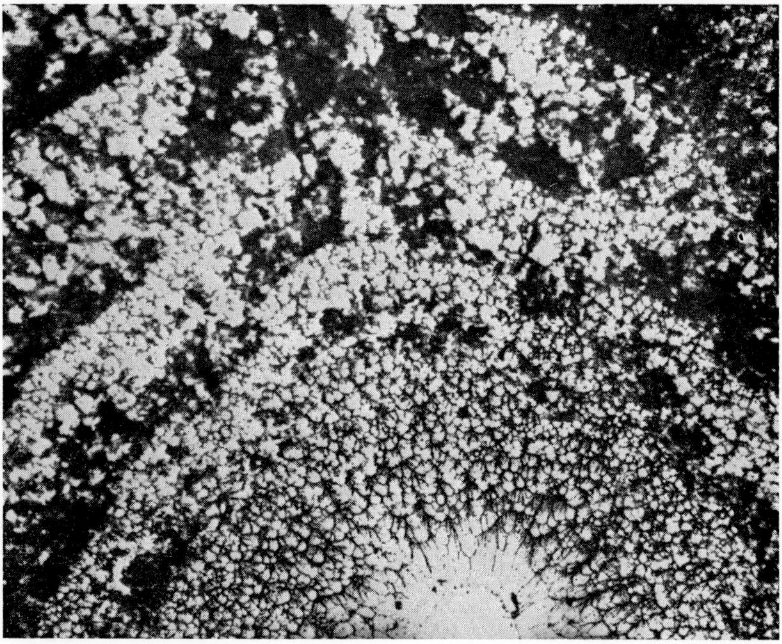

Fig. IIC-11. Fracture surface of tensile specimen of PMMA. Molecular weight $\cong$ 490,000 ($\times$30). [From (38) and (57).]

This is clearly illustrated in Figures IIC-10 to IIC-14. The hyperbolas become smaller and more concentrated. When the area immediately adjacent to the smooth glassy fracture origin in Figure IIC-13 is examined at higher magnifications, the hyperbolas appear to have lost their identity. They are also lacking in the specimen illustrated in Figure IIC-14.

The center or origin of fracture is always a smooth surface largely devoid of structure; indicating that which Zandman calls "decohesion" is responsible for the beginning of fracture. He has suggested that the increase in the size of the glassy area after a prolonged application of load is due to the orientation of the macromolecular chains along the axis of tension. While this hypothesis gains some support from the literature (1,21,22), it should only be considered as a suggestion for further investigation.

As each hyperbola contains a secondary fracture origin, the increase in the number of hyperbolas per unit area suggests the pres-

Fig. IIC-12. Fracture surface of tensile specimen of PMMA. Molecular weight $\cong$ 200,000 ($\times$30). [From (38) and (57).]

ence of larger numbers of defects or inhomogeneities where these "subfractures" originate in the lower molecular weight matrices. The number of defects might be expected to increase as the molecular weight decreases, possibly because of the larger number of chain-ends.

IIC2c. Temperature

Polymers which are hard and brittle at room conditions may become tough and ductile if the temperature is raised. Similarly, polymers displaying leathery behavior at room conditions may become brittle if the temperature is lowered sufficiently. Similar transitions occur in the fracture topography with changes in temperature. This is readily demonstrated by fracturing a number of specimens of poly(methyl methacrylate) under the same rate of loading but maintaining a different temperature for each test. The

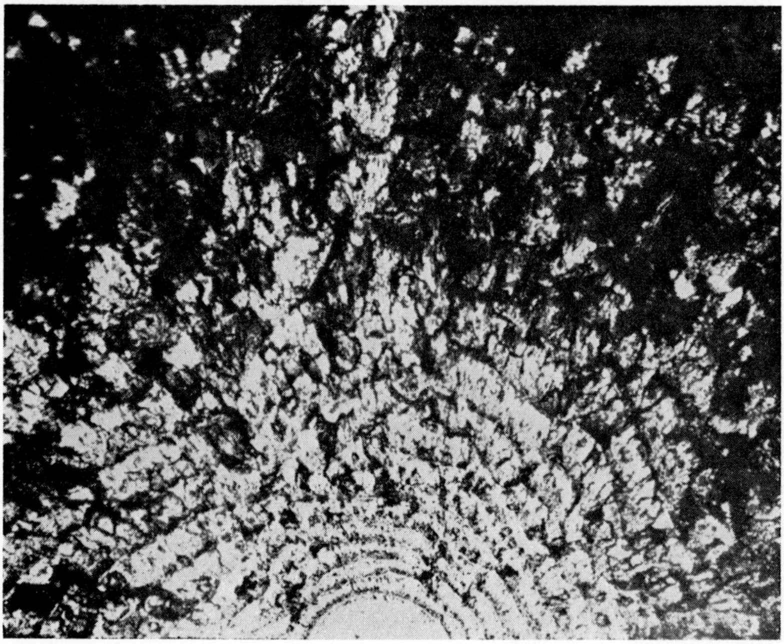

Fig. IIC-13. Fracture surface of tensile specimen of PMMA. Molecular weight $\cong$ 120,000 (×30). [From (38) and (57).]

results are shown in Figures IIC-15 and IIC-16. On the left is a specimen broken at room temperature, and adjacent to it is a tensile fracture produced at 40°C. There is little difference in the two fractures beyond a slight increase in the size of the glassy area. However, when the temperature is raised to 50°C., the whole nature of the surface changes. There is a total absence of recognizable geometric figures, and the mirror area itself is furrowed by radial markings (Fig. IIC-17). These latter appear to be the trajectories of the fracture-front. At 60, 70, and 80°C., the elongation is greater but, in general, the mechanism of fracture as reflected in the surface features is similar. This marked change in the appearance of the fracture surface between 40 and 50°C. has been related (57) to the sharp increase in elongation in the same temperature range (Fig. IIC-18). The tensile strength and modulus of elasticity decrease only gradually under these conditions. There is apparently a change in the *mechanism of deformation* which is reflected in the

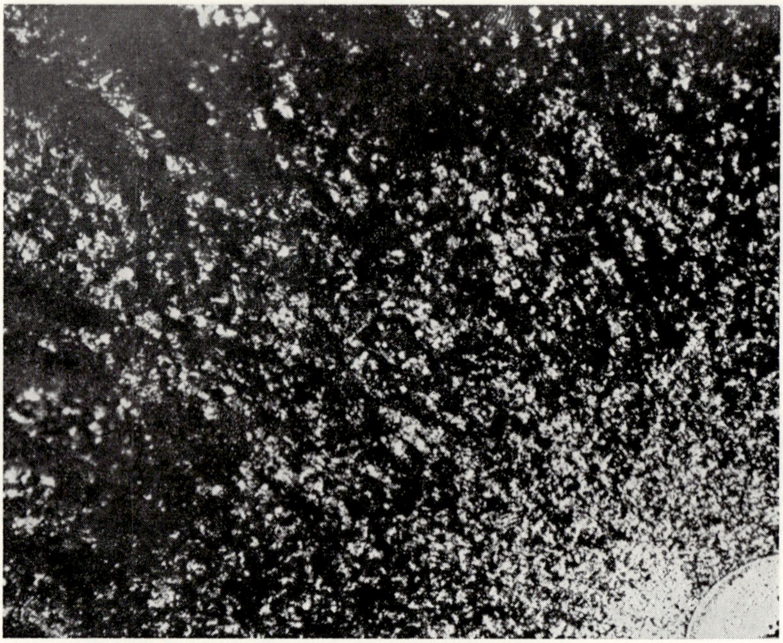

Fig. IIC-14. Fracture surface of tensile specimen of PMMA. Molecular weight $\cong$ 90,000 ($\times$30). [From (38) and (57).]

appearance of the fracture surface, but a more exact definition of the nature of this transition has not been given.

Cheatham and Dietz (16) reported the effect of temperature on the tensile fracture of annealed, molecularly unoriented polystyrene broken under static conditions at 0, 23, 52, and 82°C. All primary fractures originated at the specimen surface. At 0°C., the fracture is essentially smooth with a large mirror area and few, if any, hesitation marks. At 23°C. the mirror area is reduced, and the concentric ribs, which were prominent in Zandman's rate of loading sequence make an appearance. A further reduction of the mirror area occurs at the higher temperature along with the disappearance of the characteristic markings of brittle behavior.

The effect of temperature on centrally notched, plate specimens of poly(methyl methacrylate) has been discussed (57). The appearance of the fracture surface does not change over the temperature range of −40°C. to +49°C. (Fig. IIC-19), nor does the resistance to

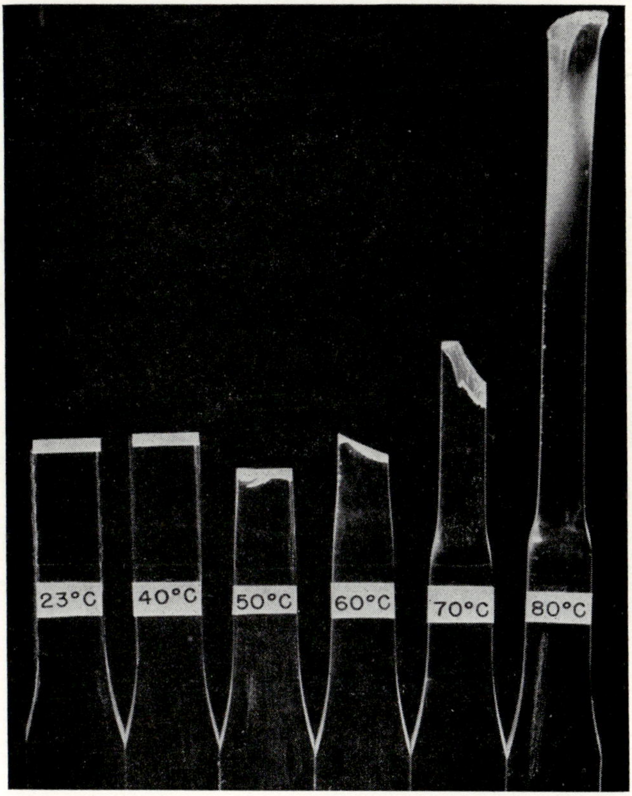

Fig. IIC-15. Tensile specimens of PMMA broken at various temperatures [From (57).]

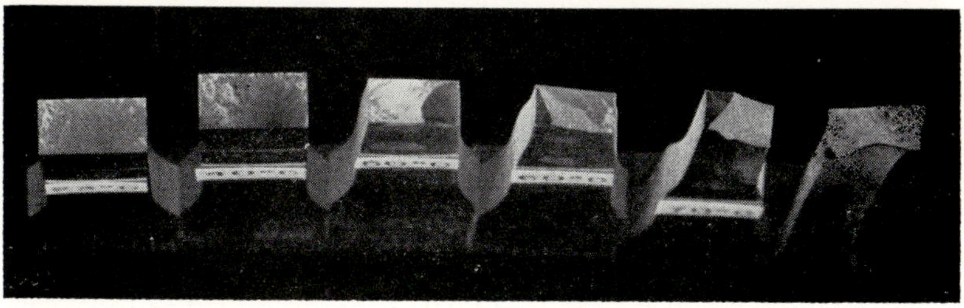

Fig. IIC-16. Tensile specimens of PMMA broken at various temperatures. See Fig. IIC-15. [From (57).]

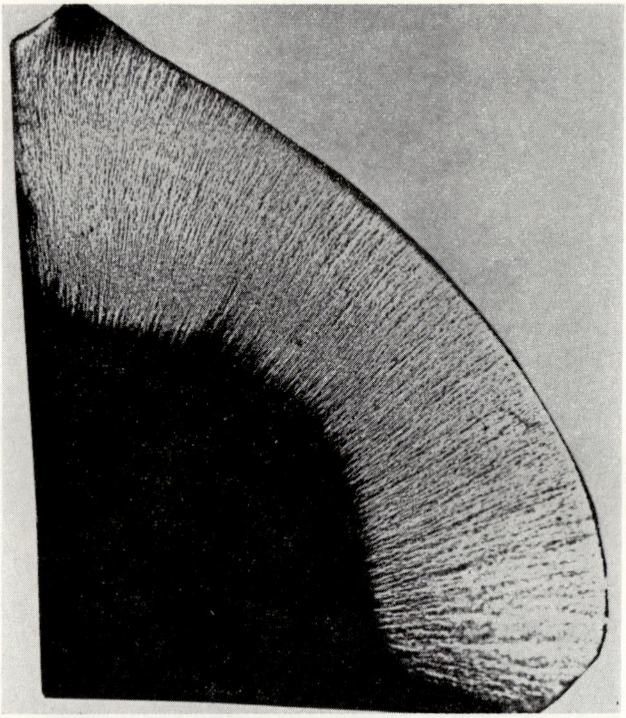

Fig. IIC-17. Fracture surface of tensile specimen of PMMA broken at elevated temperatures (×20).

crack propagation. However, for biaxially oriented poly(methyl methacrylate), the fracture surface tends to become rough as the test temperature is increased above 4°C. (Fig. IIC-20). Concurrently, the resistance to crack propagation starts to increase. The fracture surface thus reflects the change in toughness as the temperature is changed.

When specimens of well-fused vinyl chloride plastisol, which is leathery at room temperature, are broken in simple tension, the fracture surfaces are found to have structures similar in some respects to those of the poly(methyl methacrylate) broken at elevated temperature. Figure IIC-21 is a typical fracture surface of a fused plastisol broken at room temperature. The radial trajectories can be seen emanating from one central flaw; there is no evidence of

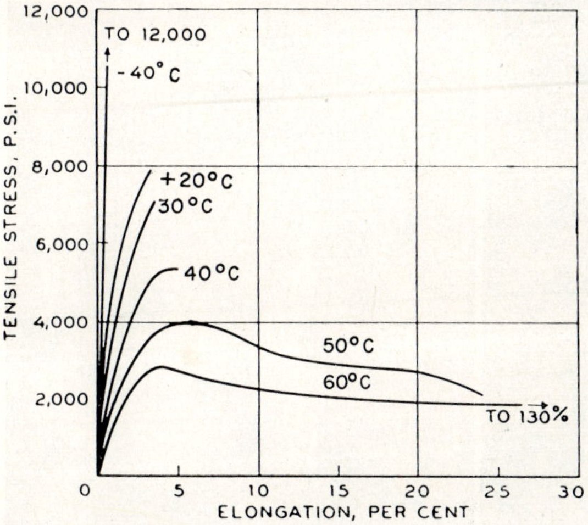

Fig. IIC-18. Effect of temperature on tensile stress-strain behavior of PMMA. [After Alfrey (1).]

geometric figures. If another specimen of the same type of plastisol is cooled to −40°C. and loaded to failure, the fracture surface again changes radically. However, in this case the fracture surface is roughly similar to the poly(methyl methacrylate) of ~120,000 molecular weight, as can be seen in Figure IIC-22.

Some highly filled rubbers, when broken statically at room conditions, will produce fracture surfaces similar to those obtained by the fused plastisols. Andrews (5), in studying fracture surfaces of notched, glassified rubber flexural specimens, obtained typical brittle failure characteristics, complete with mirror zones and Wallner lines.

IIC2d. Molecular Orientation

Considerable interest in hot-stretched, transparent polymers has resulted from the recognition of the significant enhancement of the tensile strength and crazing resistance of such materials (31). Sheets are heated to a temperature above the softening point but below the degradation temperature, stretched, and are then allowed to cool. Fracture morphology for a number of highly elongated polymers has been reported.

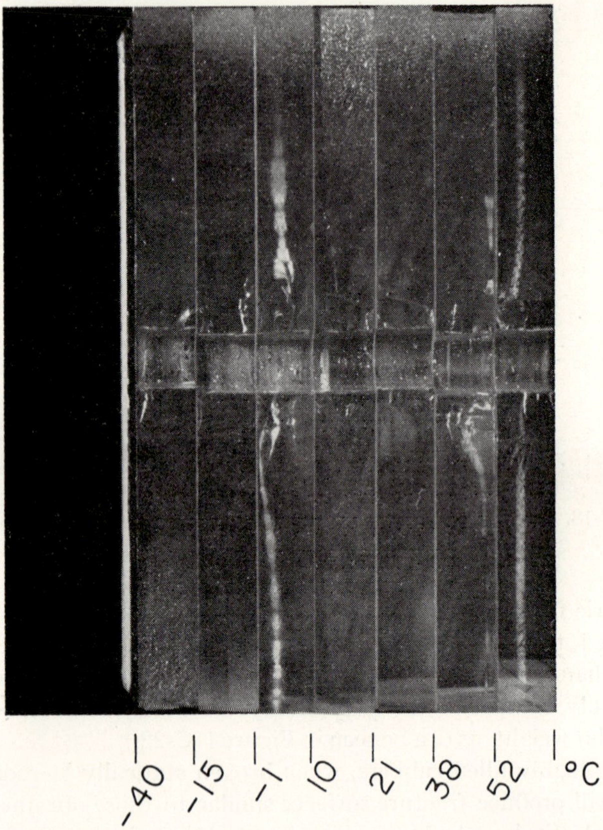

Fig. IIC-19. Effect of temperature on the fracture appearance of PMMA. [From (57).]

The fracture behaviors of uniaxially stretched polystyrene rods were studied by Cheatham and Dietz (16). Unoriented polystyrene, broken under static tensile conditions at room temperature, forms a fracture surface at approximately 90° to the tensile stress. There is a large mirror area and the characteristic markings of brittle failure. In specimens hot-stretched 580% and broken at 0°C., there was little change in the surface morphology. When a specimen stretched to 744% was broken at 23°C., the sharp fracture began to develop a fibrous periphery, but the center was much the same as found in unoriented rods. With specimens stretched to 989 and

Fig. IIC-20. Effect of temperature on fracture appearance of PMMA multiaxially stretched 90%. [From (57).]

1313% and tested at 23 and 0°C., respectively, the fracture surface develops into a ragged *fibrous* mass.

For biaxial stretch-forming in the plane of the sheet, the effects on the properties of poly(methyl methacrylate) were investigated by Axilrod et al. (8). Biaxial stretching to 50% strain did not affect the tensile strength or secant modulus of elasticity in tension. However, the total elongation, and the stress and strain at the onset of crazing, were greatly increased by the stretching. Tensile specimens of unstretched material commonly failed at 5–10% elongation with the fracture approximately flat and normal to the applied tension.

Fig. IIC-21. Fracture surface of a tensile specimen of a PVC plastisol broken at room temperature (×20).

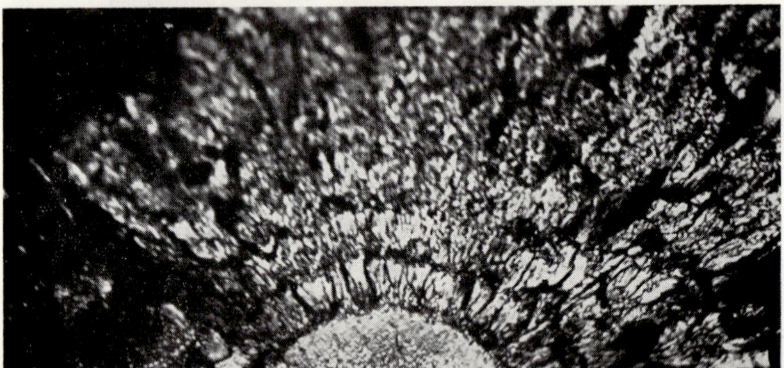

Fig. IIC-22. Fracture surface of a tensile specimen of a PVC plastisol broken at −40°C. (×100).

The surface markings have been described previously. Occasionally a specimen failed at greater elongation, sometimes associated with a fracture at an angle to the direction of the tension. Stretched specimens, which had much greater rupture strains, formed a largely laminar fracture surface. The general appearance is shown in Figure IIC-23. Almost invariably, a small mirror area was found at one

Fig. IIC-23. Fracture surfaces of tensile specimens of PMMA with varying degrees of multiaxial stretching. *A*, unstretched; *B*, stretched 50%; *C*, stretched 100%; *D*, stretched 140%.

corner of the section. When stretch-forming was increased to 100 and 150% (55), the mirror area developed on only a few of the specimens. The laminar nature of the fracture surfaces was more pronounced in these highly stretched specimens. Essentially the same behavior has been reported (56) for poly(methyl α-chloroacrylate).

As the degree of biaxial orientation is increased, the increase in resistance to crack propagation in poly(methacrylate) is reflected in the fracture surfaces of centrally notched, plate specimens (Fig. IIC-24). The surface becomes rougher at the higher degrees of stretching, indicating a greater energy for crack propagation.

It has been suggested that the beneficial effects of stretch-forming are the result of opening and elongating surface flaws, thereby reducing their ability to operate as foci of stress concentrations. In the interior of the specimen, a higher degree of order is also initiated by eliminating centers of heterogeneous arrangement of chains.

IIC2e. Type of Loading

(1) Tension

Surface markings resulting from failures induced by simple tension have been discussed in several previous sections. The effects of certain variables on the tensile fracture characteristics have also been noted.

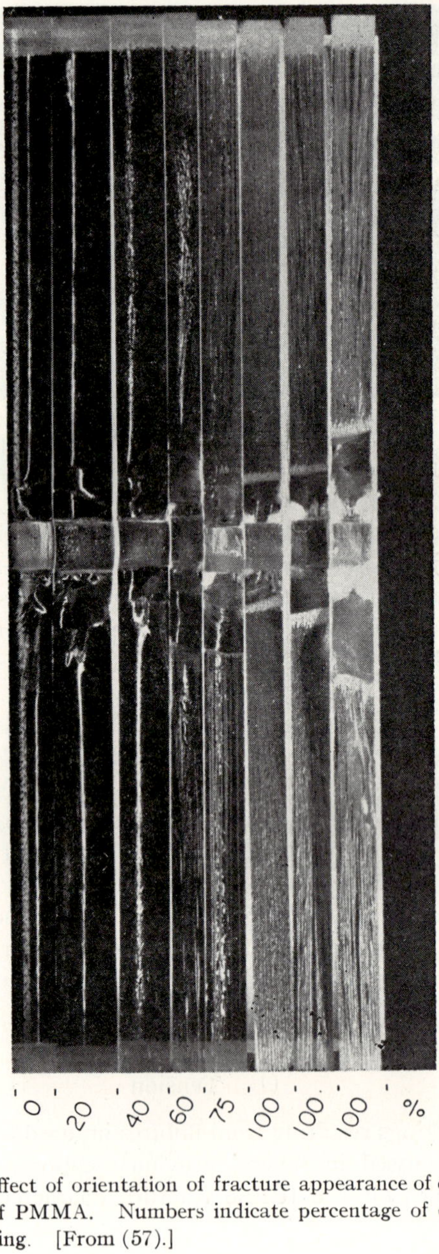

Fig. IIC-24. Effect of orientation of fracture appearance of centrally-notched plate specimens of PMMA. Numbers indicate percentage of elongation in the multiaxial stretching. [From (57).]

(2) Flexure

Zandman (59) has attempted to pinpoint the effects of various types of loading on fracture topography. Prismatic specimens were loaded in simple flexure but little data on other test conditions are given.

In general, the fracture always begins on the tension side. The fracture origin (center), as well as the concentric ribs, bears a strong resemblance to those produced in simple tension. The final stages are characterized by relatively fine ribs with numerous, well defined river systems radiating from them. These river systems result from the consolidation of level-differences as faster elements shoot ahead and stop slower ones.

IIC2f. Dynamic Fatigue Fracture

Zarek (60), using a rotating beam testing machine of the dead weight type, produced flexural failures in unplasticized poly(methyl methacrylate) rods. He used a loading frequency range of 95–2250 cycles/min. and bending stresses of ±2000 to about ±6500 lb./in.^{2}.

The types of fracture varied greatly. Under conditions of high stress–low number of reversals, the fracture surface consisted almost entirely of glassy mirror area. For low stress–high number of reversals, the fracture consisted of coarse ribs radiating from the periphery so as to intersect at the center of the section. The general conformation was suggestive of an iris diaphragm. Intermediate values of stress and reversals showed rather large, central glassy areas with river systems extending from it to the outer rim of the cross section.

The author assumed that failure occurred by the development of fine radial cracks which increase in size until rupture takes place. Each radial crack starts quite independently of the others and they all eventually join up by fracturing. These "radial cracks" are, in effect, river systems which lead away from the mirror area. It would appear that the rupture actually originates in an internal flaw which developes radially.

Because of the high strain energy losses and the low heat conductivity of poly(methyl methacrylate), the rapid cyclic stressing produces an appreciable temperature rise which undoubtedly affects the fracturing process at high rates of cycling.

IIC3. Velocity of Crack Propagation and Fracture Morphology

As stated above, fracture usually starts with a low velocity and gradually increases up to a limit which is the constant limiting fracture velocity found in brittle materials. In general, it is found that the roughness of the surface is related to the fracture velocity. As described previously, the area of slow growth surrounding the initiating flaw is relatively smooth and is referred to as the mirror area. As one progresses outward from the fracture source, the velocity of crack propagation increases, and as shown previously (Figs. IIC-1 to IIC-4), the surface roughness also. However, there is some indication that at very high speeds, there is a reduction in the level-differences and surface roughness, but this has not been studied thoroughly.

Schardin (46) has measured the fracture speed in poly(methyl methacrylate). He found that the velocity increased with increasing distance from the fracture origin. Maximum velocities were found to be between 400 and 700 m./sec. These values were dependent on the static load at the moment of rupture, with the velocity increasing with load. Schardin also showed, as one might deduce from the above, that a certain velocity does not necessarily cause the same appearance of the fracture surface. The appearance depends also on the rate of loading or the loading history of the specimen.

Considerable work has been done on the rupture of rubber, mainly at the British Rubber Producers' Research Association, in connection with the development of a tear criterion for rubber (see Section IIIB). During the course of this work, several studies have reported on the velocity of crack propagation and the appearance of the fracture surface. Greensmith and Thomas (20) reported that the torn surface of a GR-S (Polysar-S) rubber vulcanizate changed from a rough, irregular appearance at low rates of propagation to a smoother one at higher rates. The rates studied were 10^{-4} to 25 cm./sec. This change was attributed to a decrease in the effective diameter of the tip of the tear. A smaller diameter would result in a smaller volume of material being strained at the tip of the crack so that the path of fracture would be confined to a narrower section with a resultant decrease in surface roughness.

Thomas (51) reported on additional tear tests on a GR-S gum rubber at propagation speeds of 10^{-4} to 10 cm./sec. He found that the manner of tearing underwent a fairly abrupt change at a critical rate, r_c, of about 10^{-2} cm./sec. The appearance of the torn surfaces correlated with the rate changes. Below r_c, the surfaces were rough and irregular, but above it they were much smoother and reflected light specularly in most of the region.

Greensmith (19) also conducted tear tests on GR-S rubbers at rates of tear propagation of approximately 10^{-5} to 1 cm./sec. In tests on a GR-S gum vulcanizate at 25°C., he found that as the rate increased, the surfaces first became progressively more jagged in appearance and then became somewhat smoother. At 90°C., however, the roughness of the torn surface did not vary appreciably with rate of propagation over the range of 10^{-3} to 10 cm./sec. The surfaces were similar to those obtained at the lower rates at 25°C. In tear tests at 25°C. on an HAF (High Abrasion Furnace) carbon black-filled, GR-S vulcanizate, it was found that the overall roughness of the surface did not vary appreciably.

Finally, Mason (34) conducted tear tests on GR-S and natural rubber at propagation rates of up to 13 m./sec. He found that the appearance of the torn surface varied with the speed of crack propagation. At lower speeds, the surfaces were coarser; at higher speeds, smoother. He proposed that the modes of fracture in solids could be classified as: (*1*) at low speeds, a mode of quasi-static propagation, generally with smooth surfaces, obtained by careful control of boundary conditions; (*2*) propagation at intermediate rates with rough surfaces, involving a correspondingly greater energy consumption; and (*3*) fast propagation with smooth surfaces, the rate of propagation being limited by the speed of elastic waves in the material.

The results reported for rubber, in which the fracture surfaces become smoother as the rate of propagation increased, appear to be opposite to those reported for brittle polymers. These differences may be due largely to the different types of tests used. The tearing tests on rubber were conducted mainly at a constant force and the decrease in surface roughness at higher propagation velocities was attributed to a decrease in the volume of material being strained at the tip of the tear. However, tension tests were conducted on poly(methyl methacrylate). In these tests, the stress increased as failure proceeded through the specimen, due to the decrease in area. The increasingly higher stresses would tend to activate flaws over a wider region and result in a rougher fracture surface, as observed.

In addition, the *physical state* of the materials was different in the two cases, being rubbery in one instance and glassy in the other. This might also, in some yet undetermined fashion, have contributed to the differences in the behavior observed.

IIC4. Energy for Crack Propagation and the Appearance of Fracture Surfaces

The major energy requirements in the fracture process are for plastic deformation at the fracture origins and at the level-differences. These processes manifest themselves as visible surface markings. As previously mentioned, the fracture origins are usually visible as parabola foci at which considerable deformation occurs. The level-differences, also called tear lines or lag lines, represent the boundaries between fracture elements propagating on different levels. These markings tend to produce a rough surface and one can state that, in general, the rougher the fracture surface, the greater the energy required for crack propagation.

Examining the fracture surface of a typical specimen of poly(methyl methacrylate), one can postulate the energy absorption characteristics, as the fracture progresses across the specimen, from the appearance of the surface. At the initiating flaw, there is a great deal of energy absorption due to plastic deformation. As emphasized above, the mirror or slow-growth area is relatively smooth, with some evidence of tear lines on a very fine scale, indicating that there is relatively little energy required for crack propagation in this area once the crack has started to propagate. The next region consists of irregularly shaped geometric figures, the boundaries of which probably represent tear lines, suggesting a higher toughness than in the mirror area. Beyond this region, large numbers of parabolas can be seen, indicating a still greater energy required for crack propagation.

Many of the variables that affect the appearance of the fracture surface also affect the energy for crack propagation, and the latter effect is reflected in the former. For example, Figures IIC-10 to IIC-14 show the fracture surfaces of specimens of poly(methyl methacrylate) of different molecular weights. In Figure IIC-14 there are no parabolas apparent but merely indications of a shattering type of failure typical of brittle materials failing with low energy absorption. There is a faint trace of parabolas in Figure IIC-13 which becomes more obvious in Figure IIC-12, a token of greater energy for crack propagation. The large number of small parabolas might well be due to a concentration of flaws caused by the increase in the number of small molecular chains. In Figures IIC-10 and IIC-11, the parabolas are more apparent and larger, indicating greater energy for crack propagation and probably fewer flaws due to the smaller number of larger chains.

This same relationship between the appearance of the fracture surface and the energy for crack propagation can be shown with other examples. As discussed above, Figure IIC-24 shows the increasingly rough appearance of the fracture surfaces of a series of specimens of poly(methyl methacrylate) with increasing degrees of orientation. The toughness of this material, as measured by resistance to crack propagation, increases as the orientation increases and this is reflected in the fracture surfaces.

Similarly, there is a fairly sharp increase in crack propagation resistance of biaxially oriented poly(methyl methacrylate) at 4°C., and this is reflected in an increase in surface roughness, as shown in Figure IIC-20. On the other hand, unoriented poly(methyl methacrylate) shows no change in toughness in this region, and this is reflected as no change in the appearance of the fracture surface (Fig. IIC-19).

In the work previously referred to on tear behavior of rubber (19,20,51), tearing energy was determined as a function of tear propagation rate, and the appearance of the torn surfaces observed. Contrary to the behavior observed for brittle polymers, the torn surfaces tended to become smoother as the energy for tearing increased, i.e., as the tear propagation rate increased. This may be an inherent difference between rubbery and glassy materials. In some cases, the effect of available energy on the fracture surfaces produced by tearing rubber vulcanizates is quite marked (20). When fluctuations in the force versus time plots are small, the surfaces are relatively smooth and unmarked, but when fluctuations are appreciable, the path of the fracture is rough and tortuous. Some tear specimens show large cyclic variations in the tearing force and associated rate of propagation. Each large variation shows up as a level-difference mark in the fracture surface. This phenomenon has been compared to the "stick-slip" in frictional situations but is probably only another case of the discontinuous mode of fracture propagation. The transitions may be due, at least locally, to a sudden expenditure of stored elastic energy and its associated stress relief. Propagation is then delayed while stored energy reaches the required level.

IIC5. Effect of Chemical and Physical Structure

Little work on the relation of fracture morphology in polymers to chemical structure has been reported. However, it has been shown that by changing some of the test variables, such as temperature, rate of loading, or molecular weight, similar fracture surfaces may be obtained for polymers differing widely in chemical structure and composition (Figs. IIC-13 and IIC-22). Thus the evidence to date is that fracture morphology gives no indication of the specific *chemical structure* of polymers.

The *physical state* of the polymer under the precise conditions of the failure greatly affects the appearance of the fracture surface. It is true, however, that changes in the chemical structure of an amorphous polymer, through the resulting shifts in viscoelastic contribution, can *change* its physical state.

IIC5a. Thermoplastic versus Thermosetting Materials

Thermosetting materials, i.e., thermally crosslinked glasses, are generally more brittle than thermoplastics, i.e., uncrosslinked glasses. This can be inferred from the fracture surfaces of "thermosets" which are usually less featured than those of thermoplastic materials. The thermosets' fracture surfaces are often almost silica-glasslike or may be rather uniformly rough, in some cases with marked indications of

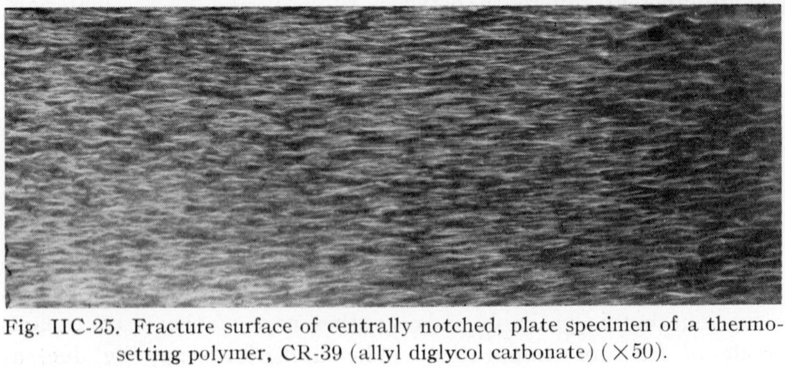

Fig. IIC-25. Fracture surface of centrally notched, plate specimen of a thermosetting polymer, CR-39 (allyl diglycol carbonate) (×50).

level-differences or tear lines oriented in the direction of crack propagation (Fig. IIC-25). However, the geometric figures and other easily distinguishable markings such as those found in poly(methyl methacrylate) are usually not observed in thermosetting materials. Feltner (18), in studies of epoxy resin, has observed some parabola markings and other features observed in poly(methyl methacrylate), but they were not as clearly defined. This entire area has not been extensively investigated as yet.

IIC5b. Glassy versus Leathery and Rubbery Materials

As shown previously (Figs. IIC-15 and IIC-16), as the temperature of a polymer increases from the glassy region and approaches the leathery region, the mode of failure changes from the typical sharp brittle separation to a tendency toward shear and separation by ductile processes. This change alters the appearance of the fracture surface. Thus the difference in morphology of materials that are glassy and leathery at room temperature is not an inherent difference. A leatherlike failure can be made to resemble a glasslike failure by changing the temperature, the rate of loading, and in some cases the size or type of specimen. It has already been shown (Figs. IIC-17 and IIC-21) that the fracture surface of poly(methyl methacrylate) at elevated temperatures resembles that of a leathery plastisol at room temperature. When the vinyl plastisol is fractured at a low temperature to produce a brittle fracture, the surface is similar to that of poly(methyl methacrylate) at room temperature.

Mason (34) reports concerning tensile failures that "... were not produced by exceptionally high-speed elongation . . ." in natural rubber and GR-S rubber at room temperature. He indicated that the fracture surfaces of the rubber showed similarities to glass surfaces. The natural rubber in particular showed a typical glassy break with some evidence of conchoidal markings. The GR-S showed similar but less pronounced effects. The speed of testing may actually have been high enough to result in a brittlelike failure whereas a somewhat lower speed would probably have produced a more ductile or tearing-type of failure, more typical of leathery or rubbery materials.

In general, then, one can only differentiate between a glassy, leathery, or rubbery response of a material during fracture by examination of the fracture surface. The "physical state" of the material as defined, for instance, by its glass transition temperature, need not be sufficient.

IIC5c. Effects of Filler

Andrews and Walsh (6) conducted an electron microscopical investigation of the rupture process in carbon-loaded rubbers in an effort to resolve some of the questions of the exact mechanism by which fillers influence the rupture process. They conducted tear tests at various rates of rupture on natural rubber and GR-S vulcanizates containing 30% by weight of carbon black. Replicas were made of the fracture surfaces for examination with the electron microscope.

They found that in a filled rubber the rupture surface may be expected to show the following characteristics: (*1*) surface roughness; (*2*) larger fractions of the fracture surface covered by the filler than the average for the rest of the material; (*3*) dependence of both on rate and mode of rupture. The rupture was not a simple plane of surface through the material but tended to travel from one filler unit to an adjacent one. Since the path followed is that requiring least energy, it indicated that filler particles or their immediate environment are zones of relative "weakness" or high stress concentration. In an unfilled material, rupture occurs as a result of high stress concentration around the advancing tear or fracture, and only the material at the very tip is at the breaking stress. If, however, there are inclusions in the vicinity, the additional stress concentrations which occur around them will be superimposed, and the breaking stress will be reached at points removed from the tip. This will result in the deviation of the rupture path through these points of internal failure, producing a rough fracture surface with excessive amounts of filler.

IIC5d. Relation of Synthetic Polymers to Other Materials

Kies et al. (30) pointed out that "the detailed markings on fracture surfaces of metals, plastics, and ceramics show many common characteristics." Since the fracture processes are generally similar for these materials, the similarity in details of the fracture surfaces is not actually surprising. In some materials, certain fracture mechanisms may be restrained as compared to other materials, with a resulting modification in the appearance of the fracture surface. In addition, the particular conditions at failure, such as temperature, rate of loading, and specimen geometry, also will tend to alter the appearance of the fracture surface. The basic *mechanism* of failure remains the same, however, and by a proper control of the conditions of failure similar fracture markings can be produced in materials having *widely* differing compositions, properties, and molecular or atomic structures.

Kies et al. (30) showed that the structural components of the parabola in poly(methyl methacrylate) have analogs in the comparable portions of the cup-cone fractures in ductile steel. The parabola in steel has a volcano-like focal crater, indicating shearing at the focus. Surrounding the shear is a bright granular region with streamers analogous to those observed in poly(methyl methacrylate). This area surrounding the focus in steel is also compared to the mirror area in a silica-glass fracture. In experiments conducted by these same authors, microscopical observations were made during the progress of cleavage fractures in cellulose acetate and in a large crystal of potassium bromide. They found that scalloped leading edges of similar appearance were produced in both materials.

In addition, these authors modeled in cellulose acetate, the chevron or herringbone pattern fracture surface observed in ship plate. In this type of fracture, the concentration of gross fracture origins is in the center of the specimen, and the fracture-front leads at the center and lags at the surface.

Irwin and Kies (25) indicated that the parabola-like fracture element observed in cellulose acetate, usually observed on a finer scale in metals, may be gross and easily observed in copper, lead-tin solders, and some steels. Parabola-shaped fracture elements are shown to exist on a cleavage facet of molybdenum.

Silica-glass fracture surfaces show such features as hackle and a linearly grooved surface often found in the more brittle of the organic polymers. In these polymers, parabolas may be modeled in some cases (18), but this appears to be more difficult in silica-glass.

IIC6. Color Patterns on Fracture Surfaces

IIC6a. Historical

It is difficult to ascertain when the strong colors on the fracture surfaces of polymers were first observed. In 1956 Busse, Orowan, and Neimark (15) exhibited some surfaces produced by fracturing large notched plates of poly(methyl methacrylate). About a year before, somewhat more diffuse colors had been observed on tensile specimens broken under static conditions (3) at the National Bureau of Standards (41). Although Berry (13) and Wolock et al. (57) both reported observing colors associated with specific fracture markings, the first detailed discussion of the phenomenon was published by Higuchi (23). Many of the same observations were made independently and reported by Berry (12) and Wolock et al. (40,57, 58).

It is entirely possible, of course, that colored fracture surfaces were observed earlier than the period mentioned. Higuchi, for example, indicates that the experiments of Benbow and Roesler (11) must have produced such surfaces. Thus far, almost all colored surfaces that have been studied have been produced on poly(methyl methacrylate).

IIC6b. Types of Color Patterns

The colors arising from interference in subsurface cracks are not considered in this discussion. Although they produce an optical effect which often appears quite close to the surface, its origin is obvious and easily ascertained.

There is considerable variety in the color patterns obtained by various fracturing techniques. Those found on statically tested tensile specimens are weak, usually located about the periphery of the mirror area and are found on relatively few specimens. They are not usually associated with an individual structure, such as parabolas, but cover an area which usually encompasses a number of figures or structures without any specificity of color or tint for each element. Fractures exhibited by Busse and his co-workers were more brilliant and formed bands transverse to the direction of fracture. Higuchi used notched sheets 3–5 mm. thick and 40 mm. wide. The exact conditions of test are not described. His figures show a variety of patterns. Most of the conic sections are differentiated in color from their surroundings and occasional transverse bands cause a reversal of colors in the banded area. The colors on the two fracture surfaces are generally complementary and the red areas are raised, while the green are depressed relative to the specimen matrix.

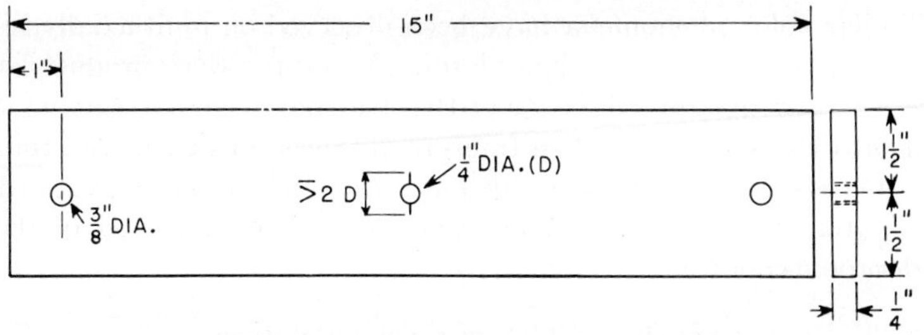

Fig. IIC-26. Schematic diagram of crack propagation specimen used for producing colored fracture surfaces. Small variations in overall dimensions and position of loading pinholes do not significantly affect fracture characteristics.

Wolock et al. (57) produced similar colors on fracture surfaces of poly(methyl methacrylate) specimens which had been broken by rapid overstress. Films of diffuse color were found on surfaces of cast polystyrene broken in the same fashion. Specific colors, however, were not related to the individual structures on the polystyrene fracture faces. Somewhat later, Wolock noted the intense differential color produced on the relatively smooth fracture surfaces of specimens of poly(methyl methacrylate) tested for resistance to crack propagation. Many of the observations to be discussed were made on fractures produced in these specimens (Fig. IIC-26) which were broken by loading through a cross-head moving at 0.02 in./min. These contain four types of color patterns or characteristics (Figs. IIC-2, IIC-3, IIC-4, IIC-7). The first is observed in the slow-growth area and actually consists of two types of patterns, namely, multicolored areas and single-colored areas. The next pattern is composed of irregularly shaped, elongated figures. These figures develop in the third pattern into oriented forms that lead, in many cases, into parabolas. The last area is darker (less reflective) with little or no color and contains many crowded parabolas in which only the foci are very distinct.

Benbow (10) reported color patterns on the surfaces of stable cracks (i.e., those propagated at an arbitrarily slow speed) produced in cast poly(methyl methacrylate). These were second-order interference colors and when viewed by reflected light, the colors on opposite sides of the crack were identical. Complementary colors, however, were obtained with extruded sheet.

Similar color phenomena have been observed on multiaxially hot-stretched poly(methyl methacrylate). Fractures were produced by inserting a wedge into the edge of the stretched sheet and forcing the tool into the specimen. The fracture surfaces formed in the thickness planes almost invariably displayed a fairly intense, first-order blue interference color. Fading took place over a period of days under room conditions.

IIC6c. Optics of Color Formation

It can be demonstrated that the strongly colored areas of the fracture surfaces produced in crack propagation specimens are covered with a thin film having properties different from that of the matrix. The presence of this film is a pre-requisite for the development of color. Under room conditions the color will disappear from crack propagation specimen surfaces over a period of months. The time for color disappearance varies greatly, however. Colors on ordinary tension specimens faded in the course of a few weeks, while those of Busse et al. (15) were still intense after almost a year. Heating to temperatures of 50–60°C. causes rapid loss of color in crack propagation specimens in well under an hour. Loss of color under these circumstances is not necessarily associated with complete loss of the film. Even when the colors have completely disappeared, an oily film still occupies the surface. The disappearance of color by aging or heating takes place by fading of the original tint without any pronounced color shift (40,58).

Concentrated vapors of a number of organic solvents such as toluene and methyl chloride will often eliminate colors in a matter of minutes. In this case a change in tint is involved. The first color changes seem concentrated along level-difference lines and a few other structures on the fracture face.

Another method of affecting color intensity of parts of the film is by exposure to low pressure. This can be demonstrated by placing one of the matching surfaces of a crack propagation specimen in a container maintained at approximately 1×10^{-5} mm. Hg pressure. After removal, it will be found that almost all of the color is removed from the slow-growth area. The color intensity over the remainder of the fracture is apparently the same as that of the matching surface which had been stored outside of the vacuum chamber. Microscopic examination and micromanipulation of the film on the slow-

Fig. IIC-27. Multiple-beam interferogram of parabola-bearing portion of a PMMA crack propagation specimen's fracture surface. The numbers identify components of the same fringe. Source was the 5461 A. mercury line. [From (40).]

growth area of the fracture surface indicates that this portion is thicker and less viscous than the rest of the film.

The fact that much of the surface is little affected by high vacuum makes it possible to apply the methods of multiple-beam interference microscopy to the study of the color phenomenon. Figure IIC-27 is a multiple-beam interferogram of a portion of a fracture produced in a crack propagation specimen. The field shows a portion of the first parabola-containing area. A layer of silver had been deposited over the surface film and the fringe displacement over the composite surface is shown. It would appear that the level-differences between adjacent secondary fractures are consistently displaced by about 150 mμ. When the color is removed by aging or heating before the specimen is prepared for interference microscopy, this displacement remains about the same.

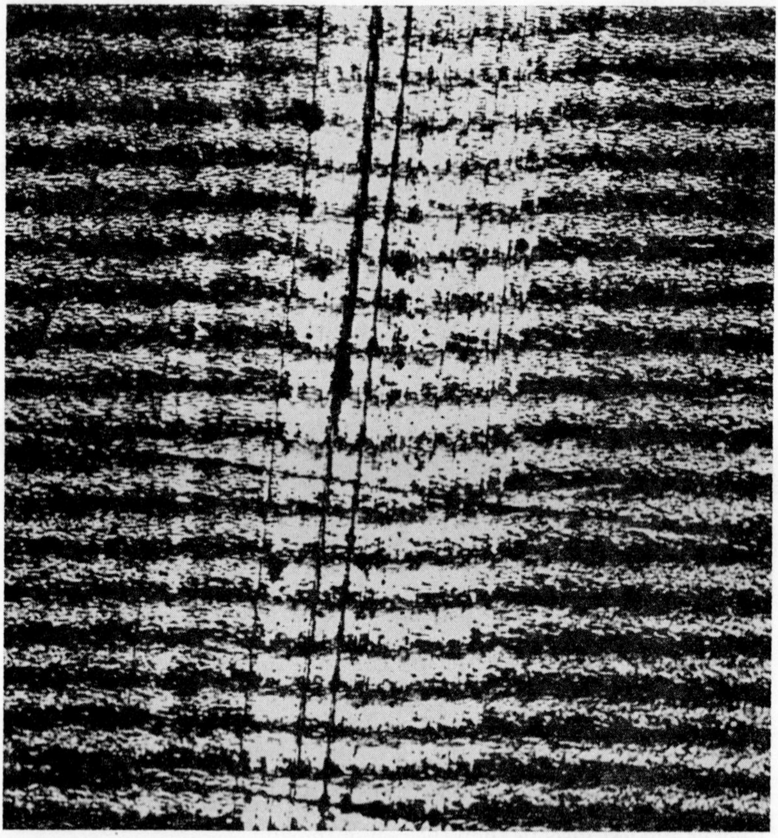

Fig. IIC-28. Multiple-beam interferogram of the swabbed area of the fracture surface of a crack propagation specimen of PMMA. Source was the 5461 A. mercury line. [From (40).]

Assuming first-order interference, Higuchi (23) has estimated a film thickness range of 400–700 mμ from the observed colors. Using a soft swab to remove the film produces a swath with a depth of less than 200 mμ in the surface as measured by multiple-beam techniques. (Fig. IIC-28). This experiment was repeated by focusing the radiation from a zirconium arc lamp on a small area of the fracture surface until all of the color had disappeared and continuing the treatment until the film was no longer detectable in the irradiated portion by microscopical examination. The depth of the circular depression was again found to be somewhat less than 200 mμ. Although the film in the slow-growth area may exceed this figure, its sensitivity to vacuum prevents direct measurement. Figure IIC-28 also shows the increased reflectivity of the treated area. Apparently swabbing or heating tends to produce a surface which is smoother than the original face of the film.

IIC6d. Possible Nature of the Color Phenomena

The presence of the film led Higuchi (23), Berry (12), and Newman et al. (40,58) to postulate its formation by the pulling-out of chain-ends and the scission of chains. These segments would have been extended in the fracture process and thus could have a somewhat different density than the matrix. The chain segments in only a relatively small thickness of material would be extended during the fracture process. The molecularly oriented layer and matrix would differ sufficiently in refractive index to lend themselves to the de-

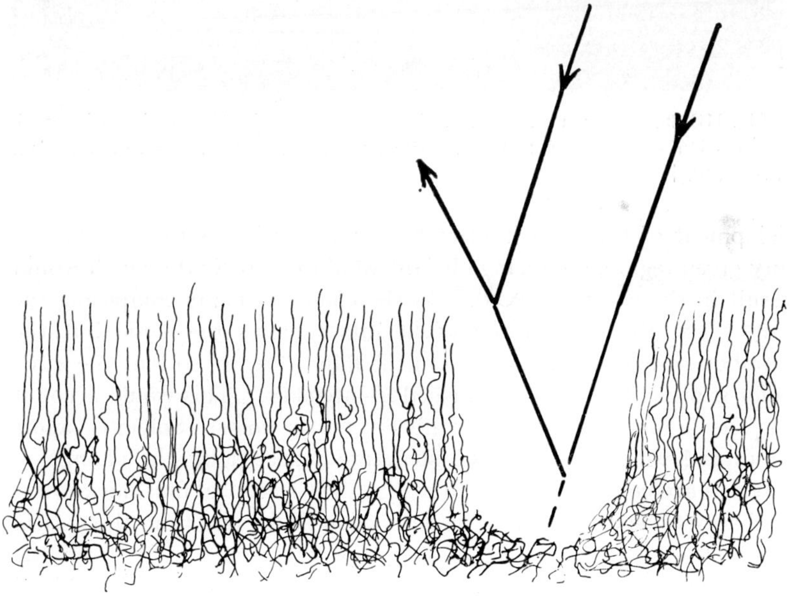

Fig. IIC-29. Diagram showing interference in the postulated molecularly oriented film on the fracture surface of PMMA. [Modified from Higuchi (23).] [From (40).]

Fig. IIC-30. Electron micrograph of a replica of the fracture surface of a crack propagation specimen of PMMA. The area is from the slow-growth region. [From (40.)]

velopment of thin film interference as shown in Figure IIC-29. In any given region, an average height would predominate which would result in the color observed. As the chain segments composing the film relax, the distribution of heights remains the same but the number of segments remaining in the film decreases so that the same color is present—but is diminished in intensity. This would account for the fading observed with no change in color. Solvent vapors produce level changes in the structure reflected by the change in interference colors. Failure to realize the theoretical film thickness is probably to be expected. The derivation of the film from the matrix would lead to a structure whose properties would differ from the matrix to the greater degree at the exposed face and tend to blend into the matrix at the other limit. While the blending section close to the matrix could contribute to the optical effect, it would be difficult to establish its boundary by the methods described.

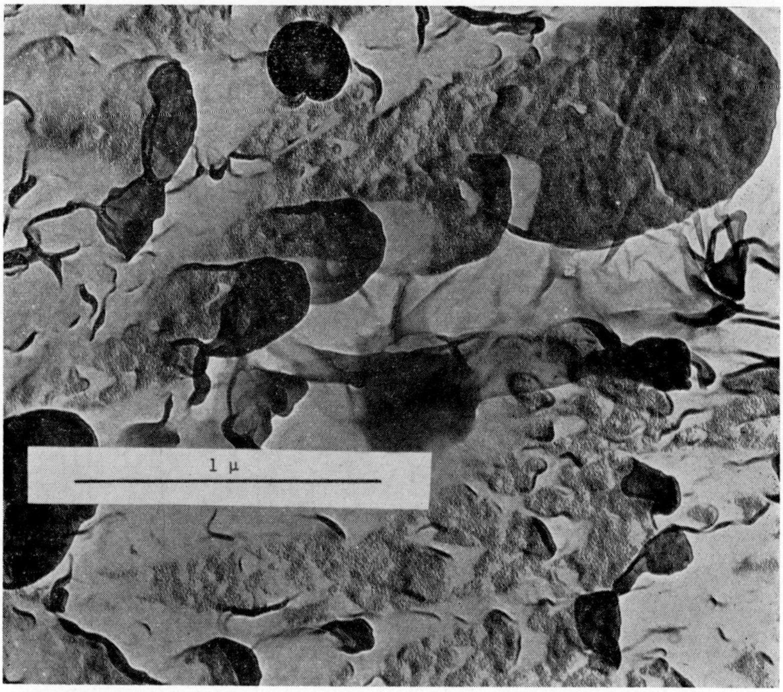

Fig. IIC-31. Electron micrograph of a replica of the fracture surface of a crack propagation specimen of PMMA. The area shown is from the crowded parabola region. [From (40.)]

Tolansky's thin film techniques (52) permit the construction of a simple model to demonstrate the optics of the system. A fracture surface with the colors removed, by aging or by some other method, is covered with a totally reflecting film of silver *in vacuo*. A drop of a dilute solution of Canada balsam in xylene is placed on the silvered surface and allowed to drain by tilting the specimen. The surface of the film of Canada balsam in contact with the fracture surface replicates the topography of the surface. The upper surface of the film, however, does not contour the fracture surface but stretches itself into a single plane for considerable distances. On illuminating such a surface, two-beam interference fringes will be formed. Wrinkling, probably caused by solvent attack on the silver and polymer substrate, is noticeable, but otherwise there is considerable similarity to the fracture-produced colors. Even the slight coloring observed in the rough, low-reflecting zone is almost identical with that observed on the original fracture surface.

Considerable fringe-broadening is apparent in the interferogram in Figure IIC-27. This is at a minimum in the slow-growth area and increases toward the crowded parabola portion. Electron micrographs of these two areas (Figs. IIC-30 and IIC-31) show the change in surface granularity responsible for the fringe-broadening. Although the surface is sufficiently smooth up to the first parabolas to permit the use of multiple-beam interference, the final surface is so rough that the method fails. Figure IIC-31 reveals this surface to be sculptured with thin leaflike projections. It is doubtful that a fragile film in the millimicron thickness range could be formed there. Only isolated fragments of color are found in this area.

IIC7. Crazing

IIC7a. General Description

When transparent thermoplastics, such as poly(methyl methacrylate) or polystyrene, are exposed to a sufficiently high tensile stress with or without solvents, deterioration of their optical clarity occurs. This deterioration results from the formation of cracks that impair light transmission and cause optical aberrations. Moreover, these cracks reduce the strength properties of the material. Commonly known as "crazing," this phenomenon is encountered in a variety of commercial products ranging from aircraft glazing and canopies to watch crystals, and is obviously of considerable economic importance to manufacturers and users of these glassy items. A number of investigators have reported the results of work on the mechanism, characteristics, and elimination or prevention of crazing (2,7,9,24,35,36,44,45,48). However, crazing remains a serious consideration in thermoplastics applications and many aspects of the problem require intensive study.

Photomicrographs and descriptions of the craze cracks can be found in the literature (38,44). The cracks vary greatly in their length and fineness. In tension specimens crazed by the application of stress or by a combination of stress and solvent, the cracks are almost invariably normal to the applied stress. Individual cracks can usually be observed by use of a low-power microscope; a 16-mm. objective and 10 × ocular is generally more than adequate. The length and number of the cracks can often be determined in this fashion. Some crazed surfaces contain cracks that appear to approach the limits of resolution of the microscope, and these require a more powerful tool for the measurement of crack dimensions.

Although the length and fineness of crazing cracks can be studied with the bright-field microscope, the determination of the depth of cracks and how they affect surface topography requires other techniques and instruments.

Recent work on poly(methyl methacrylate), polystyrene and polycarbonate has radically changed concepts of the physical form of craze cracks and added considerably to our knowledge of the phenomenon.

IIC7b. Types of Crazing

Russell (44) has classified crazing on the basis of the contributing variable as "stress crazing," "solvent crazing," and combinations of

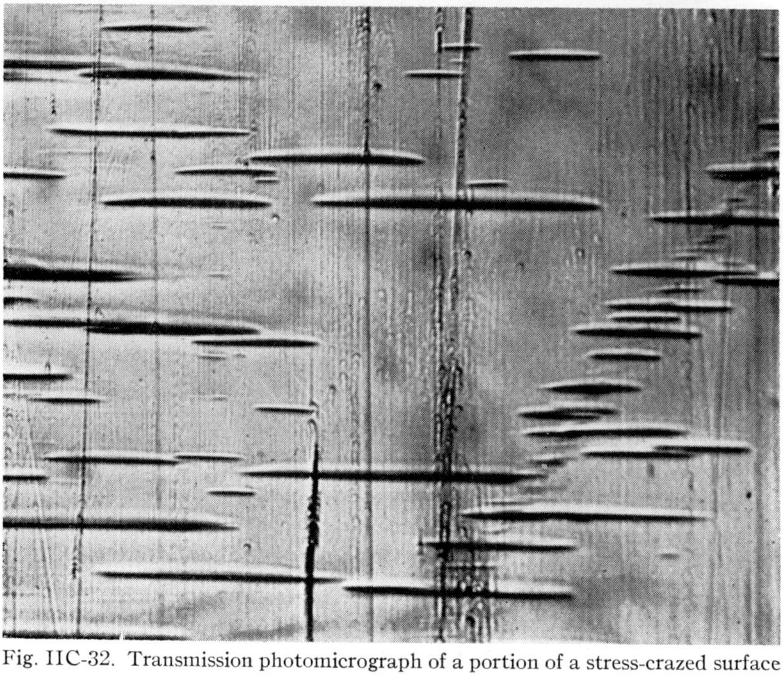

Fig. IIC-32. Transmission photomicrograph of a portion of a stress-crazed surface of a PMMA tensile specimen (×200).

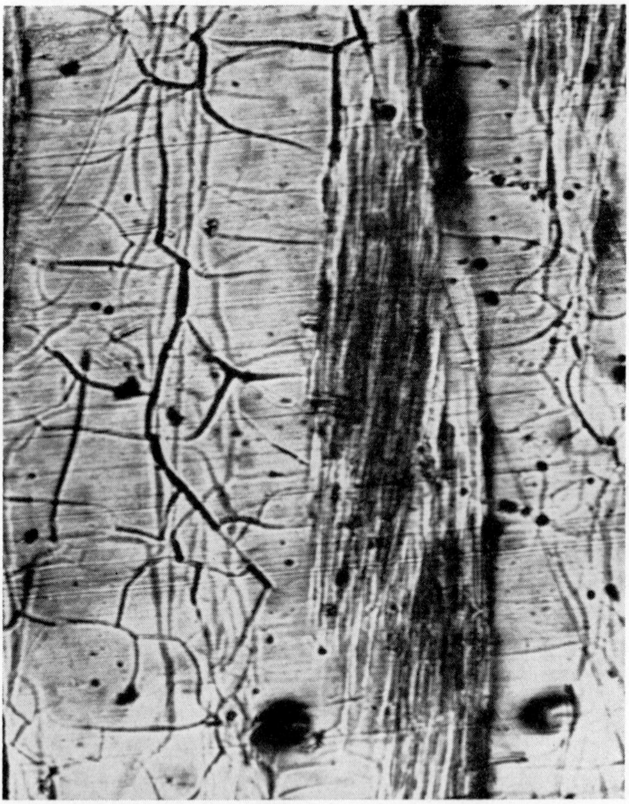

Fig. IIC-33. Reflection photomicrograph of the surface of a chlorine-containing polyester laminate. The surface shows the random cracking characteristic of crazing due primarily to the action of solvents (×60).

these. However, in addition to these categories, crazing may also be considered as divided into patterns wherein all cracks are internal or all are superficial or a combination of these characteristics. Although "solvent crazing" might often connote that no stress is involved in this type of crazing, actually this term should imply that no external force is applied. However, there is sufficient internal stress to cause crazing when the solvent is applied.

In stress crazing of linear chained high polymers, the observed cracks are always found to be at right angles to the direction of the applied tension (Fig. IIC-32). Crazing due primarily to the action of solvents is apt to be extremely random in direction (Fig. IIC-33). Combinations of stress and solvent produce patterns of larger cracks which are also oriented. The size, density, and depth of penetration of the cracks depend not only on stress and/or solvent used and the period of time employed, but also on the material under study. For comparable stress–time conditions, poly(methyl methacrylate) specimens will develop larger but fewer cracks than those of polystyrene. Specimens of polystyrene will develop some cracks which are entirely internal and, at the same time, others which intersect the surface plane. Poly(methyl methacrylate) crazes, however, tentatively appear to be always on the surface. Some experimental poly(methyl α-chloroacrylate) sheets were also found to exhibit internal crazing with all the cracks limited to a central core of the specimen. Whether this behavior is a characteristic of the material's composition or a reflection of the quality of production and fabrication is not yet clear.

IIC7c. Variables Affecting Crazing Behavior

All of the transparent glassy polymeric materials currently available are susceptible to crazing in varying degree. In some cases, stress and temperature must be adjusted to extreme conditions to obtain significant crazing. Crazing is attributed to a wide variety of causes (61). Among the more common are residual stresses produced in forming operations, exposure to solvents, and applied tensile and flexural stresses. However, recent work (50) has indicated that residual stresses, etc., might not be initiators of crazing, but merely serve to increase the severity of defects.

Sherman and Axilrod (48) have studied the crazing behavior of poly(methyl methacrylate) intensively. They found that the stress at the threshold of crazing decreases with increasing temperature (23–70°C.) very nearly at the same rate as the tensile strength; that is, the ratio of the threshold crazing stress to the maximum stress, $\sigma_c/\sigma_{\max}$, remains fairly constant. In the temperature range of 23–50°C., the strain at the threshold of crazing decreases significantly. For modified poly(methyl methacrylate) there is a further decrease in the range of 50–70°C.

Wolock et al. (55a) studied the molecular weight dependence of crazing behavior of poly(methyl methacrylate). Tensile tests were conducted on specimens from five samples of cast poly(methyl methacrylate) in which the molecular weight varied from ~90,000 to ~3,200,000. Both stress crazing and stress-solvent crazing were studied. In the stress crazing tests, it was found that the specimens of the lower molecular weight ruptured at lower strains without crazing. The stress and strain at which crazing initiated increased for samples of higher molecular weight. In the stress-solvent crazing tests conducted, using benzene and isopropanol, the lower molecular weight materials ruptured immediately upon crazing; usually at an observed craze crack. For the samples with the two highest molecular weights, there was a slight indication of increased craze resistance.

Two groups of polystyrenes have been used by Rudd (43a) to illustrate the molecular weight dependence of stress cracking. One series (anionically polymerized) having a narrow molecular weight distribution, and another (isothermally polymerized) having a broad distribution, were compared. The molecular weight of the anionic polymers varied from ~120,000 to ~270,000 while the isothermic material ranged from ~180,000 to ~450,000. Resistance to stress crazing in terms of stress decay was found to decrease with decreasing molecular weight. Reduced resistance to stress crazing at low molecular weights has sometimes been attributed to the existence of voids and imperfections due to chain-end groups. The polystyrenes of narrow molecular weight distribution used in this work, however, were found to be sensibly lacking in molecular components smaller than M.W. ≅ 75,000.

The number of solvents that are known to decrease the "craze resistance" of polymers would make a formidable list. "Good" solvents have been mainly used in solvent crazing studies. However, Ziegler and Brown (61) have shown that even "poor" solvents, such as kerosene, *n*-heptane, dairy products, and propylene glycol, depress the stress level apparently required for the onset of crazing. In fact, at sufficiently high stresses, water can be shown to have the ability to affect crazing behavior in poly(methyl methacrylate) and polystyrene. Small amounts of chemically unreacted monomer present in the polymer also act as a "solvent" in the crazing process.

When nylon plates are bent so as to exceed a critical strain and a drop of concentrated hydrochloric acid is applied to the surface in tension, catastrophic rupture results. Similar observations have been made on poly(methyl methacrylate) when bent and treated with

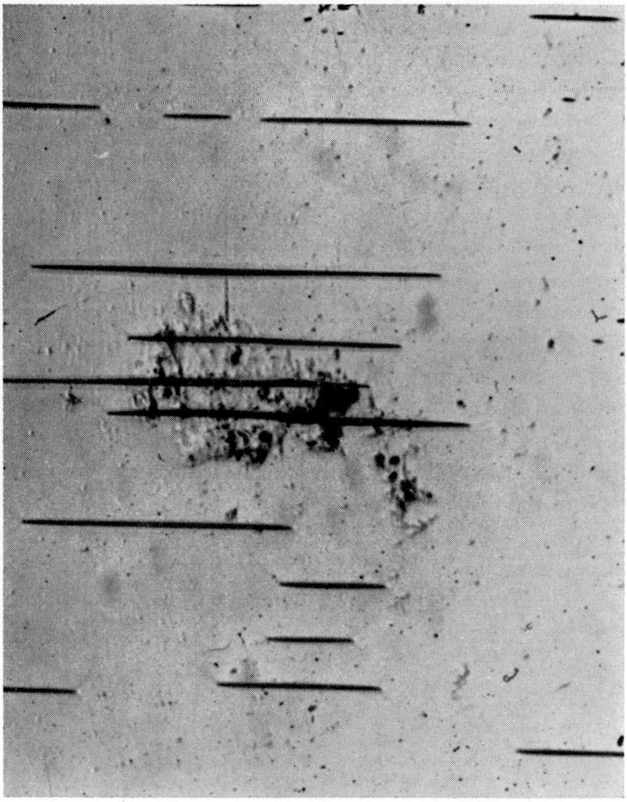

Fig. IIC-34. Transmission photomicrograph of a portion of the surface of a stress-crazed tensile specimen of PMMA. The largest cracks are associated with particles of adhesive from the original covering of pressure-sensitive paper (×200).

a variety of organic solvents (17). The fracture surfaces were found to consist of a smooth area extending about two-thirds of the way across the section from the tensed surface, the remainder being torn roughly apart. A large craze apparently served as the fracture origin. Sheldon (47) has demonstrated similar behavior in poly(ethylene terephthalate), cf. Mylar.

It has been shown (38) that the largest cracks produced by stress crazing in poly(methyl methacrylate) are usually associated with adhesive particles from protective masking paper (Fig. IIC-34). Wolock (54) designed experiments which provide further support for attributing a role in the crazing process to the masking adhesive. The masking paper was removed from tension specimens of commercially available sheet just before test. A length of pressure-sensitive cellulose tape was firmly applied to half of the reduced section of the tension specimens and then stripped off. Specimens were slowly loaded to failure or until crazing occurred. It was found that crazing was slight or absent over that part of the reduced section which had been in contact with the pressure-sensitive cellulose tape, while the area which had not been treated displayed the usual concentration of cracks. Apparently the cellulose tape strips residual masking adhesive from the surface. The adhesive would otherwise act as an ineffective solvent crazing agent which becomes more effective at elevated stress levels.

Large increases in resistance to stress crazing and stress-solvent crazing can be obtained by multiaxial stretching of thermoplastic polymers (8,31,55,56). A sheet of the polymeric glass is heated to an rubbery state and then stretched by vacuum forming or by some other technique. Some materials may be stretched considerably in excess of 100%. The crazing resistance increases with increasing degree of stretching.

IIC7d. Micromorphology of the Craze Crack

The earliest investigators (35,36,44) assumed the craze cracks to be slit-like failures in the matrix. However, this simple concept did not provide a completely satisfactory basis for crazing behavior, and Sauer, Marin, and Hsiao (45) were probably the first to express the belief that crazing is not merely a mechanical crack phenomenon although it may be closely associated with visible mechanical cracks.

Most of the early work on crack morphology utilized bright-field microscopy and limited itself to the length dimension of the crack. Newman and Wolock (38) extended this approach to the electron microscope and found that in stress-crazed specimens the minimum crack length observed was about 2 μ, which is considerably above the resolution of this instrument. This would imply the existence of some threshold condition for crack formation. Bright-field transmission microscopy is also useful for indicating the relationships of cracks and inhomogeneities in the matrix. Individual cracks in polystyrene can be shown (37, 38) to be frequently associated with such defects (Fig. IIC-35).

Multiple-beam interference microscopy (37,38) was used to observe the influence of crazing on surface topography. Each crack in poly(methyl methacrylate) specimens was found to be surrounded by an elevation which might be as high as 270 mμ under stress-solvent crazing but was only a few hundred angstroms in stress-crazed cracks. Interference microscopy is also useful in demonstrating the flow lines initiated at the ends of cracks in some specimens of poly(methyl methacrylate). The trace of these lines on the surface makes an angle of

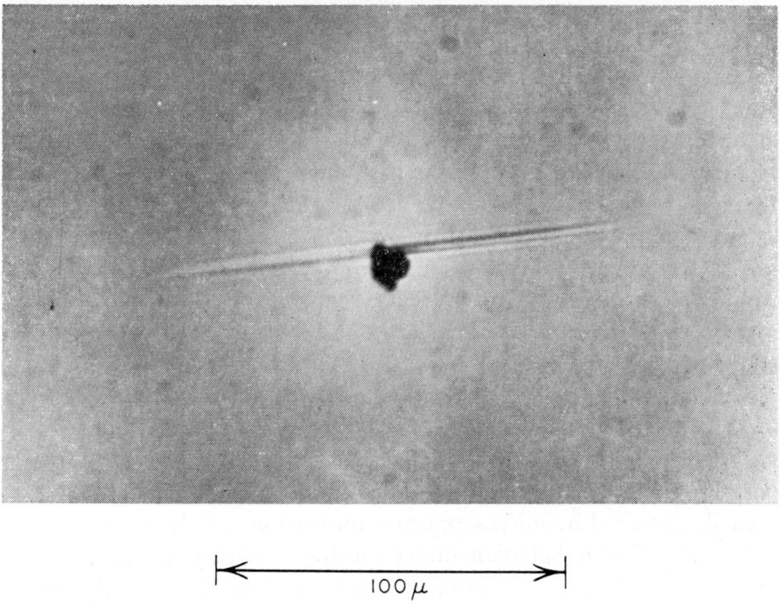

Fig. IIC-35. Photomicrograph of an internal craze crack in a PS specimen associated with a visible inhomogeneity. [From (57).]

35–45° with the cracks. These flow lines appear to be an alternate mechanism of stress relaxation.

Two-beam interference was effectively used by Bessonov and Kuvshinskiĭ (13a) to determine the internal configuration of cracks in plasticized poly(methyl methacrylate). The cracks were wedge-shaped, being wider at the surface than at the mesial edge. Growth takes place in all directions into the matrix so that initially they are roughly circular in shape. Subsequently, they become parabolic in form with the vertices directed toward the center of the section. With growth, the vertex of the parabola is often blunted.

These investigators found that cracks in polystyrene formed more easily than in poly(methyl methacrylate). The form of cracks in polystyrene was more complex than that in poly(methyl methacrylate). The surfaces were uneven, and clear interference patterns were not observed. At room temperature, polystyrene cracks formed narrower wedges.

IIC7e. Observations on Craze Matter

Loss in mechanical strength due to crazing appears to be less than would be predicted for a complete loss of the section involved in a true crack. This fact, in addition to other anomalous behavior, such as an overly pronounced crack "healing" at elevated temperatures, has caused some dissatisfaction with the concept of the crack as a wedge-shaped void. Bessonov and Kuvshinskiĭ (14) postulated the strengthening of the cracks by systems of filaments. These filaments would not prevent liquids from penetrating the cracks nor would they affect the interference fringes developed. These filaments would shorten on heating and thus heal the cracks rather efficiently.

Spurr and Niegisch (50) convincingly demonstrated the presence of oriented material in craze cracks produced in long-time tests of polycarbonates. This type of polymer crazes with the formation of very large cracks in which oriented material joining the crack surfaces is easily detected by microscopical examination. Polycarbonate does not exhibit brittle behavior under the test conditions, rupture strain being about 100%. Craze matter was also reported in polystyrene and poly(methyl methacrylate), though it was more difficult to detect and record in these polymers. The contents of the cracks are visible through difference in refractive index probably coupled with some scattering from micropores. Using the critical incident angle of a beam on the specimen surface, Kambour (26–28) has also demonstrated the presence of a solvent and stress-sensitive substance in polycarbonate craze cracks. His work indicates a porous structure in these cracks consisting of 50–60% of polymer. The porosities in crazes appear to have a large size range (27, 28). Void content in the crazes of polystyrene was calculated as 40%, in styrene–acrylonitrile copolymer 60%, and in poly(methyl methacrylate) about 40% (27).

Additional evidence of this type of structure has been found in large internal cracks in specimens of polystyrene or styrene–acrylonitrile copolymers (40). These specimens had developed high skin stresses produced by heating to the softening point followed by quenching in liquid nitrogen. The stresses may also be introduced by injection molding of the specimens. Static tension loading of the specimens causes the spontaneous appearance of numerous large craze cracks just before the rupture. Many of these occupy more than half of the cross section. The interior of these large cracks contains a thin film which increases in thickness from the crack periphery to the center. Films in the styrene–acrylonitrile copolymer adhered firmly to the crack surfaces, though the film was often found to be on one surface for a considerable area and the remainder on the other surface. Strong interference fringes were developed by the films and permitted estimates of their thicknesses. Some of these reached a maximum of 2–3 μ in the center portions of the cracks.

The quenched polystyrene specimens also contained craze matter. It was reported by Bessonov and Kuvshinskiĭ (14) that the craze films could be removed by manipulating the edges of the film. In larger tensile specimens the film was found to be detached and pieces with dimensions up to several millimeters could be picked out of the cracks (40).

The refractive index of these films appears to be significantly lower than the specimen matrix.

IIC7f. Craze Matter and Strength of Polymers

Spurr and Niegisch (50) have suggested a mechanism for the evolution of the craze crack structure. This has been divided into three steps: initiation, propagation, and termination. The induction period represents the time necessary for sufficient energy to become available at the site of the defects and inhomogeneities. Molecular distortions or rearrangements take place at the defect sites and it is assumed that they involve the movement of polymer chains in the direction of the applied tensions. Duration of the initiation period is, of course, shorter as stress and temperature increase. The crevice starts to develop as molecular segments attain some critical predisposition to the direction of the tensile force.

Propagation of the crack is maintained by the continued yield and flow of resin at the periphery of the craze. During the induction period the resin surrounding the defect has adjusted to a state favorable for its subsequent rapid orientation. The ability of the resin to undergo this deformation is assumed to depend on its lubricity and molecular weight. Relief of the stress at the initiation site redistributes excess stress to the neighboring resin in a plane perpendicular to the applied tension. This neighboring resin, in turn, is converted to craze matter. The craze tip is composed of matter in the process of conversion to craze matter, and small incremental stresses are sufficient for this conversion. The path and shape of the craze periphery then depends on the randomness or orientation of the surrounding resin being acted on by these stresses.

Termination of crack growth occurs after long stressing times. The specimen, however, continues to support the applied tensile load for additional long periods. Any excess local stress has been reduced to a negligible level over the craze crack surfaces and the incremental macroscopic creep can no longer sustain significant craze propagation. In addition, the resin undergoes oblique flow (38), which permits redistribution of the stress.

The ultimate rupture may or may not occur through a true craze crack. In the large internal cracks of the surface-compressed specimens, it usually does break through a craze. The failure appears to be in the bond between the craze matter and the crack surface. Characteristically in polystyrene, the failure shears both surfaces; in the copolymer only one interface is broken. Fracture may take place through cracks devoid of craze matter; true voids which arise from larger defects than the craze cracks.

Detailed examination of the interior surfaces of craze cracks reveal the conic sections characteristic of brittle failure. The craze matter is fairly stable. Exposure in an oven at 70°C. for 48 hr. caused little change in the film produced in the styrene–acrylonitrile copolymer. Prolonged heating (several weeks) caused some pitting in the thinner parts. Films in intact cracks are unaffected by such treatment for even longer periods.

IIC8. Summary

The study of fracture topography is an important tool in efforts to understand the nature of fracture processes. The individual events involved in failure are sculptured in the fracture surface, and a proper interpretation of these markings yields valuable information concerning the mechanism of failure.

The basic or classic fracture marking observed in polymers is the parabola, and the mechanism of formation of this figure exemplifies the entire fracture process, i.e., the initiation of fracture at flaws, the discontinuous nature of the process, the advance nucleation of flaws, the initiation of secondary fractures, and their joining up with the main crack-front, propagating usually at a different level, forming the geometric figures. The fracture origins and the level-difference lines formed are the major deformation processes for which energy must be supplied for the fracture process. In practice, the fracture path and subsequently the fracture markings are modified somewhat by imperfections, inhomogeneities, localized stress conditions, specimen geometry and other elements, so that the markings are often not the exact geometric figures described above.

Variables that affect the mechanical behavior of polymers also affect the appearance of the fracture surface. These include rate of loading, molecular weight, temperature, orientation, and type of loading. By changing these variables, one can produce duplicate fracture surfaces in two materials having widely different chemical structure.

For glassy polymers, the roughness of the surface is a direct indication of the energy required for crack propagation. In addition, results to date indicate that the surface roughness also increases with crack velocity. However, tear tests on rubber show that the torn surface becomes smoother as the rate of tear propagation and the energy for tearing increase.

Colors have been observed on the fracture surfaces of glassy polymers, mainly on poly(methyl methacrylate). In the most interesting case, these colors are associated with specific fracture markings. In this instance, the colors on the two halves of the same fracture surface are found generally to be complementary. They fade with time, exposure to solvent, and in some cases, with exposure to vacuum. The colors have been attributed to a surface layer that is molecularly oriented during the fracture process, having a density and a refractive index different from that of the matrix.

Another phenomenon of interest in fracture topography is "crazing," which is the formation of relatively fine structural discontinuities in a polymer under tensile loading. The stress may be applied externally or may be purely internal. Solvents tend to accelerate the process. In some cases, the crazes occur only on the surface and in others, are located throughout the matrix. Molecular orientation of the polymer by multiaxial stretching has been found to greatly reduce or eliminate this phenomenon. Recent studies have indicated that the crazes, in many cases, contain oriented material.

The fairly limited studies of fracture topography of polymers have proved extremely useful in contributing to an understanding of the fracture process. Further studies should prove equally fruitful in developing knowledge of the mechanism of failure.

References

1. Alfrey, T., *Mechanical Behavior of High Polymers*, Interscience, New York, 1948.
2. ANC-17 Bulletin, Plastics for Aircraft, Part II, Transparent Glazing Materials, Departments of Air Force, Navy, and Commerce (1957).
3. American Society for Testing and Materials, Method of Test for Tensile Properties of Plastics D-638, 1961.
4. American Society for Testing and Materials, *Symposium on Resinography (1963)*, Special Technical Publication, in press.
5. Andrews, E. H., *J. Appl. Phys.*, **30,** 740 (1959).
6. Andrews, E. H., and A. Walsh, *J. Polymer Sci.*, **23,** 39 (1958).
7. Axilrod, B. M., and M. A. Sherman, *Modern Plastics*, **30,** 130 (June 1953).
8. Axilrod, B. M., M. A. Sherman, V. Cohen, and I. Wolock, *J. Res. Natl. Bur. Std.*, **49,** 331 (1952).
9. Bailey, J., *Modern Plastics*, **24,** 127 (1946).
10. Benbow, J. J., *Proc. Phys. Soc. (London)*, **78,** 970 (1961).
11. Benbow, J. J., and F. C. Roesler, *Proc. Phys. Soc. (London)*, **B70,** 201 (1957).
12. Berry, J. P., *Nature*, **171,** 91 (1960).
13. Berry, J. P. (comment) in *Fracture*, B. L. Averbach, D. K. Felback, G. T. Hahn, and D. A. Thomas, Eds., Wiley, New York, pp. 263–264.

13a. Bessonov, M. I., and E. V. Kuvshinskiĭ, *Fiz. Tverd. Tela*, **1,** 1441 (1959).

14. Bessonov, M. I., and E. V. Kuvshinskiĭ, *Vyskomol. Soedin.*, **1,** 1561 (1959).
15. Busse, W. F., E. Orowan, and J. E. Neimark, paper presented to American Physical Society, Philadelphia, Pa., March 1957; also Symposium on Ultimate Strength of Polymers, Brooklyn, Nov. 1956, Polytechnic Institute of Brooklyn.
16. Cheatham, R. G., and A. G. H. Dietz, *Modern Plastics*, **29,** 113 (1951).
17. Ensanian, M., *Nature*, **193,** 161 (1962).
18. Feltner, C. E., Univ. Illinois, Theoretical and Applied Mechanics Report 224 (August 1962).
19. Greensmith, H. W., *J. Appl. Polymer Sci.*, **3,** 183 (1960).

20. Greensmith, H. W., and A. G. Thomas, *J. Polymer Sci.*, **18,** 189 (1955).
21. Hermans, P. H., *Kolloid-Z.*, **89,** 345 (1939).
22. Hermans, P. H., Kratky, O., and Platzek, P., *Kolloid-Z.*, **86,** 245 (1939).
23. Higuchi, M., *Repts. Research Inst. Appl. Mech., Kyushu Univ.*, **6,** 173 (1958).
24. Hsiao, C. C., and J. A. Sauer, *J. Appl. Phys.*, **21,** 1071 (1950).
25. Irwin, G. R., and J. A. Kies, *Welding J., Res. Suppl.*, **31,** 95-S (1952).
26. Kambour, R. P., *Nature,* **195,** 1299 (1962).
27. Kambour, R. P., General Electric Research Laboratory Report No. 63-RL-(3407c), Schenectady (August 1963).
28. Kambour, R. P., *Polymer,* in press.
29. Kerkhof, R., *Naturwissenschaften,* **40,** 478 (1953).
30. Kies, J. A., A. M. Sullivan, and G. R. Irwin, *J. Appl. Phys.*, **21,** 716 (1950).
31. Kline, G. M., I. Wolock, B. M. Axilrod, M. A. Sherman, D. A. George, and V. Cohen, Development of Craze and Impact Resistance in Glazing Plastics by Multiaxial Stretching, National Advisory Committee for Aeronautics, Report 1290 (1956).
33. Leeuwerik, J., and F. Schwarzl, *Plastica,* **8,** 474 (1955).
34. Mason, P., *J. Appl. Phys.*, **29,** 1146 (1958).
35. Maxwell, B., and L. F. Rahm, *Ind. Eng. Chem.*, **41,** 1988 (1949).
36. Maxwell, B., and L. F. Rahm, *SPE J.*, **6,** 7 (1950).
37. Newman, S. B. in *Symposium on Microscopy,* ASTM Special Technical Publication No. 257, Philadelphia, 1959, p. 132.
38. Newman, S. B., and I. Wolock, *J. Res. Natl. Bur. Std.*, **58,** 339 (1957).
39. Newman, S. B., and I. Wolock, *J. Appl. Phys.*, **29,** 49 (1958).
40. Newman, S. B., and I. Wolock, in *Adhesion and Cohesion,* P. Weiss, Ed., Elsevier, New York, 1962, p. 218.
41. Newman, S. B., and I. Wolock, unpublished data.
42. Rochow, T. G., and E. G. Rochow, *Science,* **111,** 271 (1950).
43. Rochow, T. G., and F. G. Rowe, *Anal. Chem.*, **21,** 461 (1949).
43a. Rudd, J. F., *J. Polymer Sci.*, **B1,** 1 (1963).
44. Russell, E. W., *Nature,* **165,** 91 (1950).
45. Sauer, J. A., J. Marin, and C. C. Hsiao, *J. Appl. Phys.*, **20,** 507 (1949).
46. Schardin, H., *Kunststoffe,* **44,** 48 (1954).
47. Sheldon, R. P., *Nature,* **194,** 767 (1962).
48. Sherman, M. A., and B. M. Axilrod, National Advisory Committee for Aeronautics, Technical Note 2778 (1952); also *ASTM Bull.* **No. 191,** 65 (July 1953).
49. Smekal, A., *Ergeb. Exakt. Naturw.*, **15,** 173 (1936).
50. Spurr, O. K., and W. D. Niegisch, *J. Appl. Polymer Sci.*, **6,** 585 (1962).
51. Thomas, A. G., *J. Appl. Polymer Sci.*, **3,** 168 (1960).
52. Tolansky, S., and M. Omar, *Nature,* **170,** 81 (1952).
53. Wallner, H., *Z. Physik,* **114,** 368 (1939).
54. Wolock, I., unpublished data.
55. Wolock, I., B. M. Axilrod, and M. A. Sherman, National Advisory Committee for Aeronautics, Research Memorandum 53D14 (1953); also *Modern Plastics,* **31,** 128 (1953).

55a. Wolock, I., M. A. Sherman, and B. M. Axilrod, National Advisory Committee for Aeronautics, Research Memorandum 54A04 (1954).
56. Wolock, I., and D. A. George, *SPE J.*, **12**, 20 (1956).
57. Wolock, I., J. A. Kies, and S. B. Newman, in *Fracture*, B. L. Averbach, D. K. Felbeck, G. T. Hahn, and D. A. Thomas, Eds., Wiley, New York, 1959, pp. 250–262.
58. Wolock, I., S. B. Newman, and S. G. Weissberg, paper presented at Meeting of the Division of High Polymer Physics, American Physica Society, Detroit, Mich., March 1960.
59. Zandman, F., *Etude de la Deformation et de la Rupture des Matieres Plastiques*, Publications Scientifique et Technique du Ministere de l'Air, No. 291, Paris, 1954.
60. Zarek, J. M., *Brit. Plastics*, **30**, 399 (1957).
61. Ziegler, E. E., and W. E. Brown, *Stress Cracking* (*Crazing*) *of Polystyrene*, Plastics Technical Service Bull., Dow Chemical Co., Midland, Mich., 1954.

CHAPTER III

Time-Dependent Fracture Processes

Sections

IIIA. PHENOMENOLOGICAL ASPECTS OF PROGRESSIVE WEAKENING IN THE HIGHLY VISCOELASTIC GLASSES

B. Rosen

IIIA1. Introduction

Studies of fatigue mechanisms or progressive weakening in the softer or more inelastic of the transparent glasses are particularly rewarding for three main reasons: (*a*) optical phenomena (whether in birefringence, interference, reflectance or in scattering) generate revealing as well as visually striking effects, (*b*) the inelastic contributions to crack stabilization (and even to liquidlike cavitation) are pronounced, and (*c*) the high order of flaw insensitivity in these noncrystalline bodies act to enhance the validity of structural (i.e., molecular) perspectives of the phenomena involved.

Within these general considerations, "fatigue" takes on a far more versatile connotation with these and other varieties of highly viscoelastic bodies, having implications which go well beyond the ultimate mechanical responses to dynamic loading patterns. Thus *static* fatigues, whether on creep or relaxations in tension or shear at *room temperatures,* fall fully within this terminology. Indeed, this section is mainly concerned with static rather than with dynamic fatigue.

We employ the term "progressive weakening" in our title so as to imply precursory fatigue processes rather than macroscopic ruptures, this latter aspect being treated by Landel and Fedors in Section IIIB which follows. Section VC offers a good starting line from which to develop the more formalistic approaches to the appreciation of the efficiency of stress concentrators in linear viscoelastic bodies under infinitesimal strain.

The presentation employed in the present section, however, is descriptive or phenomenological in approach. Moreover, it offers no attempt at a comprehensively thorough review of this dynamic, yet embryonic, area of study; the major objective being a condensed outline of the salient factors in a loose but tangible framework of *conceptual,* as well as of phenomenological, continuity.

IIIA2. Microfailures

IIIA2a. Some Varieties of Microfailures

As illustrated in Figure IIIA-1, and as also was considered briefly in Sections IIB and IIC, failure is readily stabilized at an optical microscopic size level with respect to its depth and breadth. As this photograph shows, these failures may frequently constitute a pattern which can be utilized in mechanical analyses by employing a critical stress or strain criterion for the more brittle of the transparent organic glasses which are also suitable for firmly adhering uniform coatings onto the elastic substrate under study (13). A careful examination of the fracture surfaces, propagation rates, and interfailure spacings in a resin such as this would reveal virtually "ideal" behavior in the

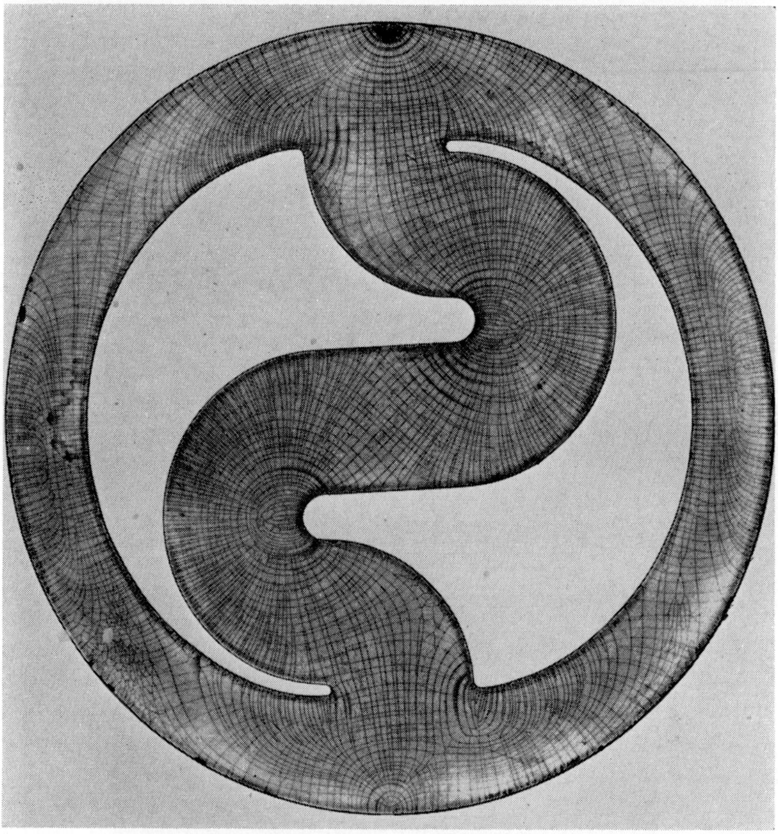

Fig. IIIA-1. Two families of isostatics obtained from a brittle lacquer coating test on a CR39 model diametrically loaded in compression. After (13). Courtesies of A. J. Durelli and McGraw-Hill Book Co.

sense of brittle fracture habits (13,45). Indeed, a two-dimensiona strain recording of this variety can be favorably compared with grid, strain gage, or photoelastic approaches to stress or strain analyses (13).

Thus these quite *brittle* glasses are clearly not the "highly viscoelastic" bodies to be treated here, but these same materials can be brought into the pertinent category at slightly higher temperatures or when sparingly softened with soluble fluids. Since these brittle lacquers are usually not composed of extremely long-chained mole-

(*a*)

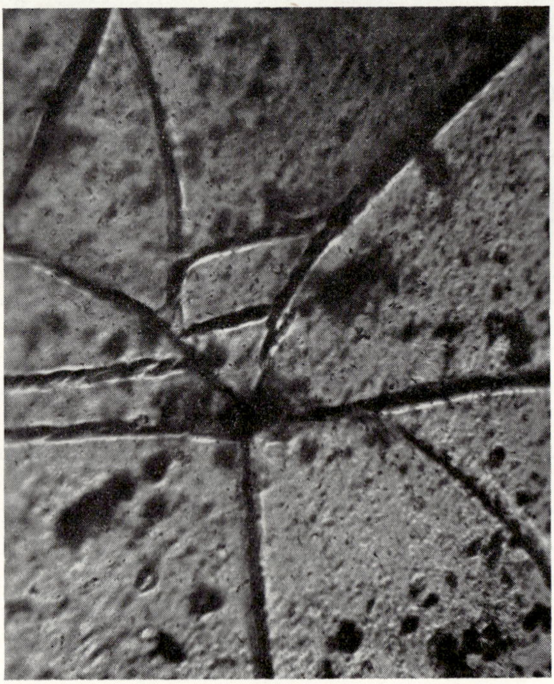

(*b*)

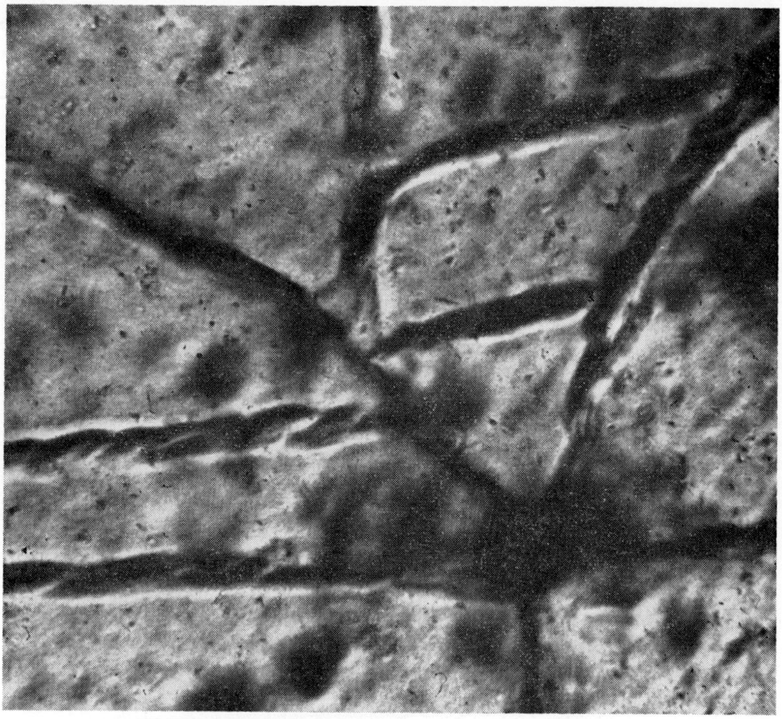

(c)

Fig. IIIA-2. Tensile microfailures in a film of a glasslike, methyl-etherified phenolic resin. The film photoelastically revealed residual biaxial tensions after a shrinkage on a chemical curing that was partly restrained by its adhesion to an aluminum foil substrate. These failures were slowly developed during the film's subsequent submersion and swelling in acetone. (*a*) Photomicrograph, somewhat reduced from 32×. (*b*) An approx. 3× magnification of the region in the upper-left corner of photo *a*. (*c*) An enlargement of the plate from photo *b*. Note the general tubular appearance of these failures in *a*, and the details of the failure loci in both *b* and *c*. Cf. Figs. IIIA-28 and IIIA-31. From Rosen (39).

cules, they usually cannot experience an appreciable "leatherlike" or a tangible "rubberlike" contribution, but pass rather abruptly into a pronounced "flow" contribution with an over-softening by heat or solute. (See Subsection IC4, p. 71, for a clarification of these terminologies on a basis of linear viscoelastic contributions.)

The situation, however, is not always as clearly "brittle" with certain varieties of microfailures such as shown (39) in Figure IIIA-2.

These were aggravated by biaxial, transient solute stresses (34) as combined with biaxial, residual "curing stresses" (38). Under a low magnification, such as shown in Figure IIIA-2*a*, these failure loci do not appear "abnormal" *per se*, yet higher magnifications reveal an apparently twisted, ropelike path; an interesting phenomenon which is not immediately anticipated in an amorphous body of this glasslike

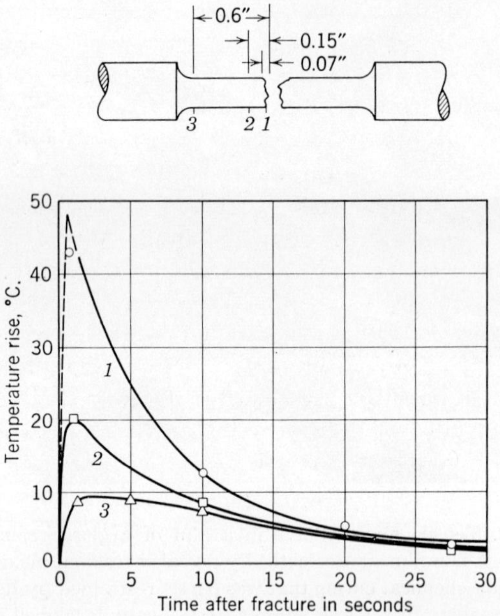

Fig. IIIA-3. Localized heating generated in a steel specimen fractured in a necking rupture during a 2 msec. test in uniaxial tension. The functions shown here reveal the general trend of the belated temperature rise on thermal diffusion away from the fracture surface and through the specimen as detected by three thermocouples located at the positions indicated in the accompanying drawing. After Nadai and Manjoine (29). Courtesy of *Journal of Applied Mechanics*.

variety. This complication, especially in view of the failures' slow and intermittent kinetics of propagation, strongly suggested highly ductile contributions which were somewhat masked by the failures' general appearance as a brittle microfracture (39). For this particular case, it was not even entirely certain that these *failure* loci were genuine *fracture* loci; optical density variations, as discussed in detail later, being frequently blamed for such apparitions as inelastically generated by structural anisotropies along a dynamic, peak stress-front (perhaps further complicated by a finite brittle contribution). In any event, there was considerable room here for doubting that these failures were that which they *appeared* to be upon an overly brief glance.

One is frequently faced with the realization that a solid's mechanical behavior at a crack-front can be quite different than that associated with the mechanical response of the macroscopic body (43). Hence, a sensibly "elastic" glass may fail by microscopically active, ductile processes insofar as the order of its viscoelastic contribution may be enhanced in the vicinity of an inhomogeneity which acts as a "stress raiser" or stress concentrator, the peak stress of which may oftentimes be orders greater than the nominal stress. This omnipresent consideration becomes particularly pertinent when adequate attention is focused on the heat generated during an inelastic deformation, i.e., that generated on an inefficient release of strain energy during the formation of new surfaces, especially if the body is a poor thermal conductor and viscoelastic in mechanical response. This adiabatic temperature effect (14), as evidenced in the relatively efficient heat-conducting metals (Fig. IIIA-3) (29), may alter the viscoelastic "state" of a noncrystalline body. When dealing with the *more inelastic* of the glasses, one can be reasonably well-assured that both fracture initiation mechanisms and decelerated crack propagation are greatly encumbered by extremely localized ductile processes.

Dissipative processes at failure-fronts, however, also imply contributions towards highly localized self-transports in the body. Thus "crack blunting" with the consequent loss in the failure efficiency therewith implied (22,43), can readily occur in the kinds of solids under consideration here. A limiting case of a ductile response in a highly stressed area is the phenomenon of *cavitation,* which is quite common in the failure of liquids (9,24) or other highly ductile bodies (1,12) in flowlike responses to shear components in biaxial or triaxial deformations. A stabilized cavitation can readily occur during the progressive weakening of the softer of the glasses (as characterized in Subsection IIIA3) in regard to the *submicro*scopic level, just as it can occur *micro*scopically in ductile metals (12). Merely to illustrate this general contribution in a polymeric body on the optical microscopic size level, however, Figure IIIA-4 shows a rubber which is

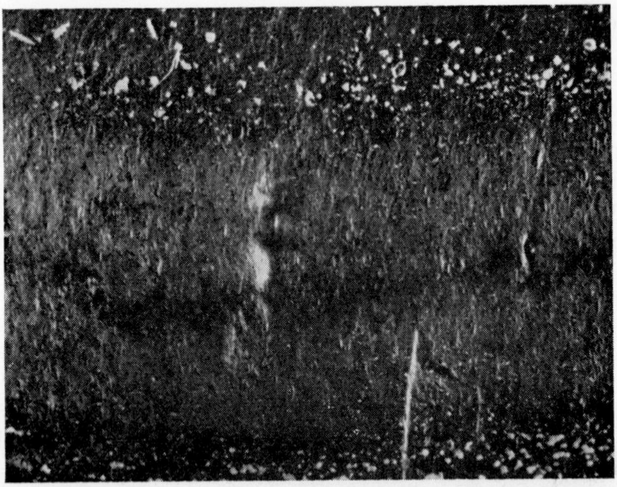

(*a*)

(*b*)

Fig. IIIA-4. Cavity-formation in a rubber generated on a dynamic notch flexing in uniaxial tension at 20 cps. Dicup-cured HAF hevea, approx. 12×. (*a*) Precursory fatigue effects; (*b*) 1 min. later. Note the drawn strands in the interior of the center cavity in (*b*) and the highly strained appearance of the blunted fracture fronts. Cf. Fig. IIIA-31*b*. From Akkerman (1). Courtesy of F. H. D. Akkerman.

gradually being made porous during dynamic notch stressing in uniaxial tension (1).

IIIA2b. "Crazing" and Associated Phenomena

In general, the failure phenomena which can be witnessed on a microscopic level do not greatly differ from those which occur on a *larger size level* in other, *more inelastic,* systems. Yet, by way of an abbreviated phenomenological orientation of the reader, let us briefly illustrate how confusing, or even how *mystifying,* this shift in magnification actually can be when we are dealing with the highly viscoelastic, transparent bodies.

Using a razor's edge, we cut a single fracture locus through the upper surface of a platen or film of our relatively soft, transparent glass such as polystyrene or poly(methyl methacrylate). We first note that this body is quite easy to cut, since the process is facilitated by a pronounced stress-relieving at the blade's edge, so that the action is closer to a well-controlled "slicing" or inelastic shear than to a purely brittle failure. Hence the depth and length of this cut can be made approximately as we choose (see Section IVA). Then let this single cut be so short and so superficial that it would escape detection were it not for a tiny ray of light which is reflected from the air–solid interface at the fracture surfaces, even when the cut is viewed through the entire transparent piece.

Although the source of the light which is being reflected may be rigidly fixed, the slightest unintentional motions of our eyes can make this light ray abruptly disappear or reappear like the twinkling of a star. A haphazardly or randomly misaligned assembly of thousands of these tiny cuts, such as obtained in biaxial tensile failure, would indeed be a virtual galaxy which may be dazzling or *crazing* to our metaphysical senses. But alas; our stars now appear to fade in brilliance as these microfailures "weld" or "heal" by self-diffusional processes at their leading edges!

Let us now superimpose the oftentimes confusing role of transient stresses on this dazzling affair. (Here we imply transient solute, i.e., swelling or shrinking, stresses as well as transient thermal stresses.) With an invisible vapor pressure or thermal gradient our stars, generated by these migrating stresses, may only twinkle at the center, at least for a while (prior to a welding or an equilibrium absorptivity), or only at the surfaces on the initial desorption of the heat or solute, etc.

The true fascination of the many possible phenomena takes firm roots when residual stress is brought into play. Let us say that the sample had been calendered, extruded, or otherwise nonhomogeneously and residually stressed so that our stars become *aligned* parallel or perpendicular to the sample's principal axis. Here, without applied or even transiently active stresses, our cooperatively oriented stars can die on welding as mysteriously as they can be born on static fatigue.

Now, as stated in connection with Figure IIIA-2, there frequently can be some *doubt* as to whether our stars are truly micro*fractures*. Therefore, we also need to superimpose the frequently made contention that these same general kinds of light reflections can be caused by birefringence generated by extremely localized cold-flowings; the hopelessness of the situation being maximized when both occur simultaneously.

The author has thus briefly attempted to convey some of the salient phenomenological sources for mystification, confusion, and consequent controversy which unfortunately widely persist in regard to the behaviors and natures of "crazing" in these highly viscoelastic, transparent glasses. With far less imagery, he will now attempt to describe some of the pertinent phenomenological findings in this embryonic branch of microfatigue study, for static stressing and straining in tension.

IIIA2c. Progressive Behavior of Tensile Microfailures in Static Fatigue

(1) Light Reflectance

In view of the above discourse, it is rather clear that photometric approaches to the study of "crazing intensity" are indicated. The experiments of Maxwell and Rahm (26,27), presented here in highly abbreviated form, are illustrative of the virtues of this rather simple technique; some of the experimental details of which are revealed in Figure IIIA-5. The highly viscoelastic glass employed was the common polystyrene, a linearly chained, fully amorphous polymeric body which can relax approximately 15% of its nominal stress in 10^4 sec. under small strains in uniaxial tension at room temperatures, and

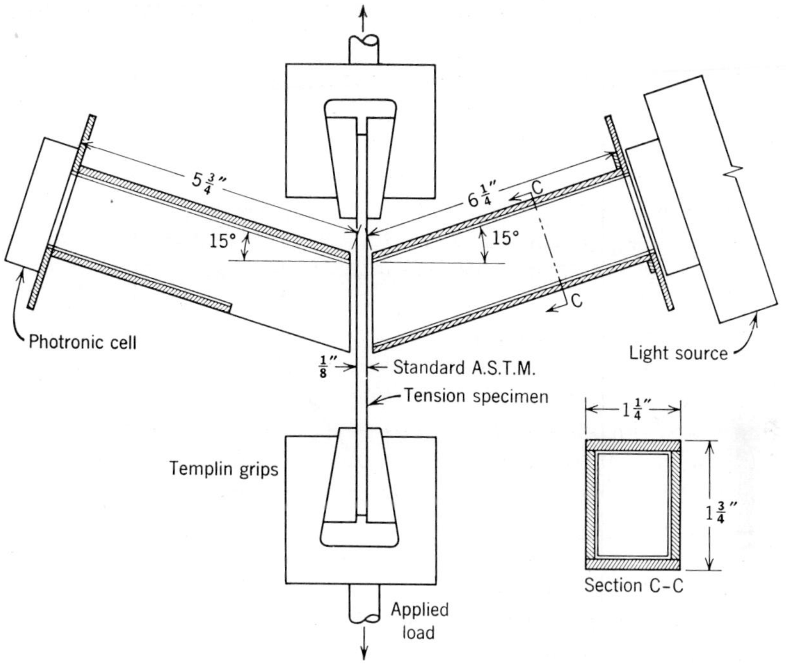

Fig. IIIA-5. Apparatus employed for the measurement of the intensity of light reflections from microfailures in polystyrene. After Maxwell and Rahm (26). Courtesy of *Industrial and Engineering Chemistry*.

whose glass temperature T_g is 90–100°C., depending on the rate of strain employed in its evaluation. Thus these are quite similar to the polystyrene specimens previously discussed in connection with Figure IC-10, p. 71.

Figure IIIA-6 indicates the general time-dependent enhancement of the crazing intensity under various loads in uniaxial tensile creep, while Figure IIIA-7 reveals the equally important role of temperature elevation in a series of otherwise controlled creep experiments (26).

In a different series of studies on polystyrene, Maxwell and Rahm investigated the roles of strain level (27). Figure IIIA-8, Curve *2*, shows the crazing intensity in uniaxial tensile creep under that particular load required to achieve an *initial* elongation of 0.82%, while Curve *1*, Part A–B, was obtained during a stress *relaxation* experiment in uniaxial tension at that same *fixed* strain level (0.82%). The re-

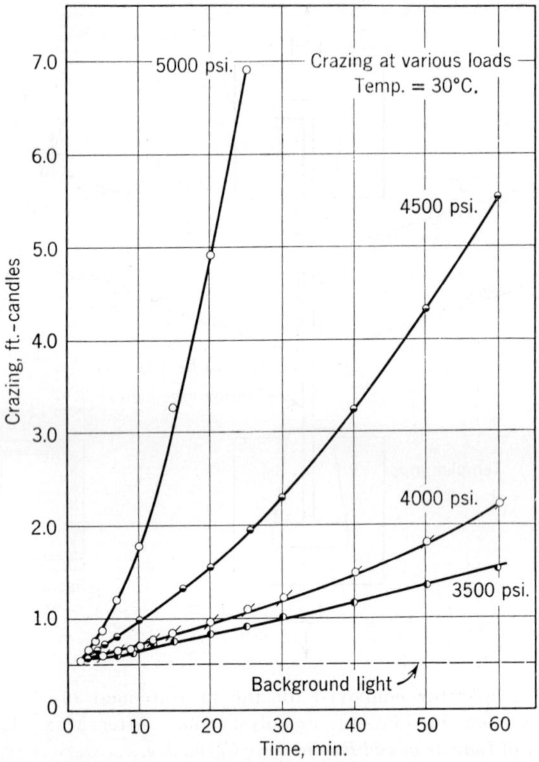

Fig. IIIA-6. Intensity of light reflections obtained during creep in uniaxial tension at various levels of nominal stress in polystyrene. After Maxwell and Rahm (26). Courtesy of *Industrial and Engineering Chemistry.*

maining portions of Curve *1* were obtained during relaxations at successively higher strain levels.

Photomicrographs showed that each higher strain level, such as those of Curve *1*, apparently generated a new population of superficial failures, the slow growth (in length) of a given failure being achieved during *relaxation* of the nominal stress at a given strain level. During the creep experiment, however, new microfailures were apparently initiated and propagated continuously at the gradually increasing strain level (27).

These experiments indicate that the general or *net* contribution of these microfailures as regards light reflections, whether due to the

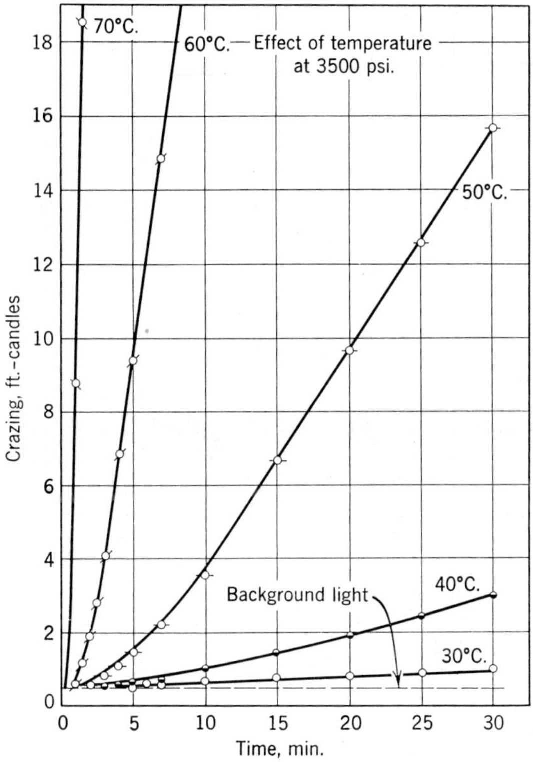

Fig. IIIA-7. Intensity of light reflections obtained during creep in uniaxial tension at various temperatures; polystyrene. After Maxwell and Rahm (26). Courtesy of *Industrial and Engineering Chemistry*.

apparent number of microfailures or their growths in length, depth, and width, is sharply enhanced with both load and temperature during uniaxial creep in tension. Moreover, these microfailures also can propagate in length, if not in number, during a gradually relaxing nominal level of uniaxial tensile stress.

Let us then focus further attention on the microfailure population and loci, and the kinetics thereof.

(2) Microcinematography

Regel (33) has employed periodic photomicrographs of microfailures in poly(methyl methacrylate) under creep in uniaxial tension. This

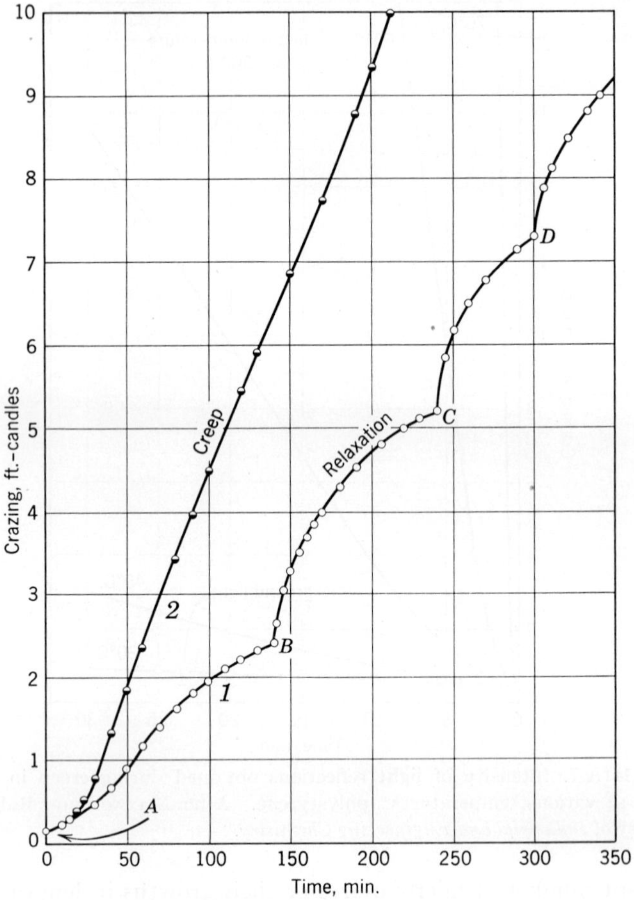

Fig. IIIA-8. Intensity of light reflections obtained during (*1*) uniaxial stress relaxations and (*2*) uniaxial creep, in tension; polystyrene. Curve *2;* at a fixed load sufficient for an initial strain of 0.0082. Curve *1*, Part *A-B;* same initial (here, fixed) strain as for Curve *2;* Parts *B-C* and *C-D*, etc., at increasing strain levels. After Maxwell and Rahm (27).

procedure was to permit the detection of flaws >1 μ in length, and studies were made of the apparent kinetics of microfailure generation, population, and propagation. This highly viscoelastic glass was further softened by addition of the solute, dibutylphthalate, which is a

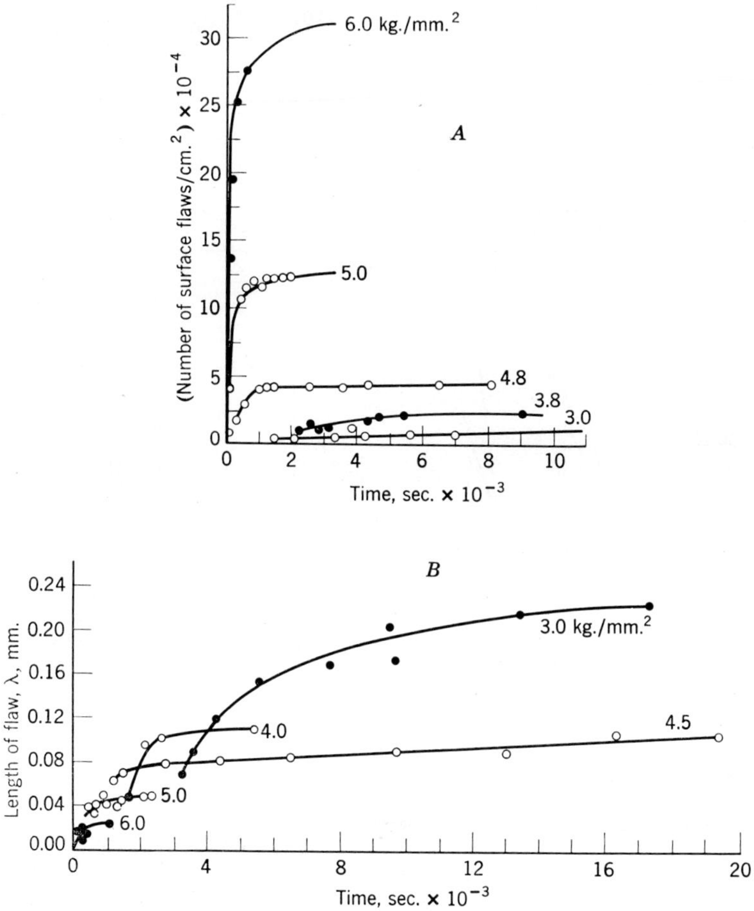

Fig. IIIA-9. Kinetics of superficial microfailure population and propagation in poly(methyl methacrylate) plasticized with 6% DBP during creep in uniaxial tension at various levels of nominal stress. (*A*) Density of failure population; (*B*) typical failure lengths in plane of surface. After Regel (33).

fluid "plasticizer" (2) of low volatility rate. Like the polystyrene employed by Maxwell and Rahm, this variety of poly(methyl methacrylate) was a synthetically prepared, glasslike and transparent, linearly chained, polymeric body.

Figure IIIA-9*A*, shows the remarkably large apparent density of superficial failures that are possible with such systems under various

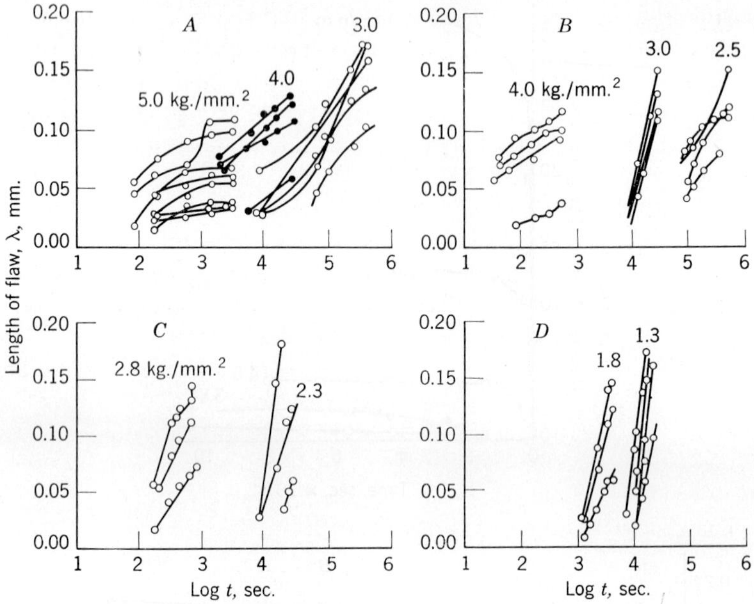

Fig. IIIA-10. Kinetics of superficial microfailure propagation in poly(methyl methacrylate) at various levels of nominal stress and concentration of fluid softening agent during creep in uniaxial tension. The individual curves imply the readings on a random sampling of microfailures found under the conditions indicated. (*A*) No plasticizer; (*B*) 10% DBP; (*C*) 20% DBP; (*D*) 30% DBP. After Regel (33).

levels of nominal stress, and their time-dependent enhancements in population density during the tensile creep. While this plot indicates that an increased stress level greatly increases both the apparent total number and the apparent rate of initiation of the microfailures, Figure IIIA-9*B* shows that the *longest* microfailures are actually formed at the *lower* stress levels where fewer microfailures were generated (33). In terms of a specimen's appearance in reflected light, this implies that while the highly stressed samples exhibited a virtual "haze" of microfailures, individual failure loci were more readily discernible at the lower stress levels.

Figure IIIA-10 displays some of the contributions of an increased quantity of the fluid softening agent (33). Each curve represents the kinetics of propagation of a randomly chosen microflaw under the plasticizer concentration levels and nominal stress levels shown. A study of these semi-logarithmic plots will reveal that: (*a*) a microflaw becomes detectable sooner at the higher stress- and plasticizer concentration levels, and (*b*) added plasticizer, like a decreased nominal stress level, enhances the rate of microfailure propagation.

It was found here, just as in the observations by Maxwell and Rahm, that the lengths of these superficial failures tend to propagate normal to the uniaxially applied load; a point that becomes entirely unambiguous in properly annealed, isotropic specimens. We will note here that both the propagation rates and apparent population rates of these tensile microfailures follow a *creeplike* pattern. Indeed, the softer the glass, the more *fertile* it becomes for that propagation and apparent population. This same general point can be surmised in reference to Figure IIIA-7 as regards a thermal softening of these viscoelastic bodies. *Crazing is not as readily generated in the harder or less viscoelastic of the glasses* such as the inorganic varieties.

The general situation is further clarified in the following.

(3) Interferometry

A two-beam, light interference fringe method was profitably employed by Bessonov and Kuvshinskiĭ in a further study of the progress of microfailures in dibutylphthlate-plasticized poly(methyl methacrylate) (5,6). The tensile creep specimens were cylindrically shaped dumbbells of 8 mm. gage length, and usually of 2 mm. diameter

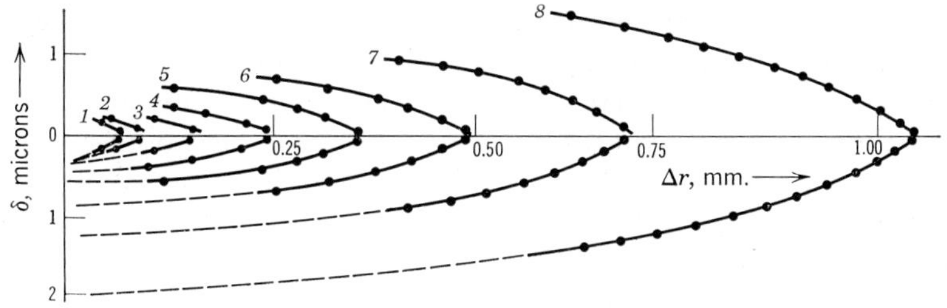

Fig. IIIA-11. Profile of a microfailure front advancing towards the center of a sample of poly(methyl methacrylate) plasticized with 6% DBP at 20°C. under 4.0 kg./mm.^{2} nominal stress during uniaxial tensile creep. (*1*) After 4 min., (*2*) 17 min., (*3*) 1.5 hr., (*4*) 5.5 hr., (*5*) 12.2 hr., (*6*) 23.5 hr., (*7*) 35 hr., and (*8*) 47.7 hr. Results of interferometry studies. After Bessonov and Kuvshinskiĭ (5).

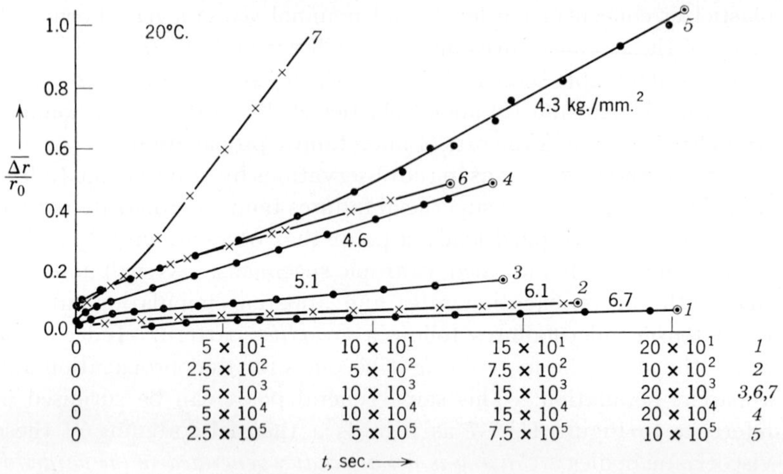

Fig. IIIA-12. Kinetics of microfailure penetration into a sample of poly-(methyl methacrylate) plasticized with 6% DBP at 20°C. under various levels of nominal stress during uniaxial tensile creep. The ordinate denotes the mean advance of the entire population of microfailures normalized to the radius of the cylindrical sample. The abscissa varies with the different stress levels. The terminal circles indicate the times for creep rupture. Curve *7* represents the advance of a single "fast" microfailure taken from the population represented in Curve *6*. Results of interferometry studies. After Bessonov and Kuvshinskiĭ (5).

eter. The microfailures grew in the plane of the transverse cross section, more rapidly along the radii than along the circumferences. Photographs of the interference patterns were made periodically and the distances between the bands were employed in a calculation of the profile of the microfailure (6). The points of Figure IIIA-11 correspond to the bright interference bands (5). The abscissa here is the sample's radius, so that when $\Delta r = 0$, we are at the sample's surface. The δ axis runs along the sample's length and thus is parallel to the uniaxially applied tension.

The ratio, $2\delta_{max}/\Delta r$, fell from ca. 8×10^{-3} to ca. 4×10^{-3}, later becoming sensibly constant for this penetration. Thus there is an initial contribution towards "blunting" of the microfailure on its advance into the specimen (5).

Figure IIIA-12 summarizes a kinetic study of the averaged, fractional advance of the entire population of microfailures along the

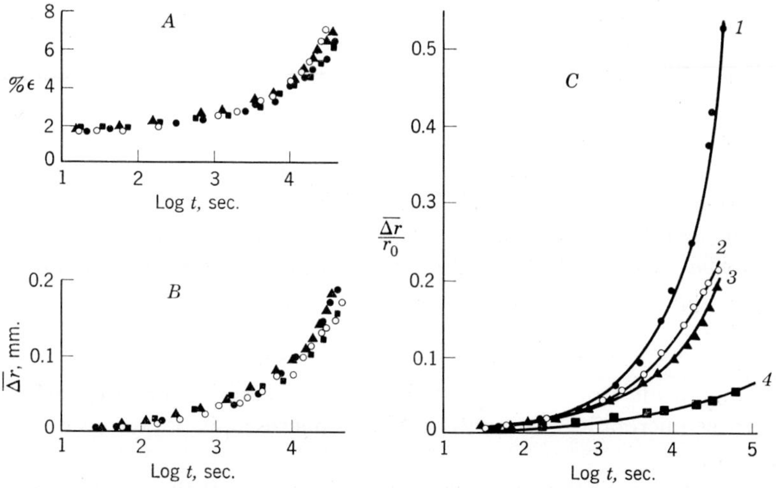

Fig. IIIA-13. Comparison of the rates of uniaxial tensile creep of the specimen and the microfailure propagation into samples of poly(methyl methacrylate) plasticized with 6% DBP when under a nominal stress of 4.6 kg./mm.^{2} at 20°C. for different sample diameters (cylindrical specimens). (*A*) Creep strain; (*B*) average *absolute* penetration of microfailure population along the radii of the specimens; (*C*) averaged *relative* or proportionate advance along the radii. Diameters of specimens: (*1*) 1.0 mm., (*2*) 1.4 mm., (*3*) 2.0 mm., (*4*) 3.0 mm. After Bessonov and Kuvshinskiĭ (5).

samples' radii at different levels of nominal stress, each stress having a different time scale, and static fatigue (delayed "brittle") *rupture* occurring at the times corresponding to the positioning of the terminal circles along the abscissae (5). Thus these are all slowly advancing microfailures, only Curve *5* reaching the center of the sample, and even there, only after $\sim 10^6$ sec. The final penetration rates here, as with Regel's studies of the length of *superficial* microfailures, appear to be steady, much like a residual rate of creep. Here again, the higher stress levels produce a *lesser progressive population* of microfailures in the consequently *more abbreviated lifetime* of the specimen. Curve *7* is an individualistic microfailure from the population represented in Curve *6*. While slightly faster-moving than its neighbors, in itself, it appears to accomplish very little as regards rupture and, moreover, does not even appear to distort the *averaged* progress of the general population of microfailures.

An interesting comparison of the average radial propagation rates of the microfailures and the *macroscopic* creep rate of the specimens is given in Figure IIIA-13 for specimens of different diameters but under the *same* nominal stress level (5). The lifetime of all specimens virtually coincided despite the percentagewise improved efficiency of the penetration of microfailures in the specimens of smaller diameter. *The general propagation of these microfailures appears to be controlled by the sample's creep and not by stress redistributions resulting from their own progress.* In brief, these microfailures are far more threatening in appearance than in macroscopic contribution.

(4) Apparent Macroscopic Contributions

Thus, aside from the alteration in its appearance, precisely what do these microfailures contribute to the macroscopic properties of the sample?

The answers to this question would be far easier if the microfailures under consideration were micro*fractures* of purely brittle origin and habit. The inelastic *contributions to* microfailure initiation and propagation, however, turn any attempted answer into a source of controversy over the inelastic *contributions of* microfailure initiation and propagation. For example, when a sample creeps in tension, microfailures apparently may be formed and propagated. When microfailures "creep" in length, width, or depth, however, the macroscopically experienced creep may conceivably become enhanced. Yet a microfailure may also "creep" in length while the sample is maintained at a *fixed* strain (see the above discussion in reference to Figure IIIA-8). Thus *macroscopic* creep, in itself, is not a necessary *contribution of* all inelastically controlled microfailures, while a localized or *microscopic* creep, flow, or stress relaxation necessarily is.

There are, however, some unambiguous evidences that samples which are badly microfailured are *also* "damaged" in the macroscopic mechanical sense, and while these same *mechanical* methods cannot be employed in defining their macroscopic contributions succinctly, they nevertheless can serve to describe the *apparent* contributions of these microfailures.

Perhaps the most convincing (but perhaps also the most deceptive) apparent contribution of microfailures is for something to happen which is mechanically irregular at the *same time* as these microfailures are made visible by significant light reflections (the real problem, of course, being whether the microfailures are actually the *cause,* or merely the concurrent *effect* of the mechanical irregularity). Thus if the viscoelastic body should simultaneously exhibit an enhanced viscoelasticity, e.g., become more "nonlinear," then one's suspicions are better fed, even if at the expense of one's logic.

It was mentioned in connection with Figure IC-20, p. 94, that a macroscopic nonlinearity in viscoelastic behavior occurred simultaneously with the appearance of crazing in polystyrene (26). Figure IIIA-14 is an attempt by Schmitt and Keskkula (42) to define a

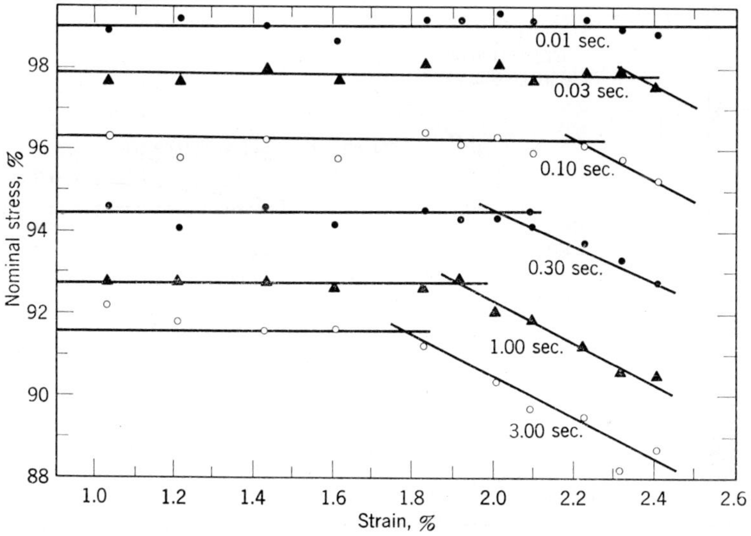

Fig. IIIA-14. Per cent nominal stress in uniaxial tension at various times during relaxation experiments with polystyrene specimens under varying strain levels. After Schmitt and Keskkula (42).

quite similar situation. Here the relative nominal stress is plotted against strain level (rather than log time) for a series of stress relaxation experiments in uniaxial tension. The isochronals should all be horizontal if the viscoelasticity is macroscopically linear. Yet these isochronals appear to break in a systematic manner so that the nonlinear viscoelasticity is achieved at the lower strains for the longer experimental times. Despite the short times involved here, the authors associated this with the onset of crazing, and thus confirmed the general remarks of Maxwell and Rahm in regard to their earlier data (26).

[We wish to pause here in order to emphasize the peculiarity of this variety of fatigue microfailure. While still a precursory aspect of a static fatigue rupture, the nominal stress level is gradually being lowered. It is suggested that delayed "brittle" ruptures in static *relaxation* experiments (a quite prevalent phenomenon with polymeric bodies, see Section IIIB) be termed "relaxive fatigue," and its microfailure analog be termed "relaxive fatigue microfailure."]

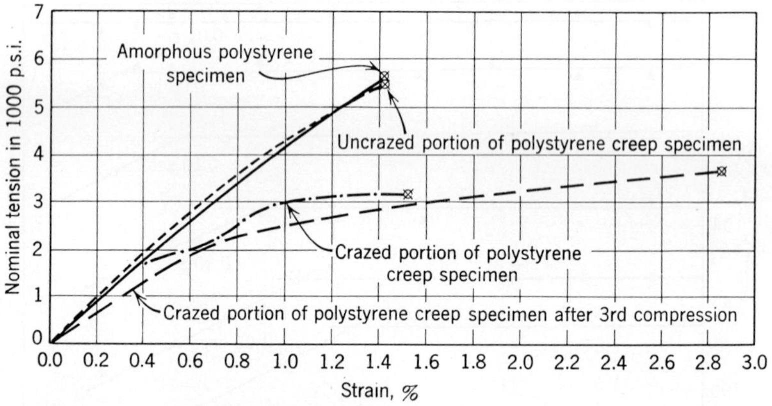

Fig. IIIA-15. Results of uniaxial tensile stress-strain studies under a fixed strain rate with polystyrene both as uncrazed and as previously crazed during uniaxial tensile creep; as discussed in the text. After Hsiao and Sauer (18). Courtesies of American Institute of Physics and J. A. Sauer.

Figure IIIA-15, after Hsiao and Sauer (18), shows a composite of several stress–strain measurements with polystyrene samples, all at a fixed temperature and strain rate. A sample was first subjected to a long-term, uniaxial tensile creep and the crazed portions near the surfaces were machined-off and the remaining portion was compared with a fresh sample for relative stress–strain behaviors in uniaxial tension. These were found to be sensibly similar relative to the strong contrast afforded by the salvaged, crazed fragments. The abrupt break in the curve for this isolated crazed portion of the creep specimen unambiguously corresponded to abrupt *fractures* in the micro*failured* areas.

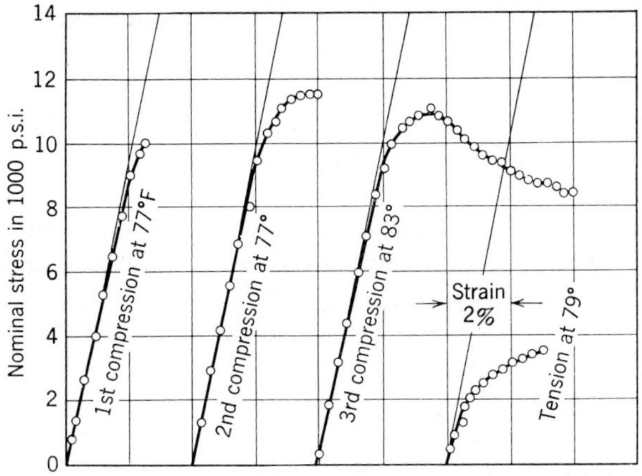

Fig. IIIA-16. Successive stress-strain experiments in uniaxial compression of a polystyrene sample which was previously crazed during creep in uniaxial tension. Curve on the right, which also appears on the right in Fig. IIIA-15, was obtained in uniaxial tension. After Hsiao and Sauer (18). Courtesies of American Institute of Physics and C. C. Hsiao.

Figure IIIA-16 shows the successive stress–strain curves in three uniaxial compressions of a sample which had previously been crazed under uniaxial tensile creep. After the third compression the sample was more ductile, yet free of the usual light reflections associated with a "crazing." A subsequent stress–strain curve in tension, shown at the right in both Figures IIIA-15 and IIIA-16, produced a far more "inelastic" response than the original specimen; an observation which was complicated by the reappearance of the microfailures themselves (18).

Regel (33) found that the time required for the apparent initiation of the crazing reflections varied systematically with the nominal stress level for uniaxial creep in tension (Fig. IIIA-17). Moreover, two further points are displayed in this interesting plot.

The general role of temperature in this time for the first detectable reflection, of. Curves *2* and *3*, appears to be appropriate for a time-temperature superposition, i.e.:

$$t_{25^\circ} \cong A \exp\{-k\sigma\}$$

$$t_{45^\circ} \cong B \exp\{-k\sigma\}$$

and

$$(t_{25°}/t_{45°}) \cong A/B$$

for most of the experimental range. As mentioned previously, the softening achieved by the higher temperature hastens the apparent microfailure initiation, so that $A/B > 1$.

A comparison of Curves *1* and *2* reveals a closely similar situation when considering the time required for creep rupture t_b and the time

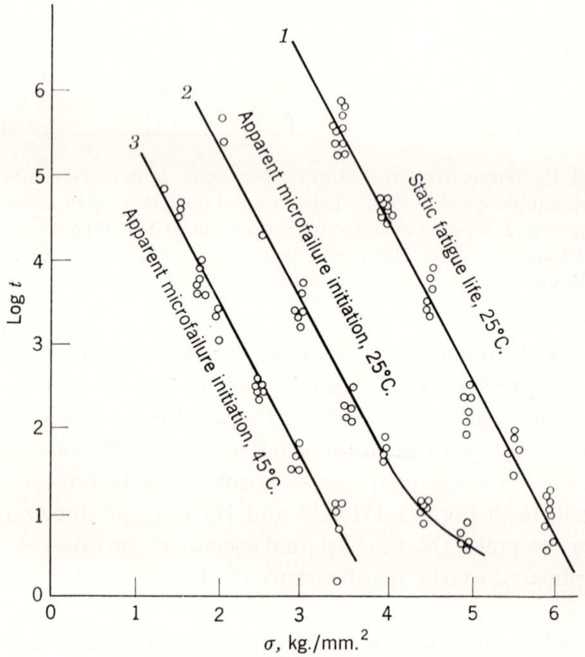

Fig. IIIA-17. Time–nominal stress functions on creep in uniaxial tension with poly(methyl methacrylate) plasticized with 6% DBP. After Regel (33).

for the apparent initiation of the microfailures; both at 25°C. Here, $(t_b/t_{init})_{25°} > 100$ for a given nominal stress. Thus the microfailures appear to be initiated in less than 1% of the time required for the delayed "brittle" rupture (33). Since all three of these functions are grossly parallel, however, further time–temperature–stress–event interrelations are, of course, possible.

Bessonov and Kuvshinskiĭ, in the same study as that described above (5), found that a relationship existed between the average residual penetration rate W of the microfailures and the time for static fatigue rupture:

$$\log t_b = f(\log W^{-1})$$

Similarly,

$$\log t_b = f(\log v^{-1})$$

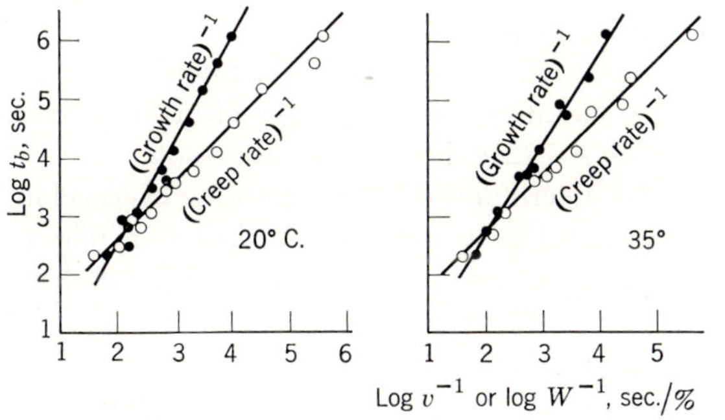

Fig. IIIA-18. Comparison of the temperature dependency of the relations between creep rupture life t_b, average residual creep rate v, and average residual microfailure penetration rate W in poly(methyl methacrylate) plasticized with 6% DBP. After Bessonov and Kuvshinskiĭ (5).

where v is the mean, residual creep rate of the specimen itself. Thus not only can one interrelate the sample's residual creep rate with its time for rupture (see Section IIIB) but also with its residual rate of failure propagation.

Beyond merely these considerations, Figure IIIA-18 displays the role of temperature in this affair (5). An increased temperature level forces an approach to the limiting condition:

$$[f(\log W^{-1})/f(\log v^{-1})]_{t_b} = \text{const.}$$

where these lines become parallel. When not parallel, one must use: const. $+ f(T_0 - T)$, where T_0 refers to the temperature for the parallel relationship. [It is interesting that a relationship of the approximate type, $t_b v^m$ = const., where $m = 1.0 \pm 0.2$, is valid for several alloys and pure metals (15).]

The remarkable simplicity of the general relationships implied in Figures IIIA-17 and IIIA-18, unambiguously establishes that the tensile microfailure is a *reliable* precursory aspect of static fatigue, regardless of how little or how much these microfailures contribute, of themselves, to the rupture of the sample.

IIIA2d. Microfailure Initiation and Healing

It has been emphasized by Newman and Wolock (31) that microfailures of the general variety herewith considered do not arise progressively from a submicroscopic level, but are generated *abruptly*. Those microfailures which could be visually discerned by their reflected light were the smallest, and no new ones were found upon improving the perspective of the affair by employing electron microscopic techniques. Thus if a 1–2 μ lower-level threshold can be universally employed for the microfailure length, the data of Regel in Figure IIIA-17 can be raised to the status of a description of microfailure initiation *per se*, rather than one of "apparent" microfailure initiation.

We have seen in connection with Figure IIIA-16, however, that when the microfailures' reflections heal totally out of view, the sample still retains a readily detectable weakening. Thus the topic of whether or not we can *see* these failures is not always the most pertinent consideration as regards progressive weakening or fatigue processes.

Fatigue processes not only *might* exist, but indeed *do* exist prior to, as well as after, the optical discernability of the microfailures. This point is amplified in the following where the history of progressive weakening and progressive strengthening can be traced in a fully *continuous* manner, with or without the presence of light reflections or "microfailures."

McAfee (28) has employed a strain-enhanced penetrant diffusion approach in the study of the precursory aspects of the static weakening of Pyrex; a sensibly brittle glass which did not "craze" during the experimental procedures, whether placed in biaxial tension, biaxial compression, or shear. By employing pressure differences of helium and other gases across the walls of a Pyrex bulb, the level of tangential, biaxial tensions or compressions could be controlled while the transport of the gases through the glass bulb could be followed simultaneously by mass spectrographic analyses. The observations in

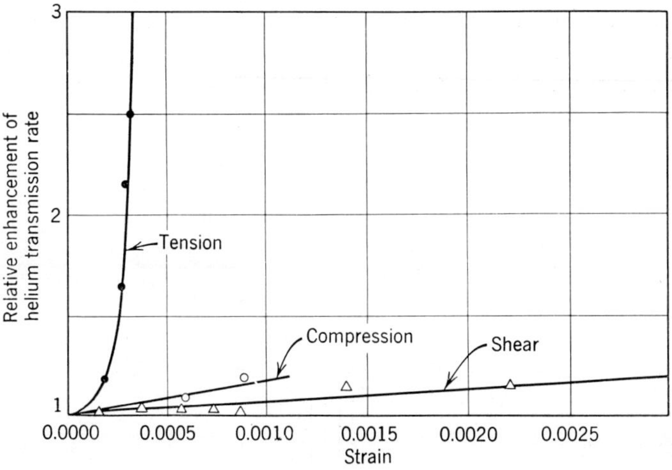

Fig. IIIA-19. Strain-enhanced, residual transport rates of helium through Pyrex glass. After McAfee (28). Courtesies of American Institute of Physics and K. B. McAfee, Jr.

shear were performed by an applied torque acting on a firmly secured glass membrane. Figure IIIA-19 shows the strain-induced enhancement of the helium flux under these procedures. An extremely careful analysis of his data led McAfee to the inescapable conclusion that the findings were generated by the *reversible* opening of *submicroscopic* fissures or channels in the random glass network.

Using much softer glasses, such as polystyrene, Rosen (35) repeated several of the general procedures of McAfee and found remarkably similar effects which differed only in the greater *order* of the enhancements observed, and not in their salient character. Crazing, however, occurred in many of the samples employed by Rosen.

The experimental arrangement (35) of the diffusion cell for both supported and unsupported films of highly viscoelastic glasses is shown in Figure IIIA-20. The unsupported films experienced small, but detectable, creep under the tangential, biaxial tensions generated by the pressure differences, while a recording system (36) was employed to ascertain the dynamic quantities of the gases being transmitted through the film during that biaxial creep. Figure IIIA-21 shows the role of time in the flux of helium and nitrogen through both supported and unsupported films. This order of enhancement

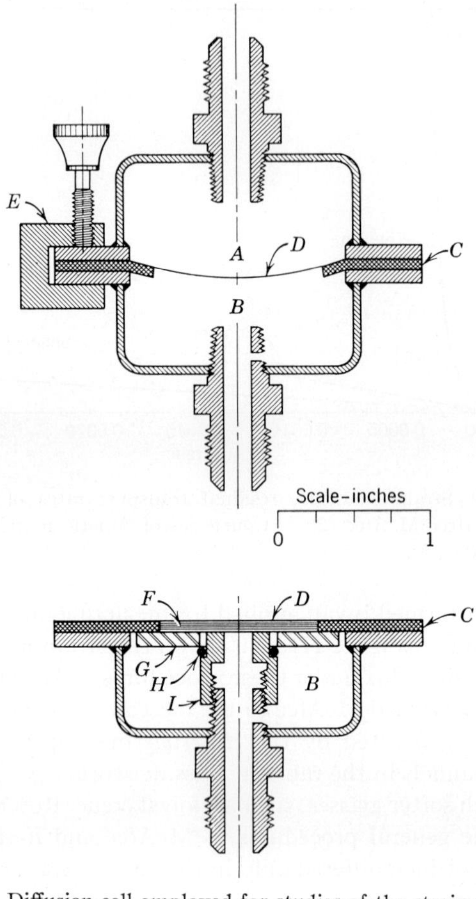

Fig. IIIA-20. Diffusion cell employed for studies of the strain-enhanced transport of gases through films of soft polymeric glasses. Above: unsupported film. Below: supported film, lower half of cell. (*A*) High pressure side; (*B*) vacuum or low pressure side; (*C*) Neoprene gasket; (*D*) circular film sample; (*E*) clamp (one of six); (*F*) layers of filter paper; (*G*) metallic washer; (*H*) O-ring; (*I*) metal fitting. Vacuum grease at interfaces. After Rosen (35).

was inexplicable on the basis of the creep-thinning of the film itself (35). Note that the enhanced transmission appears to be fully *continuous*, the whole curves being automatically recorded from the very start of the experiments. Figure IIIA-22 shows the general role of the level of pressure difference (hence the level of biaxial tension) on

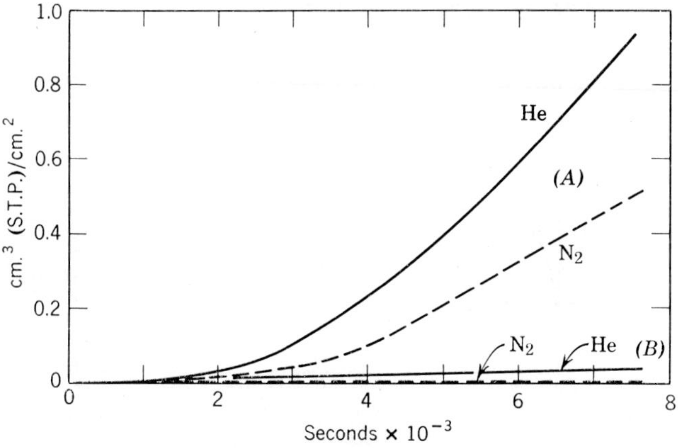

Fig. IIIA-21. Time-dependent, strain-enhancement of the transport rates of He and N_2 through 25 μ films of polystyrene at 26°C. and under a pressure difference of 12.6 cm. Hg. (*A*) Unsupported (biaxially tensed) films; (*B*) supported films. After Rosen (35).

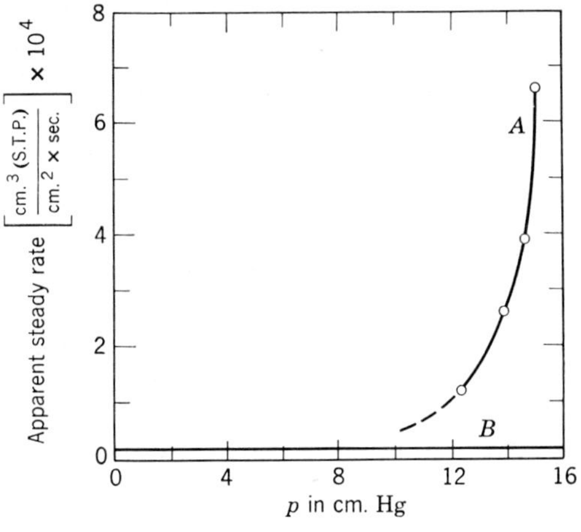

Fig. IIIA-22. Enhancement of the apparent steady rate of transport of He through 25 μ films of polystyrene at 27°C. under different pressure differences. (*A*) Unsupported film; (*B*) supported film. These films experienced 1.5 hr. exposure at each pressure difference shown. Pressures were increased successively. These are pressure-reduced, relative plots adjusted to a pressure difference of 1 cm. Hg. Hence, despite the increase in pressure, the line for the supported film is both horizontal and finite for all pressures shown. After Rosen (35).

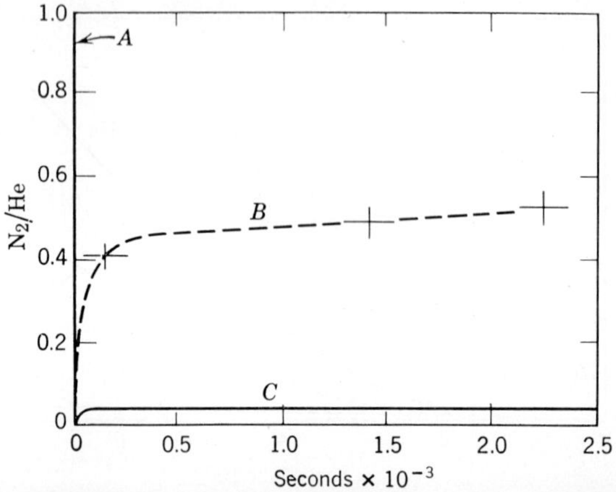

Fig. IIIA-23. Time dependency of the relative selectivities of He and N_2 transmissions through a 25 μ film of polystyrene during a strain-enhanced diffusion experiment at 26°C. under a total pressure difference of 13.0 cm. Hg in the N_2/He mixture. (*A*) Ratio of the partial pressures of N_2/He in the gas mixture placed on the high-pressure side of the diffusion cell. (*B*) Ratios of gases being transmitted, as analyzed in a mass spectrometer during the experimental intervals shown by the solid horizontal lines. (*C*) The results with the supported film (ratios calculated from the individual permeability coefficients which are unaffected by the presence of a non-interacting gas). If the film were significantly ruptured, the gas mixture indicated at Point *A* would have been transmitted through that path of least resistance. Hence the process was still highly selective as regards gas-type, despite the enhanced transmission rates observed in Figs. IIIA-21 and IIIA-22. This same point can be surmised from Fig. IIIA-21. After Rosen (35).

a single unsupported film progressively held at these increasing levels while the helium was being transmitted. Figure IIIA-23 shows that this general situation was not caused by capillary holes in the film, since the transmitted mixture of helium and nitrogen strongly favored the transmission of the helium; thus this was a highly *selective* process, and even on a possible "hole" level of a few Angstrom units. Figure IIIA-24 revealed that the healing of a damage which was rapidly induced, was sensibly complete after several weeks. This degree of healing, as grossly ascertained by the net transport of helium molecules, suggested that the process was gradually approaching perfection, even on a level which is competitive with the thermally activated transports of helium molecules. The conclusion that self-diffusion

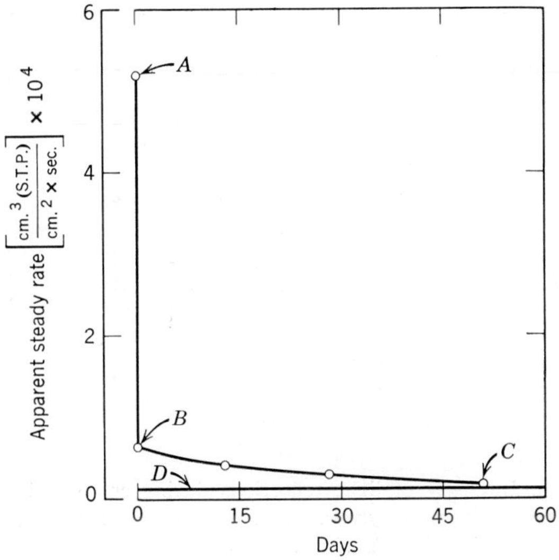

Fig. IIIA-24. Time-dependent reduction in the enhancement of He transmission rate through a 25 μ film of polystyrene at an average temperature of 27°C. subsequent to a brief period of damage. (*A*) Apparent steady transmission rate when the film was unsupported for 15 min. at a pressure difference of 37.5 cm. Hg. (*B*)-(*C*) Transmission rates while film was supported at this same pressure difference. (*B*) 1 hr. later. (*D*) Supported, fully undamaged film. After Rosen (35).

processes of the polystyrene are responsible for the latter stages of this healing process, becomes rather inescapable (35).

Most significantly, Rosen investigated the distribution of these submicrofissures over the surfaces of the films by placing a liquid at the lower-pressure side of the (inverted) membranes, and found that the thousands of tiny bubbles of the transmitted gases were sensibly homogeneously dispersed. Indeed, the film appeared to be failing *homogeneously*, not merely in one or two places where the stress was to be abnormally concentrated.

If one carefully compares the fuller findings of McAfee and Rosen, the question will arise as to whether these very similar situations in both Pyrex and polystyrene, i.e., in both a hard and a soft glass, cannot be reconciled within considerations of time or temperature, hence of the order of viscoelastic response experienced on a submicroscopic size level (35).

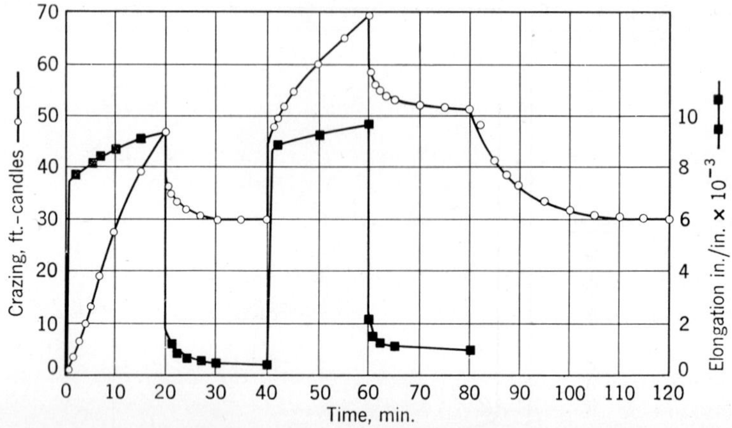

Fig. IIIA-25. Intensity of light reflections during creep and creep recovery in uniaxial tension with polystyrene. After Maxwell and Rahm (26). Courtesy of *Industrial and Engineering Chemistry*.

Figure IIIA-25 shows a few more of the fine studies of Maxwell and Rahm on light reflectance in polystyrene (26). Here the creep strains are plotted along with the crazing intensity for both loading and unloading in uniaxial tension. Note that the reflections recover (i.e., decrease) much in the manner of a retarded elastic deformation. The crazing is readily aggravated on repeated loadings. The final curve on the right shows a partial thermal recovery.

Judging from the studies of McAfee (28) and Rosen (35), as coupled with our knowledge of crazing, there apparently are *submicroscopic* fatigue processes occurring in viscoelastic bodies when they are subjected to static stressing and straining which are progressing in a *creep-like* manner, whether or not we can see them or readily detect their macroscopic consequences. Even if crazing is to be a *by-product* of these submicroscopic, creeplike processes, if we can ignore this crazing as a significant contributor to the rupture of the body, the body will probably rupture by these submicroscopic, creeplike events. Even if crazing *were* a significant contributor, since crazing also "creeps," the static fatigue of the body should still be macroscopic relaxation- or creep-dependent. Thus, it would appear that *we are forming a closer link between the net viscoelastic behavior of the macroscopic body and its time- and temperature-dependent static fatigue.* This situation becomes more compatible, the softer the body or the greater its viscoelastic contribution.

IIIA2e. Submicroscopic Analyses of Microfailures

Crazing failures in the highly viscoelastic glasses should not be confused with optical density fluctuations arising in the directions of maximized shear, such as with Lüder's lines (30). These "microfailures" are highly stress-dissipated tensile micro*fractures* which are profoundly encumbered, not merely with crack blunting, but also with micro*cavitation* processes. Crack blunting, while initially important, as seen by Figure IIIA-11, later becomes inconsequential in comparison to the strikingly large role for the liquidlike notion of cavitation. These points are clarified in the following.

The contention that "crazing" phenomena were not simply due to blunted cracks was first raised by Sauer, Marin, and Hsiao in 1949 (41). They spoke of a molecular orientation-type phenomenon which was to generate an anisotropically arranged material in the areas in, or near, the zone of reflectance of the otherwise isotropic glass. Photographs such as those shown in Figure IIIA-2 gradually appeared with much other direct and indirect evidence pointing more firmly to this same general contention (4,18,39,40). The notion that a crack was *also* there, however, persisted but with a somewhat decreasing emphasis, so that they have now come to be viewed as cracks or cracklike cavities which are *filled* with something that is anisotropic relative to the macroscopic body in which they reside. Perhaps the fullest reali-

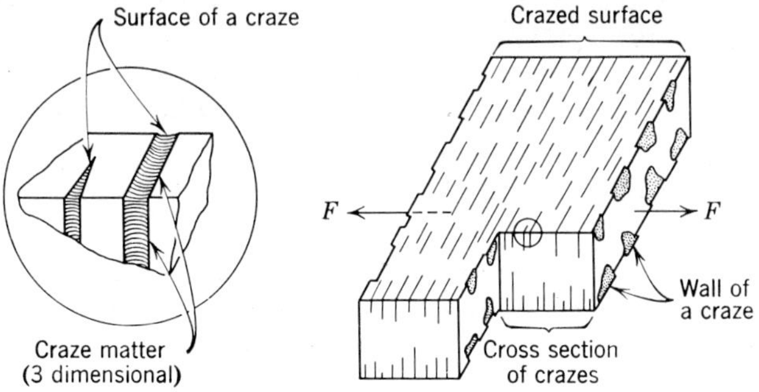

Fig. IIIA-26. An illustration defining the terms employed for a microfailured sample. After Spurr and Niegisch (44). Courtesy of W. D. Niegisch.

Fig. IIIA-27. Electron micrograph of a crazed surface (negative replica). After Spurr and Niegisch (44). Courtesy of O. K. Spurr.

zation of this view was due to the excellent studies by Spurr and Niegisch in 1962 (44), an overly brief review of which is presented below.

In reference to Figure IIIA-26, the three-dimensional aspects of the tensile microfailure can be designated as "crazed surface" (along its topographical features), "crazed cross section" (along its transverse depth), and "crazed wall" (along its longitudinal depth). Figures

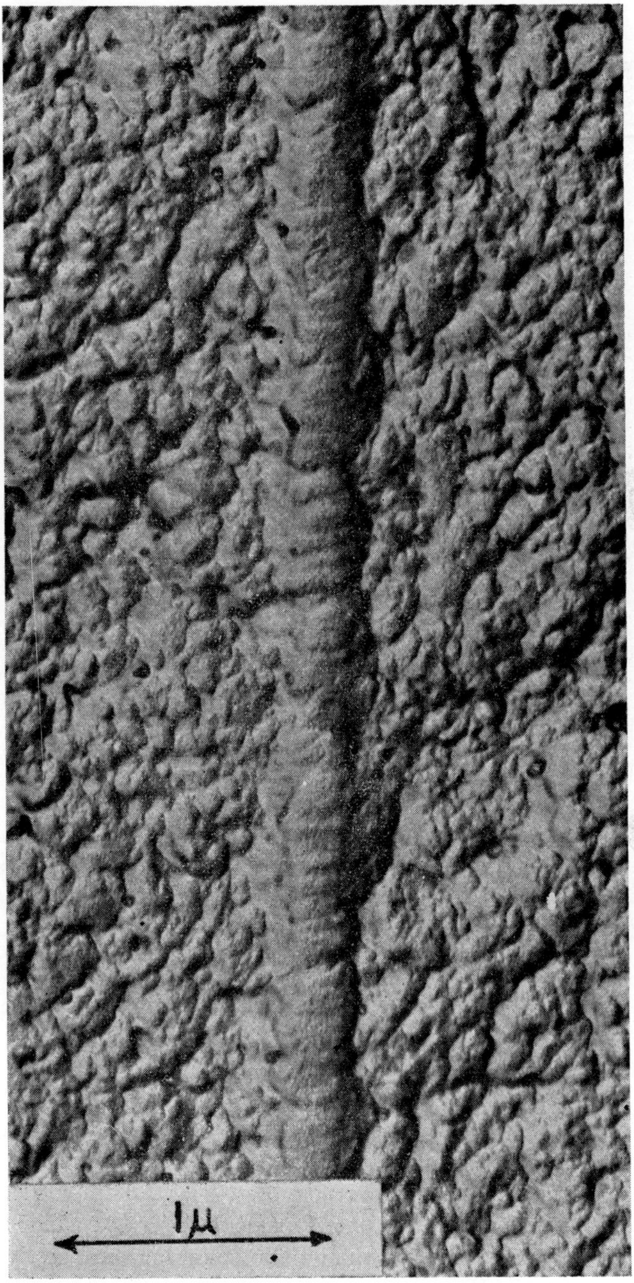

Fig. IIIA-28. Electron micrograph of a craze's cross section. After Spurr and Niegisch (44). Courtesy of W. D. Niegisch.

Fig. IIIA-29. Electron micrograph of a craze's wall. The "craze matter" is in the upper portion of this micrograph. After Spurr and Niegisch (44). Courtesy of O. K. Spurr.

IIIA-27 to -29 show electron micrographs of these three perspectives, but with different glassy materials and different patterns of loading, so that no cross comparisons are possible (44). All three views, however, show that there is a material inside of the cracks which is grossly different in replica texture than their bounding surfaces. This interesting material has been termed "craze matter" (44).

With the use of a diamond knife, microtomed sections were taken of the craze matter from the wider crazes such as obtained with polycarbonate at elevated temperatures (44). This matter is coherent, having a striated texture which runs along the craze's width, i.e., parallel to the initially localized tension which generated the craze itself. This point can also be appreciated by a study of Figure IIIA-29, where we are looking at the ends of the *bundles* of the craze matter.

X-ray microdiffraction studies of highly crazed regions of polycarbonate films produced patterns which supported the contention that there was an orientation of some type occurring along the direction of the applied tension. This observation, however, pertained to a general region of 50 μ in diameter in the film and not to the craze matter alone. Noncrazed regions did not show this effect (44).

Unlike brittle cracks, it was possible to produce crazes that crossed over each other in a perpendicular fashion without termination. Presumably the craze matter *supports* a subsequent crack propagating through it (44).

In a further study of craze matter, Kambour (21), using crazes aggravated by a localized solute softening of polycarbonate, produced unusually large crazes of 0.2 mm. width as generated by the tensile components in bending. Dimensional distortions along the transverse depth of the craze indicated that the craze matter had a 40% decrease in density relative to the uncrazed areas.

The coefficient of mutual interdiffusion of alcohol and craze matter was found to be rather large, $\sim 10^{-5}$ cm.^{2}/sec., by studying the movement of a critical refractive index boundary within the craze as the alcohol evaporated from the cavity. With some of the crazes, calculations of the refractive index of craze matter in the presence of alcohol indicated a 50–50 air–glass mixture (21).

Kambour finally indicated that if this were merely a cold-flowing, free of hydrostatic components in tension, the *rarefaction* aspect would be difficult to reconcile.

If the reader turns back to observe Figure IIIA-4 carefully, the hole formation in this rubber is seen to be not quite empty, but contains predominantly drawn strands in that cavitationlike process. A vaguely similar situation probably occurs in crazes, even if they are found on a more *microscopic* level in the *harder*, but highly viscoelastic glasses.

As pointed out by Newman and Wolock (31) in connection with

the electron microscope and light interference studies of crazed surfaces, frequently the apparent crack has "lips" projecting to an order of 100 mμ above the surface. This same situation can be viewed *macroscopically* if one slits a sheet of rubber and stretches it transversely to the direction of the slit. Here again, *we are forced to think of the behavior of a soft rubber rather than that of a relatively hard glass.*

IIIA3. Cavitation

IIIA3a. Cavities and Fractures in Liquids

A cavity in a large volume of liquid subjected to a hydrostatic tension p will expand or collapse, depending upon whether p is greater or less than the value (19)

$$p = (2\,\gamma/r) - p_v$$

where γ is the surface tension, r is the radius of a spherical cavity, and p_v is the vapor pressure. Experimental observations with liquids such as CCl_4 have approached nearly half of this theoretical strength (7), while most solids (cf. Berry, Section IIB) usually fall short of this criterion by two or three orders of magnitude.

It is clear that homogeneous liquids have a far better chance of approaching their theoretical strengths than do the inhomogeneous polycrystalline or semicrystalline bodies. The hard, homogeneous, supercooled liquids such as the silica glasses also become more predictable in the *solidlike* sense of failure. *Granting sufficient experimental patience,* however, the soft, homogeneous supercooled liquids should also be quite predictable, but only if we are to anticipate *liquidlike manifestations* of that time-dependent failure. Moreover, since failure processes necessarily focus upon microscopic or submicroscopic events, the order of experimental patience required for that predicability of fatigue in the highly viscoelastic glasses is far less than anticipated from the body's macroscopic mechanical behaviors.

"Flaws" in the sense of inclusions and exclusions become less important as the body is permitted the opportunity of becoming more liquidlike on a microscopic or submicroscopic level, and this situation adds to the body's more effectively homogeneous mechanical response. When we become more justified in *treating* a body as homogeneous, we simultaneously become more justified in *respecting* the macroscopic mechanical response of that body for its reflection of those averaged submicroscopic or even those averaged molecular processes involved in its breakdown.

In order to approach its theoretical strength, a homogeneous body which is subjected to an isotropic tension must fail *homogeneously*, not merely in one or two places where the stress is to be abnormally concentrated. Thus as we approach the homogeneous status, the population of breakdowns increase enormously under the dilative components of the stress system.

It has been shown that for studies employing stress wave loadings on liquids such as water, olive oil, glycerol, and syrup, whose viscosities range from approximately 10^{-2} to 10^3 poises (9), the tensile "strength" may, in practice, be represented by

$$P_c' = (3 \times 10^7)\eta^{1/5} \qquad \text{IIIA3-(1)}$$

in c.g.s. units, where the "strength" here relates to the onset of an easy detection of a turbidity due to an extremely large population of cavities in the transparent liquids.

It has been suggested by Kolsky (24) that the large population of cavities is replaced by a few cracks when eq. (1) is exceeded in a time $t \ll \tau$, so that the shear strain energy becomes efficiently stored at certain special localities rather than "relaxed" by forming new surfaces on a gross cavitation. He has succeeded in producing cracks in a supercooled liquid for which $\tau \cong 10$ msec. by detonations active during ~ 1 μsec.

As was mentioned in Subsec. IC2, p. 45, shear strain energy is partly stored when the body cannot efficiently dissipate the potential energy contribution of the applied shear. This situation is aggravated as $\tau = \eta/G$ is increased so as to conflict with our experimental patience. For the case of cavitation versus fracture, however, the τ to which we need refer is necessarily associated with a *microscopic* region in the body and should not be regarded in terms of the macroscopic value of τ or our patience for it.

The microscopic value of τ in a soft, noncrystalline polymeric body can be influenced by adiabatic heating, steric molecular factors such as polymer chain entanglement or crosslinks, the existence of "flaws" in the sense of foreign matter or discontinuities, and other contributions, such as a flow-induced hardening of that localized area.

If the molecules of the liquid are polyatomic, a "fracture" should not be regarded as a running cavity, if this term "cavity" is to imply new

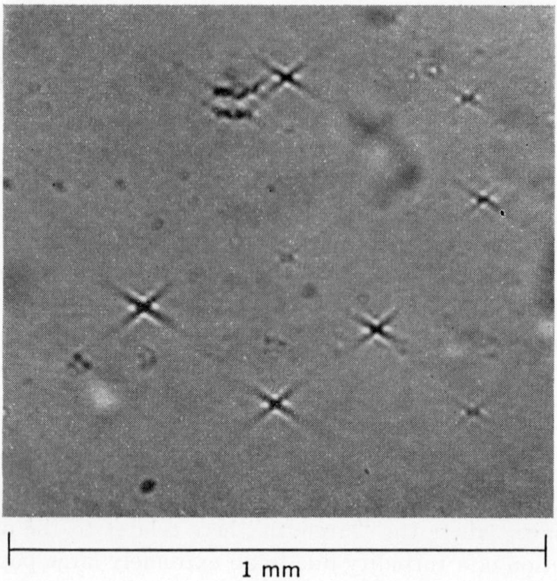

(*a*)

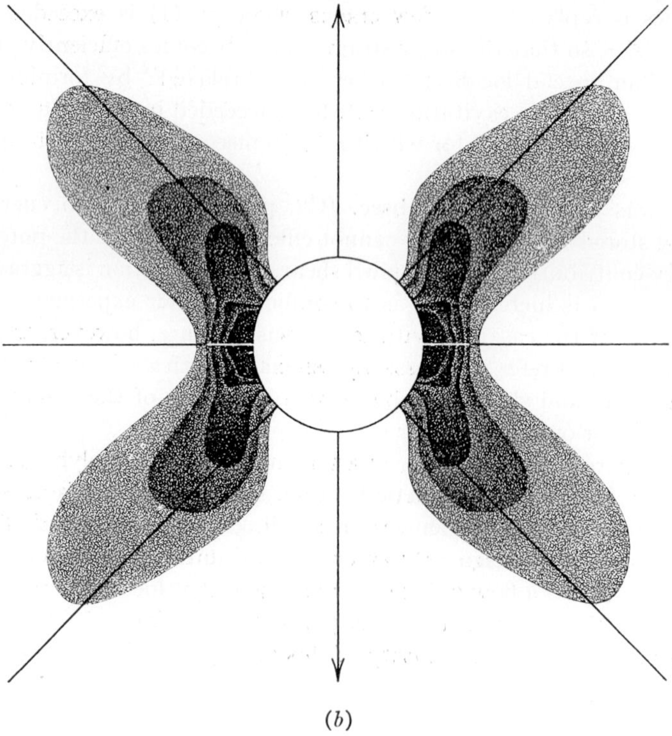

(*b*)

surfaces solely generated by shear stress-biased, intermolecular transports. *Intra*molecular dissociations, i.e., primary bond scission, must also play a key role. If the molecules of the fluid are highly asymmetric, the τ of pertinence as regards primary bond scission is related to the time for the molecule to transport itself into a certain configuration or location so that strain energy is not efficiently stored in it.

IIIA3b. Localized Strain-Hardening in Noncrystalline Polymeric Bodies

Hsiao, in Section IVC, treats the role of "chain orientation" in strengthening polymeric bodies. As emphasized in Subsection IC5d, p. 98, there is a mesomorphic state which can be generated when dealing with bodies containing large asymmetric molecules. During a pronounced inelastic shear, by a mechanism of chain segmental self-diffusion, there can be a net alignment of polymeric chains in a direction that is normal to the shear gradient. This can be accomplished concurrently with a "cold-flowing" of the body below T_g or a "hot-stretching" above T_g, even when the final substance has no tangible trace of crystallinity. A *birefringence,* however, is generated during the process, and this is a sign of *a new state of aggregation* of noncrystalline matter which is due to this chain orientation (see Subsection IIIA4b). A maximum strengthening is produced here by a factor which varies from case to case in the approximate range 2–6; hence it unambiguously is a source of considerable strain-hardening, even though fully noncrystalline in origin. Let us consider the extreme pertinence of this contribution.

Figure IIIA-30*a*, after Newman and Wolock (31), shows a photomicrograph of some former areas of stress concentration, i.e., the

Fig. IIIA-30. (*a*) Photomicrograph of the surface of a specimen of poly(methyl alpha-chloroacrylate) that had been loaded to failure. After Newman and Wolock (31). Courtesy of S. B. Newman. (*b*) Schematic drawing of the general areas to be strain-hardened by macromolecular orientation on inelastic shear in the vicinity of a planar hole. The successively contoured lines represent the fixed-incrementally spaced enhancements of the maximum shearing stress as they close-in upon the hole. The darker regions thus crudely represent those general areas where the birefringence-generating macromolecular orientation would be pronounced on a uniformly applied, uniaxial tension field acting upon a highly viscoelastic glassy polymer. The macromolecules would tend to become oriented normal to the gradient of this shear.

Lüder's lines, demarking the inelastically generated birefringence within the general areas of maximized shear, persist long after the sample is ruptured. It is quite safe to presume that there is a macromolecular orientation-induced strain-hardening existing in the immediate vicinity of these flaws, which generates this Lüder birefringence, as very approximately illustrated by the present author in the schematic drawing of Figure IIIA-30*b*.

(This general situation is important to remember when Radok discusses the role of stress concentration in linear viscoelastic bodies in Section VC. A more realistic approach in noncrystalline polymeric bodies would need to superimpose this variety, if not also other varieties, of strain-hardening. This problem, however, obviously involves highly "nonlinear" viscoelastic contributions; see Section VD.)

Thus, *areas of stress concentration in noncrystalline polymeric bodies frequently are sources of strengthening as well as weakening.* There probably are two extremes to be balanced against each other. If the τ which is associated with localized chain realignments is large in comparison to the time of the experiment, then the area of stress concentration is more efficient as regards fracture. Yet, if the chain can realign so as to reinforce that area (much like the spontaneous healing of a wound), then the area can be made less effective.

Note that while this is a flow contribution in the vicinity of a stress-raiser, it is not a fully dissipative process. These chains, once aligned, are storers of strain energy themselves. Thus one can imagine a transient case of chain alignment where the failure is initiated near a zone of maximized shear, since this area is neither overly strong nor overly weak.

IIIA3c. Initiation and Propagation of Strain-Hardened Cavities

Let us consider the situation implied by the crazing marks of Figure IIIA-31. In the photomicrograph of Part *A* of this figure, the tensile microfailures appear to be sensibly stabilized once generated and terminated in the regions of maximized inelastic shear as denoted by Lüder's lines. Only rarely have these microfailures propagated beyond the reasonably *unitized* distance. It appears that *it is far easier to generate a new failure than to propagate an old one.*

There are probably two major components of this microfailure stabilization: (*a*) failure blunting and (*b*) strain-hardening; in the present case, these are not mutually exclusive contributions. Moreover, the stabilization of the failure over such short distances would act to generate a stress redistribution toward a more hydrostatic tensile straining of the initial failure.

We are now in a better position for an attempt to put the pieces of this complex puzzle together. It tentatively appears that a genuine micro*fracture* is abruptly generated at a region of shear strain energy storage, but is stopped or retarded long enough to gather hydrostatic contributions, so that the crack is torn in both simple shear and cavitationlike events (Fig. IIIA-31*b*). The opposite walls of the crack become cold-drawn* so that the net effect is *a microfracture which is partly filled with its own old wall* and partly empty due to (*a*) the enhanced separation of the fracture surfaces in the sense of a "lipping," (*b*) the liquidlike cavitation contributions from flow processes in the sense of a bundling of the chains on cooperative interchain slip, and (*c*) primary chain scission; the last two generating entirely new surfaces of their own.

Propagation, here, might well be viewed as a *repeated initiation* which statistically favors certain, once-initiated, cracklike cavities. Note carefully in Figure IIIA-31*a* that there only are a very few cases of a failure-extension, and even most of these few are not actually continuations of the preceding "segment," but slightly displaced from the direction of maximized inelastic shear. These principally sheared areas necessarily are stronger. Thus the propagation would require a reinitiation in a zone which, while strain-hardened enough to support a significant storage of shear strain energy, is weaker than the original, now hardened, crack-front. If the reader now turns back to the photos of Figure IIIA-2, the general situation, as represented schematically in Figure IIIA-31*c*, can be re-appreciated if we assume that this apparent ropelike "twisting" is actually a series of staccato-like cracking, cavitation, and cold-drawing processes which are *impeded* at the maximized shearing of the old crack-fronts. Indeed, this interrupted and randomly erratic style of crack propagation was observed by the present author to take place in those samples shown in Figure IIIA-2.

In terms of crack propagation kinetics, a situation of this general type suggests that the old cavity has little influence on the new one

* "Cold-flown" would be more correct.

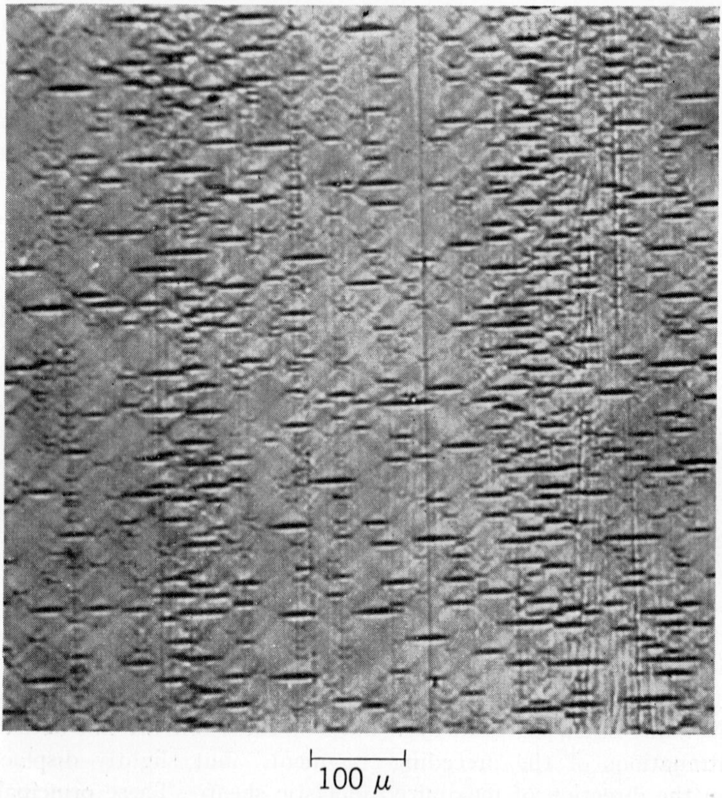

(a)

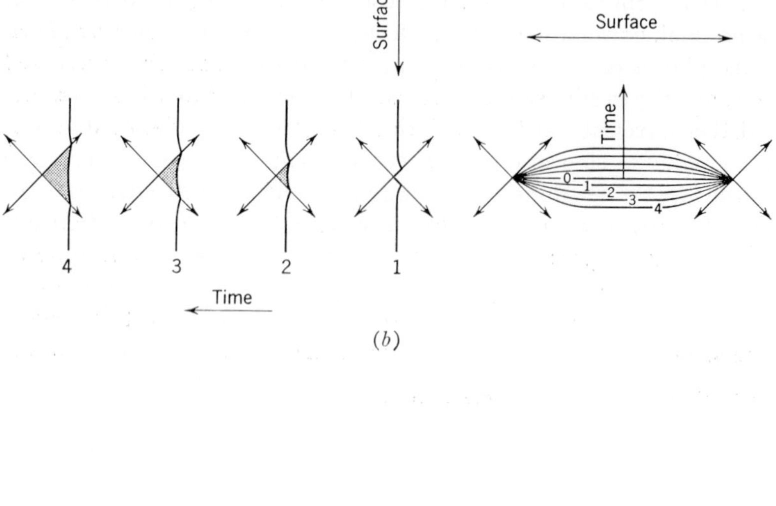

(b)

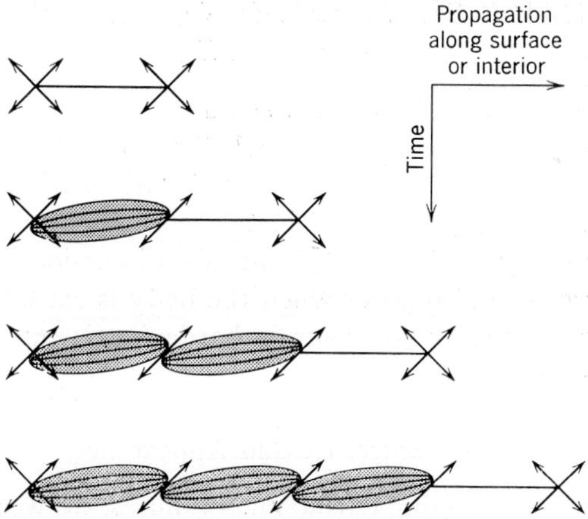

(*c*)

Fig. IIIA-31. (*a*) Photomicrograph of a crazed specimen of poly(methyl methacrylate) in uniaxial tension (vertical). After Newman and Wolock (31). Courtesy of S. B. Newman. (*b*) Cavitational enlargement of a stabilized microfracture; schematic. Cf. Fig. IIIA-4*b*. (*c*) Propagation sequences in the growth of cavitated microfractures; schematic. Cf. Figs. IIIA-2 and IIIA-28.

which is being "segmented" to it. The cavity grows as stress is redistributed throughout the whole sample by *macroscopically* reflected inelastic processes. Hence *macroscopic flow rate and microfailure growth rate would exhibit some direct relationship to each other.*

Various other aspects of tensile microfailures are treated at several places within and following the next discussion concerning a slightly different situation also met in highly viscoelastic bodies. It should be clear, however, that we have attempted, within the above, to characterize an extreme case so as to caricature the general situation with those particular varieties of tensile microfailures in polymeric bodies which exhibit a decelerating rate of propagation. Moreover, it also should be clear that the transition between formal microfracture and cracklike cavities, both of which can reflect light, is not discontinuous in conceptual approach, but should rather be viewed as extremes of the continuous spectrum of possible tensile microfailure types in glassy polymeric bodies.

IIIA3d. Submicrocavitation in Noncrystalline Polymeric Bodies

It would be incorrect, however, to assume that cavitation processes in the highly viscoelastic glasses are wholly confined to the consideration of tensile microfailures. A "purer" variety of cavitation, i.e., one which does not "propagate" in the sense of a running cracklike cavity (as readily appreciated from the discussion in Subsection IIIA3a, above) can also occur when the body is considerably softer, the stressing slower, or when strain-hardening is more effective in restraining its progress.

(1) Occurrence and Appearance

When we were discussing tensile microfailures, we were continually referring to a species of progressive failure which readily reflected light in these transparent bodies. A distinctly different optical phenomenon frequently occurs when some of the extremely soft glasses (or perhaps "glassy leathers") are made to experience a pronounced inelastic deformation. Here the body becomes cloudy, even to the extent of an opaque whitening (16,37,46,48). This scattering of light is frequently confused with that obtainable by the generation of a randomized assembly of spherulitic crystalline growths on the pronounced cold-drawing of an initially transparent, soft glass or leather (32).

This diffusion of light is not nearly as "dazzling" to our metaphysical senses, but is actually quite ugly in otherwise transparent bodies; frequently being termed "blush" or by other expressions which connote a colloidlike turbidity.

Even if the blushed sample is free from signs of crystallizations of any kind, this phenomenon is frequently complicated by the production of large and sensibly brittle-like fractures which travel their superficial paths through the whitened and obviously weakened substrate.

The systematic study of blushing is tentatively not nearly as far advanced as that of crazing. It appears that the chemical engineer is more interested in avoiding it than in understanding it, since this phenomena clearly ruins the piece by either its loss in transparency or by the pronounced weakening which accompanies its occurrence. Crazing, however, is far more interesting from an engineer's point of view, since it is more visually intriguing than unattractive as regards a lowered transparency, and does not destroy the piece for most of its purposes. Thus only a few systematic studies of blushing have been made, and only two of these are treated in the following.

(Whenever blushing is generated in the presence of a soluble fluid, the contention arises that the phenomenon is caused by the localized aggregation of the fluid in the body. It is unclear, however, as to how this fluid is to aggregate in specific domains which are large enough for a light-scattering contribution without the body being at least somewhat discontinuous in those isolated domains.)

Aside from a whitening and a pronounced weakening, blush is also not as fickle as crazing; once it decides to develop, it simply stays without offering signs of a serious attempt at a healing (37). As with any ultimate mechanical phenomena, however, blush is indeed highly fickle regarding the matters of state of stress, rate of strain, and the localized conditions of the body under examination. Like crazing, blush will usually first appear at the surface of a residually unstressed or unstrained sample when acted upon by a sensibly uniform tensile field, uncomplicated by thermal or solute stresses; later to be discovered in the sample's interior. Frequently, however, superficial fractures will relieve some of the surface stresses, and then the blush will be first found in the vicinity of the fracture-front or at other equally obvious regions of stress concentration prior to its more uniformly internal appearance. Essentially nothing, however, is tentatively known about its kinetics of appearance or the role of temperature or stress level in those kinetics. (The present author, however, has studied the diffusion-controlled migration of blushing with the transient solute stresses generated by the transient concentration distribution of acetone during a deswelling of polystyrene films, and this will be published elsewhere.)

(2) Accelerated Tensile Creep

In those substances which exhibit a crazing on creep in uniaxial tension, this tensile microfailure is a precursory aspect of that body's static fatigue or "delayed brittle" rupture. While it is true that these microfailures can generate a subtle enhancement of the "nonlinear" viscoelastic contribution of the body (see Figure IIIA-14) they do not contribute to a "delayed ductile" rupture or "accelerated tensile creep" which is terminated by a *necking* rupture of the sample. When

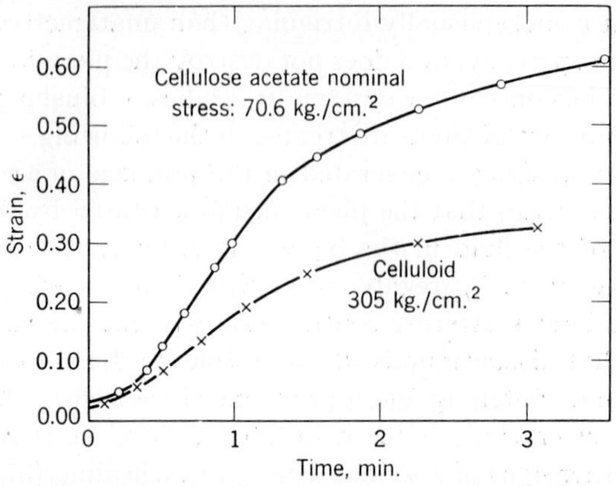

Fig. IIIA-32. Accelerated creep in uniaxial tension with cellulose acetate and cellulose nitrate. After Haward (16).

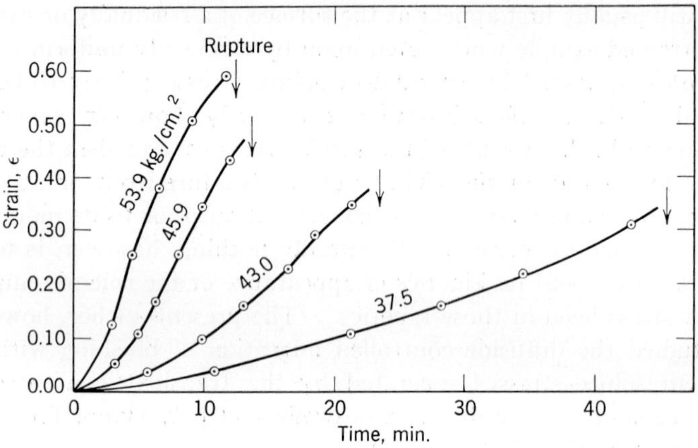

Fig. IIIA-33. Time-dependent portions of creep curves of cellulose acetate samples at different levels of nominal stress in uniaxial tension. After Haward (16).

a delayed ductile rupture is achieved in those soft *metals* which exhibit a pronounced acceleration in tensile creep, readily detectable cavities and a concurrent density-lowering occurs in the necked zone (12,22). When a glassy-leathery, transparent *polymeric body* experiences a delayed ductile failure, the necked region whitens or blushes (16,17) and the density of that region is lowered (48). The cavities in the latter

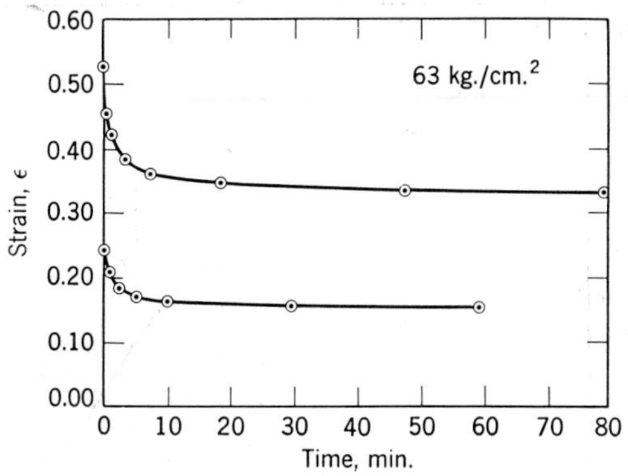

Fig. IIIA-34. Creep recovery and residual strains in cellulose acetate samples which had experienced accelerated creep for different periods. After Haward (16).

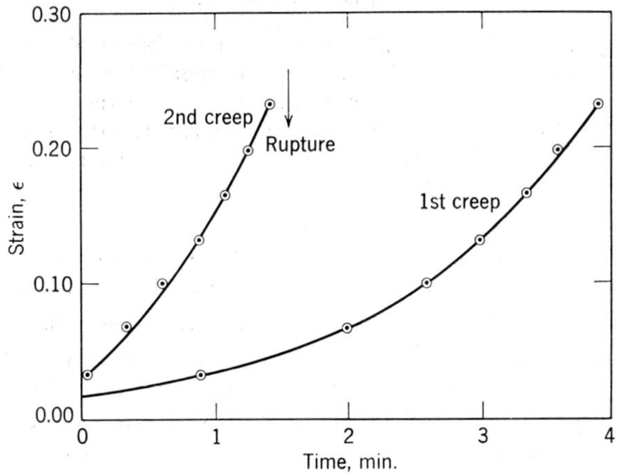

Fig. IIIA-35. Progressive weakening during the accelerated creep of cellulose acetate. After the first creep experiment, the sample was taken above its T_g, yet produced a second creep (under the same nominal stress and temperature) which was more accelerated. After Haward (16).

case are readily sustained at a submicroscopic level; rarely, if ever, detectable on an optical microscopical examination of the piece.

Figure IIIA-32, after Haward (16), shows the accelerated creep of cellulose acetate and cellulose nitrate films of 0.05 cm. thickness under two levels of nominal stress in uniaxial tension. These soft, hydro-

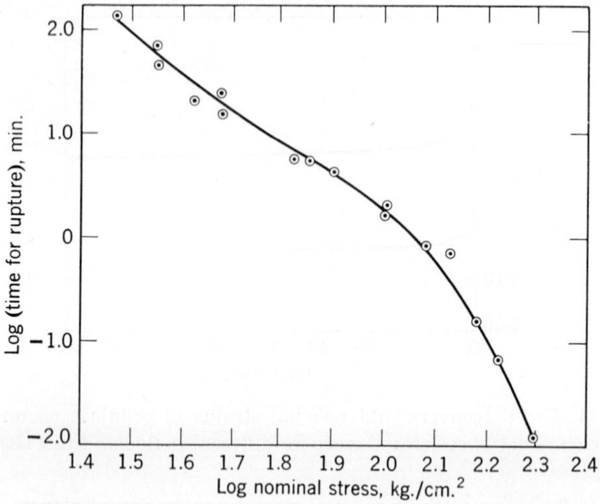

Fig. IIIA-36. Time for creep failures as a function of nominal stress in uniaxial tension with cellulose acetate. Accelerated creep rupture is more readily achieved at the lower stress levels and longer times. Thus we are viewing the transition from accelerated creep rupture to short-termed creep rupture here. The upper portion of this curve would shift to the right if the "true stress" were employed. After Haward (16).

philic glass samples were further softened by a saturation with water vapor at 33% R.H. ambient at 26°C. Figure IIIA-33 shows a set of four accelerated creeps with cellulose acetate where the time-independent, or apparent hookean portions of the curves are removed so that only the time-dependent extensions are given. By a shifting along the time axis, Haward found that these curves were superimposable to a limited degree. Just as with delayed "brittle" ruptures, the higher stress level generates a quicker rupture here. Figure IIIA-34 shows the high level of "permanent set" or residual strain achieved during creep experiments of this type. The sample's T_g was very close to room temperature, so that when the body was sparingly heated to 70°C. its original shape and appearance was sensibly reattained within a few minutes (16).

Progressive weakening of a sample at the higher extensions is shown in Figure IIIA-35. After the first extension at a nominal stress of 60 kg./cm.^{2}, the sample was brought to 65°C. for 15 min. so that the residual elongation was only 1%. On the second extension under the same load and ambient conditions, the sample's permanent weakness was reflected in a fast rupture at a far greater rate of creep acceleration. This situation became so intriguing that Haward studied a whole series of specimens and trials in experiments, such as shown in Figure IIIA-35, and constructed a table which revealed that it did not greatly matter how long a period of time was employed for the first extension, since $t_1 + t_2 = 6.5$ min. with an average deviation from the mean, ± 0.5 min. In other words, *the damage was sensibly permanent and cumulative*, and the sample's life in this type of experimental procedure was *predestined* to be approximately 6.5 min.; the heating between the first and second extensions did not repair that time-cumulative weakening (a situation which was also independent of the totaled or maximum creep strain).

As he proceeded to plot this lifetime in creep at *different* levels of nominal stress at room temperatures, Haward found the relation shown in Figure IIIA-36. He mentioned that "as the film stretches it gradually becomes white and opaque, a fact that could be explained if the material were becoming inhomogeneous."

(3) Analyses of Submicrocavities

Submicroporous turbidities have been produced in cellulose nitrate films during studies at 110–130°C. at nominal stress levels in uniaxial creep, or on transient stressing at 1–3 kg./mm.^{2} by Zhurkov, Marikhin, and Slutsker (48). Analyses were performed with both light scattering and small-angle x-ray scattering, together with dilatometric measurements. Glasses such as polystyrene, poly(vinyl chloride), and cellulose di- and triacetate could be made turbid under selective circumstances of loading at 60–80°C.

Scattering of light was studied by three methods: (*a*) the scattering indicatrix method, (*b*) measurement of the angular dependence of the degree of polarization of scattered light, and (*c*) measurement of transparency, i.e., of the dependence of the scattering coefficient on the wavelength of light. The first two optical methods mentioned above established that the turbidity was due to formation of inhomogeneities (i.e., regions with a refractive index different than that of the surrounding medium) with dimensions of several hundred Angstrom units. Neither of these two methods, however, yielded any information on the nature of the scattering regions. Hence, the "transparency" approach was needed. They concluded that the turbidity was due to formation of submicroscopic discontinuities in the form of voids. The radii of these cavities were ~1000 A. in cellulose nitrate, ~640 A. in poly(methyl methacrylate), and were found to be ~375 A. in some of the other glasses. Their enormous concentrations were found to be 10^{11}–10^{12}/cm.^{3}, the lower values pertaining to the larger voids (48).

This general situation was closely confirmed with the small-angle x-ray scattering approach. The concentration data was based upon the *assumption* of voids which was later confirmed, however, by direct measurements of density changes of the order, $-\Delta d/d \cong 10^{-3}$–$10^{-2}$.

Here again, as we saw with crazing, we have an extremely large number of flaws, except that their numbers in this case are relatively enormous. In addition, stabilization appears to become both more pronounced and more localized as the polymeric glass becomes increasingly soft. With these very large strain levels we can anticipate a pronounced noncrystalline strain-hardening contribution. (Note both Figure IIIA-30*b* and the severely strained areas at the potential cavity-fronts in Figure IIIA-4*b*.)

The inability of Haward's samples to regain their strength on an "annealing" (i.e., on taking the sample above its T_g) suggests that the uncrystallized molecules themselves were damaged. Thus this would not be a *perfect* liquidlike cavitation where self-diffusion processes can heal all damage in time or in temperature. Despite this, these cavities apparently cannot transport themselves so as to coalesce into a readily visible hole. They are pinned-down by something which also prevents a gross closing-up of these cavities, short of measures such as taking them above T_g.

Thus one probably can anticipate here both liquidlike cavitation and polymer chain scission as highly complicated by an overpowering role for a strain-hardening of some type which stops their growth, prevents their coalescence, and keeps the cavities open after the stress is removed and which thus strongly contributes to the residual strain in the body. Cold-flowing within the cavity itself would not appear to be as pertinent as a cold-flowing around the cavity in this particular case.

The term "cold-flowing," however, loses most of its colloquial meaning here since *some of these cavities are small by comparison to the macromolecules involved when fully extended.* Thus we are unavoidably thrown into a rather selective view of *chain stretching* and consequent chain scission. Perhaps we are speaking here of the collective scission and slip of isolated groupings of chains. Moreover, with concentrations of 10^{12}, i.e., a trillion, cavities for each cubic centimeter and only a few hundred Angstrom units in the radius of each, the question immediately arises: "What is *concentrating* the stress here?" Perhaps it is the macromolecules themselves!

Moreover, are we actually viewing here an exaggerated case of what is progressively happening in the harder glasses on a sub-light-scattering level? Indeed, *is this the clue we need for a macroscopically homogeneous variety of breakdown in homogeneously strained viscoelastic bodies?* As pointed out previously, this kind of situation seems to be what is required in interpreting the observations of McAfee and Rosen.

IIIA4. Macromolecular Contributions

Thus with the softer of the polymeric glasses, we cannot avoid the increasingly important role of the chainlike molecule in the progressive weakening of the body. This is clear since the enhanced softening of the body gradually forces our attentions toward rheological, or even hydrodynamic, processes and this general situation places a far greater order of sensitivity on the size and shape of the entities being transported. If these entities have a degree of *certainty* in their configurations, such as at a point of "chain entanglement" or at a cross-link in a fully noncrystalline and macroscopically isotropic body, their localized resistance to cavitational flow, hence their localized storage of strain energy in shear, is reflected in both a macroscopic contribution to viscoelasticity (see Subsec. IC4, p. 68) and in *a remarkably localized contribution to stress concentration.* (This point is further clarified by Landel and Fedors, Section IIIB.)

IIIA4a. Chain Scission

Now, if one were to attempt to sever chainlike molecules, he might choose to undergo a cutting of the sample with a razor's edge. This rather direct approach has its severe limitations, however, since in order to reduce the average chain length to half of its original value, each molecule would have to be cut on an average of one time; a thing which would demand profound orders of both patience and dedication to this destructive mission. More effective approaches to this challenging problem are discussed by Jellinek in Section IVE, where the physico-chemical methods such as irradiation and thermally induced chemical changes are employed. Any process that employs a sufficiently high level of extremely localized shear which is experienced in a period $t \ll \tau$, will also be effective. Thus the shearing components of ultrasonic waves serve very nicely here (20), while even creep at sufficiently large rates of flow, as seen from the above discussion, also may serve this end. To do this rather thoroughly, however, one should employ techniques based on a rapid dynamic shearing since the chains missed on the first shear might well be severed on the next, etc.

This *dynamic fatiguing* of amorphous polymer bodies, in which there is some direct evidence for the scission of polymer chains or an otherwise inexplicable, profound loss in strength, has been termed "mastication." On experiencing this, the noncrystalline body usually exhibits a greatly shortened relaxation time or other indications of a more pronounced flow contribution. This need not always be the case, however, since the scission of primary bonds generates free radicals which may subsequently *repolymerize* to form a different macromolecular system on these secondary chemical changes. In fact, this mechanically-induced type of repolymerization in the solid body is frequently employed in the commercial preparation of certain new varieties of polymeric substances, especially the block copolymers (11).

The measurement of the average chain size of a polymeric body constitutes a large sector of the domain of the polymer physicist or chemist (25). If the chains are linear or only mildly branched they usually can be dissolved to form a "macromolecular colloid." In the limit of infinite dilution, the thermodynamic and hydrodynamic solution methods yield various kinds of average molecular weights, depending upon the order of sensitivity of the method as regards a certain order of macromolecular size in a given size distribution (or size and shape in the case of the hydrodynamic methods).

As shown in Figure IIIA-37, the role of mastication might be followed by intermittently dissolving an aliquot of the sample and measuring its solution viscosity (11). The "limiting" term here implies an extrapolation of a certain series of solution viscosities towards infinite dilutions. Thus this limiting solution viscosity $[\eta]$ would have some relation to molecular weight M such as $[\eta] = KM^a$, where the two constants might be evaluated (25) from a previous osmotic pres-

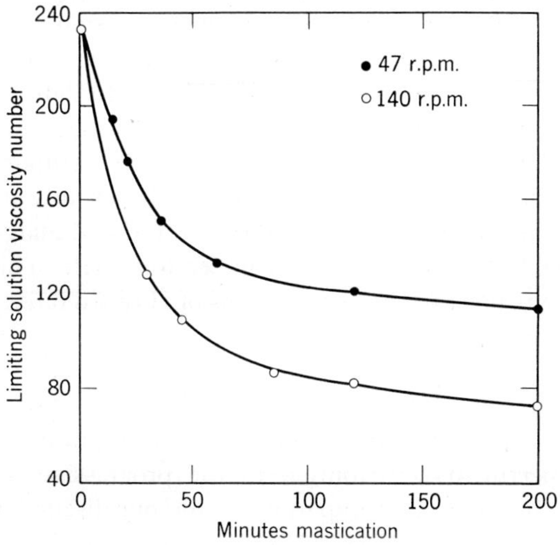

Fig. IIIA-37. Lowering of the solution viscosity of polystyrene as progressively worked at different rates at 130°C. See discussion in text. After Ceresa (11).

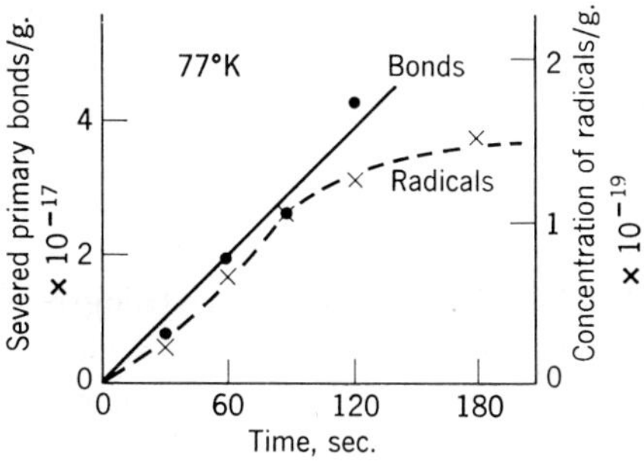

Fig. IIIA-38. The progressive destruction of primary bonds and the generation of free radicals in poly(methyl methacrylate) by action of a vibratory mill. See text. After Butyagin (10).

sure or light-scattering approach to determining the number of particles present in the colloid, etc. In any event, this figure shows that the solution viscosity, at least, is sharply lowered on a dynamic shearing of polystyrene; the specific mechanical method being best described in the original paper (11).

Figure IIIA-38 shows the kinetics of chain scission and free radical generation from poly(methyl methacrylate) in a vibratory mill (10). The solution viscosity approach was employed to detect the shift in the distribution of chain sizes, while electron paramagnetic resonance was used in the detection of the free radicals. It is important to note that the number of primary bonds severed is sensibly linear in this dynamic fatigue, since this suggests a cumulative damage law which will be treated by Landel and Fedors in Section IIIB. This consideration also reflects back to the observations of Haward on the additivity factor in the progressive weakening of his samples.

IIIA4b. Chain Orientation

That the localized redistribution of chain configurations in a shear field has a pertinent contribution in the progressive weakening of polymeric bodies was made apparent during our discussions of microfailures and cavitation above. However, it is possible to bring the entire polymeric body into this directionally biased or anisotropic status of chain configuration by a uniaxial flowing of the body near or above its T_g by a remarkably pronounced inelastic strain followed by a cooling which makes this strain residual. Such anisotropic, transparent, noncrystalline bodies are birefringent as regards the different velocities of light of a given wavelength transmitted parallel and perpendicular to the plane of the reorganized chains. As discussed in detail by Hsiao, Section IVC, this frequently produces a considerable strain-hardening.

The salient roles for birefringence and strength are illustrated in Figure IIIA-39 for samples of polystyrene which were enormously elongated when above their T_g (3). It is clear, however, that the order of chain orientation achieved depends on both the strain rate and the rate of the subsequent cooling to room temperatures, as well as the absolute residual elongation.

A net chain orientation in the noncrystalline polymeric body, however, does not greatly affect a subsequently measured T_g, its density, coefficient of thermal expansion, or even its velocity of waves in the ultrasonic range, relative to those of the truly amorphous or isotropic body (3); yet its tensile strength and ultimate elongation are greatly altered. As seen from Figure IIIA-39, the body becomes weaker in uniaxial tension when the orientation of the sample is positioned perpendicular to the tensile field. This presumably implies that it is far easier for a tensile fracture to be initiated, or perhaps propagated, parallel to the direction of the net chain alignment than perpendicular to it. It is clear that the body would also be mechanically anisotropic with respect to brittle-ductile contributions, relaxation times, etc. It is also clear that a uniaxial chain orientation would affect the precursory aspects of their time-dependent strength, i.e., their fatigue processes. This has been frequently shown (18,40).

As mentioned in Section IIC, it is also possible to generate anisotropic polymeric glasses by a *biaxial* hot-stretching of a film or plate

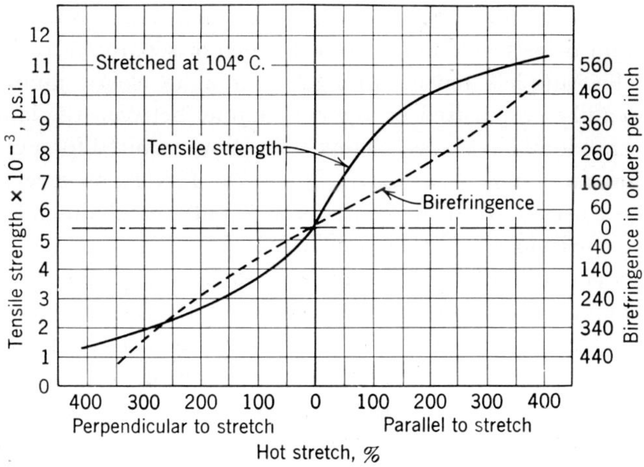

Fig. IIIA-39. Effect of per cent elongation of polystyrene above its T_g on the tensile strength and birefringence of the glass at room temperature. After Bailey (3).

such as by drum-head, vacuum procedures (23). The chains' situations are somewhat different here, however, since while the average end-to-end chain length is extended in the plane of the residual biaxial elongation, there is no directionally preferred orientation of the chains in that plane. There is, however, an anisotropy developed perpendicular and parallel to the planes of that stretching; the molecules being drawn into laminarlike strata in those parallel planes. If tensile specimens are cut from a plate of this type, as seen in Figure IIC-23, p. 257, a uniaxial tension generates a laminarlike fracture surface. But despite the difference in the *appearance* of the fracture surface, the uniaxial tensile *strength* of the biaxially stretched sample is sensibly unchanged (23).

A peculiar situation occurs here concerning crazing, however, since these bodies are remarkably *stable* as regards microfailure generation and propagation (23).

Presumably, a cavity is more difficult to generate if it must be confronted by chains which are more *elongated* (even if not more aligned) in its paths of propagation. Perhaps the subsequent strain-hardening of the cavity is more easily achieved if *some* of the molecules are already stretched in the proper sense for that strain-hardening; less *flow* orientation being required for those certain chains, and more flow required for others which need to be aligned or separated by a cavitation. The sample's tensile strength is a more elastic affair which can be readily propagated in a *brittle* manner between the laminated strata of randomly extended chains. The flow required for a cracklike cavitation is a highly selective *rheological* factor, however, which would be sensitive to the average distance between chain-ends during their extremely localized realignments.

In this connection, it is perhaps pertinent to mention that those glasses which are composed of *crosslinked* molecules may also have a moderately high resistance against crazing (39). Here, too, we can expect a considerable difficulty in moving certain of the subchains (i.e., chainlets between juncture points) into new positions for a cavitational flow, while others can perform the strain-hardening function in their unmodified positions.

Thus it would appear that the more difficult it is to transport *some* of the chains, whether they are tied down to certain juncture points or whether they are unusually efficient as regards generating a hydrodynamic friction during their realignment, the more difficult it becomes to generate a *cooperative* chain flow-dependent cavitation, and the easier it concurrently becomes for a strain-hardening role on the parts of those chains already properly aligned.

Crazing, hence cavitation, does not readily occur in bodies whose localized flow is extremely hampered or limited within the range of macromolecular dimensions, as a pronounced cold-drawing or flow on the parts of more than half of the molecules involved is apparently required. Thus *a whole localized area of the body must be transformed in a highly cooperative unison,* not merely a small portion of it, i.e., not merely some of its molecules. This is probably the reason for the occurrence of crazing on high speed impacts (see, for example, that shown in Figure IIIA-28) (44) which are performed in periods several orders of magnitude less than the macroscopic relaxation times of the whole body, despite the high order of *localized* flow fully required for that cavitation process. Thus the sketches in Figures IIIA-31*b* and IIIA-31*c* should frequently be regarded as having very brief time axes; this period depending on the fuller circumstances of microfailure generation.

IIIA4c. Chain and Segmental Relaxations

The role for extremely localized transport processes becomes more important as regards the progressive weakening of the body the more inelastic that body becomes. This is as true for the more viscoelastic of the metals as it is for the more viscoelastic of the glasses. In polymeric bodies, as clarified in Subsec. IC2, p. 51, the whole focus of attention on the *coherent* transport of polymer chains is relegated to the domain of influence of the individual segmental motions; the chain itself responding to the net contribution of all segments of that highly cooperative polysegmented aggregate.

Relaxation processes imply at least four concurrent events: (*a*) a highly localized transport process, (*b*) a highly localized stress relief, (*c*) a highly localized heat evolution and, (*d*) a highly *selective* stress redistribution.

If the body is maintained at a fixed strain, item (*b*) encourages the nominal stress level to be lowered. This does not imply, however, that item (*d*) is not active. If item (*d*) were not active here, it would be literally impossible to explain the extension of microfailures as discussed in regard to Figure IIIA-8, Curve *1*. Moreover, as readily seen from the discussion in Section IIIB, static fatigue rupture can also occur under progressively relaxing levels of the nominal stress.

Since highly pertinent stress redistributions thus are to occur even under relaxing nominal stress levels, they assuredly must occur during creep under a fixed load. Delayed "brittle" rupture, i.e., static fatigue, can generally be regarded as a macroscopic consequence of extremely localized stress relaxations, the force fields of which are selectively transferred to certain areas of the body which cannot relax as rapidly, hence are more efficient storers of strain energy in shear. Zener (47) has likened this situation in polycrystalline metals to a jigsawlike arrangement of the crystalline grains which selectively ex-

Fig. IIIA-40. A photoelastic jigsaw model. With the cooperation of M. M. Leven.

perience plasticity within the individual grains until its load is transferred progressively to certain complex grain/grain-boundary blocks which, for steric reasons, cannot relax. Bueche (8), in regard to the static fatigue of the highly viscoelastic glassy polymers, suggested that segmental relaxations cause a redistribution of stress towards certain statistically favored primary bonded chains which were stronger than their neighbors since they were, by pure chance, to be oriented parallel to the tensile field.

Figure IIIA-40 shows a photoelastic jigsaw model constructed of an epoxy resin which, at room temperatures, is one of the more brittle, hence more elastic, of the polymeric glasses. Let us say, however, that we raised the temperature of this model considerably so that immediately after the load was applied we could see the fringe distribution under crossed polarizers as shown in this figure. But now that the elevated temperature has transformed this model into one which has a greater viscoelastic contribution, we can witness a *gradual* shift in the fringe patterns so that certain intersection points of the jigsaw model would develop a relatively enhanced order of fringes during the model's uniaxial *creep*. If the "nominal" stress level were sufficiently high, we would gradually see the generation of a fracture at one of the interlinks where the fringe order is exceptionally high. Depending on the specific *location* of that fractured portion of the model, the whole model may distort severely; (*a*) but continue to bear the load, (*b*) and completely rupture the model simultaneously, (*c*) and *immediately* rupture one or more of the other interlinks without a complete failure, (*d*) and *slowly* rupture one or more of the other interlinks without a complete failure, etc. Thus, such processes are rather *selective*, and by no means are they any less complex in real bodies.

Microfailures might be *initiated* at a creeplike rate of population by this kind of stress redistribution during the body's *normal* creep rate, just as it might *propagate* a given microfailure at a creeplike rate under these same circumstances.

Accelerated tensile creep, and the cavitation phenomena associated therewith, might be generated from a once-*normal* creep rate which tends to enhance the stress at those parts of the body which tentatively are stronger, yet locally fail on that stress redistribution; only to cause subsequent stress-redistributed failures at a continuously accelerating pace until the sample either ruptures or newly discovers that it has certain very strong points (which may be or may not have been generated as indirect consequences of those localized failures and the stress redistributions therefrom).

A microfailure might propagate at a creeplike rate under *relaxing* levels of nominal stress by the enhanced *creep* generated on energy storage in those special areas which cannot relax as rapidly, etc.

It appears to this author that the general notion of a sensibly *homogeneous*, submicroscopic breakdown in a volume element of a sensibly isotropic viscoelastic body can have no possible cause other than a fully stabilized variety of intermolecular and intramolecular failures as generated by *randomly selective stress redistributions* on the multifarious, highly localized, relaxation processes. In this same highly general sense, *a macroscopic creep or a macroscopic relaxation is automatically the precursor of homogeneous failure.*

IIIA5. Significant Roles for Passive Flaws in Highly Viscoelastic Bodies

Perhaps the only important point raised in this treatment is the required emphasis on the gradual loss in pertinence of the accidentally induced "flaw." Scratching or placing a dust particle in a liquid simply does not greatly impair its strength, and that is sensibly the *submicroscopic* situation in the softer of the glassy polymeric solids. Moreover, these solids act to strengthen the region of the flaw and thus further stabilize its influence.

A quite logical, yet still interesting, role is played by cavitation processes in the highly viscoelastic glassy polymers. The transition from tensile microfractures, to slowly propagating microcavitated failures, thence to stabilized cavities, is a natural consequence of an increased flow contribution in the viscoelastic character of the bodies under study; chain orientation-induced strain-hardening, however, also playing a vital role.

As the body became softer, the population of failures increased and we had to shift our attentions to successively lower levels of inelastic contributions until we became intimately concerned with submicroscopic events such as polymer chain- and even polymer chain-segmental transports. Indeed, we were forced to question whether the molecular chains themselves were those stress concentrators of pertinence during an accelerated tensile creep and a progressive weakening in the glassy-leathers. Breaking a homogeneous body is a completely different affair than breaking one which is inhomogeneous. The inhomogeneous body fails in only a few places when under *an isotropic* stress, while a homogeneous body fails *homogeneously*. If we are to deal with the ultimate mechanical behaviors of extremely soft bodies, we will necessarily become more reminded of the character of its liquidlike, rather than its solidlike, contributions.

The required shift in perspective concerning the peculiarities of flaws in soft solids became rather marked when the tensile microfailures were seen to be inconsequential concentrators of stress and whose propagation rates were fairly reliable reflections of the body's macroscopic creep rates. It is difficult to imagine a more inefficient crack than one which cannot redistribute its stress as it propagates!

As we extrapolate this peculiar behavior further towards the microscopic responses of the even softer bodies such as the leathers and finally to the rubbers, we enter a remarkably strange domain for flaws—one where it becomes a commercial practice to *add* insoluble impurities such as carbon blacks and silicas to, indeed, *increase* the strength of these highly unorthodox solids. Developing a *conceptually unmodified*, Griffith criterion for rubbers would appear to be almost as futile as a study of the notch-sensitivity of a gas.

Another realization is met insofar as the body's *macroscopic* viscoelastic behavior, in translating information to us on the roles of the averaged contributions of strains or stresses over an isotropic area, becomes a more *reliable* connection with our net observations on strength since the body is actually approaching a dual situation: (*a*) the flaw is becoming more ineffective, and (*b*) the body is progressively weakened in a creeplike manner due to locally selective, yet macroscopically random, redistributions of the applied stress. The important implication here is simply that a sensibly isotropic viscoelastic body which is flaw-insensitive and capable of weakening *homogeneously* in a manner which reflects its macroscopic, inelastic deformation or stress relaxation is one which is ideal for *fatigue predictions*. If one can predict the first appearance of a microfailure from a knowledge of the body's residual creep rate, he has gained something that is interesting. If one can predict the lifetime of the body in creep rupture from knowledge of the time for the first appearance of a microfailure or from the creep rate of the body, then he has gained something more that is both interesting and useful.

Moreover, the bodies with which we are dealing here have macroscopic mechanical behaviors which are not too far removed from that anticipated by *linear viscoelasticity theory* (see Subsec. IC4, p. 68). Thus, as shown by Landel and Fedors in Section IIIB, the ineffective flaw situation as coupled with a homogeneous submicroscopic weakening, permits us to employ a time–temperature superposition approach to the close empirical prediction of ruptures under different times and temperatures. Furthermore, the subsequent use of Boltzmann's superposition principle then permits us to make amazingly good guesses concerning the happenings under different *patterns* of stressing and straining; whether static, transient, or dynamic.

Beyond merely this, when we are approaching a type of body such as the *rubber* within the above considerations of flaw insensitivity and a homogeneous submicroscopic weakening generated by "randomly selective" stress redistributions, we are simultaneously approaching another type of elasticity; one based on configurational entropy considerations, not on internal energy. A rather unique theory for the contributions of the flexible motions of the chainlike molecules has evolved (see Subsec. IC3, p. 64). It is, indeed, a *molecular theory* for configurational elasticity. Together with it, and its fusion with linear viscoelastic interpretations of the deviations introduced by the retarded modes of chain motion and their consequent stress redistributions, we are gradually approaching an entirely new situation in the field of fracture—an extremely useful and meaningful *molecular theory for homogeneous fatigue processes*, i.e., one which intentionally neglects the accidental flaw except as a point of *strengthening*.

IIIA6. Summary

A literature review and perspective of the salient failure mechanisms in the softer of the glassy polymeric solids has been presented, especially as regards the optical phenomena which are generated in the transparent varieties during creep rupture.

Tensile microfailures in these bodies can exhibit a creeplike rate of population and propagation and are found to exhibit marked cavitational contributions as well as a crack blunting. These localized tensile failures are unique insofar as they become filled with a striated and porous substance that is strongly "worked" by a ductile coldflowing. They appear to be internally, as well as externally, hardened by noncrystalline polymer chain alignments in shear. While the body is somewhat weakened by these tensile microfailures, they are more interesting for being reliable harbingers of the forthcoming delayed "brittle" rupture in static fatigue. Possible mechanisms for the strain-hardening and propagation of these tensile microfailures are suggested.

Upon moving towards the extremely soft end of the spectrum of supercooled glasses, i.e., toward the "glassy-leathery" viscoelastic state, the transparent bodies can be brought to a colloidlike turbidity under conditions of pronounced inelastic tensile straining such as an accelerated tensile creep. This is caused by stabilized, i.e., noncoalescing, cavities on the order of a few hundred to several hundred

Angstrom units in radius. Their numbers range to the trillions per cubic centimeter, as determined by light- and x-ray scattering studies. These generate a density-lowering and a pronounced, permanent and cumulative weakening in the body, and the cavities are probably strain-hardened by macromolecular orientation at their potential fracture fronts. These cavities, however, are on the size level of macromolecular dimensions, and suggest chain scissions during or following pronounced chain realignments.

Other considerations which are made include treatments of the roles for polymer chain scission and free radical generation; Lüder birefringence due to polymer chain realignments; locally selective, stress redistribution processes; and cross-comparisons with the metals and the harder glasses.

A general perspective is finally presented for the far-reaching roles for inefficient flaws as coupled with a progressive and homogeneous submicroscopic breakdown in the highly viscoelastic and rubberlike bodies; whether regarding the possibilities for phenomenological, viscoelastic, or even molecular theories for fatigue in these unusually interesting and experimentally convenient varieties of real solids.

References

1. Akkerman, F. H. D., *J. Appl. Polymer Sci.*, **7**, 1425 (1963).
2. Alfrey, T., Jr., *Mechanical Behavior of High Polymers*, Interscience, New York, 1948.
3. Bailey, J., *India Rubber World*, **118**, 225 (1948); see also R. D. Andrews and J. F. Rudd, *J. Appl. Phys.*, **28**, 1091 (1957); J. F. Rudd and E. F. Gurney, *ibid.*, **28**, 1096 (1957); J. F. Rudd and R. D. Andrews, *ibid.*, **31**, 818 (1960).
4. Berry, J. P., *S.P.E. Trans.*, **1**, 109 (1961).
5. Bessonov, M. I., and E. V. Kuvshinskiĭ, *Soviet Phys.-Solid State*, **3**, 950 (1961).
6. Bessonov, M. I., and E. V. Kuvshinskiĭ, *Soviet Phys.-Solid State*, **1**, 1321 (1959); see also *Vysokomolek. Soedin.*, **1**, 1561 (1959); *ibid.*, **2**, 397 (1960).
7. Briggs, L. J., *J. Chem. Phys.*, **19**, 970 (1951).
8. Bueche, F., *J. Appl. Phys.*, **29**, 1231 (1958); see also N. W. Taylor, *ibid.*, **18**, 943 (1947).
9. Bull, T. H., *Brit. J. Appl. Phys.*, **7**, 416 (1956); *Phil. Mag.*, **1**, 153 (1956).
10. Butyagin, P. Yu., *Dokl. Akad. Nauk SSSR*, **140**, 145 (1961).
11. Ceresa, R. J., *Polymers*, **1**, 477, 488 (1960).
12. Crussard, C., and J. Friedel, In *Creep and Fracture of Metals at High Temperatures*, National Physical Laboratory, Philosophical Library, New York, N. Y., 1957; also see several other contributions to this volume.
13. Durelli, A. J., E. A. Phillips, and C. H. Tsao, *Analysis of Stress and Strain* McGraw-Hill, New York, 1958.

14. Freudenthal, A. M., *The Inelastic Behavior of Engineering Materials and Structures*, Wiley, New York, 1950.
15. Grant, N. J., in *Fracture*, B. L. Averbach, D. K. Felbeck, G. T. Hahn, and D. A. Thomas, Eds., Technology Press–Wiley, New York, 1959, p. 17, as stated in relation to a comment by E. S. Machlin.
16. Haward, R. N., *Trans. Faraday Soc.*, **38,** 394 (1942).
17. Haward, R. N., *Strength of Plastics and Glass*, Cleaver-Hume Press, London; Interscience, New York, 1949.
18. Hsiao, C. C., and J. A. Sauer, *J. Appl. Phys.*, **21,** 1071 (1950).
19. Irwin, G. R., in *Handbuch der Physik*, S. Flugge, Ed., Vol. VI, Springer-Verlag, Berlin-Gottingen-Heidleberg, 1958, p. 551.
20. Jellinek, H. H. G., *Degradation of Vinyl Polymers*, Academic Press, New York, 1955.
21. Kambour, R. P., paper presented at 142nd Meeting, American Chemical Society, Atlantic City, N. J., 1962; See *Reprints, Polymer Div.*, **3,** 132 (1962); *Nature*, **195,** 1299 (1962); G. E. Res. Lab. Rept. #63-RL-(3407C), Schenectady, Aug. 1963; *Polymer*, in press.
22. Kennedy, A. J., *Processes of Creep and Fatigue in Metals*, Wiley, New York, 1963.
23. Kline, G. M. et al., "Development of Craze and Impact Resistance in Glazing Plastics by Multiaxial Stretching," National Advisory Comm. for Aeronautics, Rept. #1290, 1956; N.B.S.
24. Kolsky, H., in *Fracture*, B. L. Averbach, D. K. Felbeck, G. T. Hahn, and D. A. Thomas, Eds., Technology Press-Wiley, New York, 1959, p. 281.
25. Mark, H., and A. V. Tobolsky, *Physical Chemistry of High Polymeric Systems*, Interscience, New York, 1950.
26. Maxwell, B., and L. F. Rahm, *Ind. Eng. Chem.*, **41,** 1988 (1949).
27. Maxwell, B., and L. F. Rahm, *SPE J.*, **6,** 7 (1950).
28. McAfee, K. B., Jr., *J. Chem. Phys.*, **28,** 218, 226 (1958).
29. Nadai, A., and M. J. Manjoine, *J. Appl. Mech.*, *A* 77 (1941).
30. Nadai, A., *Theory of Flow and Fracture of Solids*, Vol. 1, McGraw-Hill, New York, 2nd Ed., 1950.
31. Newman, S. B., and I. Wolock, *J. Res. Natl. Bur. Std.*, **58,** 339 (1957).
32. Raff, R. A. V., and J. B. Allison. *Polyethylene*, Interscience, New York, 1956.
33. Regel, V. R., *Soviet Phys.-Tech. Phys.*, **1,** 353 (1956); see also *J. Tech. Phys.* (*U.S.S.R.*), **21,** 3, 287 (1951); V. R. Regel and Yu. N. Nedoshivin, *ibid.*, **23,** 8, 1333 (1953), the latter.
34. Rosen, B., *J. Polymer Sci.*, **58,** 821 (1962); *ibid.*, **B1,** 1472 (1963).
35. Rosen, B., *J. Polymer Sci.*, **47,** 19 (1960); see also *ibid.*, **44,** 546 (1960); C. H. Klute, *J. Appl. Polymer Sci.*, **1,** 340 (1959); C. H. Klute and P. J. Franklin, *J. Polymer Sci.*, **32,** 161 (1958).
36. Rosen, B., and J. H. Singleton, *J. Polymer Sci.*, **25,** 225 (1957).
37. Rosen, B., *J. Polymer Sci.*, **49,** 177 (1961).
38. Rosen, B., and A. E. Fornof, Symposium on Standards for Filament-Wound Reinforced Plastics, U.S. Naval Ordnance Lab., American Society for Testing and Materials; Special Technical Publication, No. 327, 1962, p. 40.

39. Rosen, B., "The Tensile Microflaw," delivered at Westinghouse Res. Labs., Pittsburgh, Nov. 1959; also paper presented to 141st Meeting American Chemical Society, Boston, April 1959; paper presented at 142nd Meeting, American Chemical Society, Atlantic City, N. J., Sept. 1959.
40. Sauer, J. A., and C. C. Hsiao, *ASME Trans.*, **75,** 895 (1953).
41. Sauer, J. A., J. Marin, and C. C. Hsiao, *J. Appl. Phys.*, **20,** 507 (1949).
42. Schmitt, J. A., and H. Keskkula, *J. Appl. Polymer Sci.*, **3,** 132 (1960).
43. Shand, E. B., *J. Am. Ceramics Soc.*, **37,** 52, 559 (1954); F. Forscher, *Welding J. Res. Suppl.*, p. 579-s (Nov. 1954); S. Bateson, *Phys. Chem. Glasses*, **1,** 139 (1960); J. P. Berry, *J. Mech. Phys. Solids*, **8,** 194, 207 (1960).
44. Spurr, O. K., and W. D. Niegisch, *J. Appl. Polymer Sci.*, **6,** 585 (1962); see also W. D. Neigisch, *ibid.*, **5,** S9 (1961).
45. Wells, A. A., and D. Post, *Proc. Soc. Exptl. Stress Anal.*, **16,** No. 1, 69 (1958).
46. Williams, J. L., and K. J. Cleereman, in *Styrene*, R. H. Boundy and R. F. Boyer, Eds., Reinhold, New York, 1952, p. 493.
47. Zener, C., in *Fracturing of Metals*, F. Jonassen, W. P. Roop, and R. T. Bayless, Eds., American Society of Metals, Cleveland, 1948, p. 7.
48. Zhurkov, S. N., V. A. Marikhin, and A. I. Slutsker, *Soviet Phys.-Solid State*, **1,** 1060 (1960).

IIIB. RUPTURE OF AMORPHOUS, UNFILLED POLYMERS

R. F. LANDEL AND R. F. FEDORS

IIIB1. Introduction

Despite the many studies on the rupture of unfilled amorphous polymers, an integrated concept of this phenomenon is only now emerging. Rupture depends on many parameters simultaneously, and until their interactive as well as their separate effects are known, our understanding of it will be incomplete. Thus, while studies have been made of the effects of time, temperature, degree of crosslinking, details of chain structure, and concentration of diluent, seldom has a sufficiently wide range and combination of such variables been covered in any given experimental program so as to provide an indication of the interactions. Moreover, one of the peculiarities of rupture studies has been the emphasis on the rupture *stress*. Because of the more complex response of rupture *strain*, e.g., to time or temperature, this parameter should be more sensitive to the changes in the primary variables, and thus offers a more definitive test of theoretical developments. In this section therefore, we shall attempt a synthesis of the experimental results into a more coherent whole, and then examine the extent to which various theories proposed to explain some individual feature, such as the crosslink dependence of the rupture stress in uniaxial tension σ_b, can be extended to describe other features such as the time for rupture t_b.

In order to obtain this synthesis of a coherent picture, rupture in viscoelastic bodies such as polymers should not be treated as separate phenomena, but rather as part of the larger problem of describing their stress–strain–time properties. If these properties are considered as a three-dimensional surface, then rupture represents some limiting value or discontinuity on this surface, or boundary to it. We shall call this a *physical property surface*. Any change in magnitude of one of the parameters required to describe this surface, such as an elastic modulus, will be shown to change the response at the boundary, which represents rupture. For example, factors which affect a rubber's modulus, such as crosslink density and swelling, will effect an accompanying change in σ_b. Similarly, if the small-strain modulus is time-dependent, then the failure stress must also be time-dependent, although the time dependencies will differ if the relaxation mechanisms are strain-dependent. From this perspective, the topic of rupture becomes much more tractable. Once a generalized concept of the surface is established, then the effect of changes in magnitude of the primary variables on the ultimate properties can be estimated from studies of their effects on localized regions far from the boundary of the surface. Thus a comprehensive study of the ultimate mechanical properties under a given state and pattern of stressing would imply a comprehensive study of the property surface and not just of the boundary region itself.

It should be emphasized that such surfaces will be nonconservative because of the viscoelastic nature of polymers and consequent energy dissipation. Therefore the state, i.e., the values of the stress σ and strain ϵ at a given time, will depend on the path or the experiment performed.

Such a comprehensive study would necessarily include both the development of a mathematical description of the surface and the establishment of a set of rupture criteria. The σ,ϵ relationship for such a surface for an amorphous polymeric body can be derived either from molecular theories or from theories of continuum mechanics. In the first case, the parameters of the relationship can be given a molecular significance and the associated rupture criteria will be given in terms of the energy stored in a polymer chain, the number of chains per unit volume, bond strengths, etc. In the second case, the σ,ϵ relationship can be interpreted in terms of coefficients of the constitutive equations. These coefficients cannot be assigned a molecular significance, except by analogy. The corresponding rupture criteria will simply be critical values of terms in the constitutive equations.

It should be recognized that a σ,ϵ,t surface implies a particular type of stress- or strain-loading. If bi- or triaxial stresses are applied, then each of these test modes may have its own time or temperature history and a three-dimensional, visualizable surface is inadequate. In such cases, the rupture behavior may be conveniently described in terms of an isothermal, isochronal rupture stress or rupture strain surface, called a *geometric rupture surface*, which gives the relationship between the three principal stresses or strains, whether tensile or compressive, at rupture. Such geometric surfaces can also be constructed for modes of failure other than simple rupture, e.g., yielding or cold-drawing. Several geometric failure surfaces have been proposed for metals, such as the octahedral shear stress, the constant elastic stored energy of distortion, the maximum shear stress, and the maximum principal elongation. However, very little information on such surfaces is available for elastomers.

Since the majority of rupture work on unfilled elastomers has been in uniaxial tension, we shall consider in this section only the tensile σ,ϵ,t surface. This, however, will suffice to account for many of the gross features of the response in geometric stress space.

Although a unified outlook is afforded by considering either the property or the geometric surface, it must be recognized that rupture stresses and strains are not well-defined material constants in the sense of having unique values, but rather they invariably exhibit a statistical distribution of values. As a result, the boundary in the first case, and the whole surface in the second, should be understood to represent the mean or the mode of such statistical distributions of rupture points. Very little experimental work has been done on establishing the stress, strain, and time distributions, their interrelations, and their dependence on the type or temperature of test.

In Chapter II of this book which concerns classical approaches to brittle fracture, the concept of a flaw (macroscopic or microscopic) as a stress raiser has been extensively discussed as the site of failure initiation. In crystalline solids these can usually be traced back to the existence and motions of dislocations, grain boundaries, etc. In crosslinked rubbers, on the other hand, the atomic arrangement is not that of a quasi-rigid lattice but that of flexible, disoriented chains linked in a three-dimensional network. The strands in this network are not all of the same length, so that the shortest ones will generally break first when the "net" is extended. Thus, the shortest subchain may frequently be associated with the site for the initial failure processes.

Single chain scission will not lead immediately to specimen rupture, however. It will be shown in Subsection IIIB3 that there is a unique expression for the breaking of an average chain, essentially independent of its chemical nature, but rather strongly dependent on its length. Since, in general, the shortest chains will be joined to somewhat longer ones, the additional stress (or strain) that the neighboring chains bear as a result of the first break may not be sufficient to cause *them* to break. The net result is to weaken the structure progressively, without rupturing it. In this sense, the effect is that of the *passive* flaw concept as developed phenomenologically in the previous section by Rosen. The *active* or pertinent flaw, on the other hand, is then controlled *by the average chain breaking properties*. Once the combination of stresses and strains builds up to the critical value, then the specimen will either rupture in a delayed, brittle-like manner or else it will tear. Brittle-like failure in a rubber usually will occur if the stress (or strain) distribution is macroscopically *homogeneous;* otherwise, the specimen will tear to a site where the stress (strain) values are below the critical conditions. This present section is principally concerned with the *initiation* of breaking in the *average* chain, i.e., *the progressive molecular initiation of active, homogeneously distributed flaws*. As a result, *considerations of crack propagation velocity have been excluded,* and fracture is taken to imply the rupturing or breaking of the specimen into two pieces in a time-dependent, yet brittle-like fashion, as generated by averaged molecular motions. As such, the fracture time or fracture strain becomes amenable to molecular approaches to viscoelasticity with these particular kinds of solids.

IIIB2. The Tensile Property Surface

IIIB2a. Time–Temperature Superposition

The tensile properties for an elastomer are given in terms of stress σ, strain ϵ, temperature T, and time t (or its equivalent $\dot{\epsilon}_{\text{const}}^{-1}$ for constant strain rate tests). Time and temperature are related under certain conditions, so that temperature effects can be incorporated with the time scale. To show this, let us first examine the interrelationship for behavior under small deformations. Here the interrelation can be treated by the Leaderman-Tobolsky-Ferry time–temperature superposition principle (30,72,100), first formalized by Ferry (31) in terms of a phenomenological model and subsequently given a molecular interpretation by the Rouse (84), Bueche (14), and Zimm (114) theories. These are discussed amply elsewhere (12,31) and need not be considered in detail here. See also Section IE.

It is important to subsequent discussions, however, to understand the implications of time–temperature interrelationships, and the method of superposing data. The dynamic shear storage modulus G' will be used as an illustration. Figure IIIB-1 shows the behavior of G' for the NBS polyisobutylene (34,76). At any temperature the

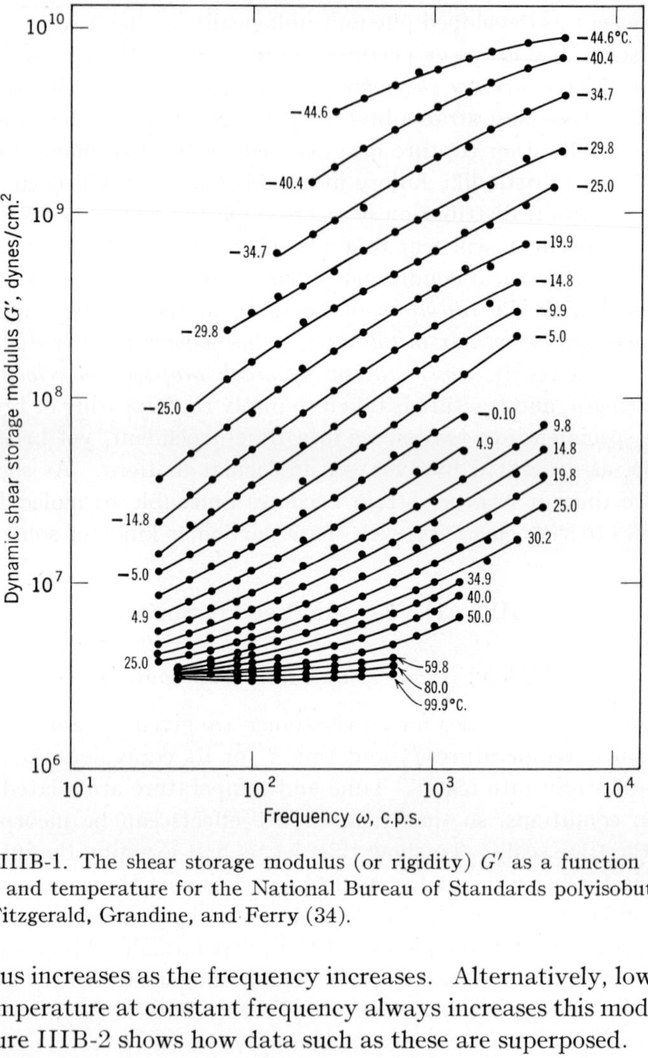

Fig. IIIB-1. The shear storage modulus (or rigidity) G' as a function of frequency and temperature for the National Bureau of Standards polyisobutylene. After Fitzgerald, Grandine, and Ferry (34).

modulus increases as the frequency increases. Alternatively, lowering the temperature at constant frequency always increases this modulus.

Figure IIIB-2 shows how data such as these are superposed. First, a small temperature and density correction factor is introduced in the ordinate (Subsec. IC4e, p. 83). The data at some temperature T_0 are then picked as a standard and the curve representing the next higher temperature is shifted horizontally to the left until the two curves coincide. The process is then repeated at progressively higher and lower temperatures until an extended master curve, repre-

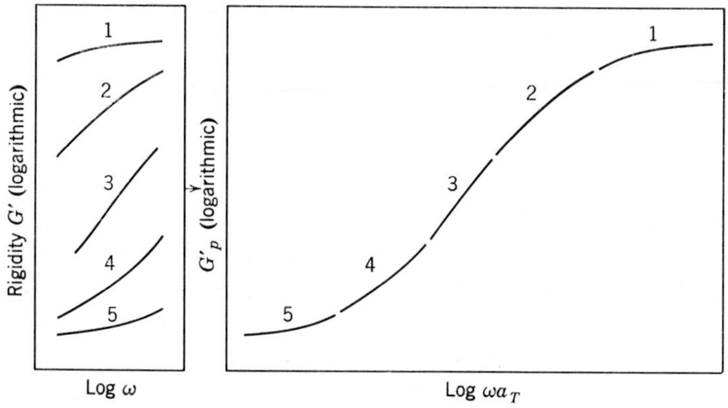

Fig. IIIB-2. Schematic illustration of the superposition of data from Fig. IIIB-1. The numbered curves in the figure on the left represent data obtained at different temperatures as a function of the frequency, ω. Curve *1* represents data obtained at the lowest temperature and Curve *5*, data at the highest temperature. In the figure on the right, the individual isothermal G' curves have been corrected for temperature and density to G_p', and then shifted horizontally, Curve *3* being taken as the reference curve, to effect superposition.

senting the response over decades of time, is constructed. Figure IIIB-3 gives the final result for the data of Figure IIIB-1. The extent to which any curve has to be shifted along the log time scale is $\log a_T$. The temperature dependence of this shift factor is described (110) by the Williams, Landel and Ferry (WLF) equation:

$$\log a_T = -C_1^0\,(T - T_0)/(C_2^0 + T - T_0) \qquad \text{IIIB2-(1)}$$

Equation (1) can be converted to an equation in which the glass temperature T_g is taken as the standard temperature, instead of the arbitrary reference temperature T_0. In this case

$$\log a_T = -C_1^g\,(T - T_g)/(C_2^g + T - T_g) \qquad \text{IIIB2-(2)}$$

Cf. eq. IE2-(1), p. 135. If $T_0 = T_g + \delta$, the constants are related as:

$$C_1^g C_2^g = C_1^0 C_2^0 \qquad \text{IIIB2-(3)}$$

$$C_2^g = C_2^0 - \delta$$

When T_g is taken as the reference temperature, it is found that C_1^g and C_2^g are nearly independent of the polymer system. These "universal" constants have the values 17.44 (dimensionless) and 51.6°K., respec-

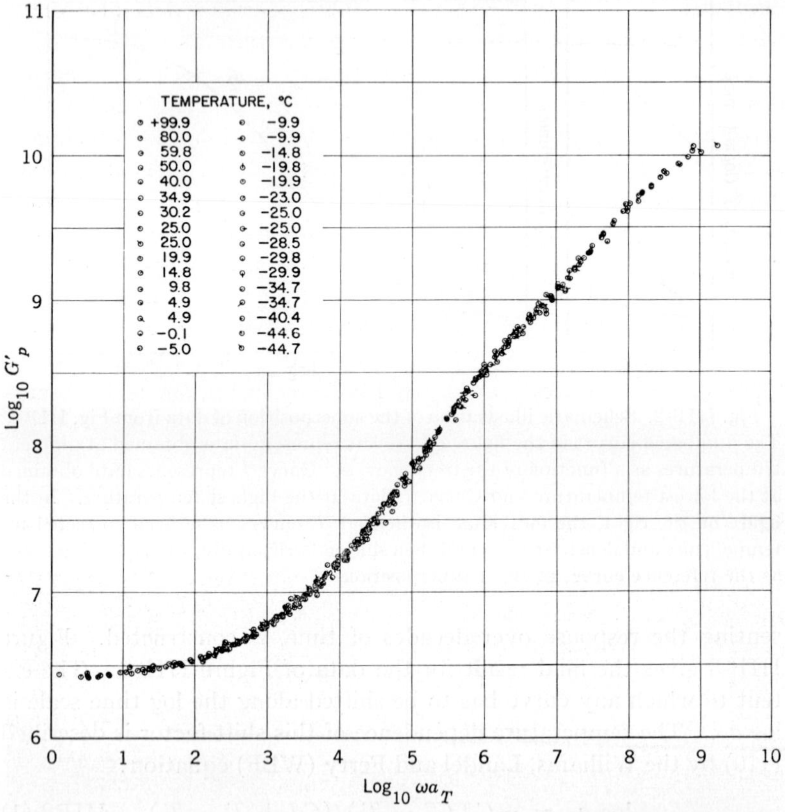

Fig. IIIB-3. Superposition of the data of Fig. IIIB-1. The individual curves from Fig. IIIB-1, representing isothermal data, have been corrected for temperature and density. The curve for $T = 25°C$. has been selected as the reference temperature and the remaining curves have been shifted horizontally to effect superposition. The amount of the shift is log a_T. We are indebted to Prof. J. D. Ferry for the detailed tables of original data on which this figure is based.

tively. Thus once T_g has been obtained, the time–temperature interrelationship is known.

To predict the effect of temperature then, it is only necessary to know T_g. A somewhat better prediction is obtained if the volumetric expansion coefficients above and below T_g are known. $C_2^g \cong 0.025\Delta\alpha$, where $\Delta\alpha$ is the difference in expansion coefficients; the *experimental* value of C_2^g can be used in place of that implied by the "universal" C_2^g, 4.8×10^{-4} deg.$^{-1}$. Log a_T does not depend strongly on C_2; therefore for most purposes the correction is not necessary.

The physical meaning of this superposition and the WLF equation is not always clear. Figure IIIB-4 shows a three-dimensional surface depicting the interrelationships between the modulus G_p', frequency ω, and temperature T for the data of Figure IIIB-1. The black lines indicate the experimental data which give rise to Figure IIIB-1 when projected to the G_p',ω plane. This can be visualized in the view of Figure IIIB-5*a*, a view of the model perpendicular to this plane.

To illustrate the superposition principle, the lowest temperature is taken as the standard, and the data at higher temperatures are translated horizontally to achieve superposition. This is illustrated in Figure IIIB-5*c*, a view of the partially superposed model; Figures IIIB-5*e* and IIIB-5*f* depict the surface after the shifting has been completed. The former, corresponding to Figure IIIB-3, represents the time dependence of the modulus at any temperature, and is the form used in constructing the individual planes or leaves of the model.

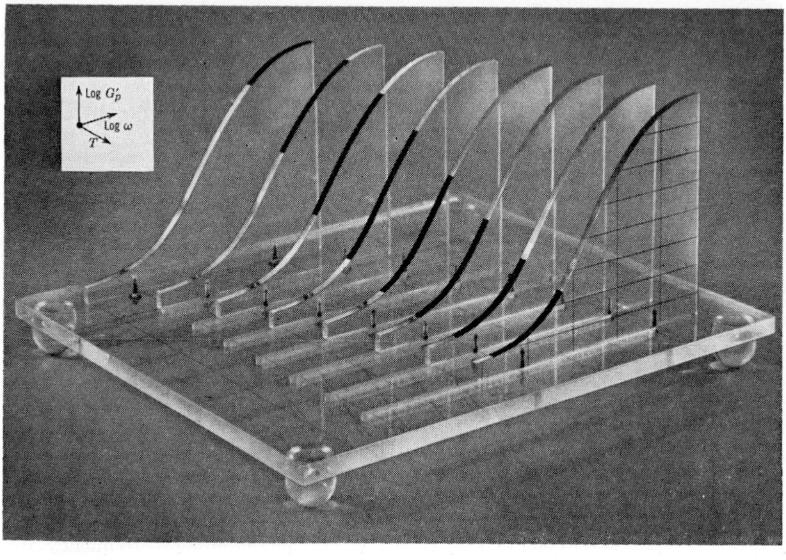

Fig. IIIB-4. View of a three-dimensional model depicting modulus (rigidity) G_p', frequency ω, and temperature T. The model was constructed from the data of Fig. IIIB-1. The black segments of the individual leaves indicate the extent of the surface scanned in Fig. IIIB-1.

(*a*)

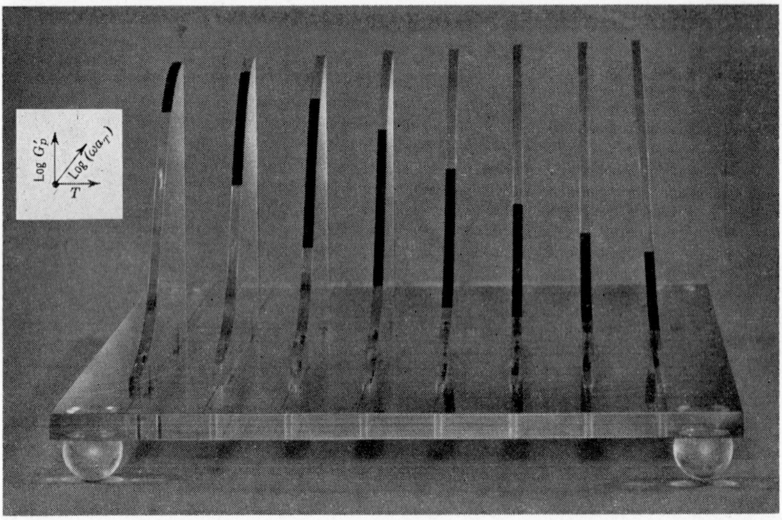

(*b*)

Fig. IIIB-5. The model of Fig. IIIB-4 viewed in the modulus-frequency and modulus-temperature planes. The three pairs of views (on this and the following two pages) show the curves unsuperposed (*a*, *b*), i.e., equivalent to the representation in Fig. IIIB-1; half superposed (*c*, *d*); and fully superposed (*e*, *f*), i.e., equivalent to the representation in Fig. IIIB-3.

(c)

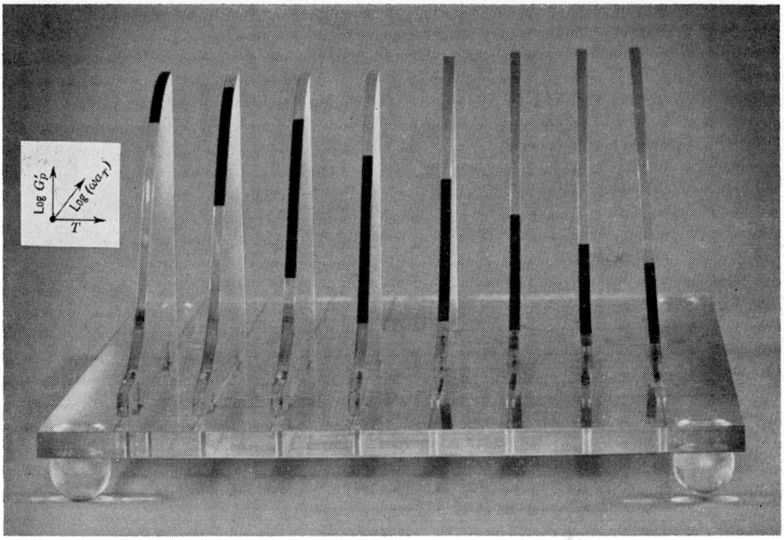

(d)

Fig. IIIB-5 (*continued*). See caption on page 370.

In an overhead view of the model (Fig. IIIB-6), the ends of the leaves represent projections on the log ω, T plane of the extent of the shift, or $-\log a_T$. In other words, these locations depict a plot of the WLF equation. Since this is true of *any* given log G_p' value, the

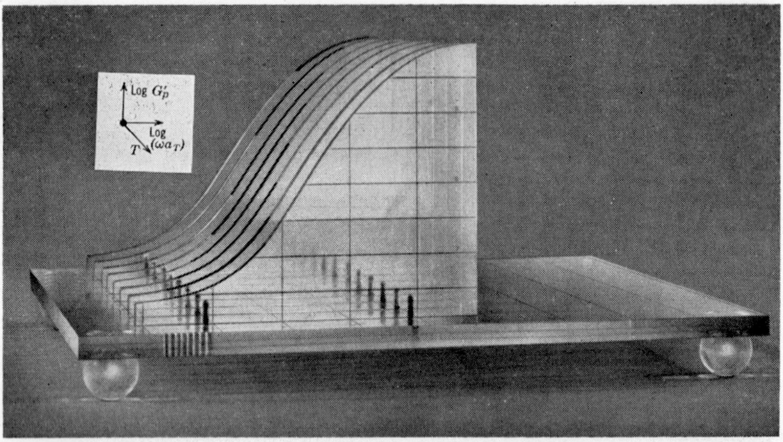

(*e*)

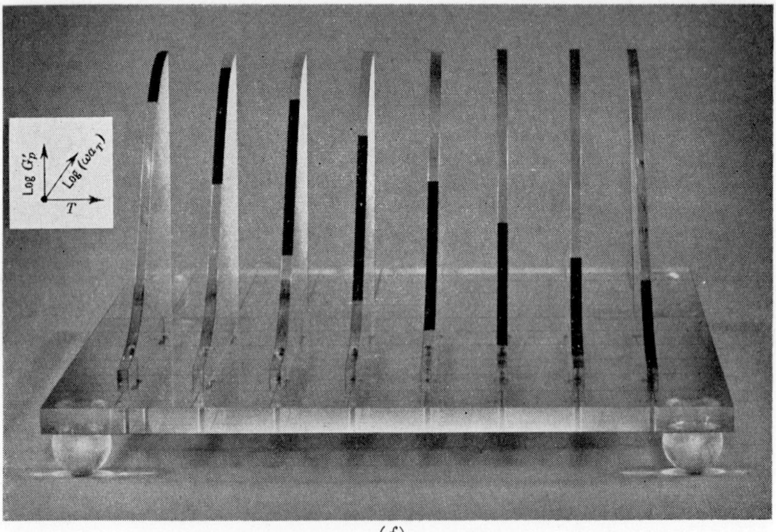

(*f*)

Fig. IIIB-5 (*continued*). See caption on page 370.

WLF equation therefore represents contour lines of this surface. *To the extent that the equation with the "universal" constants apply, all amorphous polymeric materials have the same contour line shape,* the individuality of the material being evidenced only by the separation

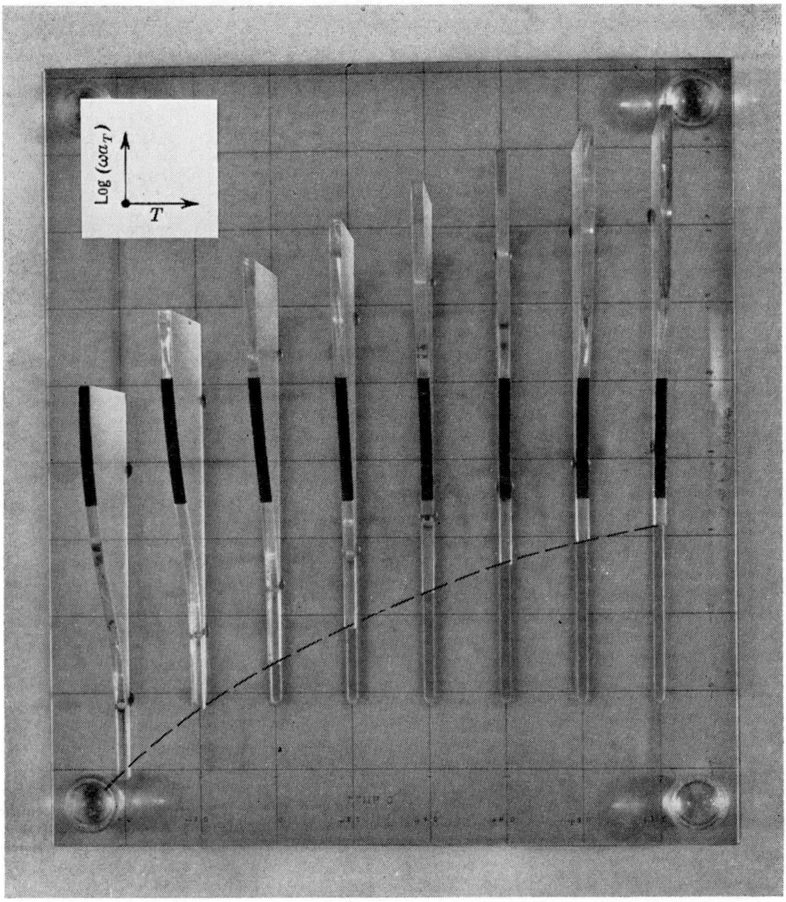

Fig. IIIB-6. View of the model of Figs. IIIB-4 and IIIB-5 from the ω, T plane. The ends of the individual leaves represent the WLF equation (dotted line), and show $-\log a_T$ as a function of the temperature.

of these lines, i.e., by the steepness of the log G_p',log ω curves equivalent to Figures IIIB-3 and IIIB-5*e*.

The two projections, to the log G_p',log ω and the log ω, T planes, separate and describe the time and temperature dependence of the properties clearly and succinctly. It should be emphasized that the third projection (to the log G_p', T plane) does not separate them clearly. The shapes of curves generated by taking cuts through the

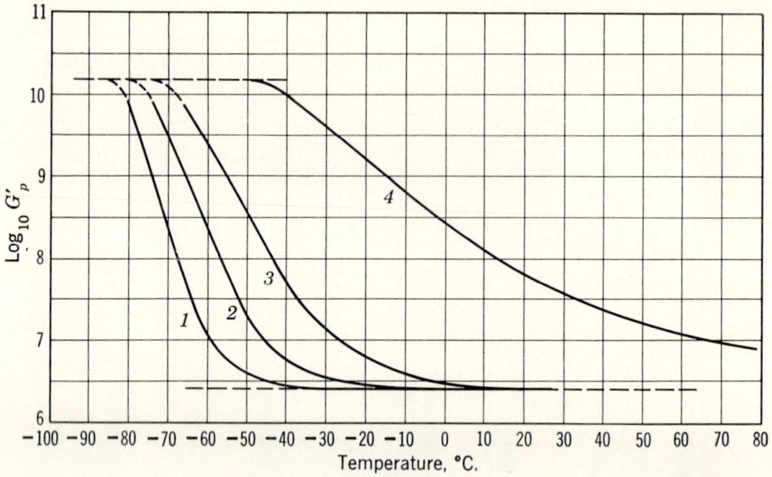

Fig. IIIB-7. Isochronal cuts through the property surface of polyisobutylene, Fig. IIIB-4, i.e., cuts parallel to the T,log G_p' plane taken at selected values of ω. As the frequency increases, the transition becomes broader and less well defined. Extrapolation of the curves to the glassy state modulus gives an indication of T_g. The frequencies represented in the figure are (*1*) 10^{-2}, (*2*) 10^0, (*3*) 10^2, and (*4*) 10^4 cycles/sec.

surface and projecting them on this plane depend on the time scale, or the cut chosen. This is depicted in Figure IIIB-7, where four isochronal cuts through the surface are shown. As can be seen, the higher the frequency (shorter the time scale), the smaller the rate of response to increasing temperature (cf. Figure IC-10, p. 71). For properties with a simple sigmoidal temperature dependence, such as modulus and compliance, the differences between the different curves is not too marked. However, the isochronal curves for parameters which pass through a maximum, such as the loss tangent or rupture strain, suffer a change in shape (e.g., see also Figs. IIIB-33*a* and *b*). It is this change in shape which makes isochronal data less desirable than isothermal data. Although either is suitable for representing the response under a given set of conditions, when the temperature is changed isothermal curves are simply shifted along the time scale without a change in shape, but the isochronal curves are not only translated, but also undergo a change in shape which is difficult to interpret readily.

IIIB2b. Stress–Strain–Time Surface

Having shown how temperature effects can be incorporated in the time scale in accordance with eq. (2) or eq. (3), let us now examine the $\sigma,\epsilon,t,$ response. Figure IIIB-8 shows the stress–strain behavior of the rubber, Viton B, at six temperatures. These are isochronal

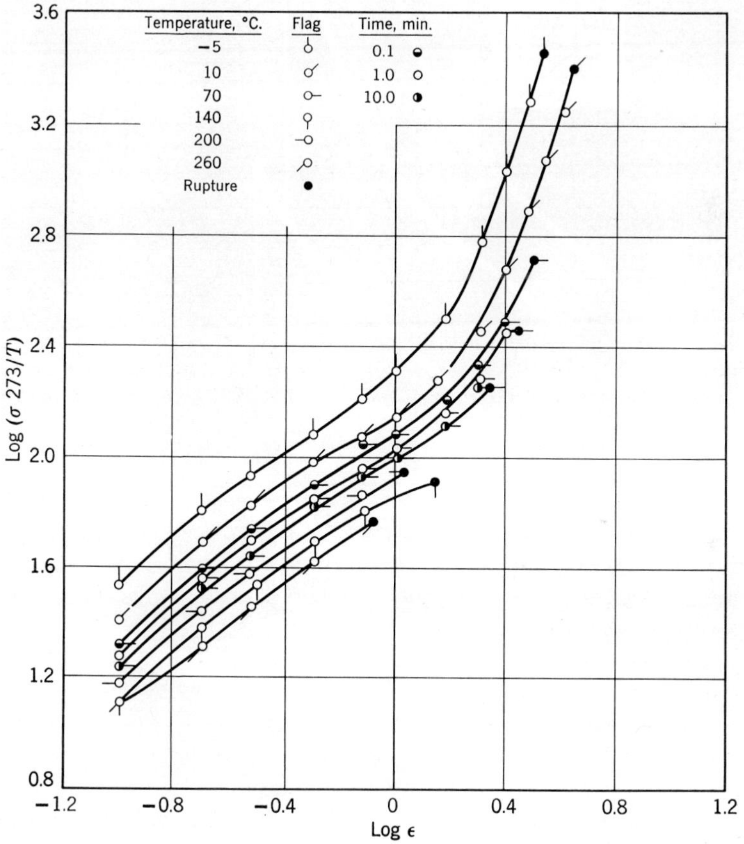

Fig. IIIB-8. One-minute, isochronal stress-strain curves for a Viton B gum vulcanizate as a function of the temperature over the range −5 to 260°C. At 70°C., the 0.1- and 10-min. isochronal curves are also shown for comparison. Filled circles indicate rupture. The stress has been reduced to a reference temperature of 273°K. The detailed data for this figure were kindly furnished by Dr. T. L. Smith.

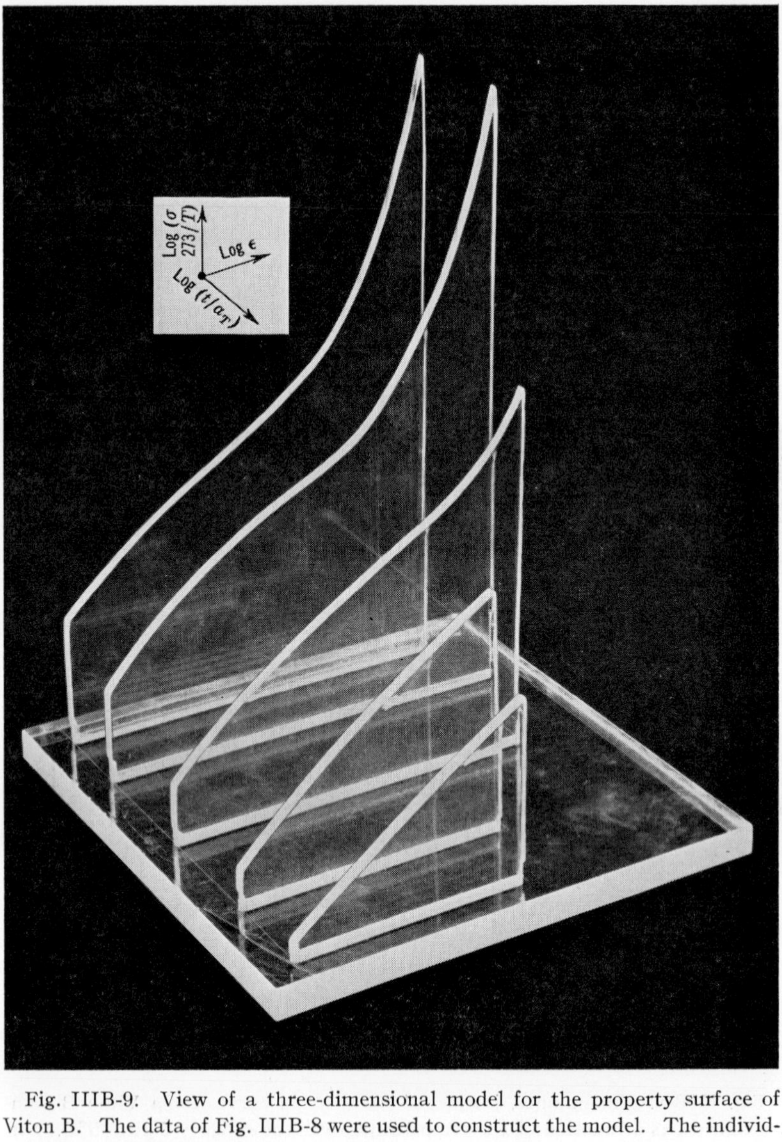

Fig. IIIB-9. View of a three-dimensional model for the property surface of Viton B. The data of Fig. IIIB-8 were used to construct the model. The individual leaves represent one-minute, isochronal stress-strain curves which have been shifted by $-\log a_T$. The ends of the leaves represent the occurrence of rupture and correspond to the filled circles of Fig. IIIB-8.

(1 min.) data obtained from an analysis of the original constant strain-rate data, in the manner of Smith (88). If these data are depicted as a three-dimensional model, the resulting surface gives a complete description of the behavior of the system (Fig. IIIB-9). The proper mathematical description of such surfaces is one of the key tasks confronting the polymer physicist today.

Rupture, on this surface, is a terminating line on the right. If this rupture line is projected to the σ,ϵ plane, as shown in Figure IIIB-10, it represents what Smith has called the "failure envelope" (86–88); if it is projected to the σ,t or ϵ,t planes it gives the time dependence of the rupture stress or rupture strain, also first extensively investigated by Smith (90). The time dependence of the stress and strain at rupture is shown more clearly in Figures IIIB-11 and IIIB-12. *These three plots represent the basic rupture description for amorphous rubbers.* Of these, the failure envelope is deemed by far the most important because its shape and location in the σ,ϵ plane are *independent of time and essentially independent of temperature.*

A change in temperature would merely shift this surface along the time axis in accordance with eq. (2) or eq. (3). This is illustrated in Figure IIIB-13, where the transparent leaves of the model are the same as those seen in Figure IIIB-9, while the opaque leaves represent the location of the surface at 100°C. The response of all time-dependent mechanisms has been shifted toward shorter times by a factor of 3,000.

The basic descriptions of rupture could have been presented without considering the property surface (Fig. IIIB-9). However, this is too restricted a viewpoint and a keen appreciation of the implication of such a surface should be developed by the reader before continuing, since it will form a background for the ensuing discussion. *The response of the polymer to any tensile experiment will simply be the path traced out on this surface.* Thus, isochronal, isotonic (constant stress), and isometric (constant strain) experiments represent cuts through the surface perpendicular to the time axis, the stress axis, and the strain axis, respectively. This is illustrated in Figure IIIB-14. If a Viton B specimen (of this ν_e) was brought to point A and held at constant stress, then the response would follow path AB. The projection of this isotonic cut to the ϵ,t plane thus corresponds to the data normally recorded in a *creep* experiment, since the stresses here are based on the initial cross-sectional area. Rupture might be expected

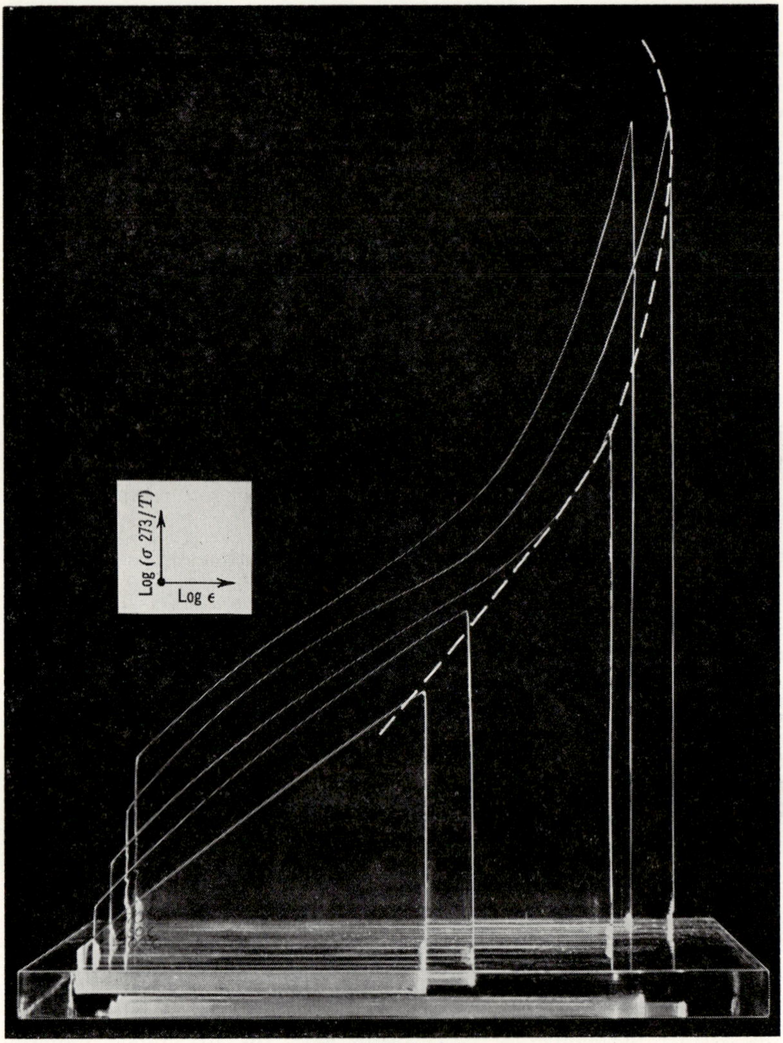

Fig. IIIB-10 (a).

after twelve decades of time have elapsed. For a creep experiment at constant true stress, $\sigma\lambda$, the specimen would break sooner. On the other hand, if the specimen at A was held at constant strain, then the response path would be curve AC. The projection of this isometric to the σ,t plane represents the course of a *stress relaxation* ex-

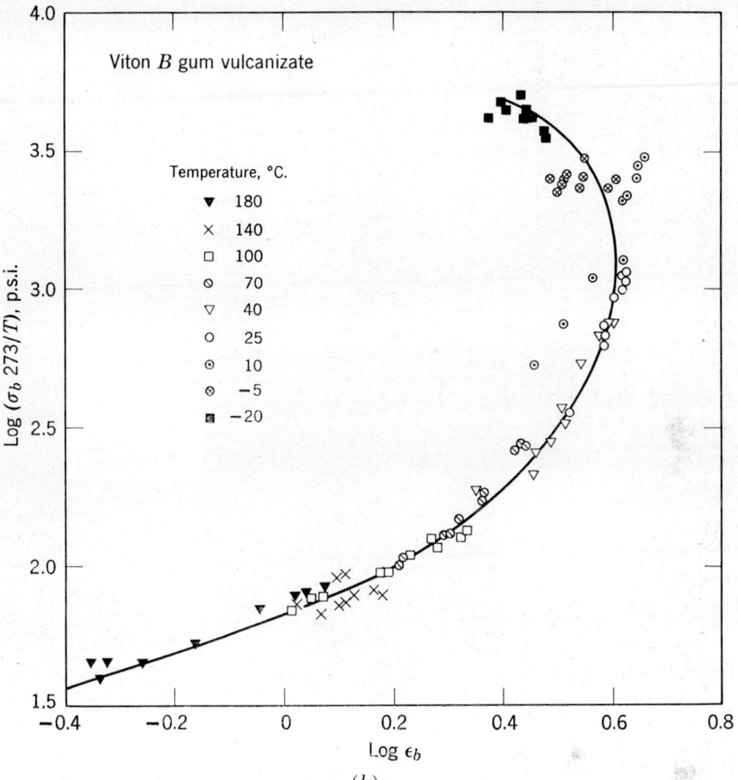

(*b*)

Fig. IIIB-10. Projection of the rupture points of the Viton B vulcanizate of Fig. IIIB-9 to the σ,ϵ plane to generate the failure envelope. (*a*) Schematic, based on Fig. IIIB-9. (*b*) Experimental results obtained from constant strain-rate tests at varying rate and temperature. From Smith (86).

periment and again the specimen could be expected to break when the path reached the limiting boundary of the surface, i.e., the failure envelope. In this experiment the rupture time required would be eighteen decades of time. Note that, starting from a given point on the surface, *rupture will always occur much faster for a creep than for a stress relaxation experiment.*

A constant strain rate test, e.g., with ring-shaped specimens tested at constant crosshead speed, does not give isochronal data. In a logarithmic representation of the surface, such as shown in Figures

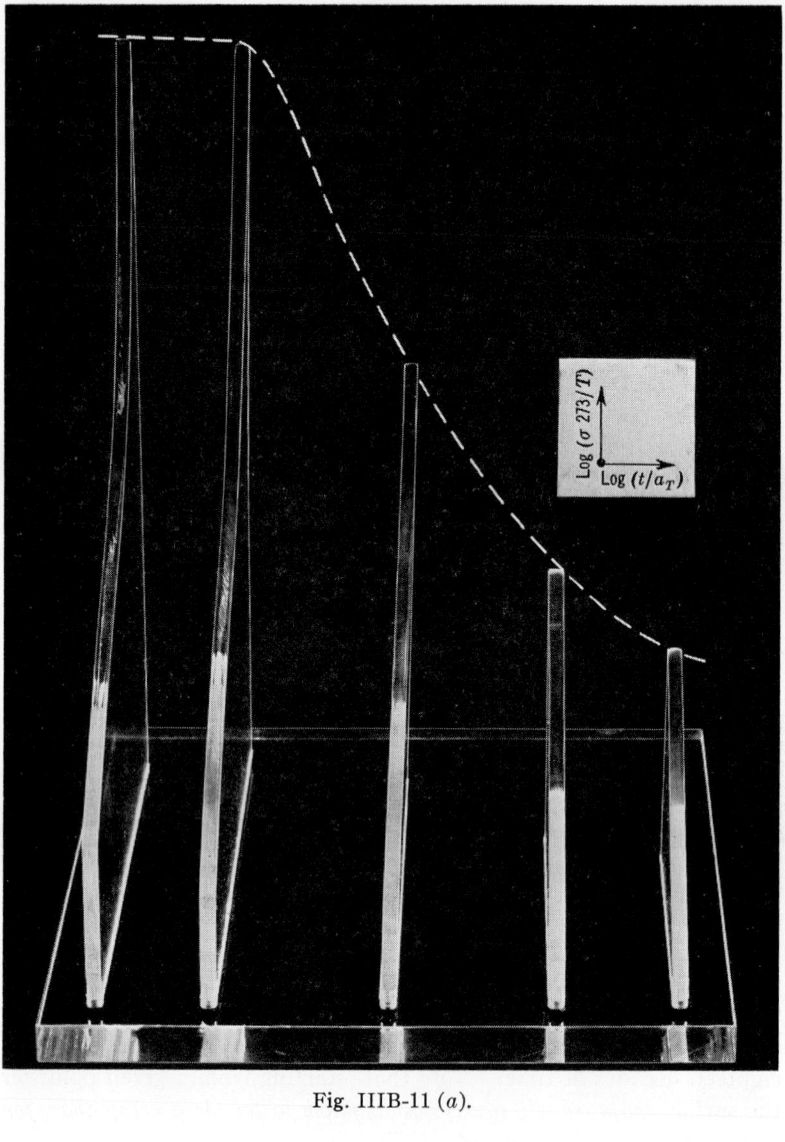

Fig. IIIB-11 (a).

IIIB-9 to 14, a constant strain rate test represents a cut through the surface with a unit slope rather than a cut perpendicular to the time axis in the ϵ,t plane as required for isochronal data.

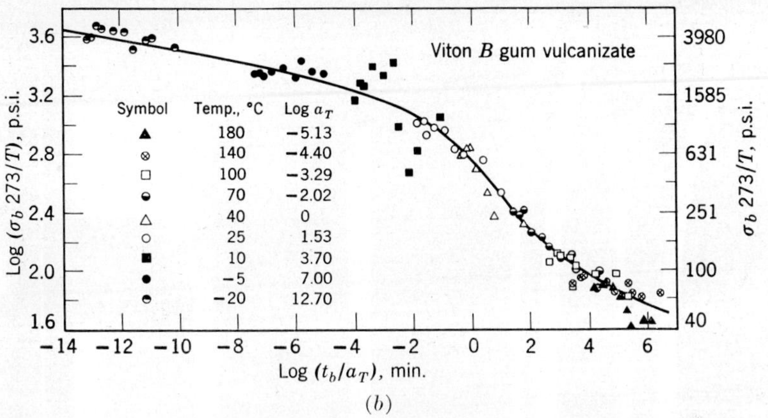

Fig. IIIB-11. Projection of the rupture points of Viton B to the σ,t plane. (*a*) Schematic, based on Fig. IIIB-9. (*b*) Experimental results obtained from constant strain-rate tests at varying rates and temperature. From Smith (86).

IIIB2c. Mathematical Description of the Surface

It is important to have a mathematical description of the property surface for several reasons: (*1*) to be able to infer the proper rupture criteria by relating the rupture properties to some characteristic feature of the σ,ϵ relationship, (*2*) to be able to manipulate the data properly in any attempt to normalize or linearize the σ,ϵ,t curves, or (*3*) to provide a rational basis for empirical descriptions of the rupture behavior, which are often required. We shall therefore consider each of these points before discussing the experimental results.

(1) The Stress-Strain Relationship and Rupture Criteria

(*a*) *Molecular Theories.* In Subsec. IC3, p. 65, it was shown that the kinetic theory of rubberlike elasticity leads to a simple tensile stress-strain relationship:

$$\sigma = (E/3)(\lambda - \lambda^{-2}) = \nu_e RT(\lambda - \lambda^{-2}) \qquad \text{IIIB2-(4)}$$

where $\lambda = \epsilon + 1$, is the principal extension ratio, and E is Young's modulus, which is proportional to ν_e, the number of effective chains per unit volume of rubber. The factor 3 stems from the assumption of incompressibility. At high extensions the theory fails to account for the experimentally observed rapid upturn in the stress-strain curve.

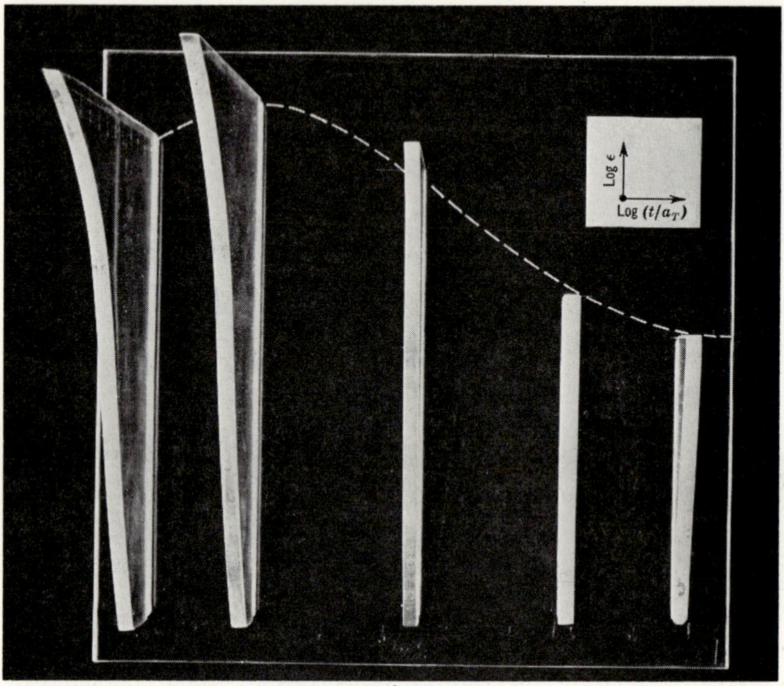

(a)

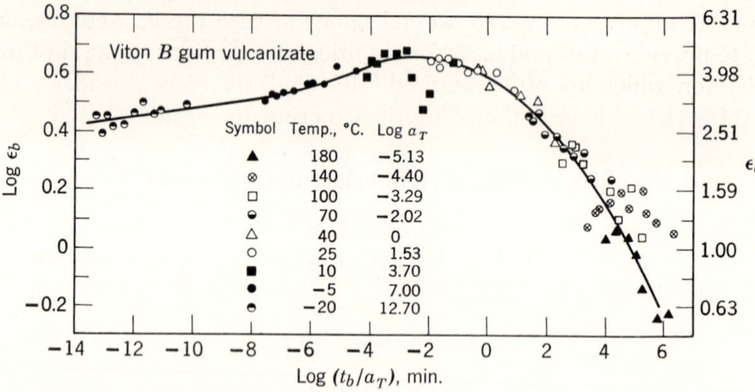

(b)

Fig. IIIB-12. Projection of the rupture points of Viton B to the ϵ,t plane. (a) Schematic, based on Fig. IIIB-9. (b) Experimental results obtained from constant strain-rate tests at varying rate and temperature levels. From Smith (86).

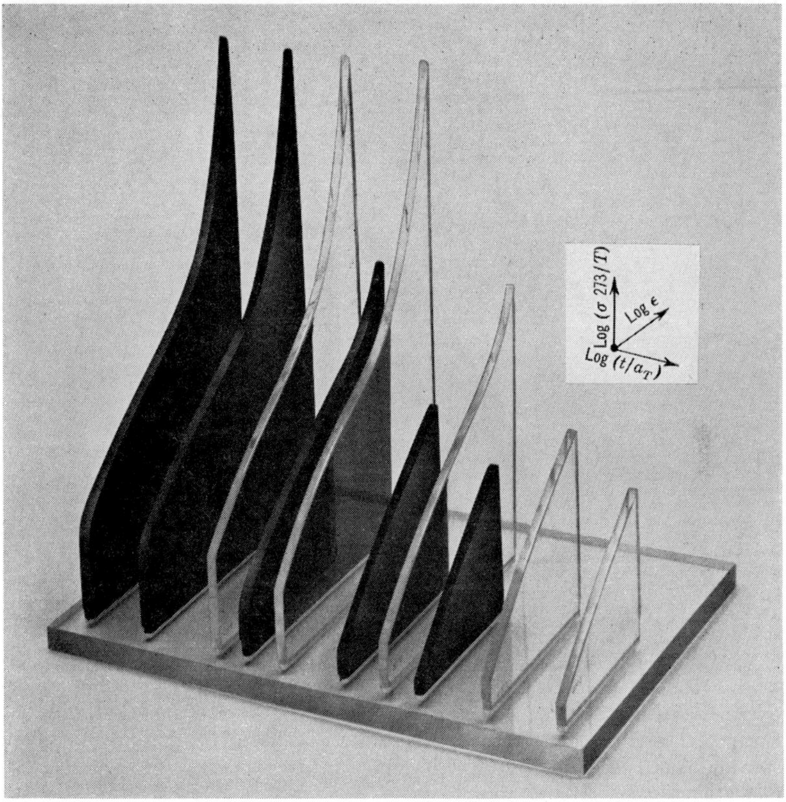

Fig. IIIB-13. Effect of a change in temperature on the tensile property surface of Viton B, showing the sample translation along the time scale as the temperature is raised from 40°C., represented by the transparent leaves (equivalent to Fig. IIIB-9), to 100°C., represented by the black leaves.

In this region the assumed gaussian distribution of chain lengths is no longer valid and a more general expression of the following type could be used instead (101):

$$\sigma = (E/9)n^{1/2}[\mathcal{L}^{-1}(\lambda/n^{1/2}) - \lambda^{-3/2}\mathcal{L}^{-1}(1/n^{1/2}\lambda^{1/2})] \quad \text{IIIB2-(5)}$$

where $\mathcal{L}^{-1}$ is the inverse Langevin function and n is the number of statistical segments for each chain. If l is the length of a segment, then nl is the length of the fully extended chain. Equation (5) appears to account for some of the deviation between experiment and eq. (4),

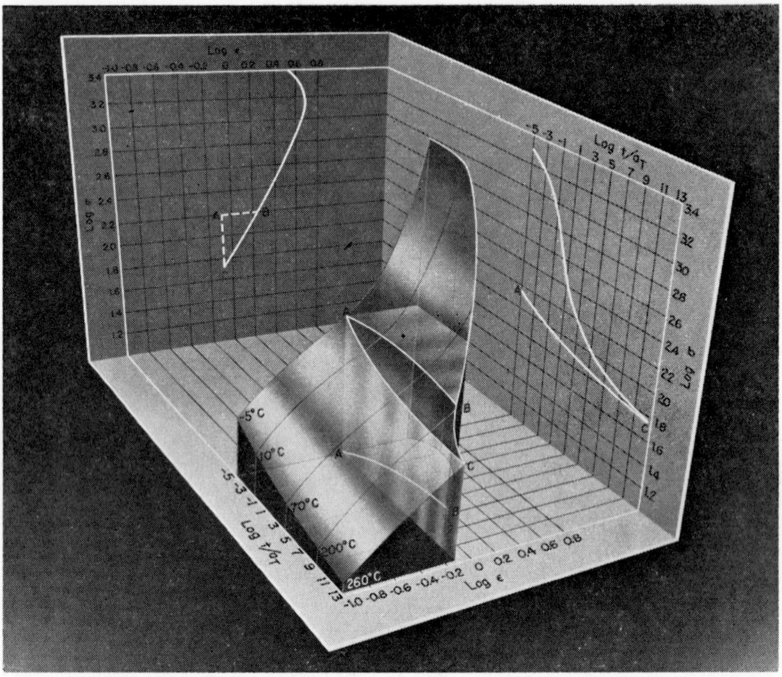

Fig. IIIB-14. Tensile property surface of Viton B showing the relationship between the paths traced on the surface in two experiments: (*1*) creep-to-rupture, Curve *AB*, and (*2*) stress relaxation-to-rupture, Curve *AC*. If the same initial starting point is used in both experiments, the figure indicates that t_b for stress-relaxation > t_b for creep-to-rupture.

insofar as it permits a rapid increase in the stress at high extensions. In addition, it provides an explicit rupture criterion in that it sets an upper bound on λ; $\lambda_{max} = n^{1/2}$. At this point, $\lambda/n^{1/2} = 1$, and $\mathfrak{L}^{-1}(1) = \infty$, so that it would take an infinite force to elongate the chain completely. Instead, the chain breaks at some extension less than this maximum possible value. Furthermore, the latter value depends on the degree of crosslinking since $n^{1/2}$ is proportional to $\nu_e^{-1/2}$, to a first approximation. Equation (5) also predicts that the ultimate properties depend on the details of the chain structure, because a flexible chain would have a larger value of n than a stiffer chain of the same extended length.

Treloar (101) has pointed out that eq. (5) suffers from certain inherent limitations. Examination of his results indicates that σ rises too rapidly beyond the region of $\lambda \cong 0.8n^{1/2}$.

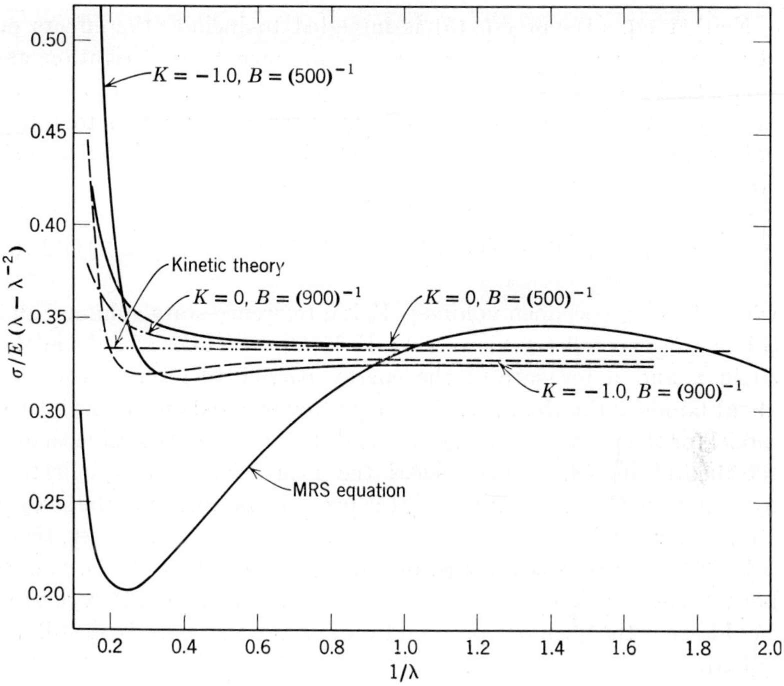

Fig. IIIB-15. Comparison of the Krigbaum-Kaneko theory (64) with the kinetic theory of rubber elasticity, eq. (4), and with the Martin-Roth-Stiehler (75) (MRS) equation, taking the MRS parameter A to be 0.40. Two values of the energy barrier parameter K and (effectively) two values of the crosslink density are shown for the Krigbaum-Kaneko theory. The latter theory predicts a minimum in the correct region of $1/\lambda$, but the minimum is much too shallow in comparison with the experimental results.

A more serious difficulty arises if the detailed behavior of an experimental stress-strain curve is considered. With an appropriate choice of both E and n, eq. (5) can be made to give a better fit than eq. (4) to experimental data for natural rubber in tension. However, Treloar has noted that if his uniaxial compression (biaxial extension) data are fitted by eq. (4), the theoretical curve lies *above* the experimental tensile data for moderate strains greater than about 40%. Since the inverse Langevin function, eq. (5), always predicts a greater stress equal to or greater than eq. (4) for the same modulus at a given λ, this discrepancy would presumably be even worse if eq. (5) were fitted to both the tension and compression data.

Neither eq. (4) nor eq. (5) is intended to include the effects of changes of internal energy on extension, since their derivation assumes that such changes are zero. Krigbaum and Kaneko (64) have attempted to account for these effects in terms of barriers to internal rotations along the chain. The general form of the uniaxial stress–strain curve thus becomes

$$\sigma = \nu_e RT(V/V_0)^{2/3}[\lambda - (1/\lambda^2)]\, \phi[\lambda, n, V/V_0, \exp\{\pm K\}] \qquad \text{IIIB2-(6)}$$

where V is the specimen volume; V_0 is a reference-state volume; and ϕ is a complicated function of λ; V/V_0, the number of links in the chain n and a measure of the energy barrier to internal rotation about bonds in the chain, K. This equation is based on a cubic lattice model for the network which is less realistic but more tractable than a tetrahedral model. Nevertheless the expression is in qualitative agreement with experiment in that it predicts an upturn in the stress–strain curve at large strains. At intermediate strains however, there is little difference between eqs. (4) and (6) (see Fig. IIIB-15). Since both equations therefore predict too high a stress in this region, it would seem that a more profound modification of eq. (4) is still required.

The maximum strain at rupture, and thus $(\lambda_b)_{max}$, is again associated with the number of links in the chain, just as it is with eq. (5). Note that in all three cases the relationship between the rupture stress σ_b and the rupture strain ϵ_b will depend on ν_e for a given polymer type. The Krigbaum-Kaneko theory in addition shows that there may be small differences between polymer types with different K's. Such factors have yet to be explored.

(*b*) *Continuum Mechanical Theories.* The stress–strain relationship can also be based on the purely formal development of continuum mechanics without invoking molecular parameters. According to this approach (29,40,41,82,92,101) the energy per unit volume, W, stored in a body on deformation can be expressed in terms of three strain invariants I_i. The form of the W,I relationship cannot be specified except for certain limitations on $\partial W/\partial I_i$. *Assuming* W to be a power series we have

$$W = \sum_{i,j,k=0}^{\infty} C_{ijk}(I_1 - 3)^i (I_2 - 3)^j (I_3 - 1)^k \qquad \text{IIIB2-(7)}$$

where, for present purposes,

$$I_1 = \lambda_1^2 + \lambda_2^2 + \lambda_3^2 \qquad \text{IIIB2-(8a)}$$

$$I_2 = \lambda_1^2\,\lambda_2^2 + \lambda_2^2\,\lambda_3^2 + \lambda_3^2\,\lambda_1^2 \qquad \text{IIIB2-(8b)}$$

$$I_3 = \lambda_1^2\,\lambda_2^2\,\lambda_3^2 = (V/V_0)^2 \qquad \text{IIIB2-(8c)}$$

and λ_1 is the ratio of the deformed to the undeformed length along one of the principal coordinates. Here V/V_0 is the ratio of the deformed to the initial volume, so that V_0 is not the reference-state volume of eq. (6). The stress–strain representation for uniform extension then becomes, in this notation:

$$\lambda_1\sigma_1 = \frac{2\lambda_1^2}{I_3^{1/2}}\left(\frac{\partial W}{\partial I_1}\right) + \lambda_1^2(\lambda_2^2 + \lambda_3^2)\,\frac{2}{I_3^{1/2}}\left(\frac{\partial W}{\partial I_2}\right) + 2I_3^{1/2}\left(\frac{\partial W}{\partial I_3}\right)$$

IIIB2-(9a)

$$\lambda_2\sigma_2 = \frac{2\lambda_2^2}{I_3^{1/2}}\left(\frac{\partial W}{\partial I_1}\right) + \lambda_2^2(\lambda_3^2 + \lambda_1^2)\,\frac{2}{I_3^{1/2}}\left(\frac{\partial W}{\partial I_2}\right) + 2I_3^{1/2}\left(\frac{\partial W}{\partial I_3}\right)$$

IIIB2-(9b)

$$\lambda_3\sigma_3 = \frac{2\lambda_3^2}{I_3^{1/2}}\left(\frac{\partial W}{\partial I_1}\right) + \lambda_3^2(\lambda_1^2 + \lambda_2^2)\,\frac{2}{I_3^{1/2}}\left(\frac{\partial W}{\partial I_2}\right) + 2I_3^{1/2}\left(\frac{\partial W}{\partial I_3}\right)$$

IIIB2-(9c)

Thus in uniaxial tension $\sigma_2 = \sigma$; $\sigma_1 = \sigma_3 = 0$; $\lambda_1 = \lambda$, $\lambda_2 = \lambda_3$, and hence

$$\lambda\sigma = \frac{2\lambda^2}{I_3^{1/2}}\left(\frac{\partial W}{\partial I_1}\right) + 4\lambda\left(\frac{\partial W}{\partial I_2}\right) + 2I_3^{1/2}\left(\frac{\partial W}{\partial I_3}\right) \qquad \text{IIIB2-(10a)}$$

$$0 = \frac{2}{\lambda}\left(\frac{\partial W}{\partial I_1}\right) + 2\left(\frac{I_3^{1/2}}{\lambda^2} + \lambda\right)\left(\frac{\partial W}{\partial I_2}\right) + 2I_3^{1/2}\left(\frac{\partial W}{\partial I_3}\right) \qquad \text{IIIB2-(10b)}$$

and eqs. (9a,b,c) reduce to the following stress-strain law:

$$\sigma = 2\left(\lambda - \frac{I_3^{1/2}}{\lambda^2}\right)\left[\left(\frac{\partial W}{\partial I_1}\right) + \frac{I_3^{1/2}}{\lambda}\left(\frac{\partial W}{\partial I_2}\right)\right] \qquad \text{IIIB2-(11)}$$

Rubbers are usually considered to be incompressible, so that $I_3^{1/2}$ is taken to be unity and $\partial W/\partial I_3 = 0$. However, small volume changes do take place, and terms involving $\partial W/\partial I_3$ cannot be neglected in any attempt to evaluate the proper form of W. Thus from eq. (8c), $\partial W/\partial I_3 = (V_0^2/2V)(\partial W/\partial V)$, and since $\partial W/\partial V$, the rate of change of elastically stored energy with volume is very large, then $\partial W/\partial I_3$ must also be large. Moreover, the complete evaluation of W will require the careful measurement of the stress–strain response and the volume change under several types of deformation in order to separate the contributions of the three partial derivatives. Thus, uniaxial σ,ϵ data, even including a determination of I_3, are insufficient to provide an adequate characterization of W unless some assumption is made about one of the three partials, since there are only two equations but three unknowns.

The kinetic theory stress-strain expression implies that

$$W = E(I_1 - 3)/6 \qquad \text{IIIB2-(12)}$$

which is the simplest possible form for W. If

$$W = C_{100}\,(I_1 - 3) + C_{010}\,(I_2 - 3) \qquad \text{IIIB2-(13)}$$

the Mooney-Rivlin stress–strain equation results (80,82):

$$\sigma = 2[\lambda - (1/\lambda^2)]\,[C_{100} + (C_{010}/\lambda)] \qquad \text{IIIB2-(14)}$$

and the molecular significance of the coefficients is that

$$C_{100} + C_{010} = E/6 = \nu_e RT \qquad \text{IIIB2-(15)}$$

Experimentally, $C_{100}/C_{010} \cong 10$ for unfilled rubbers. Equation (14) is the best known modification of eq. (4) which provides a better representation of experimental observations.

For eq. (6), W has the form

$$W = (E/3)\,[C_{10}(I_1 - 3) + C_{20}(I_1 - 3)^2 + C_{30}(I_1 - 3)^3$$
$$- C_{01}(I_2 - 3) - C_{11}(I_1 - 3)\,(I_2 - 3) + \ldots] \qquad \text{IIIB2-(16)}$$

where terms in I_3 have been omitted in accordance with the assumption that the material is incompressible and, of course, C_{10} and C_{01} are not the same constants as those appearing in eqs. (13)–(15). The coefficients again have molecular significance, being related to the Krigbaum-Kaneko parameters V/V_0, n, and K.

Little can be said at present regarding failure criteria in terms of W. The simplest criterion would evidently be a critical value of W itself, though limiting values of I_i or λ_i could be taken as well. Thus, eq.

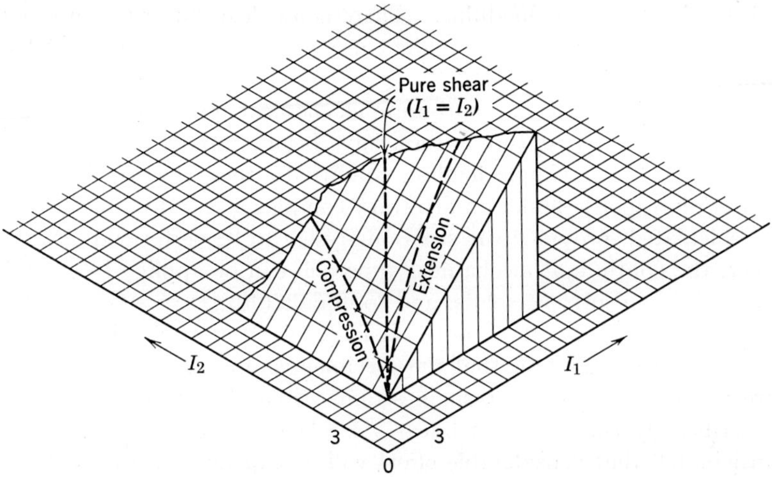

Fig. IIIB-16. The stored energy surface corresponding to the kinetic theory expression, eq. (4), plotted as a function of the strain invariants I_1 and I_2. The dotted lines indicate the paths traced on the surface for three experimental test modes—uniaxial tension, pure shear, and uniaxial compression. (From Dr. H. Leaderman, private communication.)

(12) represents a wedge-shaped figure in $W(I_1 - 3),(I_2 - 3)$ space with a slope of zero in the I_2 direction and of $E/6$ in the I_1 direction (Fig. IIIB-16). A critical value of W would therefore establish a maximum height for this figure, while a critical value of $I_1 - 3$ would fix its length.

In any event, it is clear that W and the criteria will have to be tested via multiaxial loading experiments, and the criteria will be closely allied with the geometric rupture surface. In some cases it may prove to be easier to discern criteria from the form of the σ,ϵ relationship than from the W,I equation. For example, Blatz (7) has reported that for a polyurethane foam containing 47% voids

$$W = \frac{G}{2}\left(\frac{I_2}{I_3} + 2I_3^{1/2} - 5\right) \qquad \text{IIIB2-(17)}$$

or

$$\sigma_i\lambda_i = G(I_3^{1/2} - \lambda_i^{-2}) \qquad \text{IIIB2-(18)}$$

where G is the shear modulus. There is no clear-cut criterion in eq. (17), but it is immediately evident in eq. (18) that the product $\lambda_i \sigma_i$ will always be less than $GI_3^{1/2}$.

The importance of the continuum mechanical approach is that once W and associated rupture criteria have been obtained for a material by testing under simple loading conditions, then (in principle) the material response under an arbitrary complex stress distribution can be calculated, provided the stress analysis can be performed. Moreover, the continuum mechanical approach can be applied to crystalline solids, anisotropic solids, and to liquids (40).

(*c*) *Time Dependence.* Both the molecular and mechanical models discussed above assume that the specimen is at *equilibrium* and there are no time effects. For small strains, time effects are adequately described by the theory of linear viscoelasticity (30), and it can be anticipated that considerable effort will be expended in the next few years to develop the theory of viscoelasticity at finite strains (see Section VD). For our purposes it will be sufficient to note that as an excellent approximation the time dependence of the stress can be factored out of the strain dependence over a considerable portion of the reduced time scale (3,50,70,75,88). Thus a general stress-strain equation can be written as:

$$\sigma = E(t)\epsilon f(\epsilon) \qquad \text{IIIB2-(19)}$$

which reduces to a time-dependent Hooke's Law at small strains, i.e., $f(\epsilon) \rightarrow 1$, as $\epsilon \rightarrow 0$. In addition, for sufficiently high temperatures, $E(t) \rightarrow E_e$ which is independent of time. Hence, the limiting case for low strains and sufficiently high temperatures would be Hooke's Law, $\sigma = E_e\epsilon$. In addition, it is important to note that $f(\epsilon)$ and $E(t)$ approach their limiting values independently.

Table IIIB-I gives several equations proposed to describe tensile behavior and the corresponding $f(\epsilon)$. If the analytic form of $f(\epsilon)$ did not vary with ϵ or time, eq. (19) would imply that all stress-strain curves would be parallel on log-log plots such as Figure IIIB-9. This is not true over the whole transition region. The σ,ϵ curve becomes increasingly concave upward as the temperature is lowered or the test rate increased. Bernstein, Kearsley, and Zapas (4) have derived an expression for W containing two time-dependent terms, though one of these was found to be very nearly independent of time. This is in qualitative accord with the observations of Martin, Roth, and

TABLE IIIB-I

Summary of Equilibrium Stress–Strain Relationships for Elastomers in Uniaxial Tension

Eq.	Equation	$f(\epsilon)$ In terms of λ	$f(\epsilon)$ In terms of ϵ
A[a]	$\sigma = E\epsilon\lambda^{-1}$	λ^{-1}	$(\epsilon + 1)^{-1}$
B[b]	$\sigma = (E/3)(\lambda - \lambda^{-2})$	$1 + \lambda^{-1} + \lambda^{-2}$	$(\epsilon + 1)^3 - 1/\epsilon(\epsilon + 1)^2$
C[c]	$\sigma = (E/9)n^{1/2}[\mathfrak{L}^{-1}(\lambda n^{-1/2}) - \lambda^{-3/2}\mathfrak{L}^{-1}(\lambda^{-1/2}n^{-1/2})]$	—	—
D[d]	$\sigma = 2C_1(\lambda - \lambda^{-2}) + 2C_2(1 - \lambda^{-3})$	$1 + \lambda^{-1} + (C_1 + C_2)^{-1}(C_1\lambda^{-2} + C_2\lambda)$	$(\epsilon + 1)^{-1}[1 + (\epsilon + 1) + C_2(C_1 + C_2)^{-1}(\epsilon + 1)^2 + C_1(C_1 + C_2)^{-1}(\epsilon + 1)^{-1}]$
E[e]	$\sigma = (E/3)(\lambda\beta)^{-1}\{\exp \beta(\lambda - \lambda^{-1}) - \exp[-\beta/(\lambda^{1/2} - \lambda^{-1/2})]\} - B\lambda^{-1}[\lambda^2 + \lambda^{-2} - \lambda - \lambda^{-1}]$	—	—
F[f]	$\sigma = E(\lambda^{-1} - \lambda^{-2}) \exp\{A(\lambda - \lambda^{-1})\}$	$\lambda^{-2} \exp\{A(\lambda - \lambda^{-1})\}$	$(\epsilon + 1)^{-2} \times \exp\{A\epsilon(\epsilon + 2)(\epsilon + 1)^{-1}\}$
G[g]	$\sigma = (\lambda^{-1} - \lambda^{-2})[B_1 \exp\{A_1(\lambda - \lambda^{-1})\} + B_2 \exp\{-A_2(1 - \lambda^{-2})\}]$	$[\lambda^2(B_1 + B_2)]^{-1}\left[B_1 \exp\left\{A_1 \frac{\lambda^2 - 1}{\lambda}\right\} + B_2 \exp\left\{-A_2 \frac{\lambda^2 - 1}{\lambda^2}\right\}\right]$	—

[a] Hooke's Law modified for changes in cross-sectional area during extension, assuming no volume change.

[b] Kinetic theory of rubberlike elasticity, assuming gaussian distribution of chain lengths.

[c] Kinetic theory, based on nongaussian distribution of chain lengths.

[d] Mooney-Rivlin expression based on continuum mechanics; assumes incompressibility; $C_1 + C_2 = E/6$.

[e] Carmichael-Holdaway (20) based on continuum mechanics; assumes incompressibility; B, β = constants.

[f] Martin-Roth-Stiehler (MRS) expression (75), empirical, $A \cong 0.38$.

[g] Becker-Rademacher (3) empirical, $B_1 + B_2 = E$; A_1, A_2 = constants.

Stiehler (75), and of Landel and Stedry (70), who found an indication that A of eq. (F), Table IIIB-I, could have been very weakly time-dependent.

(2) Normalization Procedures

Let us now examine the results of some possible normalization schemes which might be attempted in order to provide a rational basis for comparisons between different systems and a fresh avenue of

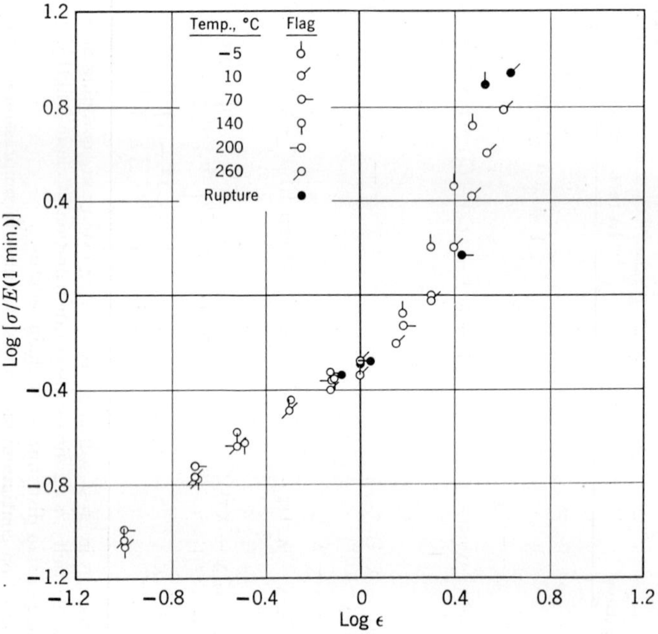

Fig. IIIB-17. The one-minute, isochronal stress-strain curves of Fig. IIIB-8 reduced to unit one-minute modulus, E(1 min.), i.e., these curves represent log $f(\epsilon)$ vs. ϵ. The filled circles denote rupture. The figure shows that the stress strain curves superpose except at low temperatures, so that $f(\epsilon)$ is a temperature-dependent function for temperatures below 10°C.

approach for future investigations. According to eq. (19), the normalization could be done by reducing the stress to unit modulus, taking either the equilibrium or time-dependent value, or by linearizing the stress-strain curve to an equivalent Hooke's Law.

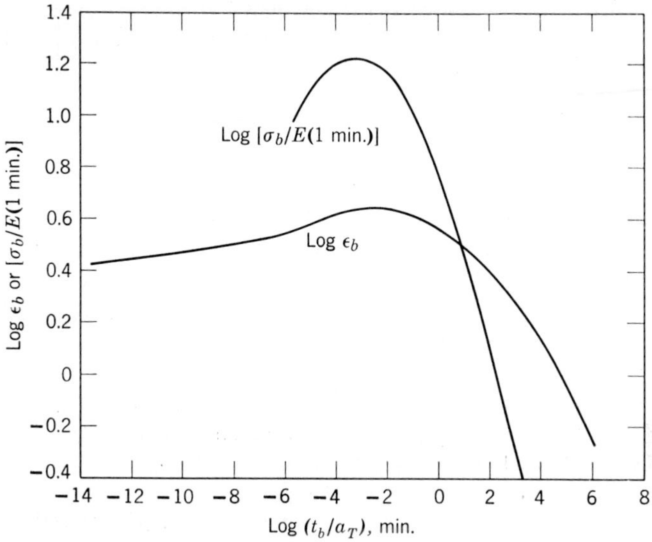

Fig. IIIB-18. Comparison of rupture stress data of Fig. IIIB-11*b*, after normalizing to unit one-minute modulus, $\sigma_b/E(1\text{ min.})$, with rupture strain ϵ_b as function of the reduced rupture time. The difference between the two curves is equal to the factor log $f(\epsilon)$.

(*a*) *Normalization to Unit Modulus.* In reducing to unit modulus note that the stress relaxation modulus $E(t)$ reaches a quasi-equilibrium value at long times, E_e, which is proportional to ν_e [cf. eq. (4)]. Therefore plots of σ/ν_e versus ϵ would allow ready comparison of $f(\epsilon)$ at various crosslink densities. On the other hand, reducing the stress-strain curve to unit modulus *throughout* the transition zone by taking the $\sigma/E(t)$ ratio would give a single σ versus ϵ curve, if $f(\epsilon)$ were independent of time. If $f(\epsilon)$ were a constant, i.e., independent of ϵ, the curve would be a straight line of unit slope. Since the rupture points would all lie on this curve, it would also represent the failure envelope. Actually $f(\epsilon)$ becomes time-dependent at low temperatures, so that the log $[\sigma/E(t)]$ versus log ϵ curves fan out in the region of the maximum in ϵ_b, as shown in Figure IIIB-17.

When this normalization procedure is applied to the log σ_b versus log t_b plot of Figure IIIB-11*b*, the latter would become simply a log ϵ_b versus log t_b plot if $f(\epsilon)$ were equal to unity, otherwise, it would be a plot of log $[\epsilon_b f(\epsilon_b)]$ versus log t_b. This is illustrated in Figure

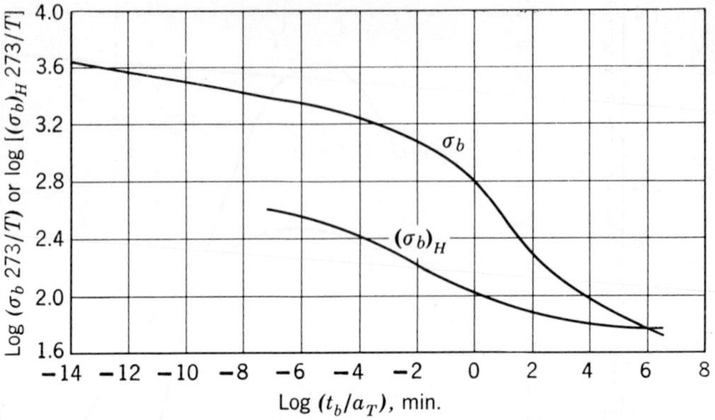

Fig. IIIB-19. Comparison of rupture stress, σ_b, with the strain-reduced rupture stress $(\sigma_b)_H$ as a function of rupture time, t_b for a Viton B gum vulcanizate. Data from Figs. IIIB-8 and 11*b* were used to construct the plot.

IIIB-18, where log $[\sigma_b/E(1 \text{ min.})]$ versus log t_b is compared with log ϵ_b versus log t_b taken from Figure IIIB-12*b*. The calculated curve maximizes at the right point on the time scale, but of course has a much higher value. The ϵ_b, t_b plot is, of course, unaffected by the normalization.

(*b*) *Normalization to Reduced Stress or Reduced Modulus.* An alternative reduction could be in terms of the strain-reduced stress σ_H (70), where $\sigma_H = \sigma/f(\epsilon)$, so that the reduced stress-strain curve is linearized to time-dependent Hooke's Law. In this case the log σ versus log ϵ curves of Figure IIIB-9 become straight lines of unit slope. The σ,ϵ,t surface becomes a curved surface of unit slope in the σ,ϵ plane, while the σ,t relationship along any isometric cut is parallel to the small-strain stress relaxation modulus $E(t)$. The log σ_b,log t_b relation becomes less time-dependent (Fig. IIIB-19), while the failure envelope in the region of $(\epsilon_b)_{\max}$ is moved toward the ϵ axis, so that the envelope becomes approximately linear up to $(\epsilon_b)_{\max}$ (see Fig. IIIB-20).

The advantage to this method of reduced variables is that first, $E(t)$ is readily measured by a number of techniques and can be given a straightforward molecular interpretation in terms of crosslink density and the chain mobility (17,30,100). Moreover, the results of the many studies on linear viscoelastic behavior can immediately be brought to bear in suggesting interesting areas of research. For example, the presence of temperature-dependent entanglement cross-links and side-chain relaxation mechanisms have been shown to give rise to characteristic responses in which the dynamic modulus, especially the loss component, is markedly different. Then, since

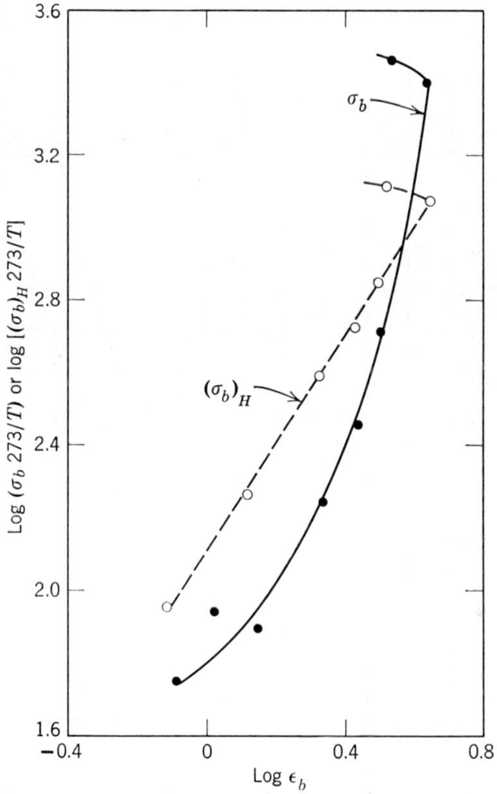

Fig. IIIB-20. Comparison of the failure envelope of Fig. IIIB-10*b* (full line) with an envelope based on the strain-reduced rupture stress $(\sigma_b)_H$.

$E(t)$ will be "anomalous" compared to the behavior of systems in which these effects are absent, anomalous rupture behavior can also be anticipated. There is some experimental evidence for this superposition in the behavior of a carboxy-terminated butadiene rubber (58,61). As a further example, since the location of the transition zone from rubbery to glassy behavior is determined by the monomeric

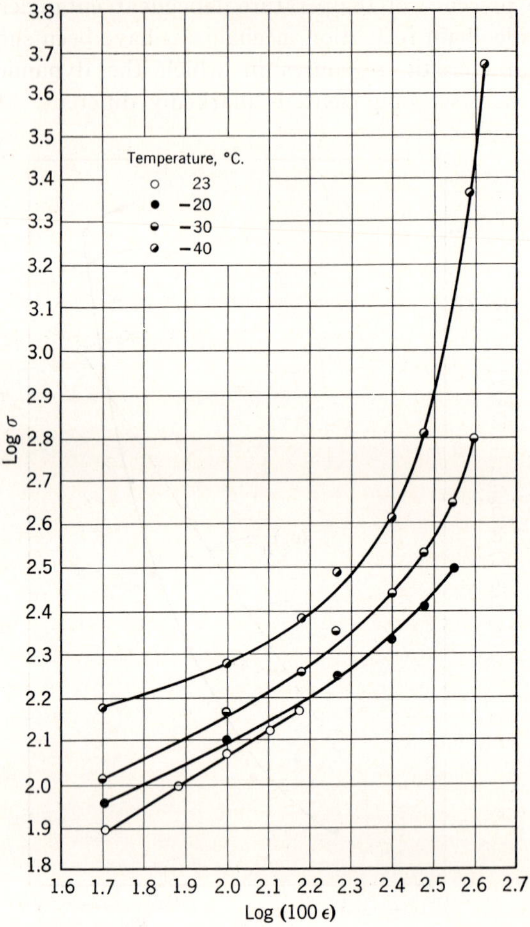

Fig. IIIB-21. Stress-strain curves for an SBR-dicumyl peroxide vulcanizate obtained as a function of temperature at a constant strain rate, showing that $f(\epsilon)$ is temperature dependent. $\dot{\epsilon}_{const} = 2.32$ min.$^{-1}$, $\nu_e = 154$ μmole/cm.^{3}.

friction coefficient ζ_0, then the location of $(\epsilon_b)_{\max}$ on the time scale should also be related to this parameter.

Secondly, the linearized form simplifies mathematical manipulations and permits the direct application of much of the mathematical formalism developed for linear viscoelasticity. The simple measures of stress and strain required for the linear case are replaced by measures appropriate to the nonlinear case at hand. Thus, Smith (88) has shown that the product $\lambda\sigma$ may be employed in treating constant strain rate data for elongations of up to about 100%, and Landel and Stedry (70) have shown that $\sigma\lambda^2/\exp\,[0.40(\lambda - \lambda^{-1})]$ may be employed for stress relaxation experiments when ϵ is used as the measure of strain. Similarly, Leaderman (71) has shown that creep and recovery data at moderate strains can be analyzed by using either nominal stress σ and the kinetic theory measure of strain, i.e., $(\lambda - \lambda^2)$ or $\lambda\sigma$ as the stress and $(\lambda^2 - \lambda^{-1})$ as the measure of strain. Finally, Mason (78) has shown how this approach may be utilized to analyze the effects of small dynamic strains superimposed on large static strains. He based his analysis on the Mooney-Rivlin equation, but the analysis should be valid for any form of $f(\epsilon)$ so long as the latter is not time-dependent. Therefore, it appears from experimental observation that a simple change of variables will suffice in many cases, and so the existing linear solutions to various problems such as creep, constant strain rate or stress relaxation experiments, which are readily available, e.g., in Section IC, may be employed to treat the nonlinear case.

Of course nature is not really so cooperative, and $f(\epsilon)$ is time- and temperature-dependent in some part of the time scale, so that a unified function does not exist over the whole transition region. This is illustrated in Figure IIIB-17 and more clearly in Figure IIIB-21.

(3) Empirical Representation of Rupture Behavior

If the property surface could be described by some general equation, rupture could then be defined in terms of a limiting surface, i.e., a surface normal to one of the three planes σ,t; ϵ,t; or σ,ϵ, and generated by the rupture points in that plane. The intersection of this limiting surface with the property surface would then define the rupture line in σ,ϵ,t space. The projection of this line to the remaining two planes would then give the curves such as seen in Figures IIIB-10, IIIB-11, or IIIB-12.

Thus three rather simple equations would suffice to describe the rupture behavior in uniaxial tension at a given temperature: one for the stress–strain curve, one for the time-dependence of the modulus, and one for the chosen limiting surface. Since each of these can be reasonably approximated with equations containing only two or three constants, only a few measurements are required to establish the values of these constants. The three equations can then be solved for the relationship between σ_b, ϵ_b, and t_b, i.e., the rupture line.

The stress-strain curve could be any suitable form such as one of those given in Table IIIB-I or eqs. (4)–(6) and (14). Equation (F) of Table IIIB-I would often be a suitable choice.

The time dependence of the modulus is known from small deformation theory to be (30):

$$E(t) = E_g + \int_{-\infty}^{\infty} H_l \exp\{-t/\tau\}\, d \ln \tau \qquad \text{IIIB2-(20)}$$

where E_g is the glassy modulus and H_l is the distribution function of tensile relaxation times τ, so that an auxiliary expression for H_l must be given. Alternatively, $E(t)$ could be given as (97,98,100)

$$2 \log E(t) = (\log E_g + \log E_e) - (\log E_g - \log E_e) \times \text{erf}\,[h \log (t/K)] \qquad \text{IIIB2-(21)}$$

where erf is the error function, h is a numerical coefficient related to the steepness of the transition and K is the value of t corresponding to $E(t) = (E_gE_e)^{1/2}$. Another empirical representation which could be used is (91a,109)

$$[E(t) - \text{E}_e]/[E_g - E_e| = 1/(1 + at)^c \qquad \text{IIIB2-(22)}$$

where a and c are constants. The corresponding distribution function is a modified power law distribution

$$H_l = (E_g - E_e)\,(a\tau)^{-c}/\Gamma(c)\,[\exp\{(at)^{-1}\}] \qquad \text{IIIB2-(23)}$$

where Γ is the gamma function.

For the limiting surface, the failure envelope is the most interesting choice, but the σ_b,t_b relationship will be taken as an example because it is the simplest to describe empirically. The breaking stress is a monotonically decreasing function of t_b, thus if a linear dependence were to be assumed for the purposes of our example then

$$\log \sigma_b = \log A - B \log t_b \qquad \text{IIIB2-(24)}$$

Combining eqs. (19), (22), and (24) gives the time dependence of the breaking strain, i.e., if $f(\epsilon) = \lambda^{-1}$

$$\epsilon_b(\epsilon_b + 1)^{-1} = A(1 + at_b)^c/t_b[E_g - E_e + E_e\,(1 + at_b)^c] \qquad \text{IIIB2-(25)}$$

IIIB3. Crosslink Dependence of Rupture

According to Subsection IIIB2c, the σ,ϵ,t surface will vary as the modulus. In the quasi-equilibrium zone, the modulus is proportional to ν_e, the number of effective chains per unit volume of whole rubber. In the glassy zone, however, the modulus is essentially independent of ν_e so that there should be little or no dependence of the rupture behavior on chain concentration. Let us therefore examine the failure envelope for the effect of a change in ν_e and then turn to the effect to be expected in the σ_b,t_b and ϵ_b,t_b relationships.

Smith (86) has noted that the high temperature end of the failure envelopes of five rubbers (Viton B, natural rubber and styrene-butadiene (SBR) both vulcanized with sulfur, and butyl rubber vulcanized both with sulfur and resin) differed by a factor approximately equal to the equilibrium modulus. In this study, Smith was primarily concerned with the effect of the chemical nature of the elastomer on ultimate properties and the observed modulus variation was a secondary effect brought about by this change of elastomer. Landel and Fedors (67) tested a single elastomer, SBR, vulcanized with dicumyl peroxide to ν_e values of 68, 107, 133, 136, and 154 μmoles/cm.^{3}. The ultimate stress and strain of the most highly crosslinked sample were determined by making measurements on ring-shaped specimens (mean diameter = 1.50 in.) at four strain rates and at eight temperatures from 23 to -70°C. These data were used to construct a failure envelope and to check the statistical distribution of rupture to be discussed in Subsection IIIB7. For the other vulcanizates, three to six rings from each sample were tested at a strain rate of 2.32 min.$^{-1}$, i.e., a portion of the envelope was obtained from statistical variability. (The determination of the envelope by statistical variability will also be discussed in Subsection IIIB7.) The results are shown in Figure IIIB-22. The data points obtained from the statistical variability of rupture determined at three temperatures and one strain rate define the same failure envelope as that determined by testing single specimens over a wide range of rates and temperatures. The data from the vulcanizates of lower chain concentration show that both rupture stress and rupture strain vary with ν_e such that the area of response in the log σ_b,log ϵ_b plane is shifted to the right (i.e., to higher ϵ_b values) as ν_e is decreased.

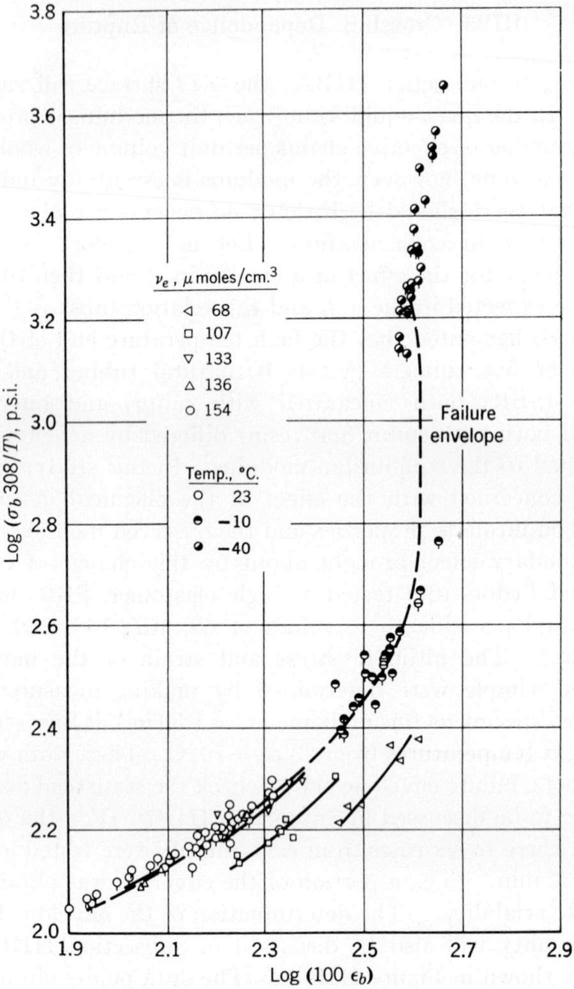

Fig. IIIB-22. Portions of the failure envelope for SBR-dicumyl peroxide vulcanizates of varying crosslink density. The portions were determined from the statistical variability of rupture by testing at a strain rate of 2.32 min.$^{-1}$ at the indicated temperatures. Each point represents the failure of a single specimen. The dotted line is the failure envelope for the sample with ν_e = 154 μmole/cm.^{3} as determined by varying test rate and temperature.

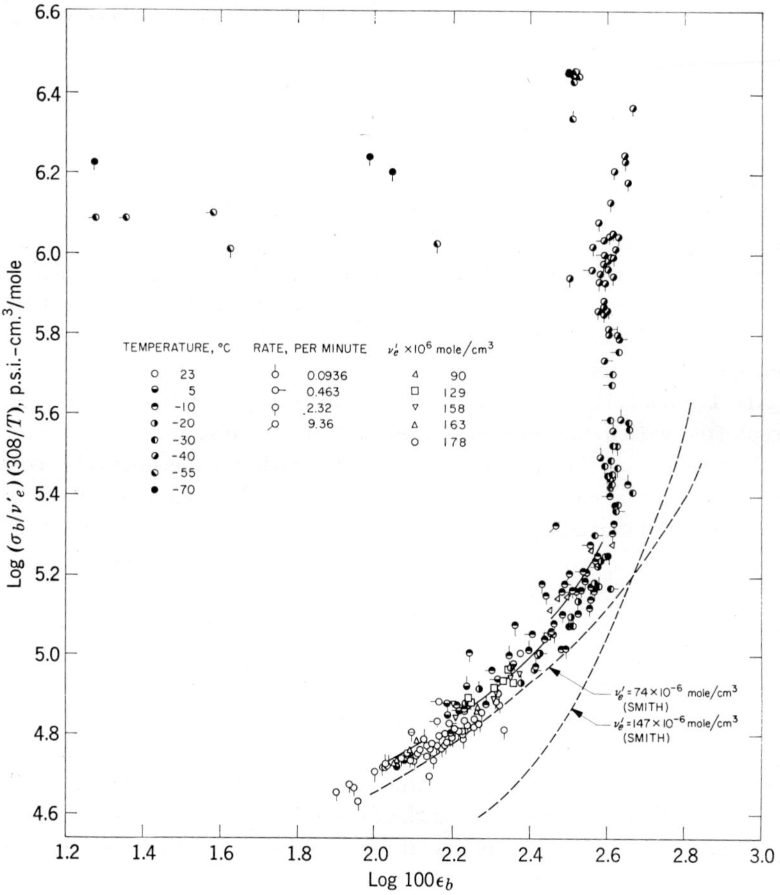

Fig. IIIB-23. The reduced failure envelope representation of the failure data for the SBR vulcanizates of Fig. IIIB-22. Also shown are the reduced envelopes for Smith's (dashed lines) two sets of data on SBR (90,91).

In attempting to analyze data such as shown in Figure IIIB-22, the results of Smith (86) suggest normalization of σ_b to unit concentration of effective chains. As we shall see in Subsection IIIB6, this result may also be derived from some of the molecular theories of tensile strength, even when this result was not of primary interest in the original derivation of the theory.

However, when comparing vulcanizates differing widely in gel fraction, the choice of the reference state is not obvious. A given specimen of vulcanized rubber contains both sol and gel (g) fractions, i.e., a portion of the material consists of a soluble fraction which is not attached to the network and hence is incapable of supporting an equilibrium load. Therefore, the concentration of effective chains can be based either on unit volume of gel, ν_e', or on unit volume of whole rubber, ν_e. The two measures of chain concentration are related by $g\nu_e' = \nu_e$. Hence, normalization of σ_b data can be attempted using either ν_e' or ν_e as the reference state.

In Figure IIIB-23, the ultimate property data for SBR have been replotted with σ_b normalized to unit ν_e'. It can be seen that the factor ν_e' superposes the envelope segments within experimental error ($\pm 5\%$). As a check on the ν_e' factor, the failure envelope data of Smith (90,91) on SBR vulcanized with sulfur were compared to these results. His ν_e' values as determined by swelling in benzene were employed and a value of $g = 0.9$ was assumed for the purposes of the comparison. Good agreement is obtained with his results for the sample with $\nu_e' = 74$ μmoles/cm.^{3}, but his data for $\nu_e' = 147$ μmole/cm.^{3} seem too low. It should be pointed out, however, that better agreement can be obtained for both sets of data by using an *average* value ν_e' as reported by Smith (91) and determined from swelling in four different solvents. For the data shown in Figure IIIB-23, normalization of ν_e superposes the failure envelope segments equally well since the value of g is essentially the same for all these vulcanizates.

To test the concept of a reduced failure envelope over a wider range in ν_e and elastomer type, data were assembled from several sources on 12 elastomer systems (15,16,26a,55,59,67,70,86). The usual test method was constant strain rate at different temperatures, though one set of data was obtained by following stress relaxation to rupture. In other cases, data were obtained by varying ν_e at constant test rate and temperature, and in one case (SBR) by the statistical variability of rupture at varying temperatures. The measured values of σ_b were reduced to both unit ν_e and to unit ν_e' using the reported, estimated, or measured values for the concentration of network chains, and to a standard temperature state of 100°C. above T_g (cf. Subsection IIIB2a).

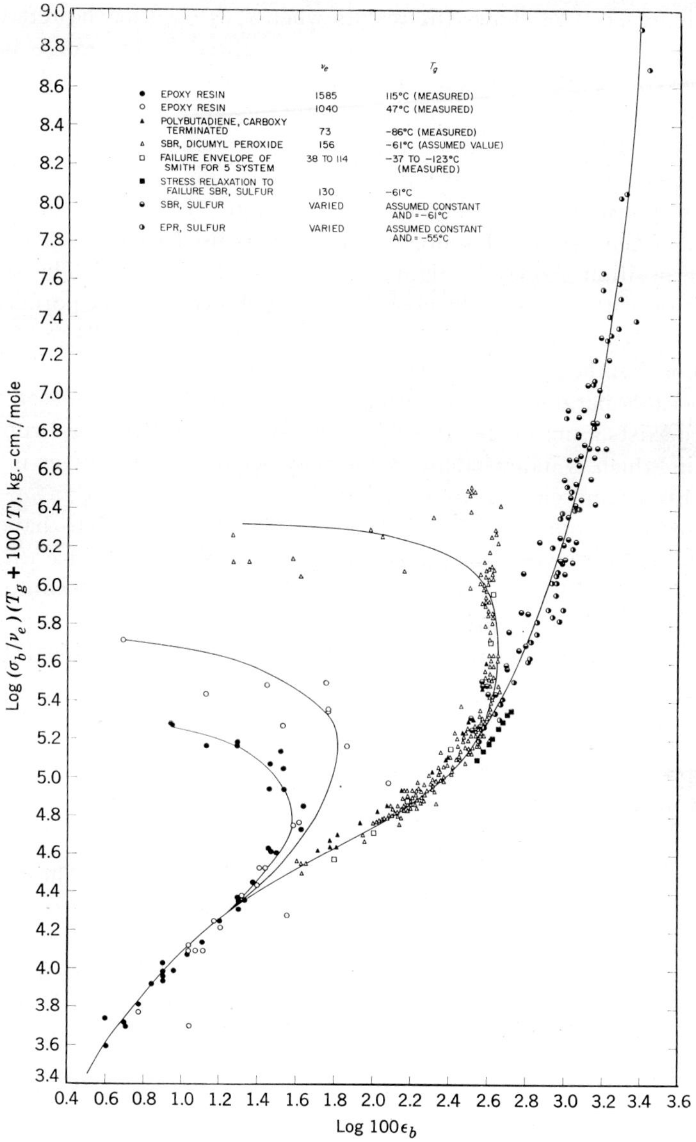

Fig. IIIB-24. Composite reduced failure envelope for 12 different systems taken from the literature (15,16,55,59,67,70,86). Included here are data obtained at constant ν_e but varying rate and temperature as well as data obtained at constant temperature but varying ν_e and rate. All data have been reduced to $T_g + 100$. In cases where T_g was not known, the value was estimated. Values of ν_e and T_g used to reduce the data are given in the key.

Figure IIIB-24 shows the results when ν_e is used as the reference state. The data depicted here cover a three-decade range in ϵ_b, 3–3000%; a three-decade range in ν_e, 2–1500; a five-decade range in $(\sigma_b/\nu_e)(T_g + 100)/T$, 10^4–10^9; and T_g values from −123 (silicone rubber) to 147°C. (epoxy). For all these systems, the high temperature portions of the envelopes superpose to define a master plot of the response which may be termed the common response curve. In view of the variety of sources and methods used to estimate ν_e, the superposition shown in Figure IIIB-24 is satisfactory. The superposition with ν_e' is equally good but is not shown since the normalized data so nearly resemble the form shown in Figure IIIB-24.

It is significant to note that these reduction schemes contain no disposable parameters. Even more importantly, it can be seen that there exists a unique relationship between rupture stress and rupture strain which is independent of the polymer over wide ranges in test conditions and chain concentration.

For many purposes, it will be convenient and useful to have an analytical expression for the common response curve. Since this expression should have the same form as that of a stress-strain curve (cf. the statistical variability results, Subsection IIIB7 and Fig. IIIB-23), a convenient choice is that of eq. (F) of Table IIIB-I. Thus

$$\log \frac{\sigma_b}{\nu_e^*}\left(\frac{T_g + 100}{T}\right) \geqq \log\left[\frac{\epsilon_b}{(1 + \epsilon_b)^2}\right] + A\left[\frac{\epsilon_b(\epsilon_b + 2)}{(\epsilon_b + 1)}\right] + 5.10 \qquad \text{IIIB3-(26)}$$

When $\nu_e^* = \nu_e$, $A = 0.50$, and when $\nu_e^* = \nu_e'$, $A = 0.40$. The data of Figure IIIB-24 are represented to within about ±10% by eq. (26).

It is evident that from given values of any two of the variables in eq. (26), the value of the third may be calculated. Thus, from the measured values of σ_b and ϵ_b, it is possible to calculate the chain concentration. A better procedure for determining ν_e from rupture data would be to compare plots of the experimental data with the master curve; the vertical separation then corresponds to $\log \nu_e$. The test temperature should be high enough to yield data in the quasi-equilibrium portion of the failure envelope; i.e., on the common response curve and not in the region of the upturn, a point which can be determined by inspection. As a rough rule, rupture points in the quasi-equilibrium region will generally be obtained when the test temperature is higher than about $T_g + 100$ for normal testing rates.

The existence of the common response curve implies that in this region, the rupture properties are independent of the chemical nature of the rubber. Presumably, therefore, the common response curve represents the "intrinsic" rupture behavior of a network. In addition, it is interesting to note that eq. (26) can be used to calculate a minimum value of σ_b at a given ϵ_b or a maximum value of ϵ_b, at a given σ_b, for any amorphous crosslinked rubber in a uniaxial test under any conditions of test temperature or rate. Figure IIIB-24 indicates also that for data which fall on the common response curve, unambiguous comparisons of the crosslink dependence of the rupture stress of elastomers should be made at constant ϵ_b, since log σ_b then varies directly as log ν_e. When tested in this manner, the rupture data should yield a linear relationship with unit slope. A semi-quantitative test of this prediction has been carried out using the 300% modulus as a measure of ν_e. It was determined that for several elastomers of widely different chemical nature, a linear relation fitted the data satisfactorily (68).

A comparison of the effect of ν_e on ultimate properties, Figures IIIB-23 and 24, indicates that decreasing ν_e at constant rate and temperature is equivalent to increasing the rate or decreasing the temperature at constant ν_e. This behavior is predicted to occur over the range of σ_b, ϵ_b values characterized by the common response curve. Hence, the common response curve can presumably be traversed in the direction of increasing σ_b and ϵ_b values by two independent paths: (*1*) increase the test rate or decrease the test temperature at constant ν_e and (*2*) decrease ν_e at constant rate and temperature. Alternatively, the common response curve can be traversed in the direction of decreasing σ_b and ϵ_b values by similar procedures.

It is implied that the common response curve, at least for low to intermediate values of ϵ_b, is the same as an "equilibrium" stress-strain curve. This would explain the superposition of the rupture data when σ_b is normalized to unit ν_e. It is evident that if the test temperature is sufficiently high and the test rate is sufficiently low (i.e., small σ_b and ϵ_b values), the common response curve would reduce to an essentially time-independent stress-strain law such as the expression predicted by the kinetic theory. Hence, theories of tensile strength of amorphous unfilled elastomers must also reduce to such a stress-

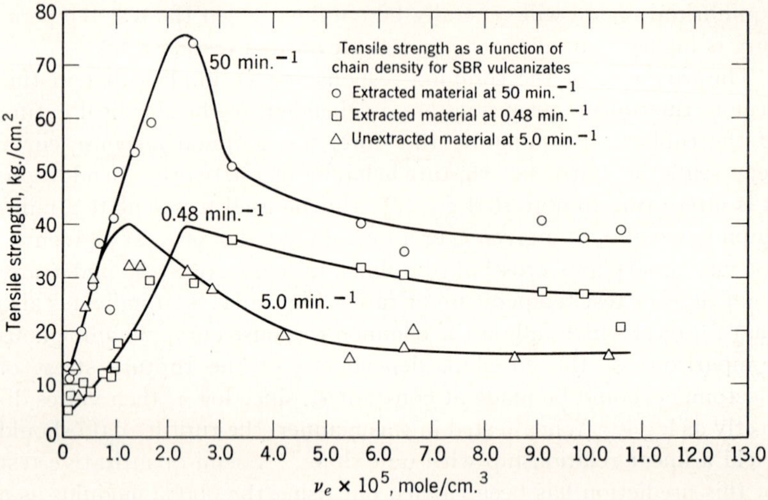

Fig. IIIB-25. The effect of rate of testing on the tensile strength of an SBR gum vulcanizate as a function of the concentration of effective network chains. After Bueche and Dudek (15).

strain law at a sufficiently high test temperature or its equivalent. This prediction is illustrated nicely in Figure IIIB-38 where the kinetic theory prediction, indicated by the dotted line, is approaching the experimental data at high ν_e values (equivalent, as mentioned above, to higher test temperatures). Hence, the correspondence of the common response curve to the Martin-Roth-Stiehler expression for stress-strain behavior is therefore not too surprising. The explanation for the generation of segments of the common response curve from statistical variability of the rupture data is also provided on the basis of the existence of an "equilibrium" stress-strain curve.

The complete failure envelopes are curves open toward the log σ_b axis. As ν_e decreases, the available experimental evidence indicates that the maximum value of log ϵ_b increases. In addition, only the lower portions of a series of envelopes representing decreasing ν_e values can be superposed by the factor log ν_e^{-1}. The upper portions of the envelopes will diverge from the common response at progressively higher values of the strain. The maximum strain attained, $(\epsilon_b)_{\max}$, i.e., the strain at which the envelope has an infinite slope, should (27) also be a measure of ν_e and data published in the literature are in qualitative agreement with this prediction (68). The latter is also implied by the ϵ_b, J'' interrelationships (Subsection IIIB4a) since the maximum in J'' is proportional to ν_e.

The effect of ν_e on strength is usually given in terms of the familiar σ_b versus ν_b plots, which are characterized by the occurrence of a maximum in σ_b at low ν_e (Fig. IIIB-25). Several explanations have been offered for this behavior, but only Epstein and Smith (27), Bartenev (2), and F. Bueche (16) have noted that the maximum is time-dependent and would apparently disappear under equilibrium conditions. Therefore attempts to provide an explanation for this maximum, whether based on the effects of crystallinity (36), chain orientation (93), or finite extensibility of chains (11), unfortunately, are in error insofar as they assume equilibrium behavior. For an amorphous system this maximum is an artifact caused by the relative rates at which σ_b and ϵ_b change with ν_e. When data such as those in Figure IIIB-25 are plotted as a reduced failure curve, log $\sigma_b T_0/\nu_e T$ versus log ϵ_b, this maximum disappears. In addition, the reduced failure curve representation of σ_b, ν_e data is time-independent over the range of strain rates covered.

The effects of a decrease in crosslink density on σ_b, t_b and ϵ_b, t_b curves, such as shown in Figures IIIB-11 and 12, follow directly from what has been stated above. The former will be unchanged at short reduced times but the quasi-equilibrium value at long reduced times will be decreased according to ν_e and will reach this quasi-limiting value at progressively longer times. The ϵ_b, t_b curves will be raised and broadened by a decrease in ν_e, and $(\epsilon_b)_{\text{max}}$ will be shifted toward longer times. The quantitative shifts should be assessable as follows. In Subsection IIIB4a it will be shown that $(\epsilon_b)_{\text{max}}$ is related to J''_{max}. Ferry (30) has shown that, according to Bueche's theory for crosslinked polymers (14),

$$J''_{\text{max}} = 0.42/\nu_e RT = 1.26/E_e \qquad \text{IIIB3-(27a)}$$

$$\omega_{\text{max}} = t_{\text{max}}^{-1} \propto \nu_e^{\,2}\, \zeta_0 \qquad \text{IIIB3-(27b)}$$

while for uncrosslinked systems, based on Marvin's (77) theory,

$$J''_{\text{max}} = 0.32\ M_e/\rho RT \qquad \text{IIIB3-(27c)}$$

$$\omega_{\text{max}} = t_{\text{max}}^{-1} \propto M_e^{\,-2}\, \zeta_0 \qquad \text{IIIB3-(27d)}$$

where M_e is the molecular weight between entanglement coupling points and ζ_0 is the average monomeric translational friction coefficient. The experimentally observed factor in eq. (27a) varies from about 0.18 to 0.30 instead of 0.42, but the time dependence, eq. (27b), is very nearly correct. Ferry's studies (30) on temperature-dependent entanglement coupling have also shown that the change in J', J'' and the frequency scale ω with a change in M_e is accurately given by eqs. (27c) and (27d). Thus $(\epsilon_b)_{\max}$ in Figure IIIB-13 would be expected to shift upward and to the right along a line with a slope of $^1/_2$ as the crosslink density decreased, just as does J''.

These relationships suggest an interesting study of comparisons between ultimate and small-strain behavior of polymers. Dynamic data could be used as a nondestructive test method for studying ultimate properties.

IIIB4. Time Dependence of Rupture

For materials which obey time-temperature superposition, studies of rupture in terms of the failure envelope represent a vast simplication because the envelope is independent of the test rate and only slightly temperature-dependent (through the factor T_0/T). Nevertheless σ_b and ϵ_b *are* time-dependent and this time dependence must also be investigated. There are a number of ways in which this can be done experimentally, though not all have been attempted. Thus, testing can be done at constant strain rate $\dot{\epsilon}_{\text{const}}$; at constant strain, i.e., stress relaxation to rupture; at constant stress, i.e., creep to rupture; and at constant stress rate $\dot{\sigma}_{\text{const}}$. In any of these tests the specimen can be repetitively loaded and the fatigue properties investigated. The primary loading mode used in fatigue is constant strain rate. Finally, a test program may be carried out involving the successive application of one or more test modes. In this way the cumulative effect of prerupture processes may be evaluated.

IIIB4a. Under Constant Rate of Strain $\dot{\epsilon}_{\text{const}}$

Villars (102), Fromandi (37), and Smith (90), have all shown that σ_b increases steadily with $\dot{\epsilon}_{\text{const}}$ at any given temperature while ϵ_b passes through a maximum. Villars was able to demonstrate this for SBR at room temperature by using a high-rate tester. The strain rates were not too reliable, so that interpretation of his results are difficult, but this was apparently the first demonstration that the breaking strains would pass through a maximum with increasing strain rate. Villars also tested natural rubber, as did Fromandi, with qualitatively similar results, but these will not be discussed because of the complications introduced by crystallinity.

A full investigation of the time dependence of rupture was first undertaken by Smith (90), who tested SBR at several rates and temperatures, and showed that the time-temperature superposition can be applied. The results were similar to those on Viton, already seen in Figures IIIB-11 and IIIB-12. The log σ_b,log t_b curve was sigmoidal in shape; though the short-time response was obscured by the fact that specimens underwent cold-drawing in this region. The log ϵ_b, log t_b curve was bell-shaped, passing through a maximum, as noted by Villars, similar to the Viton behavior indicated in Figure IIIB-12.

Kaelble (55), following this approach measured both the ultimate tensile properties and the dynamic small-strain shear behavior of two crosslinked epoxy resins having T_g's of 47 and 115°C. Although he had to carry his testing temperature up to 200°C., his results were qualitatively the same: time–temperature superposition was applicable, and σ_b increased steadily while ϵ_b passed through a maximum with increasing strain rate. The agreement between a_T derived from the dynamic and ultimate data was satisfactory. The elongations were low, $(\epsilon_b)_{\max}$ being only 70 and 40%, respectively, because of the high degree of crosslinking, but the response was rubbery.

The time dependence of the ultimate properties of even very highly filled rubbers such as solid rocket propellants are qualitatively similar to those of the unfilled binder, the response of the binder being, to a first approximation, unaffected by the presence of the filler (6,69). That is, time–temperature superposition is obeyed and the σ_b,ϵ_b, σ_b,t_b, and ϵ_b,t_b relationships resemble those of the unfilled rubber. This resemblance is maintained only if the filler is nonreinforcing, i.e., the binder forms only weak bonds with the filler. (On the other hand, the bonds should be thermally stable at the test temperature.) In the examples cited, the filler consisted of 70–80 wt.-% of an inorganic salt; the rubber, a polyether-based polyurethane. Similar results have been obtained from a wide variety of rubber types, filler types (various crystalline salts, metal powders and small glass beads), and volumetric loadings. In all cases the response is dominated by that of the rubber, and so is qualitatively similar to that of the unfilled gum stock.

Such systems can be considered as zero*th* approximations to the behavior of crystalline polymers. A high modulus material has been dispersed in the amorphous phase, (analogous to the incorrect but still useful raisin pudding concept of crystalline polymers) but the particle-rubber bond is not too strong. These particles do not deform and orient under stress as crystallites do, but instead the rubber detaches from the particles and vacuoles form around them.

Clearly the time dependence of σ_b and ϵ_b is related to the viscoelasticity of the rubber. At long reduced times it can be assumed that the specimen will break when a certain amount of energy is stored in the rubber, and the shortest chains are in the fully extended state. In this time–temperature regime, all chains can rearrange freely within the given time scale, the local viscosity being too low to affect greatly the chain motions. The response will represent equilibrium behavior and there will be no time dependence. As the test rate increases, the relaxation times for some of the longest submolecules will be greater than the time scale of the test and part of the stress will be dissipated by viscous losses as the chains move toward their equilibrium positions. Moreover, as the submolecule responds, it will act like an additional chain, so that the number of chains supporting the load will appear to have increased. Both factors will require that a greater stress must be applied before the critical amount of energy required for rupture can be stored in a chain. Note that it is the energy stored which is taken as a criterion for rupture, and not the maximum extension of the shortest chains. If the latter were adopted as the criterion, and if affine deformations were assumed, then it would be necessary to explain why this critical value increases with increasing strain rate, i.e., why chains which previously broke at the critical extension could now exceed that extension.

As the time scale grows progressively shorter the ratio of the energy dissipated to that stored will pass through a maximum. If the dissipation controls the extension, then the latter will also pass through a maximum, in accordance with observations. Dissipation is usually considered in terms of dynamic data and given as the loss modulus E'' or the loss tangent, $\tan \delta = E''/E'$. However $\tan \delta$ passes through its maximum at about the mid-point of the transition from rubberlike to glasslike behavior, while ϵ_b passes through its maximum at the very foot of the transition zone *before many of the submolecules have responded.* Rather than expecting ϵ_b and $\tan \delta$ to maximize at the same time, it would be a better approximation to say that ϵ_b had maximized just at the time scale where the modulus *enters* the transition zone. The reason for this apparent discrepancy is that the dissipation controlling ϵ_b is weighted by the longest relaxation times. As a result the loss compliance D'' rather than modulus E'' will provide a better indication of the behavior of ϵ_b.

For a linearly viscoelastic elastomer, this can be shown as follows: the stress–strain or stress–time relationship for a crosslinked network subjected to a constant rate of strain is (cf. Subsection IC4f, p. 84),

$$\sigma = E_e\epsilon + \dot{\epsilon}_{\text{const}} \int_{-\infty}^{\infty} H_l\, \tau(1 - \exp\{-t/\tau\})\, d \ln \tau \qquad \text{IIIB4-(28)}$$

where H_l is the tensile relaxation spectrum and $t = \epsilon/\dot{\epsilon}_{\text{const}}$. The total work done on the specimen per unit volume, is

$$\mathcal{E}_{tot} = \int_0^{\epsilon} \sigma\, d\epsilon = (\epsilon^2 E_e/2) + \dot{\epsilon}_{\text{const}}^2 \int_{-\infty}^{\infty} H_l\tau \times [t - \tau(1 - \exp\{-t/\tau\})]\, d \ln \tau \qquad \text{IIIB4-(29)}$$

The stored energy $\mathcal{E}_{st}$ $(=W)$ is determined by integration over the strain in the individual submolecule (or springs, if a phenomenological model is assumed), i.e.

$$\mathcal{E}_{st} = (E_e\epsilon^2/2) + (\dot{\epsilon}_{\text{const}}^2/2) \int_{-\infty}^{\infty} H_l\, \tau^2(1 - \exp\{-t/\tau\})^2\, d \ln \tau \qquad \text{IIIB4-(30)}$$

The energy dissipated $\mathcal{E}_d$ is thus

$$\mathcal{E}_d = \mathcal{E}_{tot} - \mathcal{E}_{st} = \dot{\epsilon}_{\text{const}}^2 \int_{-\infty}^{\infty} H_l\, \tau^2[(t/\tau) - (3/2) + 2 \exp\{-t/\tau\} - (1/2) \exp 2\{-t/\tau\}]\, d \ln \tau \qquad \text{IIIB4-(31)}$$

For $t < \tau$, the term in square brackets is zero; for $t > \tau$, it is just t/τ and so

$$\mathcal{E}_d = \dot{\epsilon}_{\text{const}}^2\, t \int_{\ln t}^{\infty} H_l \tau \, d \ln \tau \qquad \text{IIIB4-(32a)}$$

$$\mathcal{E}_d = \dot{\epsilon}_{\text{const}}^2\, t\eta_l' = \epsilon^2 E'' \qquad \text{IIIB4-(32b)}$$

where η_l' is the dynamic tensile viscosity, which in turn is equal to tE''.

To the extent that this expression is applicable at rupture, the rupture strain is therefore dependent on the dynamic small-strain loss modulus as

$$\epsilon_b = (\mathcal{E}_{d,b}/E'')^{1/2} \qquad \text{IIIB4-(33)}$$

or, in terms of shear loss compliance J'',

$$\epsilon_b \cong (J''\mathcal{E}_{d,b}/3)^{1/2} \qquad \text{IIIB4-(34)}$$

Since J'' is more sensitive to long relaxation (retardation) times than G'' and $t > \tau$, it is preferable to use J'' as a basis of comparison. So long as $\mathcal{E}_{d,b}$ is constant or a slowly varying function of time, then ϵ_b should pass through its maximum at the same point on the time scale as J'' does on a $1/\omega$ scale. Moreover, the two curves, log ϵ_b versus log t_b and log J'' versus log ω^{-1} should have qualitatively the same shape. This has been found to be true (65), Figure IIIB-26, for carboxy-terminated butadiene rubber (60) crosslinked with methyl bis-(azidirinyl)phosphate (MAPO), and for one of two epoxy resins studied by Kaelble (55), which are the only two systems in which both J'' and ϵ_b are directly available. It is also true for the resin-cured butyl rubber of Smith (86), where J'' was taken from the polyisobutylene data of Ferry, Grandine, and Fitzgerald (32,34).

The value of $\mathcal{E}_{d,b}/3$ evaluated at the maximum is found to be 15,000 for the butadiene rubber, 18,000 for butyl (resin-cured), and 7,000 dynes/cm.^{2} for the epoxy materials, i.e., sensibly constant in view of the range in $(\epsilon_b)_{\max}$ values, 40–600%. Further work on this point is in progress, since it may offer a convenient way to locate the time scale at which the maximum in ϵ_b will occur and to characterize a material in the region of this maximum.

It was shown in Subsection IIIB3 that $J''_{\max}$ is proportional to ν_e^{-1}, so that eq. (34) implies $(\epsilon_b)_{\max} \propto \nu_e^{-1/2}$. This is similar to the functional relationship given by the stress–strain eq. (5).

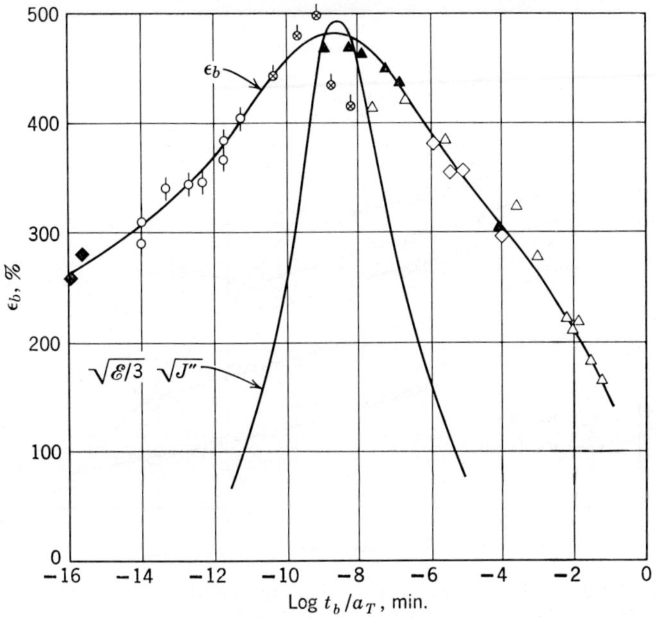

Fig. IIIB-26. A comparison (65) of rupture strain with the square-root of the loss compliance, $(J'')^{1/2}$ as functions of the reduced rupture time for a carboxy-terminated polybutadiene vulcanizate (58). In the figure, $\mathcal{E}$ corresponds to the energy dissipated at rupture; $\mathcal{E}_{d,b}$ of the text. The calculated curve has been matched to the experimental curve at the maximum.

IIIB4b. Under Constant Strain; Stress Relaxation

Instead of elongating the specimen at a constant rate until it breaks, it can be maintained at some fixed strain e.g., point A in Figure IIIB-14. The stress will relax until the failure envelope is reached, point C, at which time the specimen will break. The higher the initial strain, the sooner the specimen should break, as illustrated by the data of Smith and Stedry (91) on SBR (Fig. IIIB-27). The rupture times, indicated by the dotted line, increase with decreasing initial strain, as expected. This dotted line corresponds to a portion of the long-time side of a bell-shaped ϵ_b, t_b plot obtained at constant strain rate, e.g., to Figure IIIB-12*b* at $t_b/a_T > 10^{-2}$ min. Constant strain rate experiments were also run at the same time on the same material, but an unambiguous direct comparison between the rupture times in the

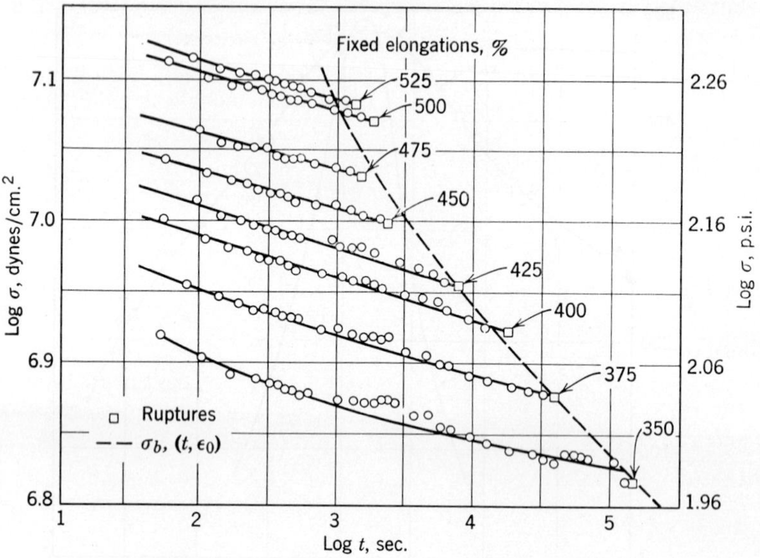

Fig. IIIB-27. Failure in stress relaxation for an SBR vulcanizate as a function of the initial strain (91). The dashed line indicates the envelope of breaking times. The numbers correspond to the initial strain. From Landel and Stedry (70).

two sets of experiments could not be made because of experimental difficulties. Nonetheless the results indicated that t_b for the former condition exceeded t_b for the latter.

Smith (87) subsequently showed that the failure envelope, rather than the ϵ_b, t_b relationship, *was apparently independent of the test method*. When this was true, and for linearly viscoelastic materials, he could show that t_b should depend on the slopes of the plot of the stress relaxation modulus versus t or the creep compliance versus t. If the time dependence of $E(t)$ is given by

$$E(t) = a(1 - b)t^{-b} \qquad \text{IIIB4-(35)}$$

then the time dependence of $D(t)$ is given as

$$D(t) = t^b \sin \pi b / [ab\pi(1 - b)] \qquad \text{IIIB4-(36)}$$

where a is a constant and b is the slope of a log $E(t)$ versus log t or a log $D(t)$ versus log t plot.

For experiments which terminate at the same point on the failure envelope, which is not the same as the paths described in Figure IIIB-14, then the rupture time $(t_b)_\epsilon$ in stress relaxation, $(t_b)_\sigma$ in creep, and $(t_b)_{\dot\sigma\text{const}}$, constant stress rate are related to those at constant strain rate $(t_b)_{\dot\epsilon\text{const}}$ by:

$$\log (t_b)_\epsilon = \log (t_b)_{\dot\epsilon\text{const}} + b^{-1} \log(1 - b) \quad \text{IIIB4-(37a)}$$

$$\log (t_b)_\sigma = \log (t_b)_{\dot\epsilon\text{const}} + b^{-1} \log [\pi b(1 - b)/\sin \pi b] \quad \text{IIIB4-(37b)}$$

$$\log (t_b)_{\dot\sigma\text{const}} = \log (t_b)_{\dot\epsilon\text{const}} + b^{-1} \log [\pi b(1 - b^2)/\sin \pi b] \quad \text{IIIB4-(37c)}$$

These relations are depicted in Figure IIIB-28. When $b \to 0$, viscous effects are small and the rupture times for the two constant rate tests or for the two static tests are the same, although the static loading conditions will give rise to rupture times which are less than those obtained in constant rate tests by a factor $\cong 2.7$. Appreciable differ-

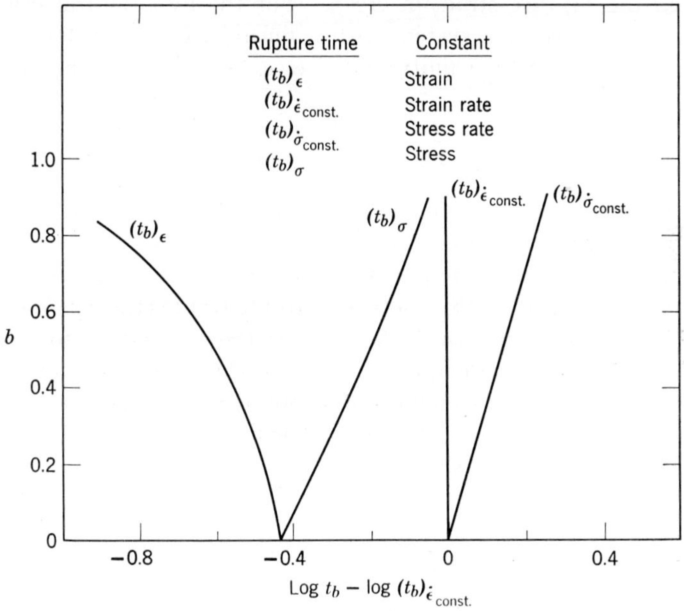

Fig. IIIB-28. A comparison of the dependence of the rupture time, relative to $(t_b)_{\dot\epsilon\text{const.}}$, on the test mode for failure in stress-relaxation, creep, constant strain rate, and constant stress rate experiments for a linearly viscoelastic material.

ences in rupture times are predicted as the specimens are taken into the time-dependent zone. Equations (36) and (37) are not expected to be exact for a real elastomer, since they were derived for a theoretical linear material. Still, it should develop that $(t_b)_{\dot{\epsilon}\text{const}} > (t_b)_\epsilon$ and $(t_b)_{\dot{\sigma}\text{const}} > (t_b)_\sigma$. An experimental test of these relations is in progress.

On the other hand, for experiments which have the same initial starting point, e.g., point A of Figure IIIB-14, different lifetime relationships are obtained (68). For the example shown, the creep lifetime will be less than the stress relaxation lifetime. The equation is

$$\log (t_b)_\sigma = \log (t_b)_\epsilon + \frac{1}{b}\left[\log\left(\frac{\pi b}{\sin \pi b}\right) - (n - 1)\log\frac{\epsilon_b}{\epsilon_0}\right] \qquad \text{IIIB4-(37d)}$$

where n is the slope of the line through the points ϵ_0, which is both the initial strain and the rupture strain in the stress relaxation experiment and ϵ_b, the rupture strain in creep. For the hypothetical example chosen by Smith of a system with a linear stress-strain law, n will always be greater than unity and if $\epsilon_b/\epsilon_0 > \pi b/\sin \pi b$, then $(t_b)_\sigma < (t_b)_\epsilon$.

IIIB4c. Under Constant Load; Creep

F. Bueche has studied the σ_b, t_b relation in creep at a series of temperatures for uncrosslinked poly(butyl methacrylate) as a test of his theory of the tensile strength of polymers above T_g (12), and for polystyrene and poly(ethyl methacrylate) to test his theory for the behavior below T_g (13). In the latter cases the data extend to the rubbery zone, though the emphasis is on the glassy state response. In all cases σ_b is linear in $\log t_b$ down to about $\sigma_b = 10^8$ dynes/cm.².

On the other hand, Bartenev (2) tried to determine whether the isothermal relationship should be

$$\sigma_b = A - B \log t_b \qquad \text{IIIB4-(38)}$$

as observed by Bueche (12,13), Busse (19) and Zhurkov (113), or

$$\log \sigma_b = \log A - B \log t_b \qquad \text{IIIB4-(39)}$$

His work was especially interesting for the number of materials studied [polybutadiene, SBR containing 10 and 30% styrene, and poly(butadiene–acrylonitrile) rubbers containing 18–40% nitrile] and for the use of 20 specimens at each stress level to establish a statistically valid t_b. The results, shown in Figure IIIB-29, indicate that eq. (39) is to be preferred. However, it is probable that the curves would flatten out at still longer times, for when his results for SBR are compared with those of Smith obtained from $\dot{\epsilon}_{\text{const}}$ tests (90), they are found to lie in the σ_b,t_b range which just covers the middle of the transition zone in σ_b. Through this region, the log-log plot is indeed linear. Note that the curves of Figure IIIB-29 are displaced according to T_g, as expected.

IIIB4d. Under Constant Stress Rate $\dot{\sigma}_{\text{const}}$

Studies of rupture at constant stress rate or constant loading rate have not, to date, been performed on elastomers. Constant loading rate equipment is customarily used for metals, but it is generally not sensitive enough to handle rubber specimens. For constant stress rate experiments on a conventional constant crosshead speed tester, the cross-sectional area of the specimens would have to be monitored and the crosshead speed adjusted accordingly.

Except for the difference in rupture times noted above, the results for the constant stress rate experiments would not be expected to be markedly different than the results obtained in constant strain rate experiments. The stress–strain curves would have the same general appearance, the principal difference being that they would be flatter, with less of an upswing at high strains.

This similarity in shape will not always prevail. For filled systems in which the load-time curve of $\dot{\epsilon}_{\text{const}}$ tests passes through a maximum, the $\dot{\sigma}_{\text{const}}$ tests will give an ϵ,t curve which will be similar in shape up to the region of the maximum found with the $\dot{\epsilon}_{\text{const}}$ test and then rise rapidly, as indicated in Figure IIIB-30. This can be deduced from a consideration of a σ,ϵ,t surface based on data obtained in $\dot{\epsilon}_{\text{const}}$ tests such as shown in Figure IIIB-31. For specimens which exhibit a yield point, a similar sharp upturn in the ϵ,t curve will occur in the region of necking. This will happen because the apparatus will try to increase the stress at a constant rate but, since the specimen elongates in an essentially discontinuous fashion, the crosshead will have to accelerate to high speed to maintain a constant stress rate.

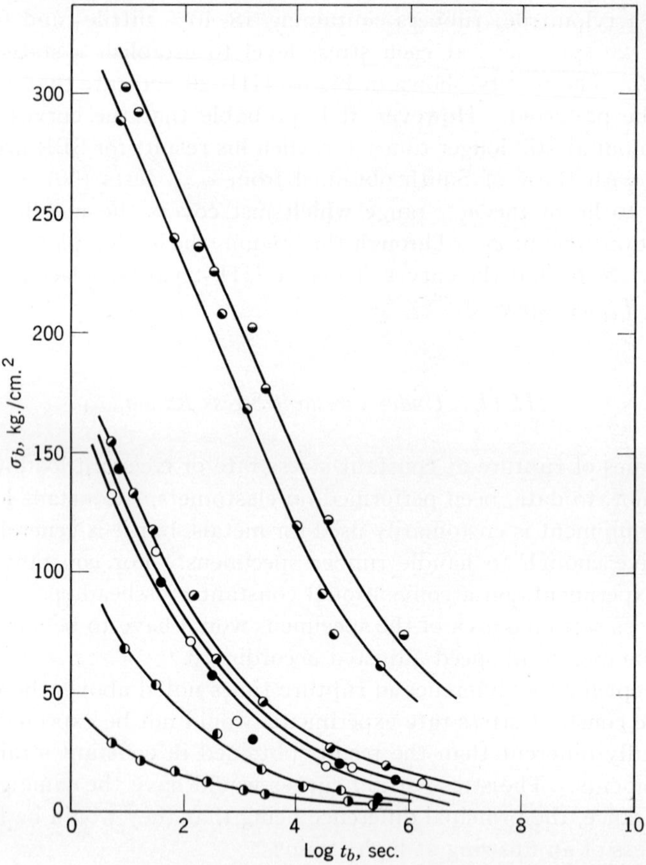

Fig. IIIB-29*a*. Rupture time in creep as a function of the initial stress for several polymer types, plotted in semi-logarithmic coordinates. ◑ SKS-10; SBR, 10% bound styrene. ◐ SKB; polybutadiene rubber. ○ SKBM; polybutadiene rubber, oil extended. ● SKS 30; SBR, 30% bound styrene. ◒ SKN-18; poly(butadiene-acrylonitrile) 18% nitrile. ◓ SKN-26; poly(butadiene-acrylonitrile) 26% nitrile. ◒ SKN-40; poly(butadiene-acrylonitrile) 40% nitrile. After Bartenev and Briuk hanova (2).

IIB4e. In Cumulative Damage Testing

One important class of tests which has apparently not been carried out on elastomers is the cumulative damage test (74,79,108). The results of this test are related to the fatigue properties, and are extensively investigated for metals. In this test, instead of determin-

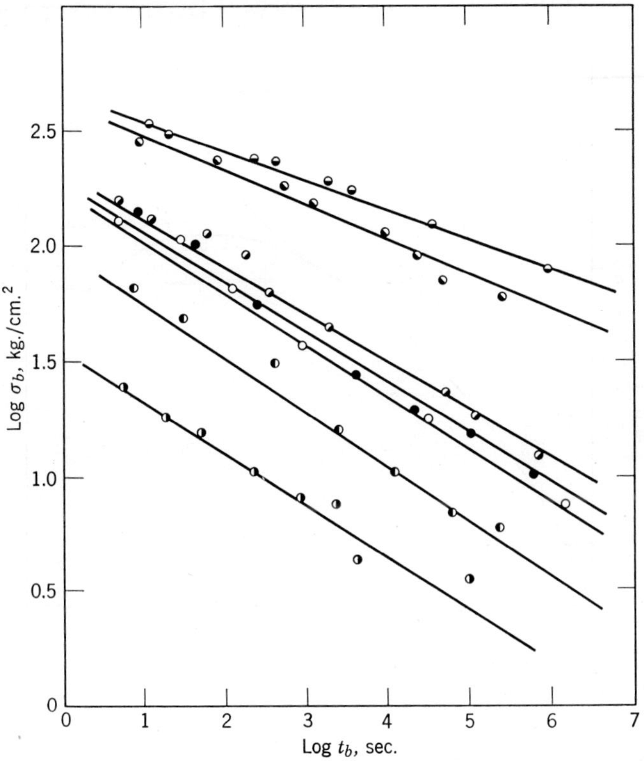

Fig. IIIB-29*b*. Data of Fig. IIIB-29*a* plotted in log-log coordinates. The symbols have the same meaning as in Fig. IIIB-29*a*. After Bartenev and Briukhanova (2).

ing the isothermal behavior over a wide range of constant strain rates, specimens could be strained at a combination of rates by applying $(\dot{\epsilon}_{\text{const}})_1$ for a short time, shifting to $(\dot{\epsilon}_{\text{const}})_2$ for a time and then shifting either back to $(\dot{\epsilon}_{\text{const}})_1$ or to a new rate, $(\dot{\epsilon}_{\text{const}})_3$. If $(\dot{\epsilon}_{\text{const}})_1 < (\dot{\epsilon}_{\text{const}})_2 < (\dot{\epsilon}_{\text{const}})_3$ then the specimen will break sooner than it would have broken under $(\dot{\epsilon}_{\text{const}})_1$ alone, but not as soon as if only $(\dot{\epsilon}_{\text{const}})_3$ had been applied. One of the simplest theories of cumulative damage assumes that the total lifetime of the specimen is the sum of the individual fractional lifetimes which have been expended. Thus, if the duration of the test is t_i under rate $(\dot{\epsilon}_{\text{const}})_i$, where the normal rupture time is $(t_b)_i$, then the lifetime expended is $(t_i/t_b)_i$ and the damage concept may be expressed as

$$\sum_{i=1} t_i/(t_b)_i = 1 \qquad \text{IIIB4-(40)}$$

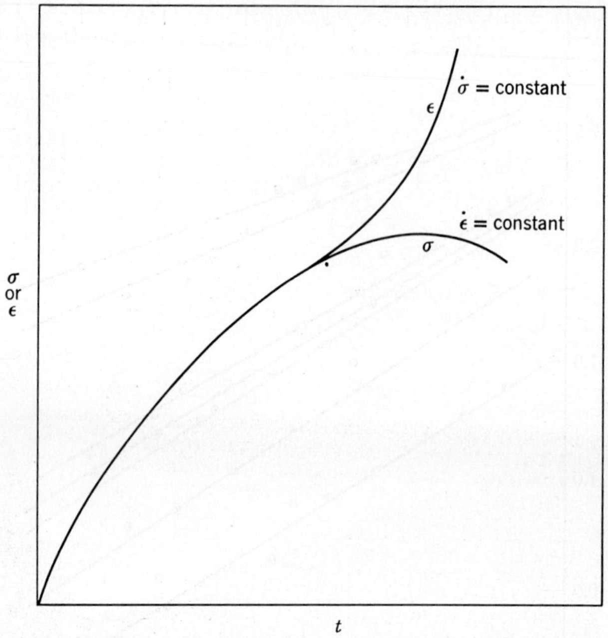

Fig. IIIB-30. The effect of a constant strain rate and a constant stress rate test on the stress-strain-time behavior of a composite solid propellant (hypothetical) (66).

The lifetime to be expected could be based on the σ_b,t_b or the ϵ_b,t_b relationships, such as shown in Figures IIIB-11*b* and IIIB-12*b* or on other relationships as well, such as total work to rupture W_b versus t_b or energy stored $\mathcal{E}_{st,b}$ versus t_b. For constant strain rate tests the ϵ_b,t_b relationship is the most convenient to employ, since the fraction of the lifetime expended in each step is equal to the fraction of the total strain available in that step.

Figure IIIB-32 shows in log-log coordinates the projection of the strain history for specimens tested at two rates differing by a factor of 100, and for specimens tested at one rate to 4% strain and then taken to rupture at the second rate. Detailed analysis (65) of such plots (and their equivalents plotted in linear coordinates) has shown that the simple cumulative strain damage theory, eq. (40), can give the correct breaking time with rupture locus *B*, i.e., when operating to the right of the maximum in the bell-shaped log ϵ_b,log t_b plot. How-

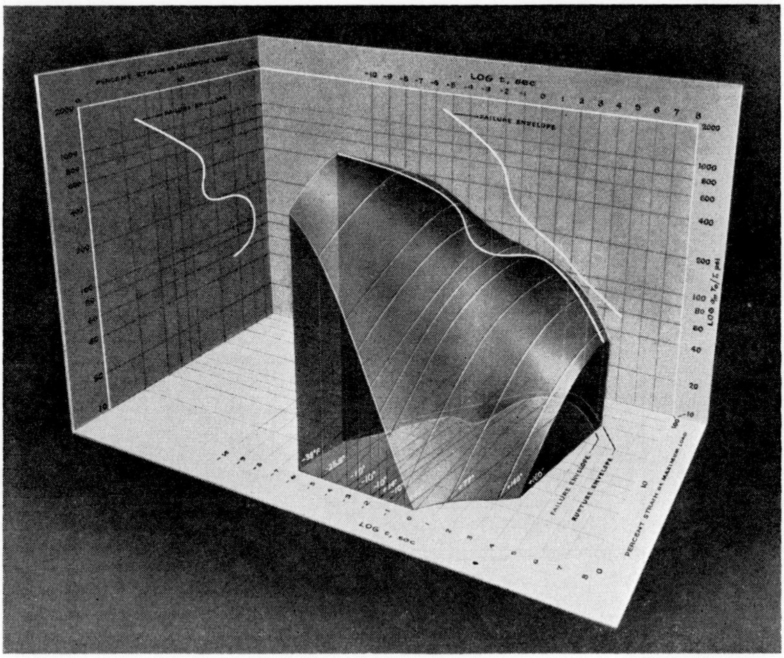

Fig. IIIB-31. Tensile property surface for a propellant system showing the relationship of the failure boundary to the surface and the projections to the σ,ϵ; σ,t; and ϵ,t planes. Note that failure does not coincide with rupture in this system.

ever, it appears that the lifetime is necessarily underestimated with rupture locus A. For an experimental test, it will be more useful, i.e., provide a more stringent test of the theory, to move to progressively higher rates when the expected rupture strain is decreasing with increasing t_b, and conversely to move to progressively lower rates when it is increasing with t_b.

If the damage depends on the order of application of stress or strain, then the simple expression of eq. (40) is no longer valid. For metals where damage depends on the order of application of the stress, Corten and Dolan (25) propose

$$\frac{t_1}{(t_b)_1} + \sum_{i=2} \frac{t_i}{(t_b)_i} \left(\frac{\sigma_i}{\sigma_1}\right)^a = 1 \qquad \text{IIIB4-(41)}$$

where a is a material constant and $(t_b)_i$ is the lifetime under stress σ_i. In addition, the damage may depend on the test mode. In polymers this is expected to be the case because it has already been pointed

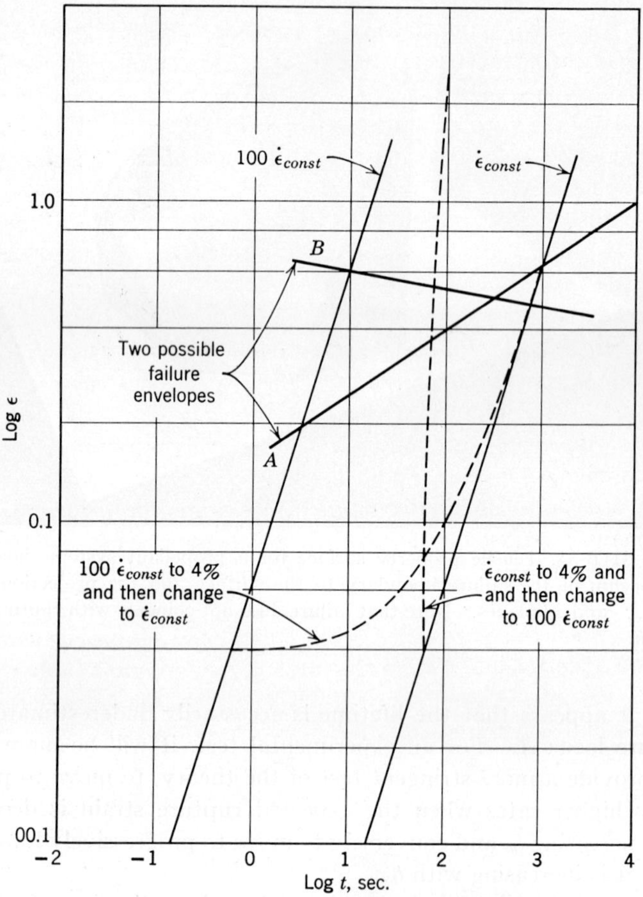

Fig. IIIB-32. Projection of the strain history for specimens tested at two rates. From Landel (66).

out that the rupture *time* will depend on the test mode [cf. eqs. (37a-d)]. Thus, rather than testing simply by changing the strain rate, the specimen could be tested part of the time at a constant strain rate and part of the time be permitted to undergo stress relaxation.

Alfrey (1) gives a qualitative discussion of cumulative damage in creep experiments (also see Section IIIA).

IIIB5. Temperature Dependence of Rupture

The temperature dependence of rupture behavior as measured at otherwise constant testing conditions is very similar to that of the time dependence. Log σ_b is still a sigmoidal function of log t_b and ϵ_b passes through a maximum. However, it is well to remember that data obtained at a fixed rate as a function of the temperature are translated *and* suffer a change in shape when the time scale is changed. This is illustrated in Figures IIIB-33*a* and 33*b* for a filled system tested at a nominally constant strain rate (69). (Failure here represents the maximum in a load-time curve as measured at constant crosshead speed, rather than literal rupture as is the case for unfilled elastomers.) The temperature effect would be qualitatively the same, however, for an unfilled rubber. The transition zones in both the σ_m, t_m and $E(t), t$ plots become more pronounced as the rate is decreased; the change in the ϵ_m curve is especially striking. As a result of such changes in shape, data from two materials tested at two different rates as a function of temperature cannot be compared, except in a qualitative fashion, unless their glass temperatures are similar or the test rates are adjusted in accordance with their respective T_g values.

Even when the test rates are the same, ultimate properties will depend on T_g, often in a straightforward manner, as noted by Borders (8) and Boyer (9). Thus Figure IIIB-34 shows the behavior of four polyurethane elastomers as a function of temperature (89). The observed differences cannot be ascribed to the diisocyanate structure without first examining the effect of the accompanying changes in T_g. Since T_g plays a role in polymer physics somewhat analogous to the critical temperature T_c for gases, comparison of mechanical properties should be made at equal values of $T - T_g$ (cf. Subsection IIIB2a). Amorphous polymers at such temperatures are then in corresponding states, just as gases are when at equal values of T/T_c. Thus when the data of Figure IIIB-34 are replotted versus $T - T_g$, single curves result, Figure IIIB-35, showing that the basic response of the polymers is independent of the diisocyanate used. Of course, the rupture behavior will not be generally independent of the chain structure and so different elastomers will have different responses when

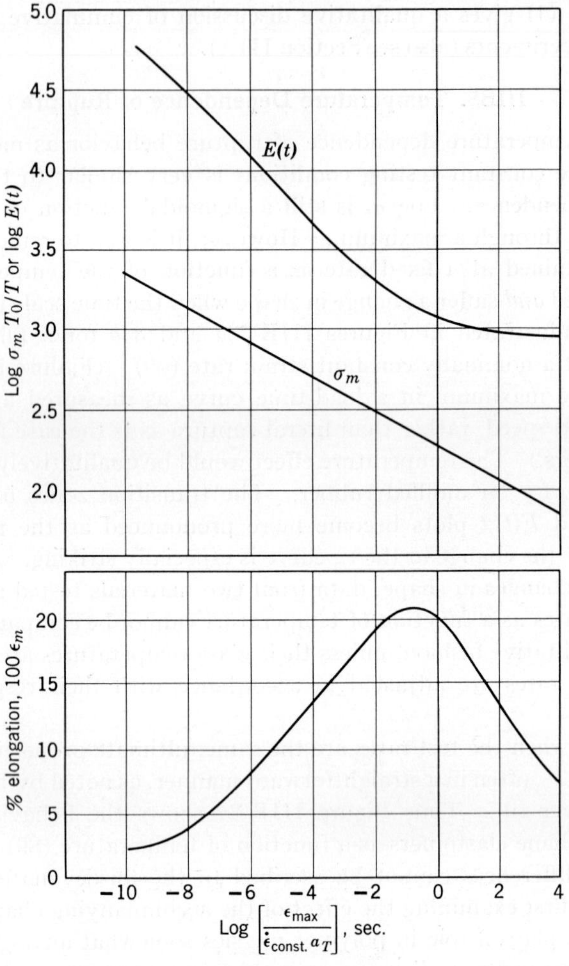

Fig. IIIB-33*a*. Time dependence of the modulus $E(t)$, the stress σ_m, and strain ϵ_m at the maximum load for a filled system as measured in a constant strain rate test. The data have been reduced to a reference temperature of 25°C. A change in temperature would translate the curves horizontally but not change their shape. From Landel and Smith (69).

compared at the same crosslink density and the same $T - T_g$ state, as illustrated in Figure IIIB-36 (89). Here the polyurethanes all have the same crosslink density, which is given as 187 μmoles/cm.^{3},

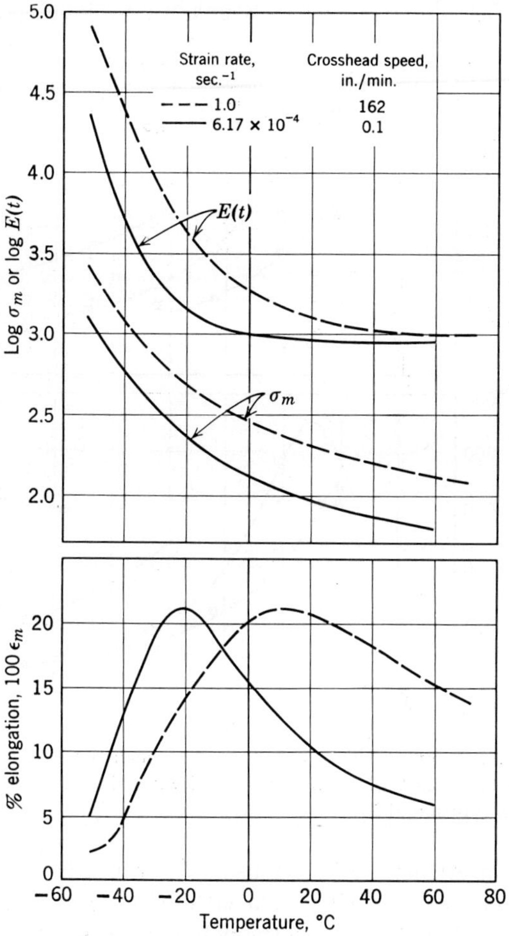

Fig. IIIB-33*b*. Temperature dependence of $E(t)$, σ_m, and ϵ_m for the same system tested at two different rates, showing the accompanying change in shape. From Landel and Smith (69).

based on stoichiometry, and the same value of $(\epsilon_b)_{\max}$ as predicted [cf. Subsection IIIB3]. Since the polyester ϵ_b curves maximize at higher $T-T_g$ values, they must also maximize at longer times in an ϵ_b, t_b plot. According to eq. (27b), this indicates that the monomeric friction coefficient ζ_0 increases through the series PPG, 80/20 polyester and 50/50 polyester. The SBR rubber has a smaller $(\epsilon_b)_{\max}$

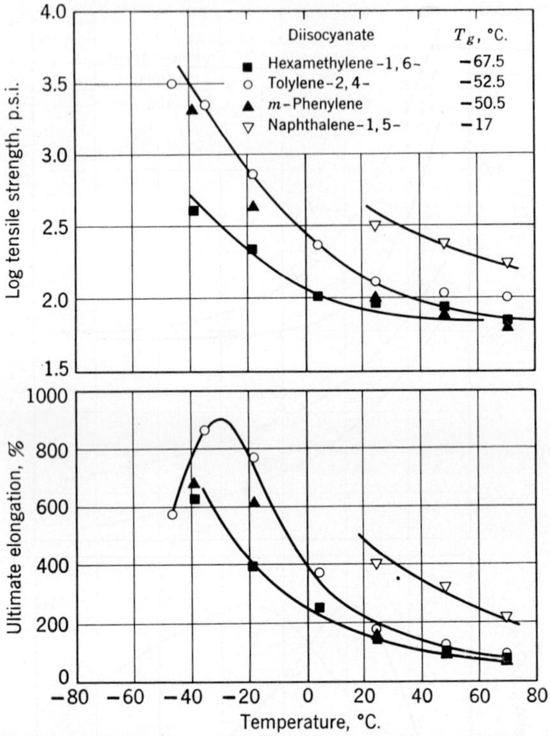

Fig. IIIB-34. Rupture stress and rupture strain as a function of the test temperature for polyurethane elastomers prepared with different isocyanates as indicated. From Smith and Magnusson (89).

than the polyurethane, 600% versus 900%, which is at variance with the reported crosslink density. The latter is 120 μmoles/cm.^{3}, so that the SBR curve should lie above the polyurethanes. The discrepancy may arise from the fact that the crosslink density for the latter systems is a calculated one and since side reactions can occur readily, the actual values could be considerably different.

In addition to the changes brought about by heating or cooling a specimen *before* testing, there is also the interesting problem of thermally induced stresses and strains. Typically, this occurs when the rubber is bonded to a metal part and the temperature is then raised or lowered from the bonding temperature. Even if the rubber is at a uniform temperature, the difference in expansion coefficient

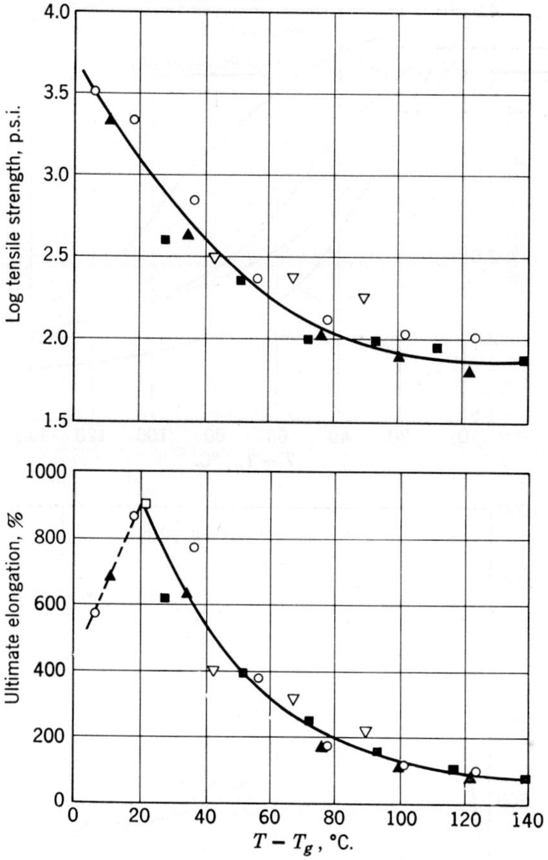

Fig. IIIB-35. Data of Fig. IIIB-34 replotted vs. $T - T_g$ showing that differences in behavior may be associated with changes in T_g and not with the changes in structure of the chain backbone. Symbols as in Fig. IIIB-34. From Smith and Magnusson (89).

between the metal and rubber will set up a strain. As the temperature changes, the strain rate will be determined by the rate of change of temperature. Since rubber is a reasonably good thermal insulator, the strain rate will be very low in a large specimen. Thus a low strain rate experiment is initiated, but as the strain increases, the path on the tensile property surface will shift along the time scale in the direction of shorter reduced times, in accordance with a_T. As a result, the

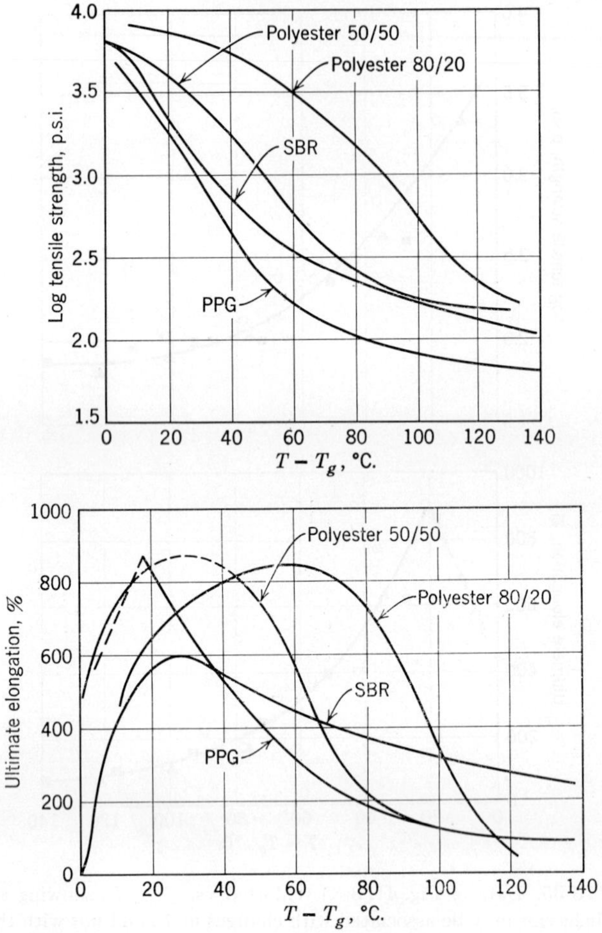

Fig. IIIB-36. Rupture stress and rupture strain as a function of $T - T_g$ for three polyurethanes and one SBR elastomer showing that changes in backbone structure can effect the response, even though the systems have the same cross-link density and are being compared at the same reduced temperature. From Smith and Magnusson (89).

experimental path across the tensile property surface is a complex one. Nevertheless, it should be possible to calculate the temperature at which breaking will occur from the WLF equation and a knowledge of ϵ_{max} versus t_{max}/a_T, as determined in constant strain rate tests (e.g., such as shown in Fig. IIIB-33*a*). Jones (54) has shown this to be true for a filled system, the calculations being performed by an iterative technique.

IIIB6. Molecular Theories of Uniaxial Tensile Strength

The polymer physicist is in a far more fortunate position than the metals or ceramics physicist in that the molecular nature of the amorphous gum polymer is relatively well defined and mathematically more tractable compared to other materials. Thus for example, with the kinetic theory of rubberlike viscoelasticity, it is possible to calculate *from first principles* the absolute magnitude of a stress under a general strain field, to within a factor of about two (101). As implied in the Introduction to this section of this book, *rupture in amorphous rubbers should be considered as simply another point on the viscoelastic stress-strain curve. Therefore, it is to be expected that, as a first approximation, theories of the breaking strength of rubbers begin at the same place as for the origins of the molecular theories for its viscoelasticity, i.e., at the molecular level. Moreover, in contrast to the behavior of classical solids, such as those of the glassy state, it is quite unnecessary to postulate the existence of flaws such as voids to account for the rupture of a rubber in the quasi-equilibrium state. Instead, the distribution of chain lengths between crosslinks provides an a priori, readily understood site for initial rupture because the shortest chains, being more quickly strained to a critical extent, will break first.*

IIIB6a. G. Taylor and S. Darin

The theory of Taylor and Darin (93) relates the tensile strength of amorphous gum rubbers to molecular parameters. It assumes that the tensile strength is proportional to the number of effective chains per unit volume ν_e which are oriented within some small angle δ to the stress axis:

$$\sigma_b = A \nu_e \omega_0 \qquad \text{IIIB6-(42)}$$

where A is a constant of proportionality and ω_0 is the orientation factor, i.e., the fraction of chains which are oriented within δ, and

$$\omega_0 \cong 1 - (1 + \delta^2 \lambda_b{}^3)^{-1/2} \qquad \text{IIIB6-(43)}$$

It is implied that δ will be independent of the crosslink density. If this is true, then eq. (42) immediately predicts that the ratio σ_b/ν_e will be a universal function of the rupture strain, i.e., it predicts the reduced failure envelope concept. However, these authors were primarily interested in studying the variation of tensile strength with crosslink density and so did not consider the implications for ϵ_b. They tried to fit their equation to a plot of σ_b versus ν_e calculated from stoichiometry, choosing the values of A and δ^2 to be 4365 and 5×10^{-5}, respectively. The extensive data of Epstein and Smith (27), coupled with direct measurements of ν_e, provided a more rigorous test of the theory. They showed that the maximum in the σ_b,ν_e plot was dependent on the degree of swelling or extraction of the specimens and on the rate of testing, a point also noted by Bartenev (2) during the course of the creep studies described in Subsection IIIB4c and subsequently confirmed by F. Bueche (16). It was concluded that the peak was due to a viscous effect rather than to some inherent property of the network structure itself. Qualitative agreement with the Taylor-Darin theory was obtained by Epstein and Smith in that the theoretical σ_b,ν_e curves had the proper shape and their location could be matched reasonably well to the experimental curves by appropriate choices of A and δ. However, the calculated curves always peaked at higher values of crosslink density and predicted too low a value of tensile strength at the highest crosslink density. Moreover, although not explicitly stated, their results would indicate that at the very lowest crosslink density, where the breaking strain ϵ_b was about 3000%, about 37% of the chains were fully oriented, and these all lay within 0.4 degrees of the stress axis. In the region of the maximum of tensile strength, where the rupture strain was about 1000%, only about 3% of the chains were oriented (within the same small angle).

According to eq. (42), the ratio of the tensile strength to the concentration of effective chains is a function of the known elongation at break, a function which contains the two parameters δ^2 and A. But σ_b/ν_e as a function of ϵ_b is just the reduced failure envelope. Both A and δ^2 can therefore be readily evaluated by comparing plots of log ω_0 versus log ϵ_b for various values of δ^2, with the reduced failure envelope. The proper value of δ^2 will be that which most nearly reproduces the experimental curve and log A will be the vertical separation of the two curves. When this is done, however, it can be seen that the theoretical curves have the wrong shape (Fig. IIIB-37). They are

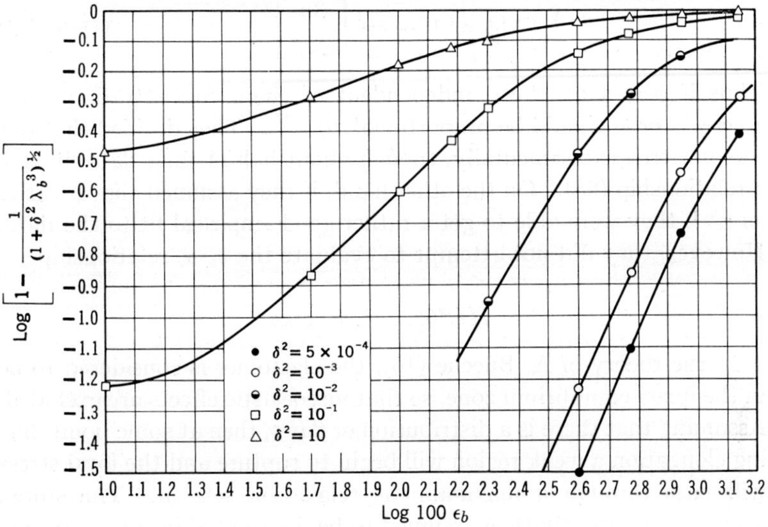

Fig. IIIB-37. Calculated failure envelopes based on the theory of Taylor and Darin (93) for five values of the parameter δ^2 as indicated.

concave downward and flatten out to a constant value, when δ^2 is large, at moderate values of ϵ_b, while the failure envelope is concave upward and rises sharply at high values of ϵ_b. This implies that either δ^2 or A or both are functions of the extension ratio.

The choice of δ^2 and A therefore depends on what portion of the reduced failure envelope it is desired to fit; as δ^2 decreases, the calculated envelopes become steeper and flatten out at higher values of ϵ_b. In the quasi-equilibrium or high temperature region of the reduced failure envelope, where the theory would be expected to be most applicable, δ^2 would have to have a value of about 10^{-1} or 10^{-2} in order to get agreement, implying that about 95% of the chains are fully elongated and oriented within about 6° of the stress axis at rupture. [These values would be changed slightly if the full Taylor and Darin formula for ω_0 had been employed in place of eq. (43).]

Epstein and Smith (27) presented an alternative rupture criterion related to the fraction, φ, of fully extended chains. However, instead of relating this fraction to σ_b as in eq. (42) they related it to ϵ_t. Assuming that a one-dimensional gaussian distribution of end-to-end chain lengths is valid,

$$\phi \approx 1 - \text{erf}\left[\frac{3n}{2\lambda_b^2}\right]^{1/2} \qquad \text{IIIB6-(44)}$$

Thus if ϕ is a constant, independent of chain concentration, then $\lambda_b \propto n^{1/2}$ or λ_b should be proportional to $\nu_e^{1/2}$. They decided that this was not true experimentally, though about half of their data fit such a relationship (68). On the other hand, if they assumed that ϕ varies as ν_e^{-1}, they were able to get a rather good empirical fit to the data. However, they did not attempt to evaluate the σ_b/ν_e relationship.

IIIB6b. A. Bueche

In the theory of A. Bueche (10), the elastomer is considered to be in the quasi-equilibrium zone, so that viscoelastic effects are excluded. Assuming that there is a distribution of flaws, then at some point during elongation a weak region will begin to rupture and the local stress will pass through a maximum as relaxation sets in. The stored energy is apparently then assumed to be distributed into neighboring regions until at some point the process becomes catastrophic and macroscopic rupture occurs.

If σ_i is the stress on the small element containing a flaw, then the rupture criterion is

$$(\partial\sigma_i/\partial t)_b = 0 \qquad \text{IIIB6-(45)}$$

where the subscript b denotes the onset of a rupture. Though not explicitly stated, the σ,ϵ curve for a macroscopic specimen would also be expected to pass through a maximum, if the response could be examined in detail during the last few milli- or microseconds before the physical parting of the ends, so that the same overall criterion can be applied to the macroscopic stress.

The macroscopic stress is then considered to be given by

$$\sigma = Cn\varphi(\lambda) \qquad \text{IIIB6-(46)}$$

where C is a proportionality constant, n is the number of bonds per unit area normal to the principal stress, and $\varphi(\lambda)$ is an arbitrary function giving the shape of the stress–strain curve.

Differentiating eq. (46) with respect to time and applying the rupture criterion gives the rate of bond breakage as

$$-(1/n)(\partial n/\partial t)_b = (1/\varphi)(\partial\varphi/\partial\lambda)_b\,(\partial\lambda/\partial t)_b \qquad \text{IIIB6-(47)}$$

The rate of bond breakage for an uncrosslinked specimen not under stress can be obtained from reaction rate theory as

$$-(1/n)(\partial n/\partial t)_b = A \exp\{-\Delta F^*/RT\} \qquad \text{IIIB6-(48)}$$

where A is the frequency factor and ΔF^* is the free energy of activation for the chain rupture process in the undeformed state. When the specimen is crosslinked and stretched, additional free energy, ΔF, is stored in the bonds. ΔF is then evaluated in a straightforward fashion, though it is assumed that there are no enthalpy changes on extension.

The final results for tetrafunctionally crosslinked systems, with no sol and free chain-ends, can be given as

$$(W_b/RT) - Z \ln [(1/\varphi)(\partial\varphi/\partial\lambda)_b] + (1/2) \ln Z - 1.62 = BZ - (\Delta H/RT) \qquad \text{IIIB6-(49)}$$

where

$$B = (\Delta F^*/RT) + \ln (\partial\lambda/\partial t)_b - \ln A \qquad \text{IIIB6-(50)}$$

$$Z \cong \rho n_0/M_0 \nu_e \qquad \text{IIIB6-(51)}$$

and W_b is the energy stored on extension per mole of chains, Z is the number of main chain bonds per network chain, ΔH is the change in enthalpy associated with the crosslinking process, n_0 is the number of main chain bonds per monomer unit, ρ is the density, and M_0 is the monomer molecular weight.

All of the terms on the left-hand side of eq. (49) may be evaluated and the left-hand side plotted against Z to evaluate B and ΔH. When this was done for a silicone rubber, ΔH was found to be somewhat too large, but ΔF^* was in good agreement with that obtained by chemical stress relaxation measurements on similar specimens if a reasonable value of A was assumed. This was as far as Bueche carried his development. However, several interesting extensions of this theory may be derived.

Equation (49) is interesting because W_b, and consequently the form of the stress–strain curve, remains unspecified. Thus any form of the stored energy function may be employed, e.g., the Mooney-Rivlin expression, eq. (13), along with its corresponding φ.

In addition, the important prediction can be made that the breaking elongation will be time-dependent even when the modulus is not, i.e., in the quasi-equilibrium region. This phenomenon has indeed been observed. Moreover, the theory can be extended to provide a useful picture of the rupture response in this region. Thus, if a kinetic theory stress–strain expression is assumed, it can be shown that the second term of the left-hand side is much larger than the other three for small ν_e. Moreover,

$$\ln\,[(1/\varphi)(\partial\varphi/\partial\lambda)_b] \cong -0.34 - 0.89 \ln \epsilon_b \qquad \text{IIIB6-(52)}$$

for $1 < \epsilon_b < 10$. Thus eq. (49) becomes

$$\ln \epsilon_b = 1.12\,[K - (\Delta H/RTZ) + \ln\,(\partial\lambda/\partial t)_b] \quad \text{IIIB6-(53)}$$

where K includes the constant factors and slowly varying terms. At constant crosslink density

$$\epsilon_b = k(\partial\lambda/\partial t)_b{}^{1.12} \qquad \text{IIIB6-(54)}$$

The theory predicts that there is no *equilibrium elongation,* the breaking strain being almost directly proportional to the test rate if $(\partial\lambda/\partial t)_b$ is identified with the macroscopic test rate.

At a constant strain rate, but variable ν_e, log ϵ_b is predicted to decrease linearly as ν_e increases, the slope depending on ΔH when Z is large and λ_b is small. The corresponding dependence of stress on ν_e may be estimated by employing the kinetic theory stress–strain expression to relate σ_b and ϵ_b. This implies that $\ln\,(\sigma_b/\nu_eRT) = \ln\,[\lambda_b - (1/\lambda_b{}^2)]$, and the results would only be valid over the high temperature portion of the failure envelope. Over the range $0.5 < \epsilon_b < 8.0$,

$$\ln RT + \ln\,(\sigma_b/\nu_e) = \ln\,(\lambda_b - \lambda_b{}^{-2}) \cong 0.54 + 0.77 \ln \epsilon_b \quad \text{IIIB6-(55)}$$

Substituting this into eq. (53) to evaluate the dependence of σ_b on ν_e at constant rate yields

$$\ln\,(\sigma_b/\nu_e) = -0.84\,\Delta H \nu_e M_0/n_0\rho RT + \text{constant} \quad \text{IIIB6-(56)}$$

For a suitably large ΔH, exemplified by the value of 140 kcal. for the silicones studied by Bueche, σ_b is predicted not to be directly proportional to ν_e. Finally, an increase in ν_e will shift the location of the failure envelope in the σ_b,ϵ_b plane toward lower σ_b and lower ϵ_b values. This is qualitatively correct.

It is interesting to note that most of the factors deemed important in evaluating ΔF prove to have no bearing on the final analysis. Thus, for example, the amount of energy stored on elongation turns out to be inconsequential in comparison with the ΔH term.

IIIB6c. F. Bueche

In his initial theory (1955), F. Bueche (12) showed how the viscoelastic response of a polymer network could account for the time dependence of uniaxial tensile rupture behavior. In a polymer network containing ν chains per unit volume, each containing N segments, the rate of rupture of a segment in the unstressed state is assumed to be

$$dn/dt = \omega \nu N \exp\{-E/kT\} \qquad \text{IIIB6-(57)}$$

where ω is the rate of segmental motion or the "jump" frequency, N is the number of submolecules per chain, and E is the activation energy for bond scission. In the presence of an applied force the exponential term is modified by the elastically stored energy per segment, V_i $(= W/N\nu)$. V_i is evaluated from the force required to displace one end of a submolecule relative to the other end. The displacement Δx_i will be time dependent and, for segments of uniform length, can be obtained from an earlier theory on viscoelastic behavior of polymers (14) as

$$V_i = (\text{force} \times \text{displacement})/2 = (3kT\Delta x_i/l^2)(\Delta x_i)/2 \qquad \text{IIIB6-(58)}$$

where

$$\Delta x_i = [4F\langle \overline{R_0^2}\rangle/(3\pi^2\nu kT)] \sum_{j\ \text{odd}}^{N} j^{-2}[1 - \exp\{-j^2t/\tau_1\}] \times [\cos\{(i+1)(j\pi/N)\} - \cos(ij\pi/N)] \qquad \text{IIIB6-(59)}$$

plus a term involving the glassy modulus to account for the instantaneous deformation. Here F is the applied load, $\langle \overline{R_0^2}\rangle$ is the mean square end-to-end separation of a chain and τ_1 is the longest relaxation time. Since there will be a distribution of chain lengths, the shorter ones will contribute a larger stress for the same value of Δx_i. This increased stress is taken into account by considering the real chains to store q^2 times as much energy as the average chain. Because of the size distribution of segment lengths, the overall rate of segment rupture is taken as the sum of the rates for the individual submolecules

$$dn/dt = \omega\nu \exp\{-E/kT\} \sum_{i=0}^{N} \exp\{q^2V_i/kT\} \qquad \text{IIIB6-(60)}$$

Rupture is postulated to occur when

$$\sum_{i=0}^{N} \exp \left\{ q^2 V_i / kT \right\} = \exp \Phi^2 \qquad \text{IIIB6-(61)}$$

which defines the rupture parameter Φ.

Since the time dependence of V_i is given by the series in eq. (59), this also accounts for the time dependence of the breaking strength. Therefore, this series was numerically evaluated and the sum plotted as a function of t/τ_1 for various values of N. As a rough approximation, the results were found to be independent of N for $N = 20$ and 60. The plot should have the same form as a σ_b,t_b plot measured in creep. By an extension of this argument, a similar result can be obtained for constant strain rate experiments. Note that the inclusion of τ_1 in eq. (59) automatically insures time-temperature superposition, because τ_1 follows the WLF equation.

When the experimental results for the breaking times of poly(butyl methacrylate) subjected to constant load were compared with the theoretical curve, a good fit was obtained over about five decades of time; a better fit was obtained by retaining the term in Δx_i accounting for the glassy state response.

On the other hand, the theory applied to constant strain rate experiments predicts a much stronger time dependence than is actually observed. The theoretical expression can fit experimental data over only about two decades of reduced time in the long-time end of the time scale. The difficulties probably arise from the mathematical approximations made in the derivation. In spite of this lack of agreement with constant strain rate experiment, the theory does provide a direct explanation for the applicability of time-temperature superposition to ultimate property data.

The dependence of ϵ_b on t_b was not explicitly considered, although it appears that it would be possible to calculate such a dependence by employing an experimental stress-strain response in conjunction with the theoretical calculation for σ_b. The location and breadth of the maximum in ϵ_b will depend strongly on the choice of the glassy state contribution. In addition, the dependence of σ_b and ϵ_b on ν_e cannot be readily assessed.

In his next theory (1957), F. Bueche (13) attempted to provide a theoretical basis for the dependence of σ_b on ν_e, specifically the maximum which occurs in σ_b,ν_e plots. Although the results were reasonably successful, later work showed the maximum to be an artifact, as noted in Subsection IIIB3, since it tends to disappear as the time scale of the experiment is increased. This led to a later reappraisal of the problem by Bueche and Dudek (15,16). In this work Bueche confirmed the earlier reports of the time dependence of $(\sigma_b)_{\max}$ and derived theoretical relationships to describe this phenomenon as follows.

In the rubbery region, it is assumed that rupture will occur when the average equilibrium stored energy per network bond becomes equal to a critical value $\mathcal{E}_0$. On the assumption that the equilibrium stress-strain curve is given by the kinetic theory expression, $\mathcal{E}_0$ is found to be:

$$\mathcal{E}_0 = [\tfrac{1}{2}\lambda_b^2 + (1/\lambda_b) - \tfrac{3}{2}]\nu_e kTm/N\rho g \qquad \text{IIIB6-(62)}$$

where ν_e is the number of effective network chains per unit volume of rubber, m is the average molecular weight per chain backbone bond in the gel, N is Avogadro's number, ρ is the density of the rubber, and g is the gel fraction. Equation (62) was applied to data obtained from SBR and EPR (ethylene–propylene rubber) tested at different rates by plotting log (g/ν_e) versus log $[(\lambda_b^2/2) + (1/\lambda_b) - \tfrac{3}{2}]$. These data were roughly in accord with eq. (62) for about $\lambda_b < 15$, and so $\mathcal{E}_0$ was taken to be a constant. Experimental evidence indicates also that λ_b is rate-dependent, which Bueche ascribes to the growth of cracks or flaws in the specimen. To the extent that this rate dependence exists, eq. (62) will be an oversimplification of the true situation.

In order to obtain a relation between σ_b and ν_e, Bueche notes that stress relaxation continues to occur even at times sufficiently long for the average chain to have relaxed. Bueche interprets this as follows. In the initial elongation, all junction points move affinely, with the result that the shorter chains bear a disproportionately larger share of the stress. These highly elongated chains can then redistribute the load to their neighbors by the motion of junction points. Because these highly elongated chains can relax, they will offer a markedly time-dependent contribution to the overall stress.

Under this assumption, σ_b is the sum of two terms, one an equilibrium component given by the kinetic theory expression and the other a time-dependent component which is taken to be proportional to the number of fully extended chains.

For the equilibrium component,

$$(\sigma_b)_{eq} = \nu_e kT[\lambda_b - (1/\lambda_b^2)] \qquad \text{IIIB6-(63)}$$

and for the time-dependent component,

$$(\sigma_b)_{\text{time depend.}} = \psi(t)\nu_e[1 - \text{erf}\,\{[\beta\nu_e/\nu g(2 - g)]^{1/2}\}] \qquad \lambda_b \gg 1 \qquad \text{IIIB6-(64)}$$

where

$$\beta = (3kT/4\mathcal{E}_0)(m/M_0)$$

Here $\psi(t)$ is an unspecified function of the test rate which is assumed to be independent of ν_e, the expression in square brackets represents the fraction of fully extended chains (analogous to φ in the work of Epstein and Smith), M_0 is the molecular weight of a statistical segment, and ν is the total number of network chains per unit volume. The assumption made in obtaining eq. (64) is that the rate-dependent component of σ_b depends only on the product of $\psi(t)$ and the number of fully extended chains.

Hence, the observed stress at break is given by,

$$\sigma_b = \nu_e\{kT[\lambda_b - (1/\lambda_b^2)] + \psi(t)\ [1 - \text{erf}\ (x^{1/2})]\} \qquad \text{IIIB6-(65)}$$

where x is the argument of the error function as given in eq. (64). Equation (65) was applied to experimental values of σ_b and ν_e obtained at different strain rates with SBR and EPR. It was found that the maximum value of σ_b as a function of ν_e was dependent on the rate of testing, with higher values of $(\sigma_b)_{max}$ obtained at the higher rates, while the value of ν_e at $(\sigma_b)_{max}$ was independent of test rate for the range investigated. The parameter β was found to be essentially independent of the test rate, while the value of $\psi(t)$ was found to be rate-dependent, as expected. For EPR, $\beta = 7.9$ and $\psi = 1.78 \times 10^{-10}$ ergs at $\dot{\epsilon}_{const} = 50$ min.$^{-1}$; for SBR, $\beta = 7.1$ and $\psi = 3.8 \times 10^{-11}$ ergs at $\dot{\epsilon}_{const} = 5$ min.$^{-1}$. The values of $\psi(t)$ for the other test rates were not reported.

Figure IIIB-38 shows the fit of the theoretical curve, eq. (65), to the experimental data obtained with SBR at a rate of 5 min.$^{-1}$ presumably using the values of β and $\psi(t)$ quoted above. The dotted line represents the equilibrium component of σ_b, and the time-dependent component is represented by the difference between the full line and

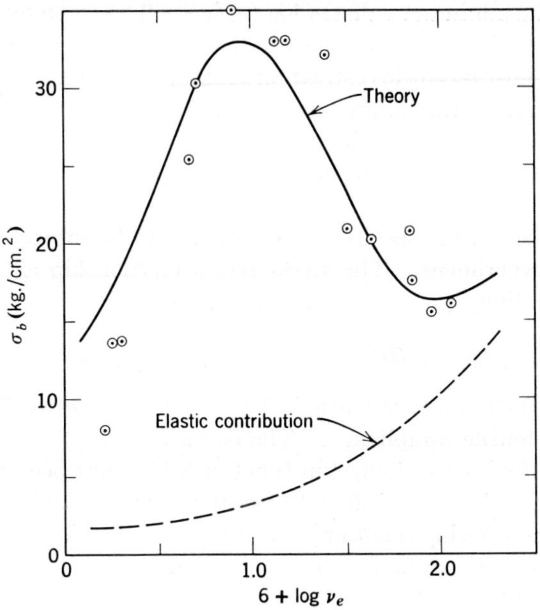

Fig. IIIB-38. Experimental rupture stress as a function of crosslink density at constant strain rate for an SBR rubber. The full line is the theoretical prediction of the Bueche-Dudek theory and the dotted line is the kinetic theory contribution to σ_b. The difference between theory and kinetic contributions represents the viscous contribution to σ_b. From Bueche and Dudek (16).

the equilibrium component. It can be seen that the elastic contribution to σ_b is by far the more important factor at high ν_e values.

Equation (65) can be written

$$\log \sigma_b/\nu_e = \log \{kT[\lambda_b - (1/\lambda_b^2)] + \psi(t)\ [1 - \mathrm{erf}(x^{1/2})]\} \quad \text{IIIB6-(66)}$$

The right hand side of this equation can be written so that ν_e does not occur explicitly, and hence the reduced failure envelope seems to be predicted. However, at a given ν_e, both λ_b and x are fixed, according to eq. (62) and (64). Therefore a change in test rate can only lead to a change in σ_b, the magnitude of which is determined by $\psi(t)$. Hence, for a series of ν_e values, eq. (66) predicts the reduced failure envelope, log σ_b/ν_e versus log ϵ_b, to be a set of vertical lines spaced, according to eq. (62), as $(\log \lambda_{b_2} - \log \lambda_{b_1}) = \frac{1}{2}(\log \nu_{e_1} - \log \nu_{e_2})$. In addition, the minimum value of log (σ_b/ν_e) will be given by the kinetic theory expression, eq. (63), when $\psi(t) = 0$.

In their most recent investigation (1964), Bueche and Halpin (17a) set out to obtain the failure envelope and the time dependence of ϵ_b. Since the latter is more difficult to describe than the σ_b,t_b relationship, a greater overall measure of success is assured, as noted in Subsection IIIB1.

In this theory, the specimen is considered to be subjected to a tensile creep experiment. The stress–strain relationship is taken to be (in our notation)

$$\sigma = [3D(t)]^{-1} (\lambda - \lambda^{-2})(1 - \lambda^2 n)^{-1} \qquad \text{IIIB6-(67)}$$

where n is the number of statistical segments in a network chain and $D(t)$ is the tensile compliance. The factor $(1 - \lambda^2 n)^{-1}$ is an approximation to the inverse Langevin function which has been used to describe finite chain extensibility (Subsection IIIB2c). The actual form of the σ,ϵ relationship is not important to the derivation so long as the time dependence can be factored into $D(t)$. Therefore, eq. (67) can be generalized to

$$\sigma = [3D(t)]^{-1} \epsilon f(\epsilon) \qquad \text{IIIB6-(68)}$$

During elongation of the macroscopic specimen, small tears are considered to form and grow. At the tip of a crack, the stress is increased over the nominal value by a stress concentration factor S and the local elongation is correspondingly greater. The criterion for break is a critical elongation of the local region or filament which is independent of rate and temperature. Thus the σ,ϵ relationship for such a filament at the instant of break is

$$\sigma_{b,r} = \epsilon_{b,r} f(\epsilon_{b,r}) = SK \qquad \text{IIIB6-(69)}$$

where K is a constant determined by the appropriate value of $\epsilon_{b,r}$ and the form of $f(\epsilon_{b,r})$. The filament breaking time t_b' is then taken to be proportional to the breaking time for the macroscopic sample t_b, i.e.,

$$t_b = q\, t_b' \qquad \text{IIIB6-(70)}$$

so that the macroscopic rupture stress σ_b is

$$\sigma_b = \sigma_{b,r}/S = K/[3D(t_b/q)] \qquad \text{IIIB6-(71)}$$

and

$$\epsilon_b f(\epsilon_b) = KD(t_b)/D(t_b/q) \qquad \text{IIIB6-(72)}$$

Equation (71) gives the time dependence of σ_b. Since $D(t_b/q)$ is a sigmoidally increasing function of time for an elastomer, σ_b is a sigmoidally decreasing function. Equation (72) gives the time dependence of ϵ_b as the ratio of the compliance observed at two times a fixed distance apart on the log time scale. [$\epsilon_b f(\epsilon_b) \cong KE(t_b/q)/E(t_b)$.] If $D(t)$ becomes time-independent, a limiting breaking elongation would be obtained. If it continues to increase slowly with time, the limiting elongation will be higher, the stronger this time-dependence. Since q is taken as 10^4, then a limiting elongation can only be sought at a time scale 10^4 (or q) greater than the time at which quasi-equilibrium behavior is approached. Moreover, at very short times, the ϵ_b,t_b curve is predicted to have the same shape as the $D(t),t$ curve.

Hence, according to eqs. (70)–(72), the measured creep response of an elastomer in conjunction with the evaluation of the constants q, n, and K is sufficient to predict the failure envelope. For an elastomer with a small Mooney-Rivlin C_2 term [cf. eq. (14)], exemplified by butyl rubber, they were able to obtain reasonable agreement with the experimental failure envelope using $q = 10^5$, $n = 56$, and $K = 0.051$. For rubbers with a large C_2, such as EPR, the simple theory using eq. (67), did not fit the experimental envelope. Bueche and Halpin (17a) postulate that a finite value of C_2 implies a dependence of $D(t)$ [as defined in eq. (67)] on stress and hence, in this case, various $D(t)$ curves will be required to fit different segments of the failure envelope. In view of this complication, the kinetic theory form factor $(\lambda_b - \lambda_b^{-2})$, in eq. (67) is replaced by the Mooney-Rivlin form factor, $(\lambda_b - \lambda_b^{-2})(1 + C_2/C_1\lambda_b)$. With this modification, the fit to the experimental EPR failure envelope was improved, but there still remains a systematic deviation; the value of the constants required were $q = 10^7$, $n = 68$, $K = 0.926$, and $C_2/C_1 = 5.12$.

Note that uncrosslinked specimens are predicted to have a maximum in ϵ_b versus t_b once the molecular weight is high enough to obtain a sigmoidal creep compliance curve. While it is difficult to assess the time dependence of ϵ_b on crosslink density from eq. (72), σ_b is predicted to be directly proportional to ν_e at times long enough so that $D(t_b/q)$ will be at the quasi-equilibrium, rubbery zone value.

IIIB6d. W. Knauss

The theory of Knauss (58–60) is perhaps the most comprehensive developed to date. It embodies many of the elements of prior theories—the establishment of rupture criteria based on a balance between stored energy and the surface energy developed by a crack (47) and a rate of bond breakage which is affected by the applied force (10,12). It also indicates, without development, how statistical effects, stress concentration factors and multiaxial stress fields may be handled as part of the future development of the theory.

The theory assumes that any fully amorphous, isotropic polymer has statistically distributed weak regions, each consisting of a set of polymer chains with a distribution of bond strengths. Failure is initiated at the weak bonds and propagates through the weak regions to form submicroscopic cracks or flaws. These in turn interact to form a macroscopic cavity which will propagate as a crack once the stored energy becomes equal to the energy which would be dissipated in the formation of new surface.

In the absence of stress the number of unbroken bonds N_1 is in equilibrium with the number of broken ones N_2

$$N_1 \underset{k_{21}}{\overset{k_{12}}{\rightleftharpoons}} N_2 \qquad \text{IIIB6-(73)}$$

The rates of bond breakage and formation, respectively, are given by

$$dN_2/dt = -\bar{k}_{21}N_2 + \bar{k}_{12}N_1 \qquad \text{IIIB6-(74)}$$

$$dN_1/dt = -\bar{k}_{12}N_1 + \bar{k}_{21}N_2 \qquad \text{IIIB6-(75)}$$

and the rates are equal in the absence of stress which implies that $N_1 = N_2$ or half the bonds are broken in the unstressed state. In the presence of stress, the average rate constants can be given by

$$\bar{k}_{12} = Ba(T) \exp\{-W/NkT\} \qquad \text{IIIB6-(76)}$$

$$\bar{k}_{21} = Ba(T) \exp\{W/NkT\} \qquad \text{IIIB6-(77)}$$

where $a(T)$ is a temperature-dependent function which arises in an averaging process of the distribution of bond strengths, W is the average stored energy per unit volume, N is the number of bonds per unit volume, and B is a normalization factor. It is then assumed that

$$a(T) = (a_T)^{-1} \qquad \text{IIIB6-(78)}$$

where a_T, as before, is the time–temperature shift factor. The only justification for this assumption is that the ultimate properties of many amorphous polymers are known to obey the superposition principle. The net rate of increase of the fraction of bonds broken due to the stored elastic energy is obtained by subtracting eq. (75) from eq. (74), and employing eqs. (76) and (77) to obtain

$$\frac{a_T}{2B}\frac{d}{dt}\left(\frac{N_2 - N_1}{N_0}\right) = -\left(\frac{N_2 - N_1}{N_0}\right)\cosh\frac{W}{NkT} + \sinh\frac{W}{NkT}$$

IIIB6-(79)

where $N_0 = N_1 + N_2$ is the total number of bonds in the weak region.

The relative number of bonds broken $(N_2 - N_1)/N_0$, due to the applied force can then be taken to be proportional to the relative size of the defect. Knauss expresses the latter in terms of an area ratio, since the rupture criterion will be expressed in terms of a surface energy, and takes the proportionality constant to be unity. If any other value were assumed, it would simply be absorbed into the constant B. Thus

$$(N_2 - N_1)/N_0 = A/A_m \qquad \text{IIIB6-(80)}$$

where A is the area of the defect and A_m is the maximum area which the defect can attain.

On the basis of dimensional analysis, the rupture criterion is set up so that rupture will occur when the stored energy becomes equal to or exceeds some critical value, Γ

$$W(A/A_m)^{1/2} \geqslant \Gamma \qquad \text{IIIB6-(81)}$$

Alternative forms of eq. (81) can be given (58,106), but this equation is more tractable and leads to quite satisfactory results.

By solving for $d(A/A_m)/dt$ and A/A_m in terms of dW/dt and W, eqs. (79), (80), and (81) can be combined to give

$$\Gamma^2 W \cosh(W/NkT) - W^3 \sinh(W/NkT) = a_T B^{-1}\Gamma^2(dW/dt)$$

IIIB6-(82)

The integrated form of this transcendental equation, combined with the appropriate time-dependent expression for W, is thus Knauss' basic description of rupture.

To calculate the behavior at break, it is then necessary to know W as a function of the loading history. For a constant strain-rate test on a linear viscoelastic body, eq. (30) gives

$$W = (1/2)E_e \dot{\epsilon}_{\text{const}}^2 t^2 + (\dot{\epsilon}_{\text{const}}^2/2) \int_{-\infty}^{\infty} H_l \tau^2 (1 - \exp\{-t/\tau\})^2 \, d \ln \tau \qquad \text{IIIB6-(83)}$$

In a real body, with a nonlinear stress–strain law, the nonlinearity could be incorporated into eq. (83) directly by replacing the linear elements of a Wiechert model with appropriate nonlinear ones, except that an analytical solution would not be obtainable in general. Alternatively, the equivalent expression involving σ_H, and thus H_H, defined as $H_H = H/f(\epsilon)$, could be employed [cf. Subsection IIIB2c(2)]. Knauss bypasses these mathematical difficulties by noting that the energy stored in a real body is smaller than, but approximately proportional to, the amount stored in a hookean body. For computational purposes he takes the factor to be $1/2$. A better value would have been $1/3$, but the actual value is eventually incorporated into N, which must be evaluated empirically; thus the choice is unimportant.

To test the theory, Knauss evaluated both the small strain viscoelastic response of a carboxy-terminated butadiene rubber by a combination of dynamic shear and stress-relaxation tests, and the ultimate properties in constant strain rate tests at a series of temperatures. The distribution function of relaxation times, usually considered as continuous, was then fitted to an eighteen-element Wiechert model using a numerical evaluation technique. Employing this value for H_l in eq. (83), he was then able to numerically fit his reduced σ_b,t_b data to obtain the three unknown constants B, Γ, and N. Once these were evaluated he could then calculate the rupture strain in these same constant strain rate tests, the velocity of crack propagation in a running crack, and the mean rupture time in a creep experiment at room temperature. The three-parameter curve fit to the σ_b,t_b data is shown in Figure IIIB-39. He did not attempt a prediction of the failure envelope, although the agreement with the ϵ_b,t_b data indicates that a reasonable fit could have been obtained.

It is important to note that, just as in the theory of A. Bueche (10), there is no equilibrium rupture stress or rupture strain. Once a sustained biasing force is applied, damage is done to the system, to the extent is given through eq. (79). According to eq. (73), however, if the stress is relieved, the specimen may "heal" if given sufficient time.

Actually, Knauss is able to calculate the amount of damage done as a function of the test modes. He predicts, for example, that in a test of the cumulative damage done at constant strain rate, in which two strain rates are employed, with the second being faster than the

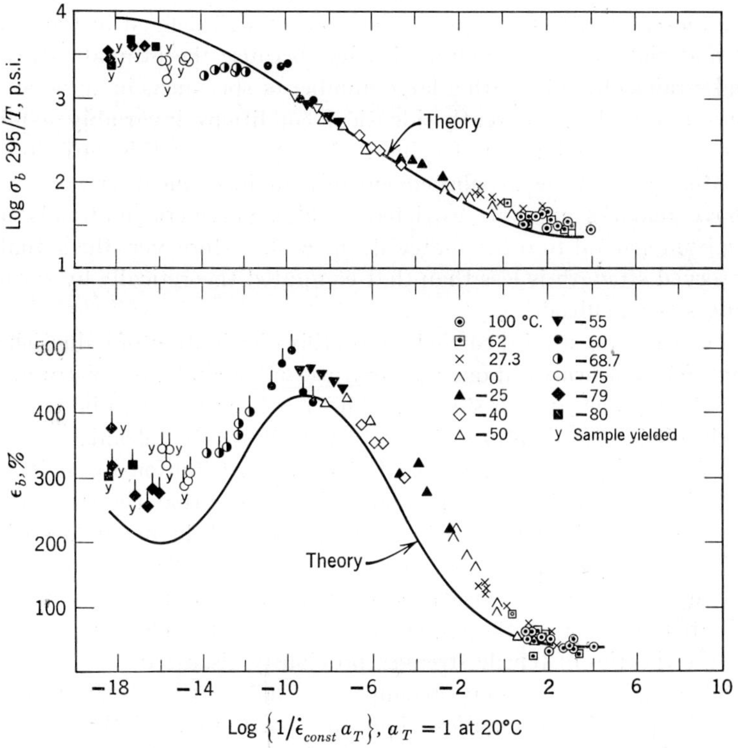

Fig. IIIB-39. Comparison between experimental ultimate stress and strain with the theoretical prediction of Knauss. The σ_b data were used to calculate the parameters of the Knauss theory, from which the solid lines were calculated. From Knauss (60).

first, the rupture time will be less than that obtained using the slower rate alone. Conversely, if the second rate is lower than the first, the rupture time will be increased. However, for strain rates differing by a factor of 100 the effect is small and would probably be masked, if not lost entirely, in the statistical distribution of rupture times.

IIIB7. An Introduction to Statistical Aspects of Rupture in Polymers

IIIB7a. Preliminary Remarks

The ultimate mechanical behaviors in uniaxial tension are probably the most frequently measured properties of polymeric materials. Characteristically, these properties are not well defined in the sense of material constants such as density, because the ultimate stresses and strains obtained with a large number of specimens in any one experiment under apparently identical conditions invariably exhibit a statistical distribution of values. It appears that this variation or scatter can not be attributed entirely to instrument errors. The above remarks are not restricted to high polymeric materials but apply in general to other materials, as well. Moreover, the actually observed strength is less than that computed theoretically by several orders of magnitude.

For rubbers, studies carried out on this distribution of values show that, for σ_b measured at one temperature and one rate, the distribution is ordinarily negatively skewed. This has been found to be true for both filled natural rubber (57,107) and SBR (57) vulcanizates and, in addition, it has been reported that the distribution for ϵ_b is also negatively skewed (107). It would also appear that the strength is dependent on the volume of the specimen, with higher values of σ_b being associated with specimens having a smaller volume (51). These effects can be considered to originate from imperfections distributed throughout the specimens and are not explicitly predicted by the theories of tensile strength previously discussed.

There have been many attempts to analyze this distribution of flaws, especially as manifested in the distribution of the ultimate stress of solids, by employing statistical methods. Rather than discuss these various efforts in detail and provide an exhaustive review, an attempt will be made to consider the underlying principles involved. Furthermore, the emphasis will be directed primarily to those aspects of particular interest to rupture behavior of elastomeric materials.

IIIB7b. Extreme Value Statistics

Kase (57) was apparently the first to attempt to derive theoretically the distribution function for rupture stress of a rubber, although the approach used by Kase, extreme value statistics, had been applied previously by others to the strength of brittle materials such as glass (33).

In the application of extreme value statistics, it is assumed that the material contains a distribution of flaws or defects of density n per unit volume, which weaken it (28). Each specimen of the material thus contains a sample of size nV where V is the specimen volume taken from the population of flaws. These samples of flaws are not identical and in each sample there will be a largest and a smallest value for the flaw size. The method assumes also that some measure of the effectiveness of the flaw in reducing the strength can be associated with each flaw and that rupture will occur when the applied load on a given specimen becomes just equal to the smallest value of the critical strength associated with the weakest flaw. In a statistical study of rupture by means of extreme value statistics, two related approaches can be considered. For example, if a sheet of a material is considered to be represented by a table of random numbers and the latter is cut into equal sized pieces, each containing several numbers, then each piece (specimen) will have a smallest number. If these numbers are taken to be proportional to tensile strength, then the specimen would break when this smallest value is reached. In this case the distribution of tensile strength is related to the distribution of the smallest numbers. On the other hand, flaws can be considered to have varying areas. Then, the physical picture is one of a sheet with holes punched out, the holes having a distribution of areas. When the sheet is cut into specimens, each piece will have a hole of largest area. In this case the specimen would be expected to break at the cross section containing the largest hole, and hence the distribution of largest values is required. These two methods of assessing the effectiveness of the flaw in reducing the strength of a material lead to essentially similar mathematical results. The term "weakest flaw" will be used to denote either or both of these concepts.

It is implied that once the applied force generates the critical stress associated with the weakest flaw, a chain of events occurs which leads to immediate rupture. If the frequency function of flaws, $f(x)$, where x is the pertinent variable, i.e., either strength or area, is assumed or is calculable for the particular model of the system chosen, then a general method exists whereby the distribution of the smallest or the largest values can be calculated (48,49).

In order to follow the work of Kase, it will be convenient to discuss some aspects of extreme value statistics (48,49). Let $f(x)$, assumed to be continuous, represent the underlying frequency function for the flaws of size x where x represents a critical stress. Then the frequency function $g_{nV}(x_1)$ for the *smallest* value of x, x_1, which is present in a sample of size nV is given by (48):

$$g_{nV}(x_1) = nVf(x_1)\ [1 - F(x_1)]^{nV-1} \qquad \text{IIIB7-(84)}$$

where $F(x)$ is the cumulative distribution function corresponding to $f(x)$. Since the specimen is assumed to rupture when the smallest value of the critical stress associated with the weakest flaw is reached, a sample value corresponds to the observed tensile strength of the specimen. Alternatively, let $f(x)$, again assumed to be continuous, represent the frequency function for an area measure of the flaw. Then the frequency function, $g_{nV}(x_m)$, for the *largest* value of x, x_m, is given by

$$g_{nV}(x_m) = nVf(x_m)\ [F(x_m)]^{nV-1} \qquad \text{IIIB7-(85)}$$

From eqs. (84) and (85), the mode or the most probable value of x_1 or x_m can be calculated. The cumulative distribution functions corresponding to eqs. (84) and (85) are respectively:

$$G_{nV}(x_1) = 1 - [1 - F(x_1)]^{nV} \qquad \text{IIIB7-(86)}$$

$$G_{nV}(x_m) = [F(x_m)]^{nV} \qquad \text{IIIB7-(87)}$$

It can be shown that when nV becomes very large, eqs. (86) and (87) approach the limiting forms (49)

$$_1\Phi(x) = 1 - \exp\left\{-nVF(x)\right\} \qquad \text{IIIB7-(88)}$$

and

$$_m\Phi(x) = \exp\left\{-nV[1 - F(x)]\right\} \qquad \text{IIIB7-(89)}$$

respectively. The distributions $_1\Phi(x)$ and $_m\Phi(x)$ apply in general since they are valid for any $f(x)$. The general eqs. (88) and (89) have three possible parametric forms (asymptotes) subject to $F(x)$ satisfying some initial conditions (49). For example, when $f(x)$ is of the exponential type, i.e., $f(x) = \delta \exp\left\{-\delta x\right\}$, the right-hand side of eqs. (88) or (89) becomes a double exponential function of x, and $\Phi(x)$ is referred to as the first asymptotic type. Another important case occurs when $f(x)$ is a limited distribution. In this case, the distribution for the smallest value of x is obtained from eq. (88) as

$$_1\Phi(x) = 1 - \exp\{-(x/x_0)^\beta\} \qquad \text{IIIB7-(90)}$$

where x_0 and β are constants. Equation (90) refines a cumulative distribution function which is referred to as the Weibull distribution or the distribution of the third asymptotic type (49). This function has been extensively applied by Weibull to studies of ultimate strength and fatigue behavior (105,105a). The frequency function corresponding to eq. (90) is:

$$_1\varphi(x) = (\beta/x_0)(x/x_0)^{\beta-1} \exp\{(x/x_0)^\beta\} \qquad \text{IIIB7-(91)}$$

It has been tacitly assumed in the above derivations and in the applications to follow that the flaws act independently and furthermore that the distribution itself or its parameters do not vary during the experiment, e.g., with strain. However, it has been shown (103, 104) that the asymptotic forms [eq. (88) or (89)], are preserved even if compliance to the first of these conditions is not strictly held. The question of the compliance to the second condition leads to perhaps the most serious objection to this method (36a). It will not be discussed in detail here.

Kase assumes that the rupture stress σ_b (in our notation) is given by:

$$\sigma_b = \sigma_0\,(1 - \gamma x) \qquad \text{IIIB7-(92)}$$

where σ_0 is the strength of a perfect specimen, x is the original area of a flaw, and γ is a positive constant. The frequency function for x is assumed to be given by:

$$f(x) = \delta \exp\{-\delta x\} \qquad \text{IIIB7-(93)}$$

for $x > 0$, where δ is a constant. Equation (93) implies that the probability of observing a flaw size x decreases monotonically with the size. Since the area is the measure of the effectiveness of the flaw in reducing the strength, Kase is looking for the distribution of the largest value of the flaw area, because the specimen would be expected to break there. The cumulative distribution function for this case in terms of σ_b is obtained by substituting x from eqs. (92) and (93) into eq. (89):

$$\Phi_\nu(\sigma_b) = 1 - \exp[-\exp\{(\delta/\gamma\sigma_0)(\sigma_b - \sigma^*)\}] \qquad \text{IIIB7-(94)}$$

where σ^*, the most probable value of the strength, is related to the specimen volume V and the density of flaws n, according to

$$\sigma^* = \sigma_0 \, [1 - (\gamma/\delta) \ln nV] \qquad \text{IIIB7-(95)}$$

The corresponding frequency function or differential distribution of breaking strengths is given as,

$$\varphi_V(\sigma_b) = (\delta/\gamma\sigma_0) \exp \{ (\delta/\gamma\sigma_0)(\sigma_b - \sigma^*) \} \times \exp [-\exp \{ (\delta/\gamma\sigma_0)(\sigma_b - \sigma^*) \}] \qquad \text{IIIB7-(96)}$$

Equation (95) predicts that the most probable value of the tensile strength should decrease as the logarithm of the specimen volume, and eq. (96) predicts that the distribution of rupture stress will be negatively skewed.

Kase tested 200 dumbbell-shaped specimens of both natural rubber and SBR vulcanizates loaded with carbon black at a constant cross-head speed of 20 in. per minute, presumably at ambient temperature. Histograms of the data for both vulcanizate types are shown in Figure IIIB-40, along with the theoretical frequency function for comparison. The values of σ^* for natural rubber and SBR were found to be 187.2 and 108.3 kg./cm.^{2}, respectively, and the corresponding values of $\sigma_0\gamma/\delta$ were found to be 10.7 and 8.9. In addition, a plot of his data for the SBR vulcanizate on extreme value probability paper (49) is shown in Figure IIIB-41. It is evident from both figures that the agreement between theory and experiment is good. Schwarzl and Staverman (92), however, report that Kase's data can also be represented by a distribution of the Weibull type with a value of $\beta = 16.6$.

This ambiguity brings out the fact that the fit of the experimental data to a theoretical distribution is not sufficient to justify the use of a particular assumed form of $f(x)$. However, one method of checking on the validity of $f(x)$ is to determine the dependence of changes in specimen volume on the mode of the tensile strength since, from eq. (84) or (85), the form of the $f(x)$ chosen automatically determines this dependence. Thus the $f(x)$ chosen by Kase leads to a predicted variation of σ^* as log V [eq. (95)]. Other forms of $f(x)$ would lead to different dependencies of σ^* on V. For a Weibull distribution, for example, σ^* varies as $V^{-1/\beta}$, while if $f(x)$ is assumed to be normal, then σ^* varies approximately (28) as $(\log V)^{1/2}$. Hence, any justification for the use of a particular form for $f(x)$ must depend on an investigation of the effect of specimen volume, or size.

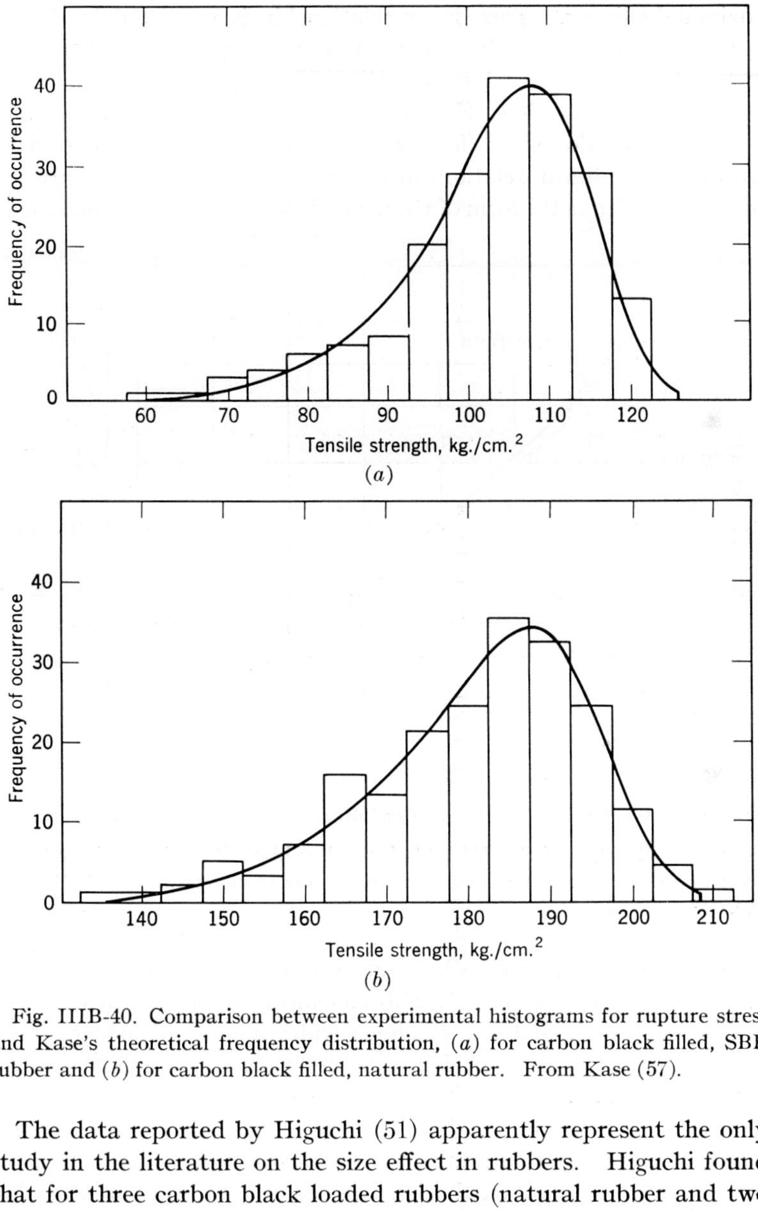

Fig. IIIB-40. Comparison between experimental histograms for rupture stress and Kase's theoretical frequency distribution, (*a*) for carbon black filled, SBR rubber and (*b*) for carbon black filled, natural rubber. From Kase (57).

The data reported by Higuchi (51) apparently represent the only study in the literature on the size effect in rubbers. Higuchi found that for three carbon black loaded rubbers (natural rubber and two SBR types) a plot of the mean tensile strength versus log V was approximately linear. This linear relationship is predicted by Kase, if it is assumed that the mean strength is proportional to the mode. However, the data of Higuchi can be replotted either as log σ_b versus log V (for the Weibull distribution) (28) or as σ_b versus $(\log V)^{1/2}$ (for the normal distribution) (28) to yield approximately linear relationships also. Hence, these data are inconclusive as far as the form of the dependence of σ_b on V is concerned.

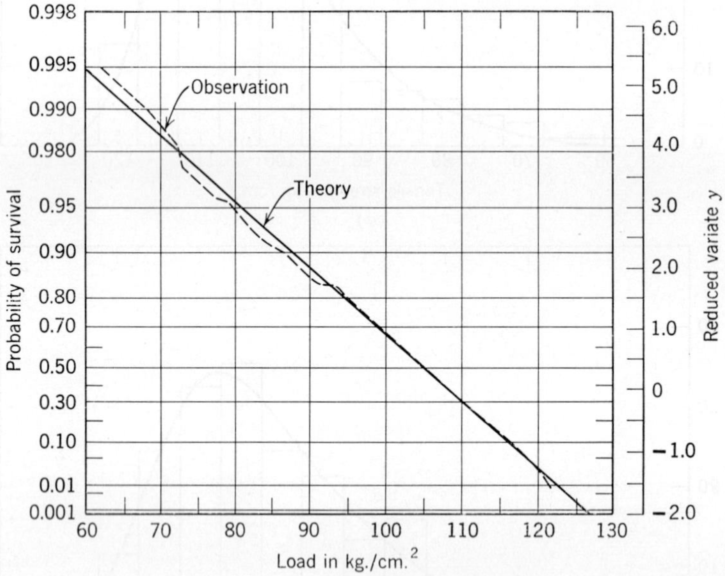

Fig. IIIB-41. Data of Fig. IIIB-40*a* for carbon black filled, SBR rubber tested for fit to a double exponential cumulative distribution function. From Gumbel (49).

Moreover, there appear to be some internal inconsistencies in Higuchi's data. For the SBR (X-243) vulcanizates, the highest values of σ_b associated with constant specimen width and length occur at the highest modulus values, which indicates that the crosslink density may have varied. In addition, in all cases where the increase in volume was accomplished by increasing the specimen length, no compensating change in strain rate was made, so that part of the σ_b decrease may be associated with the decrease in strain rate. The question of the dependence of σ_b (and ϵ_b) on volume should be carefully re-examined, taking precautions to maintain constant strain rate and crosslink density.

In spite of these difficulties the study of Kase can account for the statistical variability observed with rupture data and also provide a rational explanation for the volume effect. Note that if a size effect does exist, then this should be reflected in a difference between strength as measured in a test exerting a homogeneous stress distribution and a test exerting a heterogeneous distribution. Thus Weibull (105a) has shown that for materials which obey a Weibull distribution and subject to certain conditions, the ratio of bending to tensile strength should be $(2\beta + 2)^{1/\beta}$. Such volume dependencies could complicate studies of the geometric rupture surface. On the other hand, if rubbers tend to have β values as large as the data of Kase imply, then the volume effect would be small.

IIIB7c. Effect of Temperature on the Distribution of Rupture Values

Generally, the applications of extreme value statistics considered in the literature do not treat the time and temperature dependence of the rupture process although there is, in fact, a strong dependence. Therefore, in order to examine these contributions one could simply determine how the parameters in the distribution, eqs. (88) or (89), depend on or vary with rate or temperature. An alternative and more fundamental approach would postulate the existence of a time-dependent distribution for the flaws. This would say that a sequence of events in time must occur before the critical flaw can fail. For rubbers the time dependence would be associated with viscoelastic responses of the network.

Some preliminary experiments have been carried out on the statistical variability of both ultimate strength and ultimate strain as a function of the temperature at a constant strain rate of 2.32 min.$^{-1}$ (67). Twenty-five ring specimens of peroxide-cured SBR vulcanizate with $\nu_e = 154$ μmoles/cm.^{3} were tested at 23, -10, and -40°C. These data are shown in Figures IIIB-42, IIIB-43, and IIIB-44, plotted according to eq. (94). The ultimate tensile stress was obtained for each ring by extrapolating the force–time curve in order to calculate that stress which corresponds to the maximum strain occurring on the inside dimensions of the ring. This procedure is discussed by Smith (86). At low temperatures, where the stress–strain curves exhibit a sharp upturn, this extrapolation procedure is difficult to

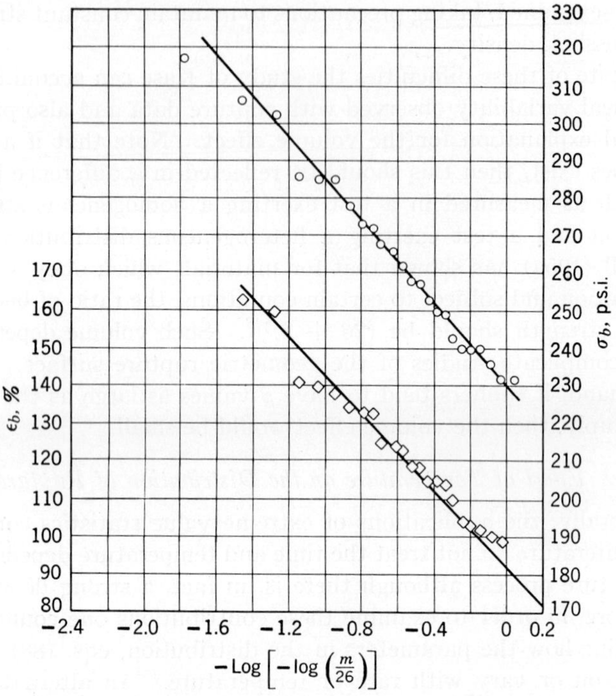

Fig. IIIB-42. The fit to a double exponential cumulative distribution function of both rupture stress (O) and rupture strain (◇) for an SBR dicumyl peroxide gum vulcanizate tested at 23°C. The parameter m is the rank of the data point in decreasing magnitude. From Landel and Fedors (67).

apply and this effect is probably responsible for the scatter in σ_b at high stress values shown in Figures IIIB-43 and IIIB-44. This is indicated also by the fact that the plots for the ϵ_b data do not exhibit the same deviations as do the σ_b values.

Although the sample size is too small to define a distribution, the data are consistent with a double exponential type distribution. The data show that both the slope and intercept of the σ_b plots increase with decreasing temperature. For the ϵ_b data, the intercepts also increase with increasing temperature, while the slopes are approximately the same at the lowest and highest temperatures. However, at -10°C., the slope is about twice that at the other two temperatures.

One interesting result of this study was that segments of a failure envelope could be defined from the statistical variability of the break points. These segments corresponded to an envelope which was the same as the one determined by changing the rate and temperature in the manner of Smith (87); these data are shown in Figure IIIB-22. The effect of varying the crosslink density was also studied as noted

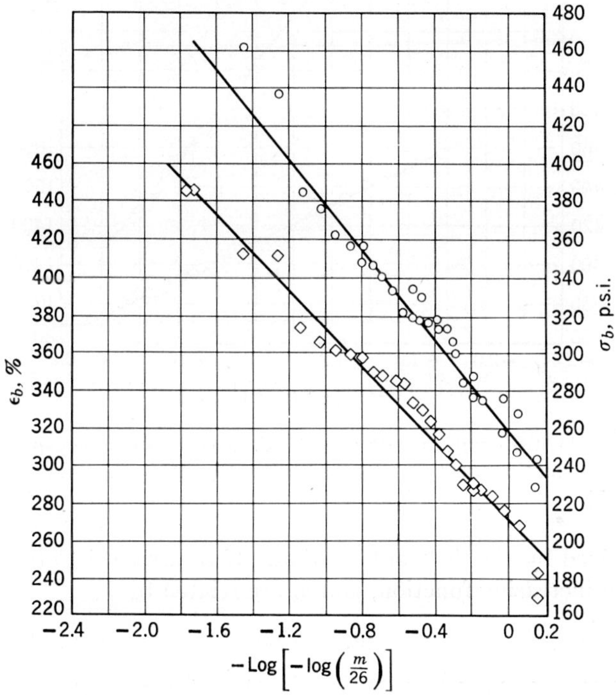

Fig. IIIB-43. SBR rupture data at −10°C. fitted to a double exponential cumulative distribution function. Symbols as in Fig. IIIB-42. From Landel and Fedors (67).

in Subsection IIIB3. Here too, portions of the failure envelope could be constructed from the statistical variability, but the sample sizes were too small to estimate how the distribution varied with crosslink density.

If, as seems plausible, the failure envelope is accepted as defining a deterministic equation between $\sigma_b T_0/T$ and ϵ_b, then the distribution

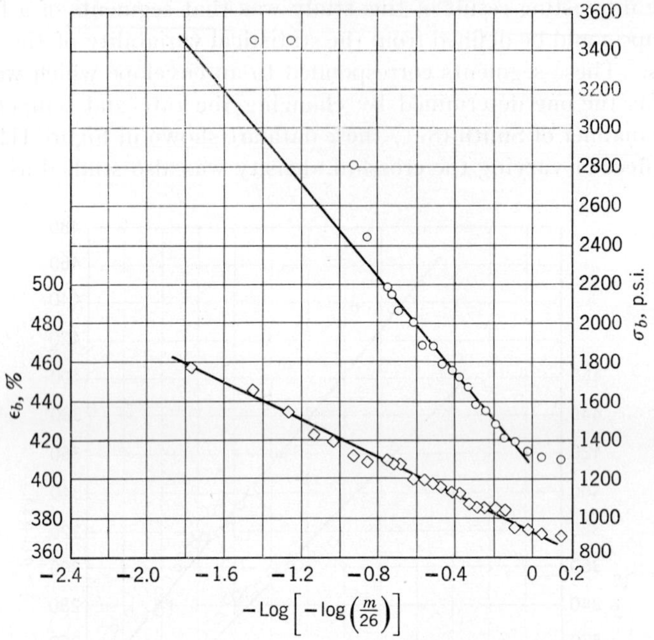

Fig. IIIB-44. SBR rupture data at −40°C. fitted to a double exponential cumulative distribution function. Symbols as in Fig. IIIB-42. From Landel and Fedors (67).

of ϵ_b is fixed by the distribution of σ_b. If ϵ_b, assumed to be a monotonically increasing function, and σ_b are related by

$$\sigma_b = s(\epsilon_b) \qquad \text{IIIB7-(97)}$$

and if the frequency distribution function for σ_b is $p(\sigma_b)$, then the corresponding distribution function for ϵ_b is given by

$$h(\epsilon_b) = p(\sigma_b)\,\frac{d\sigma_b}{d\epsilon_b} \qquad \text{IIIB7-(98)}$$

In the region of interest, σ_b is a monotonically increasing function of ϵ_b which can be approximated for small intervals in σ_b by a linear relation with slope k. Consequently, the distribution of ϵ_b will have the same form as the distribution in σ_b, the only difference being a scale change and a translation. Thus, if $p(\sigma_b)$ is negatively skewed, then $h(\epsilon_b)$ will be also.

This provides an explanation for the results obtained by Wiegand and Braendle (107), who found that the distribution for ϵ_b was of the same shape as that for σ_b, but much narrower. They studied the properties of a natural rubber vulcanizate (inner tube stock) at room

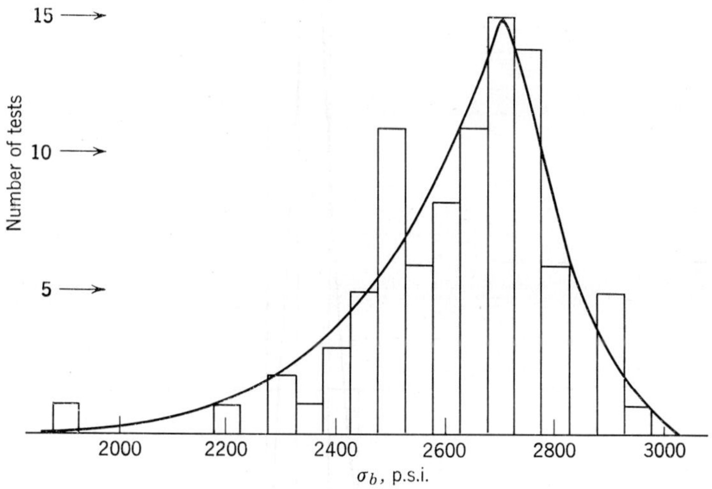

Fig. IIIB-45. Histogram for rupture stress obtained from testing 200 specimens of a natural rubber vulcanizate. The curve shown represents an approximate, continuous distribution defined by the histogram. After Wiegand and Braendle (107).

temperature. In their case, $s(\epsilon_b)$ is the stress–strain curve and not the failure envelope. Their data are shown in Figures IIIB-45 and IIIB-46.

IIIB7d. Stochastic Processes

An alternative approach to incorporating the time dependence of the rupture process is based on the assumption that rupture can occur at any time during which the load acts. In extreme value statistics, i.e., the weakest link concept, rupture can occur only when the smallest critical stress associated with a flaw is reached. The former approach is termed the stochastic process theory of rupture (52) or the theory of breaking kinetics (24). Coleman (24) has shown that the two approaches lead to equivalent results when rupture follows first-order kinetics. Stochastic process theory has been applied to various types of failure in polycrystalline metals by Yokobori (112) and in polymeric fibers by Coleman (24).

If $N(t)$ is the number of specimens that have survived to time t,

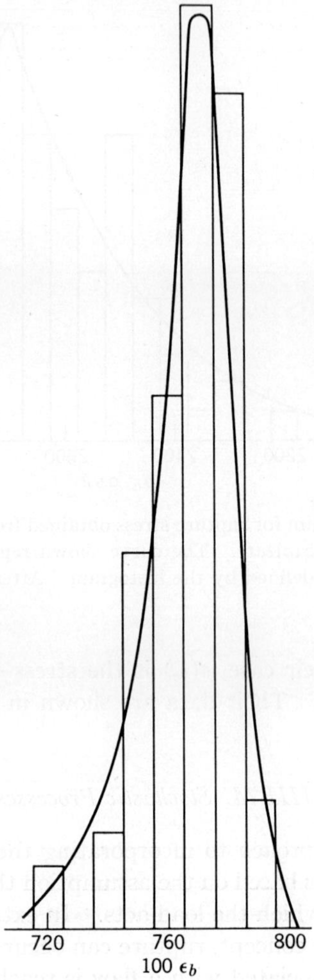

Fig. IIIB-46. Histogram for the rupture strain for the same specimens for Fig. IIIB-45, showing that the distributions in σ_b and ϵ_b are of the same form. The curve represents an approximate, continuous distribution defined by the histogram. After Wiegand and Braendle (107).

the transition probability, $\mu(\sigma,t)$, that a specimen known to be intact at time, t, ruptures in the next time interval dt, is defined as:

$$-d \ln N(t) = \mu(\sigma,t)\, dt \qquad \text{IIIB7-(99)}$$

The transition probability will in general be a function of both the time and the stress, which may itself be a function of time, i.e., $\sigma = \sigma(t)$ and $\sigma_b = \sigma(t_b)$. Equation (99) may be integrated directly to yield

$$N(t_b)/N_0 = \exp\left\{-\int_0^{t_b} \mu(\sigma,t)\, dt\right\} \qquad \text{IIIB7-(100)}$$

where N_0 is the number of specimens (assumed large) at $t = 0$. The frequency density function is given by

$$-dN(t_b)/N_0 = \mu(\sigma_b,t_b) \exp\left\{-\int_0^{t_b} \mu(\sigma,t)\, dt\right\} dt_b \qquad \text{IIIB7-(101)}$$

where $-dN(t_b)/N_0$ is the probability that a specimen will rupture in the interval dt_b. If $\sigma(t)$ is known, then eq. (101) provides the theoretical distribution of rupture times; if $t = t(\sigma)$ is known, then the equation can be reformulated to give the distribution of rupture stresses. As an example, in a constant loading rate experiment $\sigma = \dot{\sigma}_{\text{const}}\, t$ and eq. (101) becomes

$$-\frac{dN(\sigma_b/\dot{\sigma}_{\text{const}})}{N_0} =$$

$$\mu\left(\sigma_b, \frac{\sigma_b}{\dot{\sigma}_{\text{const}}}\right) \exp\left\{-\int_0^{\sigma_b/\dot{\sigma}_{\text{const}}} \mu\left(\sigma, \frac{\sigma}{\dot{\sigma}_{\text{const}}}\right) \frac{d\sigma}{\dot{\sigma}_{\text{const}}}\right\} \frac{d\sigma_b}{\dot{\sigma}_{\text{const}}} \qquad \text{IIIB7-(102)}$$

for the distribution of the breaking stress. Equation (102) thus indicates how the frequency function depends on the rate of loading. It is apparent that in order to calculate the theoretical distribution function, the functional form of μ must be known. This form may be evaluated from the experimental data, or calculated theoretically from an assumed model for the system.

Another test mode of interest is that of a sinusoidal stress imposed in addition to a constant load. This test mode is of importance in fatigue studies, and the distribution of break times has been calculated for fibers by Coleman (24).

Lacking an example of the application of the theory to fracture in rubber, the study of Hori (52) on the rupture of concrete in creep and constant rate of loading experiments will be used to illustrate the basic ideas. For relatively low stress levels, his experimental data suggested that μ depended on the stress level but was independent of the time, i.e., a semilogarithmic plot of $N(t_b)/N_0$ versus t_b, as suggested by eq. (99), was approximately linear with a slope depending on the stress level. The data could be described equally well by assuming either

$$\mu = A \exp \{c\sigma_b\} \qquad \text{IIIB7-(103)}$$

or

$$\mu = B\sigma_b{}^d \qquad \text{IIIB7-(104)}$$

Hence, eqs. (100) and (101) could be integrated directly to obtain a calculated distribution function for rupture. It is interesting to note, as pointed out by Hori, that the exponential form for μ can also be derived by an application of absolute reaction rate theory to rupture process.

Using the constants A, B, c, and d experimentally determined from creep experiments, Hori calculates both the distribution of rupture stresses and the effect of rate of loading to be expected for rupture in constant rate of loading experiments. The experimental distribution for the one rate studied was in good agreement with the theoretical prediction. For the constant rate experiment, Hori points out that the exponential dependence of μ on σ_b, eq. (103) leads to a distribution function of rupture stress which is equivalent to eq. (94) derived from extreme value theory, while the use of eq. (104) leads to a Weibull distribution for the strength [eq. (90)]. In a more extended study of the creep rupture of cement under higher stress levels, Hori (53) observed that the total creep curve can be considered to be made up of the superposition of three separate processes or stages. Based on a modified four-element Voigt model containing a non-newtonian dashpot in parallel with a plastic element having a yield value, he is able to use the theory to calculate a theoretical distribution for the rupture time in creep which is in good agreement with his experimental results.

Coleman has made extensive applications of the theory of stochastic processes to the rupture of polymeric fibers in test modes such as creep, constant rate of loading, and fatigue. These results have been adequately reviewed elsewhere (22,23).

It appears that no application of the theory has been made to rupture of rubber. This suggests a fruitful area for further research.

IIIB8. Tear and Fatigue

IIIB8a. Introduction

In addition to the test modes discussed previously, i.e., the constant rate experiments and creep and stress relaxation to rupture, other test modes of interest are those of tear and the closely related test, fatigue. In the former case the specimen, usually containing a small cut or having an appropriate geometry to give a region of high local stress concentration, is elongated until crack growth is initiated. In the latter case the specimen is subjected to repeated deformation, the amplitude (not necessarily constant) of each cycle being generally much less than that required to cause rupture in a single application. Under these conditions the lifetime of the specimen, as measured by the number of cycles to failure N, can be quite high. Megacycle lifetimes are common, though the value depends of course on the amplitude and the frequency of the excitation.

Tearing failure can be adequately characterized in only a few cases of simple specimen geometry; many of the test pieces commonly employed are not readily amenable to the stress analysis at the crack tip which is necessary to separate the material properties from the specimen geometry (56). Failure of rubber in fatigue not only suffers from the same difficulties of choice of test-piece geometry but is also operationally ill-defined. Often, after a major portion of the fatigue life has been expended, cracks appear in the material. These continue to grow until the specimen finally becomes severed (18). For some applications of rubber, "failure" can be considered as the point where the cracks are first detected, or when visible cracks reach a certain size; for others, it can be defined to occur when actual rupture of the specimen occurs.

The statistical variability of fatigue lifetime has been considered more extensively than the statistical variability of σ_b or ϵ_b as measured in tensile experiments, probably because in engineering applications it is sufficient to work at safe loads and the rupture variability can be ignored. In fatigue applications, however, the interest lies in providing as high a lifetime as possible, as for example in the flexing of a tire. Since, for rubbers, fatigue studies are therefore usually carried out in the σ_b,t_b or ϵ_b,t_b region where t_b is large, σ_b or ϵ_b are only slightly dependent on t_b [cf. Figs. IIIB-11 and IIIB-12]. As we have seen in Subsection IIIB7, this implies that small variations in σ_b or ϵ_b will lead to large variations in t_b. Hence a statistical description must be emphasized in fatigue studies, and it is clear that the fatigue life should be defined in terms of a frequency distribution $p(N)$ at each level of the imposed excitation. Consequently, more meaningful terms such as the most probable lifetime or "the expected lifetime" can be defined. However, the number of specimens needed or the time involved in obtaining the required data for the complete determination of $p(N)$ is usually prohibitive, and various methods of economizing both time and the number of specimens have been proposed (105).

Studies of fatigue failure of rubbers are generally performed using machines, such as the duPont, the DeMattia, or the Scott testers, which were designed to simulate different types of service conditions (18,21,26). The complex loading history and complex specimen geometry, in conjunction with excessive heat build-up in some specimens has made fundamental studies of fatigue mechanisms difficult to carry out. However, it was early established that cracking, initiated either artificially or from naturally occurring flaws, led to ultimate failure in fatigue experiments (21). It was also observed that specific oxidation effects were present, since the fatigue life could be prolonged by the presence of antioxidants (18). The cracking process was thought to consist of two stages: crack initiation, which occurs in weak regions in the material, and crack growth, which ultimately leads to failure. It was also suspected that a correlation should exist between tear and fatigue but the early experimental comparisons were inconclusive (18). Recent work has shown that crack growth is related to tear behavior, so it will be informative to examine this phenomenon first and then see what the results imply for fatigue.

IIIB8b. Tear

Tear can be considered as rupture propagating from a stress raiser. As such it should be possible, in principle, to give a stress or strain criterion for tear rupture. However, the analysis of the stress distribution in the neighborhood of the crack tip is too difficult to permit an exact solution at present. However, Rivlin and Thomas (83) pointed out that a detailed analysis could be circumvented by making an energy balance on a small region of a strained specimen before and after the tear had passed through it. The energy concept is similar to that employed by Griffith (47) in the analysis of crack growth in brittle materials, but because of the large strains involved and the viscoelastic nature of a rubber, the analogy cannot be pursued too far.

Consider a stretched specimen of volume V and an initial thickness a, having an elastically stored energy per unit volume, W. If a crack now propagates through the specimen this energy will be dissipated in the formation of new surface, in the kinetic energy of motion of the cracked edges, and in the heat required for molecular motion of the chains as the specimen retracts. For a rigid specimen which will crack at small strains the latter two contributions are negligible and the stored energy is primarily spent in creating new surface area. For an elastomer all contributions will have to be considered. In either case, however, if the criterion for tear is assumed to be the existence of a critical energy of dissipation $\mathfrak{T}$, then a specimen cannot tear until this amount of energy has been stored in the specimen. In terms of an energy balance

$$-V(\partial W/\partial c)_l \geqslant \mathfrak{T}a \qquad \text{IIIB8-(105)}$$

where c denotes the length of the tear or cut, V the specimen volume and the partial derivative indicating a process at constant specimen length l, emphasizes that the boundaries are fixed. $\mathfrak{T}$ has the dimensions of energy per unit area of the crack and represents the work expended irreversibly per unit specimen thickness per unit increase in cut length, c.

One of the difficulties with tear measurements is that the results generally depend on the test specimen geometry. In order to show that $\mathfrak{T}$ was indeed a material constant, independent of the test piece, Rivlin and Thomas (83,94) analyzed and employed several types of specimens. Two of these are shown in Figure IIIB-47: (*a*) shows the "trousers" test piece where, if the specimen is wide enough,

$$\mathfrak{T} = 2F/a \qquad \text{IIIB8-(106)}$$

where F is the applied force and (*b*) shows the simple tensile specimen, where

$$\mathfrak{T} = 2KcW \qquad \text{IIIB8-(107)}$$

Here K is a slowly varying function (46) of the extension ratio λ. At room temperature $\mathfrak{T}$ was found to have a value of about 1.3×10^7 ergs/cm.^{2} for natural rubber, independent of specimen geometry.

In an extended series of experiments based on this tear criterion (42–45,46a,46,81,95), it was then found that in the trousers test

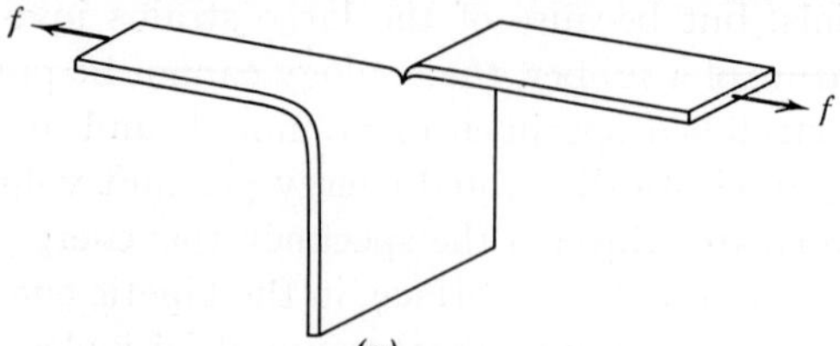

(a)

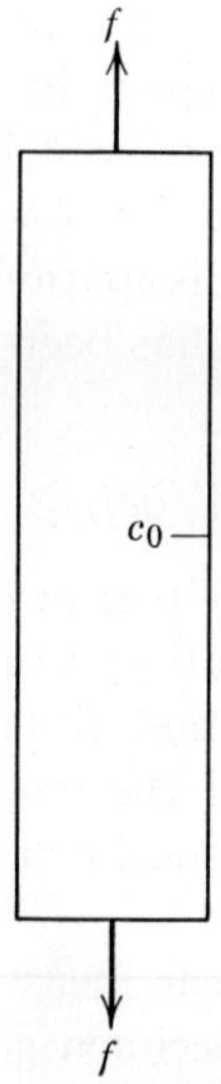

(b)

Fig. IIIB-47. Two test pieces for which the tearing energy $\mathfrak{I}$ can be evaluated from the applied force and extension. (a) The trouser test piece and (b) the simple extension test piece. The arrows indicate the direction of the applied force. c_0 is the original cut length before testing. After Gent, Lindley, and Thomas (39).

piece the cut tip is under a substantially uniaxial tension. If crack tips of varying diameter d are employed then it should turn out that

$$\mathfrak{I} = Wd \qquad \text{IIIB8-(108)}$$

Replacing W by $\mathcal{E}$, the integral under the σ, ϵ curve, Thomas (94) was able to make a direct comparison between $\mathfrak{I}/d$ and $\mathcal{E}$, using natural rubber specimens with tip diameters ranging from 0.090 to 0.294

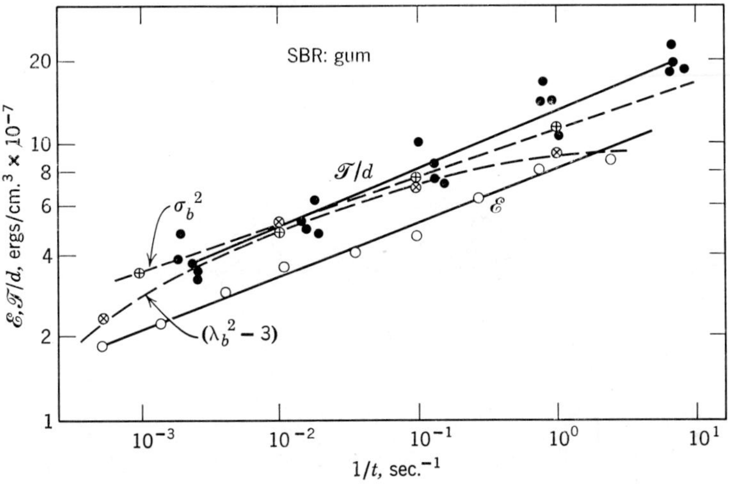

Fig. IIIB-48. Comparison of the time dependence of the energy to initiate tearing $\mathfrak{T}/d$ with the work for rupture per unit volume $\mathcal{E}$ for an SBR gum vulcanizate at 25°C. (43). Also shown are the time dependencies of σ_b^2 and $\lambda_b^2 - 3$, i.e., an estimation of the tearing energy based on the kinetic theory stress-strain law calculated from Greensmith's tensile data for the same system. After Greensmith (43).

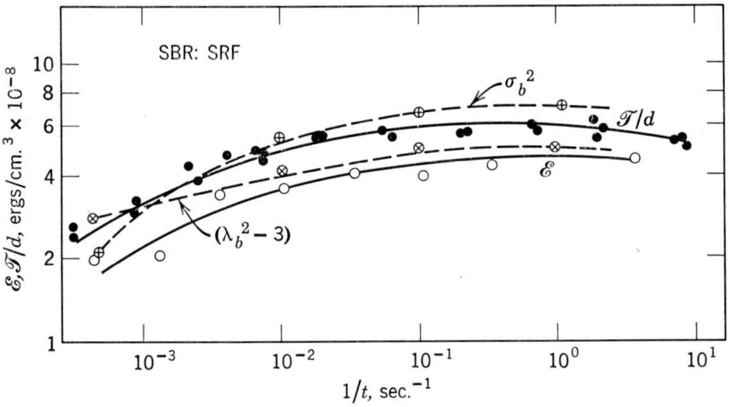

Fig. IIIB-49. Comparison of $\mathfrak{T}/d$ and $\mathcal{E}$ for a carbon black filled SBR at 25°C. As in Fig. IIIB-48, the estimations from kinetic theory, σ_b^2 and $\lambda_b^2 - 3$ are based on Greensmith's tensile data (42) on the same system, and plotted to coincide with the $\mathfrak{T}/d$ curve at $t^{-1} = 10^{-3}$ sec.$^{-1}$. After Greensmith (43).

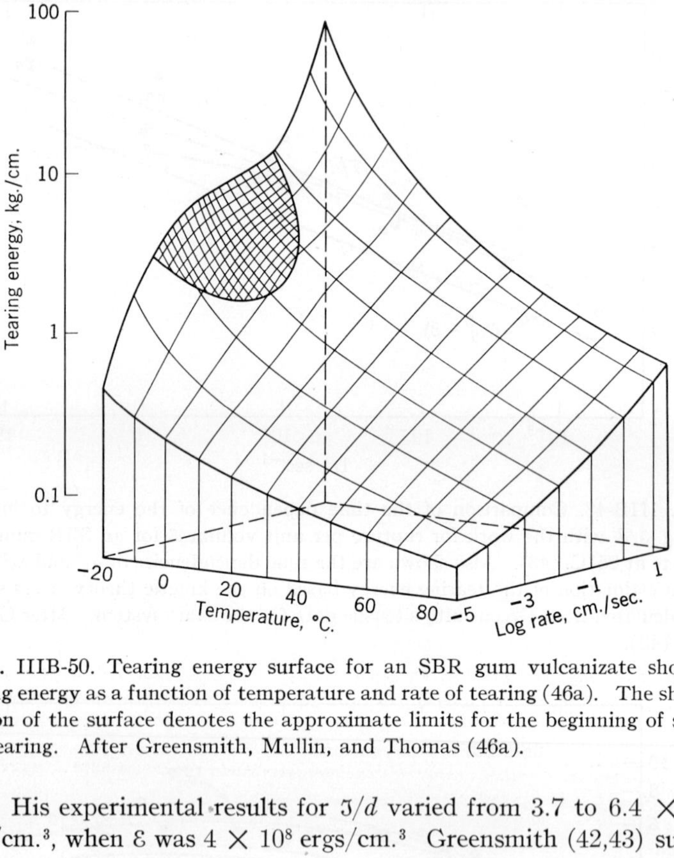

Fig. IIIB-50. Tearing energy surface for an SBR gum vulcanizate showing tearing energy as a function of temperature and rate of tearing (46a). The shaded portion of the surface denotes the approximate limits for the beginning of stick-slip tearing. After Greensmith, Mullin, and Thomas (46a).

cm. His experimental results for $\mathfrak{T}/d$ varied from 3.7 to 6.4×10^8 ergs/cm.^{3}, when $\mathcal{E}$ was 4×10^8 ergs/cm.^{3} Greensmith (42,43) subsequently repeated this in a more thorough fashion, comparing $\mathfrak{T}/d$ to $\mathcal{E}$ over a range of test rates and temperatures for SBR, both gum stock and carbon black filled. When $\mathfrak{T}/d$ was evaluated from the force required to initiate tearing, he obtained the results shown by the solid lines in Figure IIIB-48, i.e., the discrepancy is the same as that observed by Thomas. Figure IIIB-49 shows a similar set of data for SBR containing 50 parts of SRF black.

Also, it was found that the appearance of the torn surface depends on the test speed and temperature, being characterized by a transition from a smooth to a rough surface. The rough-surface tearing was described as "stick-slip" by analogy to a sliding friction response. In certain cases there was a regular fluctuation in the force required to

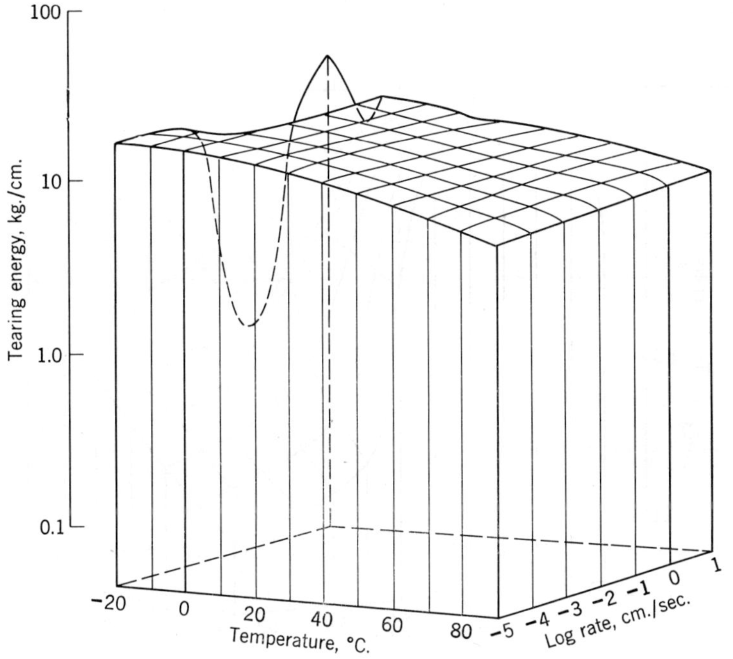

Fig. IIIB-51. Tearing energy surface for a gum natural rubber vulcanizate After Greensmith, Mullins, and Thomas (46a).

tear the specimen and large discontinuities in the tear path appeared. This intermittent tearing was called "knotty" tear. In addition to this change in surface appearance, $\mathfrak{T}$ itself was found to be a function of the test temperature and rate. The basic response of $\mathfrak{T}$ in smooth tearing is exemplified by results for unfilled SBR (Fig. IIIB-50). The shaded area indicates the beginning of stick-slip tear at lower temperatures. Natural rubber, on the other hand, has a tearing energy which is well-nigh independent of test rate and temperature (Fig. IIIB-51). This type of response is associated with the development of crystallinity at higher strains, so that the stress at the tip of the crack is distributed throughout a larger volume of specimen. When SBR is reinforced with FT carbon black, knotty and stick-slip tear occur at certain temperatures and rates. The enhancement of $\mathfrak{T}$ appears as a mesa jutting out from the basic property surface of Figure IIIB-50, as can be seen in Figure IIIB-52. This enhanced resistance

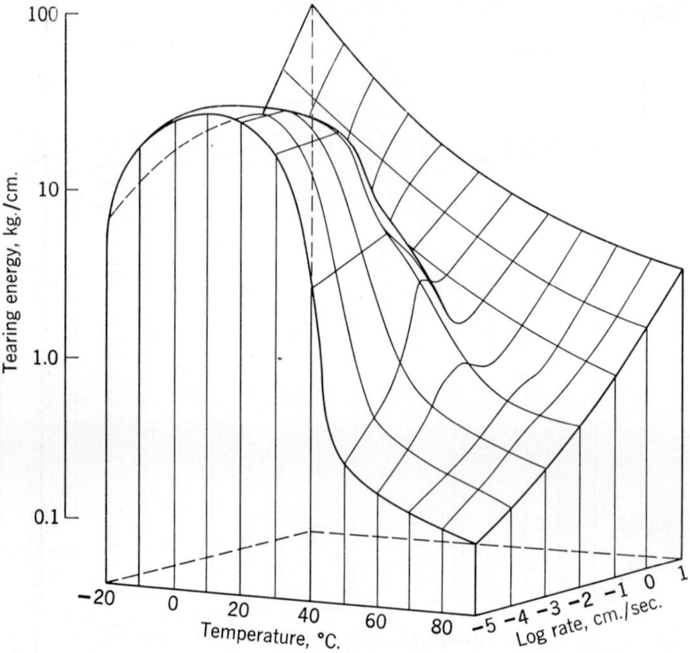

Fig. IIIB-52. Tearing energy surface for an SBR vulcanizate containing 30% by weight of FT carbon black. After Greensmith, Mullins, and Thomas (46a).

to tear is again due to the anisotropy developed in the presence of the particles and the corresponding redistribution of stress through a larger volume of specimen. The shape of the basic surface (Fig. IIIB-50) suggests that time–temperature superposition is applicable, over the range of smooth tearing. Figure IIIB-53 shows the superposed results (81) for the data of Figure IIIB-50. Also shown are data obtained with several other copolymer rubbers of butadiene–styrene and butadiene–acrylonitrile of varying monomer composition.

We would now like to obtain an indication of the meaning of the tear criterion in terms of the σ,ϵ,t surface of Figure IIIB-9. Recalling eqs. (107) and (108), it is immediately obvious that there is no direct analogy because the tear criterion depends not on a critical value of the stress or strain, but rather on a critical fraction of the rupture energy, $\mathcal{E}_b$, given by a critical value of the products $\mathcal{E}_\tau c$ or $\mathcal{E}_\tau d$. The subscript τ is used to denote that this value of the work at a

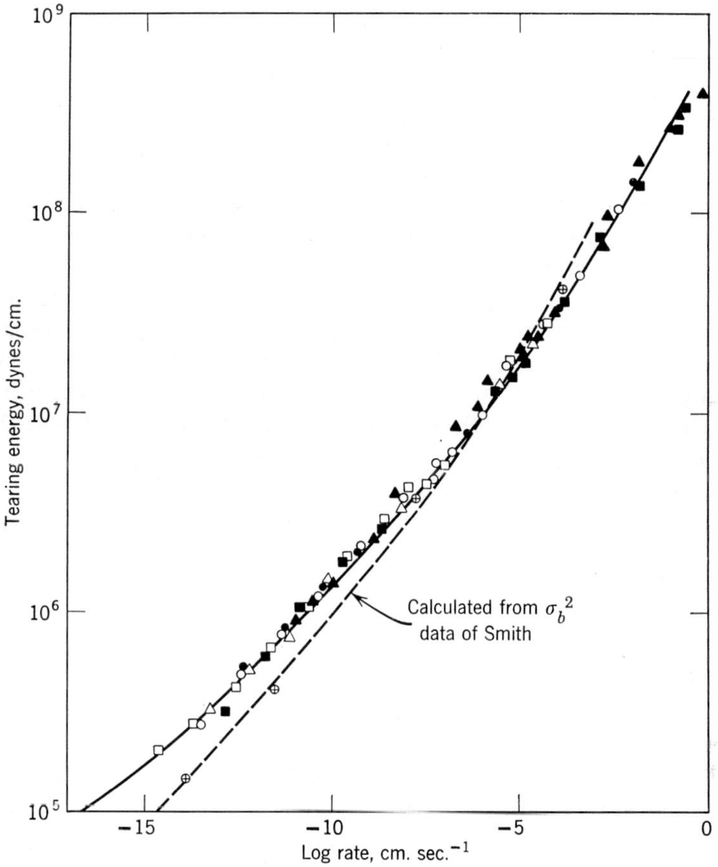

Fig. IIIB-53. Master plot for SBR and acrylonitrile-butadiene elastomers showing that tearing obeys time-temperature superposition. Also shown is the comparison of the time-dependence of σ_b^2 for SBR rubber taken from the data of Smith (90) (dashed line). After Mullins (81).

given λ is not the same as the work in the absence of the crack. Thus the product of the area under the stress–strain curve and the initial crack length should be a "constant." In general, it is not constant because $\mathcal{E}$ is time-dependent. Hence $\mathfrak{I}$ is also automatically time-dependent and should vary with the strain rate, as observed. Note that $\mathcal{E}_b$ will pass through a maximum with increasing test speed just as λ_b does, and $\mathfrak{I}$ should be expected to do so also. This has not been tested yet, since Greensmith's SBR studies were carried only to -40°C. and ϵ_b data for a similar material (67,90) are just at the $(\epsilon_b)_{\max}$ at this temperature. Presumably, the same lack of data at sufficiently low temperatures also holds for the butadiene–acrylonitrile rubbers of Figure IIIB-53.

If the stress–strain curve follows the kinetic theory, then $\mathcal{E}_b \cong (E/6)(\lambda_b^2 - 3)$ and $\sigma_b \cong (E/3)\lambda_b$ for $\lambda_b \gg 1$. Thus $\mathfrak{I}/d$ or $\mathfrak{I}$ itself should be proportional to λ_b^2 or to σ_b^2. These relationships can be subject to an approximate test, since Greensmith (43) has also reported ultimate property data on these same systems. Unfortunately the tensile modulus was not given. Nevertheless, log σ_b^2 and log λ_b^2 were calculated and compared to $\mathfrak{I}/d$ by setting the calculated curve equal to the $\mathfrak{I}/d$ values at a time scale of 10^{-3} sec.$^{-1}$. The results are also shown in Figures IIIB-48 and IIIB-49. It can be seen that the two *ad hoc* plots bracket the $\mathfrak{I}/d$ results. As an additional check, σ_b^2 from the data of Smith (90) is also plotted in Figure IIIB-53. Again the agreement is satisfactory, indicating that an estimate of tear behavior can be readily obtained from ultimate property data.

It can be shown by the same procedure employed in Subsection IIIB-2b for a linearly viscoelastic material, that $\mathcal{E}_b$ is approximately proportional to E''. Since $\mathcal{E}_b \propto \mathfrak{I}$ and $E'' = \omega\eta'$, where η' is the dynamic viscosity at the test frequency ω, this explains Mullins' observation (81) that $\mathfrak{I}$ is proportional to $\omega\eta'$. Thus, as with ultimate properties, determined at $\dot{\epsilon}_{\text{const}}$, it should prove to be possible to gain information on the tear behavior of simple systems via nondestructive, small-strain dynamic test methods.

IIIB8c. Dynamic Fatigue

Probably the most clear-cut and fundamental experiments on fatigue are the recent studies carried out by Gent, Lindley, and Thomas (39,73). These studies are based on the tearing energy concept described above. Lindley and Thomas (73) have shown, based on this concept, that fatigue is essentially a cut-growth process initiated from flaws in the specimen.

It is found for natural rubber, containing artificially induced razor cuts, that the cut growth per cycle, dc/dn, for a specimen subjected to repeated uniaxial extension of constant amplitude is given by

$$dc/dn = \mathfrak{I}^2/A \qquad \text{IIIB8-(109)}$$

where A is the cut growth constant for the material. For the simple tensile specimen (Fig. IIIB-47*b*), the cut growth per cycle is given by

$$dc/dn = (2K\mathcal{E}c)^2/A \qquad \text{IIIB8-(110)}$$

Integration of eq. (110) gives an expression for the number of cycles for the cut to grow from c_0 to c as

$$n = [A/(2K\mathcal{E})^2][(1/c_0) - (1/c)] \qquad \text{IIIB8-(111)}$$

If the length of the cut c_0 at time zero is small compared to c, then the fatigue life N will be given by

$$N = [A/(2K\mathcal{E})^2](1/c_0) \qquad \text{IIIB8-(112)}$$

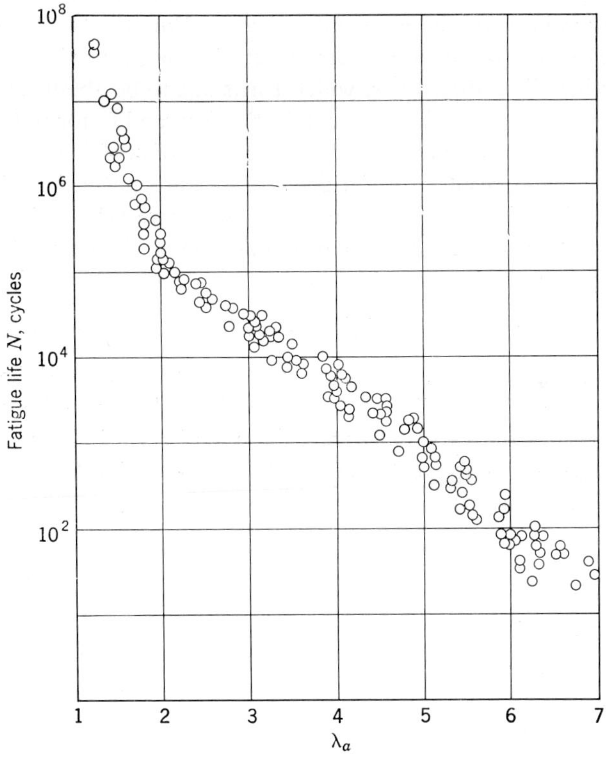

Fig. IIIB-54. Fatigue life for a natural rubber gum vulcanizate as a function of the amplitude of the extension ratio, λ_a. The data were obtained using three types of test pieces, i.e., the two shown in Fig. IIIB-47 and a "pure shear" test piece. After Gent, Lindley, and Thomas (39).

When cut growth is investigated for specimens having an artificially induced smooth razor cut, it is found (73) that eq. (112) is followed only after the specimen has been subjected to some hundreds of cycles, during which time there is a transition from smooth cut growth, which is rapid, to rough cut growth which is slower. Both types of cut growth follow eq. (112), A for the smooth cut growth process being larger than the rough cut constant by about a factor of 10.

In addition, eq. (112) was found to describe the fatigue behavior of natural rubber both with and without an artificial (razor) cut, provided $\mathcal{E}$, and thus $\mathfrak{J}$ were less than the value required for catastrophic tearing. Black-loaded stocks of natural rubber also conform to equation (112), but the values of c_0 and A differ somewhat from the corresponding values for the gum vulcanizate. Equation (112) was used to estimate the size, c_0, of defects present in the rubber in the absence of an artificially induced cut, which turns out to be about 2.6×10^{-3} cm. Gent (38) was able to calculate the fatigue life for natural gum vulcanizates to within about 20% for a range of c_0 of 5:1 and for N of 3000:1. These results for fatigue life of gum natural rubber as a function of strain amplitude are shown in Figure IIIB-54.

For SBR gum stocks, similar studies have shown that the cut growth per cycle is given by

$$dc/dn = \mathfrak{J}^4/B \qquad \text{IIIB8-(113)}$$

In this case, the cut growth depends on the fourth power of the tearing energy. In addition, B was found to depend on the frequency f, for f values less than about 10 cycles/min. and essentially independent of f for $f > 10$. The fatigue life for SBR, using simple tensile specimens (Fig. IIIB-47*b*) is given by

$$N = [B/48(K\mathcal{E})^4](1/c_0^3) \qquad \text{IIIB8-(114)}$$

The presence of carbon black in SBR vulcanizates profoundly affects the fatigue mechanism, and in fact the fatigue behavior tends to follow the cut growth relation exhibited by natural rubber [eq. (112)]. This is attributed to the fact that the presence of carbon black in SBR is similar, in its effect, to the presence of crystallites in natural rubber. Note that eqs. (112) and (114) predict that in contrast to the behavior of metals (74) there is no endurance limit for rubber, i.e., $N \rightarrow \infty$ only as $\mathcal{E} \rightarrow 0$, provided $c \neq 0$. This is analogous to the prediction in Subsection IIIB4 that there is no limiting ϵ_b or σ_b in a constant strain rate test. On the other hand, eqs. (112) and (114) imply a more severe restriction, because even if there were a limiting safe stress or strain in a single test, e.g., a $\dot{\epsilon}_{\text{const}}$ test, $\mathcal{E}$ has a finite value for any deformation used in a fatigue study and this value determines the fatigue life.

The fatigue studies described above were carried out in air. Gent has recently extended this work to studies of chemical effects in fatigue of rubber (38). He has shown that the fatigue life of natural rubber is greatly increased by performing the studies *in vacuo*. The relative improvement was found to depend on the strain amplitude, with the greater improvement being observed at the lower strains. The presence of known antioxidants in the vulcanizates increased the fatigue life in air by a factor of about 3–4; however, *in vacuo* the fatigue life was essentially unchanged by the presence of antioxidants. Irradiation with ultraviolet light during fatigue in both air and *in vacuo* increased fatigue life. Gent attributed this to further crosslinking brought about by the irradiation. He considered two possible processes whereby the lifetime in air could be reduced below that *in vacuo:* (*1*) the occurrence of chain scission reactions whereby oxygen contributes to the cut growth by promoting chain breaking; (*2*) the absence of recombination reactions whereby chains broken by the purely mechanical cut growth step are prevented from recombining to "heal" the cut by the intervention of oxygen, an efficient free radical trap. These alternative proposals can be tested, for example, by observing the effect of the presence of a free radical trap on the fatigue life *in vacuo*. If (*1*) is correct, the life time *in vacuo* will be unchanged, while if (*2*) is correct, the lifetime *in vacuo* will be reduced to a value similar to that found in the presence of air. Gent (38) concluded from his experimental data that mechanism (*1*) is probably the dominant factor.

Gent, Lindley, and Thomas (39) pointed out that the stress–strain curves obtained during the fatigue cycles were essentially constant. On this basis, and assuming that the kinetic theory expression is a good approximation to the actual stress–strain curve, the present authors have shown that eq. (112) can be written as

$$N = A/c_0\{[2K\nu_e RT[(\lambda_a^2/2) + (1/\lambda_a) - \tfrac{3}{2})]\}^2 \qquad \text{IIIB8-(115)}$$

For the given A value and several experimental points, an average value of the quantity $\nu_e T$ was calculated. If T is assumed to be

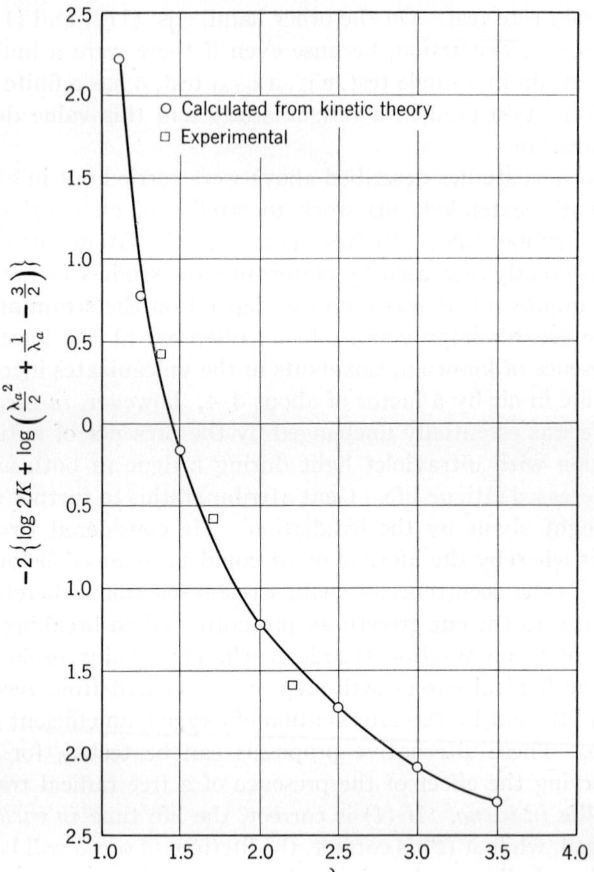

Fig. IIIB-55. Comparison of the form of the experimental N,λ_a curve of Fig. IIIB-54 with the form calculated from eq. (115) based on the kinetic theory expression for the stored energy. The value of $\nu_e T = 7.3 \times 10^{-2}$ and $c_0 = 1.5 \times 10^{-3}$ cm. were required for the fit.

310°K., $\nu_e = 236 \times 10^{-6}$ μmoles/cm.^{3}. For this value of $\nu_e T$, a theoretical N,λ_a curve was calculated and found to be in excellent agreement with the experimental one. The calculated curve based on kinetic theory is shown in Figure IIIB-55. Also shown are the average experimental values taken from their data. The agreement is satisfactory up to λ_a about 2.5. The c_0 value is calculated to be 1.5 ×

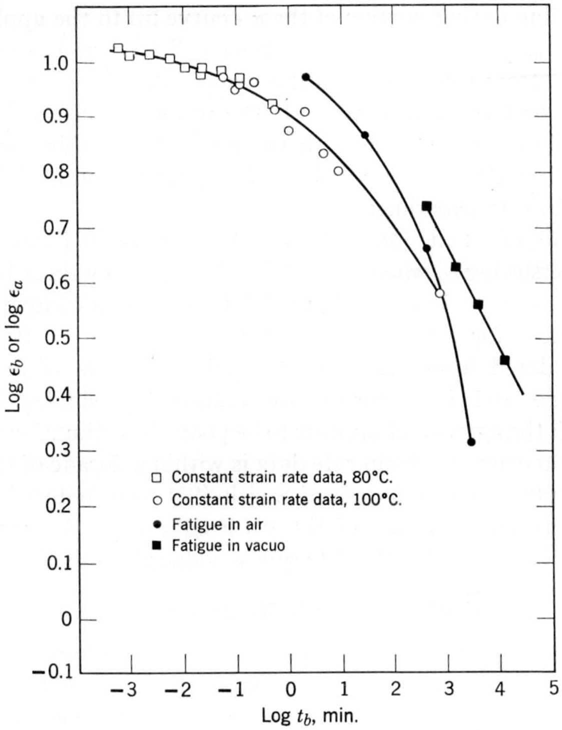

Fig. IIIB-56. Estimation of fatigue life of natural rubber from constant strain rate rupture data by the method indicated in the text. ϵ_b and t_b are the breaking strains and rupture times for specimens tested at constant strain rate; ϵ_a is the strain amplitude in fatigue and here $t_b = N/f$ (38). The results for fatigue in air are the same as those presented in Fig. IIIB-55. The standard reference temperature is 30°C.

10^{-3} cm., which compares favorably with the value of about 2.0×10^{-3} cm. reported by Gent (38). If this procedure is generally valid, at least for small strain amplitudes, it should be possible to obtain portions of the fatigue life curve by measuring the lifetimes of specimens of varying ν_e and superposing the data, provided A and c_0 do not depend on ν_e. The shorter fatigue lives of specimens of higher crosslink density would thus allow an N,λ_a curve to be obtained more quickly.

In terms of the property surface, the experiments of Gent, Lindley, and Thomas (39) correspond approximately to a trace on the surface corresponding to that portion of the σ,ϵ curve up to the applied strain amplitude ϵ_a. In the ϵ_a,t_b plot, the trace will have a saw-tooth waveform with a front that moves until it reaches the log ϵ_a,log t_b envelope, at which point rupture occurs. If the property surface is approximately independent of the path or test method, this log ϵ_a,log t_b plot should be the same as the log ϵ_b,log t_b curve obtained from constant strain rate experiments.

The data of Gent (38) are shown replotted in Figure IIIB-56, as log ϵ_a versus log t_b, where $t_b = N/f$. The data obtained both in air and *in vacuo* are included. Figure IIIB-56 shows a comparison with the ϵ_b,t_b curve for a natural rubber vulcanizate ($\nu_e = 127$ μmoles/cm.^{3}) calculated from the constant strain rate data of Smith (86). The fatigue and $\dot{\epsilon}_{\text{const}}$ curves are remarkably similar in shape, and though the agreement appears to be poor, the estimation of fatigue life based on constant strain rate data is within a decade of the experimental results. The ϵ_b,t_b data of Smith (86) were reduced from 100 and 80°C. to 30°C. by use of the WLF equation. An approximate correction was made for the differing ν_e values by use of eq. (27b).

IIIB9. Geometric Rupture Surface

In Subsection IIIB1 it was pointed out that a σ,ϵ,t property surface implies a particular type of stress or strain loading. If bi- or triaxial stresses are applied, the rupture behavior may be conveniently described in terms of the geometric rupture surface (7,62,74,81a,85). This depicts the interrelationship between the isothermal, isochronal rupture stresses and rupture strains, such as indicated in Figures IIIB-57 and IIIB-58 (7). Taking stress as an example, an arbitrary stress may be resolved into contributions along any of three orthogonal coordinates σ_1, σ_2, σ_3, and the coordinates of the rupture stress can be considered to locate a point in stress-space. Thus the stress axes define eight octants, of which the first octant (+, +, +) represents pure tensile rupture and its counterpart in the (−, −, −) octant represents pure compressive rupture. Uniaxial tensile rupture σ_b thus represents some limiting value along one of the three axes. For an isotropic body, rupture space is symmetric, so that $(\sigma_b)_1 = (\sigma_b)_2 = (\sigma_b)_3$, and a single test gives the intersections of this surface with all these axes. If a biaxial load is applied, then the response lies in one of the planes σ_1,σ_2; σ_2,σ_3; or σ_3,σ_1. Again, the response is identical in each plane. Thus a single measurement of rupture in pure shear ($\sigma_1 = \sigma_2$)

gives rise to three points on the surface. By continued testing under various combinations of tensile stresses a complete surface delineating the rupture characteristics may be constructed. This geometric rupture surface then represents a rupture criterion for the first octant.

Figures IIIB-57 and IIIB-58 illustrate such surfaces in terms of the true stress $\bar{\sigma}_b(= \lambda_b\sigma_b/I_3^{1/2})$ and the extension ratio λ_b for a polyure-

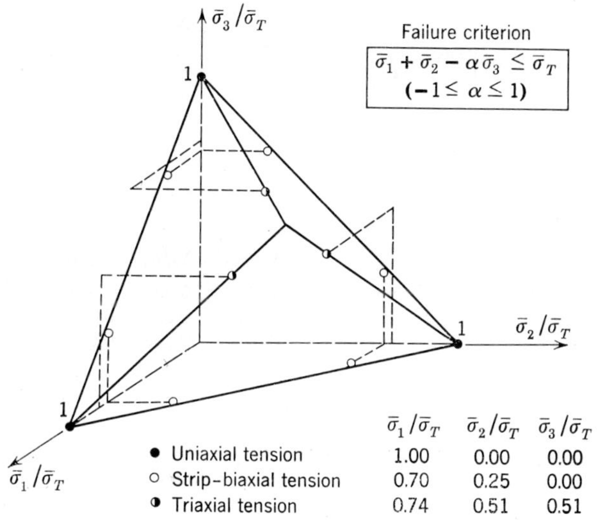

Fig. IIIB-57. Geometric failure surface in principal true stress space for a polyurethane rubber at room temperature normalized to unit tensile behavior ($\bar{\sigma}_T = \lambda_b\sigma_b$). After Ko (62).

thane rubber. In both cases these geometric rupture surfaces have been normalized to unity according to the pure tensile values. The rupture stress surface, Figure IIIB-57 is a triangular pyramid mathematically described by

$$\bar{\sigma}_1 + \bar{\sigma}_2 - \alpha\bar{\sigma}_3 \leqslant \bar{\sigma}_T; \qquad -1 \leqslant \alpha \leqslant 1 \qquad \text{IIIB9-(116)}$$

which is therefore the rupture criterion for this particular system. The rupture strain surface (Fig. IIIB-58) is an especially simple one, the criterion for rupture being a plane in λ-space given by

$$\sum_i \lambda_i = \text{constant} \qquad \text{IIIB9-(117)}$$

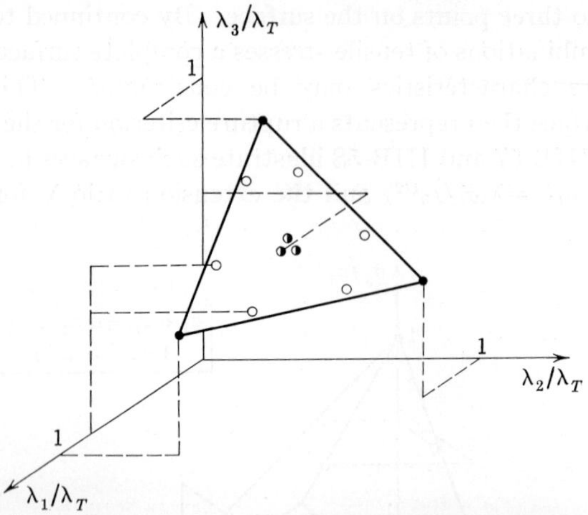

	λ_1/λ_T	λ_2/λ_T	λ_3/λ_T
● Uniaxial tension	1	0.427	0.427
○ Strip-biaxial tension	0.822	0.593	0.428
◑ Triaxial tension	0.600	0.593	0.593

Failure criterion

$\sum_i \lambda_i = \text{const.}$

Fig. IIIB-58. Geometric failure surface in principal extension ratio space for the same polyurethane rubber of Fig. IIIB-57. The data are again normalized to unit tensile behavior. After Ko (62).

Note that the relationship between such stress and strain rupture surfaces is fixed by the appropriate σ,ϵ law. For the rubber represented here, the kinetic theory expression is applicable.

These geometric rupture surfaces are important in practical applications, where complex loading conditions may be found, and in the analysis of test results in cases where the test piece is not subjected to a uniform stress field.

IIIB10. Summary

In this section we have attempted to show that there is a steady progression from small strain (preferably dynamic) to large strain and then to rupture behavior. Therefore considerable insight into the molecular origin of rupture phenomena is afforded by drawing on available knowledge of the viscoelastic response of polymers on the one hand, and by recasting the mathematical description in terms appropriate to finite strain elasticity theory on the other. Because of the paucity of data on rupture in other than uniaxial tension, little can be done at present with the latter approach. However, it is implied that such studies are required as a prerequisite to a complete understanding of the behavior of both pure and filled rubbers and crystalline polymers. In the latter systems the nonhomogeneous stress distribution in test pieces, on a microscopic scale, would seem to require a detailed stress analysis and a knowledge of the stored energy function for the amorphous portion of the matrix before the true nature of rupture phenomena can be assessed.

Thus the results presented can be considered as basic to an understanding of the behavior of the more complex systems including partly crystalline polymers, and many of the correlations indicated should carry over to such systems. Little attempt has been made to bridge the gap, however, in favor of a more intensive survey of the response of amorphous systems. Nevertheless, several references have been made to the behavior of simple filled systems and the effect of fillers, and these discussions could provide the nucleus for a future analysis.

The close connection between small deformation behavior and rupture is portrayed by considering a property surface depicting the relationships between stress, strain, and time. Temperature effects can be accounted for, in the simple systems discussed, by the WLF equation. Since the stored energy functions and the distribution functions of relaxation times for different polymers are similar, rupture behavior should also be expected to be similar, provided the comparison is made at some corresponding state. Different test modes can be considered to represent different paths across this surface. The usual representations of rupture consist of projections of the rupture points to the stress–strain, stress–time, and strain–time planes. The first of these is called the failure envelope and is deemed the most important one because it is essentially independent of test rate and temperature, is directly related to the crosslink density, and appears to be independent of the test mode.

Various theories proposed to explain certain features of rupture in amorphous systems have been examined for their ability to describe other features of rupture as well. The theories of Knauss and of Bueche and Halpin are the most general, with the former having been subjected to a wider range of experimental tests.

The statistical nature of rupture has not been extensively studied. A brief investigation revealed that the statistical distributions of breaking properties were temperature-dependent but did not permit the evaluation of the underlying flaw distribution. The functional relationship between the distributions of rupture stress, strain, and time is indicated. The statistical nature of rupture is most often encountered in dynamic fatigue studies. The results of the earlier parts of this section invited a comparison between dynamic fatigue and constant strain rate data, with the result that a close correlation was shown to exist.

IIIB11. List of Key Symbols

a_T	Time-temperature shift factor; the ratio of relaxation times at two different temperatures
A,B,C,D	Constants
$D(t)$	Tensile creep compliance
E	Tensile modulus
$\mathcal{E}$	Work indicated by the area under a stress-strain curve *or*, with subscript d, energy dissipated per unit volume
$f(x)$	Differential distribution function for x
$f(\epsilon)$	Strain function describing the nonhookean contribution to a stress-strain equation
$F(x)$	Cumulative distribution for x
G	Shear modulus
H_l	Distribution function of tensile relaxation times
I	Strain invariant
J	Shear compliance
n	Number of equivalent statistical segments per chain
N	Number of cycles to rupture in dynamic fatigue
t	Time *or*, with subscript b, rupture time
T	Absolute temperature
$\mathfrak{T}$	Characteristic tearing energy
V	Volume
W	Elastically stored energy per unit volume
ϵ	Tensile strain *or*, with subscript b, breaking strain
ζ_0	Average monomeric frictional coefficient
$\Phi(x)$	Cumulative distribution function for extreme values of x
λ	Principal extension ratio

ν_e	Concentration of effective network chains (per unit volume of whole rubber)
σ	Tensile stress *or*, with subscript b, breaking stress
τ	Relaxation time
$\varphi(x)$	Differential distribution function for extreme values of x

References

1. Alfrey, T., Jr., *Mechanical Behavior of High Polymers*, Interscience, New York, 1948.
2. Bartenev, G. M., and L. S. Briukhanova, *J. Tech. Phys. USSR* (Eng. Transl.), **3,** 262 (1958).
3. Becker, G. W., and H. J. Rademacher, *J. Polymer Sci.*, **58,** 621 (1962).
4. Bernstein, B., E. A. Kearsley, and L. J. Zapas, *Trans. Soc. Rheology*, **7,** 391 (1963).
5. Bischoff, J., E. Catsiff, and A. V. Tobolsky, *J. Am. Chem. Soc.*, **74,** 3378 (1952).
6. Blatz, P. J., *Ind. Eng. Chem.*, **48,** 727 (1956).
7. Blatz, P. J., and W. Ko, *Trans. Soc. Rheology*, **6,** 223 (1962).
8. Borders, A. M., and R. D. Juve, *Ind. Eng. Chem.*, **38,** 1066 (1946).
9. Boyer, R. F., *J. Appl. Phys.*, **22,** 723 (1951).
10. Bueche, A. M., *J. Polymer Sci.*, **19,** 275 (1956).
11. Bueche, F., *J. Appl. Phys.*, **28,** 784 (1957).
12. Bueche, F., *J. Appl. Phys.*, **26,** 1133 (1955).
13. Bueche, F., *J. Polymer Sci.*, **24,** 189 (1957).
14. Bueche, F., *J. Chem. Phys.*, **22,** 603 (1954).
15. Bueche, F., and T. Dudek, ASD-TDR 62-902 Report, Wright Patterson Air Force Base, December 1962.
16. Bueche, F., and T. Dudek, *Rubber Chem. Technol.*, **36,** 1 (1963).
17. Bueche, F., *Physical Properties of Polymers*, Interscience, New York, 1962.

17a. Bueche, F., and J. C. Halpin, *J. Appl. Phys.*, **35,** 36 (1964).

18. Buist, J. M., and G. E. Williams, *India Rubber World*, **124,** 320, 447, 567 (1961).
19. Busse, W. F., E. T. Lessing, D. L. Loughborough, and L. Larrick, *J. Appl. Phys.*, **13,** 715 (1942).
20. Carmichael, A. J., and H. W. Holdaway, *J. Appl. Phys.*, **32,** 159 (1961).
21. Carpenter, A. W., *The Chemistry and Technology of Rubber*, C. C. Davis and J. T. Blake, Eds., Reinhold, New York, 1937.
22. Coleman, B. D., *J. Appl. Phys.*, **29,** 968 (1958).
23. Coleman, B. D., *J. Mech. Phys. Solids*, **7,** 60 (1958).
24. Coleman, B. D., *J. Appl. Phys.*, **27,** 862 (1956).
25. Corten, H. T., and T. J. Dolan, *International Conference on Fatigue of Metals*, Institution of Mechanical Engineers and American Society of Mechanical Engineers, 1956.
26. Dillon, J. H., in *Advances in Colloid Science*, Vol. 3, H. Mark and G. S. Whitby, Eds., Interscience, New York, 1946.

26a. Dudek, T., unpublished results.
27. Epstein, L. M., and R. P. Smith, *Trans. Soc. Rheology*, **2**, 219 (1958).
28. Epstein, B., *J. Appl. Phys.*, **19**, 140 (1948).
29. Eringen, A. C., *Non-Linear Theory of Continuous Media*, McGraw-Hill, New York, 1962.
30. Ferry, J. D., *Viscoelastic Properties of Polymers*, Wiley, New York, 1961.
31. Ferry, J. D., *J. Am. Chem. Soc.*, **72**, 3746 (1950).
32. Ferry, J. D., L. D. Grandine, Jr., and E. R. Fitzgerald, *J. Appl. Phys.*, **24**, 911 (1953).
33. Fischer, J. C., and J. H. Holloman, *Trans. AIME*, **174**, 546 (1947).
34. Fitzgerald, E. R., L. D. Grandine, Jr., and J. D. Ferry, *J. Appl. Phys.*, **24**, 650 (1953).
35. Flory, P. J., *Chem. Revs.*, **35**, 51 (1944).
36. Flory, P. J., *Principles of Polymer Chemistry*, Cornell Univ. Press, Ithaca, N. Y., 1953.
36a. Freudenthal, A. M., *The Inelastic Behavior of Engineering Materials and Structures*, Wiley, New York, 1950.
37. Fromandi, G., R. Ecker, and W. Heideman, *Kautschuk Gummi*, **13**, WT25 (1960).
38. Gent, A. N., *J. Appl. Polymer Sci.*, **6**, 497 (1962).
39. Gent, A. N., P. B. Lindley, and A. G. Thomas, *J. Appl. Polymer Sci.*, **8**, 455 (1964).
40. Green, A. E., and F. E. Adkins, *Large Elastic Deformations and Non-Linear Continuum Mechanics*, Oxford Univ. Press, 1960.
41. Green, A. E., and Zerna, W., *Theoretical Elasticity*, Oxford Univ. Press, 1954.
42. Greensmith, H. W., *J. Appl. Polymer Sci.*, **3**, 175 (1960).
43. Greensmith, H. W., *J. Appl. Polymer Sci.*, **3**, 183 (1960).
44. Greensmith, H. W., *J. Polymer Sci.*, **21**, 175 (1956).
45. Greensmith, H. W., and A. G. Thomas, *J. Polymer Sci.*, **18**, 189 (1955).
46. Greensmith, H. W., *J. Appl. Polymer Sci.*, **7**, 993 (1963).
46a. Greensmith, H. W., L. Mullins, and A. G. Thomas, *Trans. Soc. Rheol.*, **4**, 171 (1960).
47. Griffith, A. A., *Phil. Trans. Roy. Soc. London*, **A221**, 163 (1920).
48. Gumbel, E. J., *Statistical Theory of Extreme Values and Some Practical Applications*, National Bureau of Standards, Applied Mathematics Series 33, Feb. 1954.
49. Gumbel, E. J., *Statistics of Extremes*, Columbia Univ. Press, New York, 1958.
50. Guth, E., P. E. Wack, and R. L. Anthony, *J. Appl. Phys.*, **17**, 347 (1946).
51. Higuchi, T., H. M. Leeper, and D. S. Davis, *Anal. Chem.*, **20**, 1029 (1948).
52. Hori, M., *J. Phys. Soc. Japan*, **14**, 1444 (1959).
53. Hori, M., *J. Phys. Soc. Japan*, **17**, 228 (1962).
54. Jones, J., Lockheed Propulsion Co. Report No. 578/536-F-3, May 1963.

55. Kaelble, D., paper presented to Division of Paint, Plastics and Printing Ink Chemistry, 136th Meeting of the American Chemical Society, Atlantic City, N. J., November 1959.
56. Kainradl, P., and F. Handler, *Kautschuk Gummi*, **12**, WT239, WT284, WT336 (1959); *Rubber Chem. Technol.*, **33**, 1438 (1960).
57. Kase, S., *J. Polymer Sci.*, **11**, 425 (1953).
58. Knauss, W. G., SM No. 62-28, Guggenheim Aeronautical Laboratory, California Institute of Technology, Pasadena, California, July 1962.
59. Knauss, W. G., SM No. 63-1, Guggenheim Aeronautical Laboratory, California Institute of Technology, Pasadena, California, Jan. 1963.
60. Knauss, W. G., Ph.D. Thesis, California Institute of Technology, Pasadena, California, June 1963.
61. Knauss, W. G., and R. F. Landel, unpublished results.
62. Ko, W. L., Ph.D. Thesis, California Institute of Technology, Pasadena, California, 1963.
63. Krigbaum, W. R., and R. J. Roe, *J. Polymer Sci.*, **61**, 167 (1962).
64. Krigbaum, W. R., and M. Kaneko, *J. Chem. Phys.*, **36**, 99 (1962).
65. Landel, R. F., unpublished results.
66. Landel, R. F., TM No. 33-52, Jet Propulsion Laboratory, California Institute of Technology, Pasadena, California, Sept. 25, 1961.
67. Landel, R. F., and R. F. Fedors, *J. Polymer Sci.*, **B1**, 539 (1963).
68. Landel, R. F., and R. F. Fedors, unpublished results.
69. Landel, R. F., and T. L. Smith, *American Rocket Society Journal*, **31**, 599 (1960).
70. Landel, R. F., and P. J. Stedry, *J. Appl. Phys.*, **31**, 1885 (1960).
71. Leaderman, H., *Trans. Soc. Rheology*, **6**, 361 (1962).
72. Leaderman, H., *Elastical and Creep Properties of Filamentous Materials*, The Textile Foundation, Washington, D. C., 1943.
73. Lindley, P. B., and A. G. Thomas, *Proc. Rubber Technol. Conf., 4th Conf.* London, 1962, p. 428.
74. Marin, J., *Mechanical Behavior of Engineering Materials*, Prentice-Hall, Englewood Cliffs, N. J., 1962.
75. Martin, G. M., F. L. Roth, and R. D. Stiehler, *Trans. Inst. Rubber Ind.*, **32**, 189 (1956).
76. Marvin, R. S., in *Proceedings of the 2nd International Conference on Rheology*, V. G. W. Harrison, Ed., Oxford, 1953.
77. Marvin, R. S., *Viscoelasticity:Phenomenological Effects*, J. T. Bergen, Ed., Academic Press, New York, 1960.
78. Mason, P., *Trans. Faraday Soc.*, **55**, 1461 (1959); *Physical Properties of Polymers*, Macmillan, New York, 1959, p. 262; *J. Appl. Polymer Sci.*, **1**, 63 (1959).
79. Miner, M. A., *J. Appl. Mech.*, **A12**, 159 (1945).
80. Mooney, M., *J. Appl. Phys.*, **11**, 582 (1940).
81. Mullins, L., *J. Appl. Polymer Sci.*, **2**, 257 (1959).

81a. Nadai, A., *Theory of Flow and Fracture of Solids*, McGraw-Hill, New York, 1950.
82. Rivlin, R. S., *Phil. Trans.*, **A241,** 379 (1948).
83. Rivlin, R. S., and A. G. Thomas, *J. Polymer Sci.*, **10,** 291 (1952).
84. Rouse, P. E., Jr., *J. Chem. Phys.*, **21,** 1272 (1953).
85. Schwarzl, F., and A. J. Staverman, in *Die Physik der Hochpolymeren*, H. A. Stuart, Ed., Springer Verlag, Berlin, 1956, pp. 165–214.
86. Smith, T. L., ASD–TRD 62-572 Report, Wright Patterson Air Force Base, June 1962.
87. Smith, T. L., *J. Polymer Sci.*, **A1,** 3597 (1963).
88. Smith, T. L., *Trans. Soc. Rheol.*, **6,** 61 (1962).
89. Smith, T. L., and A. B. Magnusson, *J. Appl. Polymer Sci.*, **5,** 218 (1961); *Rubber Chem. Technol.*, **35,** 753 (1962).
90. Smith, T. L., *J. Polymer Sci.*, **32,** 99 (1958).
91. Smith, T. L., and P. J. Stedry, *J. Appl. Phys.*, **31,** 1892 (1960).
91a. Smith, T. L., unpublished results quoted in ref. 30, p. 267.
92. Staverman, A. J., and F. Schwarzl, in *Die Physik der Hochpolymeren*, H. A. Stuart, Ed., Springer Verlag, Berlin, 1956, Chap. II, p. 126.
93. Taylor, G. R., and S. R. Darin, *J. Polymer Sci.*, **17,** 511 (1955); *Rubber Chem. Technol.*, **30,** 863 (1957).
94. Thomas, A. G., *J. Polymer Sci.*, **18,** 177 (1955).
95. Thomas, A. G., *J. Polymer Sci.*, **31,** 467 (1958).
96. Thomas, A. G., *J. Appl. Polymer Sci.*, **3,** 168 (1960).
97. Tobolsky, A. V., and E. Catsiff, *J. Polymer Sci.*, **19,** 111 (1956).
98. Tobolsky, A. V., *J. Appl. Phys.*, **27,** 673 (1956).
99. Tobolsky, A. V., and H. Eyring, *J. Chem. Phys.*, **11,** 125 (1943).
100. Tobolsky, A. V., *Properties and Structures of Polymers*, Wiley, New York, 1960.
101. Treloar, L. R. G., *The Physics of Rubber Elasticity*, Oxford Univ. Press, London, 1958.
102. Villars, D. S., *J. Appl. Phys.*, **21,** 565 (1950).
103. Walsh, J. E., *J. Am. Statistical Assoc.*, in press.
104. Walsh, J. E., *Skandinaziak Aktuarietidskrift*, in press.
105. Weibull, W., *Fatigue Testing and the Analysis of Results*, Pergamon Press, New York, 1961.
105a. Weibull, W., *Ing. Vetenskaps Akad. Handl.*, No. 151 (1939).
106. Westergaard, H. M., *J. Appl. Mech.*, **A6,** A49 (1939).
107. Wiegand, W. B., and H. A. Braendle, *Ind. Eng. Chem., Anal. Ed.*, **1,** 113 (1929).
108. Williams, M. L., in *Progress in Astronautics and Rocketry*, Vol. I, M. Summerfield, Ed., Academic Press, New York, 1960.
109. Williams, M. L., P. J. Blatz, and R. A. Schapery, SM No. 61-5, Guggenheim Aeronautical Laboratory, California Institute of Technology, Pasadena, California, ASTIA No. AD 256 905, Feb. 1961.
110. Williams, M. L., R. F. Landel, J. D. Ferry, *J. Am. Chem. Soc.*, **77,** 3701 (1955).

111. Williams, M. L., R. A. Schapery, A. R. Zak, and G. H. Lindsey, SM No. 63-6, Guggenheim Aeronautical Laboratory, California Institute of Technology, Pasadena, California, Feb. 1963.
112. Yokobori, J., *Kolloid-Z.*, **166,** 20 (1959).
113. Zhurhov, S. N., and B. N. Narzullaev, *J. Tech. Phys.* (*USSR*), **23,** 1677 (1953); S. N. Zhurkov and E. E. Tomasherskii, *J. Tech. Phys.* (*USSR*), **25,** 1 (1955); S. N. Zhurkov and T. N. Sanfirova, *Doklady Akad. Nauk.* (*USSR*), **101,** 2 (1955).
114. Zimm, B. H., *J. Chem. Phys.*, **24,** 269 (1956).

Addenda: See H. W. Greensmith, L. Mullins, and A. G. Thomas in *The Chemistry and Physics of Rubber-like Substances,* L. Bateman, Ed., Maclaren-London, Wiley-New York, 1963, for a recent review on the strength of rubbers. Also see the article by M. L. Williams in *Fracture of Solids,* D. C. Drucker and J. J. Gilman Eds.. Interscience, New York, 1963, on the fracture of viscoelastic materials.

CHAPTER IV

Failure Processes

Sections

IVA. FRACTURE PHENOMENA IN CUTTING

A. Kobayashi and K. Saito

IVA1. Introduction

Many of the synthetic polymeric substances have taken their places as engineering materials by virtue of their novel properties. Parts made from them are mainly produced by molding techniques since this approach is facilitated by their rheological behaviors. Precision machining techniques, however, have also been widely employed for parts in vehicles, chemical apparatus, electronic instruments, etc.

It might well be said that cutting by a tool is a variety of "controlled failure" since the kind of chip to be removed and the fracture loci must be predetermined, the objectives being a smooth and accurately shaped piece. In order to machine a good product, one must appreciate the optimum cutting conditions in view of the cutting phenomena encountered, the mechanisms of failure, and the properties of the body.

Cutting by a tool and the deformation and fracture phenomena involved stem from the sharp, mobile wedging of the work material. With reference to Figure IVA-1, when a cutting tool advances into the piece along a given axis, a cutting force R', which is the resultant of the normal force F_{fn} and the tangential or frictional force F_f, acts on the tool at the rake surface. The reaction of R' on the work material is R, and it is this reaction which deforms and fractures the work material.

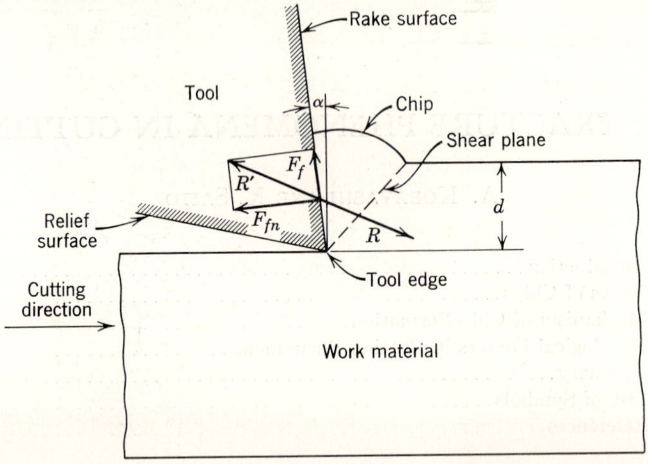

Fig. IVA-1. Forces and geometry of cutting. F_f, frictional force on rake surface. F_{fn}, normal force on rake surface. R', resultant force on rake surface. R, reaction of R' on work material. d, depth of cut. α, rake angle.

The kinds of chips obtained with a given kind of solid vary with the specific conditions of that fracture, and these include the depth and speed of cutting as well as the geometries of the tool and work material. However, it is possible to appreciate the cutting process significantly by controlling or defining these factors. Here we will attempt to clarify the salient considerations with the cutting of polymeric leathers and glasses which, as a whole, are colloquially termed "plastics."

IVA2. Types of Chips (1–3,3c)

The polytetrafluoroethylene (PTFE) samples employed were Teflon Nos. 1 and 5, obtained from E. I. du Pont de Nemours & Co., and were used for high- and low-speed cutting, respectively. These leatherlike bodies were compressed and sintered in the forms of pipe and plate. Three types of chips were observed with the cutting of PTFE. With cutting speeds, $V < 1$ m./min., continuous chips were usually produced as shown in Figure IVA-2. When compared with the other polymeric solids, described below, or with the metals, PTFE presents an unusual case, since a well-defined shear plane is not observed. Presumably, this results from a rubberlike contribution

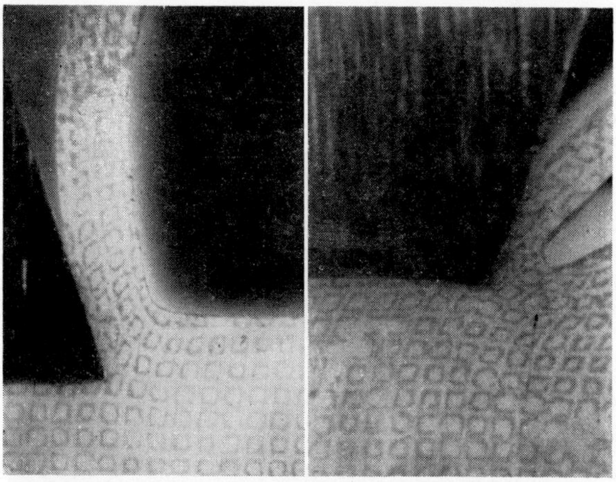

(a) (b)

Fig. IVA-2. Cutting of PTFE at low cutting speeds; the "continuous flow*" chip type. Cutting speed, $V = 0.019$ m./min. Depth of cut, $d = 0.5$ mm. (a) Rake angle, $\alpha = 20°$. (b) $\alpha = -20°$.

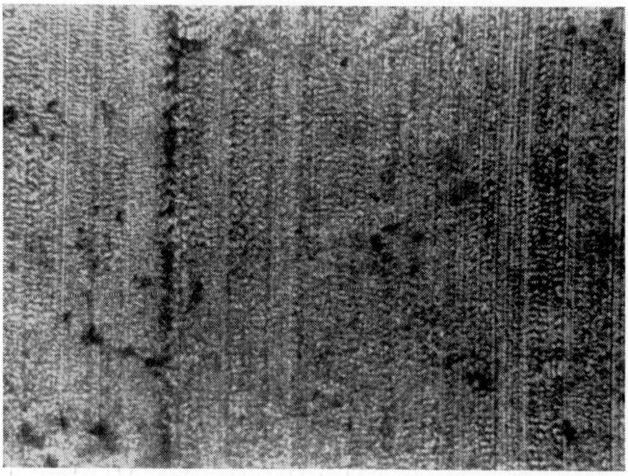

Fig. IVA-3. Topography of "continuous shear" chip type with PTFE. $V = 765$ m./min., $d = 0.046$ mm., $\alpha = 40°$.

at these low cutting speeds (see Subsection IVA3). Shear is experienced only in the zone which is closely adjacent to the rake surface. We will term this kind of chip "continuous flow.*"

* **Editor's Note:** See Tables IVA-I and -II.

When PTFE is cut with a highly raked tool at much higher speeds, V = 218–765 m./min., the more familiar "continuous shear" type of chip is obtained, just as with metals. This occurs under an extremely

Fig. IVA-4. Topography of "discontinuous shear" chip type with PTFE. V = 765 m./min., d = 0.046 mm., α = −10°.

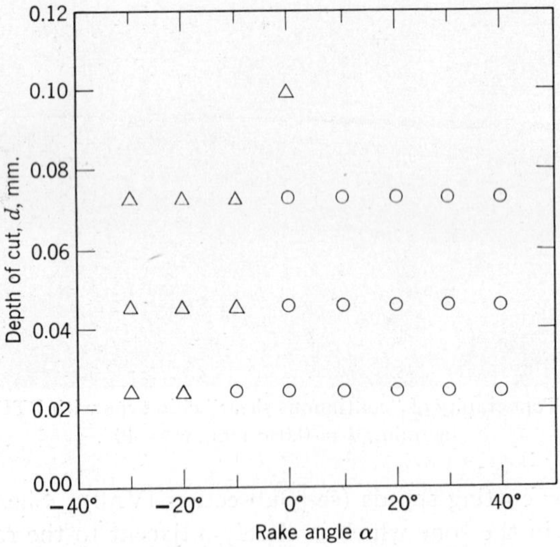

Fig. IVA-5. Summary of the cutting conditions for obtaining (Δ) "discontinuous shear" and (O) "continuous shear" chip types with PTFE. V = 218–765 m./min. Ambient work temperature, T = 25°C.

Fig. IVA-6. The "discontinuous complex shear" chip type with PS. $V = 0.2$ m./min., $d = 0.2$ mm., $\alpha = -30°$.

Fig. IVA-7. The "discontinuous crack" chip type with PS. $V = 0.2$ m./min., $d = 0.5$ mm., $\alpha = 40°$.

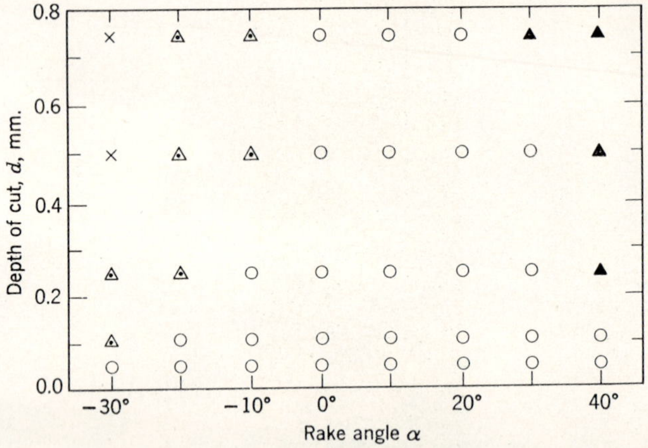

Fig. IVA-8. Summary of the cutting conditions for obtaining (○) "continuous shear," (△) "discontinuous complex shear," and (▲) "discontinuous crack" chip types with PS. (×) indicates that no chip was formed. $V = 0.2$ m./min. $T =$ 25°C.

wide range of cutting conditions. The surface topography of this kind of chip is shown in Figure IVA-3.

The third variety of PTFE chip observed, the surface of which is shown in Figure IVA-4, is obtained at the high cutting speeds with an only slightly positive, or negatively raked tool. This we will call the "discontinuous shear" type since, as may be inferred from this photograph, there was intermittent slippage.

Figure IVA-5 summarizes the cutting conditions observed for the formation of either the continuous or the discontinuous shear type of chips with PTFE No. 1 at cutting speeds, 218–765 m./min.

The polystyrene (PS) employed was a "high impact strength" variety of this glass which contained white pigment, was trade named Deck 7, and made by Kaito Chemicals Co. Here, too, three varieties of chips were identified. The continuous shear type, however, was produced under low cutting speeds, small cutting depths, and at high temperatures. Moreover, the second and third kinds of chips observed with PS differ from those mentioned above. These are shown in Figures IVA-6 and IVA-7 and are termed "discontinuous complex shear" and "discontinuous crack" types, respectively. The word "complex" implies a large contribution from compressive

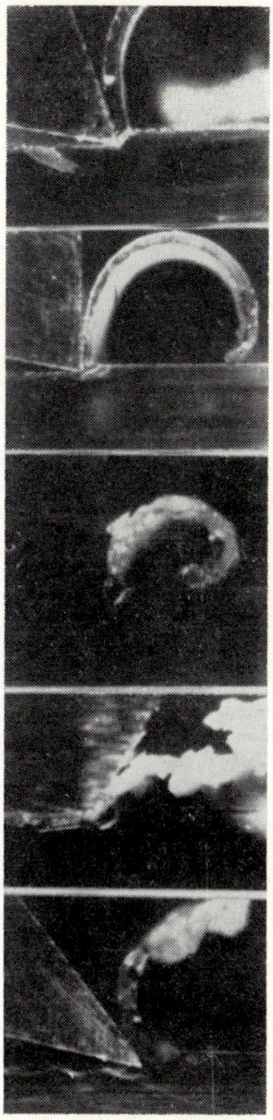

Fig. IVA-9. The five chip types obtained with PMMA. (*a*) "Continuous shear" (shear plane), (*b*) "continuous shear" (shear zone), (*c*) "discontinuous shear," (*d*) "discontinuous complex shear," and (*e*) "discontinuous crack."

or tensile stresses (see Subsection IVA3). The cutting conditions for these three varieties of chips in PS are summarized in Figure IVA-8.

The poly(methyl methacrylate) (PMMA) employed was a cast glasslike resin obtained from Fujikura Kasei Co. under the trade name Fujika Glas (cf. Plexiglas or Lucite). Here, five varieties of chips were obtained as shown in Figure IVA-9. These were not exactly the same five varieties mentioned above, however, since the continuous flow* type was missing (as mentioned before, only the

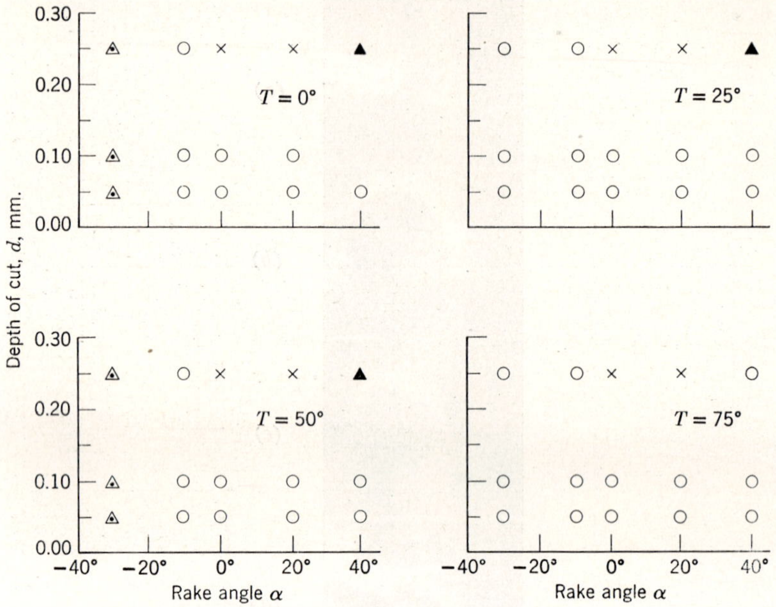

Fig. IVA-10. Summary of cutting conditions for obtaining (○) "continuous shear," (△) "discontinuous complex shear," and (▲) "discontinuous crack" chip types with PMMA. (×) = no chip. V = 0.2 m./min.

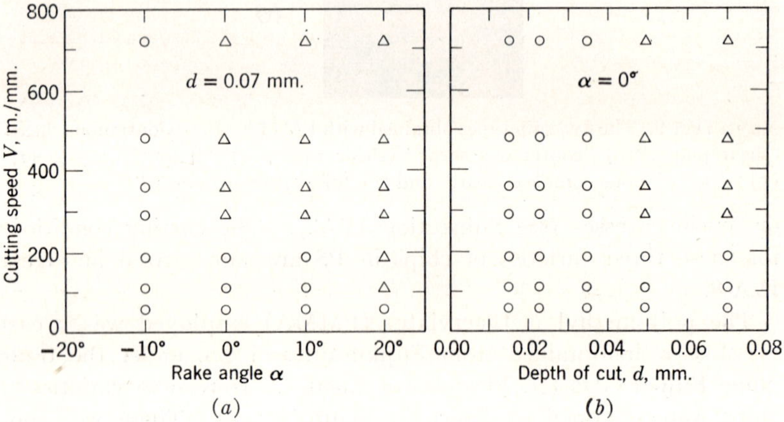

Fig. IVA-11. Summary of cutting conditions for obtaining (○) "continuous shear" and (△) "discontinuous shear" chip types with PA.

Fig. IVA-12. Topography of "discontinuous crack" chip type with PC. $V = 400$ m./min., $d = 0.5$ mm., $\alpha = 40°$.

PTFE showed this variety) and the continuous shear type is subdivided into two kinds; that generated on the shear *plane* and that generated in the shear *zone* (see Subsection IVA3). Figure IVA-10 summarizes the cutting conditions for the continuous shear, discontinuous complex shear, and the discontinuous crack types in PMMA.

The polyamide (PA) employed was nylon 6 as molded by Toyo Rayon Co. The continuous shear type was produced under most cutting conditions as seen from Figure IVA-11. However, the discontinuous shear variety was obtained when the rake angle, speed, and cutting depth were large.

The rigid poly(vinyl chloride) (PVC) employed was in the form of extruded pipes made by Sekisui Chemical Co. This sample behaved quite similarly to the PA above.

The polycarbonate (PC) employed was as extruded parts, such as pipes or plates, from Teijin Co. under the trade name Teijin Panlite. While the continuous shear type of chip predominates here also, the severer cutting conditions, i.e., large α, V, and d, generate the discontinuous crack variety as shown in Figure IVA-12.

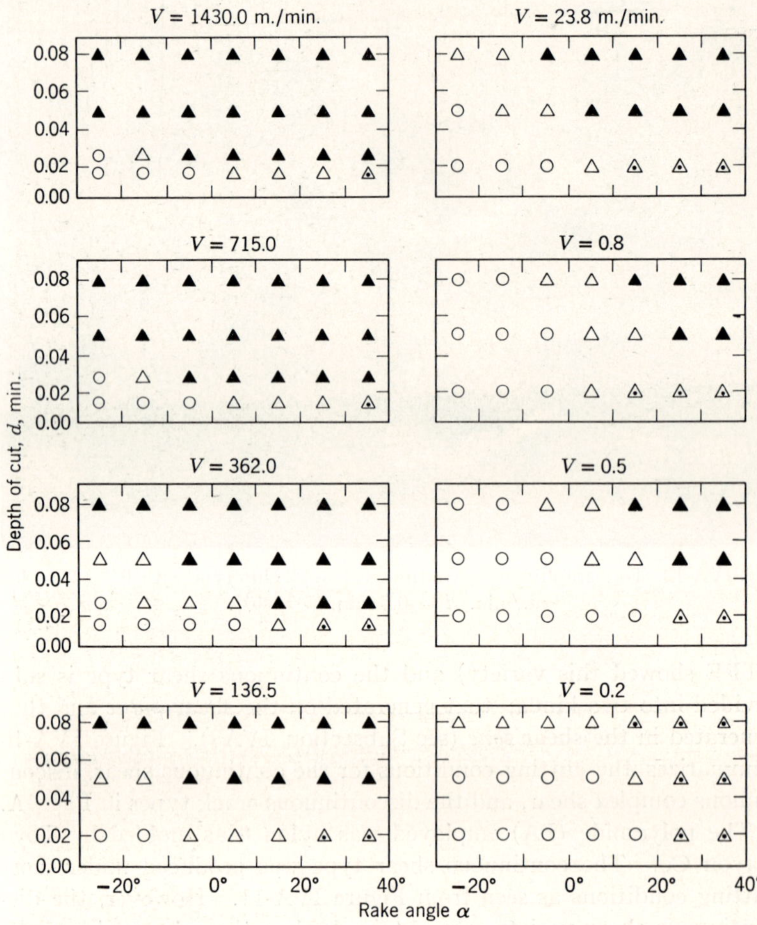

Fig. IVA-13. Summary of cutting conditions for obtaining (O) "continuous shear," (Δ) "discontinuous shear," (⨻) "discontinuous complex shear," and (▲) "discontinuous crack" chip types with CP.

The cast polyester (CP) employed was an unsaturated alkyd polyester by Toyo Koatsu Ind., Inc., with the trade name Ester C-15X. Here, too, the continuous shear type predominates at the milder cutting conditions. As summarized in Figure IVA-13, however, this shifts to the discontinuous shear, discontinuous complex shear, and then to the discontinuous crack type with increasing severity of cut-

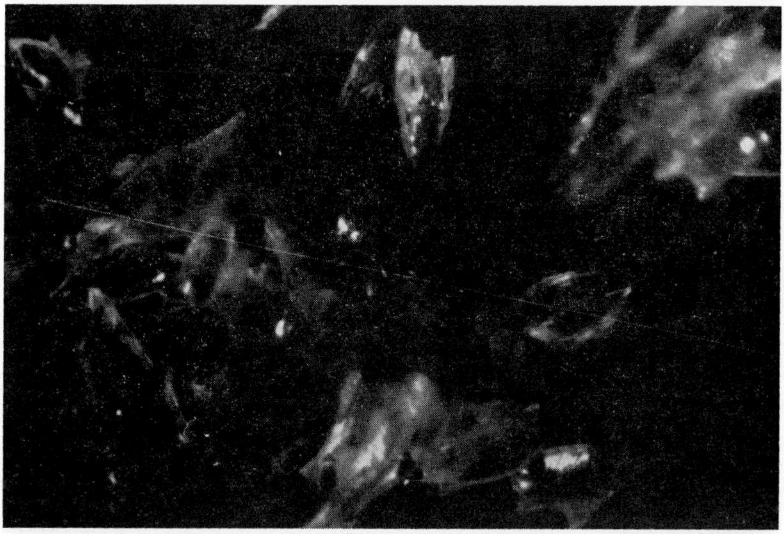

(*a*)

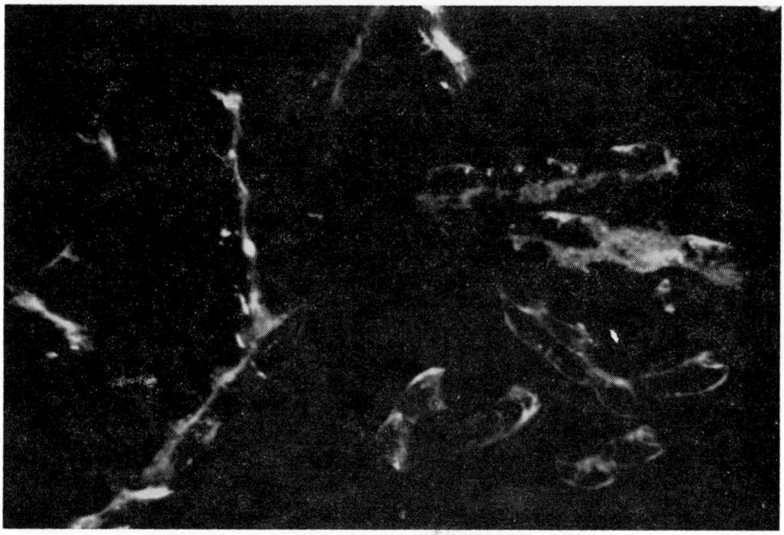

(*b*)

Fig. IVA-14. The "discontinuous crack" chip types with CP. $\alpha = 40°$. (*a*) $V = 136.5$ m./min., $d = 0.05$ mm. (*b*) $V = 1430$ m./min., $d = 0.023$ mm.

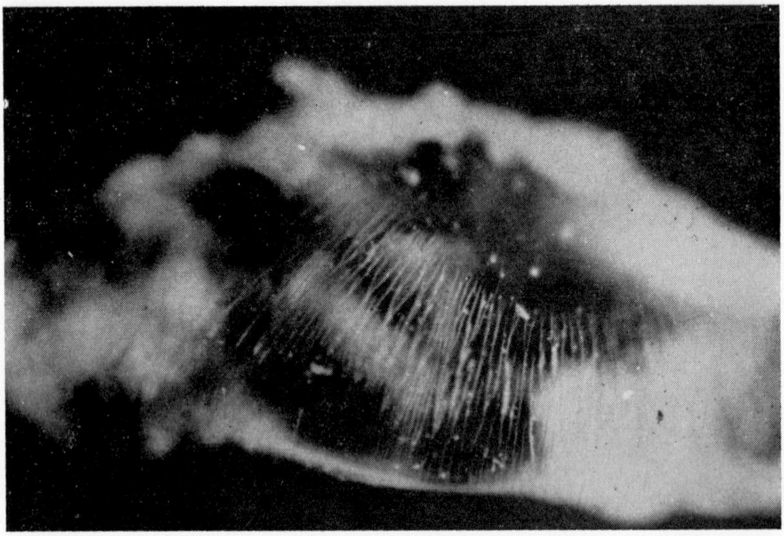

(a)

(b)

Fig. IVA-15. Hackle marks observed on surfaces of "discontinuous crack" chip type with CP. (a) and (b) correspond to the parts of Fig. IVA-14.

ting conditions. Moreover, the discontinuous crack species has a needlelike character, as shown in Figure IVA-14, with a further reflection of the sensibly brittle fracture in the hackle marks shown on the chip surfaces of Figure IVA-15.

Table IVA-I contrasts the terminologies employed here for chip type with that commonly employed for metals. (Compare with refs. 7 and 8.)

TABLE IVA-I
Terminologies for Types of Chips

For polymeric solids				For metals
Continuous	Flow*			—
Continuous	Shear			Flow
Discontinuous	Shear	Simple		Shear
		Complex	Shear with compressive stress	Tear
			Shear with tensile stress	
Discontinuous	Crack			Crack

IVA3. Mechanism of Chip Formation (3)

It was seen within the foregoing that the type of chip formed under a given cutting condition was dependent on the variety of polymeric

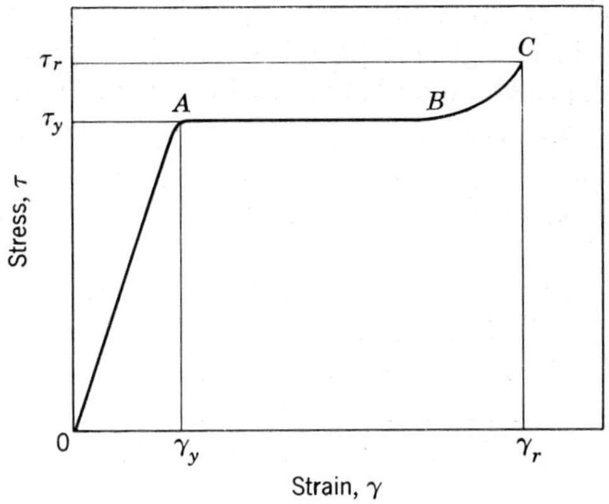

Fig. IVA-16. Highly schematic stress-strain curve for a viscoelastic body under a given strain-rate and temperature. (*A*) Yield point τ_y, γ_y. (*B*) Point for initiation of strain-hardening. (*C*) Rupture point, τ_r, γ_r. See Fig. IC-21, p. 97.

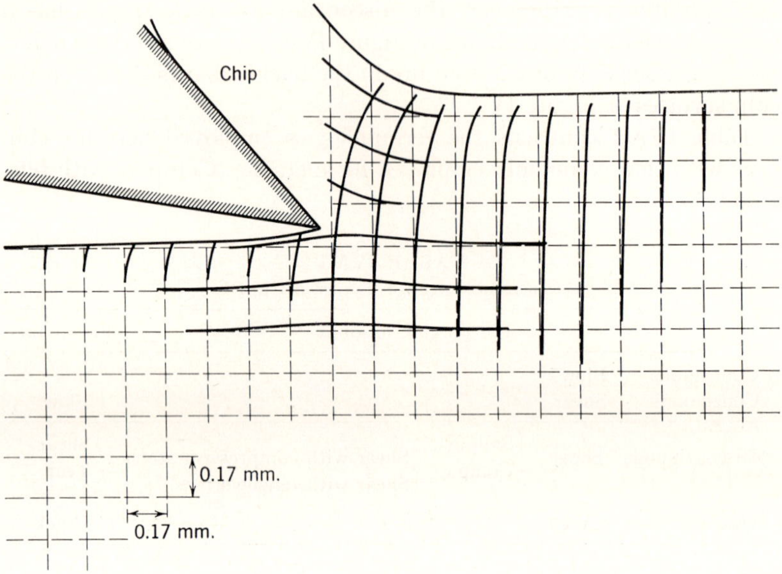

(*a*)

solid employed. The clear implication here is that the viscoelastic contribution of each of the samples varies considerably. To a first approximation, whether regarding cutting and chip formation, or deformation and fracture, the salient problem is similar.

Let us consider Figure IVA-16, which is a highly generalized stress–strain curve of a ductile body. For the present, we will intentionally ignore the viscoelastic or time- and temperature-dependent implications of this figure in regard to polymeric bodies, and later turn to those critical considerations in Subsection IVA4. Thus Figure IVA-16 is to be viewed as the deformational response of a viscoelastic body under a given strain rate at a given temperature. Since we do not permit the imposed stresses to relax, this body can "yield" at point τ_y, γ_y or at A. From points A to B the sample draws at stress τ_y, and beyond B, strain-hardens to the rupture point τ_r, γ_r or C. The region $0 \rightarrow$ A is the body's "elastic" response, while B $\rightarrow$ C implies a strengthening of the body on, say, a strain-induced crystallization (see Subsection IC5c, p. 96).

The continuous flow* variety of chip, as obtained with PTFE (but also in common with polyethylene) is formed as shown in

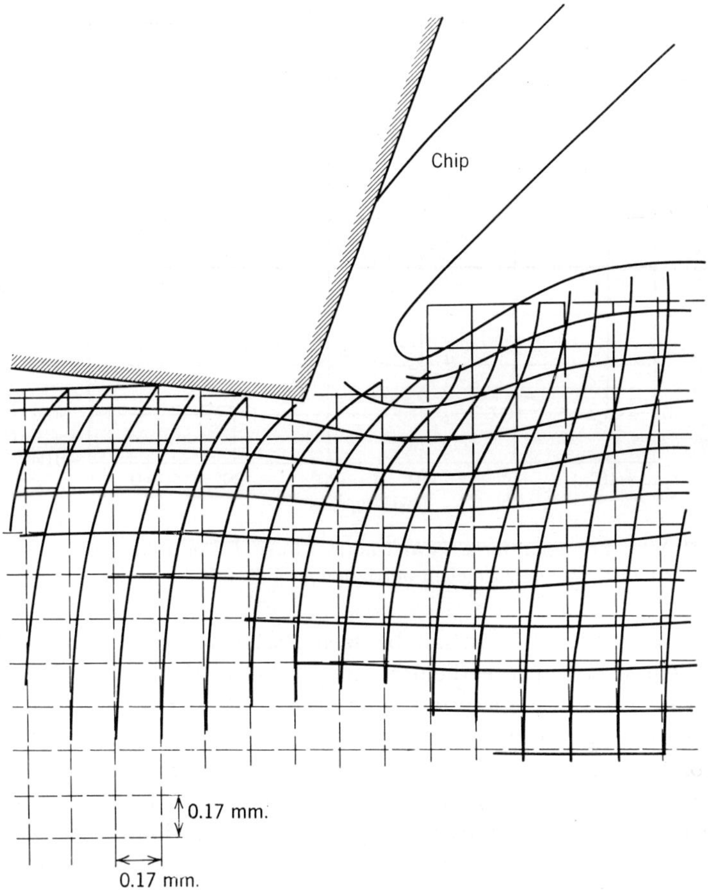

(*b*)

Fig. IVA-17. Patterns of deflection during cutting of PTFE #5. $V = 0.019$ m./min., $d = 0.5$ mm. (*a*) $\alpha = 40°$, relief angle $= 10°$. (*b*) $\alpha = -20°$, relief angle $= 10°$.

Figure IVA-17, as redrawn from actual photomicrographs. It is seen from Figure IVA-17*a* that with $\alpha = 40°$, the work material experiences a tensile strain perpendicular to the cutting direction and a compressive strain parallel to it. The chips are formed by the tensile stress accompanied by the cleavage action of the tool's edge.

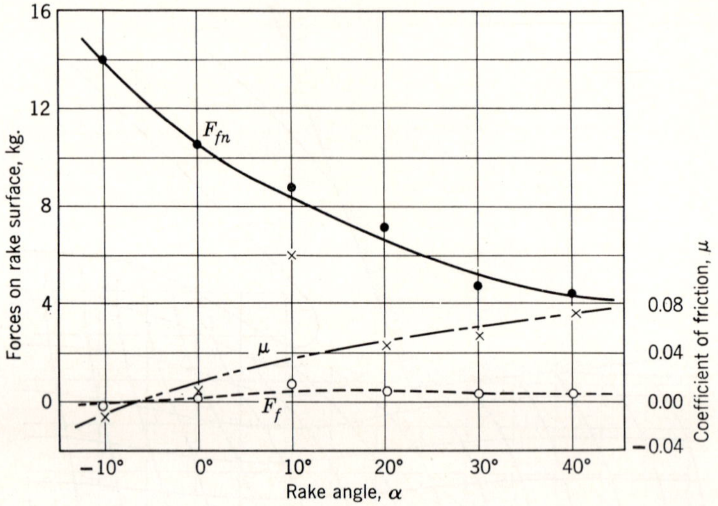

Fig. IVA-18. Frictional force F_f and normal force F_{fn} on rake surface vs. rake angle for PTFE #5, and coefficient of friction μ. $V = 0.762$ m./min., $d = 1.0$ mm.

In Figure IVA-17*b* it can be seen that compressive strains are experienced both perpendicular and parallel to the cutting direction when $\alpha = -20°$. Despite the great magnitude of these strains in the work material and chip, most of them *recover* after the cutting is completed. The volume of the system remains sensibly fixed since the ratio of d to the final chip thickness, d_c, is $r_c \cong 1$.

It is clear that this variety of cutting is not predominantly due to the ductility of the work material but to its rubberlike elasticity at these low values of V. In other words, the cutting is done in the region of $0 \rightarrow A$ of Figure IVA-16 where a rubberlike elastic contribution predominates at the large, slow deformations of these "leather-like" bodies. (It was seen in Subsec. IC4, p. 71, with regard to Figure IC-10, that leatherlike bodies exhibit both classical solid and rubberlike elastic contributions, the latter being more favored with the longer times for observation. *Ed.*)

If we are permitted, as a first approximation, to apply this system to the Merchant (6) equation for the chip's direction, where $\partial W/\partial\phi = 0$, then

$$2\phi + \beta - \alpha = \pi/2 \qquad \text{IVA3-(1)}$$

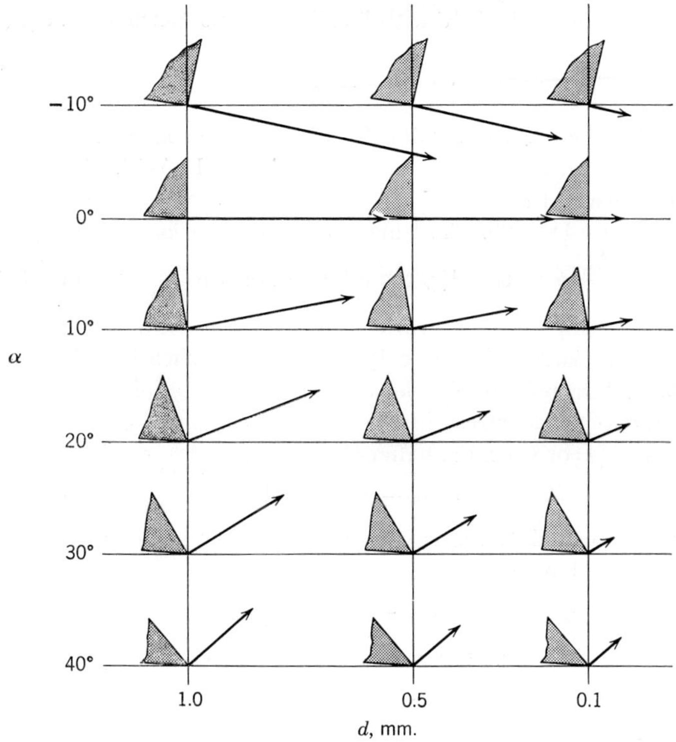

Fig. IVA-19. Cutting force vectors at various cutting depths and rake angles when the frictional angle on the rake surface β is 0° with PTFE #5. $V = 0.019$ m./min.

where W is the work done, ϕ defines the direction of the shear plane and

$$\beta = \tan^{-1} \mu \qquad \text{IVA3-(2)}$$

$$\mu = F_f/F_{fn}$$

or

$$\mu = (F_t + F_c \tan \alpha)/(F_c - F_t \tan \alpha)$$

where μ is the coefficient of friction on the rake surface. This frictional angle β on the rake surface is calculated to be nearly zero as seen from μ in Figure IVA-18 with PTFE. In this special case, eq. (1) becomes

$$2\phi - \alpha = \pi/2$$

and the direction of the resultant cutting force coincides with the normal to the rake surface as shown in Figure IVA-19. The relationship between the several components of eqs. (1) and (2) are shown in Figure IVA-20. The magnitude of ϕ can be obtained from

$$\phi = \tan^{-1}[r_c \cos \alpha/(1 - r_c \sin \alpha)] \qquad \text{IVA3-(3)}$$

For the continuous shear variety of chip, the shear plane occurs from the tool's edge, and the chip is formed by shearing along this shear plane. On tentatively viewing the work material as possessing a characteristic yield stress in shear, the shear angle ϕ is obtained after Krystof (4) or Lee and Shaffer (5) as

$$\phi = \alpha - \beta + (\pi/4) \qquad \text{IVA3-(4)}$$

One is able to calculate the values of the shearing stress τ_s and shear γ_s on the shear plane by

$$\tau_s = [(F_c \cos \phi - F_t \sin \phi) \sin \phi]/A_0 \qquad \text{IVA3-(5)}$$

$$\gamma_s = \cot \phi + \tan (\phi - \alpha) \qquad \text{IVA3-(6)}$$

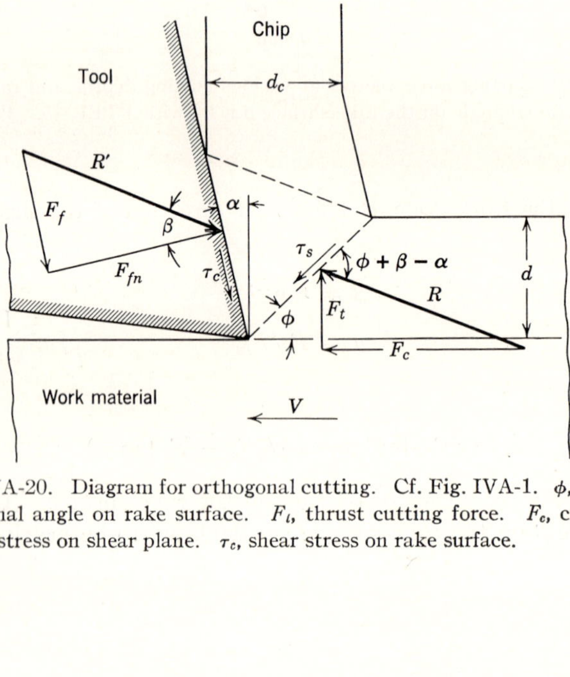

Fig. IVA-20. Diagram for orthogonal cutting. Cf. Fig. IVA-1. ϕ, shear angle. β, frictional angle on rake surface. F_t, thrust cutting force. F_c, cutting force. τ_s, shear stress on shear plane. τ_c, shear stress on rake surface.

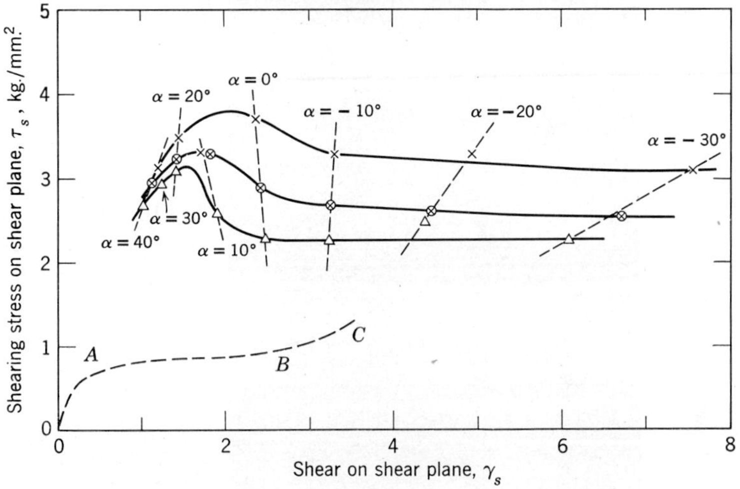

Fig. IVA-21. Shear stress vs. strain on shear plane at depths of cut (×) 0.023, (⊗) 0.046, and (Δ) 0.073 mm. with PTFE for various rake angles. V = 500 m./min. Dashed curve, from data in tension.

where A_o = cutting area = $d \times$ width of cut. An example is shown in Figure IVA-21. The dashed line represents the tensile test (under analogous strain rates). This continuous shear type of chip is produced when the shear on the shear plane is smaller than the limiting rupture strain γ_r. This is the case when $\alpha = 10° \rightarrow 40°$ for PTFE, as shown in Figure IVA-22 which presents the sectional views of the chips along the flow directions. Moreover, the values of $(\beta - \alpha)$ will characterize the mechanism of chip formation. If $(\beta - \alpha)$ becomes negative, the continuous shear variety of chip tends to be produced as clarified by eq. (4) and Figure IVA-23.

The discontinuous shear type of chip also occurs in shear along the shear plane acting upwards from the tool's edge, but here the intervals of shear are large relative to those for the continuous shear type and rupture occurs intermittently along the shear plane. The mechanism is explained by Figures IVA-21, IVA-22 and IVA-23. Here, the chip is produced when the shear exceeds γ_r as for PTFE, where α is negative as shown in Figure IVA-21. Positive values of $(\beta - \alpha)$ will give this discontinuous shear type of chip.

The discontinuous complex shear variety of chip is produced by a complex shear stress sytem combined with compressive or tensile

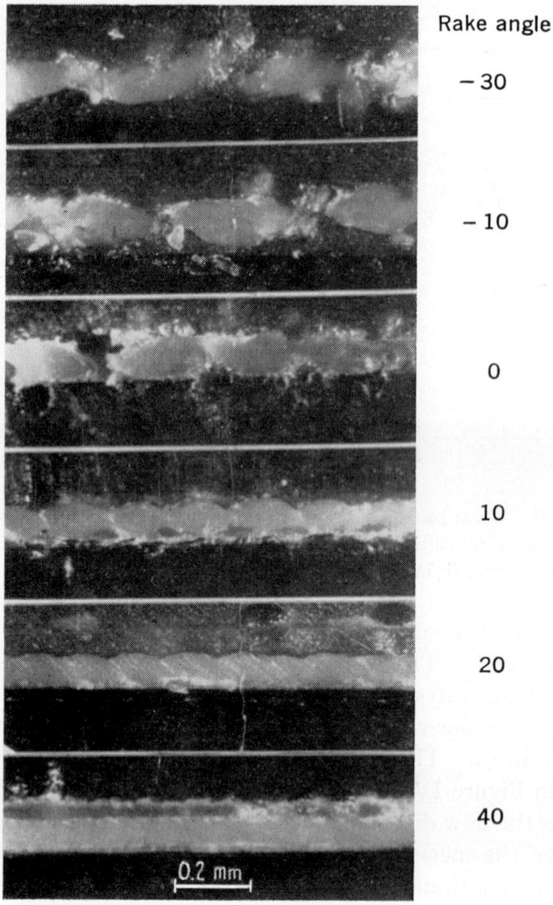

Fig. IVA-22. Sectional views of chips of PTFE formed at various rake angles. V = 765 m./min., d = 0.073 mm.

contributions. Figure IVA-24 gives examples of the relationship between the cutting forces and α. The direction of the thrust cutting force F_t turns from upwards to downwards as α goes from positive to negative; hence a negative α compresses the work material.

The discontinuous crack type of chip is generated when α and d are large. This is the brittle variety produced in the classical elastic range of stored strain energy. When the tool advances into the work material, the zone adjacent to the tool's edge deforms in-

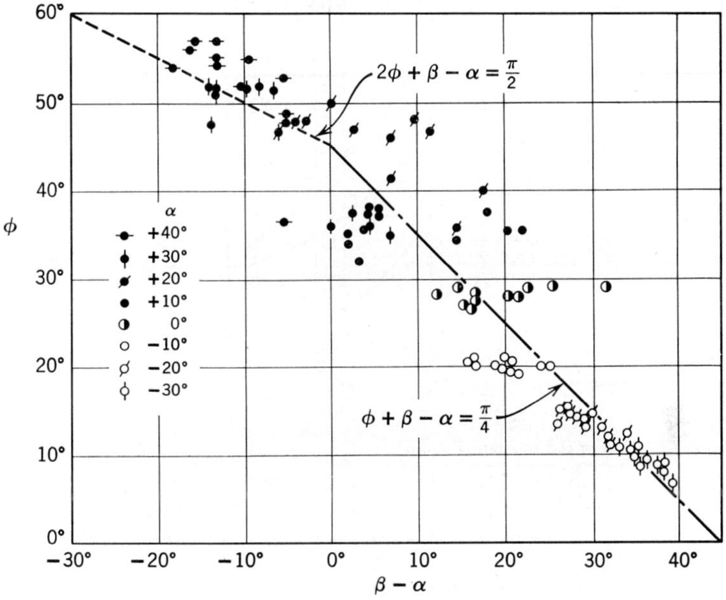

Fig. IVA-23. Relationship between ϕ and $(\beta - \alpha)$ for PTFE at cutting speeds of 218, 327, 500, and 765 m./min. and depths of cut 0.023, 0.046, and 0.073 mm.

elastically, and the mechanism of chip formation follows that illustrated in Figure IVA-25. The crack initiates in tension by the wedge action of the tool before the inelastically deformed zone can expand. The crack propagates in a periodic succession by the bending moment

TABLE IVA-II
Mechanisms of Chip Formation

Type of chip			Main cause
Continuous	Flow*		Rubberlike deformation
Continuous	Shear		Continuous slippage by shear stress
Discontinuous	Shear	Simple	Intermittent ductile rupture
		Complex	Same as above combined with compressive stresses
			Same as above combined with tensile stresses
Discontinuous	Crack		Brittle fracture

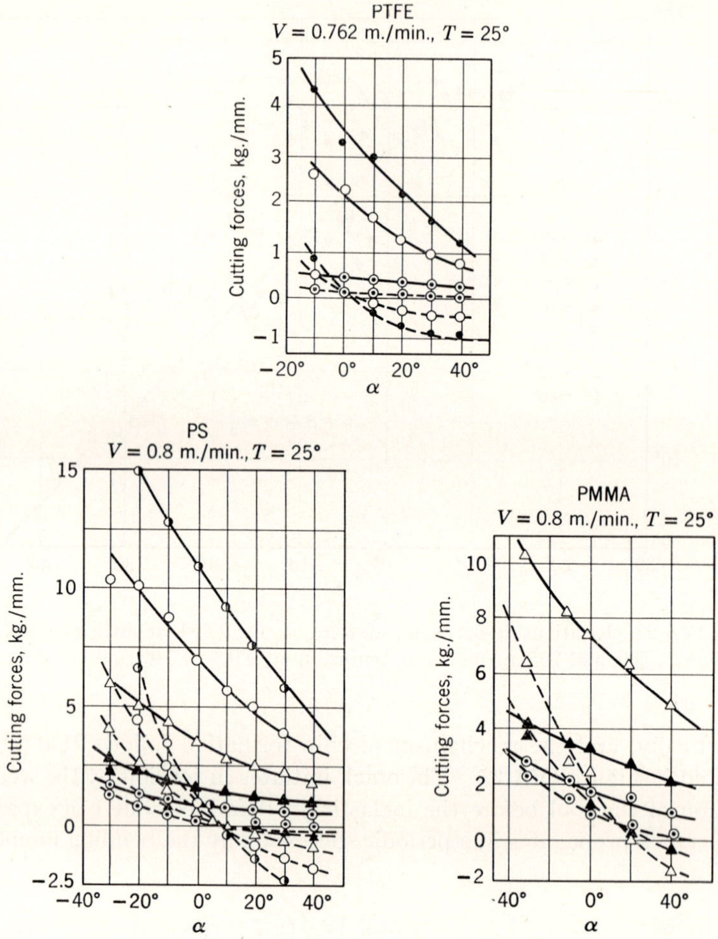

Fig. IVA-24. Relationships between cutting forces (F_t - -; F_c—) and rake angles under various cutting conditions with PTFE, PS, and PMMA. Depth of cut, *d*, is 1.0 (●), 0.75 (◑), 0.5 (○), 0.25 (△), 0.1 (▲), and 0.05 (⊙) mm.

acting on the tool's rake surface. This crack advances upward towards the free surface of the work material in *advance* of the tool.

In summary, the types of chips with the salient causes for formation are outlined in Table IVA-II (3a,3b).

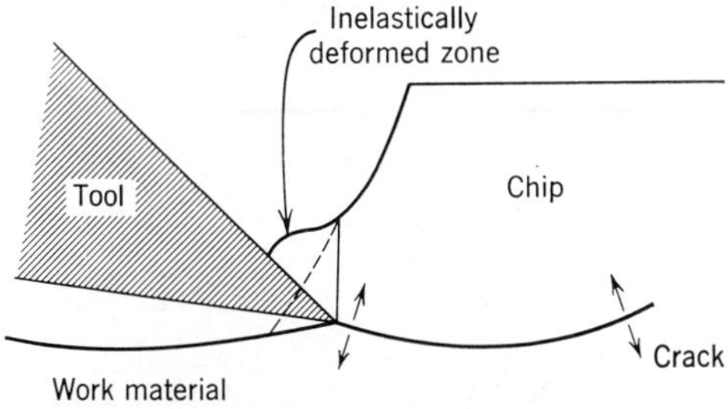

Fig. IVA-25. Schematic diagram for illustrating the mechanism of forming the "discontinuous crack" chip type.

IVA4. Rheological Factors in Cutting Phenomena (2)

Let us now turn to the extremely important roles of time and temperature in these viscoelastic bodies.

Figure IVA-26 shows how PMMA varies in chip shape with variations in V. Here one can see differences in the appearance and magnitude of the inelastically deformed zone at the tool's edge. Moreover, the type of chip progresses from the continuous to the discontinuous shear type with increasing V and the discontinuities themselves enlarge.

Figure IVA-27 shows the influence of cutting speed on the force with CP and PC. At first it increases, then decreases. But when $V > 10$ m./min., the type of chip becomes the discontinuous crack rather than the continuous shear variety at $\alpha = 0°$ for CP.

Figure IVA-28 shows the effect of ambient work temperature on the cutting forces. This temperature rise decreases these forces, as anticipated from general considerations of viscoelastic bodies (see

(a) (b) (c)

Fig. IVA-26. Effects of cutting speed on chip type with PMMA. $d = 0.1$ mm., $\alpha = -10°$. (a) $V = 0.8$ m./min., (b) $V = 0.2$ m./min., (c) quasi-static cutting.

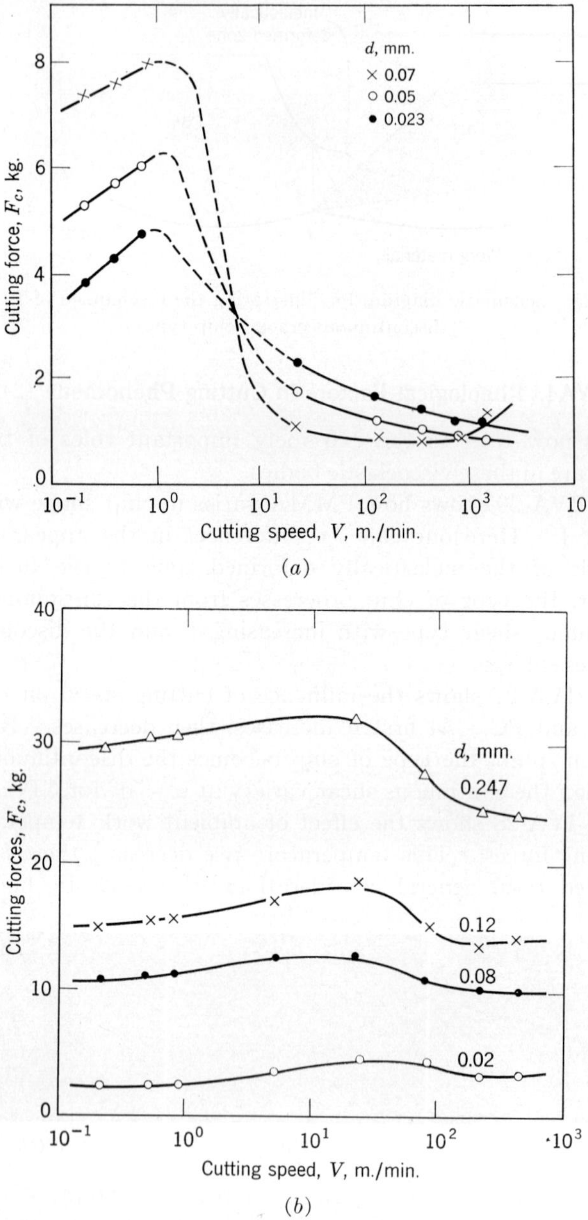

Fig. IVA-27. The effects of cutting speed on cutting force $\alpha = 0°$. (a) CP, (b) PC.

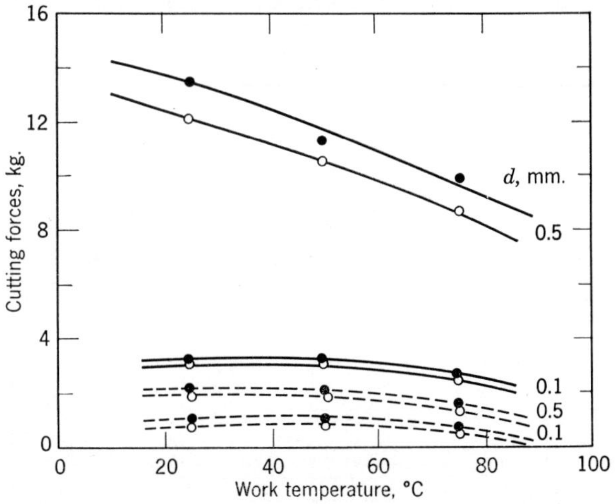

Fig. IVA-28. Effects of ambient temperature on cutting forces (F_t - -; F_c —) with PS. $\alpha = 0°$. (●) $V = 0.8$ m./min., (O) $V = 0.2$ m./min.

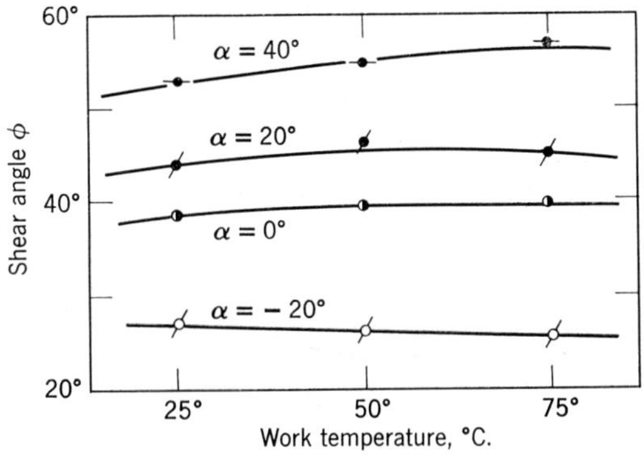

Fig. IVA-29. Effect of ambient temperature on shear angle with PS.

Section IC). The effect of work temperature on the angle of shear is only slight as shown in Figure IVA-29 for PS. Figure IVA-30 shows the reduction of τ_s with work temperature as compared to that for tension specimens.

Some preliminary insight into the important role of heat dissipation on cutting is obtained from Figure IVA-31, where the increases in temperature on the shear plane and tool–chip interface are shown

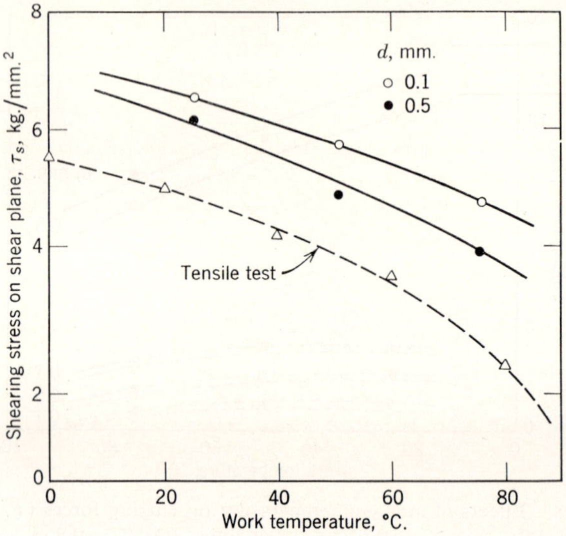

Fig. IVA-30. Effect of ambient temperature on shearing stresses on shear plane with PS.

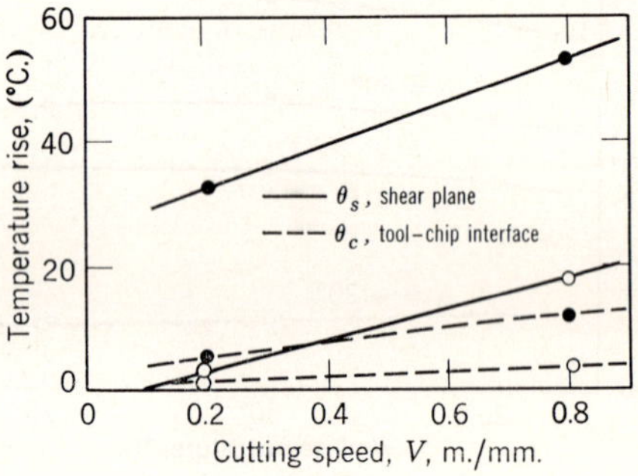

Fig. IVA-31. Effect of low cutting speeds on temperature rise, on shear plane and rake surface–chip interface. $\alpha = 0°$, $d = 0.1$ mm., (●) PMMA, (○) PS.

as being dependent on the cutting speed for PS and PMMA at low cutting speeds.

There is apparently no simple rheological interrelationship to be uncovered here, but the present authors have discussed some of these factors elsewhere (2). Moreover, further studies are in progress in regard to these considerations.

IVA5. Summary

Fracture phenomena in the cutting of several synthetic polymeric glasses and leathers are characterized in terms of cutting conditions, types of chips formed, and cutting mechanisms which include unusually strong dependencies on the rheological contributions of time and temperature. The chips formed were both continuous and discontinuous. The continuous varieties were, in part, similar to those obtained in metals cutting in shear, yet frequently exhibited a variety of chip which was generated through rubberlike deformation. The discontinuous types of chips were threefold: "discontinuous simple shear," "discontinuous complex shear," and "discontinuous crack." The "complex" variety was produced under a combined system of stresses including compressive or tensile contributions. In general, the milder cutting conditions, e.g., low cutting speeds, small rake angle of the cutting tool, and small depth of cutting, tended toward a continuous chip formation. The mechanisms of formation of discontinuous chips reflected both brittle and ductile contributions, and these are considered in some detail. Cutting speed, i.e., failure rate, plays a dominant role in these viscoelastic bodies especially when localized heating under conditions of inefficient heat dissipation is aggravated with the more severe cutting conditions.

IVA6. List of Symbols

d	Depth of cut
d_c	Chip thickness
F_c	Cutting force (component parallel to the cutting direction)
F_f	Frictional force on rake surface
F_{fn}	Normal force on rake surface
F_t	Thrust cutting force (component normal to the cutting direction)
r_c	$= d/d_c$
R', R	Resistant force on tool; reaction of R'
T	Ambient or work temperature
V	Cutting speed
W	Work
α	Rake angle of cutting tool
β	Friction angle on rake surface
γ	Strain

γ_r	Rupture strain
γ_s	Shear on shear plane
γ_y	Yield strain in shear
θ_c	Temperature on rake surface
θ_s	Temperature on shear plane
μ	Coefficient of friction on rake surface
τ	Shearing stress
τ_c	Cutting stress on rake surface
τ_r	Rupture stress
τ_s	Shearing stress on shear plane
τ_y	Shearing stress for yield
ϕ	Shear angle relative to cutting direction
CP	Cast polyester
PA	Polyamide
PC	Polycarbonate
PMMA	Poly(methyl methacrylate)
PS	Polystyrene
PTFE	Polytetrafluoroethylene
PVC	Poly(vinyl chloride)

References

1. Kobayashi, A., and K. Saito, *Bull. Electrotech. Lab.*, **25**, 433 (1961).
2. Kobayashi, A., and K. Saito, *J. Polymer Sci.*, **58**, 1377 (1962).
3. Kobayashi, A., and K. Saito, *Bull. Electrotech. Lab.*, **26**, 595 (1962).
3a. Kobayashi, A., *C.I.R.P. Annalen*, **11**, 82 (1963).
3b. Kobayashi, A., *Bull. Electrotech. Lab.*, **27**, 291 (1963).
3c. Kobayashi, A., *Modern Plastics*, **40**, 110 (1963).
4. Krystof, J., *Ber. Betr -Wiss. Arb.*, **12**, 7 (1939).
5. Lee, E. H., and B. W. Shaffer, *J. Appl. Mech.*, **18**, 4 (1951).
6. Merchant, M. E., *J. Appl. Phys.*, **16**, 267 (1945).
7. Okoshi, M., *Sci. Papers Inst. Phys. Chem. Res. (Tokyo)*, **14**, 189 (1930).
8. Rosenhain, W., and A. C. Sturney, *Engineering*, **119**, 151, 178 (1925).

IVB. A THERMODYNAMIC THEORY OF STRENGTH

MARKUS REINER

IVB1. The Huber-Hencky Criterion

In the following, the strength of a material will be understood as that property by which it resists excessive plastic deformation, which may ultimately lead to rupture; or it may refer to rupture if the material is brittle and does not suffer a plastic deformation. The theory propounded has its origin in the strength theory of Huber (4) and Hencky (2). In accordance with this theory, a material fails when a certain limit is exceeded by the maximum distortional stress-work. The mathematical expressions for both Huber's and Hencky's theories are the same, but while Huber was concerned with rupture, Hencky was concerned with plastic flow. In mathematical language, this theory can be expressed as follows.

In view of Subsection IC4h, p. 90, if the tensors of strain ϵ and of stress σ are resolved into their isotropic components ϵ_v and σ_m, where ϵ_v is the cubical dilatation and σ_m the mean tension, and their deviator components e and s, the elementary stress-work

$$dW = \sigma \cdot d\epsilon \qquad \text{IVB1-(1)}$$

can likewise be resolved into two components, of which the first is the work of dilatation or the volumetric stress-work

$$dW_v = \sigma_m \cdot d\epsilon_v \qquad \text{IVB1-(2)}$$

and the other, the distortional work

$$dw = s \cdot de \qquad \text{IVB1-(3)}$$

Huber and Hencky assume (*a*) that an isotropic pressure ($p = -\sigma_m$) may be increased beyond any limit without causing failure, and (*b*) that an isotropic tension σ_m is of no influence upon plastic flow but may cause rupture by exceeding the true molecular cohesion of the material. If the latter is not exceeded, the material may fail by w exceeding a limit of maximum distortional work.

For homogeneous or quasi-homogeneous nonporous materials, Huber's and Hencky's assumption (*a*) has been established by experiments, while assumption (*b*) is a logical consequence of the concept of cohesion. Their theory has been shown to be generally applicable to strength tests when carried out at the ordinary testing speeds. For an ideal elasto-plastic material, Hencky's criterion is identical with v. Mises' well-known flow condition (5).

IVB2. The Deficiency of the Huber-Hencky Criterion

Huber and Hencky's theory is purely statical and the velocity of strain is not considered. Improvements in the methods of testing of materials and the frequent use of high-speed tests as well as of creep tests, however, have made it increasingly evident that the velocity of strain considerably influences the result of the test. It is therefore obvious that no purely statical conception can lead to a satisfactory theory of strength. A theory taking account of the velocity is required both from theoretical and practical considerations. Such a theory was suggested by Reiner and Freudenthal (7), and developed in detail by Reiner and Weissenberg (8).

Because of the satisfactory evidence for the Huber-Hencky theory at low speeds, a good dynamic theory has to contain Huber-Hencky's static theory as a special case. There is no necessity of modifying their postulates (*a*) and (*b*). All volumetric strain is to be completely reversible, and therefore purely elastic. In this case the relation between stress and strain is accordingly one-valued, and the velocity of the strain cannot be of any influence. It is only the deviator components of stress and strain, and the distortional work, with which a more general theory must be concerned.

IVB3. Thermodynamic Considerations

The weak point of Huber's theory becomes evident when we consider a metal bar in creep. If a mild steel bar is loaded at a certain elevated temperature, it will be continuously elongated at a more or less constant residual speed. In this creep, the stress-work is not stored-up in the form of potential elastic energy, but is dissipated in the form of heat. Such dissipated stress-work is not available for overcoming the resistance of the material to brittle rupture or plastic yielding. This makes it clear that a dynamical theory of strength can only be a thermodynamic theory. Let w be the stress-work, F the free intrinsic energy, and D the bound dissipated energy (all per unit volume); then the first law of thermodynamics requires

$$\frac{d}{dt}\left(\frac{w}{\rho}\right) + \frac{d}{dt}\left(\frac{F}{\rho}\right) + \frac{d}{dt}\left(\frac{D}{\rho}\right) = 0 \qquad \text{IVB3-(4)}$$

where ρ is the density, and the free intrinsic energy is the total free energy minus the kinetic energy. The density must be introduced because the first law does not refer to unit volume, but to unit mass.

Now let us consider isothermal processes only, i.e., the heat into which the dissipated energy is converted is conducted away without delay. Then, in accordance with the second law of thermodynamics, we have

$$\frac{d}{dt}\left(\frac{D}{\rho}\right) \geqq 0 \qquad \text{IVB3-(5)}$$

where the sign (=) stands for reversible, and the sign (>) for irreversible processes.

From the fact that $d(D/\rho)/dt$ vanishes for all reversible processes it follows that the intrinsic free energy is equivalent to a generalized elastic potential Φ, which we may call strain-work (w_c); "strain' denoting the *recoverable* part of the deformation, with the subscript c indicating that it is conserved.

If we consider the volumetric strain-work separately, and the quantities in eqs. (4) and (5) refer to the distortional work only, ρ can be taken as constant and cancels out.

Using Newton's dot for differentiation in respect of time, we accordingly write

$$\dot{w} + \dot{w}_c + \dot{D} = 0 \qquad \text{IVB3-(6)}$$

where it is understood that $\dot{D}$ can only be either positive or zero.

The total distortional work can be expressed in terms of stress and velocity of strain

$$dw = -s \cdot \dot{e}\, dt \qquad \text{IVB3-(7)}$$

[The negative sign is used in eq. (7) because w is imposed by external forces.] From eq. (7) we obtain the distortional power

$$\dot{w} = -s \cdot \dot{e} \qquad \text{IVB3-(8)}$$

For purely mechanical processes Φ and $\dot{D}$ are functions of stress and strain and their differential quotients in respect to time only. Indicating by m and n, the order of the differential quotient, where m and n can have any value 0, 1, 2 . . ., we ultimately write

$$-s \cdot \dot{e} + [d\Phi(s^m, e^n)/dt] + \dot{D}(s^m, e^n) = 0 \qquad \text{IVB3-(9)}$$

The functions Φ and $\dot{D}$ are of different form for every group of materials, and the materials themselves can be distinguished by the various values of the parameters which appear in their rheological equations. These parameters can also be mechanical material constants, e.g., one group of materials is formed by all perfectly elastic bodies, which follow Hooke's Law, the different materials being distinguished by the value of the shear modulus G. Another group is formed by all simple viscous liquids, which obey Newton's Law, and the different liquids are distinguished by the value of the coefficient of viscosity η. See Subsection IC2a, p. 46.

IVB4. The Reiner-Weissenberg Criterion

On the basis of the development of the preceding subsection, Reiner and Weissenberg postulate that failure depends upon a maximum value of the intrinsic free energy, i.e., the energy $\Phi = w_c$, which can be stored-up elastically in the volume element of the material. Failure will occur at the time t_f, when

$$w_c = \int_0^{t_f} (\dot{w} - \dot{D})\, dt = R \qquad \text{IVB4-(10)}$$

where R is a material constant, which may be called the resilience of the material. Equation (10) may be called the *failure condition.* Where there is no dissipation of energy, or where the dissipation of energy is negligible, $\dot{D}$ either vanishes or can be neglected, and the condition for rupture becomes

$$\int_0^{t_f} \dot{w}\, dt = R \qquad \text{IVB4-(11)}$$

which is the condition of Huber and Hencky's theory. It can easily be shown that, for a hookean body, it is identical with the v. Mises flow condition. In an irreversible process where $\dot{D}$ is positive in accordance with eq. (5), part of the power of distortion leaks out, and only a part of it can be stored-up. At certain states of stress and deformation, when $\dot{w} = \dot{D}$ the body can be deformed without limit, as with a liquid.

In order to evaluate eq. (10) for a special case, the following have to be known: (I) the rheological equation of state of the material, and (II) the mechanical conditions of the test.

The rheological equation of state can generally be expressed as a relation between the time derivatives of stress and strain, or

$$R(s^m, e^n) = 0 \qquad \text{IVB4-(12)}$$

(It should be kept in mind that this is a tensor equation.)

The mechanical conditions of the test consist of the dynamical and kinematical boundary conditions.

From (I) the free elastic potential Φ can be calculated, and by means of (II) Φ can be integrated up to failure.

If we assume the rheological equation of state to be of the form

$$R(s, \dot{s}, e, \dot{e}) = 0 \qquad \text{IVB4-(13)}$$

this qualitatively covers a wide range. We can go on a step further by assuming R to be linear, or of the form

$$C_1 + C_2 s + C_3 \dot{s} + C_4 e + C_5 \dot{e} = 0 \qquad \text{IVB4-(14)}$$

Such a rheological equation was postulated by Hohenemser and Prager (3).

Equation (14) can be considered as composed of two equations:

$$A_1 + A_2 s + A_3 e + A_4 \dot{e} = 0 \qquad \text{IVB4-(15)}$$

and

$$B_1 + B_2 s + B_3 \dot{s} + B_4 e = 0 \qquad \text{IVB4-(16)}$$

and accordingly the material which has eq. (14) as its rheological equation is composed of two materials, which follow the rheological eqs. (15) and (16), respectively. As can be seen, the first is a solid; the second a liquid. We can neglect A_1, which can be interpreted as an initial stress, and B_1, which can be interpreted as an initial strain; then eqs. (15) and (16) can be written as

$$s - 2Ge - 2\eta_s \dot{e} = 0 \qquad \text{IVB4-(17)}$$

and

$$2\dot{e} - (s/\eta) - (\dot{s}/G_l) = 0 \qquad \text{IVB4-(18)}$$

The factor 2 in eq. (17) results from the usual definition of the material constants G and η, or what comes to the same, of shear. The constant G is an elastic modulus, and the constant η, a coefficient of viscosity. The subscripts s and l indicate the solid and the liquid states, and η_s is accordingly the measure of the viscous resistance to the deformation in a solid, while G_l is the measure of the elasticity of a liquid. Where η_s is absent, the first body is reduced to a hookean solid; where G_l is absent, the second body is reduced to a newtonian liquid. The ideal material represented by eq. (17) has been called a Kelvin solid, the other represented by eq. (18), a Maxwell material. While it is not maintained that any real material could be perfectly represented by either a Hooke or a Kelvin solid, or by a newtonian or maxwellian liquid, it may be assumed that the properties of the ideal materials represented by the above rheological equations can be encountered in varying degrees in the mechanical behavior of every real material.

IVB5. Special Cases

(*a*) In the hookean solid, the rheological equation for which is

$$s = 2Ge \qquad \text{IVB5-(19)}$$

all the stress power of the external forces is converted into elastic potential energy, and none is dissipated, or

$$d\Phi/dt = s \cdot \dot{e} = 2Ge\dot{e} = G\, de^2/dt \qquad \text{IVB5-(20)}$$

which gives

$$w_c = Ge^2 = s^2/4G; \ \dot{D} = 0 \qquad \text{IVB5-(21)}$$

(*b*) In a newtonian liquid, the rheological equation for which is

$$s = 2\eta\dot{e} \qquad \text{IVB5-(22)}$$

all the stress power of the external forces is dissipated, or

$$\Phi = 0; \ \dot{D} = s \cdot \dot{e} = 2\eta\dot{e}^2 = s^2/2\eta \qquad \text{IVB5-(23)}$$

The function $\dot{D}$ is Rayleigh's dissipation function.

(*c*) For the Kelvin solid, eq. (17) gives

$$s\dot{e} = Ge\dot{e} + 2\eta_s\dot{e}^2 \qquad \text{IVB5-(24)}$$

Comparing eq. (24) with eq. (9), we find

$$\dot{w}_c = 2Ge\dot{e} = G\, de^2/dt$$

$$w_c = Ge^2 \qquad \text{IVB5-(25)}$$

$$\dot{D} = 2\eta_s\dot{e}^2$$

(*d*) For the Maxwell material eq. (18) gives

$$s\dot{e} = (s^2/2\eta) + (s\dot{s}/2G_l) \qquad \text{IVB5-(26)}$$

Comparing eq. (26) with eq. (9) we find

$$\dot{w}_c = s\dot{s}/2G_l = G_l(ds^2/dt)/4$$

$$w_c = s^2/4G_l \qquad \text{IVB5-(27)}$$

$$\dot{D} = s^2/2\eta$$

Applying Reiner and Weissenberg's failure condition, eq. (10), to these four cases, we find the following.

(*a*) A solid which is perfectly elastic up to the yield point and is therefore an ideal plastic material, or which is perfectly elastic up to rupture and is therefore an entirely brittle material, fails through plastic deformation in the first, or through rupture in the second, case when either the stress or the strain reaches a certain limit. For such materials the present criterion is therefore reduced to Hencky's and Huber's criteria, respectively, and these in their turn are reduced to Rankine's and St. Venant's criteria, respectively, which, in this case, give identical results, provided the deviator components of the stress and strain tensors are taken, and not the complete tensors.

(*b*) A simple newtonian viscous liquid, which has no elasticity, can only be torn apart by hydrostatic tension. In laminar flow, where the mean pressure is equal to the hydrostatic pressure, a simple viscous liquid *cannot* fail by straining, because for such a material Φ vanishes and is therefore smaller than any limit which may be put to its strength. If observation of a lubricating layer points to the opposite, this shows that the lubricating oil is not a simple or newtonian liquid. Recently Barlow and Lamb (1) have actually found that lubricating oils exhibit shear elasticity.

(*c*) A Kelvin solid fails when the strain reaches a definite limit. There is no single-valued relation between stress and strain, and there is no definite limit of stress which corresponds to the limit of strain. Let e_f be the strain failure; then eq. (17) shows that

$$s_f = 2Ge_f + 2\eta_s\dot{e} \qquad \text{IVB5-(28)}$$

and the stress at which the material fails increases with the velocity of strain. (See Subsection IC5c, p. 96.) Let s_{fs} be the stress at which the material fails in a static test, i.e., with vanishing strain velocity; then there is

$$s_f = s_{fs} + 2\eta_s\dot{e} \qquad \text{IVB5-(29)}$$

As η_s can be determined from the damping of free oscillations of the material under test (see Subsection IC4g, p. 89), i.e., independently of the strength test, the dependence of the strength of the material upon the rate of deformation can be predicted in accordance with eq. (29). It is not maintained that this dependence will actually be a linear one. On the contrary, it may be expected that it will not be linear, because it is well known that eq. (17) does not correctly describe the damping of free oscillations. If, however, from experimental observations the correct form of the functional relationship which has to replace eq. (17) is determined, there is no difficulty in deriving the relationship which has to replace eq. (29). This has been carried out for the "standard linear solid" by Reiner (6). Qualitatively, however, it is shown that the thermodynamic theory gives the result that the failure stress of an elastic material which shows damping of its free oscillations, increases with increasing rate of strain.

(*d*) A Maxwell material can be regarded as a viscous liquid which shows elasticity. It fails when the stress reaches a certain limit s_f. To this limit there does *not* correspond a definite limit of the rate of deformation. Equation (18) shows that

$$\dot{e}_f = (s_f/2\eta) + (\dot{s}/2G_l) \qquad \text{IVB5-(30)}$$

Let $\dot{e}_{fs}$ be the rate of deformation at which the material fails if the stress is applied infinitely slowly; then there is

$$\dot{e}_f = \dot{e}_{fs} + (\dot{s}/2G_l) \qquad \text{IVB5-(31)}$$

The modulus of elasticity G_l can be determined by oscillation experiments, and the maximum rate of strain at which failure occurs predicted in accordance with eq. (31). This is applicable to failure of a material in creep, when it can be considered as a very viscous and, at the same time, an elastic liquid.

IVB6. General Case

If we desire to apply this theory to the strength of a polymeric body, we have to assume a rheological model representing its mechanical behavior. Sometimes the Kelvin model as shown in Figure IC-13, p. 75, is adequate. Sometimes reasoning leads to a "complex" Kelvin model as shown in Figure IC-14 of that same section. The former case has been treated here in Subsection IVB5c. The latter case can be treated on the same lines. As a "complex" Kelvin model includes both a retardation time, as does a simple Kelvin model, and a relaxation time, as does a Maxwell model, the failure stress will still increase with increasing rate of strain. On the other hand, when the load is constant, failure can take place after some time in creep.

IVB7. Summary

Strength is the ultimate resistance of a material to either plastic yielding or brittle rupture, or to rupture after plastic yielding. The criterion of strength of a material is its resilience at failure, where "resilience" is the conserved strain-work w_c. In a homogeneous, dense material, an isotropic pressure may be increased beyond any macroscopic limit without causing failure, while an isotropic tension may cause rupture by exceeding the true molecular cohesion of the material. For the deviatoric part of the stress tensor, the first law of thermodynamics supplies a relation between the distortional strain-work by measuring the resilience R of the material, the stress-power $\dot{w}$ and the rate of dissipated energy $\dot{D}$ of the form

$$w_c = \int_0^{t_f} (\dot{w} - \dot{D})\, dt = R$$

where t_f is the time at which failure takes place. This criterion is applied for the cases of hookean elastic solid, newtonian viscous liquid, and the Kelvin and Maxwell models.

IVB8. List of Symbols

D	Bound dissipated energy
$\dot{D}$	Rayleigh's dissipation function
e	Deviator of ϵ
F	Free intrinsic energy
f, l, s	Subscripts for failure, liquid, solid
G	Elastic shear modulus
m, n	Power superscripts
p	Isotropic pressure
R	Resilience
s	Deviator of σ
t	Time
W	Stress-work
W_v	Volumetric stress-work
w	Distortional stress-work
w_c	Conserved strain-work
ϵ	Strain tensor
ϵ_v	Cubical dilatation
η	Coefficient of viscosity
Φ	Elastic potential
ρ	Density
σ	Stress tensor
σ_m	Mean isotropic tension

References

1. Barlow, A. J., and J. Lamb, *Proc. Roy. Soc.* (*London*), **A253,** 52 (1959).
2. Hencky, H., *Z. Angew. Math. Mech.*, **4,** 323 (1924).
3. Hohenemser, K., and W. Prager, *J. Rheol.*, **3,** 16 (1932).
4. Huber, T., Czapismo technize, Lemberg, 1904.
5. v. Mises, R., *Göttinger Nachr. Math.-Phys. Klasse,* 582 (1913).
6. Reiner, M., *J. Mech. Phys. Solids*, **8,** 255 (1960).
7. Reiner, M., and A. Freudenthal, Congr. Appl. Mech., Cambridge, Mass., 1938.
8. Reiner, M., and K. Weissenberg, *Rheol. Leafl.*, **10,** 12 (1939).

IVC. MOLECULAR ORIENTATION-DEPENDENT FRACTURE STRENGTH†

C. C. Hsiao

IVC1. Introduction

This review article describes some specific recent developments (14,17,25) concerning the macromolecular orientation-dependent mechanical strength of some polymeric solids. These developments occupy an intermediate position between the phenomenological and molecular theories of the mechanical properties of materials.

The general analytical approach is to obtain the stress field in the vicinity of a point in a system of oriented linear spring elements. The system of elements, without preferred orientation, constitutes an ideal isotropic medium which is susceptible to analytical treatment. Upon introducing a simple homogeneous deformation, it is possible to treat mathematically the orientation of the system of elements, and thus to evaluate statistically the corresponding strength of the oriented medium. For simplicity, this article is limited to the consideration of a static system in which the microscopic constituents will not

† See pp. 100, 214, 253, and 348 for discussions relative to this consideration in the noncrystalline zones of polymeric bodies. *Ed.*

break and re-form under load so that the total number of spring elements will not change with time.

It might be of interest to point out that the concept for combining ideas concerning microscopic structure with macroscopic behavior of materials is not new. E. and F. Cosserat (3), as long ago as 1909, presented theories of oriented materials. Currently Ericksen and Truesdell (5) and others (6–8) have reported many new developments concerned with the concept of an oriented continuum with vectors or directors attached to material points. In this article an elementary analysis is considered in which the polymeric solid is regarded as a simple system of material points to each of which is attached a single vector.

IVC2. Strength Theory for Ductile Media

By considering an oriented system of linear spring elements, the stress tensor in the vicinity of a point referred to an orthogonal coordinate system may be expressed (14) as

$$\sigma_{ij}^* = \int El^2\lambda\rho^* f^* \epsilon_{mn} s_m{}^* s_n{}^* s_i{}^* s_j{}^* \, d\omega^* \qquad \text{IVC2-(1)}$$

where E is the spring modulus, l is the length of any spring element, λ is the number of springs per unit volume in the solid, ρ^* is the probability density function of the orientation of the springs, f^* is the fraction of unbroken elements, $\epsilon_{mn} s_m{}^* s_n{}^*$ is the strain of the spring whose instantaneous direction is given by the unit vector $\boldsymbol{s}^*$, $d\omega^*$ is the solid angle, and the asterisk (*) indicates that the parameter concerned is to be evaluated in the oriented system.

If the fracture mechanism is considered to be independent of time, then $f^* = f_0$ becomes a constant. In other words, the system is a static one and all the spring elements taken into account do not break and re-form during loading. However, the fracture strength is still time-dependent (10,15). Only the strengths obtained at one constant rate of straining are considered and compared. If we let $C = El^2\lambda f_0$, then the stress tensor may be reduced to the following form, since C is composed essentially of products of constants:

$$\sigma_{ij}^* = C \int \rho^* \epsilon_{mn} s_m{}^* s_n{}^* s_i{}^* s{}^* \, d\omega^* \qquad \text{IVC2-(2)}$$

Evaluation of these integrals will give us the desired strength information.

IVC2a. Isotropic Strength

First, a calculation will be made of the strength for the random or isotropic case, i.e., when $\rho^* = \rho$ is independent of θ and ϕ, the spherical coordinates and all the asterisks in the foregoing expression may be dropped as the system is without orientation. Now $\int \rho d\omega = 1$ and $d\omega = \sin\theta \, d\theta d\phi$. Thus

$$\int \rho \, d\omega = \rho \int_0^{2\pi} \int_0^{\pi} \sin\theta \, d\theta d\phi = 4\pi\rho \qquad \text{IVC2-(3)}$$

and

$$\rho = 1/4\pi$$

Assume ϵ_{mn} to be of the form

$$\begin{vmatrix} \epsilon_{11} & 0 & 0 \\ 0 & \epsilon_{22} & 0 \\ 0 & 0 & \epsilon_{33} \end{vmatrix}$$

and σ_{ij} of the form

$$\begin{vmatrix} \sigma_{11} & 0 & 0 \\ 0 & \sigma_{22} & 0 \\ 0 & 0 & \sigma_{33} \end{vmatrix}$$

and that their principal directions are coincident. Thus

$$\sigma_{11} = \frac{C}{4\pi} \int (\epsilon_{11} s_1^4 + \epsilon_{22} s_1^2 s_2^2 + \epsilon_{33} s_1^2 s_3^2) \, d\omega \qquad \text{IVC2-(4)}$$

with similar expressions for σ_{22} and σ_{33}. It is easily seen that nine integrals will arise in these expressions. Denoting the first one by $\int_{11}$, the second by $\int_{12}$, etc., then, except for omission of the appropriate strain coefficient, for σ_{11}:

$$\int_{11} = \int s_1^4 \, d\omega$$

$$\int_{12} = \int s_1^2 s_2^2 \, d\omega$$

$$\int_{13} = \int s_1^2 s_3^2 \, d\omega \qquad \text{IVC2-(5)}$$

for σ_{22}:

$$\int_{21} = \int s_2{}^2 s_1{}^2\, d\omega$$

$$\int_{22} = \int s_2{}^4\, d\omega$$

$$\int_{23} = \int s_2{}^2 s_3{}^2\, d\omega \qquad \text{IVC2-(6)}$$

for σ_{33}:

$$\int_{31} = \int s_3{}^2 s_1{}^2\, d\omega$$

$$\int_{32} = \int s_3{}^2 s_2{}^2\, d\omega$$

$$\int_{33} = \int s_3{}^4\, d\omega \qquad \text{IVC2-(7)}$$

Here

$$s_1 = \sin\theta\cos\phi$$

$$s_2 = \sin\theta\sin\phi$$

$$s_3 = \cos\theta \qquad \text{IVC2-(8)}$$

The values for the nine integrals are as follows:

$$\int_{11} = \int_{22} = \int_{33} = 4\pi/5 \qquad \text{IVC2-(9)}$$

$$\int_{12} = \int_{13} = \int_{21} = \int_{23} = \int_{31} = \int_{32} = 4\pi/15 \qquad \text{IVC2-(10)}$$

Therefore

$$\sigma_{11} = (C/5)[\epsilon_{11} + (\epsilon_{22} + \epsilon_{33})/3] \qquad \text{IVC2-(11)}$$

$$\sigma_{22} = (C/5)[\epsilon_{22} + (\epsilon_{33} + \epsilon_{11})/3] \qquad \text{IVC2-(12)}$$

and

$$\sigma_{33} = (C/5)[\epsilon_{33} + (\epsilon_{11} + \epsilon_{22})/3] \qquad \text{IVC2-(13)}$$

For simple tension in the *33* direction, $\epsilon_{11} = \epsilon_{22}$. For a fixed value of ϵ_{33} we desire ϵ_{11} and ϵ_{22} to be such that both σ_{11} and σ_{22} are zero. Setting $\sigma_{11} = \sigma_{22} = 0$ we find

$$\epsilon_{11} = \epsilon_{22} = -(1/4)\epsilon_{33} \qquad \text{IVC2-(14)}$$

Thus

$$\sigma_{33} = (C/6)\epsilon_{33} \qquad \text{IVC2-(15)}$$

The isotropic strength of the medium is thus proportional to the magnitude of the strain ϵ_{33}. For the determination of the fracture strength a corresponding fracture strain must be given.

IVC2b. Completely Oriented Strength

Next, a calculation is made for the stress in the case of complete orientation in the *33* direction. In this case the probability will be "lumped in" or divided into equal parts at $\theta = 0$ and $\theta = \pi$. Thus $\rho d\omega = {}^1/_2$ at $\theta = 0$, and $\rho d\omega = {}^1/_2$ at $\theta = \pi$, and is zero elsewhere. Therefore

$$\begin{aligned}\sigma_{33} &= C\int \rho(\epsilon_{11}s_3^2s_1^2 + \epsilon_{22}s_3^2s_2^2 + \epsilon_{33}s_3^4)\,d\omega \\ &= C\epsilon_{33}\int \rho\, d\omega \\ &= C\epsilon_{33}({}^1/_2 + {}^1/_2) \\ &= C\epsilon_{33} \qquad \text{IVC2-(16)}\end{aligned}$$

We now see that for a given strain, ϵ_{33}

$$\frac{\sigma_{33}(\text{completely oriented})}{\sigma_{33}(\text{completely random})} = 6 \qquad \text{IVC2-(17)}$$

That is, for a perfectly oriented system the fracture strength is six times that of an isotropic system. This interesting result is quite realistic, as will be confirmed later by comparison with experimental data.

IVC2c. Partially Oriented Strength

In a general case, the strength of a partially oriented medium must be obtained. To obtain such an expression we consider the stress tensor for an oriented system given before

$$\sigma_{ij}^* = C\int \rho^*\epsilon_{mn}s_m^*s_n^*s_i^*s_j^*\,d\omega^* \qquad \text{IVC2-(18)}$$

Again there will be nine integrals to evaluate. In the evaluation of these use will be made of the following equality, $\rho^* d\omega^* = \rho d\omega$, i.e., the number of linear spring elements in a small solid angle is the same before and after orientation. Thus

$$\sigma_{ij}^* = C \int \rho \epsilon_{mn} s_m^* s_n^* s_i^* s_j^* \, d\omega \qquad \text{IVC2-(19)}$$

For uniaxial orientation, if we let a finite strain ϵ indicate the degree of orientation of the system, then the new coordinates of a point x,y,z will be given by

$$x' = x/(1 + \epsilon)^{1/2}$$

$$y' = y/(1 + \epsilon)^{1/2}$$

$$z' = (1 + \epsilon)z \qquad \text{IVC2-(20)}$$

after orientation on the basis of assuming the incompressibility of the medium. Transforming the above integral by use of these relations, we obtain

$$\int_{11} = \int_{\phi=0}^{2\pi} \int_{\theta=0}^{\pi} \frac{\sin^5 \theta \cos^4 \phi}{(1 + \beta^2 \cos^2 \theta)} \, d\theta d\phi$$

$$= \frac{3\pi}{4} \left[\frac{3 + \beta^2}{\beta^4} + \frac{-3 - 2\beta^2 + \beta^4}{\beta^5} \tan^{-1} \beta \right] \qquad \text{IVC2-(21)}$$

where

$$\beta^2 = (1 + \epsilon)^3 - 1 \qquad \text{IVC2-(22)}$$

$$\int_{12} = \int_{21} = \int_{\phi=0}^{2\pi} \int_{\theta=0}^{\pi} \frac{\sin^5 \theta \cos^2 \phi \sin^2 \phi}{(1 + \beta^2 \cos^2 \theta)^2} \, d\theta d\phi$$

$$= \frac{\pi}{4} \left[\frac{3 + \beta^2}{\beta^4} + \frac{-3 - 2\beta^2 + \beta^4}{\beta^5} \tan^{-1} \beta \right] \qquad \text{IVC2-(23)}$$

$$\int_{13} = \int_{31} = \int_{\phi=0}^{2\pi} \int_{\theta=0}^{\pi} \frac{\cos^2 \theta \sin^3 \theta \cos^2 \phi}{\alpha^2 (1 + \beta^2 \cos^2 \theta)^2} \, d\theta d\phi$$

$$= \pi \left[\frac{-3}{\alpha^2 \beta^4} + \frac{3 + \beta^2}{\alpha^2 \beta^4} \tan^{-1} \beta \right] \qquad \text{IVC2-(24)}$$

where

$$\alpha^2 = 1/(1 + \epsilon)^3 \qquad \text{IVC2-(25)}$$

$$\int_{22} = \int_{\phi=\theta}^{2\pi} \int_{\theta=0}^{\pi} \frac{\sin^5 \theta \sin^4 \phi}{(1 + \beta^2 \cos^2 \theta)^2} d\theta d\phi$$

$$= \frac{3\pi}{4} \left[\frac{3 + \beta^2}{\beta^4} + \frac{-3 - 2\beta^2 + \beta^4}{\beta^5} \tan^{-1} \beta \right] \quad \text{IVC2-(26)}$$

$$\int_{23} = \int_{32} = \int_{\phi=0}^{2\pi} \int_{\theta=0}^{\pi} \frac{\sin^3 \theta \cos^2 \theta \sin^2 \phi}{\alpha^2(1 + \beta^2 \cos^2 \theta)^2} d\theta d\phi$$

$$= \pi \left[\frac{-3}{\alpha^2\beta^4} + \frac{3 + \beta^2}{\alpha^2\beta^5} \tan^{-1} \beta \right] \quad \text{IVC2-(27)}$$

and

$$\int_{33} = \int_{\phi=0}^{2\pi} \int_{\theta=0}^{\pi} \frac{\sin \theta \cos^4 \theta}{\alpha^4(1 + \beta^2 \cos^2 \theta)^2} d\theta d\phi$$

$$= 2\pi \left[\frac{3 + 2\beta^2}{\alpha^2\beta^4} - \frac{3}{\alpha^4\beta^5} \tan^{-1} \beta \right] \quad \text{IVC2-(28)}$$

Thus

$$\sigma_{11} = \sigma_{22} = \frac{C}{4\pi} \left[\frac{3\pi}{4} \left(\frac{3 + \beta^2}{\beta^4} + \frac{-3 - 2\beta^2 + \beta^4}{\beta^5} \tan^{-1} \beta \right) \epsilon_{11} \right.$$
$$+ \frac{\pi}{4} \left(\frac{3 + \beta^2}{\beta^4} + \frac{-3 - 2\beta^2 + \beta^4}{\beta^5} \tan^{-1} \beta \right) \epsilon_{22}$$
$$\left. + \pi \left(\frac{-3}{\alpha^2\beta^4} + \frac{3 + \beta^2}{\alpha^2\beta^5} \tan^{-1} \beta \right) \epsilon_{33} \right] \quad \text{IVC2-(29)}$$

and

$$\sigma_{33} = \frac{C}{4\pi} \left[\pi \left(\frac{-3}{\alpha^2\beta^4} + \frac{3 + \beta^2}{\alpha^2\beta^5} \tan^{-1} \beta \right) \epsilon_{11} \right.$$
$$+ \pi \left(\frac{-3}{\alpha^2\beta^4} + \frac{3 + \beta^2}{\alpha^2\beta^5} \tan^{-1} \beta \right) \epsilon_{22}$$
$$\left. + 2\pi \left(\frac{3 + 2\beta^2}{\alpha^2\beta^4} - \frac{3}{\alpha^4\beta^5} \tan^{-1} \beta \right) \epsilon_{33} \right] \quad \text{IVC2-(30)}$$

As before, for a simple tension test in the *33* direction, $\epsilon_{11} = \epsilon_{22}$. For a fixed value of ϵ_{33} we desire ϵ_{11} and ϵ_{22} to be such that $\sigma_{11} = \sigma_{22} = 0$. Proceeding as before we find

$$\sigma_{33} = \frac{C\epsilon_{33}}{2}\left[\left(\frac{3 + 2\beta^2}{\alpha^2\beta^4} - \frac{3}{\alpha^4\beta^5}\tan^{-1}\beta\right) - \frac{\left(\dfrac{-3}{\alpha^2\beta^4} + \dfrac{3 + \beta^2}{\alpha^2\beta^5}\tan^{-1}\beta\right)^2}{\left(\dfrac{3 + \beta^2}{\beta^4} + \dfrac{-3 - 2\beta^2 + \beta^4}{\beta^5}\tan^{-1}\beta\right)}\right] \qquad \text{IVC2-(31)}$$

It can be shown that

$$\lim_{\epsilon \to 0} \sigma_{33} = C\epsilon_{33}/6 \qquad \text{IVC2-(32)}$$

and

$$\lim_{\epsilon \to \infty} \sigma_{33} = C\epsilon_{33} \qquad \text{IVC2-(33)}$$

which check with the results obtained by the direct method for finding the limiting cases. Dividing the formula for the stress at some arbitrary ϵ by the one for $\epsilon = 0$ we obtain

$$\frac{\sigma_{33}(\epsilon)}{\sigma_{33}(0)} = \left\{3\left[\left(\frac{3 + 2\beta^2}{\alpha^2\beta^4} - \frac{3}{\alpha^4\beta^5}\tan^{-1}\beta\right) - \frac{\left(\dfrac{-3}{\alpha^2\beta^4} + \dfrac{3 + \beta^2}{\alpha^2\beta^5}\tan^{-1}\beta\right)^2}{\left(\dfrac{3 + \beta^2}{\beta^4} + \dfrac{-3 - 2\beta^2 + \beta^4}{\beta^5}\tan^{-1}\beta\right)}\right]\right\}\frac{\epsilon_{33}(\epsilon)}{\epsilon_{33}(0)} \qquad \text{IVC2-(34)}$$

In order to obtain a fracture strength theory, some assumption must be made about the ratio $\epsilon_{33}(\epsilon)/\epsilon_{33}(0)$, where $\epsilon_{33}(\epsilon)$ is the small strain at fracture when oriented an amount indicated by ϵ, and $\epsilon_{33}(0)$ is the small strain at fracture when unoriented. If this ratio is assumed to be constant, i.e., the small rupture strain $\epsilon_{33}(\epsilon)$ is not affected by orientation, we obtain a fracture theory which predicts a fracture strength at complete orientation of six times that for random orientation. The variation of fracture strength in this case, as a function of ϵ is shown in Figure IVC-1. Data of failure stress obtained for

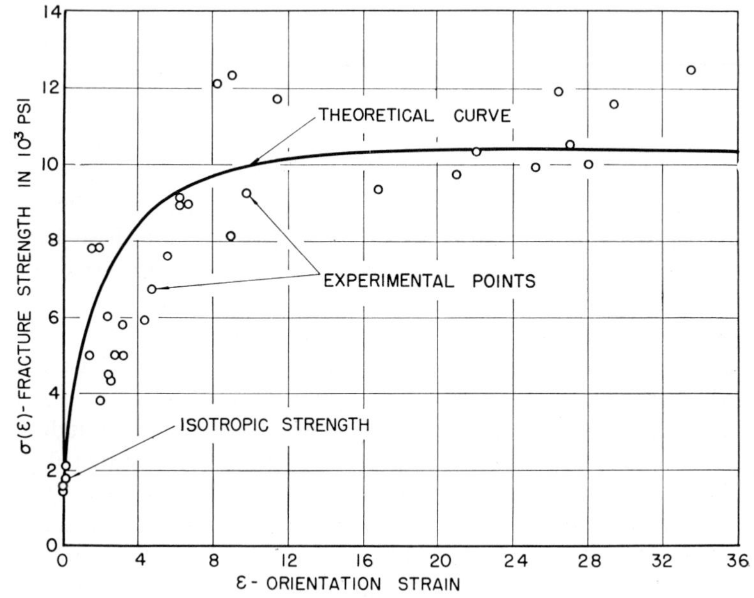

Fig. IVC-1. Theoretical and experimental strengths of polyethylene at room temperature.

polyethylene specimens tested at room temperature are also given for comparison.

IVC3. Strength Theory for Brittle Media

On the basis of the previous analysis it is seen that the fracture strength of an oriented and strained system of linear elements is not only dependent upon the orientation strain, but also on the fracture strain ratio of the medium. As given before, if σ_{ij}^* is the stress tensor, where $i,j = 1,2,3$, and the asterisk designates the quantity after orientation, then the stress tensor at the vicinity of a point may be expressed as

$$\sigma_{ij}^* = \int \lambda l \rho^* f^* F_i^* s_j^* d\omega \qquad \text{IVC3-(35)}$$

where λ is the number of linear elements per unit volume per unit solid angle, l is the length of any individual element, ρ^* is the density of probability distribution function of the elements after orientation,

f^* is the fraction of unbroken elements in the strained system, F_i^* is the force component in an instantaneous direction s_i after orientation, $\mathbf{s}^*$ is a unit vector indicating the normal direction of the fracture surface, and $d\omega^*$ is the solid angle after orientation. However, for relatively large brittle polymer samples, the mechanism of mechanical breakdown is likely to be dominated by the shape and the number of the so-called flaws or exclusions in the solid medium. Then, because of the stress concentrations developed around these heterogeneities, it may be possible to assume that the fracture strength is essentially proportional to the number and not necessarily to the direction of the linear elements per unit area penetrating the fracture surface. In other words, we may assume that f^* and F_i^* are essentially constants and are not affected by the deformation of the system. Now if we consider a unit fracture plane whose normal is indicated by the unit vector $\mathbf{s}_j^*$ and the density of probability distribution function of the oriented elements is ρ^*, then in a solid angle $d\omega^*$ there are $\lambda\rho^* d\omega^*$ linear elements penetrating the unit fracture surface. If l is the length of the elements and f_0 the fraction of the unbroken ones, then the total fraction of unbroken elements per unit area penetrating the fracture surface is

$$\int \lambda l \rho^* f_0 \, d\omega^* \qquad \text{IVC3-(36)}$$

Furthermore, if we consider that the fracture strength, i.e., the force required for rupturing all the unbroken chains in any individual linear element, is F, then the fracture strength measured at any direction after orientation is given by

$$\sigma_j^* = \int \lambda F l f_0 s_j^* \rho^* \, d\omega^* \qquad \text{IVC3-(37)}$$

But again, the number of linear elements before and after orientation should not change; therefore

$$\rho^* d\omega^* = \rho d\omega \qquad \text{IVC3-(38)}$$

where ρ is the density of distribution function without any preferred orientation, and $d\omega$ the original solid angle in an isotropic system. Instead of determining the density of the probability distribution function ρ^*, we can simply evaluate the following integral, as ρ is a constant in a random system. Therefore

$$\sigma_j^* = \lambda Fl\rho f_0 \int s_j^* \, d\omega \qquad \text{IVC3-(39)}$$

where

$$s_j^* = \sin\theta^* \cos\phi^*, \sin\theta^* \sin\phi^*, \cos\theta^* \qquad \text{IVC3-(40)}$$

$$j = x, y, z \qquad \text{IVC3-(41)}$$

and

$$d\omega = \sin\theta \, d\theta d\phi \qquad \text{IVC3-(42)}$$

In order to evaluate the integral

$$\sigma_j^* = (1/2\pi)\lambda Fl\rho f_0 \int_0^{2\pi} \int_0^{\pi/2} s_j^* \sin\theta \, d\theta d\phi \qquad \text{IVC3-(43)}$$

first of all we must determine s_j^* as functions of (θ,ϕ). This can be accomplished by introducing the conditions of flow-orientation as follows.

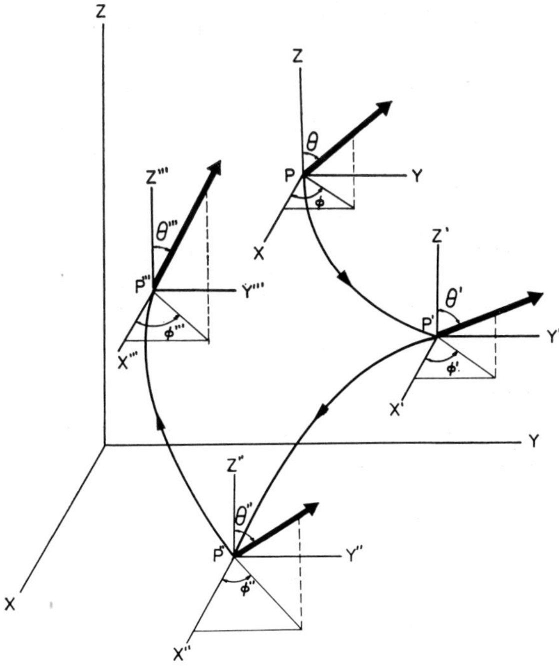

Fig. IVC-2. Successive orientation of a random vector.

Consider that incompressible flow of the idealized random system takes place during the process of orientation. If any arbitrary representative element, with its direction indicated by a random vector

$$d\mathbf{r} = \mathbf{i}dx + \mathbf{j}dy + \mathbf{k}dz$$

at $P(x,y,z)$, where **i**, **j**, and **k** are invariant unit vectors in an XYZ coordinate system as shown in Figure IVC-2, is to follow a successive orientation as a result of large deformations ϵ_x, ϵ_y, and ϵ_z along the X, Y, and Z directions, respectively, then its new position becomes

$$\begin{aligned} P'''[x''', y''', z'''] = P'''[&(1+\epsilon_x)(1+\epsilon_y)^{-1/2}(1+\epsilon_z)^{-1/2}x, \\ &(1+\epsilon_x)^{-1/2}(1+\epsilon_y)(1+\epsilon_z)^{-1/2}y, \\ &(1+\epsilon_x)^{-1/2}(1+\epsilon_y)^{-1/2}(1+\epsilon_z)z] \end{aligned} \qquad \text{IVC3-(44)}$$

Its new direction is represented by

$$d\mathbf{r}''' = \mathbf{i}dx''' + \mathbf{j}dy''' + \mathbf{k}dz''' \qquad \text{IVC3-(45)}$$

By definition, after the three successive flow orientations the direction cosine is

$$\cos\theta''' = dz'''/[(dx''')^2 + (dy''')^2 + (dz''')^2]^{1/2} \qquad \text{IVC3-(46)}$$

In terms of the random variables θ and ϕ

$$\begin{aligned} \cos\theta^* &\equiv \cos\theta''' \qquad \text{IVC3-(47)} \\ &= \frac{(1+\epsilon_z)^{3/2}\cos\theta}{\{(1+\epsilon_z)^3\cos^2\theta + [(1+\epsilon_x)^3\cos^2\phi + (1+\epsilon_y)^3\sin^2\phi]\sin^2\theta\}^{1/2}} \end{aligned}$$

Substituting eq. (47) into eq. (43) and carrying out the integration with respect to θ, then we obtain the following expression after the upper limit of ϕ is changed:

$$\sigma_z(\epsilon_x,\epsilon_y,\epsilon_z) = \frac{2\lambda Fl\rho f_0}{\pi}\int_0^{\pi/2}\frac{d\phi}{1+\left[\left(\frac{1+\epsilon_x}{1+\epsilon_z}\right)^3\cos^2\phi + \left(\frac{1+\epsilon_y}{1+\epsilon_z}\right)^3\sin^2\phi\right]^{1/2}}$$

IVC3-(48)

The value of the above integral is (11)

$$\sigma_z(\epsilon_x,\epsilon_y,\epsilon_z) = \frac{2\lambda Fl\rho f_0}{\pi}$$

$$\times \left\{ \frac{\pi(1+\epsilon_z)^3}{2[(1+\epsilon_z)^3-(1+\epsilon_x)^3]^{1/2}[(1+\epsilon_z)^3-(1+\epsilon_y)^3]^{1/2}} - \left[\frac{(1+\epsilon_z)^{9/2}}{(1+\epsilon_x)^{3/2}[(1+\epsilon_z)^3-(1+\epsilon_x)^3]} \mathbf{\Pi}(p,q) - \left(\frac{1+\epsilon_z}{1+\epsilon_x}\right)^3 \mathbf{K}(q) \right] \right\} \quad \text{IVC3-(49)}$$

where

$$p \equiv \frac{(1+\epsilon_x)^3-(1+\epsilon_y)^3}{(1+\epsilon_z)^3-(1+\epsilon_x)^3}$$

$$q = \left[1-\left(\frac{1+\epsilon_y}{1+\epsilon_x}\right)\right]^{1/2} < 1 \quad \text{IVC3-(50)}$$

and $\mathbf{K}(q)$ and $\mathbf{\Pi}(p,q)$ are, respectively, the complete elliptic integrals of the first and third kind. Some conditions of validity for using the above results are that

$$-1 < \epsilon_x,\epsilon_y,\epsilon_z < \infty \quad \text{IVC3-(51)}$$

$$\epsilon_x \neq \epsilon_z \neq \epsilon_y \quad \text{IVC3-(52)}$$

For cases when $\epsilon_x = \epsilon_z$ or $\epsilon_y = \epsilon_z$ for $-1 < \epsilon_x,\epsilon_y,\epsilon_z < \infty$, the original integral should be consulted.

The value of σ_z is that at the specific point $P'''(x''',y''',z''')$ after the successive flow-orientations. As the same condition holds for any other representative point of the system, then the macroscopic strength will be just the same as σ_z.

It might be of interest to mention that according to this analysis it appears possible for one to estimate the fracture strength in any direction if the principal large flow deformations are known.

IVC3a. Uniaxial Orientation and Longitudinal Strength

In the case of uniaxial orientation, if the strength along the direction of orientation is desired, then from eq. (48), let $\epsilon_x = \epsilon_y = 0$ and $\epsilon_z = \epsilon$.

$$\sigma_z(\epsilon) = \lambda Fl\rho f_0(1+\epsilon)^{3/2}/[1+(1+\epsilon)^{3/2}] \quad \text{IVC3-(53)}$$

When $\epsilon = 0$, the unoriented strength is given as

$$\sigma_z(0) = \lambda F l \rho f_0/2 = \sigma(\infty)/2 \qquad \text{IVC3-(54)}$$

and

$$\sigma_z(\epsilon) = \frac{(1+\epsilon)^{3/2}}{1+(1+\epsilon)^{3/2}} 2\sigma_z(0) \qquad \text{IVC3-(55)}$$

This value $\sigma_z(0)$ is only half of that when $\epsilon \rightarrow \infty$, indicating that a perfect orientation of the system will double the strength if measured in the direction of orientation. Or $\sigma_z(\infty) = 2\sigma_z(0)$, which will be the limiting magnitude of the strength as affected by flow-orientation. That is

$$\frac{\sigma_z(\text{completely oriented})}{\sigma_z(\text{completely random})} = 2 \qquad \text{IVC3-(56)}$$

For the variation of the strength as a function of large deformation and the isotropic strength of the system, Figure IVC-3 shows the general tendency. Again, experimental data obtained on samples of polyethylene are shown for comparison (11). However, in order

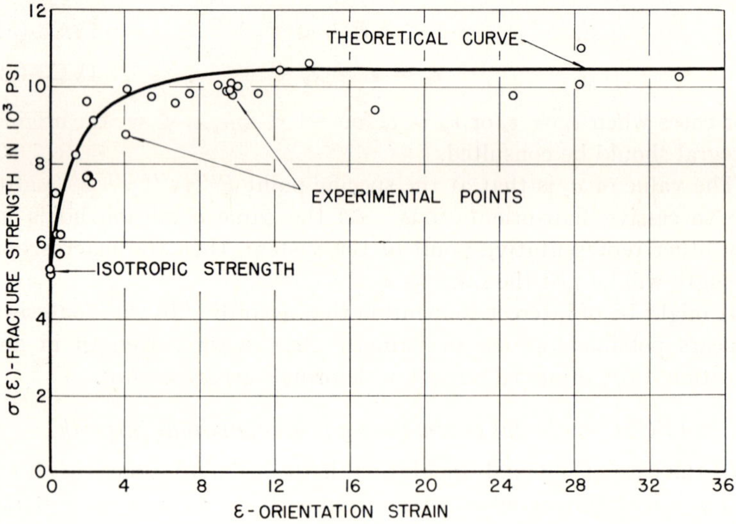

Fig. IVC-3. Theoretical and experimental strengths of oriented polyethylene at −100°C.

to make polyethylene brittle, all the tests were conducted at fairly low temperatures (~ -100°C.). Good agreement was obtained between predicted strength and measured data, as shown in Figure IVC-3.

IVC3b. Uniaxial Orientation and Transverse Strength

The appropriate analytical expressions for the transverse strength of the oriented system can be obtained by following the same general principle. Assuming that the principal stress and strain directions do not change under simple tension then

$$\sigma_x = \lambda Fl\rho f_0 \int \sin\theta^* \cos\phi^* \, d\omega \qquad \text{IVC3-(57)}$$

or

$$\sigma_x = \lambda Fl\rho f_0 \frac{1}{2\pi} \int_0^{2\pi} \int_0^{\pi/2} \sin\theta^* \cos\phi^* \sin\theta \, d\theta d\phi \qquad \text{IVC3-(58)}$$

For uniaxial orientation by a large deformation ϵ along the Z direction in tension,

$$\sin\theta^* = \frac{(1+\epsilon)^{-3/2}\sin\theta}{\{1 + [(1+\epsilon)^{-3} - 1]\sin^2\theta\}^{1/2}} \qquad \text{IVC3-(59)}$$

and

$$\cos\phi^* = \cos\phi \qquad \text{IVC3-(60)}$$

On substitution of eqs. (59) and (60) into eq. (58), the expression can be integrated. The results are

$$\sigma_x(\epsilon) = \lambda Fl\rho f_0 \frac{2}{\pi} \left\{ \mathbf{K}(h) - \frac{1}{1-(1+\epsilon)^3} [\mathbf{K}(h) - \mathbf{E}(h)] \right\}$$

$$(-1 < \epsilon < 0) \qquad \text{IVC3-(61)}$$

$$\sigma_x(\epsilon) = \lambda Fl\rho f_0 \frac{2(1+\epsilon)^{3/2}}{\pi(1+\epsilon)^3 - 1} \{\mathbf{K}(k) - \mathbf{E}(k)\}$$

$$(0 < \epsilon < \infty) \qquad \text{IVC3-(62)}$$

where $\mathbf{K}(h)$ and $\mathbf{E}(h)$ are, respectively, the complete elliptic integrals of the first and second kind. Both of them are functions of $h =$

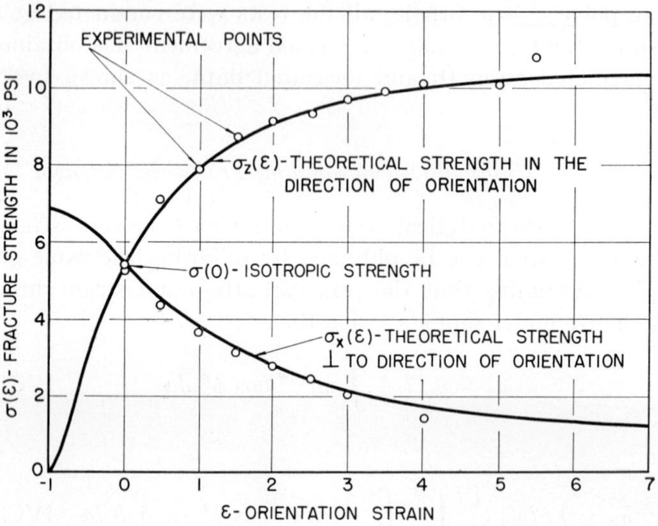

Fig. IVC-4. Comparison of theoretical and experimental strengths of polystyrene (molecular weight $\cong$ 68,000).

$[1 - (1 + \epsilon)^3]^{1/2}$. $\mathbf{K}(k)$ and $\mathbf{E}(k)$ are again the complete elliptic integrals of the first and second kind, respectively, but are functions of $k \equiv [1 - (1 + \epsilon)^{-3}]^{1/2}$.

$\sigma_x(\epsilon)$ can also be expressed in terms of the unoriented strength $\sigma_x(0)$. When this is done eqs. (61) and (62) become

$$\sigma_x(\epsilon) = \frac{4}{\pi}\left\{\mathbf{K}(h) - \frac{1}{1 - (1 + \epsilon)^3}[\mathbf{K}(h) - \mathbf{E}(h)]\right\}\sigma_x(0)$$

$$(-1 < \epsilon < 0) \qquad \text{IVC3-(63)}$$

$$\sigma_x(\epsilon) = \frac{4}{\pi}\frac{(1 + \epsilon)^{3/2}}{(1 + \epsilon)^3 - 1}[\mathbf{K}(k) - \mathbf{E}(k)]\,\sigma_x(0)$$

$$(0 < \epsilon < \infty) \qquad \text{IVC3-(64)}$$

Reasonably good agreements between the theoretical predictions and experimental results have been obtained for samples of linear amorphous polystyrene for one-directional flow-orientation. Comparison of the experimental data (11) with theoretical predictions is shown in Figure IVC-4. Amorphous polystyrene is quite hard and brittle at room temperature. Therefore all the fracture tests in simple tension were conducted at room temperature. It is interesting to note that the agreement is good for the fracture strength measured both in the direction of deformation as well as normal to the orientation.

IVC4. Strength Theory and Other Phenomena

It might be of interest to point out other areas of comparison between the experimental results and the theory. One is concerned with the molecular weight dependence of the tensile strength (12). The theory indicates an increase in strength as the molecular length increases. The experiments (12) do indeed follow the correct general trend as compared with theoretical results, as shown in Figure IVC-5. However, the exact dependence of the fracture strength upon the molecular weight and the theoretical molecular length is not known. Secondly, the theory is in general agreement with the "crazing" be-

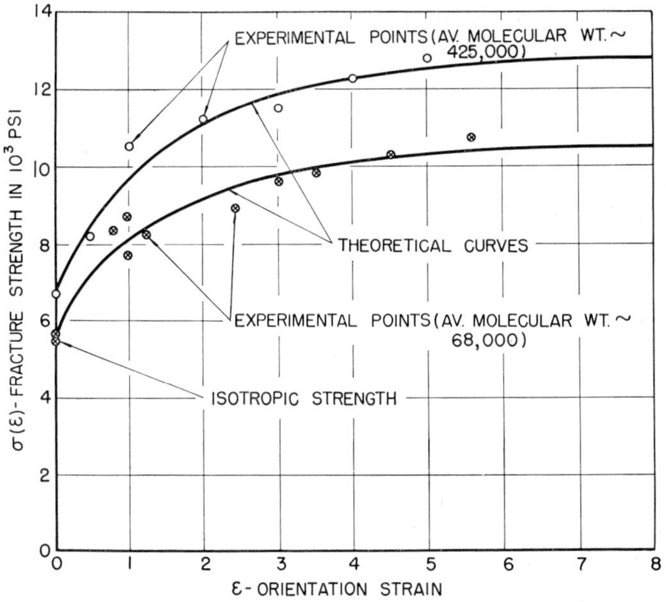

Fig. IVC-5. Theoretical and experimental strengths of uniaxially oriented polystyrene.

havior as reported earlier (13,26). Furthermore, the effect of size on the fracture strength has been studied by many scientists. For instance, Bartenev (2) has made an intensive analysis on the subject. He suggested that the strength of fine glass fibers is dependent upon the structural orientation and the orientation of heterogeneities. This treatment is, in general, somewhat similar to that in the present article where strength is considered to be dependent upon the molecular orientation and the ductility or brittleness of the material. In addition, the effect of molecular branching on strength can also be analyzed by the same general theory on a modified basis. Some results on fracture strength of poly(methyl methacrylate) (16) have already been obtained. It is hoped that further development of this topic can be continued.

On this general subject there are other areas that may be worth considering in future investigations. A natural outgrowth of the present analysis is the consideration of deformation-induced anisotropy (22,24) for various solids. This is found to be particularly important in studying polymeric materials. Another area is concerned with the fracture strength under a combined state of stress (23). As of now, to the best knowledge of the writer, the fracture criterion under an arbitrary state of combined stresses, such as biaxial and triaxial states of stress, is not yet clearly established (1,4,27). The development of the strength theories will aid not only in the future utilization of results of microscopic properties of materials in engineering analysis and design incorporating macroscopic anisotropy (18), but also in the understanding of the behavior of materials in general.

Finally, a word may be said about the orientation-dependent irradiation effect on fracture strength. It has been well established that prior molecular orientation before any atomic radiation will significantly improve the fracture strength of most linear polymers such as poly(ethylene terephthalate) (9,19), polyethylene (20), and nylon (21). However, a clear understanding of the complex phenomena encountered upon irradiation perhaps requires a good deal of additional study.

IVC5. Summary

A simplified theoretical treatment has been presented of the effect of molecular orientation on the fracture strength of polymeric solids. The theory is based on a model of the polymer consisting of ideal elastic spring elements which can be oriented under the action of homogeneous deformations. The analysis is carried out for two different sets of assumptions. The first treatment, which appears applicable to relatively ductile polymers such as polyethylene tested at room temperature, leads to the conclusion that the fracture strength of a perfectly oriented system is six times that of a completely random or isotropic system. The second treatment, which appears applicable to relatively hard, brittlelike polymers such as polystyrene at room temperature or polyethylene at $-100°$C., leads to the conclusion that the fracture strength of the completely oriented system is twice that of the random system. Other aspects of the theory are discussed such as the effect of uniaxial orientation on transverse, as well as longitudinal, strength and the effect of changes in molecular length.

Research supported in part by U.S.A.E.C. under Contract AT(11-1)-532.

IVC6. List of Key Symbols

C	Constant
E	Spring modulus
$\mathbf{E}$	Elliptic integral of the second kind
f	Fraction of unbroken spring elements
f_0	Constant
F_i	Force component in direction s_i
F	Force for rupture
h	Function of finite strain
k	Function of finite strain
$\mathbf{K}$	Elliptic integral of first kind
l	Length of spring element
$P(xyz)$	Representative point with coordinates (x,y,z)
p	Function of finite strains
q	Function of finite strains
r	Polar coordinate
s	Unit vector
*	Indicates referral to oriented system
α,β	Functions of finite strain
ϵ	Finite strain, indicating degree of orientation
$\epsilon_x,\epsilon_y,\epsilon_z$	Finite strains in directions x, y, z
ϵ_{ij}	Component of strain tensor

$\epsilon_{ij}(\epsilon)$	Fracture strain for degree of orientation given by ϵ
$\epsilon_{ij}(0)$	Fracture strain in unoriented state
θ	Polar coordinate
λ	Number of springs per unit volume
$\mathbf{\Pi}$	Elliptic integral of third kind
ρ	Probability density function of orientation of springs
σ_{ij}	Components of stress tensor
$\sigma_{ij}(\epsilon)$	Fracture stress for degree of orientation given by ϵ
$\sigma_{ij}(0)$	Fracture stress in unoriented state
ϕ	Polar coordinate
ω	Solid angle

References

1. Averback, B. L., D. K. Felbeck, G. T. Hahn, and D. A. Thomas, Eds., *Fracture,* Technology Press and Wiley, New York, 1959.
2. Bartenev, G. M., and A. N. Bovkoonenko, *Zh. Fiz. Khim.*, **29,** 508 (1955).
3. Cosserat, E., and F. Cosserat, *Theories des Corps Deformables,* Hermann, Paris, 1909.
4. Drucker, D. C., and J. J. Gilman, Eds., *Fracture of Solids,* Interscience (Wiley), New York, 1963.
5. Ericksen, J. L., and C. Truesdell, *Arch. Ratl. Mech. Anal.*, **1,** 295 (1958).
6. Ericksen, J. L., Proceedings Symposium on Structural Dynamics under High Impulse Loading, Technical Documentary Report No. ASD-TDR-63-140 (1963).
7. Ericksen, J. L., paper presented at the 4th International Congress on Rheology, Brown Univ., Providence, R. I., August 1963.
8. Flügge, S., *Encyclopedia of Physics,* 3rd Ed., Vol. I, Springer Verlag, Berlin, 1960.
9. Haynes, A., and C. C. Hsiao, *J. Appl. Phys.*, **31,** 1871 (1960).
10. Hsiao, C. C., *Nature,* **186,** 535 (1960).
11. Hsiao, C. C., *J. Polymer Sci.*, **44,** 71 (1960).
12. Hsiao, C. C., *J. Polymer Sci.*, **47,** 251 (1960).
13. Hsiao, C. C., and J. A. Sauer, *J. Appl. Phys.*, **21,** 1071 (1950).
14. Hsiao, C. C., *J. Appl. Phys.*, **30,** 1492 (1959).
15. Hsiao, C. C., Proceedings Symposium on Structural Dynamics Under High Impulse Loading, Technical Documentary Report No. ASD-TDR-63-140 (1963).
16. Hsiao, C. C., and T. S. Wu, *J. Polymer Sci.*, **A1,** 1798 (1963).
17. Hsiao, C. C., J. E. Osborn, and D. B. Rozendal, *J. Phys. Soc. Japan,* **16,** 459 (1961).
18. Hsiao, C. C., and S. R. Moghe, Tech. Rept. No. 20 AEC AT (11-1)-532, University of Minnesota (1963).
19. Hsiao, C. C., and S. T. Chow, *J. Appl. Phys.*, **31,** 1869 (1960).

20. Hsiao, C. C., and J. W. Yang, *J. Appl. Phys.*, **33,** 2870 (1962).
21. Hsiao, C. C., Y. C. Das, and A. Haynes, *Brit. J. Appl. Phys.*, **11,** 277 (1960).
22. Kao, S. R., and C. C. Hsiao, Tech. Rept. No. 22, AEC AT(11-1)-532, University of Minnesota (1963).
23. Moghe, S. R., and C. C. Hsiao, Tech. Rept. No. 17, AEC AT(11-1)-532, University of Minnesota (1962).
24. Osborn, J. E., T. S. Wu, and C. C. Hsiao, Tech. Rept. No. 12, AEC AT(11-1)-532, University of Minnesota (1960).
25. Robinson, A., J. E. Osborn, and C. C. Hsiao, *J. Appl. Phys.*, **31,** 1602 (1960).
26. Sauer, J. A., and C. C. Hsiao, *Trans. ASME*, **75,** 895 (1953).
27. Volkov, S. D., *Statistical Strength Theory,* Gordon and Breach, New York, 1962.

IVD. MORPHOLOGICAL ASPECTS OF THE DEFORMATION AND FRACTURE OF CRYSTALLINE POLYMERS

P. H. GEIL

IVD1. Introduction

This discussion will be based on the assumption that the basic structural unit of a crystalline polymer is the lamella, an anisotropic molecular crystal on the order of 100 A. thick and several microns in at least one of the lateral dimensions. They may be regarded as being somewhat similar to the grains in a metal. The molecules are oriented essentially normal to the broad faces of the lamellae (see Section IB). The large majority of the molecules are folded within the lamellae, although some, depending on the crystallization conditions, may extend from one lamella to the next. Since the material to be discussed in this section is still largely fragmentary and of a preliminary or scouting nature, it is not yet possible to correlate the morphological observations with stress–strain functions or other macroscopic assessments of the deformation and fracture of polymers. Knowledge of the morphological aspects of the deformation of polymers in uniaxial and biaxial tensile strain is needed, however, since the large asymmetry of their molecular forces is best utilized in oriented materials such as obtained by "cold-drawing" or "rolling."

In this particular section, the term "cold-drawing" will be used in regard to that process in which a sample is strained uniaxially in tension at any temperature below that at which recrystallization can take place on a short time scale (i.e., more than 30°C. below the melting point) and above that at which brittle failure occurs at small strains. The latter temperature will vary considerably with the morphology of the sample, but will be above the glass transition temperature T_g. In order to be cold-drawn, a polymer must be able to undergo a ductile deformation $\geqslant 10\%$ (usually much larger) and any polymer solid so strained will be referred to here as a "cold-drawn polymer." The deformation may take place either through the formation of a neck following yielding (Subsec. IC5c, p. 96) or *uniformly* throughout the sample. In general we will omit the term "cold"; *draw, drawing,* or *drawn* thus referring to the above process. It is the morphological changes involved in these processes with which we will be primarily concerned here. The deformation of oriented fibers is discussed by Hearle in Section ID.

IVD2. Wide-Angle X-Ray Diffraction

When a polymer is drawn, the molecules tend to become oriented with increasing strain in the draw direction. Linear polyethylene, for instance, can be strained on the order of 1000% before rupture occurs (at room temperatures and suitable strain rates). Although, as discussed below, some rather complex changes in orientation of both the crystals and the molecules occur during the initial stages of the drawing of crystalline polymers, the molecular axes in a fully drawn fiber or film (crystalline or noncrystalline) are highly aligned in the draw direction. The other two crystal axes (the molecular axis is parallel to one of the crystal axes) are randomly oriented normal to the draw direction. Thus a fiber has a cylindrical symmetry. The changes in orientation which occur in the macroscopic necking region that can be observed visually when many fibers or films are drawn are believed to be essentially the same as those that occur more gradually and throughout a "uniformly drawing" sample. Direct experimental evidence concerning this, however, appears to be lacking. The "macroscopic necking" to be discussed here should not be confused with the sharp discontinuity, on a microscopic scale, between folded and unfolded molecules which is sometimes observed when lamellae are drawn (Subsections IVD4 and IVD5). The macroscopic necking region is often many microns long (see Fig. IVD-4).*

One of the most complete studies of the change in unit cell orientation during drawing, as a function of elongation and subsequent relaxation, has been reported by Aggarwal, Tilley, and Sweeting (1). Using microcamera x-ray techniques they obtained diffraction patterns from various portions of the necking region of a low density, branched polyethylene. Optical retardation and dimensional changes were also measured. Summarizing their results, they indicate that the drawing of compression-molded initially "unoriented" polyethylene films occurs in three stages: "*1.* alignment of the **a** axis perpendicular to the stretching direction, accompanied by a sharp increase in optical retardation (measured perpendicular to the plane of the film); *2.* progressive alignment of the **c** axis, with increasing percentage elongation, but with comparatively small increase in optical retardation; *3.* slipping of crystallites aligned nearly perfectly with the **c** axis parallel to the stretching direction, attended by only a small increase in percentage elongation and optical retardation but with progressively increasing contraction in width." It should be noted that the orientations given are not absolute and that there is considerable deviation for all but the most highly oriented specimens. It would appear that macroscopic yielding is associated with the transition from stage *1* to stage *2*, but considerably more information is needed to interrelate these observations, stress–strain functions, and observable changes in morphology.

If a drawn crystalline specimen is annealed at temperatures near but below the melting point, relaxation and a contraction in length can occur. Belbeoch and Guinier (2a) have shown, for instance, that when a drawn fiber of branched polyethylene (melting point = 115°C., originally drawn 500% at room temperature) is annealed, contraction begins at about 50°C. The amount of contraction increases with temperature but remains less than 50% up to 95°C. At 105–110°C. the relaxation is complete (the sample returning to its original length). Belbeoch and Guinier showed that below 80°C. the contraction is accompanied by an increase in crystal size or perfection with little or no change in crystal orientation. Above 90°C. the **c**

* **Editor's Note:** The terminologies as specifically employed in this section in regard to "cold-drawing" and "macroscopic necking" are not entirely identical to those usually employed.

(molecular) axis begins to rotate away from the fiber axis. Upon complete relaxation, they suggest that the **b** axis is normal to the fiber axis while the **a** and **c** axes are randomly rotated about **b**. Similar changes are expected to occur in other polymers.

Several authors (8,30) have experimentally determined the elastic (Young's) modulus of polymer crystals in the direction of the molecular axis by measuring the change in lattice spacing when a highly oriented fiber is strained in tension. In general, the results indicate that for a fully elongated molecule such as polyethylene (for which the lattice extension must be due only to increases in bond angle and length, since the molecular backbone is a planar zig-zag), the elastic modulus of an idealized fiber composed of completely oriented crystallized molecules is on the order of 2.5×10^6 kg./cm.^{2}. In the case of helical molecules, such as polypropylene and polyoxymethylene, in which the molecules are not fully extended (lattice extension may occur by rotation about the single bonds of the backbone as well as by changes in bond angle and length), the elastic modulus is (30) on the order of 5×10^5 kg./cm.^{2}. In the case of a number of polyesters, whose diffraction patterns are such as also to suggest a helical conformation, the value found was an order of magnitude lower (8). All of these values were derived from the change in lattice spacing in the crystal and the overall increase in sample length under the assumptions that a model consisting of two hookean springs in series can be used to approximate the fiber; one spring characteristic of a perfect crystalline region and the other of an oriented noncrystalline region, and that no slip occurs along the backbones of the molecules. See Hearle, Section ID.

IVD3. Small-Angle X-Ray Diffraction

One or more orders of small-angle x-ray diffraction are found for drawn fibers of most crystalline polymers. Porod (28) has reviewed this subject and its interpretation in terms of a model of the fringed micelle type (see Sections IB and ID). Although this kind of structure has not yet been disproven for highly drawn polymers, a number of questions concerning its applicability have been raised (13,20,35).

The small-angle pattern from a drawn fiber (or film) generally has the form of lines or streaks normal to, and on the meridian of, the fiber pattern. Under some conditions, four-point or quadrant diffraction patterns are found (13,28). Whereas the line pattern is in-

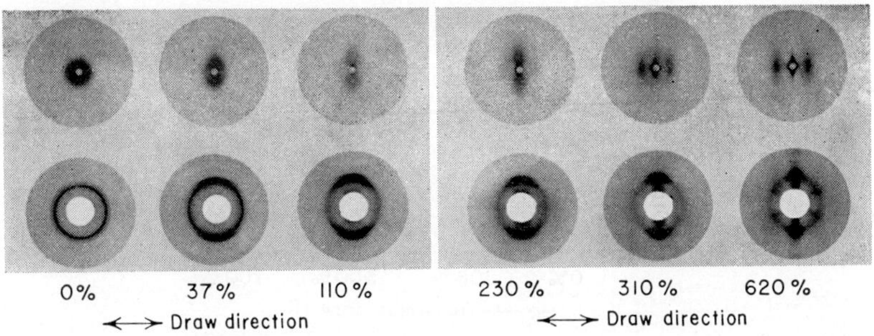

Fig. IVD-1. Small and wide-angle diffraction patterns from a linear polyethylene as a function of the amount of draw (16).

dicative of a periodic spacing or linear lattice along the fiber axis, the four-point diagram suggests that some type of three-dimensional arrangement of the diffracting structures must exist. In addition, diffuse scattering is found, presumably a result of elongated voids in the fibers.

The changes observed in the wide-angle x-ray diffraction pattern during drawing do not have any clear relationship to the changes in the small-angle pattern. Hendus (16) has obtained some excellent patterns (Fig. IVD-1) showing the changes that occur during the uniform drawing of films of linear polyethylene. The wide-angle patterns he obtained simultaneously, however, do not appear to agree with those obtained by Aggarwal et al. (1) in the necking region. This difference may be real, and a result of the difference in type of draw or polymer (Aggarwal et al. used a branched polyethylene), or the difference may be due to the limited number of reflections visible in Hendus' patterns which do not permit a complete interpretation.

At low elongations, and extending well beyond yield, the originally symmetrical, small-angle diffraction pattern takes on a V-shaped structure with no change in periodicity. A quadrant-type diffraction pattern is formed in which each reflection consists of a line inclined to the draw direction. With further elongation, an entirely new periodicity is seen to appear, the pattern consisting of streaks along the meridian. These results suggest that the structural units (probably lamellae) causing the original periodicity are retained for elongations up to about 250%, but become oriented with respect to the draw direction. At higher elongations a new morphological struc-

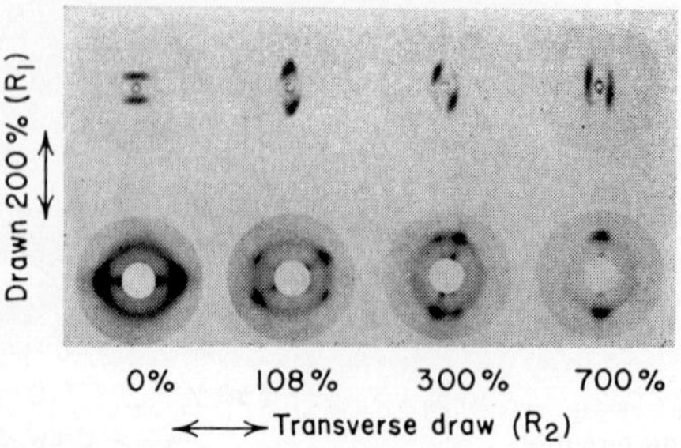

Fig. IVD-2. Changes in the wide and small angle diffraction patterns as a drawn, branched polyethylene film (originally drawn 200%) is redrawn at nearly right angles to the original draw direction (16).

ture may develop. The reorientation of the molecules within the structural units must occur independently (to some extent) of the reorientation of the structural units, since the wide-angle x-ray pattern shifts in a nearly continuous fashion.

Hendus (16) also investigated the changes in the diffraction pattern that take place when a drawn sample (in this case, branched polyethylene) is redrawn at an inclination to the original draw direction (Fig. IVD-2). Despite an almost complete realignment of the molecular axes, the small-angle pattern takes on a skewed orientation, again suggesting that the molecules reorient within morphological units which are also being oriented. Similar experiments are needed at higher and lower original "draw ratios."

IVD4. Microscopy of Drawn and Fractured Spherulitic Samples

Relatively few papers have been published which describe optical or electron microscope observations of the inelastic deformation of polymer spherulites (3,13–15,22,31,34), and in many cases the reproductions of the published micrographs have not been sufficiently clear to permit observation of the details of the drawing process. We summarize below, Keith and Padden's optical microscope observations (22) of the drawing of linear polyethylene.

(*1*) Spherulites of polyethylene in thin films (5–50 μ), grown by very slow cooling and thus having a coarse structure, undergo brittle failures at low tensions and very small elongations. The fracture locus follows an irregular course without regard to the boundaries or internal structure of the spherulites. (*2*) In more rapidly cooled thin films, brittle failure takes place at the boundaries of the spherulites. (*3*) With films crystallized at increasing rates, drawing is first found to take place at the spherulite boundaries before rupture and then a crystallization rate is reached for which the spherulites themselves begin to yield in place of, or in conjunction with, yield at the boundaries. When the spherulites yield, they often do so locally; a neck forming across the spherulite diameter (see ref. 13, Fig. VII-8). The extinction rings of the spherulite can still be seen in the drawn region at low elongations but tend to disappear at higher elongations. (*4*) The resulting drawn material is microfibrillar and readily splits parallel to the fibril axis. (*5*) They suggest that the differences in the drawing behavior on different crystallization rates are a result of differences in the number of "tie" molecules (see Section IB) holding the lamellae together.

Several authors (15,21,24,34) have commented on the susceptibility of a number of polymers to fracture when strained along spherulite boundaries if they had initially been cooled slowly from the melt [probably corresponding to those described under (*2*) above]. This is a result of a partial restrainment of the localized contraction toward the center of each spherulite that takes place when the polymer crystallizes. This potential for contraction can be sufficiently large so that a number of small voids are formed at the spherulite boundaries or in the case of polypropylene (21,24), that the sample even falls apart into individual spherulites. Radial cracks and voids are also often found in slowly cooled samples of many polymers. In samples which are cooled more rapidly [(*3*) above], one expects more "tie" molecules between adjacent lamellae within a spherulite and, at the boundaries, between interleaved lamellae from neighboring spherulites. In this case, partially restrained contraction during crystallization will result in residual stresses between the lamellae. Annealing may relieve these stresses, permit recrystallization of the tie molecules into single lamellae and permit brittle fracture during subsequent deformation, whereas the original rapidly cooled polymer would draw plastically.

We next consider optical and electron microscope observations (13) of the drawing and fracture (at liquid nitrogen temperatures) of polyethylene and polyoxymethylene and, where possible, attempt to correlate these observations with the x-ray results discussed previously.

As indicated by Keith and Padden (22), the first visible elongation of a spherulitic polyethylene film occurs at spherulite boundaries which are normal to the draw direction (Fig. IVD-3). In thin films,

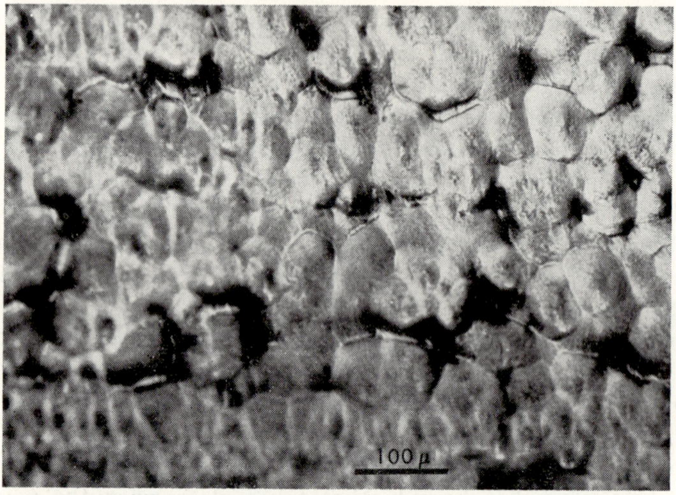

Fig. IVD-3. Slightly drawn thin film of linear polyethylene (13).

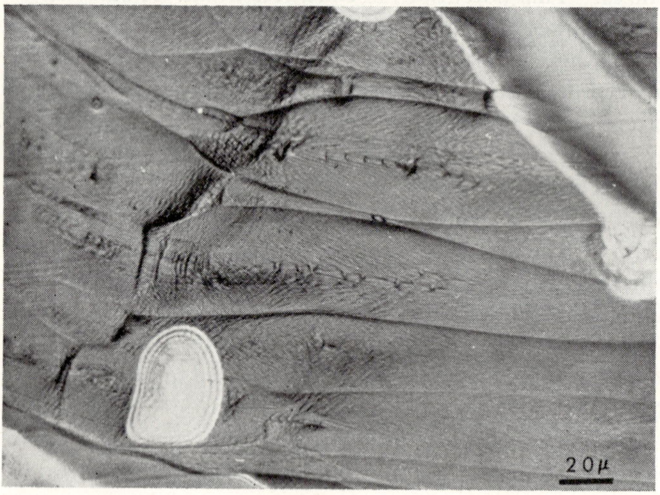

Fig. IVD-4. Replica of the necking region of a drawn film of linear polyethylene, replicated while taut. The interference patterns are between the poly(vinyl alcohol) replica and the microscope slide (13).

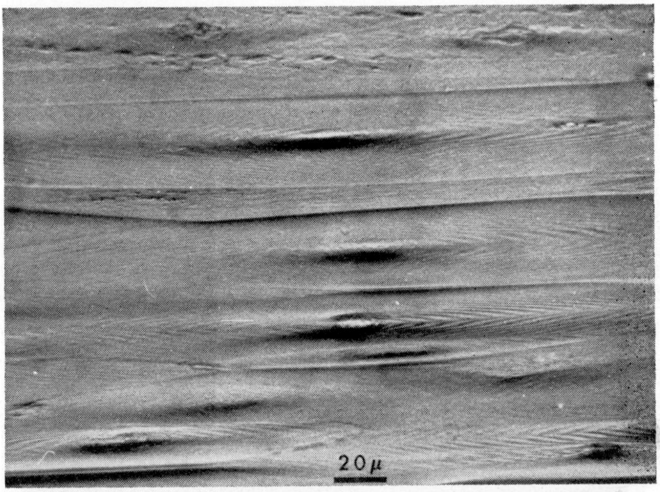

Fig. IVD-5. Drawn spherulitic film of linear polyethylene (13).

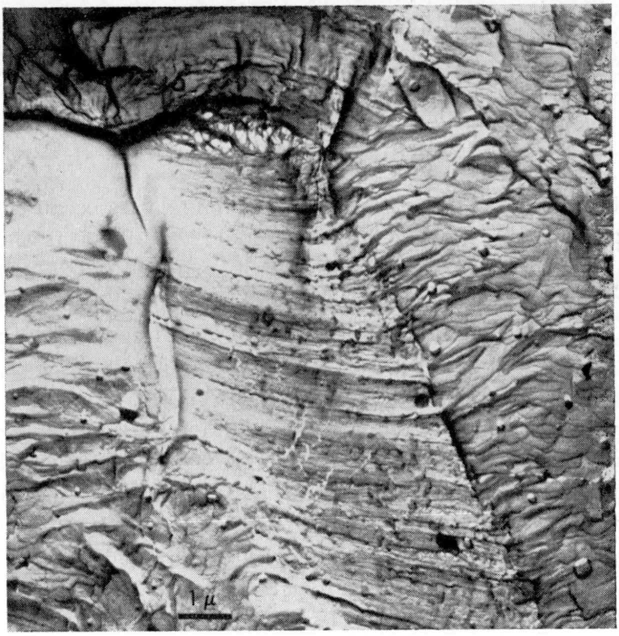

Fig. IVD-6. Electron micrograph of the boundary region between spherulites in a slightly drawn film (as in Fig. IVD-3) (13).

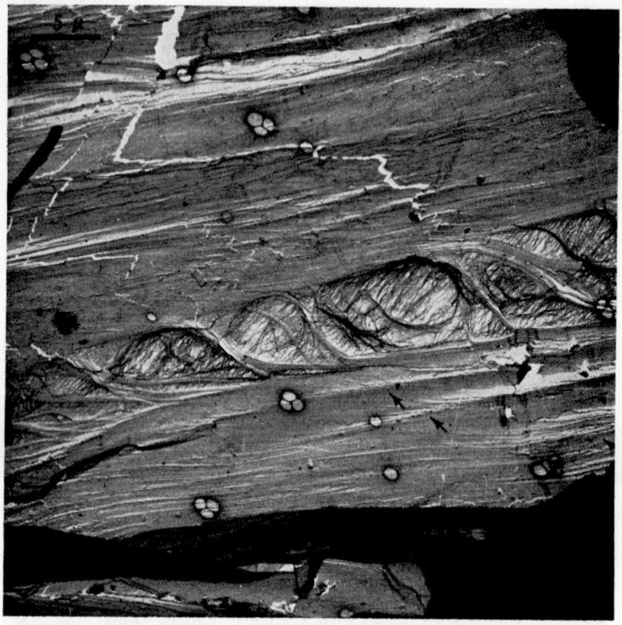

Fig. IVD-7. Low magnification micrograph of drawn linear polyethylene (as in Fig. IVD-5). Remnants of the ring structure can be seen (arrows) (13).

at least, drawing will occur first at these boundaries because they are thinner than the rest of the film. It should be noted that, since the **b** axis in polyethylene spherulites is radially oriented (23,27), the **a** and **c** axes of the portion that draws first will be perpendicular to the draw direction. While this appears to agree at least in part with the wide-angle x-ray results [**a** axis normal to the draw direction, stage (*1*)], the cause of the x-ray results is probably more subtle. As will be seen in the next subsection, the deformation characteristics of polyethylene lamellae are significantly different in the directions of the **a** and **b** axes.

With further elongation of linear polyethylene samples at room temperature, a neck usually forms extending across the entire sample. A replica of a necking region is shown in Figure IVD-4. Almost the entire elongation is restricted to a distance which is less than a spherulite diameter but is still of considerable size. The total change in orientation reported by Aggarwal et al. (1) takes place within this region.

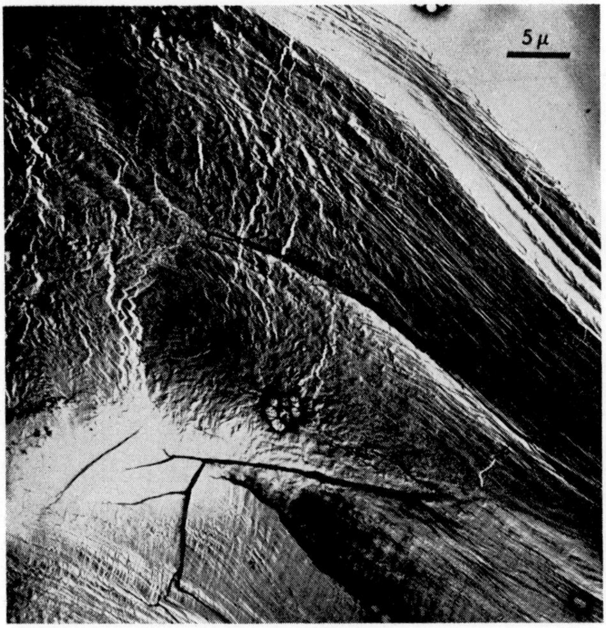

Fig. IVD-8. Electron micrograph of the necking region in a drawn linear polyethylene film (13).

The fully drawn material is shown in Figure IVD-5. Spherulites and their internal ring structure (Section IB) are still visible on the surface and, to some extent, in polarized light (13,34). When drawn to this extent, the polymer is fibrous in structure. The relationship between the (still present) ring structure and the newly formed fibrils is not known.

Electron micrographs (Fig. IVD-6) from a slightly elongated polyethylene (similar to that in Fig. IVD-3) indicate that there can be an abrupt change between drawn and undrawn material. The lamellar structure appears to be retained up to the line of draw. In some cases, in this polymer and polyoxymethylene (13), fragments of the lamellae appear to break off and remain as discrete regions in the drawn material.

Biaxial components of the tensile stress will generate a splitting of the drawn regions and the formation of voids (Figs. IVD-6 and IVD-7). Figure IVD-7 is from a fully drawn region, similar to that in

Fig. IVD-9. Replica of a film of branched polyethyene drawn 50% (13,25).

Figure IVD-5. The individual fibrils in the void area are on the order of 100 A. in diameter. Although occasionally a periodic structure of about the same size as the long period spacing (9,13) is seen on these fibrils, the significance of the periodicity is not known. In some cases, at least, it appears to be an artifact.

Electron micrographs of a necked region similar to that in Figure IVD-4 show a gradual transition from the lamellar to the fibrillar structure (Fig. IVD-8). Although the micrographs do not permit an interpretation of the deformation process in detail, at least the early portions of the process appear to involve a twisting of the lamellae. This twisting has been observed most clearly in a sample of branched polyethylene drawn 50% (Fig. IVD-9) (25). In this case the sample underwent uniform draw, at least up to this elongation. It was suggested (25) that groups of the lamellae twisted as units toward the draw direction. These results agree at least partly with the x-ray patterns obtained by Aggarwal et al. (1) if one assumes that the lamellae twist so as to bring the **b** axis toward the draw direction, the

Fig. IVD-10. Fracture surface of polytetrafluoroethylene which was drawn 100% after the surface was produced. Non-uniform shear (*A*) and uniform rotation (*B*) of the striations have taken place (32).

a and **c** axes being randomly oriented in planes tending to be normal to the draw direction. Whether or not deformation also occurs within these polyethylene lamellae through slip and twinning is not known as yet, but we believe it is likely. Slip between lamellae has been observed in slowly cooled polyoxymethylene samples (13). Twinning has been observed in drawn polyethylene single crystals, the twinning being such as to result in the alignment of the **b** axis with the draw direction (see next subsection).

Some of the types of deformation which probably occur within the lamellae apparently can be observed when polytetrafluoroethylene is drawn (32). On fracture surfaces (fractured at liquid nitrogen temperatures) of undeformed, highly crystalline polytetrafluoroethylene, bands can be seen which are on the order of 1000 A. wide and many microns long. The molecules are normal to the bands (6). A series of striations is also normal to the bands, and thus apparently parallel to the folded molecules.

It appears reasonable to assume that the bands correspond to the lamellae in other polymers and that the striations remain parallel to the molecules when the sample is deformed. By following the deformation of the bands and striations when a sample is drawn, Speerschneider and Li (32) have noted two types of deformation (see also Subsection IVD6 for the effect of rolling and compression on the band structure). They suggest that the initial stages of inelastic deformation (Fig. IVD-10) involve a relative displacement of the striations within the bands. In those bands which are oriented normal to the principal strain, a nonuniform sliding or shear of groups of the striations takes place (*A* in Fig. IVD-10) whereas if there is a component of the stress parallel to the bands, a uniform rotation or tilting and shear of the striations occurs (*B* in Fig. IVD-10).

Fig. IVD-11. A region from the same sample as in Fig. IVD-10 in which bowing (*C*) and kinking (*D*) of the striations have taken place (32).

Speerschneider and Li suggest (32) that if the deformation resulting from the shear of the striations, as pictured in Figure IVD-10, is locally insufficient to accommodate the macroscopic deformation, then the striations themselves may deform (Fig. IVD-11). In a few cases (as at *C* in Fig. IVD-11) the striations become bowed, whereas in other regions a kink develops (as at *D* in Fig. IVD-11). The shear and kinking noted by Speerschneider and Li suggest that the deformation is taking place by slip (possibly through dislocation motion) and twinning in a manner related to that observed in metallic solids (see also Subsection IVD6). In all such comparisons, however, the asymmetry of the molecular forces in polymer crystals must be considered. Slip within a lamella, normal to the molecules, for instance, would require chain scission, and is less likely.

As discussed above, many crystalline polymers can undergo large inelastic deformations, i.e., draw, when a stress is imposed, the morphology of the polymer changing from lamellar to fibrillar. Besides this form of deformation, however, it is often found that a crystalline polymer will undergo an apparent brittle fracture rather than undergoing draw when a stress is imposed. In the case of polyoxymethylene, for instance, a slowly cooled film will fracture when a tensile force is applied, whereas a rapidly cooled film may yield and draw. This is presumably related to the perfection of the lamellae and the number of "tie" molecules extending from one to the other, i.e., the morphology of the sample (13), but at present no quantitative information is available. In addition, factors such as molecular weight, branching, tacticity (which affects morphology) and relative temperature (with respect to the T_g) are significant.

In attempts to study the internal morphology of crystalline polymers, the observations of fracture surfaces have been employed (2,6, 11,13). Slowly cooled polyoxymethylene (11) and polychlorotrifluoroethylene (4) undergo brittle fracture readily when cooled to liquid nitrogen temperatures, the fracture taking place primarily between lamellae. When the failure occurs across a lamella, the micrographs suggest that it is not brittle but, instead, small areas are drawn, become fibrillar, and actually *melt* adiabatically and retract onto the fracture face. A granular structure develops along each broken edge of the lamellae (13). The melting is even more obvious in the case of polyethylene, as discussed below. Polytetrafluoroethylene, slowly cooled to liquid nitrogen temperatures (6), also can fracture in a brittlelike manner; revealing the bands and striations. However, it is not known if the fracture takes place between the bands or within them. The observations of the striations on bands on the fracture surfaces, and their absence on bands on exterior surfaces, suggest the latter. Fibers can often be seen on these fracture surfaces, but many of these are artifacts resulting from the replication techniques used for electron microscopy (13).

Whereas fractionated (i.e., narrow molecular weight distribution) linear polyethylene undergoes brittle, interlamellar fracture at liquid nitrogen temperatures (2), the whole (i.e., wide molecular weight distribution) polymer usually undergoes a quasi-brittle type of failure

(13). Similar failure takes place when rapidly cooled samples of other crystalline polymers are fractured at temperatures below T_g. Although the fracture appears grossly brittle, observation shows (13) that microscopic drawing has taken place on the fracture surface (Fig. IVD-12). In some areas of rapidly cooled samples, large drawn fibers are seen, while in others, numerous small fibrils are present. In

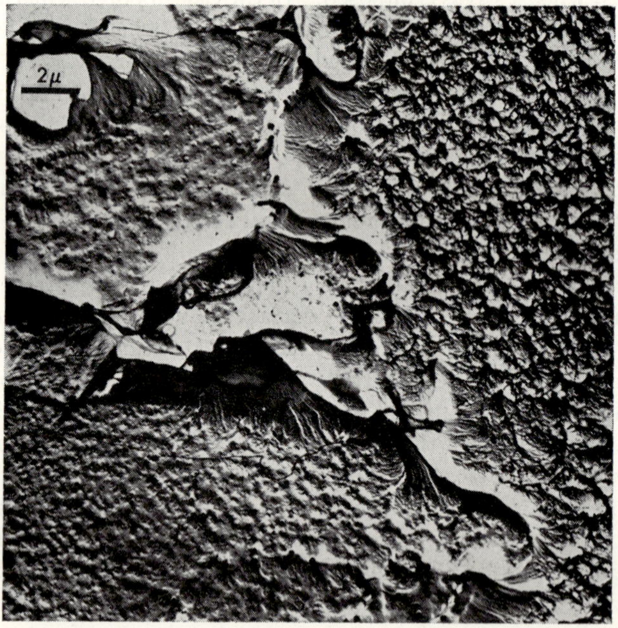

Fig. IVD-12. Fracture surface (fractured at liquid nitrogen temperature) of a 0.060-in. thick, Dry Ice–acetone quenched sample of linear polyethylene (13).

both cases the ends of the drawn fibers appear to have melted. Sometimes the adiabatic heating resulting from the inelastic deformation, or the overly rapid release of elastic strain energy, is sufficient to melt the small fibrils completely, and they contract and form small mounds on the surface. In a more slowly cooled whole polymer, although lamellae can sometimes be recognized, the appearance of the fracture surface again often suggests local inelastic deformation and melting (13).

IVD5. Drawing and Fracture of Polymer Single Crystals

To date, only a small amount of research has been devoted to the study of the deformation of single crystals of polymers; and most of this has been restricted to polyethylene. Scratching or shearing of crystals has revealed fractures taking place only between fold planes

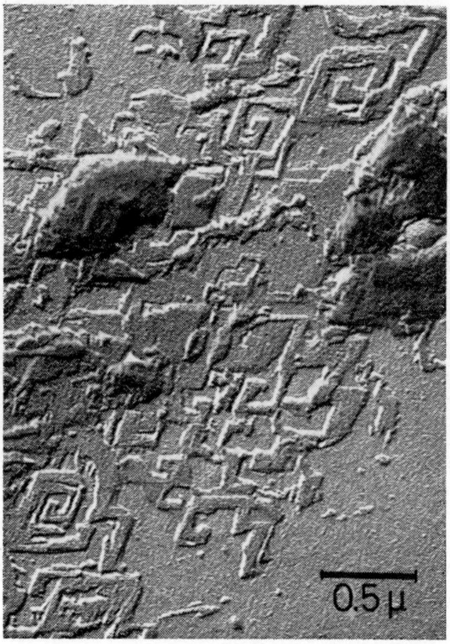

Fig. IVD-13. Polyethylene crystal deformed by shearing between glass plates (33).

and between lamellae (12,33). In a dendritic crystal, for instance, fracture planes are found parallel to the growth faces, changing direction at each fold domain boundary (12). In polymer crystals containing spiral growths, the fracture planes are also found to spiral (33) (Fig. IVD-13). In addition, each lamella in a spiral growth fractures independently, suggesting that there are few, probably no, tie molecules joining neighboring lamellae in solution-grown crystals. When fracture does not occur, fibers $\geq$ 50 A. in diameter are drawn out of the scratched crystal.

As noted in Section IB, polyethylene crystals grow in the form of hollow pyramids which collapse when deposited on a substrate and the

Fig. IVD-14. Single crystal of polyethylene drawn 100% on a Mylar substrate. The **a** axis is nearly vertical in the micrograph (13).

solvent subsequently evaporated. The collapse involves various types of deformation which have not yet been defined in detail. Molecular slip and tilting or rotation have been suggested (13). In addition, cracks sometimes develop, particularly at the apices, across which fibers are drawn (13,18). These fibers, each of which apparently results from the unfolding of molecules from a group of neighboring fold planes, are on the order of 100 A. in diameter. The number of fibers depends on the length of the crack. Where more than one lamella is involved, all the molecules within a given fiber appear to come from a single lamella (13).

Observations of polyethylene crystals drawn on soft substrates indicate that the type of deformation is dependent on the molecular weight, number of defects, type of fold domain (see Section IB), direction of draw, and on the adhesion to the substrate (13,13a). Most of the results have been obtained with the use of Mylar (E.I. du Pont de Nemours and Co., Inc.) films as a substrate. Suitable films of Mylar can be drawn as much as 200% before rupturing. Viton (E.I.

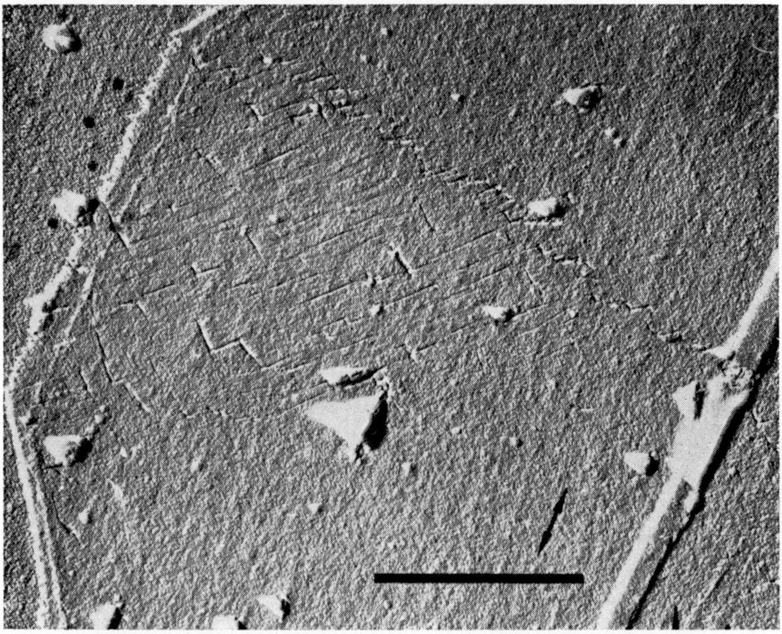

Fig. IVD-15. Portion of a truncated polyethylene crystal drawn 50%. Slip bands and cracks are observed in the {100} fold domain. Cracks are also present along the fold domain boundary. The bar's length represents 1 μ (13a).

du Pont de Nemours and Co., Inc.), which can be drawn as much as 1000%, has been used for a few experiments (13). Results available at the time of this writing are summarized below.

For diamond-shaped crystals grown at 80° C. in tetrachloroethylene from a Melt Index 5, Marlex (Phillips Petroleum Co.) linear polyethylene it was found (13a) that "brittle" fracture occurred at elongations as low as 10%. The resulting cracks appeared to be preferentially aligned along those {100}, {010}, or nongrowth face {110} planes which are nearly perpendicular to the draw direction. The edges of the cracks were irregular in path and fibers were not observed to be drawn across the cracks. With further elongation, the entire crystal was observed to break up into small pieces on the order of 100 A. in length.

On the other hand, diamond-shaped polyethylene crystals grown from Melt Index 3.3 and 4.7, Alathon (E. I. du Pont de Nemours and

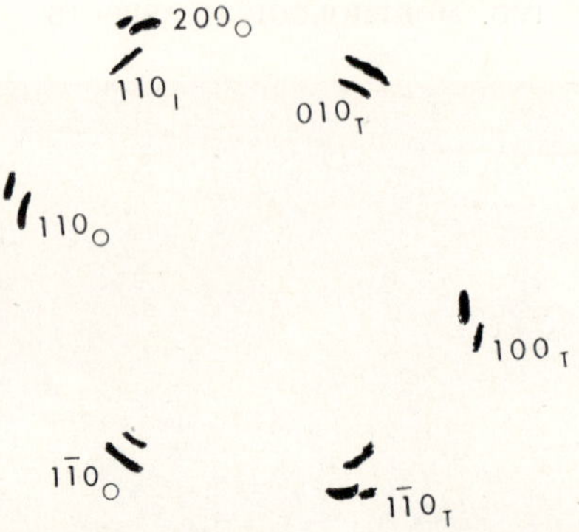

(*a*)

(*b*)

Fig. IVD-16. Electron diffraction pattern (tracing) (*a*) from a polyethylene crystal (*b*) drawn 25% on Mylar. The bar's length represents 1 μ. The diffraction pattern indicates that both {110} twinning (*c*) and a crystal transformation to a triclinic unit cell (*d*) have occurred. The reflections in the diffraction pattern are labeled according to their source, T (new, triclinic lattice), O (orthorhombic lattice after twinning), and I (initial orthorhombic lattice) (13a).

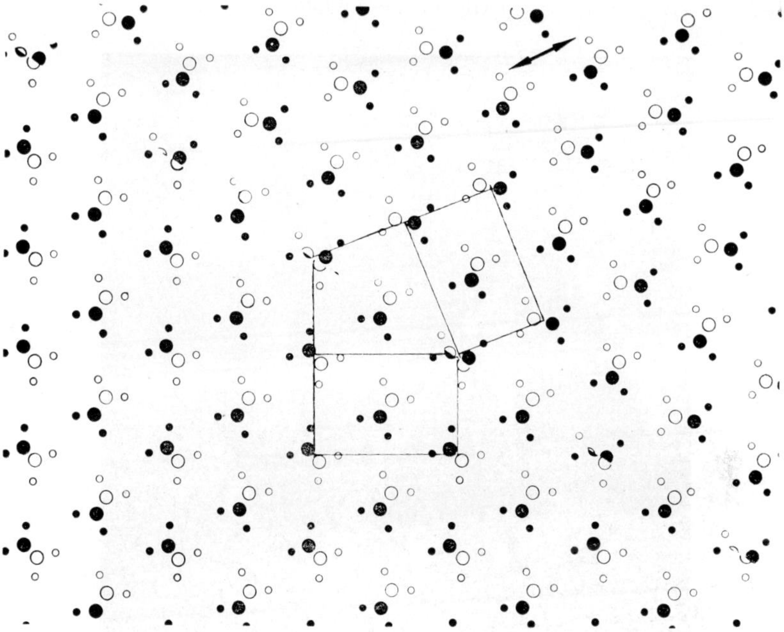

(*c*)

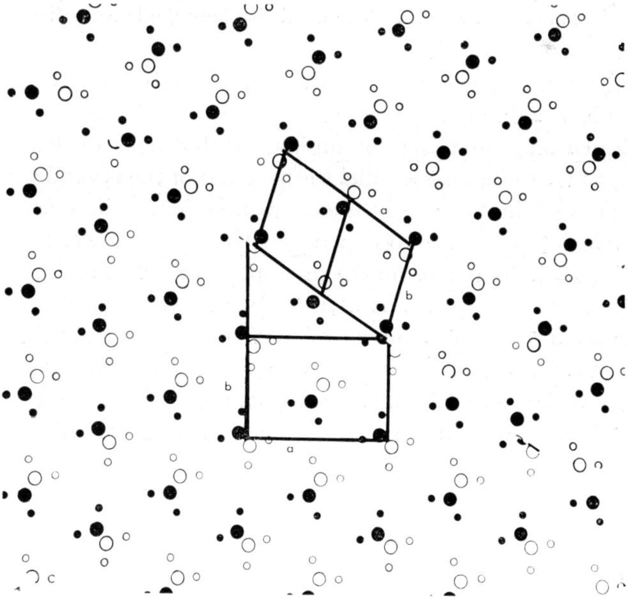

(*d*)

Fig. IVD-17. Polyethylene single crystal drawn 50% on Viton. The cause of the black spots is not known. The **a** axis is along the longest diagonal (13).

Co., Inc.) linear polyethylene (in which there probably were more defects within the crystals) were found to deform plastically (13,13a). When drawn on Mylar, sets of similar parallel lines of deformation (Fig. IVD-14) are formed parallel to the **a** axis of the crystal whenever a component of the **b** axis is in the draw direction. The separation of the lines does not appear to depend on the angle between the draw direction and the **b** axis or on the amount of draw. The lines can be observed on crystals drawn as little as 10%. Although the lines can be observed in dark field micrographs, it has not yet been possible to correlate the various portions of the lines with the Bragg reflection used to form the dark field image. A few crystals have been observed in which similar lines are formed parallel to the original $\{310\}$ planes. Although with samples drawn 50% or more, a few crystals are found in which actual tears or fractures are formed during the elongation, in general, the crystal appears to merely become thinner throughout and to take on a rippled structure.

Fig. IVD-18. Portion of the crystal in Fig. IVD-17 at higher magnification. The arrows indicate a region in which shear has taken place between fold planes (13).

No deformation lines have been observed in crystals in which the **a** axis is parallel, or nearly so, to the draw direction. In general, only a surface roughness is produced. These results appear to agree with the x-ray studies of the deformation of bulk polyethylene in which it appears that deformation is easiest parallel to the **b** axis.

The diamond-shaped crystals have only {110} fold domains. For truncated crystals grown at 85°C. in xylene from Melt Index 5.0, Fortiflex linear polyethylene (Celanese Corporation of America), it was found that the {100} fold domains differed considerably in their deformation characteristics from the {110} fold domains. Whereas, in these crystals, the {110} fold domains deformed in a manner similar to that in the diamond-shaped crystals described above, cracks and slip bands formed at elongations of as low as 10% in the {100} fold domains (Fig. IVD-15). In addition it was found that the fold domain boundaries between {100} and {110} fold domains were particularly susceptible to failure, whereas failure, in general, did not occur at the boundaries of neighboring {110} fold domains.

Electron diffraction patterns from the crystals grown from Alathon polyethylene, and drawn on Mylar, indicate that both twinning and a crystal transformation can take place. In Figure IVD-16 is shown a crystal drawn 25% in which the diffraction pattern indicates that a 110 twin boundary swept through the crystal and also that a

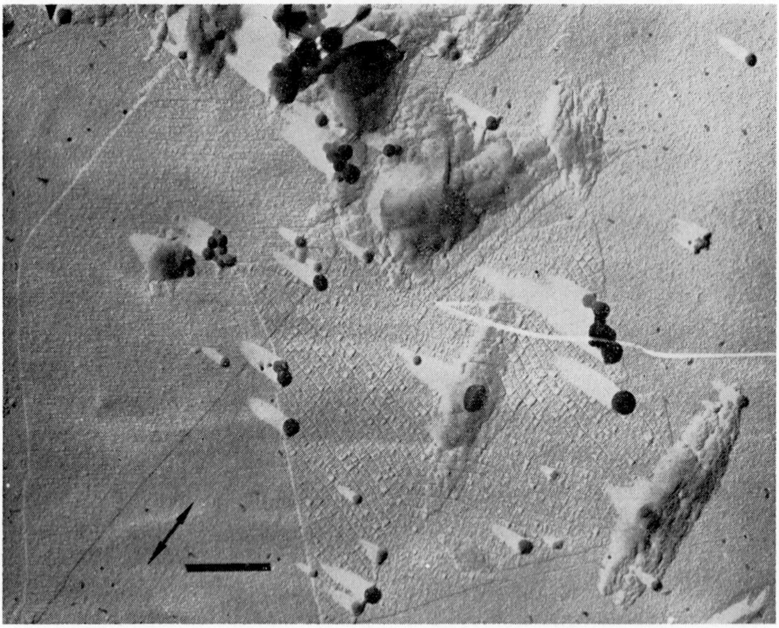

Fig. IVD-19. Nylon 6 crystals drawn 50% on Mylar. The bar's length represents 1 μ (13b).

portion of the crystal has transformed from an orthorhombic lattice to a monoclinic lattice (26,33a,33b). The corresponding changes in molecular packing are indicated in the diagrams. The twinning was such as to bring the **b** axis nearly parallel to the draw direction, in agreement with the previously described x-ray observations of the drawing process in bulk polyethylene. The crystal transformation, as suggested by Frank et al. (10), may be a martensitic type of transformation.

The deformed regions in crystals drawn on Viton differ considerably from those in crystals drawn on Mylar. It is believed that the crystals adhere to the Viton in only a few areas. When the substrate is drawn, cracks develop in the crystal, the polymer being drawn across the cracks (Figs. IVD-17 and IVD-18). Broad ribbons of drawn material are formed when the cracks in a given fold domain are not parallel to the growth face. However, if the crack is parallel (or nearly so) to the growth face, only small fibrils are formed. Thus, at a fold domain boundary there is an abrupt change in the appearance of the crack. Diffraction patterns and dark field micrographs have not yet been obtained from these crystals.

Nylon 6 and polyoxymethylene single crystals, when drawn on Mylar, develop cracks at elongations as low as 5% (13b). The cracks in the nylon 6 crystals form parallel to the hydrogen-bonded planes when the draw direction is within approximately 45° of the normal to these planes. At larger angles, the cracks cross the hydrogen-bonded planes, forming various angles with the growth faces (Fig. IVD-19). Plastic deformation, involving molecular slip and tilt, accompany the crack formation, particularly when the draw direction is within 45° of the normal to the hydrogen-bonded planes. Crack formation in polyoxymethylene crystals appears to occur preferentially either parallel to, or perpendicular to, the hexagonal unit cell faces. A limited amount of plastic deformation accompanies the crack formation. In both polymers there was no evidence of deformation twinning.

Although only a limited amount of research has been done, as of this writing, it is anticipated that the results obtained with single crystals will apply directly to the plastic deformation of the lamellae in bulk polymers. The electron diffraction results obtained when a drawn polyethylene crystal is annealed, i.e., **b** axis rotates so that it is normal to the draw direction (13c), as well as those obtained during the initial drawing, are in agreement with corresponding x-ray diffraction studies on bulk polyethylene. The initial results (13,13a, 13b) suggest that the plastic deformation takes place through a combination of twinning, crystal transformations, and slip. The latter process, when parallel to the molecular axis, may involve a dislocation motion of the type suggested by Reneker (29) to explain the molecular motion involved in recrystallization during annealing (Section IB).

IVD6. Deformation by Rolling

Rolling, often accompanied or preceded by drawing, has long been used to obtain three-dimensionally extended polymer samples to use in x-ray, unit cell determinations. The polymer is repeatedly passed between two rolls. In the cases discussed below, these rolls are turned by hand, are at room temperature, and the roll-gap is decreased slightly following each pass. In polyethylene, for instance, Bunn found (4) that in a highly rolled sample the **c** axes are parallel to the roll direction while the {110} planes are predominantly parallel to the surface. In a lightly rolled sample (only a few passes and therefore a small decrease in thickness), there is a tendency for the {100}

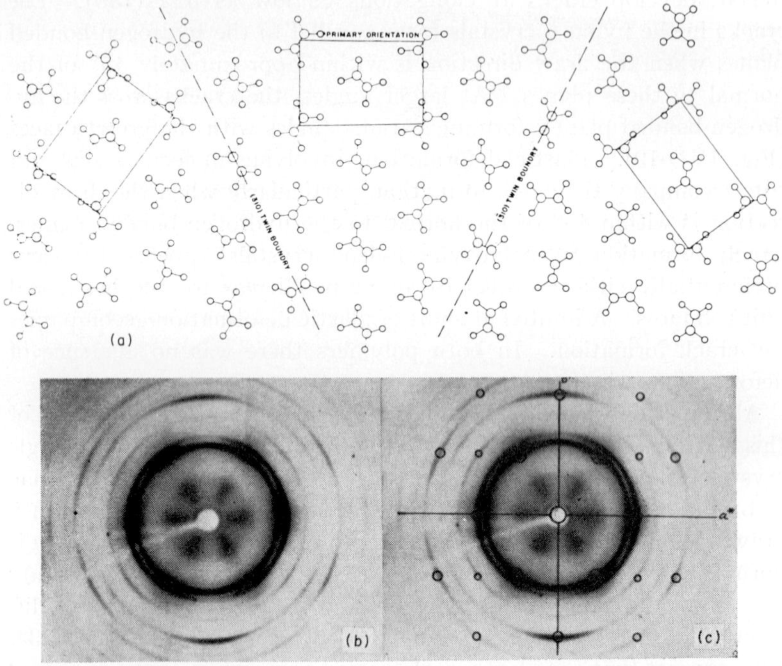

Fig. IVD-20. Twinning of the polyethylene unit cell. The composition planes, in this case, are {310} planes. A diffraction pattern from rolled linear polyethylene corresponding to this type of twinning is shown in (*b*). The reciprocal lattice due to the material with "primary orientation" is superimposed on the diffraction pattern in (*c*). The additional reflections result from rotation of this reciprocal lattice by ±53° (7,13).

planes to be parallel to the surface (10). In the nylons, the hydrogen-bonded planes become parallel to the surface during rolling (5,19).

Frank, Keller, and O'Conner have discussed (10) the expected deformation of polyethylene during rolling in terms of slip and twinning. Their wide-angle x-ray diffraction observations on samples which were previously oriented by drawing are in agreement with their expectations. They suggest that previously unoriented polyethylene will deform in a similar way when rolled, preferred orientation of the molecules taking place during the rolling.

On the basis of considerations derived from the deformation of simple solids (e.g., metals) and recognition of the long chain nature of polymers, they suggest that the most likely slip-mode will involve dislocations with a Burger's vector in the ⟨010⟩ direction, the dislocation involving extra {010} half planes of molecular segments. (See Figs. IVD-16 and IVD-20 for diagram of unit cell; the ⟨010⟩ direction is the shortest lattice translation.) The slip plane would be a {100} plane. The second most likely type of slip ⟨100⟩ {010} is not expected to play a significant role in the deformation process since (*a*) this slip system involves a considerably larger lattice displacement for formation of the dislocation and, (*b*) it is orthogonal to the ⟨010⟩ {100} slip system and therefore is subject to the same driving forces as those which can be relieved by the latter system. Frank et al. indicate (*hk*0) dislocations may also play a role in slip, dissociating into pairs of $(\frac{1}{2}h,\frac{1}{2}k,0)$ dislocations with {110} slip planes, but that the twinning modes discussed below will be in competition with it.

Frank et al. suggest that deformation twinning is most likely to occur with {110} and {310} composition planes; i.e., during the deformation the rearrangement of the molecular segments is such that the twin and the original lattice are mirror images of each other, these planes being the interfaces. The proposed {310} twins are shown in Figure IVD-20 while {110} twinning is depicted in Figure IVD-16. They indicate that both of these twins, with a slight empirical preference for {310}, should operate when there are compressive stresses near the ⟨010⟩ direction or tensile stresses near the ⟨100⟩ direction.

On the above basis, Frank et al. predict (10) the following behavior when drawn polyethylene is rolled. Those crystals with {100} and {010} planes close to 45° to the axis of compression will deform by ⟨010⟩ {100} slip, undergoing a rotation which brings {100} more

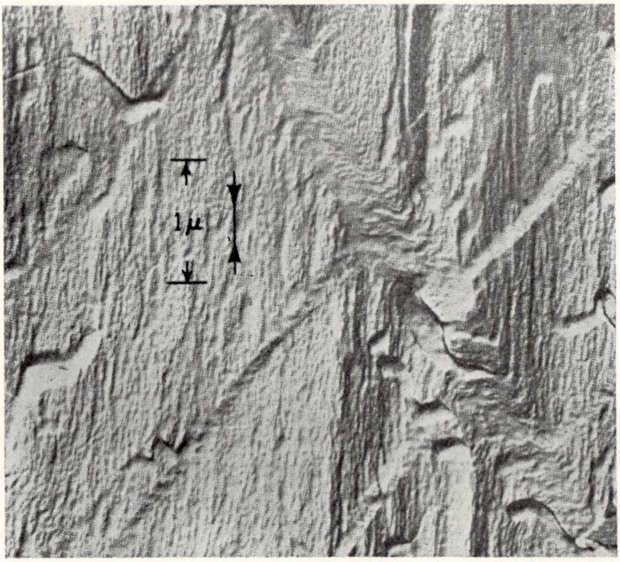

Fig. IVD-21. Electron micrograph of the cleavage surface (fractured at liquid nitrogen temperature) of a singly oriented nylon sample containing kink bands. The double arrow shows the compression direction and axis of the sample (35).

nearly parallel to the compression plane. Crystals with the {010} planes close to normal to the compression axis should twin, resulting in a moderate deformation (25% shear if {310} twin) accompanied by a large change in orientation of the unit cell. The new orientation of the unit cell is favorable for ⟨010⟩ {100} slip. Slip and twinning is expected to cease when the {100} planes are nearly parallel to the plane of compression.

They suggest (10) that with a large amount of deformation, residual stresses within the sample induce twinning following removal of the roller pressure such that the {100} planes rotate slightly and transform into {110} planes. The residual stress, closely associated with molecular segments in an amorphous phase and in the folded conformation, is assumed equivalent to a sideways compression and/or to localized tensile components parallel to the principal compression. The {110} planes, being the planes of closest packing, would preferentially be expected to lie parallel to the compression surface in highly compressed samples. (In Fig. IVD-20 the axis of compression is horizontal.)

Frank et al. were also able to describe observations on the effect of annealing and subsequent redrawing of rolled, drawn polyethylene as well as various aspects of the initial drawing process in terms of the proposed deformation mechanism. Although additional detailed observations are needed to confirm all aspects of the proposed mechanism, this paper is an excellent start toward understanding the deformation of crystalline polymers.

Zaukelies has subsequently discussed (35) the deformation of previously drawn, and drawn and rolled nylon in a similar manner, pointing out that the results can be explained most satisfactorily in terms of a defect-crystal model rather than a two-phase, crystallite-in-amorphous-matrix model. He finds that distinct kink bands can be observed both optically and in the electron microscope (Fig. IVD-21) when the samples are compressed. The formation of the bands involves slip in the molecular axis direction along $\{010\}$ planes. (These planes contain the hydrogen bonds.) Although microscope observations could be interpreted in terms of slip between, and bending of, adjacent 100 A. diameter fibrils, the fact that definite kink band angles, related to the crystallographic lattice, are observed indicates that the deformation is occurring within the fibrils. The angle between the molecular axis in the original sample and in the kink band, which varies with temperature, is found to be related to the number of adjacent molecular planes which slip as a unit when the kink band forms. Zaukelies indicates that at 25°C. two adjacent $\{010\}$ planes act as a unit, whereas at 100°C., three adjacent $\{010\}$ planes are involved.

Zaukelies suggests (35) that the slip occurs through the motion of (*1*) an edge dislocation, composed of a linear array of point dislocations related to those described by Reneker in polyethylene (29) except that the long nylon unit cell involves a considerably larger lattice distortion, and/or (*2*) several types of screw dislocations. He indicates that the formation and motion of these screw dislocations may also occur during the rolling process itself.

The tensile fracture of a notched, originally drawn and rolled nylon is found to depend on the angle between the notch and the $\{010\}$ planes. Zaukelies indicates (35) that if a drawn and rolled bar is notched so that the leading edge of the crack is parallel to the $\{010\}$ planes, considerable slip in the $\langle 010 \rangle$ direction along these planes takes place before the sample fails. However, if the notch is perpendicular to the $\{010\}$ planes, the resulting failure is brittle and no slip occurs. (In his paper Zaukelies uses the convention that the $\langle hkl \rangle$ direction is normal to the $\{hkl\}$ plane, whereas we use here the convention (17) that the $\langle hkl \rangle$ direction is the vector to the *hkl* real lattice point, thus $\langle 001 \rangle$ is parallel to the molecular axis.)

Electron microscope studies of the deformation by rolling of previously unoriented polymers have been reported only for polytetrafluoroethylene (13). These observations suggest that the lamellar, folded-chain nature of the crystals is retained up to quite high degrees of deformation (a thickness reduction of at least 6-fold) but that twinning, kinking, and slip take place on a molecular scale. Although the lamellae retain their identity during the deformation, the molecules within them become tilted with respect to the lamellae in a manner similar to that observed when polytetrafluoroethylene is drawn (Figs. IVD-10 and IVD-11).

IVD7. Summary

A number of wide- and small-angle x-ray diffraction, and optical and electron microscopical observations of the deformation of crystalline polymers during uniaxial "drawing" and "rolling" have been reported. Unfortunately, they have not been correlated among themselves or with the macroscopic features of the deformation process. The observations on both isolated single crystals and spherulitic polymers suggest that the deformation process during both drawing and rolling involves slip and twisting of lamellae at low strains, followed by slip and twinning on a molecular scale within the lamellae, unfolding of the molecules and the formation of fibrils and, finally, slip between the molecules in the fibrils at successively higher strains. Depending on the cohesion between the lamellae, rupture may occur at various stages in this process. Brittle fracture is found to occur only between lamellae which are not coherent, and between fold planes within the lamellae. Fracture of a crystalline polymer, even at low temperatures, generally involves inelastic deformation on at least a micro scale accompanied by adiabatic heating, even to the extent of localized melting.

References

1. Aggarwal, S. L., G. P. Tilley, and O. J. Sweeting, *J. Polymer Sci.*, **51**, 551 (1961).
2. Anderson, F. R., paper presented at American Physics Society meeting, St. Louis, March 1963.
2a. Belbeoch, B., and A. Guinier, *Makromol. Chem.*, **31**, 1 (1959).
3. Brown, A., *J. Applied Phys.*, **20**, 552 (1949).
4. Bunn, C. W., *Trans. Faraday Soc.*, **35**, 482 (1939).
5. Bunn, C. W., and E. V. Garner, *Proc. Roy. Soc.* (*London*), **189A**, 39 (1947).
6. Bunn, C. W., A. J. Cobbold, and R. P. Palmer, *J. Polymer Sci.*, **28**, 365 (1958).
7. Clark, E. S., personal communication.
8. Dulmage, W. J., and L. E. Contois, *J. Polymer Sci.*, **28**, 275 (1958).
9. Fischer, E. W., and G. Schmidt, *Angew. Chem.*, **74**, 551 (1962).
10. Frank, F. C., A. Keller, and A. O'Connor, *Phil. Mag.*, **3**, 64 (1958).
11. Geil, P. H., *J. Polymer Sci.*, **47**, 65 (1960).
12. Geil, P. H., and D. H. Reneker, *J. Polymer Sci.*, **51**, 569 (1961).
13. Geil, P. H., *Polymer Single Crystals*, Interscience (Wiley), New York, 1963.
13a. Geil, P. H., *J. Polymer Sci.*, **A2**, 3813 (1964).
13b. Geil, P. H., *J. Polymer Sci.*, **A2**, 3857 (1964).
13c. Geil, P. H., *J. Polymer Sci.*, **A2**, 3835 (1964).
14. Hammer, C. F., T. A. Koch, and J. F. Whitney, *J. Appl. Polymer Sci.*, **1**, 169 (1959).
15. Harris, P. H., and J. H. Magill, *J. Polymer Sci.*, **54**, S47 (1961).
16. Hendus, H., personal communication.
17. Henry, N. F. M., and K. Lonsdale, Eds., *International Table for X-Ray Crystallography, Vol. 1, Symmetry Groups*, Kynoch Press, Birmingham, 1952.
18. Hirai, N., H. Kiso, and T. Yasui, *J. Polymer Sci.*, **61**, S1 (1962).
19. Holmes, D. R., C. W. Bunn, and D. J. Smith, *J. Polymer Sci.*, **17**, 159 (1955).
20. Hosemann, R., *Polymer*, **3**, 349 (1962).
21. Inoue, M., *J. Polymer Sci.*, **55**, 443 (1961).
22. Keith, H. D., and F. J. Padden, Jr., *J. Polymer Sci.*, **41**, 525 (1959).
23. Keller, A., *J. Polymer Sci.*, **17**, 351 (1955).
24. Padden, F. J., Jr., and H. D. Keith, *J. Applied Phys.*, **30**, 1479 (1959).
25. Peck, V., and W. Kaye, *J. Appl. Phys.*, **25**, 1465 (1954).
26. Pierce, R. H. H., Jr., J. P. Tordella, and W. M. D. Bryant, *J. Am. Chem. Soc.*, **74**, 282 (1952).
27. Point, J. J., *Bull. Acad. Roy. Belg.* (*Classe Sci.*), **41**, 974, 982 (1955).
28. Porod, G., *Fort. Hochpolymer.-Forsch.*, **2**, 363 (1961).
29. Reneker, D. H., *J. Polymer Sci.*, **59**, S39 (1962).
30. Sakurada, I., Y. Nukushina, and T. Ito, *J. Polymer Sci.*, **57**, 651 (1962).
31. Sella, C., and J. J. Trillat, *Compt. Rend.*, **248**, 410 (1959).
32. Speerschneider, C. J., and C. H. Li, *J. Appl. Phys.*, **33**, 1871 (1962).
33. Stuart, H. A., *Ann. N. Y. Acad. Sci.*, **83**, 3 (1959).
33a. Tanaka, K., T. Seto, and T. Hara, *J. Phys. Soc. Japan*, **17**, 873 (1962).
33b. Turner-Jones, A., *J. Polymer Sci.*, **62**, 556 (1962).
34. Yu, Y. F., and R. Ullman, *J. Polymer Sci.*, **60**, 55 (1962).
35. Zaukelies, D. A., *J. Appl. Phys.*, **33**, 2797 (1962).

IVE. GENERAL ROLE OF CHEMICAL CHANGES

H. H. G. JELLINEK

IVE1. Introduction

Polymer molecules can be modified in a variety of ways by chemical and physical environments. The most profound changes in molecular structure, as far as the rheological properties of amorphous and partly amorphous polymeric bodies are concerned, are caused by chain scission and crosslinking. While these changes are usually quite insignificant from the chemical point of view, they give rise to major modifications of the macroscopic properties of the sample.

Chain scission is classified as a polymer degradation reaction. However, these degradation reactions also include changes of the polymer side groups, rupture of crosslinks, double-bond formation in the main chains, and reactions with a variety of chemical compounds (oxidation, ozonization, bromination, etc.).

Chain scission can follow a number of different mechanisms; indeed, there is a whole spectrum of possibilities. At one end of this spectrum, there is the purely random scission process where main chain links are broken statistically along the backbone of the polymer molecule. At the other end of the spectrum, there is the "depolymerization" reaction which, in its simplest form, is merely a reversal of a polymerization and consists of three processes: (*1*) initiation, (*2*) depropagation (i.e., a rapid production of monomer), and (*3*) termination. Between these two extremes there are chain scission reactions which represent transitions. Thus, if "chain transfer" reactions play a role, a random reaction may be superimposed on the pure depolymerization process. Of course, there also are cases where chain scission and "repolymerization" or chain scission and crosslinking occur simultaneously. Detailed treatments of such reactions are found in a number of monographs (7,12,14,22). Here, only some of the main points will be outlined.

Degradation and crosslinking can be brought about by heat; ultraviolet light; high energy radiations; ultrasonic waves and other mechanical treatments; electric discharges; and oxidation, ozonization and other chemical reactions. In this section, degradation and crosslinking reactions will be discussed generally, while the cracking of elongated rubber on exposure to ozone, and the effects of ultraviolet light and high energy radiations on polymers will be treated in more detail.

IVE2. Chain Scission

It may be noted that the "number-average chain length" at any stage of the main chain scission process, whatever the mode of breakdown and for any molecular size distribution, is always (provided the monomer itself is not broken down into smaller fragments) given by

$$P_t = \left(\sum^{i} N_i P_i / \sum N_i\right)_t \equiv P_0/(s + 1) \qquad \text{IVE2-(1)}$$

where P_t is the number-average chain length of the chain scission process at time t (including all polymer chains and free monomer), P_0 is the initial number-average chain length, and s is the average number of main chain links broken in each "original chain" of length P_0 at t (14).

The simplest case of a random degradation process is that for a polymer sample of uniform chain length, especially for small degrees of degradation, i.e., where the percentage of broken main chain links is very small. It is assumed that each main chain bond is of equal strength and accessibility, and that the rate of breaking links at any time in the degradation process is proportional to the number n of main chain links remaining in the polymer sample at time t,

$$-dn/dt = kn \qquad \text{IVE2-(2)}$$

where k is a rate constant. If the number of chain scissions is very small, n can be replaced by n_0, the initial number of main chain links. Hence,

$$-dn/dt = kn_0 \qquad \text{IVE2-(3)}$$

and

$$n_0 - n = kn_0t \qquad \text{IVE2-(4)}$$

Here $(n_0 - n)$ represents the number of broken main chain links in the system. Further, $n_0 - n = N_0s$, where N_0 is the original number of polymer chains in the system and s is the average number of scissions for each original chain. Since $s = (P_0/P_t) - 1$ and $n_0 = N_0P_0$ [more accurately, $n_0 = N_0(P_0 - 1)$] one obtains

$$N_0[(P_0/P_t) - 1] = kN_0P_0t$$

or

$$(1/P_t) - (1/P_0) = kt \qquad \text{IVE2-(5)}$$

Equation (5) is encountered in all cases where a random degradation reaction is operative and the percentage of chain scissions is small. The equation is also independent of the type of degradation (thermal, hydrolytic, oxidative, photo, high energy radiation, etc.), as long as chains are ruptured purely at random. Moreover, this equation is not very dependent on the initial size distributions as long as this distribution is sensibly random. Of course, the rate constant k is dependent on the particular polymer studied and the cause of degradation (heat, light, etc.).

For higher degrees of degradation $\alpha = s/P_0$, and a somewhat more accurate expression has to be used:

$$\ln[1 - (1/P_0)] - \ln[1 - (1/P_t)] = kt \qquad \text{IVE2-(6)}$$

The number distribution for a degree of degradation α for one base mole of polymer chains is given by,

$$\left.\begin{aligned} n_p &= (1/P_0)\,\alpha(1-\alpha)^{P-1}\,[(P_0-1)\alpha - P\alpha + 2] \\ n_{p_0} &= (1/P_0)(1-\alpha)^{P_0-1} \end{aligned}\right\} \qquad \text{IVE2-(7)}$$

Here n_p is the number of chains of length P per base mole of polymer.

If α is appreciable ($s > 5$), eq. (7) tends to

$$n_p = \alpha^2(1 - \alpha)^{P-1} \qquad \text{IVE2-(8)}$$

Equation (8) is often referred to as the random distribution. As α increases further, the type of distribution remains the same. Thus, the whole process could equally as well have been started with a sample initially having a random distribution of molecular sizes. Weight distributions are obtained by multiplying eq. (7) by P and P_0, respectively. These distributions then refer to a unit weight (1 g.) of polymer sample.

The "weight-average chain length" $P_w = \sum N_i P_i^2 / \sum N_i P_i$ is given by

$$P_w/P_0 = (2/s^2)(s - 1 + e^{-s}) \qquad \text{IVE2-(9)}$$

and the "viscosity-average chain length" P_v for $s > 6$ by

$$P_v/P_0 = (1/s^2)\ [(\gamma + 1)\ \Gamma(\gamma + 1)]^{1/\gamma} \qquad \text{IVE2-(10)}$$

where γ is the exponent in the intrinsic solution viscosity relationship $[\eta] = KP_v^{\gamma}$.

The ratio of the weight-average to number-average chain length is

$$P_w/P_n = [2(s + 1)/s^2](s - 1 + e^{-s}) \qquad \text{IVE2-(11)}$$

This ratio rapidly approaches 2 for $s > 5$. Thus a purely random degradation process is characterized by a very large decrease in chain length for small numbers of chain scissions and the effective absence of monomer or very small chains for moderate values of s.

Thus it can be seen that a very insignificant chemical reaction (very small amounts of oxygen, for instance), sufficient to break one, two, or three links on the average for each original chain, can profoundly affect the molecular size, hence the rheology, of the polymer. The percentage of the total main chain links ruptured is very small indeed. An example may illustrate this: $P_0 = 10^3$, $s = 1$, hence $\alpha = 10^{-3}$; 0.1% of the main chain links are ruptured, but even such small chemical reactions may make the polymer more inelastic. For one base mole of polymer or $1/P_0 = 10^{-3}$ moles of original chains of a polymer initially of uniform chain length, only 10^{-3} moles of O_2 are needed if one oxygen molecule is required for each scission. (If the molecular weight of the monomer is 100, there are 100 g. of polymer and 32 mg. of O_2 involved.)

At the other end of the degradation spectrum, the depolymerization process has characteristics quite in contrast with those of the random case. In its simplest form it consists, as already pointed out above, of the three following reactions.

(*1*) First there is initiation or activation of polymer chain-ends. A variation of this process would be initiation taking place at random along the chains.

(*2*) Depropagation consists of a zipping-off of monomers in rapid succession starting at the activated chain-ends, or in the case of random initiation, at the newly formed chain-ends. The "kinetic chain length" (i.e., the number of monomer units produced for each active chain-end) may be smaller, equal to, or larger than the initial polymer chain length.

(*3*) Termination can come about by mutual interaction of depropagating chains (second-order termination kinetics) or by rearrangement of activated chain-ends (first-order termination kinetics).

These three reactions can be exemplified generally for vinyl polymers as follows.

Initiation (random):

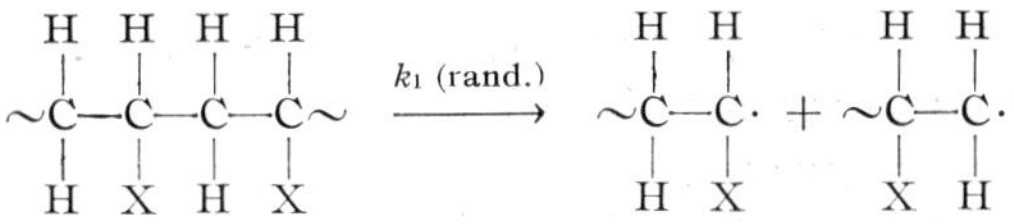

Initiation (chain-end):

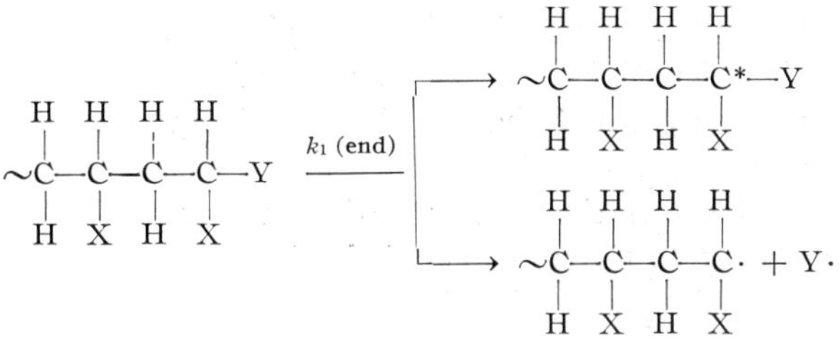

Depropagation:

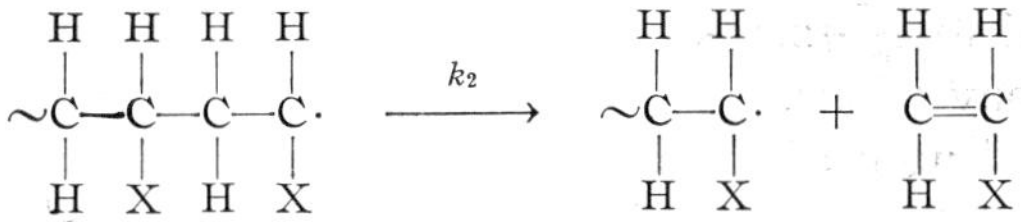

Termination:

$$2 \text{ radicals} \xrightarrow{k_3} \text{polymer; Second-order kinetics}$$

$$1 \text{ radical} \xrightarrow{k_3'} \text{polymer; First-order kinetics}$$

The simple case as exemplified above may be complicated by transfer reactions, which can be of the intermolecular or intramolecular type.

Transfer (intermolecular):

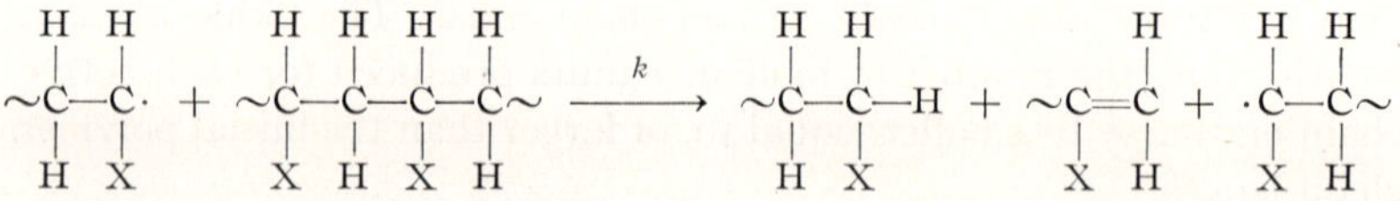

Transfer (intramolecular)

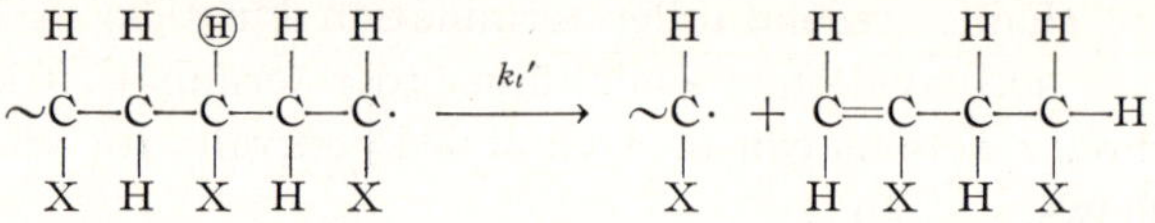

The simplest cases of degradation are illustrated in Figure IVE-1, where the yield of monomer and the number-average chain length, not including monomer, are plotted against percentage conversion of polymer into monomer (this presupposes a high vacuum system where the monomer is taken out continuously; actually, not only the monomer units are removed but polymer molecules up to the tetramer are also assumed to be removed for the examples given in Fig. IVE-1). The initial chain length in all cases, is assumed to be $P_0 = 1{,}000$; no transfer is considered, and the kinetic chain lengths ϵ are indicated on the curves.

Figure IVE-1 shows that the rate of monomer production is first order with respect to the polymer left for large kinetic chain lengths for random and chain-end initiation. The rate of monomer production is approximately zero-order over an appreciable part of the reaction with respect to residual polymers for small zips and random initiation and strictly zero-order for small zips and chain-end initiation up to high conversions.

The chain length of the residual polymer remains approximately constant over a large range of conversion for large kinetic chain lengths in either random or chain-end initiation. This is understandable,

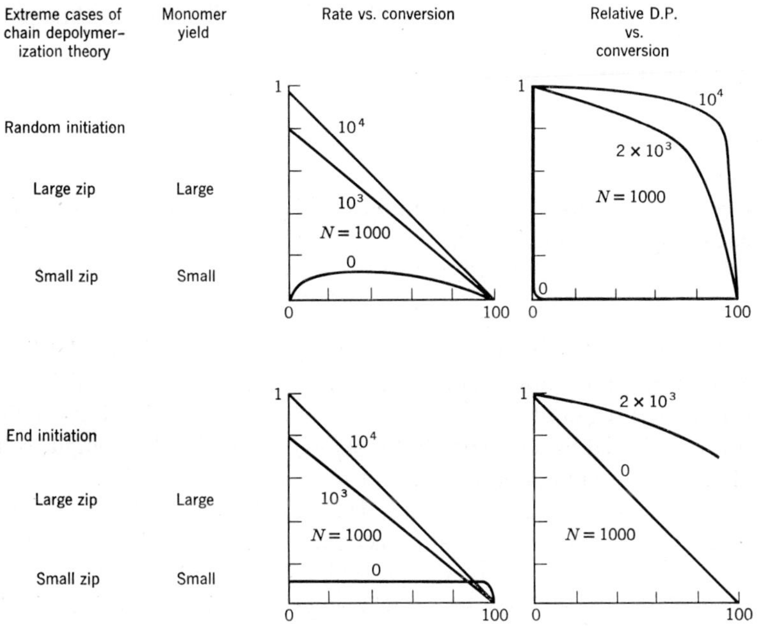

Fig. IVE-1. Schematic representation of extreme cases of degradation reactions; all curves refer to an initial chain length of $P_0 = N = 10^3$, no transfer reactions are taking place. The kinetic chain length ϵ is indicated above the curves. Rate scale for rate curves with $\epsilon = 0$ multiplied by a factor of 100 for random initiation and 50 for chain-end initiation. [Wall (22).]

particularly in the case of polymers of uniform chain length, because as soon as a chain is initiated it degrades completely to monomer, and the residual polymer consists only of chains of length P_0. A small zip, of course, approximates the purely random case, if random initiation takes place. In the case of chain-end initiation with small zip, the chains must progressively become smaller with percentage conversion. The limiting case would be stepwise production of monomer.

Thermal degradation in vacuum or in an inert atmosphere can take place by a random degradation process or by a reverse polymerization reaction. The mode of breakdown depends largely on the structure of the polymer. There is one general rule for vinyl polymers. Polymers which have tertiary hydrogen atoms usually tend towards a random breakdown, whereas polymers which have other groups in place of the tertiary hydrogen atoms tend to depolymerize (such as methyl).

Chain reactions can also take place at the side groups; thus poly(vinyl chloride) tends to split off HCl in rapid succession, forming double bonds in the main chains.

Oxidative degradation and ozonization tend to be random processes unless there are special sensitive groups in the chains not distributed at random in the polymer. However, unless these reactions proceed in polymer solutions, they are complicated by diffusion processes or formation of protecting layers.

IVE3. Crosslinking

Crosslinking can take place under the influence of ultraviolet light, high energy radiation, oxidation, etc. Those polymers which have tertiary hydrogen atoms tend to crosslink. This crosslinking reaction can be followed experimentally by changes in molecular weight, solubility, degree of bulk swelling, and mechanical properties.

There are a number of concepts which are important in this connection. The crosslinking density q is defined as the fraction of the total amount of monomer in the sample which is crosslinked. The number of crosslinked monomers in the whole system is then qm_0, if m_0 is the number of monomers in the polymer sample. The number of crosslinks is $qm_0/2$. Crosslinking usually takes place at random.

A "gel" begins to form in the reaction vessel, at least from a theoretical point of view, when an average of one monomer in each original chain is crosslinked; this is true for a polymer sample of uniform chain length. If the polymer has a molecular size distribution, the formation of a gel (or the "gel point") is reached when there is one crosslinked monomer for each weight-average chain length. At this point an insoluble fraction, the "gel fraction," appears. If crosslinking increases, this gel fraction increases while the sol fraction (the soluble part of the polymer) decreases.

The above statements can be expressed more concisely. The condition for the gel point is given by the probability $p_{c(\mathrm{crit.})}$ that one monomer is crosslinked for each original chain. For a polymer of initially uniform chain length, this implies

$$p_{c(\mathrm{crit.})} = 1/(P_0 - 1) \cong 1/P_0 \qquad \text{IVE3-(12)}$$

For a sample having any molecular size distribution, the condition for the gel point is given by

$$p_{c(\text{crit.})} = 1/(P_w - 1) \cong 1/P_w \qquad \text{IVE3-(13)}$$

where P_w refers to the original sample.

The crosslinking coefficient ϑ may be defined as

$$p_c P_w = \vartheta \qquad \text{IVE3-(14)}$$

At the gel point $p_c = p_{c(\text{crit.})}$ and the crosslinking coefficient becomes unity:

$$p_{c(\text{crit.})} P_w = \vartheta_{(\text{crit.})} = 1 \qquad \text{IVE3-(15)}$$

Another quantity of importance in this connection is the crosslinking index γ, which is defined by

$$p_c P_0 = \gamma \qquad \text{IVE3-(16)}$$

For a sample of uniform chain length, $P_0 = P_w$, hence $\vartheta = \gamma$. For a random distribution of chain lengths $P_w = 2P_0$; hence $\vartheta = 2\gamma$. It can be shown that the broader the molecular size distribution, the fewer crosslinks for each average polymer molecule are needed to reach the gel point.

Changes in average molecular weights can be calculated as functions of the crosslinking index and coefficient. Similarly, the sol (i.e., soluble) or gel (insoluble, i.e., crosslinked) fractions can be expressed as functions of these parameters. Theories have been extended to include simultaneous chain scission and crosslinking reactions.

The number-average molecular weight between crosslinks M_c, is of importance for highly crosslinked samples in the configurational elastic properties of rubberlike polymers. M_c is given by

$$M_c = m_w/q \qquad \text{IVE3-(17)}$$

where m_w is the molecular weight of the monomer.

IVE4. Ultraviolet Irradiation

Degradation by ultraviolet radiation (15) will be discussed only briefly. This is a highly specific process which requires absorption of photons of sufficient energy by particular groups or bonds present in the polymer molecules. As such groups or bonds are usually distributed at random over the chain molecules, the photodegradation process is usually a random process. Moreover, in some cases a few monomer units may be zipped-off from the newly formed chain ends.

Such a chain scission reaction, as already pointed out above, profoundly influences the mechanical properties of the polymer.

Polymer solutions may be considered first. In this case we have an equation equivalent to eq. (5):

$$(1/P_t) - (1/P_0) = kt = \phi_s I'_{abs.} t/n_0 \qquad \text{IVE4-(18)}$$

where ϕ_s = no. of links broken/no. of photons absorbed, = the quantum yield for chain scission, $I'_{abs.}$ is the average number of quanta of a particular wavelength absorbed for each cubic centimeter of the solution (the solvent would be transparent here), t is the time of irradiation and n_0 the initial concentration of main chain links in the polymer solution. $I'_{abs.}t$ is simply the average irradiation dose per cubic centimeter of solution.

If Beer-Lambert's Law is obeyed, eq. (18) can be written (for a cylindrical reaction vessel):

$$(1/P_t) - (1/P_0) = k_1(1 - e^{-k_2 n_0 l}) I_0 t/n_0 \qquad \text{IVE4-(19)}$$

where k_1 is a rate constant, k_2 is an optical constant, l is the length of the reaction vessel, and I_0 is the incident radiation intensity. If light absorption is small, eq. (19) reduces to

$$(1/P_t) - (1/P_0) = 2.3\phi_s E I_0 t/n_0 l \qquad \text{IVE4-(20)}$$

For this case, the degradation should be independent of polymer concentration, since the optical density E is directly proportional to n_0. Such photodegradation reactions may, in certain cases, be complicated by repolymerization and crosslinking.

The treatment of polymer films is more complicated. Ultraviolet radiation usually cannot penetrate deeply into such films. Therefore, maximum degradation takes place at the irradiated surface, decreasing continuously with increasing depth; exponentially, if Beer's Law is obeyed. For the overall reaction with a sample of initially uniform chain length, one has

$$(1/P_t) - (1/P_0) = (kI_0 t/m_1 x_1)(1 - e^{-ax_1}) \qquad \text{IVE4-(21)}$$

where m_1 is the number of monomers per cubic centimeter of the film, x_1 is the thickness of the film, and a is an optical constant. For small absorption this reduces to

$$(1/P_t) - (1/P_0) = kI_0 ta/m_1 \qquad \text{IVE4-(22)}$$

The molecular size distributions obtained, however, depend on the distance from the irradiated surface. The degradation of a polymer of uniform chain length, for small distances from the irradiated surface, for example, may be sufficiently advanced for the distributions to have attained the random type. At larger distances, however, the random distribution may not yet have been reached ($s < 5$), while for still larger distances no degradation may have taken place at all since the radiation may not be able to penetrate that deeply. At any one distance x, the degree of degradation α_x is given by

$$\alpha_x = s_x/P_0 = \alpha_0 e^{-ax} \qquad \text{IVE4-(23)}$$

and at the surface it is at its maximum

$$\alpha_0 = kI_0ta/m_1 \qquad \text{IVE4-(24)}$$

The weight-average and viscosity-average chain length are, of course, different from those which would have been obtained if the distribution at various stages of the degradation were the same throughout the whole sample. Equations have also been elaborated for cases of polymers initially having a random distribution (15). Polymer molecules having tertiary hydrogen atoms tend to crosslink.

IVE5. High Energy Irradiation

High energy radiation (2,8,9) is concerned with radiation of photons, α-particles, neutrons, and electrons, all having energies in the neighborhood of 10^5 e.v./photon or particle. This energy is far above that needed for breaking bonds or producing ions in polymer molecules. The energy sufficient for such processes is about 10 e.v. Hence, high energy radiation leads to a random excitation or removal of electrons and consequently to random breaking of bonds. In other words, whereas the action of ultraviolet radiation is dependent on specific groups or bonds in the molecule, high energy radiation is, to a large extent, nonspecific. If the energies are of magnitude $\cong 10^6$ e.v., then nuclei are affected and radioactivity can be induced. However, high energy radiation of polymers is primarily concerned with orbital electrons and subsequent reactions of the ions and radicals produced. Ionization does not usually take place on ultraviolet irradiation, hence radical processes are mainly involved and comparison of the reactions induced by ultraviolet and high energy irradiation may give clues as to whether a particular high energy irradiation reaction proceeds *via* ions or radicals.

A useful concept in this connection, one which does not presuppose a knowledge of mechanisms, is the "G value." This G value is defined as the number of individual acts of one kind taking place for each 100 e.v., absorbed by the irradiated system. Thus G(crosslinking) is the number of crosslinks formed per 100 e.v., absorbed by the irradiated system, G(main chain scissions) is the number of scissions produced per 100 e.v., and $G(CO_2)$ is the number of CO_2 molecules evolved for each 100 e.v. absorbed, etc. Thus one system can simultaneously have a multitude of G values.

If chain scission only takes place and the rate of breaking links is a linear function of radiation dose, an equation equivalent to eq. (5) is obtained,

$$(1/\bar{M}_n) - (1/\bar{M}_{n_0}) = r/E_d N \qquad \text{IVE5-(25)}$$

where $\bar{M}_n$ is the number-average molecular weight when a radiation dose r has been received and $\bar{M}_{n_0}$ the initial value. E_d is the average energy absorbed by the system for each main scission, and N is Avogadro's Number.

If random crosslinking only takes place and the rate of crosslinking is a linear function of the radiation dose, one has,

$$q = q_0\, r \qquad \text{IVE5-(26)}$$

q_0 is the crosslinking density for a unit radiation dose and is characteristic of the irradiated system. Extent of crosslinking can also be expressed in terms of G values,

$$G(\text{crosslinks}) = 0.48 \times 10^6\, q_0/m_w \qquad \text{IVE5-(27)}$$

Here the radiation dose is expressed in megarads. The number-average molecular weight between crosslinks is given by,

$$M_c = m_w/q = m_w/q_0 r = 0.48 \times 10^6/rG(\text{crosslinks}) \qquad \text{IVE5-(28)}$$

The crosslinking density at the gel point is

$$q = q_0\, r_{\text{gel}}$$

where r_{gel} is the radiation dose needed to reach the gel point. Since here, $qP_w = 1$, we also have

$$q_0\, P_w\, r_{\text{gel}} = 1$$

or

$$r_{\mathrm{gel}} = 1/q_0 P_w = m_w/q_0 \bar{M}_w \qquad \text{IVE5-(29)}$$

The term q_0 can be expressed in terms of G(crosslinks); hence

$$r_{\mathrm{gel}} = 0.48 \times 10^6/G m_w P_w = 0.48 \times 10^6/G\bar{M}_w \qquad \text{IVE5-(30)}$$

r_{gel} is expressed in megarads. Determination of $\bar{M}_w$, the weight-average molecular weight, and r_{gel} gives G(crosslinks) directly, or if G(crosslinks) and r_{gel} are known, $\bar{M}_w$ can be evaluated. Usually, $r_{\mathrm{gel}} \cdot \bar{M}_w \cong$ constant; hence crosslinking is sensibly independent of molecular weight.

If chain scission and crosslinking take place simultaneously, the situation becomes more complicated and reference is necessarily made to the literature (2,8,9). Polymers with tertiary hydrogen atoms tend to crosslink. Side groups are also attacked by high energy irradiation, and gases are often produced. Eventually, the accumulation of such gases in the polymer may give rise to the development of cracks in the sample.

IVE6. Oxidative Degradation

A brief outline will be given here of possible oxidation mechanisms of unsaturated polymers and polymers having tertiary hydrogen atoms.

Initially, a radical may be formed. In principle, this can be achieved by ultraviolet or high energy irradiation, catalysts, or direct oxidation:

$$\mathrm{R{-}CH{=}CH{-}CH_2{-}R' + O_2 \rightarrow R{-}CH{=}CH{-}\dot{C}H{-}R' + HO_2\cdot}$$

The radical can then react further with oxygen:

$$\mathrm{R{-}CH{=}CH{-}\dot{C}H{-}R' + O_2 \rightarrow R{-}CH{=}CH{-}CH(O_2\cdot){-}R'}$$

Reaction of this peroxide radical with another polymer molecule may lead to hydroperoxide formation and production of a polymer radical, which can propagate a chain reaction:

$$\mathrm{R{-}CH{=}CH{-}CH(O_2\cdot){-}R' + R{-}CH{=}CH{-}CH_2{-}R' \rightarrow}$$

$$\mathrm{R{-}CH{=}CH{-}CH(OOH){-}R' + R{-}CH{=}CH{-}\dot{C}H{-}R'}$$

The polymer molecules containing hydroperoxide at random along their chains may occasionally undergo chain scission reactions due to hydroperoxide decomposition, hence a random chain scission ensues. A possible chain scission reaction here is:

$$\text{R—CH=CH—}\underset{\text{OOH}}{\underset{|}{\text{CH}}}\text{—R}' \rightarrow \text{R—CH=CH—}\underset{\text{O}\cdot}{\underset{|}{\text{CH}}}\text{—R}' + \cdot\text{OH} \rightarrow$$
$$\text{R—CH=CH—CH=O} + \text{R}'\cdot$$

Crosslinking could occur between radicals either forming oxygen bridges or leading directly to carbon–carbon crosslinks. These reactions constitute termination reactions of the oxidative chain reactions. Of course, there are a number of other possible reactions which could take place (7,12,14,22). Termination may also be brought about by

$$2\text{ROO}\cdot \rightarrow \text{stable product} \xrightarrow{\text{e.g.}} \text{O}_2 + \underset{\text{O}}{\underset{\|}{\text{RCR}}} + \underset{\text{OH}}{\underset{|}{\text{RCHR}}}$$

"Saturated" polymers, such as polyethylene and polypropylene, usually have a number of double bonds which make them susceptible to oxidation.

IVE7. Ozonization

Unsaturated hydrocarbons are very sensitive to ozone, even to the very small concentrations encountered in the atmosphere. This atmospheric concentration varies from zero to a few parts in 10^7 parts of air. Ozonization is a very complex process. The initial reaction between ozone and an olefin may be formulated as follows:

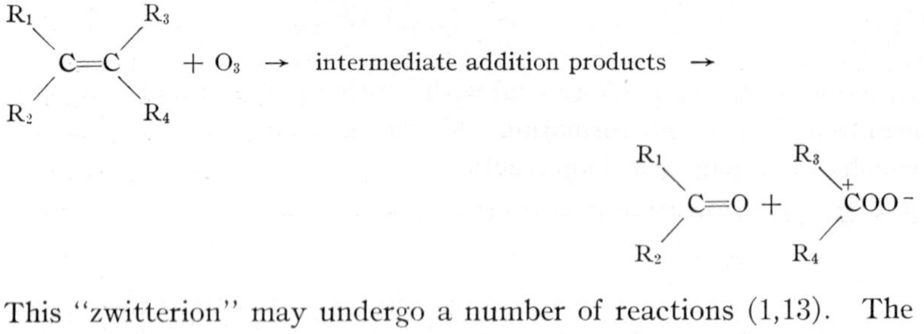

This "zwitterion" may undergo a number of reactions (1,13). The intermediate addition products are formulated in a variety of ways:

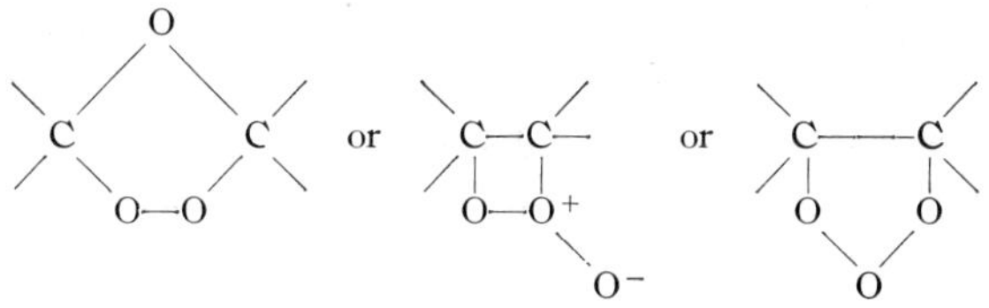

The effect of ozone on elongated polymers when leading to cracking will subsequently be discussed in more detail.

IVE8. Chemorheology

Degradation or crosslinking of polymers can be followed experimentally by various means. Thus degradation of long chain molecules can be quantitatively observed by changes in molecular weight,

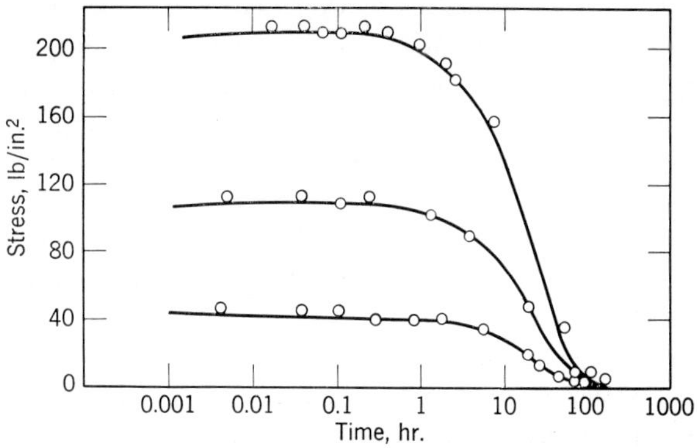

Fig. IVE-2. Continuous stress relaxation in a hevea gum vulcanizate at 100°. [Grassie (12).]

changes in bulk weight in vacuum, absorption spectra of several kinds, electron spin, uptake of oxygen, mechanical properties, etc. Similarly, crosslinking can be detected by changes in molecular weight, extent of swelling, mechanical properties, solubility measurements, spectra, etc. A very sensitive means of studying chain scission and the further crosslinking of crosslinked rubbers, which are three-dimensional networks, is measuring stress relaxation, creep, and permanent set. This field, developed by Tobolsky, is referred to as chemorheology (21). The changes in these network polymers can be brought about by heat and oxidation, vulcanization, and ultraviolet and high energy irradiation. In rubbers, chain scission and crosslinking do not appear to be strongly dependent on elongation at sufficiently low extensions.

A crosslinked rubber which undergoes only chain scission will show a chemically induced, continuous stress relaxation when held at a constant elongation. A typical example is shown in Figure IVE-2. Nominal stress is plotted as a function of the logarithm of time. In this particular case, and in a number of others, this can be represented by an exponential function,

$$S(t)/S(0) = e^{-k't} \qquad \text{IVE8-(31)}$$

where k' is a rate constant which depends on temperature and follows the Arrhenius equation permitting the evaluation of an apparent energy of activation E_a

$$k' = Ae^{-E_a/RT} \qquad \text{IVE8-(32)}$$

Thus chain scission can be measured in this way giving valuable information about the chemical process. The rate of chain scission is independent of elongation of the network up to fairly high elongations. Samples must be thin so that diffusion of oxygen, for instance, does not become rate-controlling.

However, crosslinking can take place at the same time or also on its own. Small amounts of crosslinking reactions do not interfere with this continuous stress relaxation. These crosslinks are built into the network under the prevailing conditions of elongation in a relaxed fashion. However, such crosslinking reactions can be detected by the so-called intermittent stress relaxation method. This relaxation will be affected by scission and crosslinking and is carried out as follows: The reaction is allowed to proceed at zero elongation. At various intervals very rapid extensions are carried out up to the same elongation as that used for the continuous stress relaxation. This intermittent stress relaxation is concluded during time intervals very small compared with the time of the total reaction. If crosslinking is taking place at zero elongation, then a plot of $S(t)/S(0)$ versus log t will differ from the corresponding curve obtained for continuous stress relaxation. This is shown in Figure IVE-3. If the stress increases for intermittent stress relaxation, a net crosslinking has taken place. Figure IVE-3 shows that chain scission and crosslinking take place on oxidation in the case of GR-S, whereas chain scission is predominant in the case of butyl rubber.

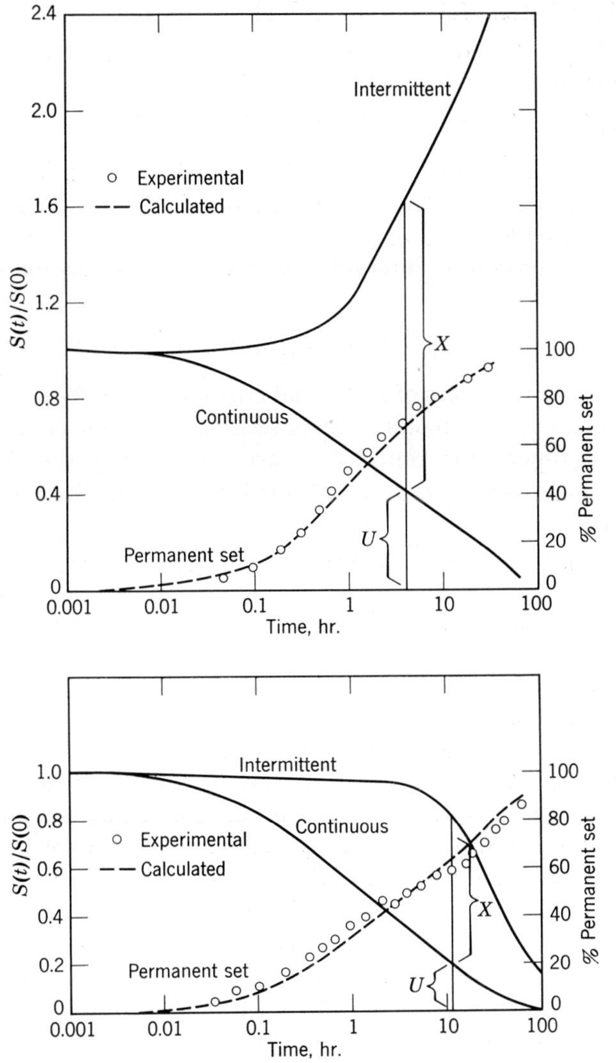

Fig. IVE-3. Continuous and intermittent stress relaxation and permanent set at 130° and 50% elongation in a GR-S vulcanizate (upper figure) and a butyl gum vulcanizate (lower figure). [Grassie (12).]

The initial stress for continuous and intermittent stress relaxation can be shown to be given by the following relationship [cf. eq. IC3-(23), p. 65],

$$S(0) = 2c(0)RT\,[(L_0/L_u) - (L_u/L_0)^2] \qquad \text{IVE8-(33)}$$

where $c(0)$ is the initial number of moles of crosslinks/cubic centimeter, the unextended length of the sample is L_u and the uniaxially extended, fixed length is L_0.

For stress relaxation intermittent at time t, one has for the corresponding stress:

$$S_i(t) = [2c(0) + 2c(t) - q(t)]RT[(L_0/L_u) - (L_u/L_0)^2] \quad \text{IVE8-(34)}$$

where $c(t)$ are the moles of additional crosslinks per cubic centimeter, at time t and $q(t)$ the moles of chain scissions per cubic centimeter at the same time. For continuous stress relaxation this becomes, (crosslinking reaction is not measured by this method unless excessive),

$$f(t) = [2c(0) - q(t)]RT[(L_0/L_u) - (L_u/L_0)^2] \quad \text{IVE8-(35)}$$

If no crosslinking takes place, $S_i(t) = S(t)$; if crosslinking occurs, $S_i(t) > S(t)$.

Many rubber samples which have been held at constant elongation under conditions where chemorheological reactions can take place, exhibit an irreversible deformation or permanent set. This occurs since the original network structure has been altered by chain scission and that a new network structure has been formed by crosslinking in a relaxed condition in the elongated state. The permanent set will reflect the relative importance of chain scission and crosslinking which have taken place. Figure IVE-3 also shows this permanent set for GR-S and butyl gum vulcanizate. U is the fraction of original network chains (chains between crosslinks) at a particular time of the process and X is the relative amount of new network chains. U/X is connected with the percentage of permanent set by

$$\begin{aligned}\%\text{ permanent set} &= 100(L_s - L_u)/(L_x - L_u)\\ &= \left\{\left[\frac{c_1}{(U/X)c_2 + 1} + 1\right]^{1/3} - 1\right\} c_3 \quad \text{IVE8-(36)}\end{aligned}$$

where $c_1 = (L_x/L_u)^3 - 1$, $c_2 = (L_x/L_u)^2$, and $c_3 = 100/[(L_x/L_u) - 1]$, where L_x is the elongated length and L_s the "set" length of the rubber. The permanent set in Figure IVE-3 has been calculated by means of eq. (36).

Rubber samples subjected to constant load show chemically induced creep under suitable conditions. If only scission takes place, the sample will show an immediate elastic elongation and will then continue to elongate indefinitely. The chemical creep function CCF is defined by:

$$\text{CCF} = \frac{[L(0)/L_u] - [L_u/L(0)]^2}{[L(t)/L_u] - [L_u/L(t)]^2} \qquad \text{IVE8-(37)}$$

In absence of crosslinking CCF $= n(t)/n(0)$, where $n(0)$ and $n(t)$ are the moles of network chains per cubic centimeter at $t = 0$ and t, respectively. $L(0)$ is the length of the specimen immediately after imposing the load.

IVE9. Exposure Cracking by Ozone

Atmospheric exposure cracking of most rubberlike polymers is to a large extent due to ozone. Most authors are agreed that this type of fracture takes place when an unsaturated polymer is exposed to ozone and is at the same time kept under tensile stress. While a minimal stress is necessary for the cracking to take place, this does not, however, mean that degradation (chain scission) or crosslinking does not always occur, but only that these reactions do not in themselves lead to crack initiation and crack propagation. It is not the intention of reviewing here all literature pertaining to crack formation and crack growth due to the influence of ozone, only some of the more recent, significant papers will be outlined.

Newton (18) gives quite a detailed review of the literature before 1945. He emphasizes that three distinct phenomena have to be considered with respect to exposure cracking: (*a*) rate of ozone attack, (*b*) rate of crack formation, and (*c*) rate of crack growth. He investigated the relationship between the "degree of cracking" (number of cracks/unit area) and percentage of strain. An example is shown in Figure IVE-4. Newton suggested that ozone rapidly attacks the surface of strained and unstrained rubber forming a thin layer of an ozonide, which reduces further adsorption of ozone. In the stretched rubber, however, cracks are formed, exposing fresh surface to the ozone.

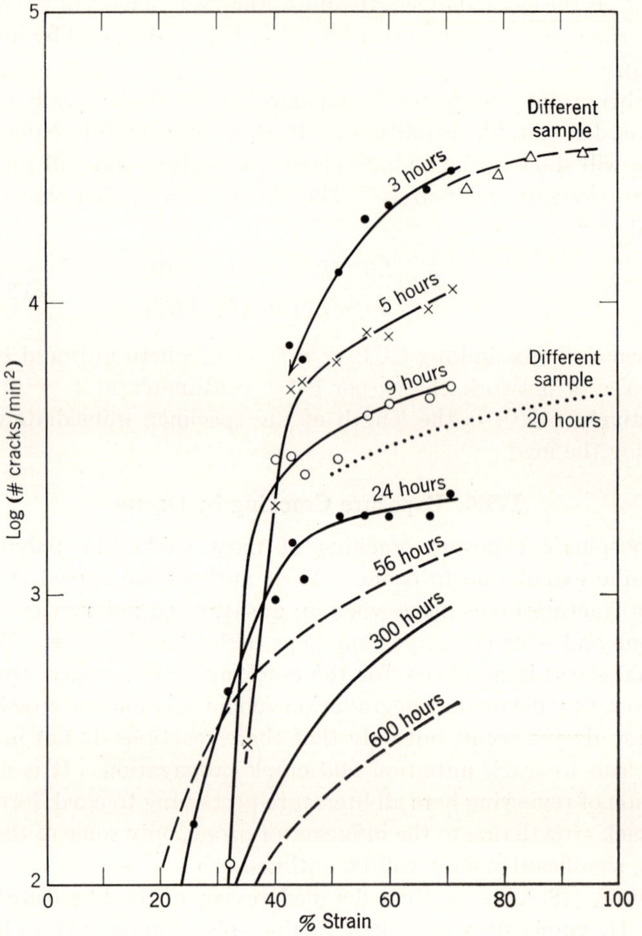

Fig. IVE-4. Logarithm of number of cracks/mm.² as a function of tensile strain after various times of exposure for rubber samples. [Newton (18).]

An interesting study of ozone cracking of rubber was carried out by Smith and Gough (20). The formation of cracks and their growth was followed with a cine-camera. The samples were kept under tension during ozone exposure. Figure IVE-5 shows the number of cracks plotted against exposure time at 15% strain. If the total number of cracks T is plotted as a function of time (some cracks heal), a curve as shown in Figure IVE-6 results. The rising part of the curve corresponds to the formation of cracks during the first 50 or 60 min. The number of cracks which were developed during exposure are

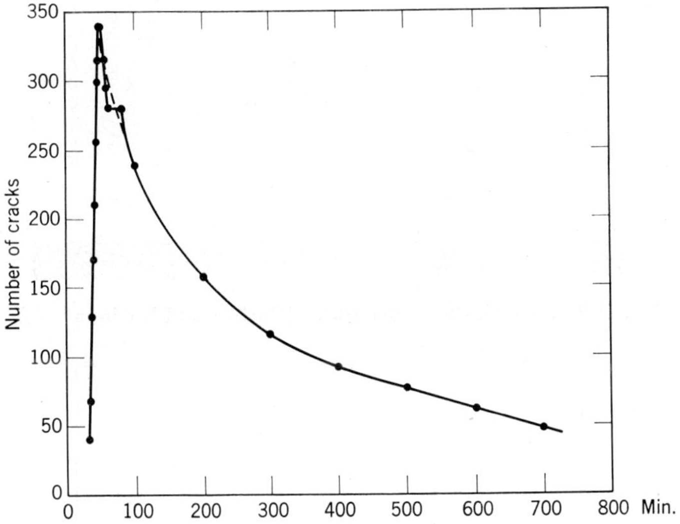

Fig. IVE-5. Number of "visible" cracks as a function of exposure time at 15% elongation. [Smith and Gough (20).]

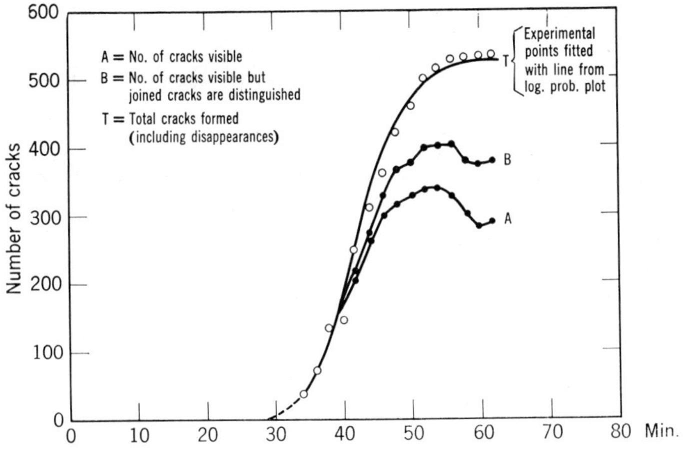

Fig. IVE-6. Number of cracks as a function of time, 15% strain. [Smith and Gough (20).]

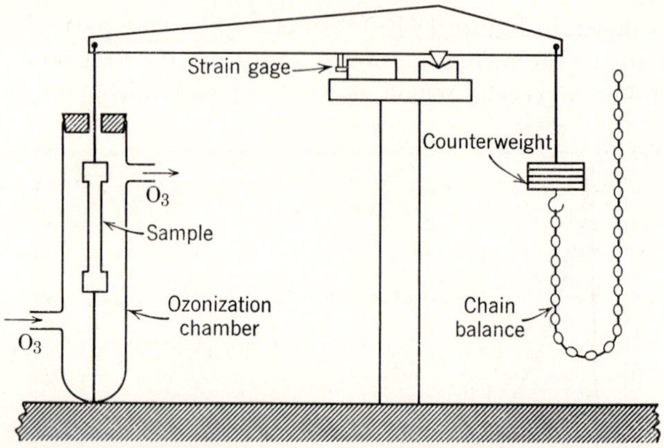

Fig. IVE-7. Ozonization apparatus. [Buckley and Robinson (6).]

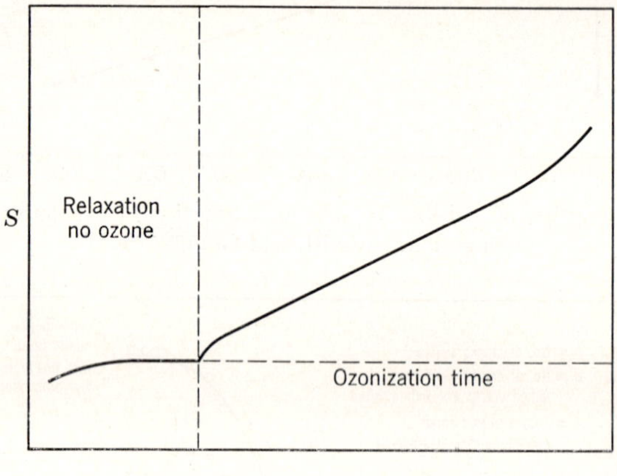

Fig. IVE-8. Typical relaxation and ozonization curve (S is the length of the sample). [Buckley and Robinson (6).]

about 10^{10} times smaller than the number of double bonds available in the surface of the rubber. These authors present a tentative mechanism endeavoring to explain the relationship between the number of double bonds and the number of crack nuclei as well as the role played by stress.

Buckley and Robinson (6) investigated the effect of ozone on butyl rubber. The rubber was stretched and was allowed to relax; subsequently the creep of the sample was observed as ozonization proceeded. Their apparatus is shown in Figure IVE-7. The samples were exposed at varying stress levels and ozone concentrations at 25°C. Length S of a typical sample as a function of ozonization time is shown in Figure IVE-8.

On exposure to ozone, the length of the sample changed immediately without an induction period. Three parts could be distinguished in these length versus time curves: (*1*) a rapid, short-termed increase in length, (*2*) a linear increase, and (*3*) an accelerated increase.

The tensile strain energy or equilibrium work for the sample before ozone exposure was represented by:

$$W = kNTA(L^3 - 3L + 2)/2L \qquad \text{IVE9-(38)}$$

where

$$N = Z_1/\{kT[L - (1/L^2)]\}$$

Here W is the equilibrium work, k is Boltzmann's constant, T is the absolute temperature, N is the number of molecular chains supporting the stress, A is the cross-sectional area of the sample, Z_1 is the measured stress, and $L = l/l_0$, l being the extended equilibrium length and

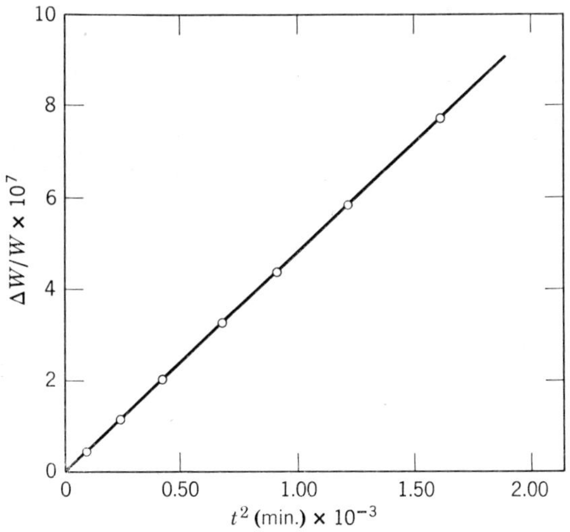

Fig. IVE-9. Change in work as a function of the square of exposure time. [Buckley and Robinson (6).]

l_0 the original length. The change in work ΔW on exposure to ozone was assumed to be given by:

$$\Delta W = \int_0^S k'S\,dS = k'S^2/2 \qquad \text{IVE9-(39)}$$

where the equilibrium length before exposure is taken as $S = 0$. Hence,

$$\Delta W/W = k'S^2L[L - (1/L^2)]/\ [Z_1(L^3 - 3L + 2)] \qquad \text{IVE9-(40)}$$

Since the second phase of stress relaxation (see Fig. IVE-8) as a function of ozone exposure time is linear ($dS/dt = K$, or $S = Kt$, strictly $S = Kt + c$, however the constant c is small), one obtains,

$$\Delta W/W = k'K^2t^2L[L - (1/L^2)]/[Z_1(L^3 - 3L + 2)]$$

or

$$\Delta W/W = k''t^2 \qquad \text{IVE9-(41)}$$

As can be seen from Figure IVE-9, eq. (41) was obeyed.

The effect of vulcanization on the reaction rate constant K of strained samples is given in Figure IVE-10. The mole percentages of

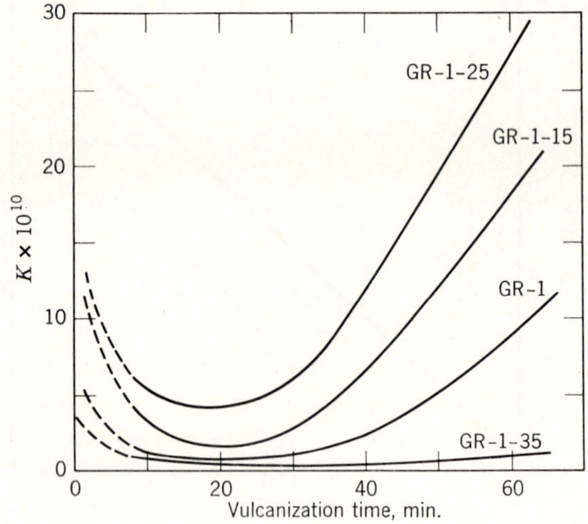

Fig. IVE-10. Cracking rate constant as a function of vulcanization time and polymer type. [Buckley and Robinson (6).]

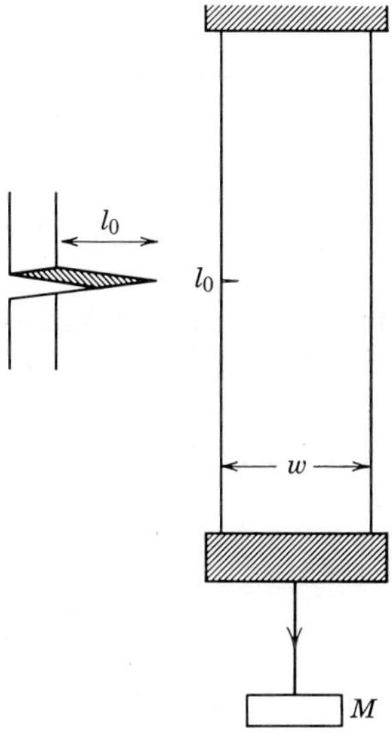

Fig. IVE-11. Testing arrangement for test pieces having an incision of length l_0 in the center of one edge. [Braden and Gent (3).]

unsaturation for GR-I-35, GR-I, GR-I-15, and GR-I-25 were 0.60, 0.89, 0.97, and 1.17, respectively. The vulcanization was carried out at 165°C. The minimum in the curves is connected with the "reversion" of the network.

The authors claim to have shown that cracking of vulcanizates in an ozone atmosphere takes place when energy is available for the formation of new surfaces after chain scission has taken place. Ozone attacks these polymers also in absence of strain, but only chain scission occurs; the rate of degradation being dependent on the gas diffusion in the respective polymer.

Braden and Gent (3,4) have studied quite extensively the attack of ozone on rubber vulcanizates. (The previous authors had investigated crack initiation and propagation over the whole surface of the sample.) However, Braden and Gent made a study of the rate of growth of a single incision in a thin sheet of rubber. This gives a much better control over the experimental conditions. Further, the basic process of crack growth without interference from surface irregularities and foreign material can be investigated.

The experimental arrangement is shown in Figure IVE-11. The whole assembly was placed in a chamber which allowed measurements to be carried out over a range of temperatures and ozone concentrations.

Several points were found to be of special importance with respect to the growth of a cut: (*1*) a well-defined value of tensile load has to be exceeded before propagation takes place; (*2*) the cut grows linearly across the rubber sheet; (*3*) the cut propagates at a uniform rate except for a short induction period (see Fig. IVE-12); (*4*) the rate of

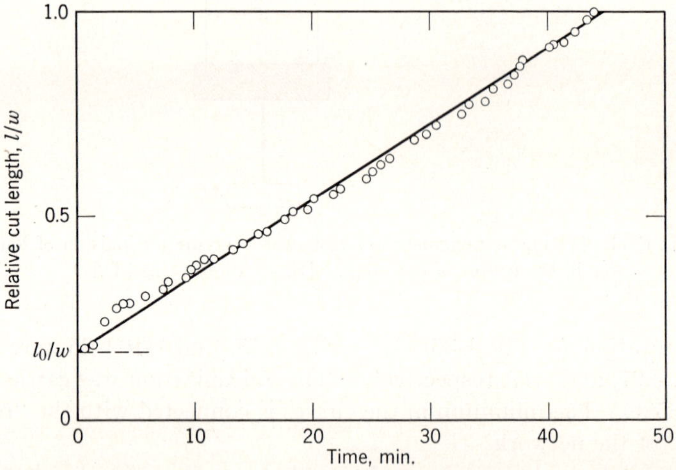

Fig. IVE-12. Cut growth as a function of time during ozone exposure. [Braden and Gent (3).]

cut growth was unaffected by the magnitude of the nominal stress except for very large stresses (minimum stress about 70 g./cm.^{2}); however, the region around the tip of the cut increased with stress and changed from a sharp incision to a wide strip; (*5*) the cut growth rate increased appreciably for high stresses ($\geq$2,000 g./cm.^{2}); (*6*) the rate of cut growth was also independent of the thickness of the test strip in the range of 0.02–0.2 cm. The effect of a number of variables, such as test piece composition, vulcanization agents, crosslinking, fillers, nature of polymer, plasticizers, temperature, and ozone concentration, were also studied.

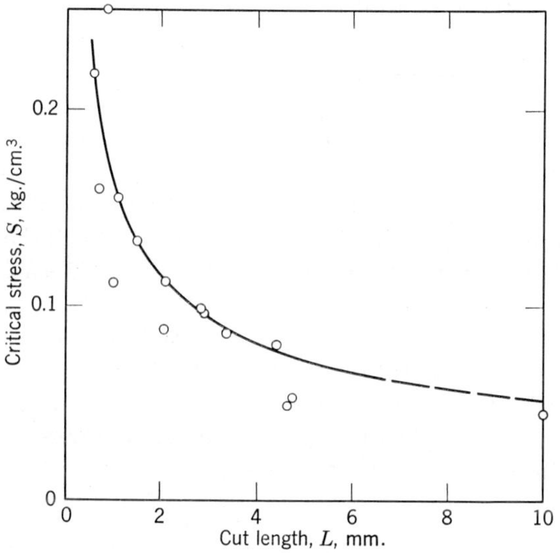

Fig. IVE-13. Minimum applied stress as a function of cut length l. The curve obeys eq. (43). [Braden and Gent (4).]

In another paper, Braden and Gent (5) describe further work on cut growth and show that the phenomenon of critical stress is not merely due to a protective oxidation layer on the surface of the incisions.

The functional relationship of the length of the cut and the critical stress is shown in Figure IVE-13. It can be calculated from classical theory that the tensile stress S_t at the tip of the cut is given by:

$$S_t = S[1 + 2(l/r)^{1/2}] \qquad \text{IVE9-(42)}$$

where S is the applied tension, l the length of the incision, and r the effective radius at the tip of the incision. This relationship should be applicable in the present case, if the assumption is made that a critical stress S_t at the tip of the incision is necessary for inducing growth while exposed to ozone. If $l >> r$, as is probably the case, eq. (42) reduces to,

$$S = S_t r^{1/2}/2l^{1/2} \qquad \text{IVE9-(43)}$$

Equation (43) is satisfactorily obeyed by the experimental results. $S_t r^{1/2}$ was chosen to fit the data. A value for S_t of 3.2 kg./cm.^{2} or larger was found. Since a classical treatment was used, it cannot be expected that the quantitative predictions are numerically correct, however, they will indicate whether the basic assumptions are reasonable.

Experiments with sheets having circular ungreased areas of different sizes showed that the cracking stress increased with decreasing area, indicating that a number of flaws are distributed over the surface of the rubber sheet. The probability of a flaw being located in the ungreased area decreases with decrease in area. Such flaws may be due to foreign matter or to imperfections in the mold surfaces.

The number of cracks formed was also investigated by performing a series of experiments where one edge of the test pieces was left ungreased and a range of stresses was applied. A number of cracks were formed, which grew at a uniform rate across the test piece. It is indicated by classical theory that the number of cracks N formed for a unit length is given by the relationship.

$$N = 1/d = 4S^2/\pi S_t^2 r \qquad \text{IVE9-(44)}$$

where d is the overall distance between adjacent cracks, S the applied stress, S_t the critical stress at each tip of a crack for growth to occur, and r the effective radius for the tip of a crack. This relationship is followed by the experimental results. A value of $S_t r^{1/2}$ similar to that obtained for one razor incision was derived from these experiments, showing that the assumption of a critical stress at the tip of a crack is justified.

It was shown by a number of experiments that the critical stress and the extensions are dependent on the stiffness of the vulcanizates. However, the stored energy was found, to a first approximation, to be independent of the stiffness.

In general, it can be shown that the stored energy W in bulk rubber is related to the energy T for cut growth at the tip of a cut of length l as follows,

$$T = 2KlW \qquad \text{IVE9-(45)}$$

K is a constant $\cong 2$. W was found to be essentially constant for a number of rubber vulcanizates and a number of polymers. This indicates that T also must have a characteristic value. Equation (45) can be transformed for small extensions of the bulk rubber into

$$S = (TE/Kl)^{1/2} \qquad \text{IVE9-(46)}$$

where S is the critical bulk stress and E is Young's modulus. In this way, a value of $T = 120$ ergs/cm.^{2} is obtained from experiments with razor cuts of 2 mm. length. The T values are somewhat larger for other polymers. These values are about twice that of the free surface energy for creating 1 cm.^{2} of new surface ($\gamma \cong 50$ ergs/cm.^{2}). Tearing would require about 10^6 ergs/cm.^{2}.

Braden and Gent (5) also investigated the effect of antiozonants. Inhibition of crack formation can in principle be achieved by reducing the energy available for such a process.

Gent (11) investigated the chemical effects which lead to cracking during fatigue experiments with soft vulcanized rubbers. These studies refer to oxygen rather than to ozone. Vulcanizates showed longer life times *in vacuo* than in air. The life time is approximately inversely proportional to the square of the elastic energy W for a unit volume of the test piece. The product $W^2 \times n$, where n is the number

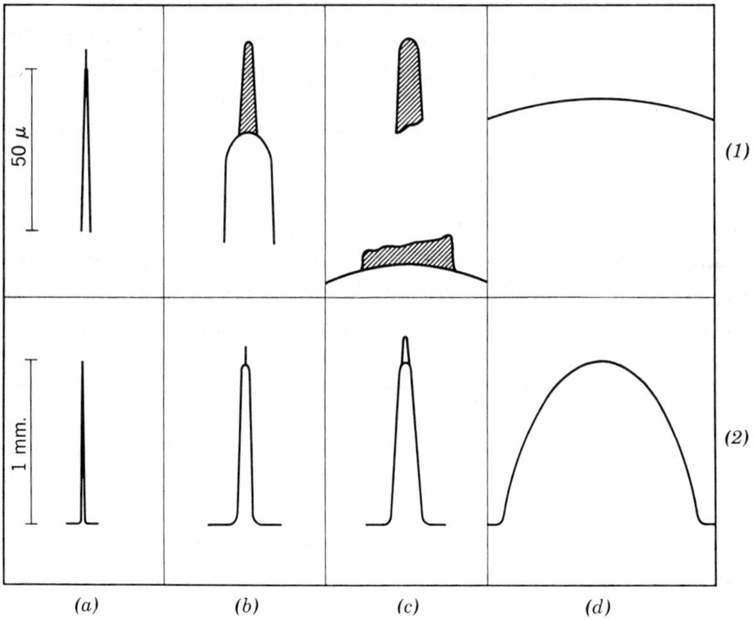

Fig. IVE-14. Shape of cracks at "minimum stress:" (*1*) view of tips, (*2*) overall view. Shaded area: crack grown in bulk of material: (*a*) plastic film, (*b*) natural rubber, (*c*) polychloroprene, (*d*) butyl rubber. [Salomon and van Bloois (19).]

of cycles for fatigue rupture, was found to be approximately constant for tests in the presence of air. Suitable additives, such as amines, prolong the life of vulcanizates in air considerably. However, fatigue lives *in vacuo* are not appreciably changed by additives.

Salomon and van Bloois (19) studied the effect of ozone on natural rubber hydrochloride. Stressed films of this material exposed to ozone ultimately ruptured. The minimum nominal stress level below which rupture did not take place was found to be 70 ± 5 kg./cm.^{2} irrespective of ozone concentration, even if the exposure was prolonged for several hours. The time for initiation of rupture decreased exponentially with stress until the hookean limit was reached; beyond this limit, rupture times increase and eventually decrease again when the film becomes oriented. Experiments showed that the rupture time is the same whether the exposure is continuous or intermittent. Cracks, according to these authors, are initiated at points in the surface where foreign microscopic particles are located. The cracks have tips characteristic of the type of polymer in Figure IVE-14.

For a brittle-type fracture the following equation holds for a two-dimensional crack at right angles to the minimum principal stress σ,

$$\sigma^2 = 4E\gamma_0\gamma_1/\pi l \qquad \text{IVE9-(47)}$$

where E is Young's modulus, γ_0 the free surface of the newly formed surface, l the length of the crack, and γ_1 a correction factor. γ_0 has a value of about 50 ergs/cm.^{2}. If primary bonds would have to be broken for the formation of cracks, the energy required would be about ten times larger than the above value. For true brittle fracture $\gamma_1 = 1$; however, in the case of rubber, energy dissipation due to inelastic processes also takes place, which makes $\gamma_1 > 1$. Taking $\gamma_0 = 50$ ergs/cm.^{2}, the value of γ_1 for natural rubber hydrochloride can be estimated from experimental data ($\sigma = 7 \times 10^7$ dynes/cm.^{2}, $E = 8 \times 10^9$ dynes/cm.^{2}, $l = 2\text{–}5 \times 10^{-4}$ cm.); hence, $2 < \gamma_1 < 5$. This value indicates that almost true brittle fracture occurred.

On the basis of all their experimental results, these authors come to the following conclusions:

Ozone is adsorbed by solid surfaces forming a thin oxidized layer (0.01–0.05 μ) which makes certain polymer surfaces resistant to further attack. If such a surface is stretched while exposed to ozone, narrow and deep capillaries are formed at the edges of foreign particles embedded in the polymers. The adsorbed film, being highly mobile, concentrates in the capillaries and cracking takes place.

The rate of fracturing is influenced by Young's modulus of the polymer and hardening of the surface; stresses at the tip of a crack increase with increase in modulus. However, the fracturing process is restricted to a narrow stress field around the tip of the crack, which keeps the energy dissipation needed for fracture small (γ_1 small). A good chemical antiozonant should soften the ozonized layer and produce blunting of the tip. It is believed by these authors that different rates of crack growth in unsaturated polymers is rather due to their different rheological properties than to the number of double bonds in the polymer.

IVE10. Miscellaneous

Crabtree and Biggs (10) observed that stretched rubbers which are irradiated by ultraviolet light show crack formation in the presence of peroxides. This cracking takes place whether oxygen is present or not. This whole process can be catalyzed by nitrogen peroxide. The authors suggest that free radicals such as tertiary butoxy, phenyl, benzoyl, acetyl, and hydroxyl are formed on photolysis of the respective peroxides leading directly to cracking. The alternative assumption is that photolysis of peroxides ultimately leads to ozone as a product and similarly photolysis of nitrogen peroxide leads to ozone (17):

$$NO_2 \xrightarrow{h\nu} NO\cdot + O\cdot$$

$$O\cdot + O_2 + M \rightarrow O_3 + M\cdot$$

$$NO. + O_3 \rightarrow NO_2 + O_2$$

For peroxide radicals, one could formulate the following reactions,

$$ROO\cdot + O_2 \rightarrow RO\cdot + O_3$$

Also,

$$2ROO\cdot \rightarrow 2RO\cdot + O_2$$

The latter reaction would then lead to ozone formation in the absence of oxygen.

Ultrasonic degradation of polymer solutions has been studied quite extensively (14). It has been shown that ultrasonic chain scission here is due to cavitation of gas-saturated liquids exposed to ultrasonic waves. This type of degradation is of a purely mechanical nature, the long chain molecules being fragmented by the frictional and impact forces generated by collapsing cavities.

Treatment of solid polymers with ultrasonic waves has been studied by Langton (16). This worker exposed ultrasonic waves to solid poly(methyl methacrylate) placed in water. The surface was investigated with an electron microscope and appreciable surface damage in the forms of pits was ascertained. Although the poly(methyl methacrylate) heated quickly on exposure, the surface of the polymer was continuously cooled by water. It is most likely that the pits are caused by the violent collapse of cavities at or near the surface of the polymer. Of course, such cavitation phenomena have previously been observed, for instance, on ships' propellers.

IVE11. Summary

Polymer degradation reactions which are relevant to the understanding of crack initiation and propagation in polymers as influenced by chemical compounds, ultraviolet light, and high energy radiation have been discussed briefly. The importance of chain scission, crosslinking, oxidation, ozonization reactions, and chemorheological processes has been indicated.

The only extensive work on the influence of chemical reactions which is available is that of ozonization of unsaturated, rubberlike polymers leading under suitable conditions to crack formation and growth. However, the field has scarcely been developed beyond the purely descriptive stage. Numerous observations on crack formation and propagation can be found in the literature; different authors emphasizing different aspects of the observed phenomena. The only quantitative formulations deal with the statistical evaluation of such data, some purely rheological formulations, and some energy relationships. The existence of critical stresses for the initiation and growth of cracks was recognized early. Only recently (see refs. 3–6 and 19) have studies been made of the growth of single cracks as a function of ozone concentration, stress, composition of material, degree of crosslinking, etc. Processes occurring at the tips of growing cracks have also been considered. The chemical and physicochemical aspects have so far received little attention, and degradation mechanisms have scarcely been applied to fracture problems of this nature. A thorough understanding of these problems can only be achieved by a concerted study of the chemical, rheological, and physical phenomena.

Some work concerning ultrasonic irradiation of polymers and also the effect of a number of chemical compounds have also been reviewed briefly.

IVE12. List of Symbols

CCF	Chemical creep function
E	Young's modulus
G	"G value" (individual processes/100 e.v. absorbed)
$I'_{\text{abs.}}$	Average absorbed light intensity/cm.^{3}
I_0	Incident light intensity
m_w	Molecular weight of monomer
M_c	Number-average molecular weight between crosslinks
$\bar{M}_n$	Number-average molecular weight
$\bar{M}_w$	Weight-average molecular weight
n	Number of main chain links at time t
n_0	Initial number (or concentration) of main chain links
n_p	Number of chains of length P in one base mole of polymer
N	Number of polymer chains
$p_{c(\text{crit.})}$	Critical crosslinking probability
P_c	Crosslinking probability
P_0	Initial number-average chain length
P_t	Number-average chain length at time t
P_v	Viscosity-average chain length
P_w	Weight-average chain length
q	Crosslinking density
r	Radiation dose
s	Average number of main chain scissions for each original chain
S	Stress (in some instances, elongation)
W	Strain energy
α	Degree of degradation
ϵ	Kinetic chain length
γ	Crosslinking index; surface energy
ϕ_s	Chain scission quantum yield
ϑ	Crosslinking coefficient

References

1. Bailey, P. S., *Chem. Revs.*, **58,** 925 (1958); see also A. D. Delman, B. B. Simms, and A. E. Ruff, *J. Polymer Sci.*, **45,** 415 (1960).
2. Bovey, E. A., *The Effects of Ionizing Radiation on Natural and Synthetic High Polymers,* Interscience, New York 1958, p. 287.
3. Braden, M., and A. N. Gent, *J. Appl. Polymer Sci.*, **3,** 90 (1960); see also M. Braden and A. N. Gent, *Kautschuk Gummi,* **14,** WT 157 (1961).
4. Braden, M., and A. N. Gent, *J. Appl. Polymer Sci.*, **3,** 100 (1960).
5. Braden, M., and A. N. Gent, *J. Appl. Polymer Sci.*, **6,** 449 (1962).
6. Buckley, D. J., and S. B. Robinson, *J. Polymer Sci.*, **19,** 145 (1956).
7. Burnett, G. M., *Mechanism of Polymer Reactions,* Interscience, New York, 1954, pp. 325–406.
8. Chapiro, A., *Radiation Chemistry of Polymeric Systems,* Interscience, 1962, p. 712.
9. Charlesby, A., *Atomic Radiation and Polymers,* Pergamon Press, London, 1960, p. 556.
10. Crabtree, J., and B. S. Biggs, *J. Polymer Sci.*, **11,** 280 (1953).
11. Gent, A. N., *J. Appl. Polymer Sci.*, **6,** 497 (1962).
12. Grassie, N., *The Chemistry of High Polymer Degradation Processes,* Interscience, New York, 1956, p. 335.
13. Houst, P. L., E. R. Stephens, W. E. Scott, and R. C. Doerr, *Atmospheric Ozone–Olefin Reactions,* The Franklin Institute, Philadelphia, Pa., 1958.
14. Jellinek, H. H. G., *Degradation of Vinyl Polymers,* Academic Press, New York, 1955, p. 315.
15. Jellinek, H. H. G., *Pure Appl. Chem.*, **4,** 419 (1962); also, *Special Lectures,* IUPAC, Physical Chemistry Section, Commission of Macromolecules, Macromolecular Chemistry, International Symposium, Montreal, 1961, Butterworths, London, 1962.
16. Langton, N. H., *Plastics (London),* **21,** 118 (1956).
17. Leighton, P. A., *Photochemistry of Air Pollution,* Academic Press, New York, 1961.
18. Newton, R., *Trans. Inst. Rubber Ind.*, **21,** 113 (1945); *J. Rubber Res.*, **14,** 27, 41 (1945) (gives a list of references prior to 1945); see also *Ind. Eng. News,* **44,** 834 (1952).
19. Salomon, G., and F. van Bloois, *Proceedings of the 4th Rubber Technology Conf., London, 1962,* W. Heffer and Sons, Cambridge, 1962.
20. Smith, D. M., and V. E. Gough, *Trans. Inst. Rubber Ind.*, **29,** 219 (1953).
21. Tobolsky, A. V., *Properties and Structure of Polymers,* Wiley, New York, 1960, pp. 223–265.
22. Wall, L. A., in *Analytical Chemistry of Polymers,* Part II, G. M. Kline, Ed., Interscience, New York, 1962, Chapter 5.

Addenda: Papers dealing with this topic which have appeared since this article was written are: E. H. Andrews and M. Braden, *J. Appl. Polymer Sci.*, **7,** 1003 (1963); G. Salomon and F. van Bloois, *ibid.*, **7,** 1117 (1963). An excellent review concerning ozone attack on rubbers has appeared; see E. H. Andrews, D. Barnard, M. Braden, and A. N. Gent in *The Chemistry and Physics of Rubber-like Substances,* L. Bateman, Ed., Maclaren, London; Wiley, New York, 1963.

IVF. SPECIFIC ROLE OF IRRADIATION

A. CHARLESBY*

IVF1. Introduction

Exposure to radiation offers an exceptionally valuable method of modifying polymers under strictly controllable conditions and of studying the effect of such modifications on the resultant physical properties. Irradiation can be carried out at any temperature, with polymers in the crystalline, rigid, or fluid states, or in an oriented molecular chain arrangement; the degree of change being controlled by the exposure dose. It must be admitted, however, that most of the radiation research on polymers has been carried out by chemists primarily concerned with the preparation of new or

* We are particularly indebted to Professor Charlesby for fully editing as well as authoring this portion of the volume. B.R.

modified chemical structures; the effect on physical properties receiving far less attention. It may be confidently expected that this situation will be remedied as experience in the use of radiation as a research tool in polymer physics becomes more widespread. In this section we shall indicate the basic reactions involved and give selected illustrations of how these have been used to study many of the physical properties of polymeric structures. Many of the more technical aspects, which have received considerable attention, will not be discussed, since they cannot be readily extended to assist our basic understanding of polymer properties. Attention will also be drawn to certain lines of research which, in the opinion of the present author, could be profitably developed.

IVF2. High Energy Radiation

The forms of radiation with which we are here concerned are variously known as high energy, atomic, nuclear, or ionizing radiations, according to preference and background of author. None of these terms is really very suitable, since the energies involved can be far lower than those involved in elementary particle physics and the sources are not necessarily nuclear. The term "ionizing radiation" is chosen in contradistinction to uv radiation which, because of its lower energy, can cause excitation but no ionization; however the observed changes are not necessarily due to ionization of the irradiated material.

The various types of radiation which may be considered are: x-rays and gamma rays; fast electrons; fast neutrons; fast charged heavy particles such as protons, deuterons, and alphas. Related radiations of low energy are uv, visible light, and slow neutrons. Suitable radiation sources are reactors, radioactive isotopes, and electrical accelerators. Nuclear reactors allow large amounts of material to be irradiated simultaneously, but have the disadvantages of lack of flexibility and mixed radiation. Radioactive sources can provide gammas (Co 60, Cs 137), betas (Sr 90), alphas (Po 210), and are convenient for many research purposes. Electron accelerators and other high voltage equipment provide very powerful radiation sources of limited penetration; their high cost and maintenance are only justified for large-scale production of irradiated material or for specialized research purposes, as when the effects of extremely high dose rates are being investigated.

In practice, due to the difficult experimental working conditions and low penetration, little work has been carried out on the alpha irradiation of thin polymer films. The major interest here lies in a comparison of alpha and electron (or gamma) radiation effects. Alpha irradiation studies of polymers may provide a lead as to the important difference in the biological effects produced by similar doses of different forms of radiation. Furthermore the energy deposition with alphas is far more localized than electrons or gammas, and this may influence such reactions as destruction of crystallinity. Experimental results on polymers are largely confined to continuous gamma irradiation (dose rates of the order of 10^{-1} megarad/hr.) and electron irradiation given in bursts, with instantaneous dose rates in the range 10^2 to 10^4 megarads/hr.

IVF3. Interaction of Radiation with Matter

When high energy radiation passes through matter, it may interact with the nucleus or with the orbital electrons; the incident particle being absorbed or continuing with diminished energy. The main processes of energy transfer are as follow:

Capture by the nucleus to give a new nucleus, which may be unstable, in which case it will subsequently emit further radiation. This process is primarily observed with slow neutrons whose energy is insufficient to cause the other reactions described below. With very few exceptions, gammas and fast electrons only produce nuclear changes at energies higher than those used in this type of research, so that induced radioactivity by these forms of radiation may be completely ignored.

Displacement of the struck nucleus, which is thereby ejected from its parent molecule or position in the lattice. This reaction predominates, e.g., between fast neutrons and hydrogen atoms. No change in the structure of the nucleus is involved in this type of reaction, although there is a change in atomic arrangement. This reaction is mainly of importance in metals, but with organic systems it is overshadowed by the changes due to electronic rearrangement.

Collision with an orbital electron, which is ejected (ionization) or raised to a higher energy level (excitation). These reactions are the primary cause of energy loss by incident charged particles, and by gamma and x-ray photons. X-rays and gamma rays, for example, lose most of all of their energy in a single collision; liberating an electron by the Compton and photoelectric effects or by pair production; the last giving an electron and a short-lived positron. The ejected electron carries high energy, and can continue to cause further ionization and excitation along its track. Fast electrons, protons, deuterons and alphas also lose their energy in smaller amounts by liberating electrons along their track and these low energy δ electrons (average energy about 100 ev.) cause further ionization and excitation. Thus a single charged particle of energy of say 1.5 Mev. passing through a polymer will leave a final track typically comprising about $1.5 \times 10^6/10^2 = 15{,}000$ ionized molecules and δ electrons; these δ electrons can produce a further 30,000 ionizations, plus an unspecified number of excitations. Since each of these ionizations and excitations may cause the disruption of a polymer molecule, of molecular weight say 10^6, a single fast electron can cause modification in chemical structure over a total molecular weight of about 5×10^{10}. It is this high ratio which renders many polymers particularly sensitive to the effects of high energy radiation. Biological systems show the same effect, but are often more sensitive to radiation.

It has generally been assumed that interaction between the incoming high energy particle or the δ electron and the irradiated material occurs at random to give rise to ionization and excitation. This assumption is now being further studied from the theoretical point of view. The energy absorbed in this way can be transferred further to a sensitive or radiation-resistant part of the molecule to produce the reactive chemical species; either ionized or excited. The ion-molecules formed in this way can then react chemically, as in ionic polymerization. More frequently, however, changes in electronic structure occur (the exact nature of which is still under discussion) which lead to the scission of primary bonds and hence the formation of radicals. These uncharged but chemically reactive groups react very rapidly with each other in the liquid or gaseous state, to reconstitute the original structure or to produce new ones. In solid systems, these radicals may be formed some distance apart and the subsequent reactions occur far more slowly. Indeed it is possible to detect these radicals using electron spin resonance techniques, and to observe the decay in their concentration as they react. In the case of polyethylene for example, the reaction may be written

$$—CH_2—CH_2—CH_2— \rightsquigarrow —CH_2—\dot{C}H—CH_2— + H$$

where $\leadsto$ denotes the change following irradiation, the dot indicates a radical on the polymer chain, and H is a mobile hydrogen atom radical. H is then believed to abstact a second H from another polymer molecule

$$H + \text{—}CH_2\text{—}CH_2\text{—}CH_2\text{—} \rightarrow H_2 + \text{—}CH_2\text{—}\dot{C}H\text{—}CH_2\text{—}$$

The two polymer radicals eventually react together to give a crosslink

$$\begin{array}{c} \text{—}CH_2\text{—}CH\text{—}CH_2\text{—} \\ | \\ \text{—}CH_2\text{—}CH\text{—}CH_2\text{—} \end{array}$$

As written the reaction is only one over-simplified explanation of the observed formation of crosslinks by irradiation. It leaves out of account the distance over which the H atom migrates before abstraction, as well as the mechanism by which the two radicals, once formed, can migrate toward each other before combining. From the present point of view, however, it is the number and distribution of these crosslinks in relation to the dose which determines the resultant physical properties. The mechanism of formation may influence the subsequent physical properties. Thus if the crosslinking reaction is a slow one, and the irradiated samples are allowed to stand in air, the radiation-induced radicals may react with oxygen rather than with each other, giving a very different product.

IVF4. Dose Units

When discussing radiation effects arising from electronic changes, it is usual to express doses in terms of the rad (radiation absorbed dose). A rad unit is defined as an energy absorption of 100 ergs/g. For polymers we are usually concerned with higher doses, conveniently expressed in megarads. One megarad equals an absorbed dose of 10^8 ergs/g. = 10 joules/g. = 2.4 cal./g. One kw.-hr. of high energy radiation, fully absorbed, can therefore give $3.6 \times 10^6/10 = 3.6 \times 10^5$ megarads to 1 g. of irradiated material, or 1 megarad to 360 kg. Electron accelerators now available have outputs of several kilowatts, and can therefore treat large quantities of polymeric materials. On the other hand 1,000 curies of Co 60 emits a total of about 15 w. of gamma power.

In earlier polymer radiation work, doses have been expressed in roentgens (r) or reps. These units are less convenient for calculation, and are now being abandoned for polymer work. They may be taken as approximately equivalent to 0.93 rads in most polymers. Other units are 10^{20} ev./g. (= 1.6 megarads) and joules/g. (= 0.1 megarad).

Changes produced by radiation are often expressed in terms of the *G* value, this represents the number of changes of a specified type corresponding to a total energy absorption of 100 ev. Not all the energy is necessarily utilized in promoting the change. Thus irradiation of air produces an average of one ion-pair for each 34.5 ev. absorbed, and the corresponding *G* value is 2.9. Values in the neighborhood of 3 are very common; much higher values (>10) are indicative of chain reactions, whereas much lower values (<1) are representative of radiation resistance as, for example, in polystyrene.

A radiation dose of 1 megarad, or 10 joules/g., is equivalent to 6.25×10^{19} ev. absorbed/g. By definition this dose produces 6.25×10^{17} *G* changes in a total molecular weight of 6.02×10^{23}, or one change of the type considered per $0.96 \times 10^6/G$ molecular weight. This simple relationship enables us immediately to translate a dose into a resultant change in structure, provided only that the relevant *G* value is known.

Attention should be drawn to a possible ambiguity in *G* values quoted in the literature. When crosslinks are formed between two chains by a junction of type ⌶, each crosslink is associated with two "crosslinked units" (one on each polymer), and the number of chain lengths between such links or junctions is increased by two for each new crosslink introduced. In referring to published *G* values for crosslinking it is therefore necessary to ascertain whether *G*(crosslinks) or *G*(crosslinked units) is implied. For such junctions *G*(crosslinks) = 0.5*G*(crosslinked units). Other forms of junction, e.g., "endlinking" ⊤ and "cyclization" ȣ, have also been considered. For example the maximum tensile strength of irradiated rubber is usually found to occur at doses of about 45 megarads. Since *G* is about equal to 2 for the production of crosslinked units [*G*(crosslinks) $\cong 1$], we immediately deduce that $M_c = 0.96 \times 10^6/(45 \times 2) \cong 10^4$, if end corrections are ignored.

IVF5. Nature of the Changes Produced by High Energy Radiation

In the previous subsection we have attempted to show in general how the number of physical and chemical changes produced by exposure to any radiation dose may be deduced if the appropriate

G value is known. Thus if a measured amount of mechanical damage is caused by a dose of *r* megarads, we can then obtain an explicit expression between the radiation-induced change in physical and chemical structure and the resultant deterioration in mechanical properties. Now we consider the general nature of the structural changes produced.

Some of the earliest work on irradiation dealt with the initiation of polymerization reactions when monomers were exposed to radiation. The polymers formed, e.g., by irradiation of many vinyl monomers at room temperature, are found to proceed by a radical reaction, the kinetics of which follow conventional chemical routes. Radiation merely replaces the catalyst. Thus apart from the greater flexibility allowed in their preparation due to the much closer control possible in the initiation stage (temperature, dose rate), in the absence of chemical catalysts, no new physical properties emerge. More recently this field has been greatly extended by the discovery of radiation-induced ionic polymerization (particularly in the presence of fine metallic oxides, etc.) and of solid state polymerization. New types of polymer have been prepared by these means, some with potentially novel characteristics. For example, the irradiation of single crystals of certain monomers may give rise directly to a uniaxially oriented crystal of polymer. The rearrangement of monomer units necessary to accommodate the polymer molecule within the crystal lattice raises interesting problems of crystal morphology. However, present interests have been largely confined to the production of such new polymers, and to the study of the kinetics of the chemical reactions involved, rather than devoted to the measurement of physical properties.

A second and quite distinct line of research arose from the irradiation of polymers already formed. Many of the physical and chemical changes induced by radiation in a wide range of polymers could be ascribed to two basic reactions: crosslinking (i.e., joining of two molecules together by a lateral covalent bond) and degradation (main chain scission). The crosslinking reaction can be readily followed by the initial increase in viscosity, and the subsequent reduction in solubility and swelling, as well as by the increase in elastic moduli. The degradation reaction can be followed quantitatively by the reduced average molecular weight, determined by viscosity, osmotic pressure, light scattering, etc. Both these reactions show a linear relation between dose and number of crosslinks between molecules, or scissions in the main chain. These two aspects will be considered in detail below.

An entirely different reaction can occur when the polymeric mixtures irradiated contain unsaturated groups, of a nature which allows a chain reaction to develop. This reaction shows some analogy with polymerization. In this case linearity no longer holds between dose and number of crosslinks (or bridges between polymer molecules). This more complex system is also discussed in Sub-section IVF9.

Many physical properties are affected by the presence in the polymer of fine particles of carbon black, silica, etc. The nature of the reinforcement produced can be studied using samples with varying amounts of such reinforcing fillers, exposed to various doses. This aspect is summarized briefly in Sub-section IVF10.

Considerable efforts have been devoted to the production of graft polymers using radiation techniques. Here again the emphasis is on preparation, and only limited information is available on the resultant physical and mechanical properties.

Finally it must be emphasized that many of the radiation-induced reactions can be profoundly modified by the addition of very small amounts of certain compounds, which either interfere with the initial physical processes, e.g., by the so-called energy transfer process, or react with the ions or radicals to give new types of chemical reactions. Thus aromatic groups are particularly radiation resistant; oxygen frequently enhances the radiation effect as in degradation; iodine and similar radical scavengers react with polymer radicals to inhibit crosslinking. Many other reactions of this nature are under close investigation in radiation chemistry, as they provide valuable information on intervening processes. They can also be compared with nonradiation induced reactions, in which the influence of physical or chemical environment changes the structure and physical properties of a polymer.

IVF6. Crosslinking and Degradation

In a wide range of linear polymers, the following major changes are observed as a result of exposing the polymer to high energy radiation.

(*a*) Crosslinking (dimerization), i.e., lateral bridge formation between polymer molecules, giving rise to higher molecular weight structures with increased branching, and eventually leading to network formation.

(*b*) Degradation (main chain scission) but without any significant degree of depolymerization down to monomer. This reduces the average molecular weight.

(*c*) Gas liberation, usually H_2, but sometimes including CH_4, CO, and CO_2, and other products of side chain decomposition. The gases may be evolved from the specimen, or congregate in the form of bubbles dispersed throughout it.

(*d*) Scission at branch points, the severed sections often remaining trapped within the polymer matrix.

(*e*) Changes of crystallinity (where present initially).

(*f*) Changes in chemical structure, e.g., by the production or elimination of unsaturation.

(*g*) Endlinking and cyclization have also been suggested as possible reactions, but evidence for their occurrence is not unambiguous. It is convenient to give some illustrations of these effects.

Most of the changes in physical properties are due to crosslinking and degradation. In most polymers one or other of these predominates. One simple rule is that vinyl polymers with unit ($—CR_1R_2—CH_2—$) degrade, while those with unit ($—CHR—CH_2$) or ($—CH_2—CH_2—$) crosslink, the distinction being possibly due in part to steric factors.

IVF7. Polymers Which Degrade

Changes in mechanical properties are most readily analyzed in those polymers which degrade when exposed to radiation. By degradation we mean a random scission of the main polymer chain, with minor chemical changes in the neighborhood of the break to stabilize it. Degradation therefore differs from the "depolymerization" observed at higher temperature in certain polymers, which break down by a chain reaction to give monomer. The ability to degrade under radiation, and to depolymerize under heat appears to be related, as is shown by the difference in the heat of polymerization of crosslinking and of degrading polymers.

The radiation induced scissions are located at random along the chain and their number is proportional to dose. Thus the reciprocal of the number-average molecular weight, $1/\bar{M}_n$, increases uniformly with dose r

$$1/\bar{M}_n = (1/\bar{M}_n)_0 + k.r$$

where k is a constant, determined by the G value for main chain scission. Here $(\bar{M}_n)_0$ is the initial number-average molecular weight and $\bar{M}_n$ its value after dose r. Exposure to a range of doses therefore enables us to produce a number of specimens of identical structure and molecular arrangement, apart from well-determined changes in molecular weight. In particular we can determine the relationship between tensile strength, elastic moduli, viscoelasticity, and molecular weight.

When the initial molecular weight of the irradiated polymer is not random, it rapidly becomes so as a result of irradiation. Thus a plot of $1/\bar{M}_v$ or $1/\bar{M}_w$ against dose (see Fig. IVF-1) will be linear only if the initial distribution is random, and the ratio of the extrapolated value of $1/\bar{M}_v$ or $1/\bar{M}_w$ $(r \to 0)$ to its true initial value is a measure of the departure of the initial molecular weight distribution from a random (exponential) type. Here again, any effect of variation in the molecular weight distribution on mechanical properties

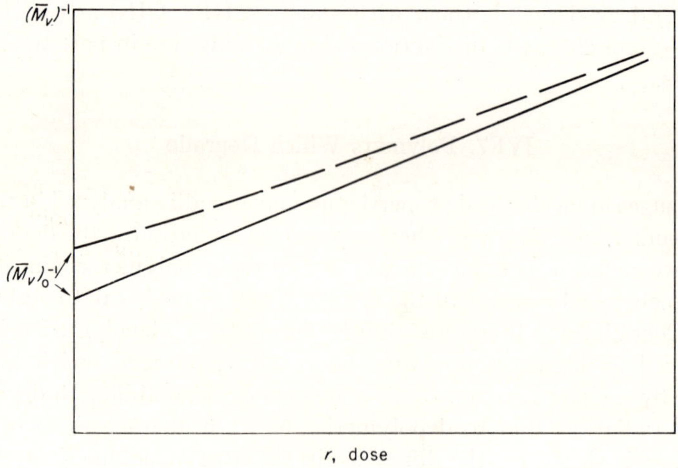

Fig. IVF-1. Variation of reciprocal viscosity-average molecular weight of a degrading polymer: (—) initial random molecular weight distribution, (– –) observed for an initial nonrandom distribution.

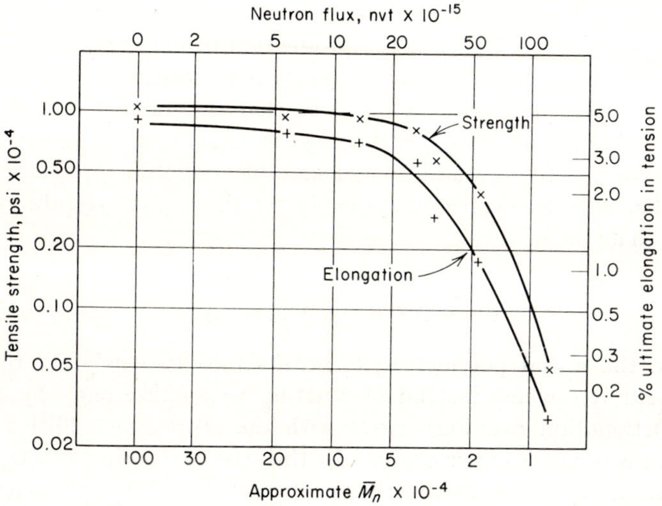

Fig. IVF-2. Effect of reactor radiation on molecular weight, tensile strength, and ultimate elongation in tension of PMMA (56).

can be readily ascertained. Of particular interest is the radiation dose (hence the molecular weight) above which the tensile strength begins to fall rapidly (see Fig. IVF-2). This is the critical $\bar{M}_n$ below which strength is largely influenced by the number of molecular chain ends, rather than by the intrinsic strength and arrangement of the chains in the specimen.

In considering examples of changes in mechanical properties we can distinguish (*1*) amorphous highly flexible (rubbery) polymers, (*2*) amorphous rigid (glassy) polymers, (*3*) randomly crystalline polymers, and (*4*) oriented crystalline polymers such as fibers.

Amorphous highly flexible polymers which are not crosslinked can be considered macroscopically as liquids whose high bulk viscosity is due to a high molecular weight and resultant entanglements. Their mechanical behavior following irradiation will differ essentially from the corresponding polymer without irradiation only insofar as they are of lower molecular weight, appropriate to the prior dose received. An excellent example of this type of degrading polymer is polyisobutylene. No general agreement has yet been reached on the chemical processes which take place following ionization and excitation, and which eventually result in chain scission, evolution of side groups (H, $\dot{C}H_3$) and the production of unsaturation. The G value for main chain breaks varies with temperature, the value at room temperatures being about 5 so that the number of radiation-induced breaks per gram, per megarad is $0.625 \times 10^{18}\, G \cong 3 \times 10^{18}$ or 5 μmoles/g. By the addition of suitable radiation protectors this G value may be decreased, possibly by the repair (resealing) of a main chain break

$$R_1R_2 \rightsquigarrow R_1\cdot + R_2\cdot$$

$$R_1\cdot + R_2\cdot + A \rightarrow R_1AR_2$$

where the initial polymer R_1R_2 breaks into two polymer radicals $R_1\cdot$ and $R_2\cdot$ which instead of reacting to stabilize (e.g., by a disproportionation reaction), react with the protecting additive A to reform a single polymer chain. In this case a chemical reaction has occurred, but the additive A has protected the polymer against radiation-induced deterioration in mechanical properties. Other forms of protection are also possible.

Even in this simple system, there remain a number of fundamental questions to which generally accepted answers have not yet been given. In the first place, we know that the G value for scission varies with temperature, this variation being small below the second-order transition point, but increasing rapidly above it. We may then inquire whether the G value increases to a maximum value (corresponding to one scission for each initial excitation or ionization) or whether it increases indefinitely, to become a chain reaction, typical of chain depolymerization. Similarly we can inquire whether the G value changes when the polymer is irradiated in a stressed condition. It can be accepted that radiation causes main chain scission, but owing to the "cage" effect which limits large molecular motions, the two broken chain ends (radicals) may link up again and restore the initial situation. It is quite probable that when the chains are under directional stress, the probability of permanent separation is increased, resulting in a higher G value for permanent scissions. Positive evidence of any such increase would provide interesting data on the vibrational properties and displacements of molecular chains. In the presence of oxygen, radiation induced scission may also lead to higher G values, due to stabilization of the broken chains. All these phenomena appear to offer potential ap-

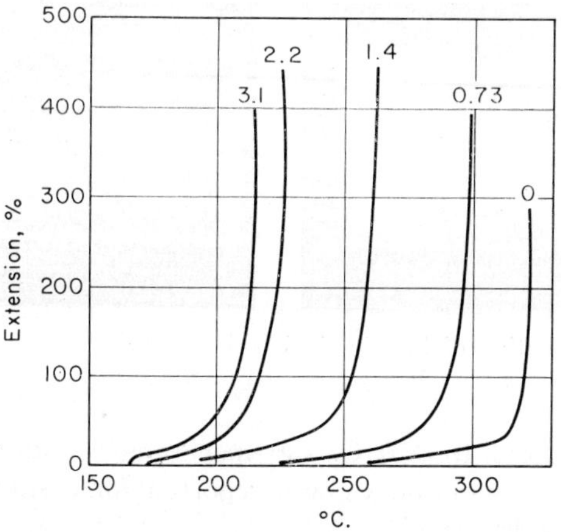

Fig. IVF-3. Elongation of PMMA under constant stress of 1.4 g./mm.² with various doses (in megarads) (8).

proaches to the study of the reactions involved during main chain scission.

As an example of a rigid amorphous noncrosslinking polymer we may choose poly(methyl methacrylate) (PMMA). Irradiated at room temperature or below, this polymer degrades with a G value which depends on temperature. At 20°, G(scission) $\cong$ 1.6, as determined by changes in viscosity in solution. Here again, the precise chemical reactions which lead to scission are still under discussion. Indeed it could be argued that the radiation-induced ionization and excitation produce the radicals observed by electron spin resonance; these radicals are relatively stable and only cause main chain scission after a lengthy period. The final main chain scission could then occur either on heating or on dissolution, when adequate flexibility is available for the scissioned molecular chains to separate. Thus the change in mechanical properties of an irradiated sample of PMMA might well depend on the time lapse between irradiation and testing, but this has not yet been tested.

Some early experiments with the uniaxial tensile creep of irradiated PMMA were made where the temperature was raised uniformly (8). Under the experimental conditions used, high elongations (20–400%) were observed at decreasing temperature depending on dose (see Fig. IVF-3). Similarly the temperature at rupture fell

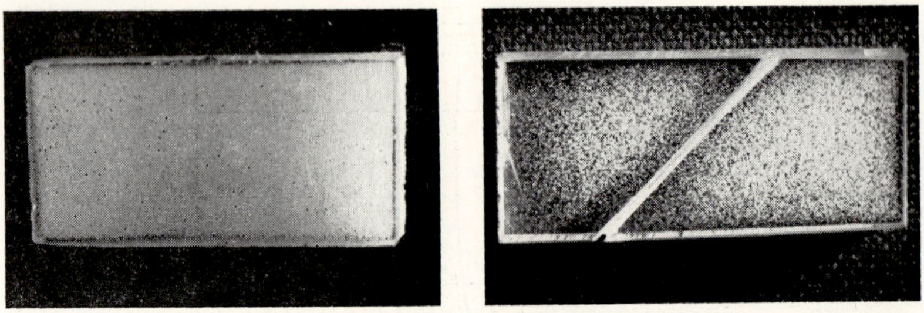

Fig. IVF-4. Formation of bubbles in irradiated PMMA due to trapped gases subsequently liberated by heating (41).

from 350 (0 dose) to 150°C. (30 megarads). Indeed a softening temperature as low as 50°C. was reported for a sample irradiated with 200 megarads.

The exact significance of these measurements is not clear, since they would depend on a combination of recoverable and irrecoverable elongation. Nevertheless these radiation doses can be readily converted to molecular weights. A dose of r megarads produces $6.25 \times 10^{17}Gr \cong 10^{18}r$ chain breaks/g., so that the number of molecules per gram, initially $N/(\bar{M}_n)_0$ becomes $N/\bar{M}_n = N/(\bar{M}_n)_0 + 10^{18}r$. If $M_n \ll (\bar{M}_n)_0$, $\bar{M}_n \cong 6 \times 10^{23}/10^{18}r = 6 \times 10^5/r$; $M_w \cong 1.2 \times 10^6/r$. Thus at the constant stress used in these experiments a decrease in molecular weight $\bar{M}_w$ to about 3×10^5 reduces the temperature at rupture from 350 to 200°C. This should be readily checked by similar measurement on a range of specimens of different molecular weight, but unirradiated. If agreement is not obtained, however, it may be surmised that the radicals initially produced by radiation not only cause immediate chain fracture but subsequently react at the higher temperatures to give enhanced breakdown at point of stress concentration. Here again is a problem in polymer physics which might well repay further study using radiation techniques.

In addition to main chain scission, irradiated PMMA may also show changes in mechanical properties due to the trapped gases. It is known that the side group $COOCH_3$ decomposes, and together with H from another part of the chain, gives rise to CO, CO_2, CH_4, and H_2. The former three gases remain trapped and dispersed, but can diffuse together to form bubbles if the specimen is heated after irradiation (see Fig. IVF-4). One interesting feature is that bubble

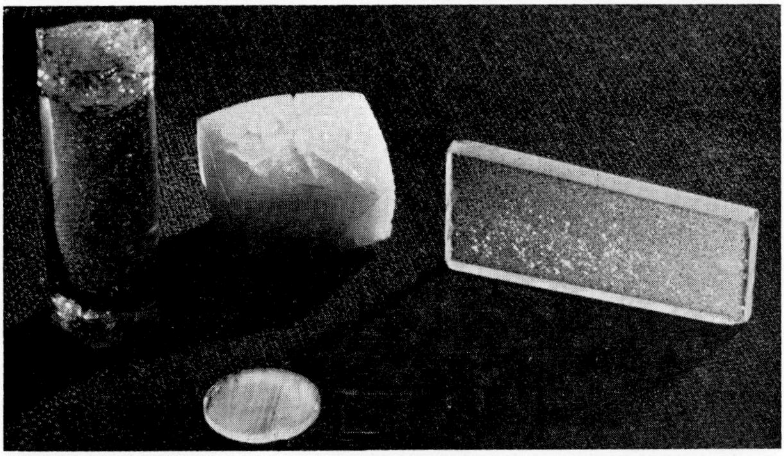

Fig. IVF-5. Internal fissures and cracks produced in irradiated PMMA standing for several weeks (54).

formation does not occur within a distance of about 0.5 mm. of the edge of the specimen, provided that this edge was present during irradiation. On this view, the gases already exist immediately after irradiation, and the role of heat is to weaken the polymer matrix, thereby allowing the gases to diffuse together. However, a very different view can be taken, whereby the radicals formed by radiation are stable at room temperature, and it is only following the application of heat that the chemical reaction takes place; resulting in the liberation of side groups. The existence of these trapped radicals within the matrix can be readily shown by electron spin resonance, although the patterns obtained cannot be readily reconciled with the expected chemical changes. The existence of reactive radicals (and possibly charged groups) within the solid matrix of the polymer may indeed have significant effects on its behavior under stress, but so far there is little relevant quantitative evidence.

When irradiated PMMA is allowed to stand for long periods, fissures can be formed within the bulk of the specimen, though less often on its surface (see Fig. IVF-5). These fissures may follow the distribution lines of peak stress within the specimen; initiation of the crack occurring due to weakening of the sample in the presence of radicals. Alternatively they may be due to stress engendered by the trapped gases, or to a contribution from the slow production of

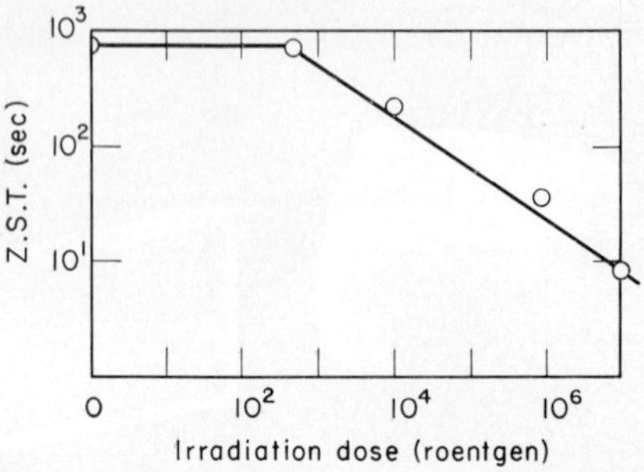

Fig. IVF-6a. Decrease in Z.S.T. of Teflon (36).

monomer (acting as plasticizer) by the radicals due to a depolymerization reaction. It is known that traces of monomer greatly weaken PMMA in tension, but little monomer is present in the gases liberated during irradiation. Nevertheless such monomer may be formed over long periods; in past irradiation studies, measurement of monomer was carried out very shortly after completion of irradiation.

It should be added that exposure to uv light produces the same type of breakdown behavior at room temperature as with high energy radiation, though with a much lower quantum efficiency. At higher temperatures the uv reaction occurs by chain depolymerization but recent results (unpublished) indicate that no such transition from degradation to depolymerization takes place with high energy radiation, which is absorbed at random along the chain.

Among the crystalline polymers which degrade under irradiation, Teflon (polytetrafluoroethylene) has been most closely studied. In spite of its very inert chemical behavior, it shows remarkably high sensitivity to radiation, and marked reductions in tensile strength and ultimate elongation occur at doses well below one megarad in the presence of oxygen. Even in the absence of oxygen, considerable reduction in mechanical properties is observed at doses which are low compared with those needed in other polymers. To explain this high sensitivity we must either assume *(1)* the G value for chain fracture is abnormally high, or *(2)* the G value is of the same order as in other polymers ($\cong 3$) but that a very high degree of polymerization is required to give the polymer its excellent mechanical properties, so that even a very limited amount of degradation is effective. It is difficult to reconcile a chain depolymerization reaction of the

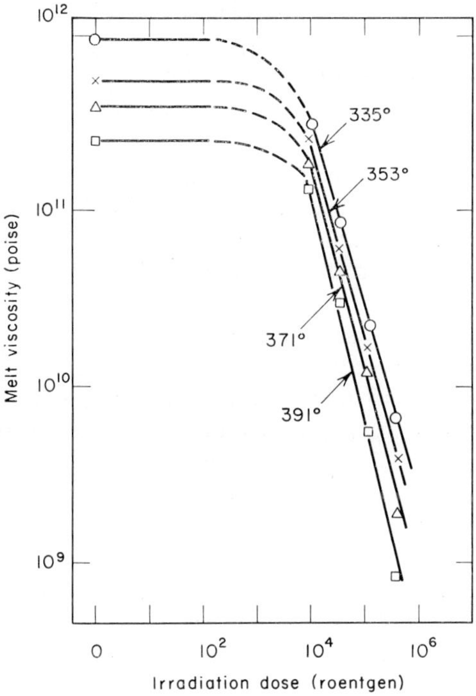

Fig. IVF-6b. Decrease in melt viscosity of Teflon at 335–391°C. as a function of dose (36).

order required with the known high degree of thermal stability of the polymers. Direct measurement of changes in molecular weight cannot be carried out due to the chemical inertness of the polymer, and to clarify the position we may consider associated changes in physical properties. Thus Nishioka and co-workers (36) claim that doses as low as 10^{-2} megarads cause a significant reduction in melt viscosity and in "zero strength time" (Z.S.T.) when exposures are carried out with gamma radiation in air (see Fig. IVF-6). Other authors report loss in shear strength, tensile strength, and ultimate elongation at relatively small doses. On the other hand Young's modulus is substantially unaltered for doses up to at least 10 megarads and the impact strength may indeed be increased due to the plasticizing effect of the fragments produced by radiation. The change at high doses is in any case obvious in irradiated film, which is converted to a friable powder. The effect of radiation on Teflon is also shown by a reduction in melting point, which is presumably related to crystal size and lattice stability.

It is surprising that x-ray diffraction studies show that irradiation causes an increased degree of crystallinity (28). Thus we must envisage a system in which a very few chemical changes cause a marked loss in mechanical properties, without removing crystallinity. The most probable site would be in the amorphous region near the crystallite surface, but the whole problem requires much closer study. This remarkable sensitivity to radiation of Teflon appears to be largely associated with the presence of oxygen; thus Wall and Florin (49) have found negligible residual tensile strength for films irradiated to 6 megarads in air, but only 50% decrease when the same dose is given in vacuum. All these experiments appear to involve problems of oxygen diffusion throughout the largely crystalline lattice, and the situation does not appear to be adequately understood.

As a second example of a crystalline polymer which degrades under radiation we may consider the polycarbonate studied in considerable detail by Golden (14,15). From the reduced viscosity at various radiation doses it is found that G(scission) = 0.09 for irradiation in vacuum, and 0.14 in oxygen. These low values are probably indicative of radiation protection by the aromatic group, and doses of the order of 10^3 megarads are needed to halve the molecular weight. From the nature of the gases evolved it is clear that the carbonate is the source of weakness. From the plot of $1/\bar{M}_v$ against dose it is deduced that the initial molecular weight is nonrandom, though it rapidly becomes so as a result of degradation. The maximum fall in elongation at break and yield strength takes place for doses in the region of 100 megarads; brittle fracture also ensues at this dose. The fracture energy F (in ft-lb.) can be related to $\bar{M}_v$; $F = 0.0044(\bar{M}_v - 13{,}500)$ (see Fig. IVF-7). It is of interest to note that these changes occur in a relatively small range of molecular weights, as deduced from the viscosity, and here again it may perhaps be useful to confirm, with unirradiated samples, that the change in physical properties is due only to the reduced molecular weight, and not to the reactivity of radicals trapped in the strained matrix.

In spite of its interest, little basic information is available on the effect of radiation on oriented degrading polymers. One outstanding

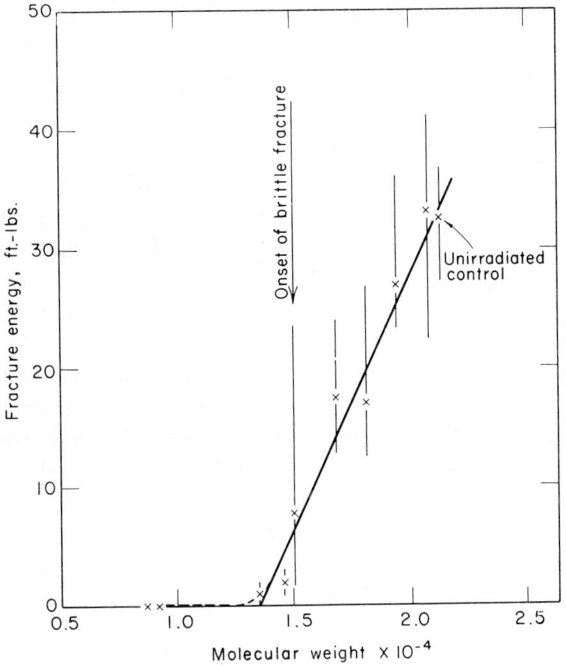

Fig. IVF-7. Fracture energy of a polycarbonate in terms of molecular weight (as modified by radiation) (14).

case is cellulose. From changes in viscosity in irradiated samples, attempts have been made to deduce the G value for main chain scission. In wood pulp, cotton linters, and cotton, the degree of crystallinity is about 70%, and we may expect different effects in the crystalline and amorphous regions, molecular chains in the latter suffering immediate scission, whereas radicals trapped in the former react very slowly, giving rise to a considerable post-radiation effect (possibly by reaction with oxygen). The presence of strongly absorbed water can greatly affect the changes resulting from radiation, both by the radicals produced in it which then react with the polymer, and mainly by the greater flexibility it permits in the molecular chains (31). The chemical modifications introduced in the crystalline regions also dislocate their structure, and hence we expect to find a reduction in crystallinity, as well as a reduction in orientation. It is worthy of note that in spite of the relatively high G value for scission, cellulose maintains much of its initial strength for doses of a few megarads. For cellulose, as for dextran, main chain scission is by no means the only form of molecular structural damage.

IVF8. Crosslinking Polymers

The formation of crosslinks, which occurs when many long chain polymers are exposed to high energy radiation, causes many profound changes in their physical and chemical behavior. The transformation from a linear (or branched) structure to a molecular network is exemplified by the vulcanization of rubber, but with radiation this transformation can occur in a very wide range of polymeric materials, to an extent which is directly proportional to dose, over a very wide range of temperatures, and even in pre-stressed systems. No additives (such as sulfur) are needed, although considerable modifications can ensue when active fillers such as carbon black are incorporated prior to irradiation. In presenting some representative examples of such modifications we shall relegate to a later stage other aspects such as network formation occurring by a chain reaction and the effect of fillers and reinforcing materials. Here we consider only flexible amorphous (such as rubber), rigid amorphous (polystyrene) and largely crystalline (polyethylene) polymers. Before doing so, it may be useful to summarize some of the main analytical results.

It is generally accepted that crosslinking (dimerization of polymer chains) occurs at random throughout the polymer. This assumption is justified by the many deductions made from it which have subsequently been confirmed experimentally. It is not valid for copolymers in which one monomer is more radiation sensitive than the other. The exact mechanisms by which such crosslinks are formed are not yet generally agreed, though they probably occur in most cases by the combination of two adjacent radicals. Various suggestions have been made to account for the mobility needed to allow the initially dispersed radicals to meet. The G value for crosslinking is of the order of unity for many polymers, but the efficiency of crosslinking is also sometimes expressed in terms of a q_0 value, this being the proportion of monomer units crosslinked per unit dose (one megarad). q_0 is related to G

$$q_0 = 2.08 \times 10^{-6}\, Gw$$

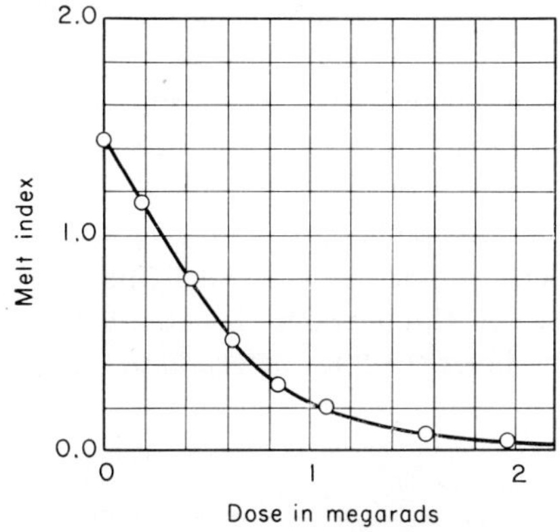

Fig. IVF-8. Decrease in melt index (increase in viscosity) of irradiated polyethylene below the gel point (16).

where w is the monomer unit weight and G is the number of crosslinks per 100 e.v. absorbed (i.e., half the number of crosslinked units). The usual M_c value (molecular weight of the chain between crosslinks) is directly related to q_0 with a small correction for the initial molecular weight. If this correction is ignored:

$$M_c = w/q_0r = 0.48 \times 10^6/Gr$$

where r is the radiation dose in megarads. Thus if $w = 100$ and $G = 1$, then $q_0 = 2.08 \times 10^{-4}$ and $M_c = 5 \times 10^5$ per megarad. To produce a crosslinked network in rubber corresponding to maximum tensile strength requires a dose of 45 megarads so that with G(crosslinks) $\cong 1$ this maximum corresponds to $M_c \cong 1.1 \times 10^4$; $q_0r = 6 \times 10^{-3}$. These relationships enable us to convert radiation dose to density of crosslinking and hence study the influence of crosslinking density on physical properties. The conversion only requires a knowledge of G, with a correction (often minor) for initial molecular weight.

For small doses, crosslinking links molecules together to give branched polymer, which is still soluble. The increase in weight-average molecular weight $\bar{M}_w$ above its initial value $(\bar{M}_w)_0$ obeys the following relationship

$$\bar{M}_w = (\bar{M}_w)_0/(1 - \delta)$$

where δ is the crosslinking coefficient, i.e., the number of crosslinked units per weight average molecule ($\delta = q_0 r(\bar{M}_w)_0/w$), Associated with this increase in molecular weight, there is an increase in viscosity and a fall in melt index (see Fig. IVF-8). At $\delta = 1$, $\bar{M}_w$ becomes infinite and the formation of closed loops or insoluble network commences. This point (gel point) represents a drastic change in the physical behavior of the polymer, since it is no longer completely soluble nor can the network be extended indefinitely without disrupting primary bonds. The corresponding gelation dose r_{gel} can be readily measured, and used to determine either q_0 or G or $(\bar{M}_w)_0$. Thus

$$r_{\text{gel}} = w/q_0(M_w)_0 = 4.8 \times 10^5\, G/(M_w)_0$$

At doses above the gelation dose, the value of M_c decreases and the network becomes tighter. If the molecular chains are flexible, the configurational elastic properties of the network can be readily determined from the statistical theory of high elasticity, although a correction must be applied for the ends of the polymer not linked into the network. For inflexible chains there is little apparent change in mechanical properties on going through the gel point, since it is the entanglements which determine the elastic moduli. For partly crystalline polymers we must distinguish between the reactions occurring in the crystalline and in the amorphous phase.

IVF8a. Crosslinking of a Flexible Polymer

To illustrate the change in behavior with increasing dose or degree of crosslinking, we may choose rubber irradiated in the absence of any of the normal vulcanizing additives. Small doses result in an increase in branching and molecular weight, and hence viscosity, until the gelation dose is reached. Beyond this dose we may speak of elastic deformation, elastic moduli, tensile strength, and elongation at break, hardness, etc. Young's modulus increases with dose, according to the approximate theoretical formula for small elongations:

$$\mathrm{E} = 3\rho RT/M_c = 3\rho RT \times 2.08 \times 10^{-6} Gr$$

when ρ is the density. To this formula must be added a correction factor to allow for the initial finite molecular weight. This can be

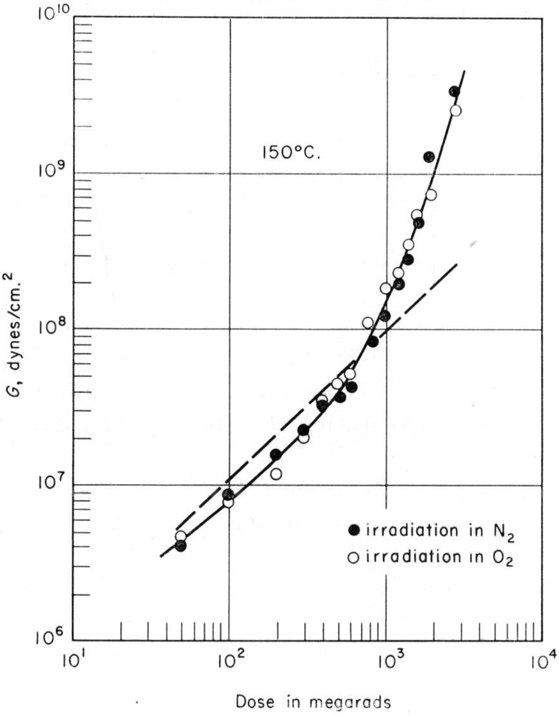

Fig. IVF-9. Shear modulus of irradiated rubber: (—) observed (22), (– –) calculated assuming $G = 0.83$ (21), $E = 3\rho RT/M_c$, effect of chain ends neglected.

approximated to by stating that E differs from zero only above the gelation dose

$$E = 3RT \times 2.08 \times 10^{-6}G(r - r_{\text{gel}})$$

This formula is found to hold satisfactorily up to high doses, with the possible exception of the proportionality factor 3, whose value depends on the particular theory of high elasticity chosen Further detailed measurements using radiation may well give a more accurate experimental value for this factor, now that the G value for crosslinking rubber has been determined with an adequate degree of accuracy. At very high doses ($>10^3$ megarads) E reaches higher values than are deduced from the formula. This departure would be expected to arise when the distance between successive crosslinks

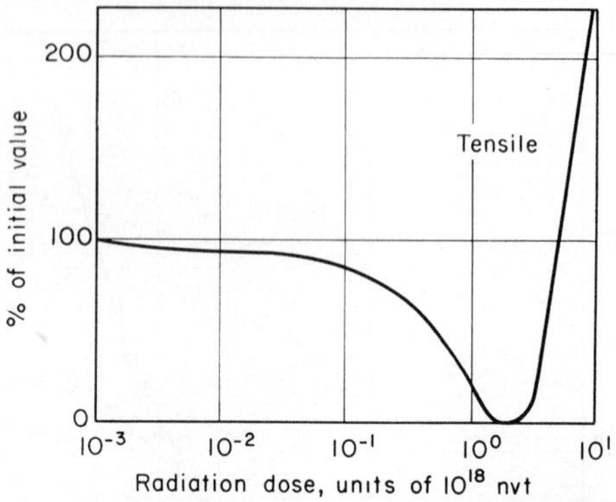

Fig. IVF-10. Tensile strength of a cured rubber, subsequently exposed to nuclear reactor radiation (10^{18} nvt $\sim$ 600 megarads) (2).

becomes comparable with segment length, so that the individual chains between crosslinks can no longer take up a randomly coiled configuration (Fig. IVF-9).

For the tensile strength the situation is somewhat more complex. This increases with dose, and reaches a maximum value in the neighborhood of 45 megarads. At higher doses it decreases to a low value at a few hundred megarads, when the mechanical behavior corresponds to that of a somewhat overvulcanized rubber (see Fig. IVF-10, where the initial chemical cure corresponds to an equivalent radiation dose of nearly 100 megarads). At even higher doses it rises rapidly as the highly irradiated rubber begins to take on many of the properties of an ebonite or glass, with low extensibility and high strength. We may therefore distinguish three deformation regimes: high elasticity ($r \cong 5$ and 100 megarads, $M_c \cong 1 \times 10^5$ to 5×10^3); a cheeselike structure which is friable ($r \cong 100$ and 600 megarads; $M_c \cong 10^3$); a dense glasslike network ($r > 600$ megarads, $M_c < 10^3$). The ability to produce a range of specimens covering these three distinct patterns of mechanical behavior, which can nevertheless be compared quantitatively (in terms of the dose), should be of considerable help in assessing quantitative theories of fracture in rubberlike and glasslike systems. The qualitative differences are already clear; if $M_c \gg$ segment length, we have high-elastic rubberlike behavior; M_c comparable to segment length corresponds to a weak, brittle, cheeselike structure which crumbles into fragments when

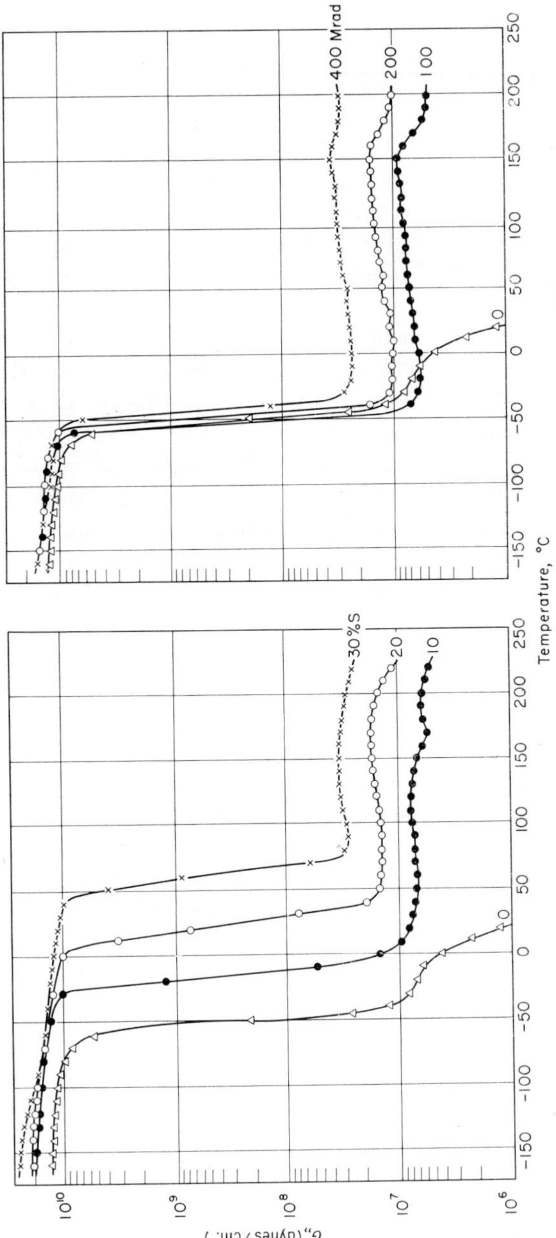

Fig. IVF-11. Shear modulus G of rubber in terms of crosslinking (22). (*a*) Sulfur vulcanization. (*b*) Radiation crosslinking. The difference in temperature between the curves in (*a*) and (*b*) shows that the transition is due to the S bridges.

overstressed; $M_c \leqslant$ segment length, i.e., a few monomer units only, leads to hard glasslike properties.

We may also compare the maximum tensile strength obtained with rubber, crosslinked by radiation, with that found with rubber vulcanized by conventional chemical techniques. Although various claims have been made of practical advantages for the radiation cure, it is fairly definite that the tensile with radiation rarely exceeds 10^8 dynes/cm.^{2}, comparable with that for a peroxide cure, but only half of that for a sulfur cure. These maximum tensile strengths all correspond to the same value of M_c ($\cong 10^4$), and the lower value for the tensile strength in the former two cases presumably shows that, because of steric strain at the crosslink, this junction rather than the main chain itself, is the point of rupture. Alternative explanations are, however, possible.

The difference between sulfur and radiation crosslinking of rubber is also found when the shear modulus is measured at various temperature. For rubber containing little or no sulfur, the elastic modulus falls from a high value ($\cong 10^{10}$ dynes/cm.^{2}) at low temperatures to the low value (about $10^{6.5}$ dynes/cm.^{2}) typical of highly elastic rubber, the transition occurring at -50°C. (see Fig. IVF-11). With increasing sulfur content this transition temperature rises from -30 to 0°C. at 10% sulfur, from 0 to 40°C. at 20% sulfur, and from 50 to 70°C. at 30% sulfur. For radiation crosslinking (doses up to 400 megarads), the transition temperature to a high elastic rubber is substantially unaffected by the radiation dose. Thus the increased transition temperature observed with sulfur is not due to the higher density of crosslinks, but to some physical (steric) hindrance due to the presence of sulfur. Measurements with dicumyl peroxide crosslinked rubber also reveal the absence of any marked temperature variation and in this respect too it is similar to radiation-induced crosslinking.

The same elastic modulus (at room temperature) is obtained with 10% sulfur as with 100 megarads of radiation. The former corresponds to 2×10^{21} sulfur atoms/g. of rubber, the latter to 6.2×10^{19} crosslinks/g. ($G = 1$). Since the elastic modulus of high elasticity rubber is determined only by the density of crosslinks, each crosslink must correspond to about 30 sulfur atoms.

The behavior shown by irradiated rubber (polyisoprene) is also shown by a wide range of polymers, some of which cannot be readily crosslinked by chemical means. One of many examples is dimethylsiloxane polymer, in which radiation can produce a transparent elastomer whose maximum tensile strength is obtained at a dose of 10 megarads. At higher doses the tensile strength decreases until a glasslike structure is approached, when it rises again. With a value of G(crosslink) of 2.4, this maximum high elasticity tensile strength corresponds to $M_c \cong 2 \times 10^4$, and is at least as high as for conventional chemical cure. Irradiated silicone fluid that has been converted to a high elastic network, however, has a very low tensile strength, and tears with ease. These weak properties can be drastically improved by the incorporation of suitable fillers. These are discussed below.

IVF8b. Crosslinking of an Amorphous Rigid Polymer

The irradiation of amorphous rigid polymers which crosslink (such as atactic polystyrene) produces little obvious change in mechanical behavior, since the presence of entanglements completely overshadow any effect of crosslinks. At higher temperatures, when the polymer softens, it is converted into a network with high elastic properties appropriate to the degree of crosslinking. Very high degrees of crosslinking may produce a more brittle structure. This may be revealed by internal fractures of a flat disklike character, which, however, do not occur near the surface.

Due to the protective effect of the benzene groups, polystyrene is remarkably radiation resistant, and the G value for crosslinks is only about 0.04. For doses up to 10^3 megarads, elastic moduli and shear strength are unaffected at room temperature, whereas tensile strength rises above this dose, which corresponds to a value of $M_c = 10^4$. This enables us to assess the importance of entanglements and end groups in the polymer. It is of interest to note that impact strength shows significant decreases even at much lower doses.

The insertion of a crosslink into a specimen causes a rearrangement in the molecular chains in its neighborhood, and they are usually drawn closer together. Given the opportunity, this results in a higher density for the material (which in the case of partly crystalline polymer is likely to be overshadowed by the reduced degree of crystallinity). The density of the crosslinked polymer can also be changed by the applied pressure, and an approximate equation of state, analogous to the equation of state of a gas, can (as a first approximation) be written for a flexible amorphous polymer (4)

$$(P + P_0)(V - V_0) = (1 - c)RT/M$$

Here P is the external pressure, P_0 the internal pressure ($\cong 10^4$ atm.), V is the specific volume, and V_0 the extrapolated minimum specific volume occupied by the polymer at low temperature, high pressure ($V_0 \cong 0.8$ cc./g.), while c is the proportion of the total polymeric system immobilized by crosslinks, and M is a quantity representing the molecular weight of the effective unit in the flexible polymer chain. Over a certain range c is proportional to dose.

IVF8c. Crosslinking in a Partly Crystalline Polymer

The properties of a crosslinked crystalline polymer are best exemplified in the case of polyethylene whose behavior under radiation has received more attention than any other polymer. Polyethylene can be obtained with a range of molecular weights, degrees of branching and of crystallinity, but the main features of the radiation-induced changes are common to both high and low density material. We find not only crosslinking, but also a smaller degree of degradation, the formation of main chain (vinylene) unsaturation, the elimination of other forms of unsaturation (vinyl, vinylidene where present), the liberation of hydrogen and to a much lesser extent of other hydrocarbons (largely from the branches), and the destruction of crystallinity. For doses up to at least a hundred megarads it is the crosslinking which is primarily responsible for the changes in mechanical properties induced by radiation. Much higher doses are required to eliminate crystallinity at room temperature.

For doses below that required to induce gelation (usually 2–5 megarads) the effect of radiation is to increase the average molecular weight and degree of long chain branching; the melt viscosity rises, and irradiated polyethylene when extruded tends to take up an irregular globular structure. It is also found that even small radiation doses largely eliminate environmental stress cracking.

Beyond the gel point and up to about 100 megarads, there is a slight rise in the tensile strength toward a maximum, but the elongation at rupture, after an initial rise, decreases fourfold for a dose of 100 megarads (Fig. IVF-12). An examination of the stress/strain curve shows that this reduction is due primarily to a reduced degree

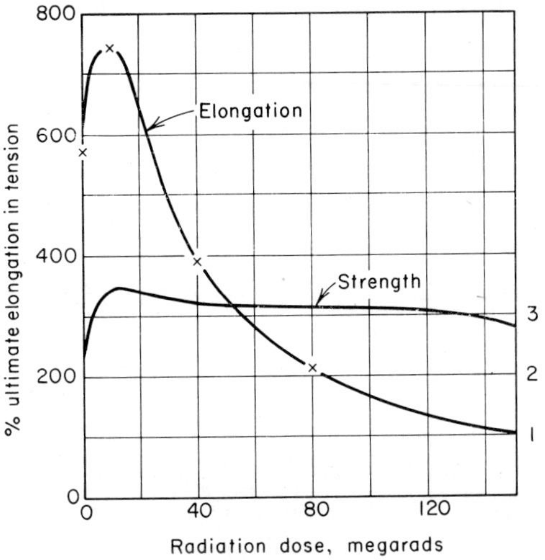

Fig. IVF-12. Tensile strength and elongation at rupture of irradiated, low-density polyethylene (30).

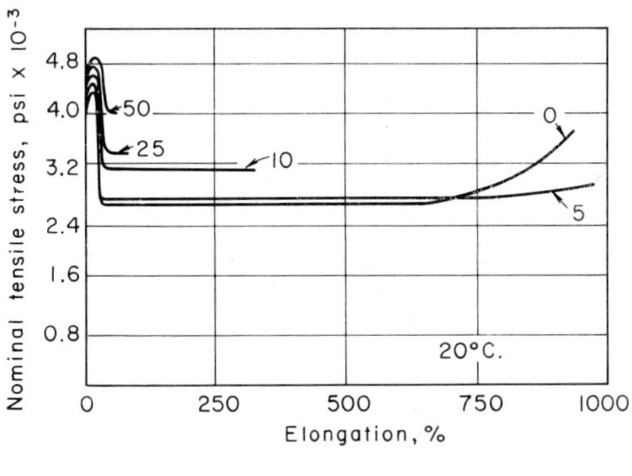

Fig. IVF-13. Cold flow, elongation at rupture of irradiated high-density polyethylene. (Doses in megarads.)

of cold flow before the point is reached at which the stress rises rapidly to a maximum and rupture ensues (see Fig. IVF-13). If we assume that this cold flow is due to reorientation of the crystals, to which the polymer chains in the amorphous regions must conform, then this reduction in extension can be ascribed plausibly to the additional constraints imposed in these amorphous regions by the presence of crosslinks.

It is not known whether crosslinks are also formed in the crystalline regions. Although the same total G value for crosslinking is observed for low and high-density polyethylene, in which the percentage of

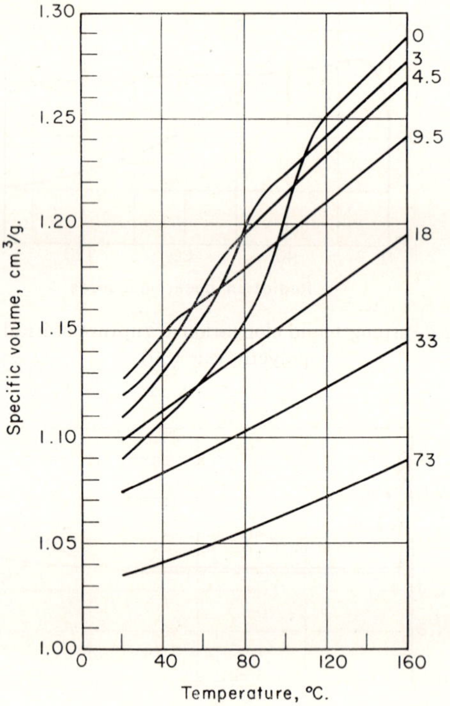

Fig. IVF-14. Effect of dose on crystalline melting point of low-density polyethylene (dose in pile units $\simeq$ 50 megarads).

amorphous structure differs by a factor of three, it is possible that the radicals formed in the crystalline region migrate to the amorphous region and form crosslinks there.

For doses of several hundred megarads, delivered at room temperature, crystallinity is greatly decreased, and this together with the high density of crosslinking gives rise to a weaker material, with some of the crumbly, cheeselike properties already mentioned in highly-irradiated rubber. At even higher doses (several thousand megarads) a hard glasslike structure is obtained, which ruptures by a conchoidal fracture. This very highly crosslinked network is transparent, apart from a dark yellow color due to the presence of chromophores in the form of conjugated unsaturation.

The relatively high stability of crystals to irradiation at room temperature is of interest. Disruptions or stresses within the crystallite may be induced by the intrusion of crosslinks; they are certainly caused by the increase in vinylene unsaturation, such that for a dose of 100 megarads we expect one vinylene group to be present for each 500 carbon atoms. For a crystallite of, say, 10^6 carbon atoms, there will be 2000 such centers, which cannot be readily accommodated within the lattice. Nevertheless the degree of crystallinity is hardly affected at room temperature. However, the melting point is successively reduced, so that for a dose of about 500–800 megarads, no crystallites remain with melting point above room temperature. This decrease in crystallite size with dose can be deduced from their melting points (as revealed by a specific volume/temperature plot; Fig. IVF-14) and enables us to correlate crystal stability with the density of irregularities within the crystal lattice.

Although the individual crystals appear to be very stable under irradiation in spite of the intrusion of unsaturation and possibly crosslinks, it must be assumed that they can only accommodate within themselves these irregularities if they are produced by irradiation in the formed crystal. If after irradiation the specimen is taken above the maximum melting point and then recooled, there is a considerable decrease in crystallinity. Thus the irregularities can inhibit recrystallization.

The ability to crosslink polyethylene by relatively small radiation doses has led to a number of useful applications. These are largely based on the fact that whereas at room temperature the mechanical strength of the polymer is determined by the binding of molecules within the crystal, at higher temperatures (above 115–135°C.) when ordinary polyethylene melts, irradiated polyethylene is converted to a rubberlike structure, whose deformability is determined by the radiation dose. If maintained in a deformed condition while cooling, the specimen retains this shape after crystallization has taken place. On subsequent heating the crystals melt, and the specimen reverts

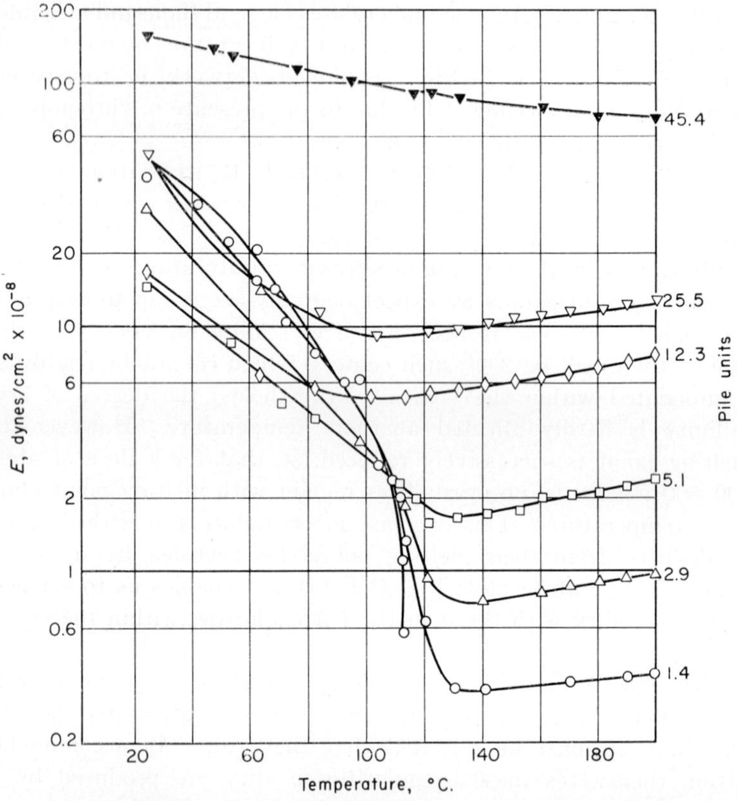

Fig. IVF-15. Variation of Young's modulus with temperature at various doses (in pile units) (5).

to the shape it held during irradiation. This is the so-called "memory effect."

Irradiated polyethylene provides an interesting system in which to study such properties as elastic deformation and speed of response, permanent set, tensile strength and elongation, since deformation, the density of crosslinking (and to a lesser extent the degree of crystallinity), can be varied at will. A number of techniques have been used to study several of these properties, including nuclear resonance, dynamic mechanical measurements, and birefringence. The change in mechanical properties of irradiated polyethylene has been followed both in static and dynamic measurements. The static measurements

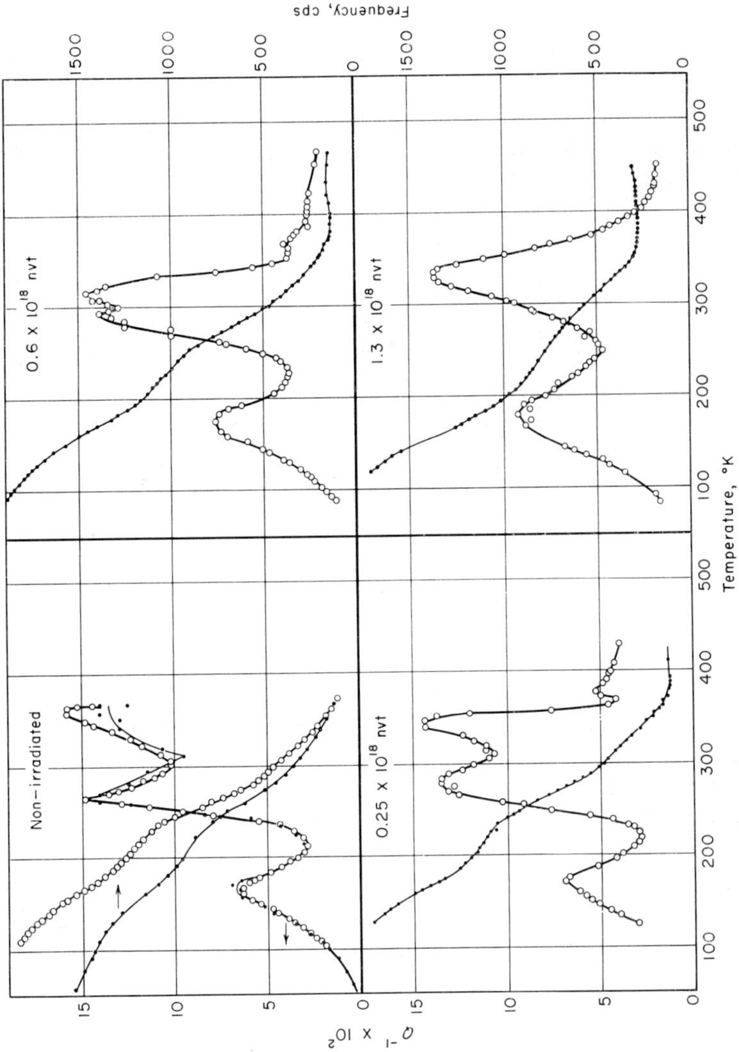

Fig. IVF-16. Damping and resonance frequency for irradiated low-density polyethylene following nuclear radiation (10).

show a fall in Young's modulus with temperature to a minimum value, which is zero for unirradiated (molten) polyethylene, but which differs from zero once a network has been formed by radiation (see Fig. IVF-15). This finite modulus (and tensile strength) corresponds to the properties of a high elasticity network. For much higher doses, when the residual crystallinity is small and the density of crosslinks high, the transition from a behavior dominated by crystallinity to one determined by the residual radiation-induced crosslinks is far less marked.

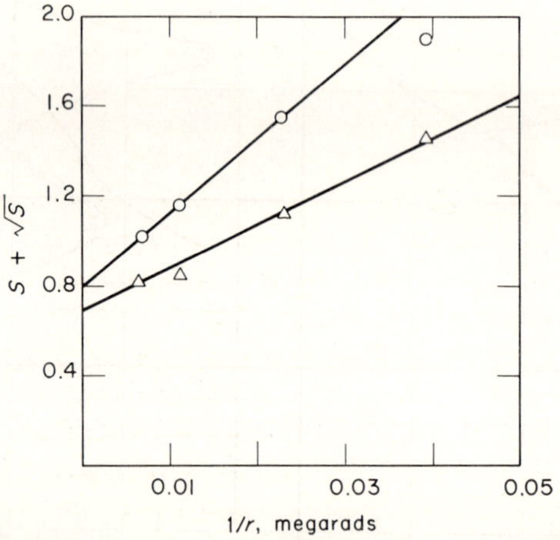

Fig. IVF-17. Effect of radiation on gel formation in polypropylene, where s is the soluble fraction: (O) irradiated at 25°C, and (△) irradiated at 25°C. and annealed at 180°C.

A number of dynamic measurements have been made over a range of temperature. The mechanical loss peaks reported at 165, 265, and 355°K. in low density polyethylene are altered. Thus the peak at 355°K. associated with crystal melting decreases in height and moved to lower temperatures. On the other hand the peaks at 265 and 165°K. both decrease in height and move to higher temperatures; these peaks have been associated with branches and segmental motion, and these changes may be due to loss of branch points and a decreasing mobility in the amorphous regions (Fig. IVF-16).

For high density irradiated polyethylene the peak corresponding to crystallinity shows a considerable decrease in temperature if the irradiated specimen is preheated before measurement. This change is to be expected if the initial crystals contain trapped radicals and other dislocations which interfere with recrystallization once the melting point has been reached.

The behavior of polyethylene may be contrasted with that of polypropylene. The latter polymer has a structure which is intermediate between that of polyethylene (which crosslinks under radiation) and polyisobutylene (which degrades), and in fact when polypropylene is irradiated it shows both crosslinking and degradation, the former being somewhat more frequent. As a result of these two opposing tendencies a relatively high dose is needed to attain the gel point, and even at very high doses, the soluble fraction does not tend to zero. A very convenient form of comparing the two reactions consists in plotting $s + s^{1/2}$ versus the reciprocal dose $1/r$. The extrapolated value to infinite dose gives the ratio of degradation to crosslinking. Figure IVF-17 shows that the ratio is about 0.8, with a gelation dose of about 25 megarads. However, the crosslinking reaction can proceed further if the irradiated specimen is subsequently annealed to allow radicals to meet and crosslink. Crosslinking and degradation have sometimes been considered as opposing tendencies insofar as they are equivalent to reducing or increasing the number of molecules, but this is an oversimplification, since a highly crosslinked network of short chains would behave very differently to one with a correspondingly smaller number of crosslinks between longer chains. The effect of simultaneous crosslinking and chain scission on mechanical properties could usefully be studied in irradiated polypropylene.

IVF8d. Crosslinking of an Oriented Crystalline Polymer

Two effects of radiation on an oriented crystalline polymer must be considered: (*a*) the effect of crosslinking and (*b*) the destruction of crystallinity.

Both these effects can be seen in the case of polyethylene terephthalate (Mylar or Terylene) which can be obtained in the form of uniaxially or biaxially oriented film. This polymer has very high radiation resistance; nevertheless, in the unoriented form it shows a considerable drop in tensile strength at doses above 10^3 megarads, presumably due to intensive crosslinking. For uniaxially oriented polymer, the reduction in tensile strength is far smaller and doses about ten times greater can be accumulated for a similar drop. Crystallinity is also more affected in the unoriented specimens. For nylon both tensile strength and elastic modulus decrease more rapidly in the case of oriented fibers and approach the value for unoriented material at high doses. The doses required to achieve these changes are so high that the chemical structure is greatly modified and one may question whether we are in fact studying the properties of these two polymers (23).

Some studies have also been made on the mechanical properties of irradiated polyethylene fibers (42). Highly elongated filaments show little of the loss of tensile strength observed with unelongated filaments after irradiation. Polyvinyl chloride fibers irradiated in vacuum show a shrinkage above 100°C. under stress and a rapid elongation at 160°C. However, when irradiated in vacuum to doses of about 40 megarads, the initial shrinkage above 100°C. is retained as crosslinks can now hold the individual molecules together.

IVF9. Network Formed by a Chain Reaction

In the polymers considered above, we have been concerned with homopolymers, which crosslink under the influence of radiation, each such crosslink being formed separately. There is a linear relationship between crosslinking density and radiation dose, which can be represented by a G value for crosslinking, typically about 1–3. Only under exceptional conditions (e.g., steric hindrance in a highly crosslinked, dense network) is there a serious departure from linearity. We can now consider a different set of conditions, in which a chain reaction is involved, so that each radical (or pair of radicals) can give rise to a number of crosslinks. This situation is somewhat analogous to a polymerization reaction, but takes place in the presence (usually) of another polymer which is crosslinked via the polymerization bridges. It is an important feature of the process that a number of such bridges are formed for every radical (or radical pair) produced by radiation so that a crosslinked network structure can be reached at much lower doses than are needed for the random crosslinking discussed earlier. We may visualize such a molecular network having the structure

```
         ⋮
         B
—AAAAAAAA—
   B     B
   B     B
—AAAAAAAA—
   B     B
   B     B
—AAAA    B
   B     B
   B     B
—AAAAAAAA—
   B     B
   ⋮     B
         ⋮
```

where A represents the initial polymer comprising a number of units A, and B are separate monomer units which under the influence of radiation constitute a polymerization chain extending through a number of polymers A and serving to bind them together. An example of such a network is the well-known unsaturated polyester–styrene mixture, often used in conjunction with glass fiber to produce a stiff, tough material. Other examples, which have been mainly studied by radiation, include polyvinyl chloride or polymethyl methacrylate linked together by allylic derivatives, and vinyl silicones, for which no additional polymerizing compound is needed.

To understand the nature of the network formed, and relate this to its mechanical properties, we need to know the number and length of the chains B, as well as the average distance between them (i.e., the number of A units between B bridges). Apart from the basic properties of polymers A and B (e.g., their flexibility, softening temperature) we would also wish to know the mechanism of termination of the polymerization chains B (since we may assume that initiation takes place either by radical formation on polymer A, or in monomer B, by exposure to radiation). Some mechanism of termination is also an essential feature, since without it a single polymerization chain resulting from a single radical could in theory link all the polymer molecules A together into a single system. Among the possible terminations there may be mentioned trapped radicals, cyclization, exhaustion of monomer B, resonance stabilization, and reaction with inhibitor or oxygen. As a result the properties of an irradiated network of this character will depend not only on the initial materials present and on the radiation dose, but also on the dose rate and the time left standing after irradiation.

It must be admitted that our knowledge of such network systems formed by radiation is far less advanced than that for simple radiation-induced crosslinking (as in polyethylene, rubber, polystyrene).

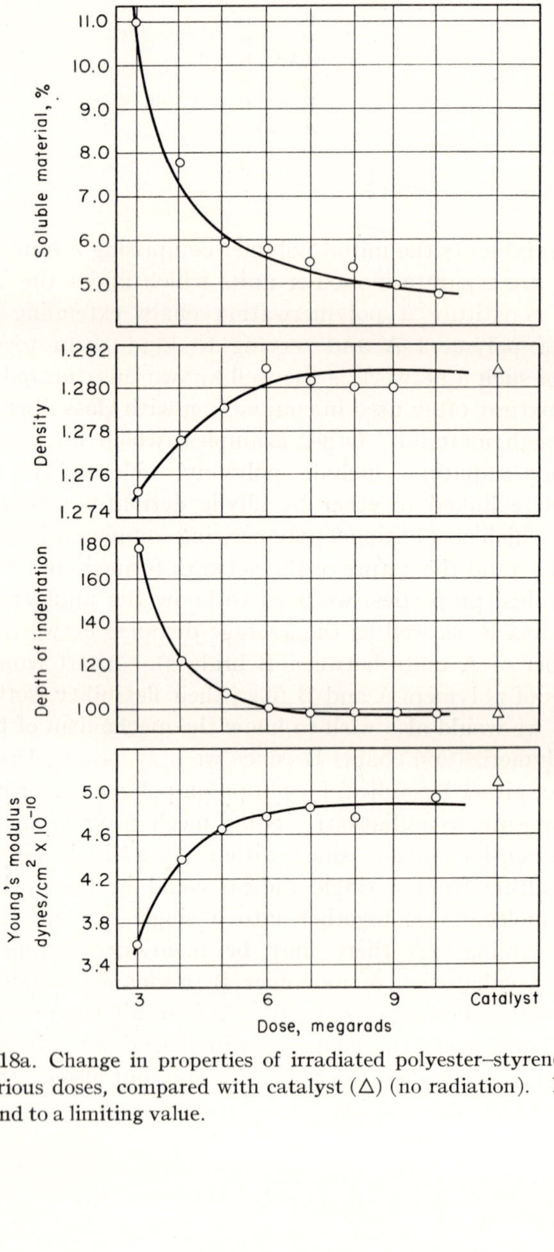

Fig. IVF-18a. Change in properties of irradiated polyester–styrene mixtures. Effect of various doses, compared with catalyst (△) (no radiation). Mechanical properties tend to a limiting value.

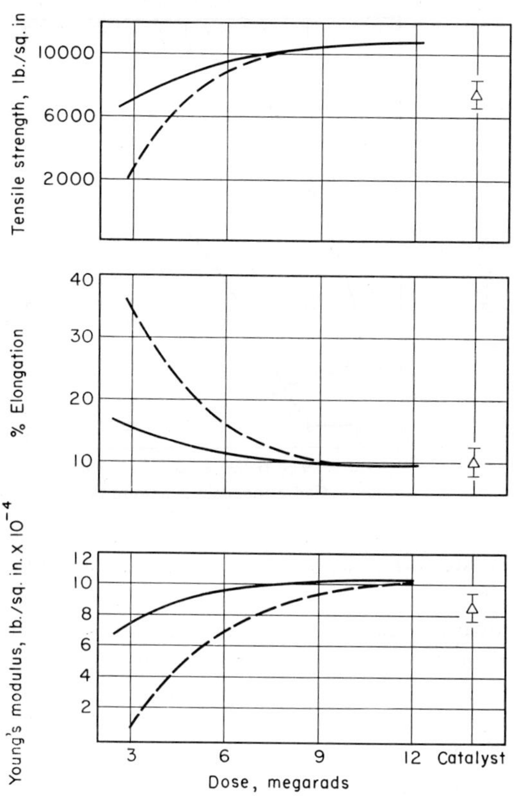

Fig. IVF-18b. Effect of standing on a partially cured polyester–styrene mixture: (- -) immediately after radiation, (——) after ageing. The reaction of trapped radicals only partly completes the cure (6).

To some extent this is due to the greater number of variables involved, as discussed above, and to the existence of a post-radiation effect. Among the techniques used have been solubility studies, measurement of residual unsaturation, dynamical mechanical studies, and electron spin resonance measurements of trapped radicals.

For a typical polyester–styrene mixture, there is a change in physical properties from gelation up to the dose corresponding to the state of full cure. Thus the solubility decreases, while the density, hardness, elastic modulus, and tensile strength rise to maximum values, which are similar to those found for chemical cure with cat-

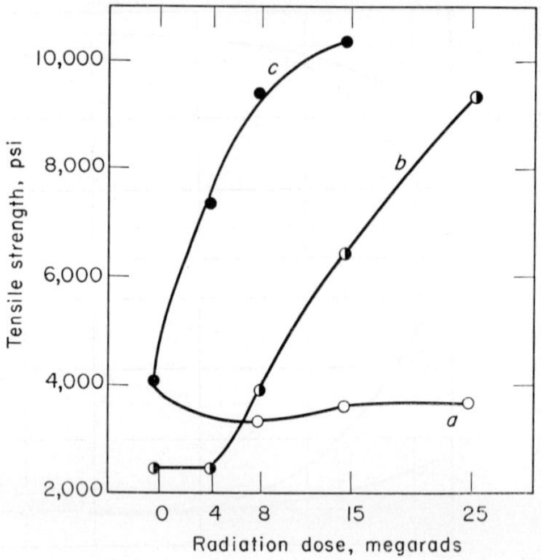

Fig. IVF-19. Network formation in PVC; effect on tensile strength (37): (*a*) + dioctyl phthalate; (*b*) + diallyl sebacate; (*c*) + triallyl cyanurate.

alysts. If, however, the radiation ceases at a lower value, the specimen will continue to cure, at least in part, due to the continued reaction of the trapped radicals, until limited by exhaustion of reactive sites, by loss of radicals, or by lack of mobility within the crosslinked structure. The last can be overcome by a rise in temperature, the second by further irradiation, and effectively it is found that heating can permanently raise the transition temperature (as deduced from measurements of elastic modulus and damping) though only in a partly cured system. In a so-called fully cured system, the introduction of further radicals by radiation does not affect the elastic properties more than can be explained by the production of individual crosslinks by direct linking together of pairs of polymer radicals in the absence of a chain reaction (Fig. IVF-18).

The comparison of crosslinking by a chain and by a linear reaction can best be illustrated in the case of polyvinyl chloride. This polymer is notoriously difficult to crosslink by radiation to an extent which allows good mechanical properties at elevated temperatures. By the use of certain unsaturated additives far better properties are achieved both at room and higher temperatures. Thus with 30 pph of triallyl cyanurate, an irradiated PVC sample shows the following tensile strengths after irradiation to 15 megarads; at 20°C. 1.1×10^8 dynes/cm.^{2}; at 70°C. 4×10^8 dynes/cm.^{2}; at 100°C. 2.3×10^8 dynes/cm.^{2}; at 150°C. 0.3×10^8 dynes/cm.^{2} or above (depending on composition). Without the cyanurate, the material is extremely weak above its usual softening point, in spite of the presence of some radiation-induced crosslinks. Figure IVF-19 shows that even at room temperature considerable increase in strength may be obtained at modest doses.

Since the curing of such unsaturated systems occurs by an exothermic reaction, the use of chemical catalysts sensitive to temperature to produce the initiating radical suffers from the disadvantage that the reaction once started is difficult to control. The use of radiation in the preparation of samples for fundamental studies of the properties of such crosslinked networks therefore offers many advantages. In particular the ability to work with identical mixtures, differing only in the concentration of radicals produced by radiation, should offer interesting prospects for elucidating the relationship between mechanical properties and network structure.

IVF10. Reinforcing Fillers

The ability to obtain higher tensile strength and moduli in polymers by incorporating fine particles, such as carbon black or silica, is retained when crosslinking is carried out by radiation. The improved properties of rubber for example can be achieved in the absence of sulfur or other compounds, showing that the reinforcement depends only on a reaction between rubber and carbon black.

The reinforcement is most obvious in the case of polydimethyl siloxane, which when crosslinked in the absence of an additive is an extremely weak material. With Aerosil (SiO_2) or similar material, present to the extent of about 40 pph of polymer, an extremely useful high temperature rubber is obtained by radiation alone, the maximum tensile strength being achieved by a dose of about 10 megarads. Reinforcement by carbon black is also possible with radiation, a reaction which is difficult to achieve by chemical means.

It is by no means certain that the nature of the interaction and reinforcement with rubber–carbon black mixes is the same as with silica. However since the most detailed work has been carried out with rubber–carbon black, this system will be considered here.

It is known that carbon blacks, as used for reinforcement, give rise to strong electron spin resonance patterns of the character expected for radicals. It therefore seemed feasible to postulate that reinforcement takes place by an attachment of rubber radicals to the surface radicals present in carbon black. In fact, recent work has thrown doubt on this interpretation of the E.S.R. patterns, and the reaction is almost certainly not due to a simple radical-radical combination. [The E.S.R. patterns are due to mobile charge carriers, whose concentration and character is in no way related to the reinforcing ability of the carbons (45).]

The increase in Young's modulus with radiation dose (i.e., crosslink density) for pure rubber is in accord with the theory of high elasticity. With carbon black the modulus is enhanced by an amount which is approximately proportional to the radiation dose and to the concentration of carbon black. Thus with no radiation there is relatively little reinforcement—even in the presence of considerable amounts of carbon. Presumably the reinforcement (expressed in terms of the increased modulus at the same dose) is proportional to the number of radicals formed in the rubber, and the reaction of such radicals with carbon black (45).

The maximum tensile strength for rubber without any additives is obtained at a dose of 45 megarads, corresponding to a value for M_c of about 10^4. A similar maximum is observed for sulfur vulcanization, again when M_c has approximately the same value. Thus the position of this maximum is determined by M_c and not by the nature of the crosslinks (which do decide the actual value of this maximum tensile strength). If therefore the carbon black reinforces by increasing the density of crosslinking (i.e., reducing M_c for a given dose), it should decrease the dose corresponding to the maximum tensile strength by the same factor as it increases the elastic modulus. In fact the dose for maximum tensile remains unaltered at about 45 megarads, independent of the amount or type of carbon black incorporated. It is, therefore, necessary to abandon the view that reinforcement consists primarily of increased crosslinking density.

Several suggestions may be made to explain these experimental results. One is based on the theory of rupture derived by Bueche, but makes the additional assumption that whereas the usual crosslink operates between two chains, the links via a carbon black particle binds about three polymer chains. Thus each carbon particle can be considered to terminate six rubber lengths;

```
      .          .
        .      .
         R    R
          R  R
           R
   ...RRRCRRR...
           R
          R  R
         R    R
        .      .
      .          .
```

Hence the scission of any one of these chains still allows its load to be carried by the five remaining ones.

A second suggestion is that the carbon-rubber links are different in nature from the rubber-rubber crosslinks, and are mainly effective in increasing modulus, but not tensile strength. This would agree with the evidence favoring the crosslink as being the weak point and the point of fracture. A third possibility is that rubber-carbon links can be readily made, so that if a rubber chain breaks, the terminal rubber radicals formed by the fracture can still be held by the carbon particles. In any case the precise nature of the bond between rubber and carbon black is still unexplained, although it may involve a link between a mobile charge and a radical.

A different problem of considerable interest concerns the properties of carbon-loaded rubber irradiated to very high doses, to produce a material similar in physical properties to an ebonite. This does not yet appear to have received any detailed attention.

IVF11. Properties of Grafted Polymers

Radiation offers a powerful technique for producing graft polymers, and a considerable amount of research has been devoted to the preparation of a wide range of grafts and to the study of the kinetics. Very little information has been published on the mechanical properties of these grafts however. Grafted polyethylene, for example, can be shown to have different maximum melting points, depending on whether the graft has been formed only in the amorphous regions into which monomer can penetrate, or in subsequently recrystallized material in which the graft intervenes in the crystal, reducing their size and number. These result in changes in the melting point of the crystallites which can be detected in specific volume/temperature curves. The properties of the graft will also depend on the reaction temperature and concentration of monomer. By control of diffusion of monomer, differential materials can be prepared, i.e., with a higher concentration of graft copolymer near the surface. In particular it becomes possible to form grafts on the surface of Teflon, with potentially useful physical properties.

IVF12. Crosslinking of Polymers in Solution

Crosslinking can occur in dissolved polymers exposed to radiation when a flexible, swollen network is formed. The physical properties of such a system may well differ from those of a similar polymer, crosslinked in the solid state. For example, there is a great difference in the properties of irradiated polymer solutions depending on their concentration during irradiation. The configuration of the polymer molecules during crosslinking is also affected by any prior dose. Unfortunately, little experimental work appears to have been published on the mechanical properties of such irradiated polymer solutions.

IVF13. Biological Macromolecules

In the wealth of published data on the effects of radiation on biological systems, very little information is available on the changes in physical properties resulting directly from the absorbed radiation.

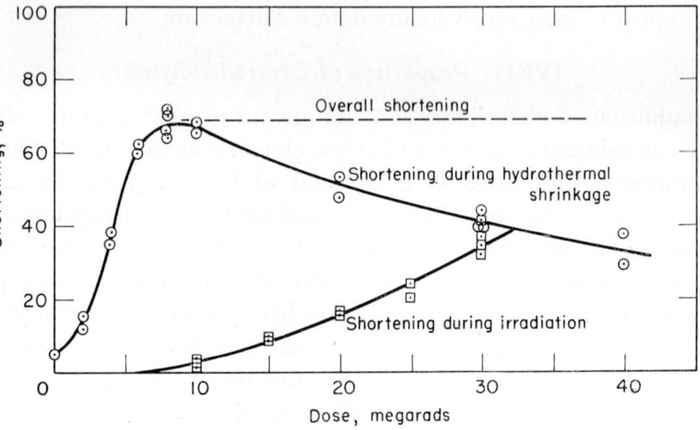

Fig. IVF-20. Shortening of rat tail fibers during irradiation and subsequent hydrothermal shrinkage (1).

This lack is not surprising in view of the remarkably small doses required to cause cell modification and death, several orders of magnitude lower than those needed to produce any modification in mechanical behavior.

One example of the type of information that can be obtained is given in recent work of Baily and Rhodes (1) on collagen, the fibrous protein of connective tissue. Collagen has an oriented, almost crystalline structure, with triple helices bound together by hydrogen bonds. At higher temperatures the collagen studied (rat tail tendon) shrinks, complete collapse occurring at a temperature of 60°C. Radiation destroys the hydrogen bonds, reducing the shrinkage temperature to 30°C. at a dose of 40 megarads (see Fig. IVF-20). Crosslinking of the collagen also occurs under radiation if the crystalline structure has been impaired by the effect of water and heat. Dry collagen shows no such change. However, the doses required indicate that the changes in mechanical properties are to be ascribed to the rupture of a high proportion of the H bonds (about 10%) rather than to the far fewer crosslinks introduced in the structure. This example emphasizes the dangers involved in a simple extension of polymer radiation studies of crosslinking and degradation to the far more complex biological systems.

IVF14. Conclusion

The aim of this subsection has been to show, in fairly general terms, the type of physical changes which can result from the chemical reactions induced by irradiation. For fuller details on any polymer, reference is made to the three volumes published in the field, as well as to review articles which have appeared in recent years. Discussion on technological materials, and factual data on changes of density (due to crosslinking) and properties such as tensile strength, elastic moduli and shear strength has been omitted; much data along these lines will be found in numerous publications, in particular the valuable reports of Bopp and Sisman (2) and of Harrington (17). Other physical changes which appear to fall outside the subject matter of this volume include the changes in electrical and optical properties (e.g., radiation-induced conductivity, thermoluminescence, optical absorption) and the formation of electrical discharges.

The emphasis on lack of experimental data on physical properties is justified by the present state of the subject; research has been largely focussed on the radiation mechanisms involved and on the new chemical structures which can be produced. Much of the detailed information at present available on changes in specific polymers is of primary interest to engineers concerned with the choice of suitable materials for use in a nuclear environment. However, radiation offers a powerful tool to those wishing to investigate basic physical properties of polymers, and it is to be hoped that the considerable gaps in our knowledge will be successively closed as polymer physicists make more use of the technique.

References

1. Bailey, A. J. and D. N. Rhodes, paper presented at 2nd International Congress of Radiation Research, Harrogate, 1962.
2. Bopp, C. D., and O. Sisman, ORNL 1373 (1953); *Nucleonics*, **13** (7), 28 (1955); **13** (10), 51 (1955).
3. Bueche, A. M., *J. Polymer Sci.*, **19**, 297 (1956).
4. Charlesby, A., *Phil. Mag.*, **44**, 578 (1953).
5. Charlesby, A., and N. H. Hancock, *Proc. Roy. Soc.*, **A128**, 245 (1953).
6. Charlesby, A., and V. Wycherley, *J. Appl. Rad. Isotopes.*, **2**, 26 (1957).
7. Charlesby, A., and J. W. C. Spackman, paper presented at the Fourth Rubber Technology Conference, London, 1962.
8. Chapiro. A., *J. Chim. Phys.*, **53**, 295 (1956).
9. Chapell, S. E., J. A. Sauer, and A. E. Woodward, *J. Polymer Sci.*, **A1**, 2805 (1963).
10. Deeley, C. W., D. E. Kline, J. A. Sauer, and A. E. Woodward, *J. Polymer Sci.*, **28**, 109 (1958).
11. Fukada, E., R. W. B. Stephens, and A. Charlesby, *J. Phys. Chem. Solids*, **16**, 53 (1960).
12. Garratt, P. G., *Polymer*, **3**, 323 (1962).
13. Glick, R. E., R. P. Gupta, J. A. Sauer, and A. E. Woodward, *Polymer*, **1** (3), 340 (1960).
14. Golden, J. H., and E. A. Hazell, paper presented at 2nd International Congress of Radiation Research, Harrogate, 1962; *J. Polymer Sci.*, **A1**, 1671 (1963).
15. Golden, J. H., *J. Polymer Sci.*, **45**, 534 (1960): *Makromol. Chem.*, **66**, 73 (1963).
16. Harper, B. G., *J. Appl. Polymer Sci.*, **2**, 363 (1959); **5**, 601 (1961).
17. Harrington, R., *Nucleonics*, **14** (9), 70 (1956); *Rubber Age, N. Y.*, **81** (9), 71 (1957); **84** (5), 798 (1959); **85** (6), 963 (1959); **86** (5), 819 (1960); **88** (3), 475 (1960).
18. Harrington, R., and R. Giberson, *Mod. Plastics*, **36** (3), 199, 314 (1958).
19. Hart, E. J., and R. L. Platzman, "Mechanisms in Radiobiology," in *Radiation Chemistry*, Vol. I, Academic Press, N. Y., 1900.

20 Haynes, A., and C. C. Hsiao, *J. Appl. Phys.*, **11,** 871 (1960).
21. Hayden, P., *Nature,* **184,** 1865 (1959).
22. Heinze, H. D., K. Schmieder, G. Schnell, and K. A. Wolf, *Kautschuk Gummi,* **14** (7), 208 (1961).
23. Hsiao, C. C., Y. C. Das, and A. Baynes, *Brit. J. Appl. Phys.*, **11,** 277 (1960).
24. Hsiao, C. C., and S. T. Chow, *J. Appl. Phys.*, **31** (11), 1869 (1960).
25. Karpov, V. L., Y. M. Malinsky, V. Serenkov, R. S. Klimanova, and A. S. Freiden, *Nucleonics,* **18,** 88 (1960).
26. Kern, E. L., G. Eierman, E. G. Bobalek, and S. M. Skinner, *J. Appl. Polymer Sci.*, **2,** 252 (1959).
27. Kline, D. E., *J. Polymer Sci.*, **A1,** 1621 (1963).
28. Kniess, F., and O. Kratky, *Monatsh.*, **90/4,** 506 (1959).
29. Kogadkin, A., Z. N. Tarasova, I. L. Gol'berg, and K. G. Kuanyshev, *Kolloidn. Zh.*, **24,** 141 (1962).
30. Lawton, E. J., J. S. Balwit, and A. M. Bueche, *Ind. Eng. Chem.*, **46,** 1703 (1954).
31. Leavitt, F. C., *J. Polymer Sci.*, **45,** 536 (1960), **51,** 349 (1961).
32. Matsumae, K., M. Watanabe, A. Nishioka, and T. Ichimiya, *J. Polymer Sci.*, **28,** 653 (1958).
33. Margolin, E. D., and T. D. Philips, *J. Appl. Polymer Sci.*, **2,** 368 (1959).
34. Merrill, L. J., J. A. Sauer, and A. E. Woodward, *Polymer,* **1** (3), 351 (1960).
35. Mullins, L., and D. T. Turner, *J. Polymer Sci.*, **43,** 35 (1960).
36. Nishioka, A., K. Matsumae, M. Watanabe, M. Tajima, and M. Owaki, *J. Appl. Polymer Sci.*, **2,** 114 (1959).
37. Pinner, S. H., *Plastics,* February 1959, January 1960.
38. Pinner, S. H., and V. Wycherley, *J. Appl. Polymer Sci.*, **3,** 338 (1960).
39. Pinner, S. H., *British Plastics,* 518 (October 1962).
40. Rady, A. A., *J. Appl. Polymer Sci.*, **1,** 129 (1959).
41. Ross, M., and A. Charlesby, AERE M/R 1067, 1953; *Atomics,* **4** (8), 189 (1953).
42. Sakurada, I., W. Tsuji, and F. Kimura, *Bull. Inst. Chem. Res. Kyoto Univ.*, **37,** 368 (1959).
43. Shinohara, Y., *J. Appl. Polymer Sci.*, **1,** 251 (1959).
44. Sobue, H., and Y. Tazima, *Nature,* **188,** 315 (1960).
45. Spackman, J. W. C., and A. Charlesby, paper presented at the Fourth Rubber Technology Conference, London, 1962.
46. Sisman, O., and C. D. Bopp, ORNL 928 (1951).
47. Szanto, I., and K. Gal, *J. Polymer Sci.*, **53,** 321 (1961).
48. Turner, D. T., *J. Polymer Sci.*, **1B,** 93, 97 (1963).
49. Wall, L. A., and R. E. Florin, *J. Appl. Polymer Sci.*, **2,** 251 (1959).
50. Wells, H., and I. Williamson, *Chem. Process Eng.*, **41,** 191 (May 1960).
51. Williams, T. F., *Nature,* **186,** 544 (1960).
52. Witt, E., *J. Polymer Sci.*, **41,** 507 (1959).
53. Yanova, L. P., and A. B. Taubman, *Deistvie Ionizirayuskchikh Izluchennii na Neorganicheskie i Organicheskie Sistemy,* Academy of Sciences USSR, Moscow, 1958, p. 314.

Full references to early papers will be found in the following volumes:

54. Bovey, F. A., *Effects of Ionizing Radiation on Natural and Synthetic High Polymers*, Interscience, N. Y., 1958.
55. Chapiro, A., *Radiation Chemistry of Polymeric Systems*. Interscience, N. Y., 1962.
56. Charlesby, A., *Atomic Radiation and Polymers*, Pergamon, Oxford. 1960.

CHAPTER V

Supplementary Contributions

Sections

VA. FRAGMENT SIZE IN SINGLE FRACTURE: THEORY

J. J. GILVARRY

VA1. Introduction

When a brittle solid is fractured, it shatters into a large number of fragments of irregular shape. The problem of specifying the distribution of the number of fragments as a function of size is an old one which has been approached almost completely on empirical or semi-empirical lines. Two general types of distribution function have been employed, of which one is a normal distribution in terms of either the particle dimension or its logarithm, and the other is characterized by appearance of the particle dimension as the argument of an exponential function. A survey and bibliography of the extensive literature on the problem is available (8).

A certain amount of energy is abstracted from the system which is applying the stress when a brittle solid fractures. The correlation of this energy with fragment size also is an old problem. The classical hypotheses in this connection are those of von Rittinger (37) (formulated in 1867), that the energy required is proportional to the new surface formed, and an alternative one of Kick (31) (dating from 1883). Repeated attempts have been made over the years to discriminate experimentally between the two assumptions or to replace both of them (5), but with no semblance of agreement among different investigators (8). In view of von Rittinger's postulate, the question of the energy relationship in brittle fracture obviously is related closely to the problem of the distribution function for the sizes of the fragments.

Geological deposits (6) and various dusts (9,29) tend to conform to a normal distribution in terms of either the particle dimension or its logarithm. Kolmogoroff (32), Halmos (28), and Epstein (11) have shown, by application of the central limit theorem to a mathematical model, that the distribution function approaches the logarithmic-normal form asymptotically, when each fragment is exposed to repetitive fracture. For this reason, such types of distribution function will be ignored in what follows since the present review is concerned primarily with the problem of single fracture.

The other general type of distribution function is represented by the empirical relation of Rosin and Rammler (16,38), in which the particle dimension appears as the argument of an exponential function. [The applicability of this function was first pointed out by Gates (13).] Special cases of the relation correspond to forms with which the names of Schuhmann (43) and of Gaudin (14) are associated. As will appear, this type of distribution applies when conditions approximate single fracture to some degree. The occurrence of the exponential function led Bennett (4) to base a heuristic interpretation of the Rosin-Rammler form on the Poisson law. The same mathematical feature has been invoked to justify an analogy with the Maxwell-Boltzmann law of statistical mechanics (26). Similarly, Poncelet (36) has suggested equipartition of energy in rupture of atomic bonds as the underlying physical mechanism determining fragment sizes. However, none of many such efforts has yielded a rigorous derivation of the proper distribution function for fragment size in single fracture, based on a closely defined physical model and deduced by the laws of probability.

In this review, the distribution function for fragment size in single fracture will be derived from the theory of probability, as first done by Gilvarry (17,18). The cases of both a three-dimensional (17) and a two-dimensional (18) solid will be treated. The physical point of departure is the Griffith theory of brittle strength (25) described by Berry in Section IIA of this volume. On this theory, the failure of a brittle solid to exhibit the high tensile strength implied by its cohesive energy is explained on the basis of the high concentrations of stress at the tips or edges of submicroscopic flaws or cracks which necessarily exist in a gross specimen. Over the years, the theory has been placed on a firm experimental basis as is evident from the older reviews of Orowan (35) and of Anderson (1). Lately, objective existence of the minute flaws postulated in the Griffith theory has been demonstrated experimentally in a most convincing fashion by Gordon, Marsh, and Parratt (23), and by Levengood (33).

At the outset, it should be emphasized that only brittle fracture is considered in this review; ductile, fatigue, and creep fracture are ignored. Discussion of the experimental results obtained by Gilvarry and Bergstrom (19,20) and by Gilvarry (21) will be deferred to Section VB.

VA2. Theory for the Three-Dimensional Case

The development of the theory proceeds from a close specification of the physical model. Physical postulates on the flaw distribution consistent with Griffith's hypothesis are made, and the corresponding mathematical distribution is identified. As will emerge, the physical assumptions made are sufficient to yield a complete and unique mathematical solution of the problem posed. Finally, some shortcomings of the model will be discussed, as reflected in the mathematical behavior of the distribution function obtained.

VA2a. Physical Assumptions

Griffith's hypothesis presupposes that the flaws or cracks ulti mately responsible for fracture exist in a brittle solid prior to appli cation of the stress system producing fracture. It does not follow, however, that every flaw or crack participates in the fracture process, since only those which can grow to macroscopic dimensions under the applied stresses produce fracture surfaces. The critical condition for crack propagation has been obtained analytically by Griffith in the two-dimensional case where a crack can be specified by its length only, in which case the criterion for propagation is that the length exceed a critical value depending on the applied stress, Young's modulus, and the specific surface energy (25). Those pre-existent cracks meeting the criterion on length propagate catastrophically until they are stopped by the boundary of the specimen or by an existent crack in the path. It will be assumed that such flaws are distributed throughout the volume of the specimen. The three-dimensional generalization of Griffith's criterion has been treated by Sack (42) and by Elliott (10); the only essential change arising from this extension is in a numerical coefficient.

The flaws which actually contribute to the fracture of the specimen will be referred to as being "activated." The use of this word does not necessarily imply the existence of an activation energy for propagation of a flaw. This question is a moot one, but Gurney (27) gives reasons for believing that a definite activation energy exists.

As is well known, the initial fracture surface in a specimen of glass (and probably in most brittle solids) starts in a flaw on the surface of the original specimen. The total number of original surficial flaws, however, is insufficient to explain the large number of fragments formed in brittle fracture, since total surface areas of fragments as measured by gas adsorption exceed original specimen surfaces by large factors (2). As an idealization, the original specimen will be considered infinitely large, so that the surficial flaws of this specimen can be disregarded. Therefore, attention will be confined to the effect of the internal flaws of the original specimen in producing fragmentation.

Further reasons for imposing this limitation can be noted. The surface of the original specimen is exposed to mechanical abrasion which affects the density of flaws on the initial surface. Further, the degree of activation of these surficial flaws may depend on the ambient medium in which the fracture takes place. Both factors are important in determining the strength of many brittle materials, including glass and rock salt; in the latter case, both effects are pronounced (24,30). However, neither factor should have a significant effect on the activation of flaws in a fragment formed in the specimen's interior.

The internal flaws of the original specimen are distributed throughout the volume, in the first instance. When such a flaw is activated, two contiguous surfaces result, each of which contains flaws distributed areally, arising from intersection of flaws by the propagating fracture. Clearly, a new fracture surface can arise by the activation of such a flaw on an antecedent surface. Similarly, the edge formed by intersection of two fracture surfaces will contain flaws distributed linearly, the activation of which can produce new fractures. For brevity, the three types of flaws in question will be referred to as volume, facial, and edge flaws, according as they do not lie on an antecedent fracture surface, they lie on one such surface, or they fall on the edge produced by antecedent fracture surfaces, respectively. The first (and basic) hypothesis made is that *fragmentation proceeds by activation of volume, facial, and edge flaws.* Griffith's criterion for an edge flaw has not been treated in the literature, but there is no reason to expect an essential difference in analytic form for this case.

Secondly, it will be assumed that the activated flaws of volume, facial, and edge type are distributed independently of each other. The hypothesis is physically reasonable, in view of the random orientation of the fracture surfaces within the specimen and the random nature of the stresses during the fracture process.

The third hypothesis required is that the activated flaws of a particular type (volume, facial, or edge) are distributed at random, independent of any stress system producing fracture. This assumption is insufficient to specify uniquely the distribution of the flaws of a given type. To be precise, it will be hypothesized that the activated flaws of any type correspond to points distributed over a domain at random, both individually and collectively in the sense of Fry (12). The former qualification implies that the position of one point is independent of that of all others. The latter signifies that the probability of a subdomain containing a particular number of points is independent of the number in any other subdomain (which does not overlap). Bateman (3) has shown that a set of points (over a single domain) characterized by these two properties conforms to a Poisson distribution.

Single fracture of a brittle solid will be defined as fracture by an external stress system which is permanently removed at the same instant the first one or few Griffith flaws begin to propagate. Subsequent flaws are activated by stress waves produced by the propagation of prior ones, as is indicated in a flash photograph of incipient fracture obtained by Poncelet (36). The removal of the external stress permits fragmentation to be controlled by the postulated random distributions of activated flaws in the original specimen. Comminution of a specimen by repetitive fracture of fragments is excluded by the definition, since there is no assurance that the flaw distributions in the fragments exposed to the repeated breakage are the same as in the original specimen.

VA2b. Poisson Law

To fix the notation and to make comparison with the subsequent results, a prefatory discussion of the Poisson law will be given.

Consider a geometrical domain which is not restricted in its dimensionality (it may be a line, area, etc.). Let a set of M points be distributed over this domain at random, individually and collectively. If the extent of the domain is T, then the *a priori* probability that a point lie in a differential subdomain of extent dt is $T^{-1}\,dt$. Assume that M and T both become infinite in such a manner that

$$M/T \to \gamma \qquad \text{VA2-(1)}$$

where γ is a finite constant. Under this condition, the probability $P(n|t)$ of n points falling in a subdomain of extent t is fixed by the Poisson distribution

$$P(n|t) = e^{-\gamma t}(\gamma t)^n/n! \qquad \text{VA2-(2)}$$

In particular, the probability $P(0|t)$ of no point lying in t is given by

$$P(0|t) = e^{-\gamma t} \qquad \text{VA2-(3)}$$

and the probability $P(1|dt)$ of one point lying in a differential element dt is

$$P(1|dt) = \gamma dt \qquad \text{VA2-(4)}$$

For the expectation $\langle n|t\rangle_{\text{av}}$ of n over the subdomain t, one finds

$$\langle n|t\rangle_{\text{av}} = \gamma t \qquad \text{VA2-(5)}$$

Accordingly, γt is the expected number of points in the subdomain, and γ represents the mean density of points over the entire domain.

Equations (3) and (4) yield directly the probability $p(t)\,dt$ of one point lying in the range t to $t + dt$ with none in the interval 0 to t, since this probability is

$$p(t)\,dt = P(0|t)\,P(1|dt) \qquad \text{VA2-(6)}$$

or

$$p(t)\,dt = \gamma e^{-\gamma t}\,dt \qquad \text{VA2-(7)}$$

If t_0 represents the upper limit on the range of t, one must have

$$\int_0^{t_0} p(t)\,dt = 1 \qquad \text{VA2-(8)}$$

for proper normalization. Accordingly, it is necessary that the product γt_0 be large enough so that

$$e^{-\gamma t_0} \ll 1 \qquad \text{VA2-(9)}$$

This requirement is in accordance with the limit, eq. (1), if t_0 and T are identified.

VA2c. General Distribution Function

In this subsection, the general form of the distribution function for fragment size under the Poisson assumption will be determined.

Let constants γ_v, γ_s, and γ_l represent the mean volume, areal, and linear densities of activated flaws of volume, facial, and edge type, respectively, for a specimen in the process of fracture. As regards the domain extent T appearing in the limit, eq. (1), the parameter γ_v is defined relative to the volume of the specimen, γ_s is defined in terms of the total area of the fracture surfaces, and γ_l is specified relative to the total length represented by the intersections of fracture surfaces. If the total volume, face area, and edge length of a fragment are denoted by v, s, and l, respectively, the mean numbers of flaws which have been activated in the volume, faces, and edges of a fragment are then given by $\gamma_v v$, $\gamma_s s$, and $\gamma_l l$, respectively, as follows from the definitions. Hence, γ_v, γ_s, and γ_l equal the mean volume, areal, and linear densities of activated flaws in the volume, faces, and edges of a fragment, respectively.

In any fragment formed in the interior of the original specimen, the faces correspond to the course of propagating fracture surfaces (or fracture planes, in the case that the faces are flat). Assume that a particular face II of a fragment was the last one formed. Then the fracture surfaces producing the other faces must have propagated until they met the sides of the original specimen or an existent fracture surface athwart their path. However, the fracture surface producing II must have propagated until it was terminated in every direction by one of the fracture surfaces corresponding to the faces immediately adjacent. Thus, II necessarily was formed by activation of a flaw interior to or on the totality of fracture surfaces creating the adjacent faces. Construct a differential element Σ consisting of the slice formed by II and a surface displaced infinitesimally from it along

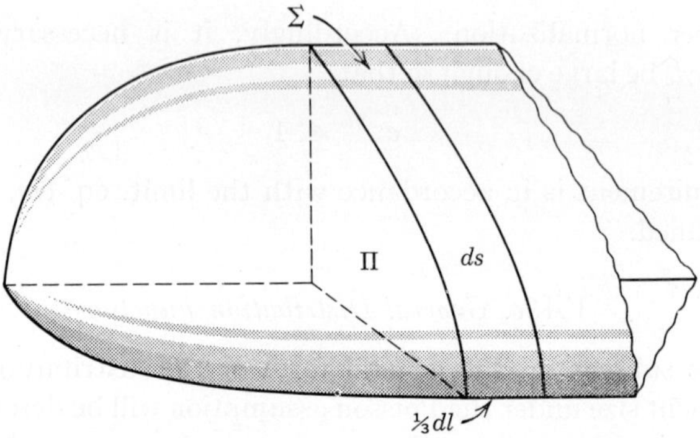

Fig. VA-1. Construction of the differential element Σ for a fragment formed by one curved and three planar fractures in a solid. Gilvarry (17).

the local outward normal, bounded in directions along the surface of II by prolongations of the adjacent faces. It is clear that the fracture surface creating II must have originated by activation of a flaw contained in Σ. Hence, the probability of formation of the fragment in question is compounded of two probabilities: the probability of at least one activated flaw in Σ, and the probability of no activated flaws in the fragment itself. The construction of Σ is illustrated diagrammatically in Figure VA-1.

In the process of constructing Σ, the total edge length, face area, and volume of the fragment increased from l to $l + dl$, from s to $s + ds$, and from v to $v + dv$, respectively. The first probability mentioned above is $P(1|dl,ds,dv)$, i.e., that one edge flaw be activated in dl of Σ, *or* one facial flaw be activated in ds of Σ, *or* one volume flaw be activated in dv of Σ. Since the three probabilities are mutually exclusive, one has

$$P(1|dl,ds,dv) = P(1|dl) + P(1|ds) + P(1|dv) \qquad \text{VA2-(10)}$$

where the right-hand side is in the notation corresponding to eq. (2). The second probability in question is the probability $P(0|l,s,v)$ that the fragment itself contain no activated edge flaw, no activated facial flaw, *and* no activated volume flaw. Hence, its value is given by

$$P(0|l,s,v) = P(0|l)\, P(0|s)\, P(0|v) \qquad \text{VA2-(11)}$$

in the notation of eq. (2), since the three Poisson distributions are assumed independent. Accordingly, the probability $dp(l,s,v)$ of formation of a fragment with total edge length, total face area, and total volume in the ranges l to $l + dl$, s to $s + ds$, and v to $v + dv$, respectively, is fixed by

$$dp(l,s,v) = P(0|l,s,v)\, P(1|dl,ds,dv) \qquad \text{VA2-(12)}$$

Using eqs. (2), (10), and (11), one obtains

$$dp(l,s,v) = e^{-Q}dQ \qquad \text{VA2-(13)}$$

where Q is the linear form

$$Q = \gamma_l l + \gamma_s s + \gamma_v v \qquad \text{VA2-(14)}$$

in l, s, and v.

It is clear that eq. (13) is independent of the choice of a particular face as the last one formed. Furthermore, the derivation requires no assumption on the shape of a fragment; in particular, geometric similarity of the fragments is not necessary. No restriction exists on the type of fracture surface, which may be conchoidal, of the type corresponding to a cleavage plane, or of other form.

Clearly, the number $dn(l,s,v)$ of fragments with total edge length, total face area, and total volume in the ranges l to $l + dl$, s to $s + ds$, and v to $v + dv$, respectively, is the product of the *a priori* number $q(l,s,v)$ of particles with these values of l, s, and v, by the corresponding probability $dp(l,s,v)$, or

$$dn(l,s,v) = q(l,s,v)\, dp(l,s,v) \qquad \text{VA2-(15)}$$

If V_0 is the volume of the initial specimen, it is obvious that q is fixed by

$$q = V_0/v \qquad \text{VA2-(16)}$$

and is independent of l and s. Accordingly, one has

$$dn(l,s,v) = V_0\, v^{-1}\, e^{-Q} dQ \qquad \text{VA2-(17)}$$

This equation represents the general form of the distribution function for fragment size in single fracture, when the distributions of edge, facial, and volume flaws correspond to separate and independent Poisson forms. It is clear from the derivation that the general distribution function is independent of particle shape.

It follows directly from a general development of Ruark (39), that the probability $P(n|l,s,v)$ of n activated flaws (unspecified as to edge, facial, or volume type) existing in an interior domain of the specimen during fracture is given by

$$P(n|l,s,v) = e^{-Q}Q^n/n! \qquad \text{VA2-(18)}$$

where l, s, and v represent the corresponding total edge length, face area, and volume, respectively. Geometrically, the domain in question is polyhedral in shape, but with faces not necessarily flat. To interpret Q physically, one can compute the expectation $\langle n|l,s,v\rangle_{\text{av}}$ of n over a definite l, s, and v from

$$\langle n|l,s,v\rangle_{\text{av}} = \sum_{n=0}^{\infty} n\, P(n|l,s,v) \qquad \text{VA2-(19)}$$

which yields

$$\langle n|l,s,v\rangle_{\text{av}} = Q \qquad \text{VA2-(20)}$$

Thus, Q physically represents the expected number of activated flaws of any type during the fracture, over an interior domain of the specimen having total edge length l, total face area s, and volume v. Accordingly, Q is the direct analog of γt of eq. (5) for the one-dimensional case. One verifies that eq. (18) yields eq. (11) for $P(0|l,s,v)$ when $n = 0$, and gives eq. (10) for $P(1|dl,ds,dv)$ when $n = 1$ and Q is replaced by the differential element dQ. Equation (13) for $dp(l,s,v)$ then appears as the direct generalization of eq. (7) for the one-dimensional form. It can also be shown that the mean-square deviation of n from its mean value equals Q, and other properties analogous to those for a Poisson distribution hold (7).

The method used to obtain $P(n|l,s,v)$ can be applied to find the allied probability $P'(n|l,s,v)$ that n Griffith flaws (pre-existent and unactivated) lie in an interior domain of the specimen before fracture, where the domain has edge length l, face area s, and volume v. The result is given by eq. (18) with Q replaced by Q', where Q' is defined analogously to Q in terms of parameters γ_l', γ_s', and γ_v' having meanings precisely analogous to γ_l, γ_s, and γ_v, respectively. Similarly, the direct analog $dp'(l,s,v)$ of $dp(l,s,v)$ is given by eq. (13) in terms of Q'. Since the flaws in question are all volume flaws of the specimen, whether they lie in an edge, a face, or in the volume of the interior domain, one must have

$$\gamma_l' = \gamma_s'^{1/2} = \gamma_v'^{1/3} \qquad \text{VA2-(21)}$$

To the extent that the orientation of the fracture surfaces is random, one expects the similar relation,

$$\gamma_l \simeq \gamma_s^{1/2} \simeq \gamma_v^{1/3} \qquad \text{VA2-(22)}$$

to hold approximately in the case of fracture. In contrast to eqs. (21), the relations, eq. (22), represent an approximation, because of the likelihood of different degrees of activation of edge, surface, and volume flaws by the stress system.

The equation for $P'(n|0,s,v)$, corresponding to omission of terms involving l, has been given previously by Ruark (40). As will emerge, however, it is precisely these terms, i.e., those which correspond to edges, which are the dominant ones in fracture.

VA2d. Properties of the Distribution

The probability Y that a fragment have total edge length, total face area, and volume less than l, s, and v, respectively, is clearly given by

$$Y(l,s,v) = \int_0^l \int_0^s \int_0^v dp \qquad \text{VA2-(23)}$$

Normalization of the probability $dp(l,s,v)$ of eq. (13) to unity requires that Y satisfy

$$Y(l_0,s_0,v_0) = 1 \qquad \text{VA2-(24)}$$

where l_0, s_0, and v_0 are values of l, s, and v, respectively, corresponding to some largest fragment. The value of the integral appearing in eq. (23) is independent of the path of integration in the space of the variables l, s, and v, since eq. (13) states that $dp(l,s,v)$ is an exact differential in Q. Hence, Y is a function of Q only, given by

$$Y(Q) = 1 - e^{-Q} \qquad \text{VA2-(25)}$$

The normalization condition, eq. (24), is fulfilled if

$$e^{-Q_0} \ll 1 \qquad \text{VA2-(26)}$$

where Q_0 is the largest value of Q for the fragments; the restriction is analogous to the inequality, eq. (9). Thus, the "largest" fragment must be interpreted as the one for which Q is a maximum.

The cumulative volume $V(l,s,v)$ of the fragments with edge length up to l, face area up to s, and volume up to v is given by

$$V(l,s,v) = \int_0^l \int_0^s \int_0^v v\, dn(l,s,v) \qquad \text{VA2-(27)}$$

Since the volume of the fragments is necessarily conserved in the fracture process, physical reasons demand that

$$V(l_0,s_0,v_0) = V_0 \qquad \text{VA2-(28)}$$

where l_0, s_0, and v_0 correspond to the largest fragment. Furthermore, it is necessary on physical grounds that eq. (28) hold regardless of the path of integration in the space of the variables l, s, and v which is used in evaluating $V(l_0,s_0,v_0)$ from eq. (27), since the condition of volume conservation is independent of particle shape. One notes, however, that the integrand in eq. (27) becomes

$$vdn = V_0\, e^{-Q}/dQ \qquad \text{VA2-(29)}$$

with use of eq. (17), and thus is an exact differential in Q. Hence V is a function only of Q, given by

$$V(Q) = V_0(1 - e^{-Q}) \qquad \text{VA2-(30)}$$

and satisfying eq. (28) by virtue of the condition, eq. (26), required for normalization of $dp(l,s,v)$. Equation (30) implies that the fracture process conserves volume independently of particle shape; thus the general distribution function, eq. (17), meets the necessary physical requirements for interpretation as such.

The cumulative fraction y of the initial volume represented by particles of edge length up to l, face area up to s, and volume up to v is defined by

$$y(l,s,v) = V(l,s,v)/V_0 \qquad \text{VA2-(31)}$$

By comparison of eqs. (25) and (30), one finds

$$y = Y \qquad \text{VA2-(32)}$$

Thus, the probability Y for formation of fragments with edge length up to l, face area up to s, and volume up to v can be evaluated experimentally as the cumulative fraction y of the volume represented by the fragments in question.

The derivation of eq. (32) depends on elimination of $1 - e^{-Q}$ between eqs. (25) and (30), and thus presupposes the particular Poisson form obtained for dp. However, one can show that eq. (32) actually is independent of an assumed model for the fracture process, and applies to plural fracture or comminution, with the understanding that $dp(l,s,v)$ represents a total probability for the overall process. With this interpretation of $dp(l,s,v)$, eq. (15) remains valid, since the *a priori* number q of fragments is a quantity independent of accidental circumstances depending on a probability, by definition. Use of the general expression, eq. (16), then yields

$$vdn = V_0\, dp \qquad \text{VA2-(33)}$$

Integrating both sides of this equation and noting the defining relations (27) and (31), one obtains eq. (32) without appeal to a model. This physical identification of the volume ratio of eq. (31) in terms of the probability of eq. (23) seems not to have been derived explicitly in the literature.

VA2e. Similarity Approximation

Experimental distributions of fragment size generally are expressed in terms of y of eq. (31) as a function (usually) of a mean linear dimension fixed by the fragment shape and the characteristics of a sieve. The integral distribution function, eq. (25), depends on three variables, l, s, and v. It can be reduced to a function of one variable by assuming that all fragments are geometrically similar.

Let x represent some mean linear dimension of a fragment and express the total edge length l, face area s, and volume v by

$$l = \lambda_l\, x \qquad \text{VA2-(34a)}$$

$$s = \lambda_s\, x^2 \qquad \text{VA2-(34b)}$$

$$v = \lambda_v\, x^3 \qquad \text{VA2-(34c)}$$

respectively, where λ_l, λ_s, and λ_v are (dimensionless) constants over the distribution, on the average. Then, the products $\lambda_l\gamma_l$, $\lambda_s\gamma_s$, and $\lambda_v\gamma_v$ can be written as

$$\lambda_l \gamma_l = k^{-1} \qquad \text{VA2-(35a)}$$

$$\lambda_s \gamma_s = j^{-2} \qquad \text{VA2-(35b)}$$

$$\lambda_v \gamma_v = i^{-3} \qquad \text{VA2-(35c)}$$

respectively, in which k, j, and i are constants, each having the dimensions of a length. The parameter Q of eq. (18) assumes the form

$$Q = (x/k) + (x/j)^2 + (x/i)^3 \qquad \text{VA2-(36)}$$

as a function of x. Physically, k in this expression represents the mean spacing of edge flaws with respect to x, as is clear from eqs. (34a) and (35a). The parameters j^2 and i^3 can be interpreted similarly, but the spacing in question is relative to area and volume, respectively.

The probability $p(x)\,dx$ of formation of a fragment with mean dimension in the range x to $x + dx$ is fixed by eq. (13) as

$$p(x)\,dx = dp(\lambda_l x, \lambda_s x^2, \lambda_v x^3) \qquad \text{VA2-(37)}$$

and thus one has

$$p(x) = e^{-Q} dQ/dx \qquad \text{VA2-(38)}$$

in terms of Q of eq. (36). Explicitly, $p(x)$ becomes

$$p(x) = \exp\left\{-\left[\frac{x}{k} + \left(\frac{x}{j}\right)^2 + \left(\frac{x}{i}\right)^3\right]\right\}\left(\frac{1}{k} + 2\frac{x}{j^2} + 3\frac{x^2}{i^3}\right) \qquad \text{VA2-(39)}$$

Similarly, the number $n(x)\,dx$ of fragments with mean dimension in the range x to $x + dx$ is given by

$$n(x)\,dx = dn(\lambda_l x, \lambda_s x^2, \lambda_v x^3) \qquad \text{VA2-(40)}$$

which yields,

$$n(x) = \omega V_0 x^{-3} p(x) \qquad \text{VA2-(41)}$$

when the parameter,

$$\omega = \lambda_v^{-1} \qquad \text{VA2-(42)}$$

is introduced for convenience. The normalization condition, eq. (26), now requires

$$(x_0/k) + (x_0/j)^2 + (x_0/i)^3 \gg 1 \qquad \text{VA2-(43)}$$

where x_0 is the largest value of x for the fragments.

The cumulative fraction y of the initial volume corresponding to fragments of dimension up to x becomes

$$y = 1 - \exp\left\{-\left[\frac{x}{k} + \left(\frac{x}{j}\right)^2 + \left(\frac{x}{i}\right)^3\right]\right\} \qquad \text{VA2-(44)}$$

from eq. (30). One obviously has $y(x_0) = 1$, in view of inequality (43). As x becomes small, the term x/k in the square brackets of eq. (44) becomes dominant over the remaining two, and the equation reduces to

$$y = 1 - e^{-x/k} \qquad \text{VA2-(45)}$$

This relation is simply the empirical Rosin-Rammler (16,38) equation

$$y = 1 - \exp\{-(x/k)^m\} \qquad \text{VA2-(46)}$$

for the case $m = 1$, but derived by a rigorous argument. In the limit when $x \rightarrow 0$, Q becomes small with dominant term x/k, and one obtains

$$y = x/k \qquad \text{VA2-(47)}$$

This result yields a theoretical derivation of the empirical Schuhmann (43) law

$$y = (x/k)^m \qquad \text{VA2-(48)}$$

for the case $m = 1$. Gaudin's (14) relation follows directly (43) from eq. (47).

Empirically, it is found that the size distribution of fracture fragments can be fitted to a fair approximation by the Rosin-Rammler or Schuhmann distributions with m reasonably close to unity, over most of the range of x. Both distributions obviously correspond to the limit where only edge flaws are operative in fracture, since these forms depend only on x/k as the variable. In general, it follows that the dominant mechanism producing fragmentation is *activation of edge flaws*. Thus, activation of facial and volume flaws plays only a secondary role in the fracture process, in the sense that such flaws are important only for larger values of x, where the number of fragments is smaller.

The derivation of the Rosin-Rammler relation given here can be compared with the discussion of Bennett (4). This author assumes that fracture proceeds by activation of random flaws in the specimen, to yield fracture planes normal to rectangular axes ξ, η, and ζ oriented parallel to the edges of a parallelepipedal specimen. The parameter k of the Rosin-Rammler equation is taken to represent the mean linear spacing of these fracture planes, as measured along the coordinate axes. By analogy with the Poisson law, it then is stated that y as defined by eq. (31) is given by

$$y = k^{-3} \int_\tau e^{-\xi\eta\zeta/k^3}\, d\xi d\eta d\zeta \qquad \text{VA2-(49)}$$

where τ designates the path of integration in the space of the variables ξ, η, and ζ, from zero up to a maximum value in each case. Correctly, it is stated that the precise value of eq. (49) depends on the selection of the path τ, which determines the correlation between fracture planes normal to different axes. Bennett, however, asserts that a path of integration exists in an average sense, by virtue of which eq. (49) yields the Rosin-Rammler relation, eq. (45).

That Bennett's assertion is not true can be shown by examining the form of eq. (49) when x is small. In this case, the exponential appearing can be replaced by unity, and the integral reduces to

$$y = k^{-3} \int_\tau d\xi d\eta d\zeta \qquad \text{VA2-(50)}$$

However, the integrand is now the element of volume in the space of the variables ξ, η, and ζ, and one obtains the limit

$$x \rightarrow 0; \quad y \rightarrow (x/k)^3 \qquad \text{VA2-(51)}$$

under the similarity approximation, independently of the path τ of integration. Thus, no path of integration exists which yields the Schuhmann line of eq. (47) as the proper limit of the Rosin-Rammler form, eq. (45), for x small. In point of fact, the physical basis of Bennett's treatment is easily identified. In eq. (44), let k and j become infinite, to yield

$$y = 1 - \exp\{-(x/i)^3\} \qquad \text{VA2-(52)}$$

This relation yields precisely the limit, eq. (51), when x is small, with k replaced by i. Hence, Bennett's approach is equivalent to assuming that activation of volume flaws of the fragments represents the dominant mode of fragmentation (at least for x small).

A difference in principle should be noted between the theoretical distributions obtained and the Rosin-Rammler and Schuhmann relations. The theoretical forms have been derived for the case of single fracture. The Rosin-Rammler and Schuhmann equations have been established empirically for the case of comminution by repetitive means (usually by a mill). This circumstance is reflected in the fact that m in eqs. (46) and (48), as fitted to the size distribution in a mill product, rarely has the precise value unity (it may range from perhaps 0.5 to possibly 1.5). Further, m is *not* a constant of the material, in general, but depends also on the type of mill. As will appear in Section VB of this volume, however, the exponent of x/k in the limiting form of eq. (47) is closely unity over a large range of x, when the experimental conditions correspond to single fracture.

VA2f. Moments of the Distribution

The volume $V(l,s,v)$ of eq. (27) can be regarded as a generalized moment of the distribution $dn(l\,s,v)$, which becomes the third moment (with respect to $x = 0$) under the similarity approximation. In this subsection, the other moments of physical interest will be considered.

The cumulative number $N'(l,s,v)$ of fragments with edge length from l to l_0, face area from s to s_0, and volume from v to v_0 is defined by

$$N'(l,s,v) = \int_l^{l_0} \int_s^{s_0} \int_v^{v_0} dn(l,s,v) \qquad \text{VA2-(53)}$$

from eq. (17), where l_0, s_0, and v_0 correspond to the largest fragment. The corresponding cumulative edge length $L'(l,s,v)$ and face area $S'(l,s,v)$ follow as:

$$L'(l,s,v) = \int_l^{l_0} \int_s^{s_0} \int_v^{v_0} l\, dn(l,s,v) \qquad \text{VA2-(54a)}$$

$$S'(l,s,v) = \int_l^{l_0} \int_s^{s_0} \int_v^{v_0} s\, dn(l,s,v) \qquad \text{VA2-(54b)}$$

respectively. For the moment, no path of integration in the space of the variables l, s, and v has been specified.

The integrals above are all of the form

$$\int_\tau f(l,s,v)e^{-Q}(\gamma_l dl + \gamma_s ds + \gamma_v dv) \qquad \text{VA2-(55)}$$

where τ designates the path of integration from the point (l,s,v) to (l_0,s_0,v_0) and the function f is proportional to $1/v$, l/v, and s/v in the case of N', L', and S', respectively. For the line integral, eq. (55), to be independent of the path τ, it is necessary that

$$\gamma_s[(\partial/\partial l)fe^{-Q}]_{s,v} = \gamma_l[(\partial/\partial s)fe^{-Q}]_{l,v} \qquad \text{VA2-(56)}$$

hold, as well as the two symmetric equations obtained by cyclic interchange of the variables l, s, and v appearing explicitly. One verifies immediately that this condition for the integrands of N', L', and S' to be exact differentials is not satisfied. Hence, the values of the three moments in question are *not* independent of particle shape, in contrast to the case with $V(l,s,v)$ of eq. (27). From a physical standpoint, this behavior is necessary, of course. Accordingly, the assumptions made are insufficient to permit evaluation of these moments in the general case. This circumstance arises because the three distributions corresponding to edge, facial, and volume flaws are assumed independent, whereas the integrals in question depend on the correlation between the variables l, s, and v. To treat the general case, it would be necessary to introduce conditional probabilities.

Under the similarity approximation, the requisite path of integration to evaluate the moments in question is specified automatically, since l, s, and v are proportional to powers of the mean dimension x. In this case the cumulative number $N'(x)$ of fragments with dimensions from x to the maximum x_0 is given by

$$N'(x) = \omega V_0 \int_x^{x_0} x^{-3} e^{-Q} dQ \qquad \text{VA2-(57)}$$

with $Q(x)$ given by eq. (36). The corresponding cumulative edge length $L'(x)$ and face area $S'(x)$ become

$$L'(x) = \tau V_0 \int_x^{x_0} x^{-2} e^{-Q} dQ \qquad \text{VA2-(58a)}$$

$$S'(x) = \sigma V_0 \int_x^{x_0} x^{-1} e^{-Q} dQ \qquad \text{VA2-(58b)}$$

respectively, in which the parameters τ and σ are defined by

$$\tau = \lambda_l/\lambda_v \qquad \text{VA2-(59a)}$$

$$\sigma = \lambda_s/\lambda_v \qquad \text{VA2-(59b)}$$

respectively.

The asymptotic behavior of $N'(x)$, $L'(x)$, and $S'(x)$ for $x \to 0$ can be determined directly by noting that, in this case, the dominant contribution to each integral arises from the region of integration close to the lower limit. One obtains:

$$N'(x) \to \tfrac{1}{2}\omega(V_0/k)\, x^{-2} \qquad \text{VA2-(60a)}$$

$$L'(x) \to \tau(V_0/k)\, x^{-1} \qquad \text{VA2-(60b)}$$

$$S'(x) \to \sigma(V_0/k) \ln (k/\gamma x) \qquad \text{VA2-(60c)}$$

for $x \to 0$, where γ in the last relation equals e^C in terms of Euler's constant C, and appears through use of the expansion of the exponential integral function. One sees that all three integrals are divergent for $x = 0$. It follows that the cumulative number $N(x)$, cumulative edge length $L(x)$, and cumulative face area $S(x)$ of fragments with mean dimension up to x, defined by:

$$N(x) = \omega V_0 \int_0^x x^{-3}\, e^{-Q} dQ \qquad \text{VA2-(61a)}$$

$$L(x) = \tau V_0 \int_0^x x^{-2}\, e^{-Q} dQ \qquad \text{VA2-(61b)}$$

$$S(x) = \sigma V_0 \int_0^x x^{-1}\, e^{-Q} dQ \qquad \text{VA2-(61c)}$$

respectively, are represented by divergent integrals.

VA2g. Discussion

For $x \to 0$, the moments $N'(x)$, $L'(x)$, and $S'(x)$ represent the total number, the total edge length, and the total face area, respectively, of all the fragments formed in the fracture process. As such, they necessarily must be finite, on physical grounds. Accordingly, it is clear that the idealization corresponding to the simple Poisson law neglects some essential element in the physical situation. The element in question is flaw depletion, arising from the fact that the total number of Griffith flaws is a bounded number. The existence of the depletion effect removes the physical difficulty raised by the divergence of the integrals in question, since a theory incorporating depletion properly must yield bounded values for the corresponding physical quantities.

The expected total numbers of activated edge and facial flaws contributing to fragmentation throughout the specimen are $\gamma_l L(x_0)$ and $\gamma_s S(x_0)$, respectively. The fact that these numbers are infinite is not inconsistent with the theoretical assumptions made, since it was postulated that the flaws are distributed *collectively* at random. As emphasized by Fry (12), only an infinite set of points can be so distributed. In particular, a set of points cannot be distributed collectively at random if their total number is fixed at a finite value beforehand; in such a case, an excess of points falling in one subdomain increases the chance of a deficit in some other subdomain, and vice versa. Thus, collective randomness of the activated flaws can obtain approximately only if their number is sufficiently large that an excess or deficit of flaws somewhere has an inconsequential effect on the total number of flaws available everywhere else. Accordingly, the assumption of a Poisson distribution for the activated flaws implies that the successive propagation of fracture surfaces throughout the specimen does not deplete the total number of flaws available.

The distribution function for the number of atoms decaying radioactively in unit time is of Poisson form, only if the number of atoms is large enough and the decay constant small enough to justify neglect of depletion of the number of atoms (12,41). A precisely analogous assumption has been made in this paper, represented by the limit, eq. (1). The present development yields correctly the asymptotic form of the general results, valid for the intermediate range of fragment size, as will be shown from experimental data in Section VB of this volume.

VA3. Theory for the Two-Dimensional Case

In the preceding discussion of the three-dimensional case, the presence of surficial flaws in the original surface of the specimen was ignored. It was possible to treat the effects of only the internal flaws of a specimen by considering the original specimen as infinitely large. The purpose of this section is to derive the distribution function of Gilvarry (18) for the sizes of the fragments in single fracture when only surficial flaws are activated in a specimen of the form of a thin plate.

VA3a. Physical Assumptions

An idealized problem is treated. The specimen is regarded as a two-dimensional manifold of continuous curvature and of infinite extent. By imposing these restrictions, one implies that no edges are present in the surface of the original specimen, and one can disregard any effect of the specimen boundary. In particular, however, the manifold in question need not be plane.

The surficial flaws of the specimen are distributed over its surface area, in the first instance. When such a flaw is activated, two contiguous fracture edges result, each of which contains flaws distributed linearly, arising from intersection of flaws by the propagating fracture. Clearly, a new fracture can arise by activation of such a flaw on an antecedent fracture edge. For brevity, the two types of flaw in question are referred to as facial and edge flaws, according to whether they do not or do lie on an antecedent fracture, respectively. The first (and basic) hypothesis made is that fragmentation proceeds by activation of facial and edge flaws.

To solve the mathematical problem, two further hypotheses analogous to the corresponding ones of Subsection VA2a are made. It is assumed that the activated flaws of facial and edge type are distributed independently of each other. Further, it is hypothesized that the activated flaws of a particular type (facial or edge) are distributed at random, individually and collectively in the sense of Fry (12).

As a basic restriction, only single fracture is considered. Thus, the external stress system must be permanently removed at the instant when the first one or few Griffith flaws begin to propagate. Subsequent flaws are activated by stress waves produced by propagation of prior ones. This physical restriction implies that fragmentation is controlled entirely by the postulated random distribution of activated surficial flaws in the original specimen.

VA3b. General Distribution Function

Let constants δ_a and δ_c represent the mean areal and linear densities of activated flaws of facial and edge type, respectively, for a specimen

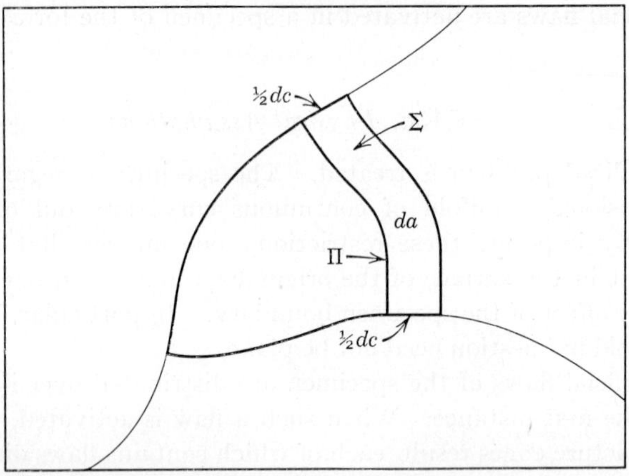

Fig. VA-2. Construction of the differential element Σ for a fragment formed by three curved fractures in a flat plate. Gilvarry (18).

in the process of fracture. If the area and perimeter of a fragment are denoted by a and c, respectively, the mean number of flaws which have been activated in the surface and perimeter of a fragment are given by $\delta_a a$ and $\delta_c c$, respectively. Thus, δ_a and δ_c equal the mean area and linear densities of activated flaws in the surface and perimeter of a fragment, respectively.

The argument yielding the probability $d\pi(c,a)$ of formation of a fragment with perimeter and area in the ranges c to $c + dc$ and a to $a + da$, respectively, is precisely analogous to that given in Subsection VA2c for the three-dimensional case. If a particular edge II of a fragment was the last one formed, construct a differential element Σ, as indicated in Figure VA-2. This process increases the perimeter of the fragment by dc and its area by da. The probability in question is then the joint probability that one edge flaw be activated in dc or one facial flaw be activated in da, with no edge or facial flaws activated in the fragment itself. The individual probabilities can be obtained from the Poisson law, which yields:

$$d\pi(c,a) = e^{-R}dR \qquad \text{VA3-(62)}$$

where R is the linear form in c and a,

$$R = \delta_c c + \delta_a a \qquad \text{VA3-(63)}$$

It is clear that the Poisson form of eq. (62) is independent of the choice of a particular edge as the last one formed. In addition, no assumption is made on the shape of a fragment, so that geometric similarity of the fragments is not required. Furthermore, it is obvious that the argument applies indifferently to a curved or plane surface of the original specimen. A basic limitation is that the curvature of this surface be continuous, so that no edges are present before fracture.

The number $d\nu(c,a)$ of fragments with perimeter and area in the ranges c to $c + dc$ and a to $a + da$, respectively, is the product of the *a priori* number of such particles by the corresponding probability $d\pi(c,a)$. The *a priori* probability in question is simply A_0/a, where A_0 is the surface area of the initial specimen. Thus, one has, for the general case,

$$d\nu(c,a) = A_0 a^{-1} e^{-R} dR \qquad \text{VA3-(64)}$$

In direct analogy with a result of Subsection VA2c, one can show that the probability $\Pi(\nu|c,a)$ of ν activated flaws (unspecified as to edge or facial type) existing in a surficial domain of the specimen during fracture is given by

$$\Pi(\nu|c,a) = e^{-R} R^{\nu}/\nu! \qquad \text{VA3-(65)}$$

where c and a represent the corresponding total perimeter and area, respectively. The physical interpretation of R is fixed by the result that the expectation $\langle\nu|c,a\rangle_{\text{av}}$ of ν over a definite c and a is given by

$$\langle\nu|c,a\rangle_{\text{av}} = \sum_{\nu=0}^{\infty} \nu\, \Pi(\nu|c,a) \qquad \text{VA3-(66)}$$

which yields

$$\langle\nu|c,a\rangle_{\text{av}} = R \qquad \text{VA3-(67)}$$

Thus R represents physically the expected number of activated flaws of any type during the fracture, over a surficial domain of the specimen having total perimeter c and total area a.

It is clear that the analogs of these results for fracture of a one-dimensional body (a long thin rod) follow directly. In this case, the pertinent expressions are directly analogous to results of the kinetic theory of gases associated with the mean-free-path of a gas molecule.

VA3c. Properties of the Distribution

The probability Z that a fragment have a perimeter and area less than c and a, respectively, is given by

$$Z = \int_0^c \int_0^a d\pi \qquad \text{VA3-(68)}$$

Normalization of the probability $d\pi(c,a)$ of eq. (62) to unity requires that Z satisfy

$$Z(c_0,a_0) = 1 \qquad \text{VA3-(69)}$$

where c_0 and a_0 are the values of c and a, respectively, corresponding to some largest fragment. The value of the integral appearing in eq. (68) is independent of the path of integration in the space of the variables c and a, since eq. (62) states that $d\pi(c,a)$ is an exact differential in R. Hence, Z is a function of R only, given by

$$Z(R) = 1 - e^{-R} \qquad \text{VA3-(70)}$$

The normalization condition, eq. (69), is fulfilled if

$$e^{-R_0} \ll 1 \qquad \text{VA3-(71)}$$

where R_0 is the largest value of R for the fragments. Thus the "largest" fragment must be interpreted as the one for which R is a maximum.

The cumulative area $A(c,a)$ of the fragments with perimeter up to c and surface area up to a is given by

$$A(c,a) = \int_0^c \int_0^a a d\nu \qquad \text{VA3-(72)}$$

Since the surface area of the fragments is necessarily conserved in the fracture process, physical reasons demand that

$$A(c_0,a_0) = A_0 \qquad \text{VA3-(73)}$$

where c_0 and a_0 correspond to the largest fragment. Furthermore, it is necessary on physical grounds that eq. (73) hold regardless of the path of integration in the space of the variables c and a which is used in evaluating $A(c_0,a_0)$ from eq. (72), since the condition of conservation of surface area is independent of particle shape. One notes, however, that the integrand in eq. (72) becomes

$$a d\nu = A_0 e^{-R} dR \qquad \text{VA3-(74)}$$

with use of eq. (64), and thus is an exact differential in R. Hence, A is a function only of R, given by

$$A(R) = A_0(1 - e^{-R}) \qquad \text{VA3-(75)}$$

and satisfying eq. (73) by virtue of the condition, eq. (71), required for normalization of $d\pi(c,a)$. Equation (75) implies that the fracture process conserves surface area independently of fragment shape, and hence the general distribution function, eq. (64), meets the necessary physical requirements for interpretation as such.

The cumulative fraction z of the initial surface area represented by particles of perimeter up to c and area up to a is defined by

$$z(c,a) = A(c,a)/A_0 \qquad \text{VA3-(76)}$$

By comparison of eqs. (70) and (75), one finds

$$z = Z \qquad \text{VA3-(77)}$$

Thus, the probability Z for formation of fragments with perimeter up to c and area up to a can be evaluated experimentally as the cumulative fraction z of the original surface area represented by the fragments in question. Equation (77) can be derived by a general argument, independently of a model, in a manner analogous to that used in the corresponding case in Subsection VA2d.

In the limit for the small fragments in a distribution, the term in R which depends on c must become dominant over the term in a. Hence, one obtains

$$z = 1 - \exp\{-\delta_c c\} \qquad \text{VA3-(78)}$$

asymptotically in this case. In the opposite limit corresponding to the large fragments, the term in R depending on a becomes dominant, to yield asymptotically,

$$z = 1 - \exp\{-\delta_a a\} \qquad \text{VA3-(79)}$$

VA3d. Similarity Approximation

The integral distribution function (64) depends on two variables, c and a. It can be reduced to a function of one variable by assuming that all fragments are geometrically similar.

Let x represent some mean linear dimension of a fragment, and express the perimeter c and area a of the fragment by

$$c = \mu_c x \qquad \text{VA3-(80a)}$$

$$a = \mu_a x^2 \qquad \text{VA3-(80b)}$$

respectively, where μ_c and μ_a are (dimensionless) constants over the distribution, on the average. Then the products $\mu_c \delta_c$ and $\mu_a \delta_a$ can be written as

$$\mu_c \delta_c = g^{-1} \qquad \text{VA3-(81a)}$$

$$\mu_a \delta_a = f^{-2} \qquad \text{VA3-(81b)}$$

respectively, in which g and f are constants, each having the dimensions of a length. The parameter R of eq. (63) assumes the form

$$R = (x/g) + (x/f)^2 \qquad \text{VA3-(82)}$$

as a function of x. Physically, g in this expression represents the mean spacing of edge flaws with respect to x; the parameter f^2 can be interpreted similarly, but the spacing in question is relative to an area.

The probability $\pi(x)\, dx$ of formation of a fragment with mean dimension in the range x to $x + dx$ is fixed by eq. (62) as

$$\pi(x)\, dx = d\pi(\mu_c x, \mu_a x^2) \qquad \text{VA3-(83)}$$

and thus one has

$$\pi(x) = e^{-R} dR/dx \qquad \text{VA3-(84)}$$

in terms of R of eq. (82). Explicitly, $\pi(x)$ becomes

$$\pi(x) = \exp\left\{-\left[\frac{x}{g} + \left(\frac{x}{f}\right)^2\right]\right\}\left(\frac{1}{g} + 2\,\frac{x}{f^2}\right) \qquad \text{VA3-(85)}$$

Similarly, the number $\nu(x)\, dx$ of fragments with mean dimension in the range x to $x + dx$ is given by

$$\nu(x)\, dx = d\nu(\mu_c x, \mu_a x^2) \qquad \text{VA3-(86)}$$

which yields

$$\nu(x) = \eta A_0\, x^{-2}\, \pi(x) \qquad \text{VA3-(87)}$$

when the parameter

$$\eta = \mu_a^{-1} \tag{VA3-(88)}$$

is introduced for convenience. The normalization condition, eq. (71), now requires

$$(x_0/g) + (x_0/f)^2 \gg 1 \tag{VA3-(89)}$$

where x_0 is the largest value of x for the fragments.

The cumulative fraction z of the initial surface area corresponding to fragments of mean dimension up to x becomes

$$z = 1 - \exp\left\{-\left[\frac{x}{g} + \left(\frac{x}{f}\right)^2\right]\right\} \tag{VA3-(90)}$$

from eq. (70). One obviously has $z(x_0) = 1$, in view of inequality, eq. (89). As x becomes small, the term x/g in the square brackets of eq. (90) becomes dominant over the other one, and the equation reduces to

$$z = 1 - e^{-x/g} \tag{VA3-(91)}$$

This relation is the two-dimensional analog of the empirical Rosin-Rammler (16,38) equation. In the limit when $x \rightarrow 0$, R becomes small with dominant term x/g, and one obtains

$$z = x/g \tag{VA3-(92)}$$

as the two-dimensional analog of the empirical Schuhmann (43) law. A similar analog of Gaudin's (14) relation can be given.

The distributions represented by eqs. (91) and (92) obviously correspond to the limit where only edge flaws are operative in fracture, since these forms depend only on x/g as the variable. In the corresponding limit $x \rightarrow 0$, the number of fragments becomes very large, as follows from eq. (87). Thus, the dominant mechanism producing fragmentation is activation of edge flaws, in general, in the sense that the majority of the fragments are engendered by such flaws.

VA3e. Moments of the Distribution

The cumulative number $N'(c,a)$ of fragments with total perimeter from c to c_0 and total area from a to a_0 is defined by

$$N'(c,a) = \int_c^{c_0} \int_a^{a_0} d\nu(c,a) \tag{VA3-(93)}$$

from eq. (64), where c_0 and a_0 correspond to the largest fragment. The corresponding cumulative perimeter $C'(c,a)$ follows as

$$C'(c,a) = \int_c^{c_0} \int_a^{a_0} c\, d\nu(c,a) \qquad \text{VA3-(94)}$$

One verifies immediately that the condition for the integrands of N' and C' to be exact differentials is not satisfied. Hence the values of these two moments are not independent of fragment shape, in contrast to the case with $A(c,a)$ of eq. (72).

Under the similarity approximation, the requisite path of integration to evaluate the moments in question is specified automatically, since c and a are proportional to powers of the mean dimension x. In this case the cumulative number $N'(x)$ of fragments with mean dimensions from x to the maximum x_0 is given by

$$N'(x) = \eta A_0 \int_x^{x_0} x^{-2}\, e^{-R} dR \qquad \text{VA3-(95)}$$

with $R(x)$ given by eq. (82). The corresponding cumulative perimeter $C'(x)$ becomes

$$C'(x) = \rho A_0 \int_x^{x_0} x^{-1}\, e^{-R} dR \qquad \text{VA3-(96)}$$

in which the parameter ρ is defined by

$$\rho = \mu_c/\mu_a \qquad \text{VA3-(97)}$$

The asymptotic behavior of $N'(x)$ and $C'(x)$ for $x \to 0$ can be determined by noting that, in this case, the dominant contribution to each integral arises from the region of integration close to the lower limit. One obtains

$$N'(x) \to \eta(A_0/g)\, x^{-1} \qquad \text{VA3-(98a)}$$

$$C'(x) \to \rho(A_0/g) \ln\, (g/\gamma x) \qquad \text{VA3-(98b)}$$

for $x \to 0$, where γ in the second relation is defined as in Subsection VA2f, in terms of Euler's constant. One sees that both integrals are divergent for $x \to 0$. It follows that the cumulative number $N(x)$ and perimeter $C(x)$ of fragments with mean dimension up to x, defined by

$$N(x) = \eta A_0 \int_0^x x^{-2} e^{-R} dR \qquad \text{VA3-(99a)}$$

$$C(x) = \rho A_0 \int_0^x x^{-1} e^{-R} dR \qquad \text{VA3-(99b)}$$

respectively, are represented by divergent integrals.

VA3f. Discussion

In the three-dimensional theory of Subsection VA2, it was found that the total number, the total edge length, and the total face area of all the fragments formed in the fracture process were represented by divergent integrals. This mathematical result conflicts with the fact that the actual quantities in question are necessarily finite, on physical grounds. One sees that a similar difficulty arises in the two-dimensional case, in connection with the total number and total perimeter of the fragments. As discussed in Subsection VA2g, the physical reason is neglect of depletion of the Griffith flaws. This element in the physical situation is neglected entirely in the idealization corresponding to the simple Poisson law (12,41).

VA4. Summary and Conclusions

The theoretical development of Gilvarry reviewed here differs sharply from prior work of others, in that it is not empirical or semi-empirical. It proceeds from a closely defined physical model, and translates physical assumptions directly into mathematical implications. The essential new theoretical concept introduced is the existence and importance of the edge flaws in fragmentation. Clearly, the presence of such flaws must be postulated to explain the linear argument in the Schuhmann equation and in the exponential of the Rosin-Rammler relation, since the facial and volume flaws ordinarily considered would imply a higher power of x. It then follows mathematically (for x small) that the edge flaws play the preponderant role in fragmentation. These facts are central to the problem, and failure to recognize them explains the lack of success in previous efforts, such as that of Bennett (4), to derive the proper distribution function from the Poisson law.

The theory predicts the existence of three distinct regions of x as regards the size distribution: a domain of constant differential probability p for small x, arising from the effect of edge flaws, followed, as x becomes large, by two regions of variable p corresponding to the effect of facial and volume flaws. The latter regime of variable differential probability had not previously been inferred experimentally. In Section VB of this volume, it will be shown from experimental results that the phenomenon in question indeed exists. It is reflected in the presence of two peaks in the experimental curve of the differential parameter p against x, corresponding to the effects of facial and volume flaws in the specimens.

In view of the fact that the theory yields rigorous derivations of the empirical Rosin-Rammler and Schuhmann laws for the first time, it obviously is correct in some intermediate range of values of x, approaching low values. However, it cannot be exact in the limit $x \rightarrow 0$, as otherwise, certain integrals over the distribution become divergent which are known on physical grounds to be finite. This deficiency has been traced to neglect of flaw depletion, a feature inherent in the Poisson law.

To incorporate flaw depletion into the theory, an essential preliminary step is the extension of the theoretical development by basing it on the binomial distribution, which is not restricted to an infinite population. Note that the Poisson distribution represents the limit of a binomial one when the number of events becomes large and the probability of an individual success is small (12). Recently, Gaudin and Meloy (15) have derived a distribution function for fragment size in single fracture, under the assumption of a binomial distribution. As a model, they assume a linear continuum depending in extent on one variable x, or essentially the case of a long thin rod. As such, their model cannot represent the effect of surface and volume flaws, whose existence in fact is conclusively indicated by results in Section VB. It should be mentioned that the results of Gaudin and Meloy differ from those of Manning (34), who previously treated the same mathematical situation by a different method.

It can be shown that the Gaudin-Meloy relation reduces properly to the forms for small fragment size given here when the number of fracturing surfaces becomes large. This behavior is consistent, of course, for the Poisson limit of a binomial distribution. For use as a point of departure in treating the depletion problem, however, the analyses of Manning and of Gaudin and Meloy require extension. The mathematical details will be given in a further publication (22).

The author wishes to thank the Editor of the *Journal of Applied Physics* for permission to use quotations and figures from articles appearing originally in that journal.

VA5. List of Symbols

VA5a. Three-Dimensional Case

C	Euler's constant
i	Mean linear spacing of volume flaws
j	Mean linear spacing of facial flaws
k	Mean linear spacing of edge flaws
l	Edge length of fragment
l_0	Value of l for largest fragment
L	Total edge length of fragments, cumulated up to particular size
L'	Total edge length of fragments, cumulated from particular one up to largest
m	Exponent of mean dimension x in Schuhmann and Rosin-Rammler laws
M	Number of points in domain, in Poisson distribution
n	Number (frequently differential) of fragments
N	Total number of fragments, cumulated up to particular size
N'	Total number of fragments, cumulated from particular one up to largest
p	Probability (differential)
$P(n\|a,\cdots)$	Probability of n points falling in domain $a,\cdots$
q	*A priori* number of fragments in fracture
Q	Linear form in l, s, and v, or x, x^2, and x^3
Q_0	Value of Q for largest fragment
s	Surface area of a fragment
s_0	Value of s for largest fragment
S	Total surface area of fragments, cumulated up to particular one
S'	Total surface area of fragments, cumulated from particular one up to largest
t	Extent of subdomain in Poisson distribution
t_0	Largest subdomain in Poisson distribution
T	Extent of total domain in Poisson distribution

v	Volume of a fragment
v_0	Value of v for largest fragment
V	Total volume of fragments, cumulated up to particular size
V_0	Volume of initial specimen
x	Mean dimension of a fragment
x_0	Value of x for largest fragment
y	Cumulative fraction of initial volume in fragments, up to particular size
Y	Probability of formation of fragments up to particular size
γ	Statistical density; exponential of Euler constant C
γ_l	Mean linear density of edge flaws
γ_s	Mean areal density of facial flaws
γ_v	Mean volume density of volume flaws
λ_l	Proportionality constant between l and x
λ_s	Proportionality constant between s and x^2
λ_v	Proportionality constant between v and x^3
σ	Ratio λ_s/λ_v
τ	Ratio λ_l/λ_v
ξ,η,ζ	Rectangular coordinates
ω	Reciprocal of λ_v

VA5b. Two-Dimensional Case

a	Surface area of a fragment
a_0	Surface area of largest fragment
A	Total surface area of fragments, from smallest to a particular one
A_0	Surface area of initial specimen
c	Perimeter of a fragment
c_0	Perimeter of largest fragment
C	Total perimeter of fragments, cumulated up to particular size
C'	Total perimeter of fragments, cumulated from particular one up to largest
f	Mean linear spacing of facial flaws
g	Mean linear spacing of edge flaws
N	Total number of fragments, cumulated up to particular size

N'	Total number of fragments, cumulated from particular one up to largest
R	Linear form in c and a, or x and x^2
R_0	Value of R for largest fragment
x	Mean dimension of a fragment
z	Cumulative fraction of surface area in fragments, up to particular size
Z	Probability of formation of fragments up to definite size
γ	Exponential of Euler constant
δ_a	Mean areal density of facial flaws
δ_c	Mean linear density of edge flaws
η	Reciprocal of μ_a
μ_a	Proportionality constant between a and x^2
μ_c	Proportionality constant between c and x
ν	Number (frequently differential) of fragments
π	Probability (differential)
$\Pi(\nu \vert a, \cdots)$	Probability of ν points falling in domain $a, \cdots$
ρ	Ratio μ_c/μ_a

References

1. Anderson, O. L., in *Fracture*, B. L. Averback, D. K. Felbeck, G. T. Hahn, and D. A. Thomas, Eds., Wiley, New York, 1959, p. 331.
2. Axelson, J. W., and E. L. Piret, *Ind. Eng. Chem.*, **42,** 665 (1950).
3. Bateman, H., *Phil. Mag.*, **20,** 704 (1910); *ibid.*, **21,** 745 (1911).
4. Bennett, J. G., *J. Inst. Fuel*, **10,** 22 (1936).
5. Bond, F. C., *Trans. AIME*, **193,** 484 (1952).
6. Dapples, E. C., *Basic Geology for Science and Engineering*, Wiley, New York, 1959, pp. 27, 120.
7. David, F. N., *Probability Theory for Statistical Methods*, Cambridge Univ. Press, Cambridge, 1949, p. 61.
8. Department of Scientific and Industrial Research, *Crushing and Grinding, a Bibliography*, Her Majesty's Stationery Office, London, 1958.
9. Drinker, P., *J. Ind. Hyg.*, **7,** 305 (1925).
10. Elliott, H. A., *Proc. Phys. Soc. (London)*, **59,** 208 (1947).
11. Epstein, B., *J. Franklin Inst.*, **244,** 471 (1947).
12. Fry, T. C., *Probability and its Engineering Uses*, Van Nostrand, New York, 1928, pp. 216, 237.
13. Gates, A. O., *Trans. AIME*, **82,** 875 (1915).
14. Gaudin, A. M., *Trans. AIME*, **73,** 253 (1926).
15. Gaudin, A. M., and T. P. Meloy, *Trans. AIME*, **223,** 40 (1962).
16. Geer, M. R., and H. F. Yancey, *Trans. AIME*, **130,** 250 (1938).

17. Gilvarry, J. J , *J. Appl. Phys.*, **32,** 391 (1961).
18. Gilvarry, J. J., *J. Appl. Phys.*, **33,** 3214 (1962).
19. Gilvarry, J. J., and B. H. Bergstrom, *J. Appl. Phys.*, **32,** 400 (1961).
20. Gilvarry, J. J., and B. H. Bergstrom, *J. Appl. Phys.*, **33,** 3211 (1962).
21. Gilvarry, J. J., *J. Appl. Phys.*, **33,** 3218 (1962).
22. Gilvarry, J. J., to be published.
23. Gordon, J. E., D. M. Marsh, and M. E. M. L. Parratt, *Proc. Roy. Soc. (London)*, **A249,** 65 (1959).
24. Greenler, R. G., and W. S. Rothwell, *J. Appl. Phys.*, **31,** 616 (1960).
25. Griffith, A. A., *Trans. Roy. Soc. (London)*, **A221,** 163 (1921).
26. Griffith, L., *Can. J. Res.*, **A21,** 57 (1943).
27. Gurney, C., *Proc. Phys. Soc. (London)*, **59,** 169 (1947).
28. Halmos, P. R., *Ann. Math. Statist.*, **15,** 182 (1944).
29. Hatch, T., and S. P. Choate, *J. Franklin Inst.*, **207,** 369 (1929).
30. Joffe, A., *The Physics of Crystals*, McGraw-Hill, New York, 1928.
31. Kick, F., *Dinglers Polytech. J.*, **247,** 1 (1883); *ibid.*, **250,** 141 (1883).
32. Kolmogoroff, A. N., *Dokl. Akad. Nauk SSSR*, **31,** 99 (1941).
33. Levengood, W. C., *J. Appl. Phys.*, **30,** 378 (1959).
34. Manning, A. B., *J. Inst. Fuel*, **25,** 31 (1952).
35. Orowan, E., *Repts. Progr. Phys.*, **12,** 185 (1949).
36. Poncelet, E. F., *Trans. AIME*, **169,** 37 (1946).
37. Rittinger, R. P. von, *Lehrbuch der Aufbereitungskunde*, Ernst und Korn Berlin, 1867, p. 19.
38. Rosin, P., and E. Rammler, *J. Inst. Fuel*, **7,** 29 (1933).
39. Ruark, A. E., *Phys. Rev.*, **65,** 88 (1944).
40. Ruark, A. E., *Phys. Rev.*, **65,** 287 (1944).
41. Ruark, A. E., and L. Devol, *Phys. Rev.*, **49,** 355 (1936).
42. Sack, R. A., *Proc. Phys. Soc. (London)*, **58,** 729 (1946).
43. Schuhmann, R., *AIME Tech. Publ. No. 1189* (1940).

VB. FRAGMENT SIZE IN SINGLE FRACTURE: EXPERIMENT

J. J. GILVARRY

VB1. Introduction

As has been emphasized, the theoretical development of Section VA of this volume for the distribution of fragment sizes corresponds to single fracture of a specimen. However, experimental determinations of size distributions in which conditions were chosen to approximate single fracture were virtually nonexistent in the literature prior to 1961. The available results corresponded to the product of a milling operation, or of an impact experiment in which the hammer was not restrained after fracture began. To fill this need, a series of experiments was carried out by Gilvarry and Bergstrom (11,12) in which the conditions for single fracture were met reasonably. A longer series meeting these conditions has been performed by Bergstrom, Sollenberger and Mitchell (5). As compared to their results, the data of Gilvarry and Bergstrom for size distributions reflect more closely controlled conditions and higher accuracy, in general.

This experimental work was performed with silica-glass spheres. An effort was made by Gilvarry and Bergstrom to extend the results to Lucite spheres, but the breaking instrument was a hydraulic press in which the travel of the piston had the low value permitted by the low compressibility of glass. In view of the much higher compressibility of Lucite, a redesign of the apparatus would have been entailed. The glass of the specimens used was homogeneous and isotropic, however, in common with most glassy polymeric solids of interest.

The data obtained with fracture of these glass spheres permits direct confrontation of the theory of Section VA with experiment in the three-dimensional case. As will appear, it also permits a test for the two-dimensional case, as shown by Gilvarry (13).

VB2. Experiment for the Three-Dimensional Case

The restriction was imposed in Subsection VA2, p. 669, that the initial specimen be sufficiently large that one could ignore the effect of the original surficial flaws in fragmentation. Such a limitation cannot be maintained in practice, since one must deal with specimens of finite size. In such a case, the fragments can be divided into two classes: the exoclastic fragments engendered by surficial flaws of the specimen, and the endoclastic fragments formed by propagation of internal flaws of the original specimen. The theory of Subsection VA2 applies only to the endoclastic fragments. [These terms are neologisms, formed from the word *clastic* (capable of being broken apart) and the combining forms *endo-* and *exo-*.]

VB2a. Experimental Method

A compressive load was applied to the spherical glass specimens through diametrically opposed platens of tungsten carbide, the motion of which was arrested by a stop after fracture started. Rates of loading varying through a factor of one thousand could be produced by the apparatus, described elsewhere by Bergstrom and Sollenberger (6). No statistically reliable evidence of an effect of the rate of loading on the form of the distribution function for fragment size has been found with this particular experimental arrangement (5). This fact constitutes a necessary prerequisite that conditions actually correspond to single fracture.

When a brittle solid fractures, the fragments necessarily fly apart at high speed. Speeds up to 10^3 cm./sec. have been determined cinematographically for the coarser fragments, but speeds for the finer particles seem to be much lower (5) than this figure. If the fragments are contained by a rigid retaining chamber around the original

Fig. VB-1. Disposition of fragments in the gelatin after fracture. It is seen that no fragments reach the walls of the Lucite retaining ring (11).

specimen, impact of the fragments on the walls induces secondary fracture which alters the distribution function of the sizes (markedly so for the larger fragments). This effect was avoided by filling the chamber with congealed gelatin to decelerate the fragments, which were removed from the gelatin matrix by washing with hot water A photograph of the retaining chamber, containing a specimen fractured in gelatin, is shown in Figure VB-1. The gelatin used was an ordinary comestible variety.

The size distributions of the fragments were determined in two ways. In one method, the usual procedure was followed: screening the fragments through a set of standard sieves and weighing the mass retained on each sieve. Cumulation of the measured masses yields y of eq. VA2-(31), p. 678, as a function of a mean dimension x. Because of the small mass (generally about 20 g.) of a specimen, this method does not give reliable results in the present application for values of x below roughly 100 μ. Tyler standard screens were used, specifications of which are given by Taggart (21). For the size range 20–45 μ, use was made of etched metal screens (manufactured by Buckbee Mears Co., St. Paul, Minn.).

For the smaller dimensions, a Coulter counter (4,7) was employed; an electrical instrument operating by virtue of the change in resistance that suspension of small particles produces in a conducting fluid flowing through an orifice. The instrument directly yields the cumulative number of particles with volume down to a particular value. The manufacturer (Coulter Electronics, Inc., Chicago, Ill.) claims accuracy to a limiting particle dimension $\sim$1 μ.

VB2b. Overall Distribution

The overall distribution neglects the distinction between exoclastic and endoclastic fragments.

Experimental values of the overall probability y' are shown as a function of x in Figure VB-2 for two specimens, as determined by means of eq. VA2-(31), p. 678, from the measured masses retained on sieves. For $x < 100$ μ, a significant error appears which arises from unavoidable loss of the finer fragments in the operations of washing

TABLE VB-I
Physical Parameters for Experimental Specimens

Specimen no.	Diameter, cm.	x_0', cm.	k', cm.
1	7.6	5.71	23.2
2	2.4	1.33	2.49
3	2.4	1.88	6.28
4	2.4	1.12	2.17
5	2.4	1.88	3.24
Mean of 2–5	2.4	1.88	3.04

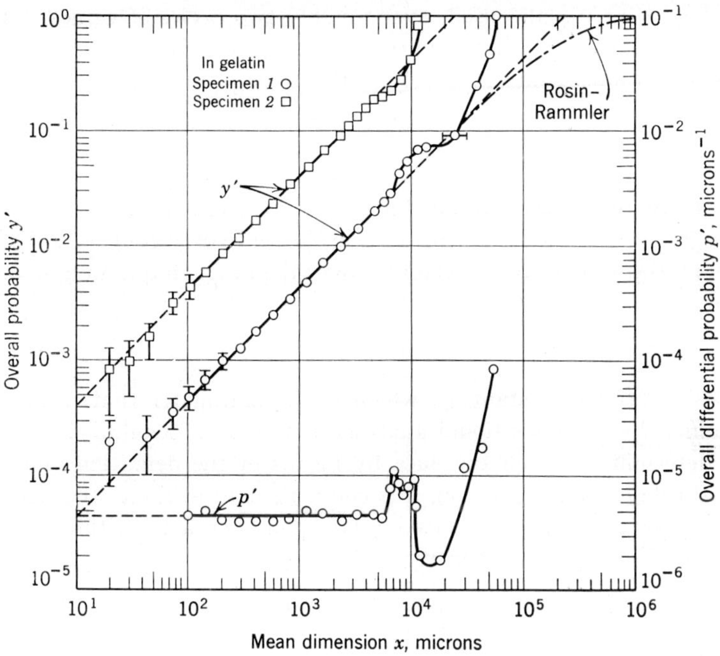

Fig. VB-2. The probability y' of single fracture for the overall distribution as a function of the mean dimension x of the fragments for two specimens. The overall differential probability p' is shown also for Specimen 1, on the right scale (11).

out the gelatin and screening the particles. This loss is measured as the difference between the original mass of the specimen and the sum of the masses of the recovered fragments. In Figure VB-2, each datum point was determined as the mean of two limits; one limit corresponds to ignoring the deficit in mass, and the other assumes the entire loss to arise in the sieve of smallest aperture (a supposition which is not necessarily true). The limits of error obtained in this way are shown in Figure VB-2, when they appear on the scale employed. Physical parameters for the specimens are given in Table VB-I. (Note that x_0' of this table is the mean dimension of the largest fragment recovered.)

In the case of both specimens, one notes from Figure VB-2 that the data for y' closely represent a straight line of inclination 45° on the logarithmic scales used, over roughly two cycles of values in the intermediate ranges of the variables. Accordingly, the function in this region must conform closely to the analog

$$y' = x/k' \qquad \text{VB2-(1)}$$

of the Schuhmann law for $m = 1$ of eq. VA2-(47), p. 681. Let $p' = dy'/dx$ represent the overall differential probability of formation of a fragment with mean dimension x, in the sense of Section VA. From eq. (1), this differential probability over the range in question has the value

$$p' = 1/k' \qquad \text{VB2-(2)}$$

in terms of the constant k', which is the parameter that would be obtained by a conventional analysis in terms of Schuhmann's law. To determine k' from the data by means of the definition and the chord rule for differentiation, it is convenient to evaluate the derivative at positions midway between data points on a logarithmic scale. By use of the operator identity

$$d/dx = (1/x)\, d/d \ln x \qquad \text{VB2-(3)}$$

one obtains,

$$p_i' = (y_{i+1}' - y_{i-1}')/[x_i \ln (x_{i+1}/x_{i-1})] \qquad \text{VB2-(4)}$$

for the differential probability p_i' at the point

$$x_i = (x_{i-1}\, x_{i+1})^{1/2} \qquad \text{VB2-(5)}$$

For the Tyler standard screens, the size ratio x_{i+1}/x_i equals $2^{1/}$ for the smaller and $2^{1/4}$ for the larger. In any one series, therefore, the point x_i for the ith sieve lies on a logarithmic scale midway between the points x_{i-1} and x_{i+1}. The value of k' for the linear portion of the curve of y' is then determined by

$$k' = 1/[\langle p_i' \rangle_{\text{av}}] \qquad \text{VB2-(6)}$$

where the average is over the linear part in question. Values of this parameter are tabulated in Table VB-I for five specimens and the mean of four of them.

Individual values of p_i' from eq. (4) are plotted against x in Figure VB-2 for Specimen 1. The value of $1/k'$ is indicated by the horizontal solid line of length delimiting the range of x used in the averag-

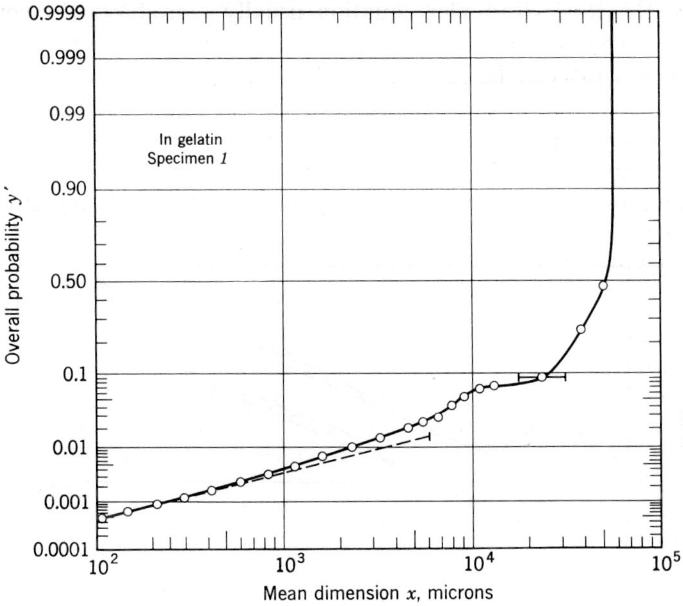

Fig. VB-3. The probability y' of single fracture for the overall distribution plotted as a function of the mean dimension x on a coordinate paper which yields a straight line for the data if it corresponds to a logarithmic-normal distribution (11).

ing process. Over this region, values of p_i' fluctuate but show no steady departure from a constant. Thus, for specimens fractured in gelatin, no evidence exists in this domain of x for a value of the Schuhmann exponent m differing from unity. Values of k' from Table VB-I have been used to construct the Schuhmann lines in Figure VB-2. The data diverge significantly from the corresponding line only for x approaching x_0'. For comparison purposes, the analog

$$y = 1 - e^{-x/k'} \qquad \text{VB2-(7)}$$

of the Rosin-Rammler curve of eq. VA2-(45), p. 681, has been shown for Specimen 1. As $x \rightarrow x_0'$, this curve lies on the side of the Schuhmann line opposite to that on which the data fall; hence the Rosin-Rammler form yields no improvement over the Schuhmann law for the overall distribution in the case of a specimen fractured in gelatin. This conclusion is at variance with the general case for a mill product, where the Rosin-Rammler equation usually reproduces the correct sense of the deviation from the Schuhmann line for large values of x.

Figure VB-2 can be compared with Figure VB-3, which shows the result of plotting y' for Specimen 1 against x on a special coordinate paper which rectifies the data points for any distribution satisfying a logarithmic-normal law (16). The dashed straight line has been drawn tangent to the curve at the point $x = 100\ \mu$; it represents a

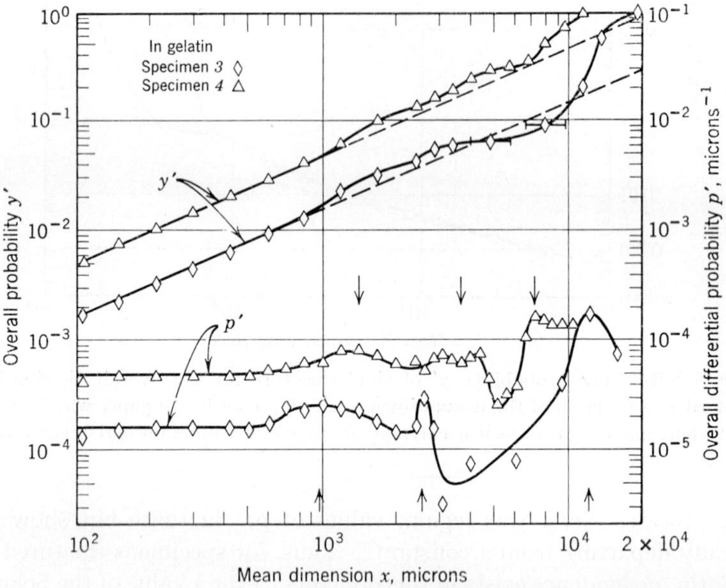

Fig. VB-4. The probability y' (left scale) and differential probability p' (right scale) of single fracture for the overall distribution as a function of the mean dimension x for two specimens. The horizontal scale is approximately twice the vertical. Data points indicated by horizontal bars correspond to sets of successive sieves where only the one of smallest aperture contained fragments (11).

logarithmic-normal distribution with parameters fitted to the curve of the data at the common initial point. The line has been extended to the value of x at which the horizontal line (solid) of Figure VB-2 terminates. At this upper limit, the logarithmic-normal distribution is in error by $\sim 50\%$, whereas the corresponding error of the Schuhmann line is essentially within the accuracy of the data at this point. The fact that the data cannot be represented satisfactorily by a logarithmic-normal distribution is in agreement with the conclusions of Section VA.

One sees from Figure VB-2 that the curve of p_i' for Specimen 1 at large x shows a structure consisting of a series of peaks, not predicted by the Schuhmann or Rosin-Rammler equations. Figure VB-4 has been constructed to display the nature of this structure in the overall distribution (the horizontal scale is roughly twice the vertical). For the two specimens appearing, the curve of p' computed from eq. (4) shows, in each case, three local maxima indicated by vertical arrows which can be correlated with points of maximum slope in the corresponding curves of y. Physically, each maximum corresponds to a value of x at which the probability $p'(x)\,dx$ of forming a fragment with dimensions in the range x to $x + dx$ is locally large for a fixed dx. The figure shows that this probability is greatest for the fragments of largest size. The existence of at least two minima, however, indicates that the probability in question is not a monotonic function of x, as one might suspect *a priori*. Further, one notes that the Schuhmann range of constant p' is not the region of maximum differential probability; this domain yields the overwhelming number of the fragments simply because of its wide extent, and not because of an inherently high differential probability.

At first sight, the conclusions from Figure VB-4 seem to be at variance with the results of Section VA. The general form of eq. VA2-(39), p. 680, predicts the existence of three distinct regions of x: a domain of constant p for small x arising from edge flaws, followed as $x \rightarrow x_0$ by *two* regions of variable p corresponding to the effect of facial and volume flaws. The results of Figure VB-4 differ in the sense that *three* local maxima appear. However, eq. VA2-(39) applies only to the endoclastic fragments, while y' and p' of Figure VB-4 correspond to the overall distribution of endo- and exoclastic fragments.

The curve of p' as a function of x for Specimen 4 is anomalous in that one local maximum consists of two less pronounced ones. A similar but less marked case occurs for Specimen 2 but not for Specimen 5 (curves not shown). This anomaly will be resolved in Subsection VB2c.

In the neighborhood of the most pronounced maximum, the individual sieves generally retained only a few fragments (occasionally, only one or none). The question arises as to whether the multiple maxima of Figure VB-4 may represent mere artifacts introduced by poor sta-

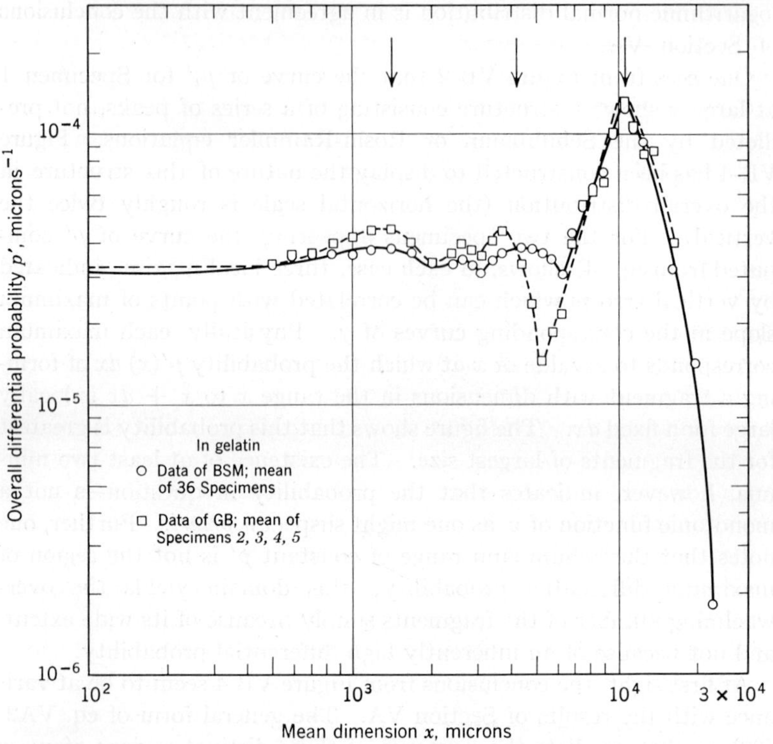

Fig. VB-5. The overall differential probability p' of single fracture as a function of x for the mean of 36 specimens (from the data of Bergstrom, Sollenberger, and Mitchell) compared to the corresponding probability for the mean of 4 specimens (12).

tistics. If such were the case, averaging the data for a sufficiently large number of specimens should cause the peaks to disappear. The value of p' from eq. (4) is shown (dashed) in Figure VB-5, as obtained from y' corresponding to an average of four specimens (Nos. 2, 3, 4, and 5) of the same initial diameter. One sees that the peak corresponding to the largest fragments is accentuated, indicating that its location is subject only to slight fluctuation, and it certainly is real. The other two maxima are broadened as compared to the case with individual specimens, but their permanence under averaging indicates their existence. An origin in sieve imperfections can be ruled out (11).

It might be argued that the evidence represented by the dashed curve of Figure VB-5 is questionable, since it is based on data from only four specimens. To meet this objection, the curve of the overall differential probability corresponding to the average of 36 specimens was constructed (12) from prior results of Bergstrom, Sollenberger, and Mitchell (5). The specimens were glass spheres of initial diameter 2.4 cm., fractured in gelatin on the same apparatus employed to break the specimens of Gilvarry and Bergstrom. As one notes from the solid curve of Figure VB-5, three peaks appear, as is consistent with the interpretation given. Moreover, the three maxima lie at approximately the same position for the mean of these 36 specimens as for the average of the other four, as indicated by the vertical arrows. In each case, the amplitudes and widths of corresponding peaks are approximately the same. These results eliminate the possibility that the peaks experimentally observed are merely artifacts arising from poor data since the averaging process should cause the peaks in the individual distributions to vanish if such were the case.

For both curves of Figure VB-5, an anomaly exists in that the middle peak exhibits a less pronounced maximum as a satellite. As already noted, this feature appears also for Specimens 2 and 4. Surprisingly, however, the location, amplitude, and width of this supernumerary peak are closely the same for the two curves of Figure VB-5. The cause of this anomaly is discussed in the following subsection, VB2c.

The fact appearing generally from these results, i.e., that for small x, y' is a linear function of x, and p' is a constant, confirms the conclusion of Section VA that edge flaws represent the dominant cause of fragmentation in the fine sizes.

VB2c. Endoclastic Distribution

Under point loading, failure of a specimen generally starts in a conical fracture surface with apex at the point of application of the force (20). In the present experiments, however, the load was distributed over a finite area as the spherical specimens deformed elastically under the pressure of the flat platens. Failure of the specimen in this arrangement starts in conical fracture surfaces with bases on the platens; the base of the cone corresponds approximately to the area of contact of specimen and platen at the instant of fracture (17). The propagating conical surfaces, however, produce a failed cylinder

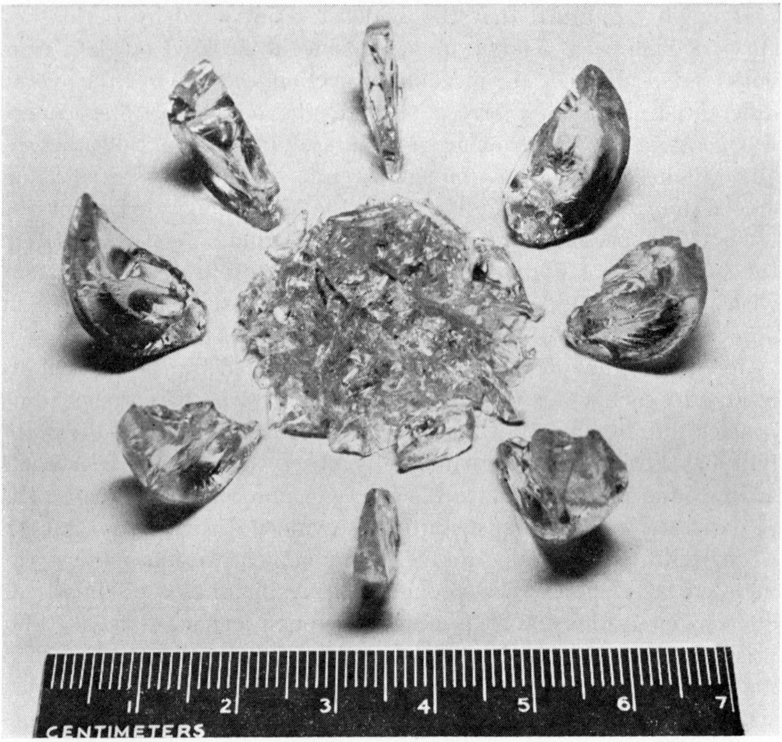

Fig. VB-6. Fragments of a specimen fractured in gelatin, as recovered from the matrix. The fragments are disposed radially in their correct relative position (11).

extending diametrically through the specimen, which yields the bulk of the fine fragments, apparently. The remainder of the specimen breaks into pieces which approximate lunes in shape. Figure VB-6 shows the fragments from a specimen, arranged in the correct relative angular position as recovered from the retaining chamber. The sequence of events in the fracture process has been illustrated diagrammatically by Bergstrom and Sollenberger (6).

The lunes of the exoclastic particles are responsible for the main mass of the fragments (of both classes). Typically, they correspond in total mass to the order of 75–90% of that of the original specimen. Considerably smaller exoclastic fragments are identifiable through the presence of original surface of the specimen. In general, few exoclastic particles were found with values of x less than that which

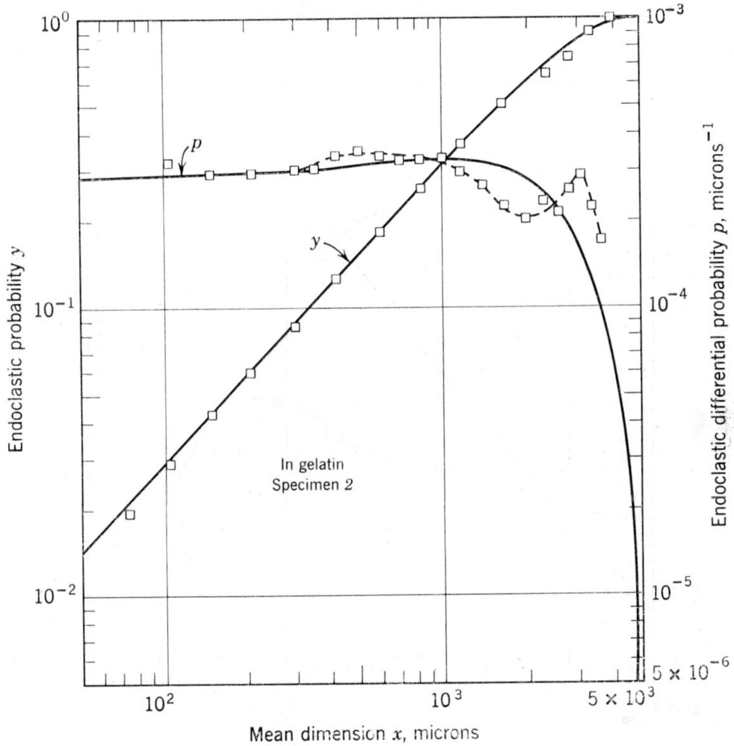

Fig. VB-7. The probability y (left scale) and differential probability p (right scale) of fracture for the endoclastic distribution as a function of the mean dimension x for Specimen 2 (11).

corresponds to the first minimum in Figure VB-4. This situation did not obtain for Specimen 4, however. In this case, 28 such fragments were recovered, extending in values of x considerably below that corresponding to the first minimum. (Throughout this article, ordinal numbers of maxima and minima are reckoned in order of decreasing x from the largest value of x for the distribution in question).

The identifiable exoclastic fragments were sieved separately and the corresponding masses were subtracted from the total originally measured to obtain the masses of the endoclastic fragments retained on each sieve. The upper limit x_0 of endoclastic dimensions was taken to correspond to one sieve size larger than the first (in order of de-

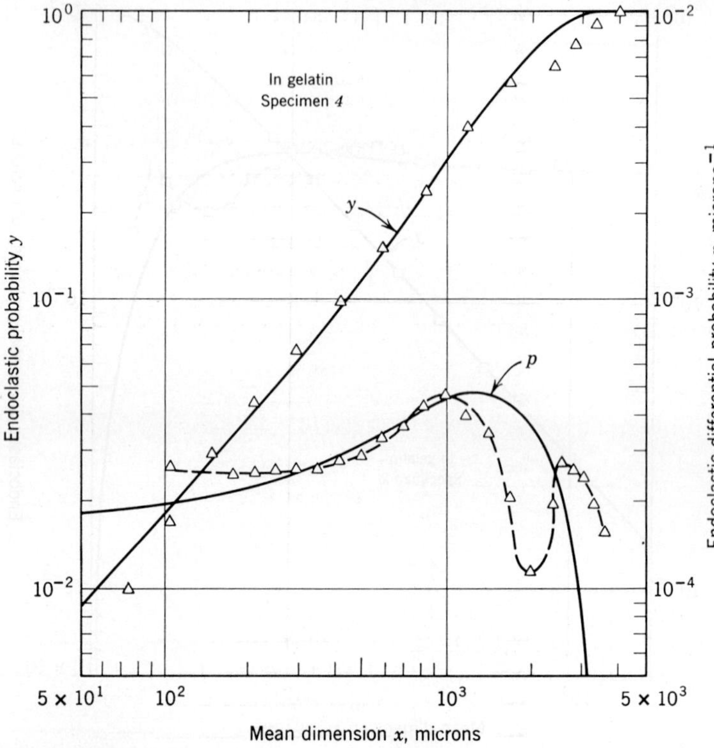

Fig. VB-8. The probability y (left scale) and differential probability p (right scale) of fracture for the endoclastic distribution as a function of the mean dimension x for Specimen 4 (11).

creasing x) for which the endoclastic mass did not vanish. Values for three specimens are tabulated in Table VB-II. Cumulation of the masses yields values of the probability y for formation of an endoclastic fragment with mean dimension between 0 and x, as shown by the plotted points in Figure VB-7 for Specimen 2 and in Figure VB-8 for Specimen 4. Values of the differential probability p from the analog of eq. (4) also appear (these points are connected by the dashed curve).

It is clear from Figures VB-7 and VB-8 that only two of the three maxima appearing in the overall differential probability p' correspond to the endoclastic particles, to which alone the theory of Section

TABLE VB-II
Parameters of the Endoclastic Distributions for the Specimens

Specimen. no.	x_0, cm.	k, cm.	j, cm.	i, cm.
2	0.40	0.36	0.36	0.44
3	0.33	0.39	0.23	0.25
4	0.40	0.59	0.27	0.24

VA applies. Thus, the main peak in p of Figures VB-2 and VB-4 stems entirely from the exoclastic fragments. In the experimental data for Specimen 4, one notes that the supernumerary peak appearing in Figure VB-4 has vanished. It arose because of the very large number, 28, of exoclastic fragments appearing from this specimen. A slight additional peak persists in the first maximum for Specimen 2,

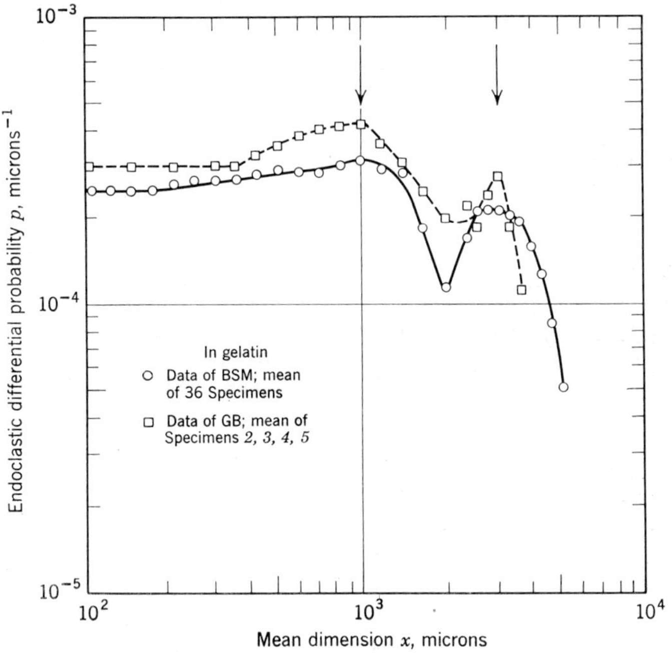

Fig. VB-9. The differential probability p of single fracture for the endoclastic distribution as a function of x for the mean of 36 specimens (from the data of Bergstrom, Sollenberger, and Mitchell) compared to the corresponding probability from the mean for 4 specimens (12).

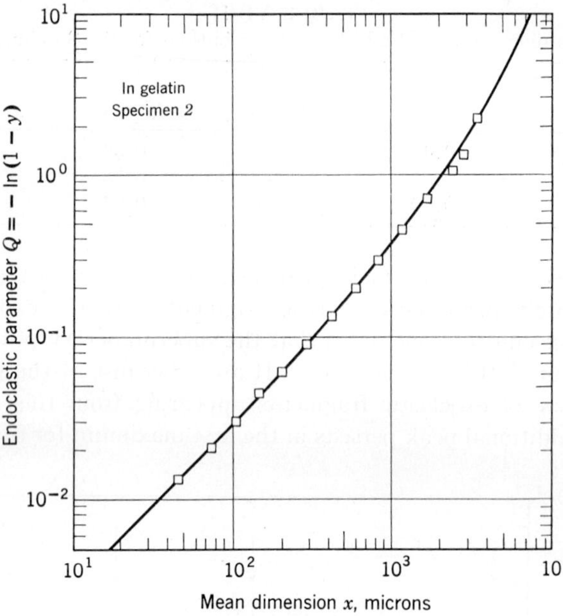

Fig. VB-10. The parameter Q for the endoclastic distribution as a function of x for Specimen 2 as given by the data and by the fitted function (11).

but its amplitude is reduced considerably as compared to the case in the overall distribution. Thus, the conclusion follows that the supernumerary peak arises from the effect of exoclastic particles.

The results to this point are subject to some statistical uncertainty since they are based on only four specimens. The endoclastic differential probability p for the mean of the 36 specimens of Bergstrom, Sollenberger, and Mitchell is shown (solid curve) as a function of the mean dimension x of the fragments in Figure VB-9 for comparison with the corresponding curve (dashed) for the mean of Specimens 2, 3, 4, and 5. One sees that the two curves show excellent agreement as regards their general course. In particular, they both exhibit two pronounced local maxima occurring at approximately the same value of x in each case, as indicated by the vertical arrows. The results of the mean of the 36 specimens confirm with good statistics the prior conclusions. Thus, the data from both sources are in agreement with the theory of Section VA, which predicts the existence of three dis-

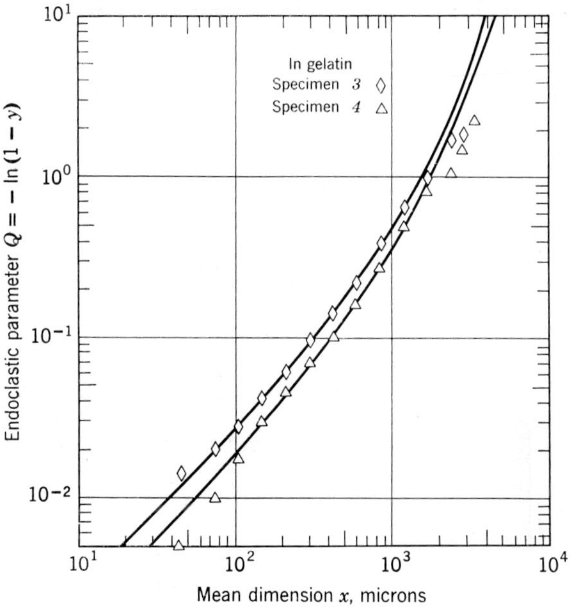

Fig. VB-11. The parameter Q for the endoclastic distribution as a function of x for Specimens 3 and 4 as given by the data and by the fitted function (11).

tinct regions of x: a domain of constant p for small x arising from edge flaws, followed, as x becomes large, by two regions of variable p corresponding to the effect of facial and volume flaws.

To discuss the quantitative agreement of theory and experiment, consider the relation

$$Q = -\ln(1 - y) \qquad \text{VB2-(8)}$$

which eq. VA2-(44), p. 681, yields. Values of Q obtained from this relation are shown in Figure VB-10 for Specimen 2 and in Figure VB-11 for Specimens 3 and 4. The parameters k, j, and i entering the cubic form Q of VA2-(36), p. 680, have been derived by the method of least squares from the data for the three specimens and are tabulated in Table VB-II. The curves of Figures VB-10 and VB-11 correspond to the fits obtained. Equation VA2-(34), p. 679, has been used to construct the corresponding curves (solid) of y in Figure VB-7 and VB-8. One sees that agreement is excellent.

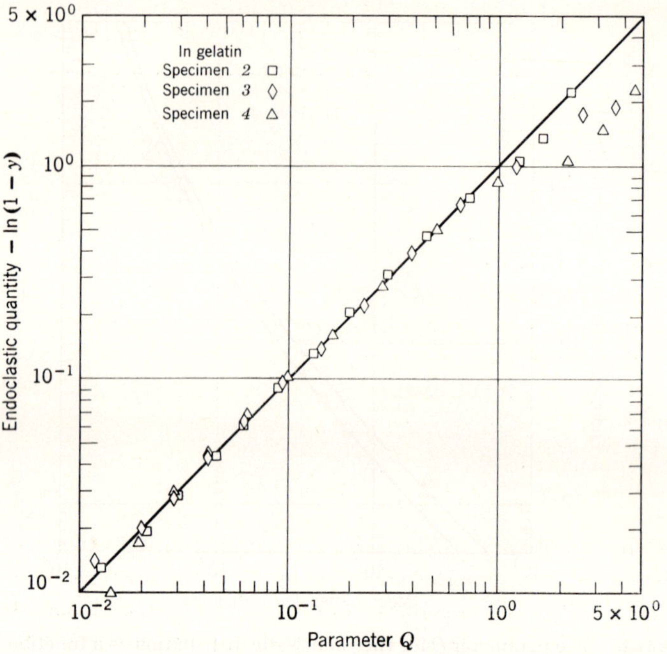

Fig. VB-12. The quantity $-\ln(1 - y)$ from the data as a function of Q from the fitted function for three specimens (11).

Independently of the specimen, eq. (8) implies that all data for $-\ln(1 - y)$ should be rectified on a line of 45° inclination when plotted against the corresponding Q of the fitted function. Figure VB-12 shows that this prediction is borne out over the major part of the data.

As compared to the case for the integral probability, agreement of theory and experiment is not as good for the differential probability p. The positions of any maxima and minima in the theoretical curve corresponding to eq. VA2-(38), p. 680, are given by the roots of $dp/dx = 0$, which yields the quartic equation

$$\left(\frac{1}{k^2} - \frac{2}{j^2}\right) + \left(\frac{4}{kj^2} - \frac{6}{i^3}\right)x + \left(\frac{6}{ki^3} + \frac{4}{j^4}\right)x^2 + \left(\frac{12}{j^2i^3}\right)x^3 + \left(\frac{9}{i^6}\right)x^4 = 0 \quad \text{VB2-(9)}$$

Since k, j, and i are necessarily positive, the terms involving x^2, x^3, and x^4 are also positive. However, the constant term and the coefficient of x can be positive or negative, depending on the relative values of k, j, and i. It follows that the largest number of variation of signs in eq. (9) is two, which is the largest number of positive real roots, by Descartes' rule of signs. Since maxima and minima must alternate in position, it follows that the curve corresponding to p of eq. VA2-(38) can display at most one peak, contrary to a statement made by Gilvarry and Bergstrom (11).

In fact, Table VB-II shows that k, j, and i are closely equal for actual specimens, as predicted in Section VA. In this case, eq. (9) shows only one variation of sign, and no more than one positive real root can appear. Identifying this root for the moment as x_2, one can obtain it analytically as

$$x_2 = \frac{1}{2}(\sqrt{2} - j/k)j \qquad \text{VB2-(10)}$$

under the approximation $(x_2/i)^3 << 1$, implying that x_2 is not near the region of x where the effect of the volume flaws is important. Evaluating d^2p/dx^2, one finds that x_2 of eq. (10) corresponds to a maximum in dp/dx. It is clear from the approximation made that this maximum arises from the effect of facial flaws, whose spacing is specified by j. Thus, the dependence of p of eq. VA2-(38), p. 680, on terms arising from volume flaws is too weak to yield a peak. It is possible for two positive real roots of eq. (9) to appear (when j is large compared to k and i), in which case a minimum appears between $x = 0$ and the position of the maximum; this situation is not of practical importance.

Values of x_2 computed from this relation with values of k and j from Table VB-II are compared in Table VB-III with positions of the second maxima as inferred from the data. One sees that agreement is reasonable as regards general magnitude, but is by no means exact. Thus, eq. VA2-(39), p. 680, with values of k, j, and i from Table VB-II certainly predicts the existence of the maximum corresponding to facial flaws, but the position is not given very accurately.

The fitted functions for Q have been used in eq. VA2-(38), p. 680, to construct the solid curves for p appearing in Figs. VB-7 and VB-8. In conformity with the analysis above, the theoretical curves for p show only one maximum. However, they do fit the data in a qualita-

TABLE VB-III
Position x_2 of the Second Maximum of the Endoclastic Distribution, as Compared with the Result of Eq. (10)

Specimen no.	x_2, μ	x_2 from eq. (10), μ
2	720	830
3	950	700
4	1300	990

tive or semiquantitative fashion, in that they reproduce the rise of the differential probability above the horizontal line $p = k^{-1}$ corresponding to the empirical Schuhmann law, in the region of x where the effect of the facial flaws should be predominant.

The failure of the maximum corresponding to the volume flaws to appear in the theoretical relation of Section VA for p can be understood on the basis of flaw depletion. It was emphasized in Section VA that depletion of the internal flaws of the specimen must be taken into account, if all moments of the distribution having physical significance are to be finite. In that particular context, depletion of internal flaws by the effect of the edge flaws was primarily under discussion. However, the internal flaws of the specimen available to produce volume flaws in a fragment can be depleted through the effect of those internal flaws yielding volume, facial and edge flaws in other fragments. Thus, eq. VA2-(36), p. 680, predicts the correct manner in which the effect of the volume flaws appears in Q only when depletion is neglected; presumably, its inclusion would steepen the rate at which the term in x^3 loses effect as x decreases from large values, to yield the maximum found in the data. It is a consequence of the considerations adduced here that the distributions of volume, facial, and edge flaws are not independent as assumed in Section VA, but are coupled to an extent through the depletion effect.

VB2d. Distribution in the Fine Sizes

The theory of Section VA states that y (or y') in single fracture becomes a linear function of x for x small, down to values at which depletion effects appear. Limitations imposed by the mechanical process of repetitive sieving and the washing operation to remove the gelatin make the error large in the region of x below 100 μ. Since the sensitivity of a Coulter counter extends down to ~1 μ, this instrument was used to determine the distribution function for single fracture at low values of x. This procedure avoids the loss of fragments in repetitive use of sieves, since the particles are suspended in a fluid during the measurement. The Coulter counter, however, is not an absolute instrument in two senses: the particle volume must be determined by calibration and, in general, measurements can be made only on a small aliquot of the total specimen. Further, different ranges of particle dimensions require use of different orifices, and the corresponding results must be matched in a region of overlap of particle dimensions. These facts are not objections in the present instance, since the interest is in the functional form of the results, not their absolute values.

Initially, it was found by Gilvarry and Bergstrom (11) that unavoidable loss of the finer particles in the operation of removing the gelatin (or the bulk of it) limited the accuracy obtainable with a Coulter counter. Later, this difficulty was overcome by means of a special technique (12). A specimen was fractured in gelatin, and the entire gelatin matrix containing the suspended fragments was transferred to a large beaker, without a sieving operation. The gelatin in the beaker was dissolved in a quantity of water sufficiently large so that any suspended remnant of the gelatin did not interfere with the operation of the Coulter counter, as verified by blank tests.

Separate aliquots were removed from the beaker of fluid containing the suspended fragments, and each aliquot was subjected to one sieving operation. This single screening was necessary to prevent clogging of the nozzle of the Coulter counter by particles too large for its range. Sieves of sizes 210, 45, and 20 μ were used for the aliquots passed through nozzles of diameters 400, 100, and 50 μ, respectively. For these three nozzle sizes, Figure VB-13 shows the results for the cumulative number $N'(v)$ of particles from the corresponding aliquots, cumulated from the highest particle volume appearing in each case down to the value v appearing as the abscissa. By multiplying the particle counts by suitable constants, the results for the different nozzle sizes were matched at the values of v where they overlap.

From eqs. VA2-(34c), p. 679, and VA2-(60a), p. 685, one finds that $N'(v)$ for the entire specimen should satisfy

$$N'(v) = \tfrac{1}{2}\omega^{1/3}(V_0/k)v^{-2/3} \qquad \text{VB2-(11)}$$

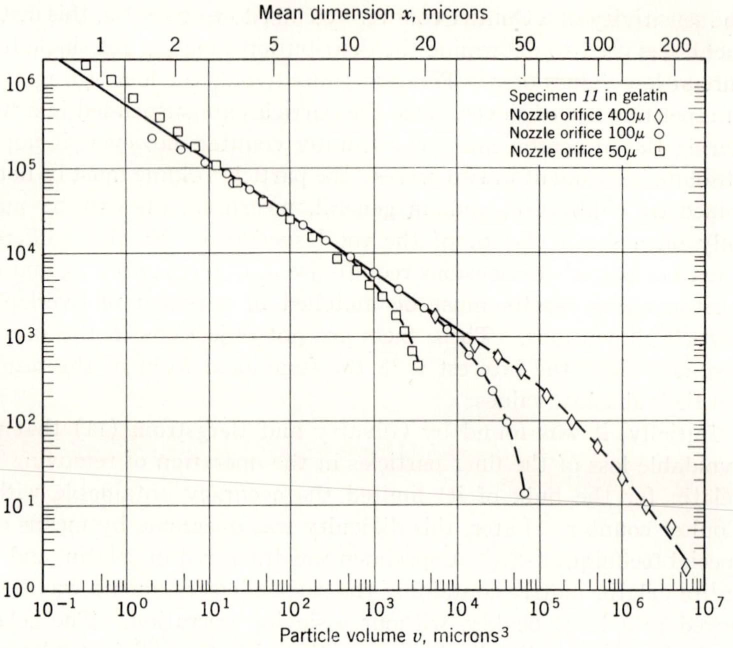

Fig. VB-13. The cumulated number $N'(v)$ of fragments with volume down to v as a function of particle volume v, determined by a Coulter counter for three aliquots from a specimen fractured in gelatin (12).

in the limiting case as $v \to 0$. In Figure VB-13, the dashed curve appearing for each aliquot runs asymptotically into a straight line (shown solid) corresponding to the power $v^{-2/3}$. The deviation of the dashed curves from the straight line arises because the proportionality to $v^{-2/3}$ implied by eq. (11) is asymptotic, derived by neglecting one limit of integration. Thus, this result does not apply until v becomes small relative to the smallest fragment volume removed from the aliquot by the sieving operation necessary to prevent clogging of the nozzle.

It is clear that the agreement of the experimental results with theory is excellent. Some deviation of the data points from the straight line appears for the smallest values of x, but this region corresponds to the limiting sensitivity (about 1 μ in particle diameter) of the Coulter counter. An approximate scale of linear dimensions is provided in Figure VB-13, based on the over-simplified assumption of spherical particles. One sees that the theory of Section VA is confirmed over a range of particle dimensions from roughly 50 μ to about 1 μ in single fracture. In conjunction with the preceding findings from sieving operations, this result implies agreement of experiment with the theory of Section VA, from the largest fragment size appearing down to about 1 μ in single fracture, at least as regards the integral distribution.

VB2e. Discussion

The parameters k, j, and i of Table VB-II represent flaw spacings referred to the mean dimension x of a fragment, and thus do not correspond to true flaw densities. To convert these quantities to proper densities one must evaluate the parameters λ_l, λ_s, and λ_v of eqs. VA2-(34a,b,c), p. 679. For this purpose it is necessary to determine the average particle shape. The histogram of Figure VB-14 shows percentage counts of the number of faces of each fragment for 155 particles arising from fracture in gelatin and 1060 fragments formed without use of this medium (as determined microscopically by E.W. Price for fragments of mean dimension down to 1680 μ). It is clear

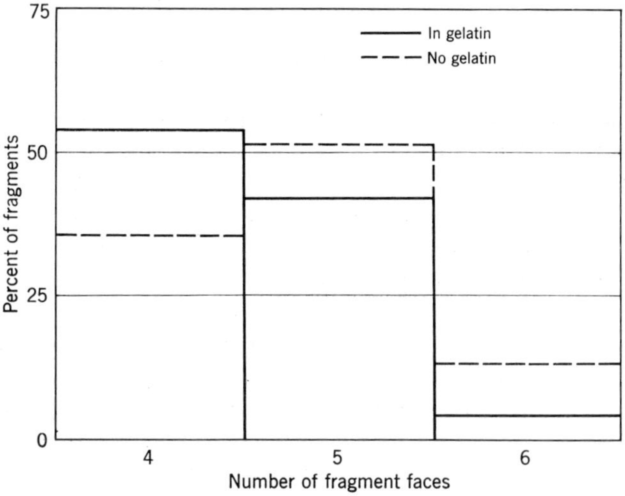

Fig. VB-14. Histogram of the percentage of fragments possessing a particular number of faces (11).

that the most probable particle shape in this range of larger sizes is tetrahedral in the case of single fracture. Qualitative examination down to a size of 74 μ for fragments formed in gelatin showed that the most probable shape is a flat tetrahedron (sphenoid) with the ratio of maximum to median dimension < 2. Columnar and cubic forms occur with decreasing relative frequency in that order. A fraction of the fragments up to a limit of perhaps 10% for any size examined showed an acicular shape in which the ratio of the maximum to the median dimension exceeded 10; the relative numbers decreased with mean dimension x for the case of single fracture. The presence of this acicular fraction implies some departure from randomness in the internal stress system producing fracture. It is generally conceded (9) that the dimension of a particle selected by a sieving operation is neither the maximum or minimum, but some intermediate value, on the average. In view of the generally flat shape of the fragments in question, however, their mean dimension was taken to correspond to the edge of a tetrahedron (assumed regular for purposes of simplicity). The corresponding values of the parameters λ_l, λ_s, and λ_v are tabulated in Table VB-IV; shown also are values of ω and of τ and σ from eqs. VA2-(42) and VA2-(59), pp. 680 and 685, respectively.

TABLE VB-IV
Parameters of the Distribution, Corresponding to a Tetrahedral Shape of Fragment

Parameter	Value
λ_l	6
λ_s	$\sqrt{3}$
λ_v	$\sqrt{2}/12$
ω	$6\sqrt{2}$
σ	$6\sqrt{6}$
τ	$36\sqrt{2}$

With use of the parameters of Table VB-IV, eqs. VA2-(35a,b,c), p. 680, yield values of the flaw densities γ_l, γ_s, and γ_v for single fracture corresponding to k, j, and i of Table VB-II, as shown in Table VB-V. One notes that the approximation, eq. VA2-(22), p. 677, is satisfied within roughly an order of magnitude, since the data of Table VB-II indicate that

$$k \approx j \approx i \qquad \text{VB2-(12)}$$

The implication of this result is that the distribution of orientations of fracture planes is not far from a random one. Furthermore, the degrees of activation of the different types (edge, facial, and volume) of internal flaws cannot differ greatly.

The flaw densities of Table VB-V refer to the internal flaws. Apparently, no results for direct comparison exist in the literature, since estimates of flaw densities made heretofore correspond to surficial

TABLE VB-V
Flaw Densities of Specimens, for Edge, Facial, and Volume Flaws of Internal Type

Specimen no.	γ_l, cm.$^{-1}$	γ_s, cm.$^{-2}$	γ_v, cm.$^{-3}$
2	0.47	4.4	75
3	0.42	11.	540
4	0.28	8.2	580

flaws. Anderson (2) has inferred a value of 10^3 cm.$^{-2}$ for the density of surficial flaws required by the theory of Fisher and Holloman (8) to explain the experimental results of Anderegg (1) on the size effect in the strength of glass fibers. The figure applies to the flaw density on the surface of a fiber drawn from a melt. From the rule of eq. VA2-(21), p. 677, this areal density implies a linear density of 30 cm.$^{-1}$. Thus, the sense of the difference from the results of Table VB-V is correct in that a surficial flaw density exceeds any corresponding internal density. In point of fact, the excess is large. Such a conclusion is necessary if glass specimens are to fail by virtue of surficial flaws, in general.

VB3. Experiment for the Two-Dimensional Case

To test the theoretical predictions of Subsection VA3, p. 686, which are the two-dimensional forms of the results of Subsection VA2, p. 669, an experimental arrangement is desirable which is the two-dimensional analog of that described in Subsection VB2. Thus, a flat plate should be compressed by forces parallel to its surface. Unfortunately, data of this kind apparently do not exist in the literature.

However, the exoclastic fragments from the spherical glass specimens used in the work of Gilvarry and Bergstrom (11,12) were formed by virtue of surficial flaws, and these fragments can be identified as the ones displaying original surface of the specimen, in general. Since the surfaces of the specimens were spherical and therefore closed, the theoretical requirement of Subsection VA3, p. 686, that no boundary exist in the surface, was met. The additional requirement of a continuous curvature of the specimen surface also was, in general, fulfilled. If specimens broken in gelatin are considered, the prerequisite for single fracture, that no secondary fracture of the exoclastic fragments on the walls of the retaining chamber occur, is met.

To the extent indicated, the exoclastic fragments from the specimens to which the results of Subsection VB2 correspond conform to the requirements of the theory. The purpose of this subsection is to display the agreement of the corresponding data with the theoretical predictions of Subsection VA3, as shown by Gilvarry (13).

VB3a. Experimental Method

The experimental work was carried out on the exoclastic fragments from Specimens 2, 3, and 4, to which the results of Subsection VB2 correspond. All specimens considered were broken in gelatin.

According to the theory of Subsection VA3, p. 686, the two-dimensional distribution function for fragment size in single fracture depends on the surface area a of a fragment and its perimeter c. In general, two types of surface area on the fragments were measured in the course of this work. One was the spherical surface corresponding to the original surface of the specimen. The other, referred to as the core surface, will be defined and discussed in Subsection VB3c. The area in question on an exoclastic fragment was measured by cementing thin lead foil to the surface, folding and trimming it to conform to the area, and weighing the foil on an analytical balance. In the case of the larger fragments, it was necessary to cut the foil into gores to conform to the spherical surface.

The perimeter corresponding to the original spherical surface on an exoclastic fragment was determined by wrapping a strip of masking tape around the perimeter in question. The trace of the spherical surface on the surface of the tape was outlined by rubbing with a lead pencil. Stripping off the tape and laying it flat rectifies in one dimension the curve corresponding to the perimeter. The length of the resulting sinuous curve was determined by means of a cartographic odometer.

VB3b. Overall Distribution

The first step in the analysis of the data is to determine the distribution function fixed by the measurements of the original spherical surface on each exoclastic specimen. This distribution will be referred to as the overall distribution. For purposes of distinction, physical quantities associated with it will be designated by a prime on the symbol. Thus a_n' denotes the original spherical surface on the nth exoclastic fragment.

The number of exoclastic fragments recovered varied from 5 in the case of Specimen 3, to 28 for Specimen 4. When this number is small, the exoclastic fragments all tend to have the shape of a lune, and some geometric similarity may exist. When the number is large, as in the case of Specimen 4, however, the smaller fragments depart appreciably from the shape of a lune, and no geometric similarity holds for all the exoclastic fragments. Hence, the similarity approximation discussed in Subsection VA3, p. 686, cannot be used in the present work.

The sieving operation of Subsection VB2 yielded a histographic representation of fragment volumes at uniform intervals of a mean dimension of the fragments. However, the measurements of surface area obviously yield a set of results at uneven intervals of a_n'. To determine z_n', the value of z of eq. VA3-(76), p. 691, corresponding to the nth fragment for the overall distribution, the fragments were ranked in order of increasing a_n' from a lowest value a_1' to a maximum. The value of A_n' for the nth fragment, corresponding to the integral A of eq. VA3-(72), p. 690, was determined from

$$A_n' = \sum_{\nu=1}^{n} a_\nu' \qquad \text{VB3-(13)}$$

with a maximum A_0', corresponding to A_0 of eq. VA3-(73), p. 690, fixed by summation over all the fragments. The probability z_n' for the nth fragment is then given by

$$z_n' = A_n'/A_0' \qquad \text{VB3-(14)}$$

In general, A_0' as determined from the recovered exoclastic fragments varied from 90 to 98% of the total initial surface of the specimen. The corresponding deficit in initial surface area does not arise significantly from failure to recover smaller exoclastic fragments, but from a characteristic feature of the fracture process in the experi-

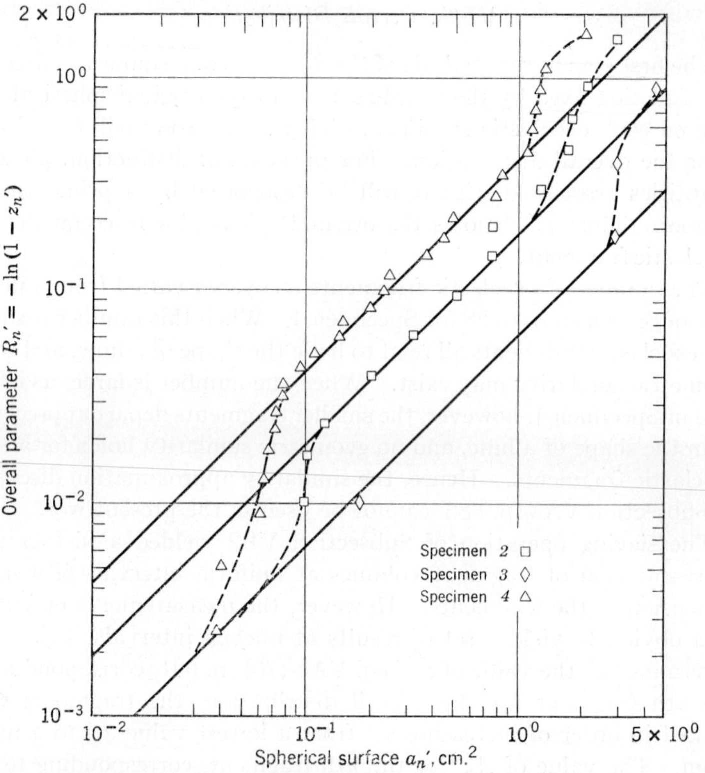

Fig. VB-15. The overall parameter R_n' from the data as a function of the original spherical surface appearing on each exoclastic fragment (13).

mental arrangement. As loading progresses, both specimen and platen deform elastically over a circular area of contact. The final extent of the contact area depends on the amount of compression required to fracture, and varies over rough limits of 2–10% of the initial spherical surface, from specimen to specimen. When fracture takes place, fragmentation over the contact area contributes to failure of the central core, as discussed in Subsection VB2c and amplified in Subsection VB3e. The trifling amount of original spherical surface that particles arising from the contact region may possess individually is below the limit of detection by the methods of this work. In this experimental arrangement, therefore, fragments possessing initial spherical surface corresponding to the interfaces between specimen and platens must be classed as endoclastic, and not as exoclastic. For this reason, z_n' of eq. (14) was normalized to the total area A_0' of the recovered exoclastic fragments, and not to the total initial surface.

Each of the larger exoclastic fragments retained a significant fraction of the total initial surface of the specimen. Thus, one expects the approximation of eq. VA3-(79), p. 691, to be valid for large a_n', corresponding to a dominant effect of the larger fragments in the distribution. If the overall parameter R_n' is defined by

$$R_n' = -\ln (1 - z_n') \qquad \text{VB3-(15)}$$

then eq. VA3-(79), p. 691, where δ_a' is a constant, yields

$$R_n' = \delta_a' \, a_n' \qquad \text{VB3-(16)}$$

The overall parameters R_n' of eq. (15) is shown in Figure VB-15 as a function of a_n', from the data for the fragments corresponding to three specimens. The data points for Specimen 3 clearly are sparse. One sees that over the intermediate range of values of a_n', the data points for each specimen fall on a straight line of 45° inclination on the logarithmic plot, as demanded by eq. (16). The straight lines appearing were determined by a least-squares fit of the data over this intermediate range. However, one notes that the data points for large a_n' and small a_n' diverge sharply from the straight lines, as indicated by the dashed curves in each case. [In this subsection, the numbering of the specimens and the coding of the data points (by squares, diamonds, and triangles) in the figures is the same as in Subsection VB2.]

It is clear that for each specimen, the distribution corresponding to z_n' is a composite of at least three distributions. In general, these correspond to the larger fragments, those of intermediate size, and the smaller fragments. In the following sections, the overall distribution will be analyzed into component distributions.

VB3c. Effect of Core Surface

In the present work, the load was distributed over a finite area as the spherical specimens deformed elastically under the pressure of the compressing platens. As was noted in Subsection VB2c, failure of

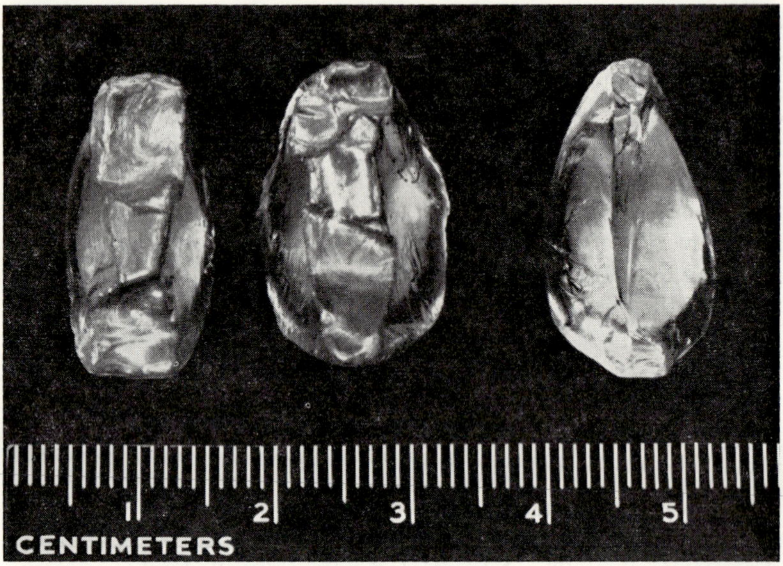

Fig. VB-16. Three typical exoclastic fragments. The two on the left show core surface (accentuated by means of a lacquer coating). The one on the right shows no such surface (13).

the specimen in this arrangement starts in conical fracture surfaces with bases on the platens; the base of the cone corresponds approximately to the area of contact of specimen and platen at the instant of fracture (17). The propagating conical surfaces produce a failed core extending diametrically through the specimen.

The exoclastic fragments responsible for the main mass of the fragments have a shape approximating a lune, as is evident from Figure VB-6. The outer surface of the lune corresponds to the original spherical specimen. For most of the larger lunes and for many of the smaller ones, an inner surface is identifiable, corresponding to the surface of the failed core which produces the endoclastic fragments. Typical lunes such as these are shown on the left in the photograph of Figure VB-16. On the right is shown a lune not exhibiting this core surface. The hypothesis will be examined that activation of surficial flaws in the core surface in question is a factor in fragmentation for the exoclastic fragments.

The core surface $a_n{}^*$ was measured with the aid of lead foil on each exoclastic fragment for which it was identifiable. These fragments were then ranked in order of increasing a_n^*, from the minimum value a_1^* found to a maximum. The value of A_n^* for the nth fragment, corresponding to the integral A of eq. VA3-(72), p. 690, is determined by

$$A_n^* = \sum_{\nu=1}^{n} a_\nu^* \qquad \text{VE3-(17)}$$

with a maximum value A_0^* fixed by summation over all the fragments exhibiting core surface. It was estimated that A_0^* as determined from the fragments in question coincided with the actual surface area of the failed core within $\sim$10 or 20%, in general. The value of z_n^* corresponding to z of eq. VA3-(76), p. 691, is then fixed by

$$z_n^* = A_n^*/A_0^* \qquad \text{VB3-(18)}$$

for the nth fragment. Physically, z_n^* represents the probability of fracture of an exoclastic fragment through activation of surficial flaws in the surface produced in the specimen by failure of the central core.

The exoclastic fragments exhibiting core surface are relatively large. Thus, eq. VA3-(79), p. 691, implies that the approximation

$$z_n^* = 1 - \exp\{-\delta_a^* a_n^*\} \qquad \text{VB3-(19)}$$

corresponding to a dominant effect of the large fragments, should be valid with δ_a^* a constant. The parameter δ_a^* represents physically the flaw density (relative to area) in the surface produced in the specimen by failure of the central core. If R_n^* is defined by

$$R_n^* = -\ln(1 - z_n^*) \qquad \text{VB3-(20)}$$

then eq. (19) requires that

$$R_n^* = \delta_a^* a_n^* \qquad \text{VB3-(21)}$$

The parameter R_n^* of eq. (20) from the data for three specimens is shown as a function of a_n^* on logarithmic scales in Figure VB-17. In each case, the data points (shown solid) corresponding to the largest values of a_n^*, lie on a straight line (shown solid) of 45° inclination, as demanded by eq. (21). For smaller values of a_n^*, the data points (shown open) deviate from the straight lines, as indicated by the dashed curves. The straight lines for each specimen were determined by a least-squares fit to the data points which appear solid. The

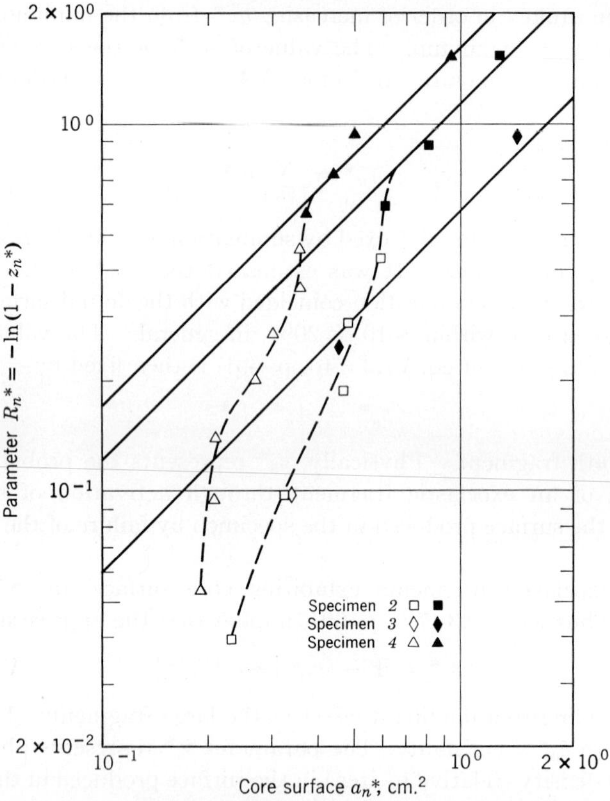

Fig. VB-17. The parameter $R_n^* = -\ln(1 - z_n)$ from the data as a function of the core surface a_n^* for every exoclastic fragment on which such surface appears. Data points fitted by the straight line are shown solid (13).

corresponding values of the flaw density δ_a^* are shown in Table VB-VI for the three specimens.

Independently of the specimen, eq. (19) implies that all data for $-\ln(1 - z_n^*)$ should be rectified on a line of 45° inclination when plotted against the corresponding R_n^* of eq. (21). Figure VB-18 shows that this prediction is borne out for the fragments having larger values of a_n^* (data points shown solid). In Figure VB-19, a direct plot of the probability z_n^* against a_n^* is shown on logarithmic scales for the three specimens. One sees that the data points (shown solid)

TABLE VB-VI
Flaw Densities δ_a^* and δ_a of Specimens, for Core and Spherical Surface of Fragments, Respectively

Specimen no.	δ_a^*, cm.$^{-2}$	δ_a, cm.$^{-2}$
2	1.2	0.50
3	0.60	0.21
4	1.7	1.0

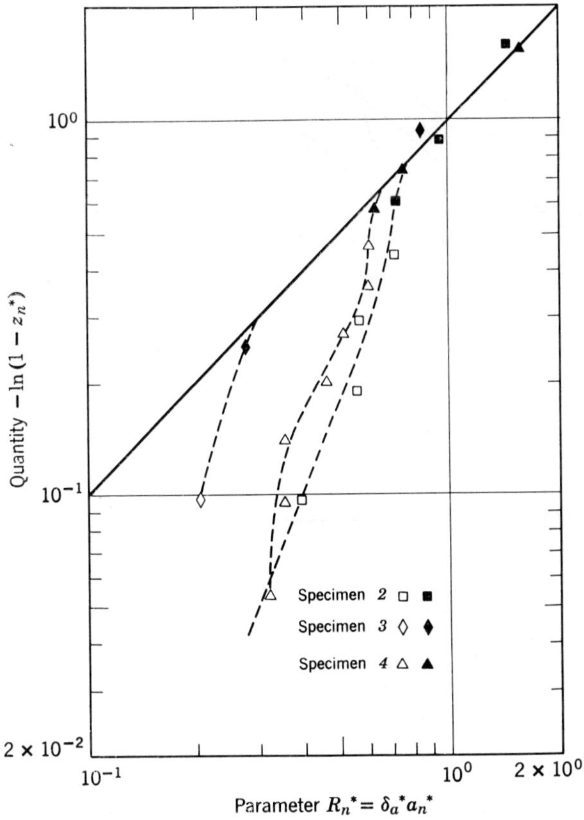

Fig. VB-18. The quantity $-\ln(1 - z_n^*)$ from the data as a function of the parameter R_n^* from the theoretical relation of eq. (21) with values of δ_a^* from Table VB-VI. Data points fitted by the straight line are shown solid (13).

for the larger values of a_n^* conform well to the exponential behavior demanded by eq. (19), as evaluated with values of δ_a^* from Table VB-VI.

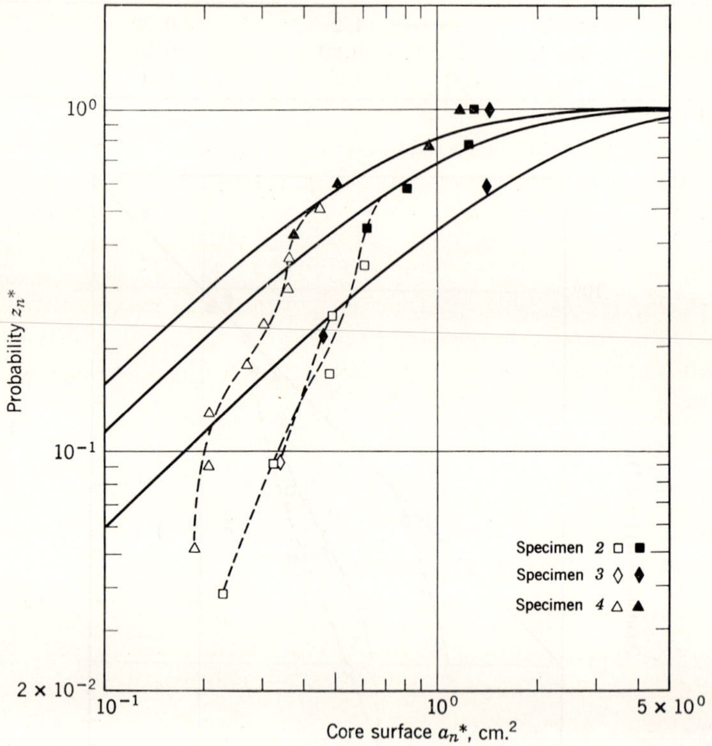

Fig. VB-19. The probability z_n^* of fracture from the data as a function of the core surface a_n^*. The curves represent the exponential relation of eq. (19), with values of δ_a^* from Table VB-VI. Data points fitted by the curve are shown solid (13).

It is clear that the distribution corresponding to the surficial flaws in the surface formed by failure of the central core is a component distribution in the overall distribution. These flaws produce the deviation from the straight lines of Figure VB-15 appearing for the larger values of a_n'.

VB3d. Effect of Spherical Surface

The foregoing discussion makes it evident that the overall distribution of Subsection VB3b does not correctly take into account the effect of surficial flaws on the original spherical surface of the specimens in producing fragmentation, since some of the fragments were formed by activation of flaws in the core surface. The latter fragments can be identified as those represented by the data points shown solid in the plots of Figures VB-17 and VB-19 against the core surface a_n^*. The data corresponding to these fragments were separated from the overall distribution, and the remaining values of a_n' were reranked in order of increasing spherical surface a_n, from a lowest value a_1 to a maximum. The set a_n differs from the set a_n' of Subsection VB3b only in the deletion from the latter of those values of a_n' for which the corresponding data fit the distribution of eq. (19) in terms of a_n^*. Because of the sparsity of data in the case of Specimen 3, one datum was regarded as belonging jointly to the set a_n' and the set a_n.

The value of A_n for the nth fragment was determined from

$$A_n = \sum_{\nu=1}^{n} a_\nu \qquad \text{VB3-(22)}$$

with a maximum A_0 fixed by summation over all the fragments in question. The corresponding value of z_n is given by

$$z_n = A_n/A_0 \qquad \text{VB3-(23)}$$

It is clear that activation of surficial flaws in the surface of the specimen produced by failure of the central core is the initial step in fragmentation for the exoclastic particles. For those particles not broken in this step, z_n physically represents the probability of fracture through activation of surficial flaws in the original spherical surface. Typically, values of A_0 were about 40% of the total original surface of the specimen.

One expects the approximation of eq. VA3-(79), p. 691, to be valid, at least for the larger values of a_n. Under this approximation, eq. (23) yields

$$R_n = \delta_a a_n \qquad \text{VB3-(24)}$$

if the parameter R_n is defined by

$$R_n = -\ln (1 - z_n) \qquad \text{VB3-(25)}$$

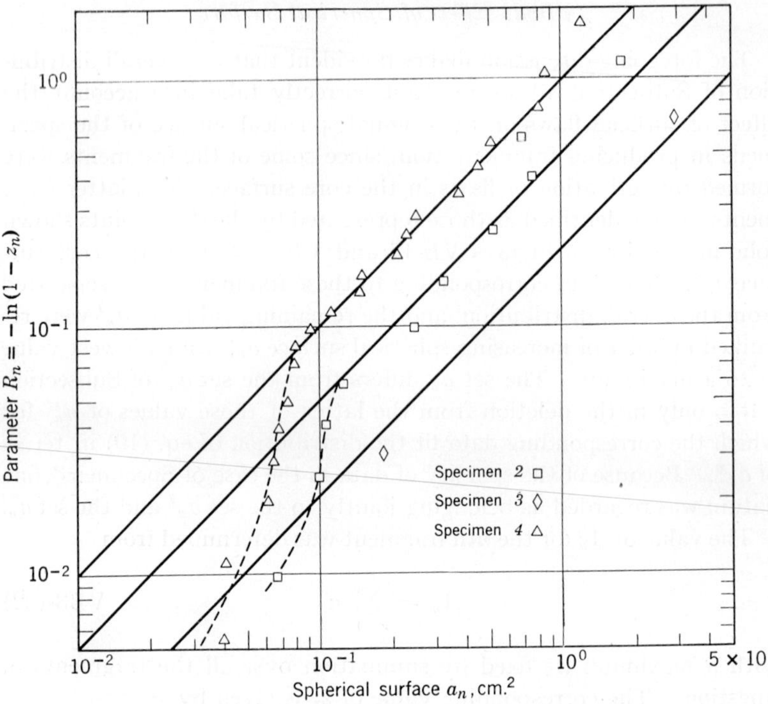

Fig. VB-20. The parameter $R_n = -\ln(1 - z_n)$ from the data as a function of the spherical surface a_n of the exoclastic fragments. The set a_n differs from the set a_n' of Fig. VB-15 by deletion of those values of a_n' for which the corresponding points in Fig. VB-17 are shown solid (13).

Physically, the constant δ_a represents the areal flaw density in the original spherical surface of those exoclastic fragments that survived fracture through activation of flaws in the core surface.

The parameter R_n of eq. (25) from the data for the three specimens considered is shown as a function of a_n in Figure VB-20, on logarithmic scales. For Specimens 2 and 4, the data points corresponding to the larger values of a_n lie on a straight line of 45° inclination, as demanded by eq. (24). These points represent the majority of the data. For the smaller values of a_n, a minority of the data points deviate from the straight lines, as indicated by the dashed curves. In the case of Specimen 3, only two data points are available, but these lie on a reasonably straight line of 45° inclination. The straight lines

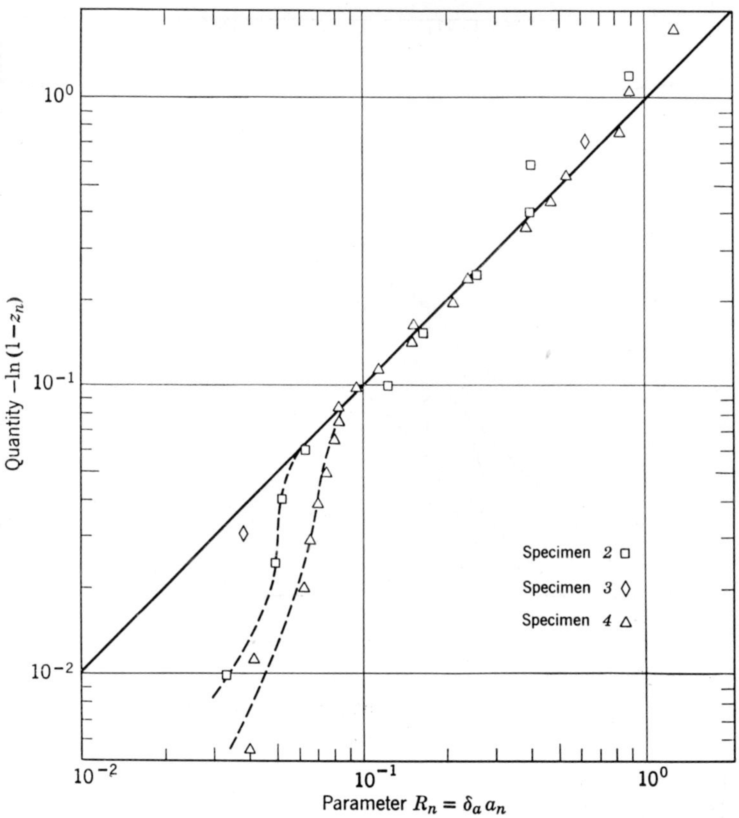

Fig. VB-21. The quantity $-\ln(1 - z_n)$ from the data as a function of the parameter R_n from the theoretical relation of eq. (24) with values of δ_a from Table VB-VI (13).

for each specimen were determined by a least-squares fit to the data points (for the larger values of a_n, in the case of Specimens 2 and 4). The corresponding values of the flaw density δ_a are shown in Table VB-VI for the three specimens.

Independently of the specimen, eq. VA3-(79), p. 691, implies that all data for $-\ln(1 - z_n)$ should be rectified on a line of 45° inclination when plotted against the corresponding R_n of eq. (24). One sees from Figure VB-21 that this prediction is borne out for the fragments with larger values of a_n in the case of Specimens 2 and 4, and for the two

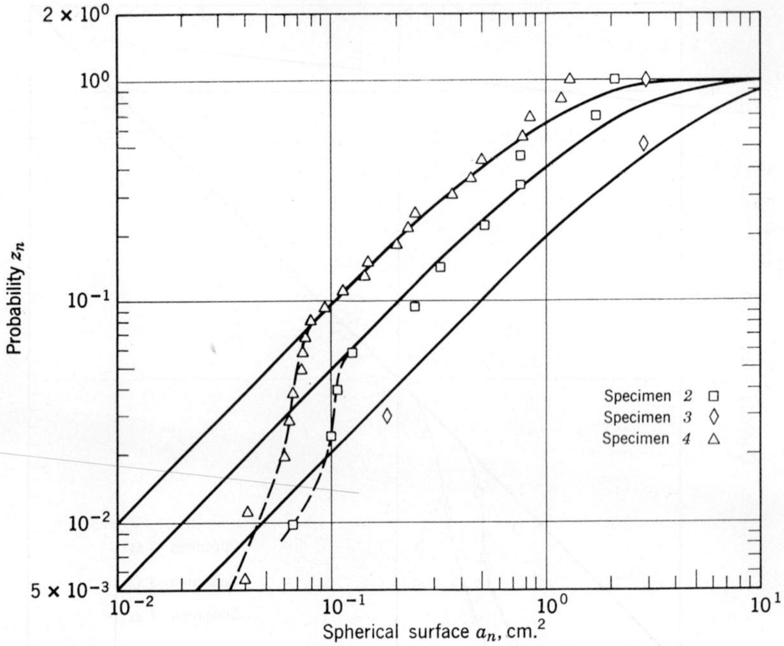

Fig. VB-22. The probability z_n of fracture from the data as a function of the spherical surface a_n. The curves represent the exponential of eq. VA3-(79), p. 691, with values of δ_a from Table VB-VI (13).

data points corresponding to Specimen 3. A direct plot of z_n against a_n is shown on logarithmic scales in Figure VB-22 for the three specimens. The data points for the larger values of a_n conform well to the exponential behavior demanded by eq. VA3-(79), as evaluated with values of δ_a from Table VB-VI. Discrepancies appear only for the very largest values of a_n.

Actually, eq. VA3-(79) represents an approximation valid only for the larger fragments. The complete expression corresponding to eq. VA3-(63), p. 688, is

$$R_n = \delta_a a_n + \delta_c c_n \qquad \text{VB3-(26)}$$

where c_n is the perimeter corresponding to the original spherical surface on a fragment, and δ_c is a constant. For the points falling on the dashed curves of Figures VB-20, VB-21, and VB-22 in the cases of Specimens 2 and 4, an attempt was made to correlate the data with an expression of the form of eq. (26), by using measured values of c_n. These attempts were unsuccessful. In view of the reasonable fit obtained by means of eq. VA3-(79), p. 691, to the data for the larger values of a_n in the cases of these specimens, one must conclude that over this range of a_n the term $\delta_c c_n$ is small relative to $\delta_a a_n$. Assuming an upper limit of 10% for the ratio $\delta_c c_n/\delta_a a_n$, one can set an upper limit to δ_c from the measured data. In this manner, one concludes that the linear flaw density δ_c corresponding to the perimeter of the original spherical surface on the fragments cannot exceed roughly 10^{-1} cm.$^{-1}$ for Specimens 2 and 4.

In the case of each specimen, the value of δ_a in Table VB-VI is less than the corresponding value of $\delta_a{}^*$. This fact is expected, since the primary mechanism of fragmentation in the exoclastic fragments is activation of surficial flaws in the core surface.

VB3e. Discussion

The exoclastic fragments for which the data are not correlated either by eq. (19) in terms of $a_n{}^*$ or by eq. VA3-(79), p. 691, in terms of a_n are less than 10 in number. It has been established that the discrepancy appearing in their case cannot be resolved by taking into account the perimeter c_n corresponding to the spherical surface on the fragments. Further, these fragments all show a relatively small amount of original spherical surface of the specimen, as compared to the remaining surface on the fragment. Accordingly, the presumption is that these particles actually were formed by activation of internal, rather than surficial flaws. Thus, they should be regarded as endoclastic rather than exoclastic particles.

If the fragments in question are separated out, the remaining fragments represent those which are truly exoclastic. All the data for the size distribution can then be correlated in one diagram. Define $a_n\dagger$ equal to $a_n{}^*$ or a_n for a fragment, according as the particle was produced by activation of surficial flaws in the core surface or spherical surface of the specimen, respectively. Correspondingly, define $z_n\dagger$ equal to $z_n{}^*$ or z_n, and let $R_n\dagger$ equal $R_n{}^*$ or R_n, where $R_n{}^*$ and R_n are given by eqs. (21) and (24), respectively. The quantity $-\ln(1 - z_n\dagger)$ from the data for the three specimens considered is shown in Figure VB-23 as a function of $R_n\dagger$, as evaluated from the flaw densities of Table VB-VI. Data points corresponding to the core surface $a_n{}^*$ are shown solid, those for spherical surface a_n appearing open. All

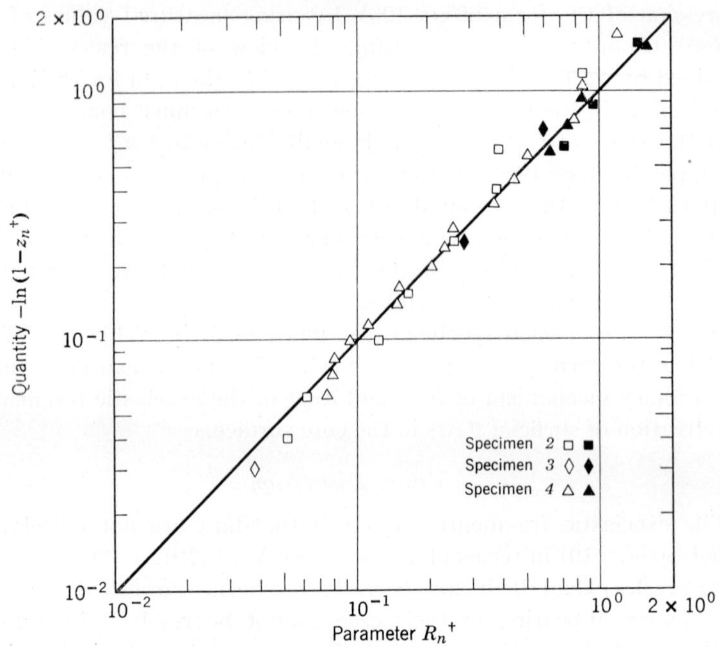

Fig. VB-23. The quantity $-\ln(1 - z_n\dagger)$ from the data as a function of the parameter $R_n\dagger$ from the theoretical relation for the true exoclastic fragments. The data points shown solid correspond to the core surface, and those shown open correspond to the original spherical surface of the fragments (13).

the data for the true exoclastic fragments lie reasonably well on a straight line of 45° inclination, as required by the theory of Section VA.

If these fragments, for which the classification is ambiguous, are indeed endoclastic, as suggested, the absence of fragmentation through activation of flaws in the perimeter of the original spherical surface of the fragments is understandable. The effect of these flaws, which should be dominant for small fragment size, is simply eliminated by the competing process of fragmentation through activation of internal flaws. Thus, the effect of the internal flaws is equivalent to a depletion effect on the surficial flaws. In this connection, one notes that the dashed curves of Figures VB-20, VB-21, and VB-22 have the form expected for a depletion effect, since in each case they lie below the curve corresponding to the surficial flaws.

Estimates of surficial flaw densities, which can be compared directly with those of Table VB-VI, are not available. It must be remembered that this work was carried out on polished spherical specimens of glass, which show high strength under compressive loading. As already noted, Anderson (2) has inferred a value of 10^3 cm.$^{-2}$ for the density of surficial flaws required by the theory of Fisher and Holloman to explain the experimental results of Anderegg on the size effect in the strength of glass fibers. The figure applies to the flaw density on the surface of a fiber drawn from a melt. It exceeds the figures of Table VB-VI by orders of magnitude. Results for these fragile fibers, however, are not directly comparable with those for spheres under compression.

The values of the surficial flaw densities δ_a^* and δ_a in Table VB-VI are less than the corresponding values of γ_s for internal flaws given in Table VB-V, whereas one, in general, expects a surficial flaw density to exceed a corresponding internal flaw density. This supposition seems reasonable if specimens are to break at the outset by virtue of surficial flaws, primarily. The initial fracture by failure of the central core, however, splits the specimen into two disjoint portions which subsequently are subject to fragmentation independently of each other. Thus, the flaw densities represented by δ_a^* and δ_a and by γ_s do not correspond strictly to flaws in the same parent body. For the exoclastic fragments, on the other hand, one finds properly that δ_a^* exceeds δ_a, since the former corresponds to the primary mechanism of fragmentation.

Presumably, failure of the central core starts by activation of a surficial flaw in the area of contact of the specimen and the compressing platens. Flaw densities referred to this surface do not appear, however, in the analysis of the data on fragment size. Thus, the development in the form given yields no direct information on the densities of the flaws responsible for inception of fracture in the specimen, but only on those creating fragmentation throughout the specimen.

VB4. Summary and Conclusions

In general, one sees that good agreement between theory and experiment obtains. In the three-dimensional case, the predicted linear variation of y with x is observed over a reasonable range of values at small x, for single fracture. The result confirms the existence and importance of the edge flaws. The physics of the situation correctly predicts the possible number of peaks in the differential probability for the endoclastic distribution. The existence of two peaks in this distribution, corresponding to the effect of facial and volume flaws, has been demonstrated from the average data for a large number, 40, of specimens. The same data confirm the presence in the overall differential distribution of a third peak arising from the exoclastic fragments. These results leave little possibility of explaining the origin of these peaks as spurious effects arising from poor data. Finally, it has been shown by means of a Coulter counter that the data display the behavior at fragment sizes approaching 1 μ which is demanded by the limiting form of the theoretical distribution function.

This distribution in its two-dimensional form has been shown to describe correctly the fragment sizes resulting from fracture of a two-dimensional body by applying it to the special case of the exoclastic fragments from a spherical specimen. In this situation, fragmentation proceeds by activation of surficial flaws not only in the original surface of the specimen, but on the core surface resulting from failure of the central cylinder. The latter mode of fragmentation of the exoclastic fragments is primary in point of time in the fracture process, but the former, in general, is predominant in regard to number of fragments. In the special case of exoclastic fragments from a spherical specimen, evidence has been presented that fragmentation by edge flaws in the perimeters of the fragments is inhibited by a depletion effect arising from internal flaws. It should be noted, however, that the data do not necessarily support the last conclusion as a general result for two-dimensional fracture.

To exemplify these conclusions, one can note from Figure VB-13 that the total number of fragments formed in fracture of one of the specimens considered must surely exceed a million, the actual figure probably being well in excess of a billion. Figure VB-12 implies that all the experimental data for the endoclastic fragments from Specimens 2, 3, and 4 can be correlated in one diagram of $-\ln(1-y)$ plotted against Q. Similarly, Figure VB-23 indicates that all the experimental data for the definitely exoclastic fragments from these specimens can be correlated in one diagram of $-\ln(1-z_n\dagger)$ plotted as a function of $R_n\dagger$. Thus, of the probable billion or more of particles formed in fragmentation of these three specimens, the theory of Section VA of this volume fails to account quantitatively for the integral size distribution only in the case of less than ten fragments, which are ambiguous as to whether they are exoclastic or endoclastic.

Both y and z_n, however, are integral parameters associated with the distribution of fragment sizes. As regards the differential parameters p and p', the theory is obviously correct in predicting that these quantities approach constants in the intermediate range of fragment sizes. However, this behavior of p and p' cannot be correct in the limit of a fragment size which is vanishing, as certain integrals over the distribution become divergent in this case which are known on physical grounds to be finite. This deficiency has been traced to neglect of flaw depletion in the theoretical development, a feature inherent in the Poisson law. For the same reason, the theory is only partially successful in reproducing the details of the peaks in the differential probability, associated with the effect of facial and volume flaws. The general trend of the data is followed, but the fit to the experimental points is only semiquantitative.

No definite experimental evidence for a depletion effect in the fine sizes has been observed down to the lowest dimension, 1 μ, corresponding to available measurements. Deviations from the Schuhmann relation for a fragment size in the neighborhood of 2 μ have been reported by Gaudin and Fuerstenau (10) for the product of a milling operation. Since they found the effect to be reproducible, it may represent a depletion effect. Experimentally, the instrument required to attack this problem is the electron microscope, since the fragment size at which such effects appear may be a relatively small multiple of molecular dimensions. A program in this direction has been undertaken by the author jointly with Dr. A. P. Welch.

To this point, no mention has been made of plural fracture (defined as single fracture of the original specimen, followed by a sequence of only a few secondary fractures of the fragments). If the experimental arrangement differs from that for single fracture only in the absence of a decelerating medium such as gelatin, no secondary fracture by the piston or hammer is possible. Any secondary fracture necessarily arises by virtue of the kinetic energy of the fragments. As already noted, the cinematographic results of Bergstrom, Sollenberger, and Mitchell indicate that the speeds can be high for the larger fragments. Hence, plural fracture has been simulated by Gilvarry and Bergstrom (12,14) in an experimental arrangement differing from the previous one in the absence of the gelatin and the presence of a steel retaining ring instead of one of Lucite (see Fig. VB-1).

The overall features of the distribution of sizes in the fragments produced by plural fracture can be predicted only qualitatively for the larger sizes. Superimposed on the normal endoclastic distribution will be that corresponding to secondary fracture of the exoclastic particles from the primary fracture, which are the only fragments expected to show a high speed of impact on the chamber walls. Residual surficial flaws in these fragments as well as the presence of internal flaws of facial and volume type should conspire to yield three local maxima in the overall distribution, instead of the single peak which the exoclastic fragments yield in single fracture. Thus, a total of five peaks in the overall differential probability at large fragment size can be expected. This prediction seems to be borne out by experiment, although the statistics are by no means as good as in the case of single fracture (12,14).

The course of the overall distribution function for small fragment size can be predicted *quantitatively* for plural fracture (12,14). The argument proceeds from the fact shown by Bergstrom, Sollenberger, and Mitchell, that the finer particles in single fracture move at relatively low speeds. Thus, particles in this range of size from the primary fracture should undergo little secondary fracture in the first instance. So long as the number of secondary fractures is restricted to only a few, such fractures will distort the distribution function for the finer fragments only slightly. Accordingly, the overall probability y' for the fine fragments in plural fracture must approach a form corresponding to the Schuhmann law of eq. (1), asymptotically as x becomes small. This result must hold down to values of x at which depletion effects appear. In conformity with these arguments, measurements on a Coulter counter show that the cumulated number $N'(v)$ of particles in plural fracture obeys the proportionality to $v^{-2/3}$ of eq. (11), established for the fine sizes in single fracture (11,14).

The problem of the distribution function in comminution of a specimen has also been discussed by Gilvarry and Bergstrom (14). In this case, only qualitative statements can be made on the basis of the present theory.

It is clear that the theory and the experimental methods reviewed yield a tool to study the distribution of Griffith flaws in a solid. This method is the only one available at present to investigate the distribution of internal flaws, since studies of the brittle strength of solids yield information primarily on the surficial flaws. Furthermore, only the latter type of flaw can be revealed by decoration techniques (3,15,19). The existence of the internal flaws can be demonstrated by stress pulses initiated by explosives (18).

The author wishes to thank the Editor of the *Journal of Applied Physics* for permission to use quotations and figures from articles appearing originally in that journal.

VB5. List of Key Symbols‡

VB5a. Three-Dimensional Case

- k' Effective value of k in an overall distribution
- p' Differential probability in an overall distribution
- x_0' Value of x_0 for largest fragment in an overall distribution
- x_1 Value of x for peak in differential probability p corresponding to volume flaws
- x_2 Value of x for peak in differential probability p corresponding to facial flaws
- y' Value of y for an overall distribution

VB5b. Two-Dimensional Case

- a' Surface area of a fragment in an overall distribution
- a^* Core surface on an exoclastic fragment from a sphere
- $a\dagger$ Defined as a^* or a, according as fragment is from core or spherical surface, respectively
- A' Value of A for an overall distribution
- A_0' Effective value of A_0 for an overall distribution
- A^* Sum of core surface on exoclastic fragments, from smallest to a particular one
- A_0^* Total core surface on all exoclastic fragments
- R' Value of R for an overall distribution
- R^* Value of R for exoclastic fragments showing core surface
- $R\dagger$ Defined as R^* or R, according as fragments are from core or spherical surface, respectively
- z' Value of z for an overall distribution

‡ For symbols not appearing in this list, consult the corresponding list in Section VA.

z^* Value of z for fragments showing core surface

$z^\dagger$ Defined as z^* or z, according as fragments are from core or spherical surface, respectively.

δ_a' Effective value of δ_a for an overall distribution

δ_a^* Mean areal density of facial flaws on core surface

References

1. Anderegg, F. O., *Ind. Eng. Chem.*, **31**, 290 (1939).
2. Anderson, O. L., in *Fracture*, B. L. Averback, D. K. Felbeck, G. T. Hahn, and D. A. Thomas, Eds., Wiley, New York, 1959, p. 331.
3. Andrade, E. N. da C., and L. C. Tsien, *Proc. Roy. Soc.* (*London*), **A159,** 346 (1937).
4. Berg, R. H., *Symposium on Particle Size Measurement*, STP **234**, ASTM, Philadelphia, 1958.
5. Bergstrom, B. H., C. L. Sollenberger, and W. Mitchell, *Trans. AIME*, **220,** 367 (1961).
6. Bergstrom, B. H., and C. L. Sollenberger, *Trans. AIME*, **220,** 373 (1961).
7. Coulter, W. H., *Proc. Natl. Electron. Conf.*, **12,** 1034 (1956).
8. Fisher, J. C., and J. H. Holloman, *AIME Tech. Publ. No. 2218* (1947).
9. Gaudin, A. M., *Principles of Mineral Dressing*, McGraw-Hill, New York, 1939, p. 51.
10. Gaudin, A. M., and M. Fuerstenau, private communication.
11. Gilvarry, J. J., and B. H. Bergstrom, *J. Appl. Phys.*, **32,** 400 (1961).
12. Gilvarry, J. J., and B. H. Bergstrom, *J. Appl. Phys.*, **33,** 3211 (1962).
13. Gilvarry, J. J., *J. Appl. Phys.*, **33,** 3218 (1962).
14. Gilvarry, J. J., and B. H. Bergstrom, *Trans. AIME*, **220,** 380 (1961).
15. Gordon, J. E., D. M. Marsh, and M. E. M. L. Parratt, *Proc. Roy. Soc.* (*London*), **A249,** 65 (1959).
16. Herdan, G., *Small Particle Statistics*, Elsevier, New York, 1953, p. 127.
17. Jaeger, J. C., *Elasticity, Fracture and Flow*, Methuen, London; Wiley, New York, 1956, p. 74.
18. Kolsky, H., in *Fracture*, B. L. Averback, D. K. Felbeck, G. T. Hahn, and D. A. Thomas, Eds., Wiley, New York, 1959, p. 281.
19. Levengood, W. C., *J. Appl. Phys.*, **30,** 378 (1959).
20. Roesler, F. C., *Proc. Phys. Soc.* (*London*), **B69,** 981 (1956).
21. Taggart, A. F., *Handbook of Mineral Dressing*, Wiley, New York, 1945, p. **19**–102.

VC. EFFECTS OF STRESS RAISERS IN LINEAR VISCOELASTIC BODIES

J. R. M. Radok

VC1. Introduction

Alfrey (1) presents a summary of the formulae of the linear theories of elasticity and viscoelasticity, largely simplified to the case of incompressible media, in Appendix III of his book. In particular, he concludes that for incompressible viscoelastic materials, in the case of prescribed loading on the boundaries (first boundary value problem) and zero initial conditions, the stress distribution is identical to that in an incompressible elastic material subject to the same instantaneous surface forces. It is readily seen that this result depends strongly on the fact that one is dealing with zero initial conditions and that body forces are absent; otherwise, it is true only for materials of the Voigt type. If this were not so, this conclusion would imply that all initial states must be possible elastic states of stress, thus it would impose a severe restriction on viscoelastic theory.

A formal study of the theory of viscoelasticity shows that one must make a distinction between basic, characteristic times of the material and pseudo-characteristic times of a body subject to nonhomogeneous deformation. The term "characteristic time" is applied to the absolute value of the inverse of the coefficient of time t in the exponents of the terms constituting the solution. The former arise irrespective of the shape and mode of loading and deformation of the body, while the latter are closely linked to the particular body under consideration.

Situations may arise where the former (characteristic) times do not enter into a particular solution, but they apparently should not be disregarded *ab initio*.

Without entering into a general discussion of the merits of a linear theory of viscoelasticity (in fact, assuming this theory to be valid for certain materials under trivial strains), it seems reasonable to look into the possibilities of realistic, quantitative stress analysis. Work published in this area exhibits artificialities in the configurations and loading conditions selected, indicating that (at the present stage) viscoelastic stress analysis is far removed from the effectiveness of its elastic analogy (7). Little attention has been given to effects of body forces or variations in the simplest modes of loading and initial conditions, even when these aspects of problems are probably more readily fixed than the boundary conditions (for example, the case of rockets).

While an effective stress analysis of viscoelastic bodies tentatively cannot be approached, a qualitative analysis of stresses and deformations is well within the reach of the existing theory (9). Such analyses throw light on fundamental aspects of the theory such as those of correlation with experiments, specification of material properties, design of specimen, influence of impurities, etc. It is indicated that studies of this kind should precede an effective stress analysis since they will help in gaining a better understanding of the behavior of viscoelastic materials.

From this point of view, the characteristic times determining the behavior of a body composed of viscoelastic material are of foremost importance since they will enter into the description of its transient stress and displacement distributions. Some of the material in this section has been discussed previously within a framework of stress analysis. It will be re-examined here in order to elucidate somewhat the influence of finite dimension, geometry, and boundary conditions on the characteristic times of a viscoelastic body.

The following results are most easily obtained by use of operator methods, a summary of which is first presented for two- and three-dimensional bodies. The salient features of the above effects are these demonstrated within a study of the viscoelastic tube externally bounded by an elastic shell. The investigation of cylindrical inclusions in a two-dimensional body, which follows, arises quite naturally as the next stage of complexity.

VC2. Notation. Basic Characteristic Times

The linear viscoelastic stress-strain law will be written (10)

$$P^p \, s_{ij} = Q^q \, e_{ij} \qquad \text{VC2-(1)}$$

$$R^r \, \sigma = S^s \, \epsilon \qquad \text{VC2-(2)}$$

where the orders p, q, r, and s of the operators, P^p, Q^q, R^r, and S^s of the form

$$P^p = \sum_{k=0}^{p} p_k \frac{\partial^k}{\partial t^k}, \text{ etc.} \qquad \text{VC2-(3)}$$

will often be omitted, unless there is a need for emphasis, and the stress and strain deviators are defined (cf. Subsection IC4h, p. 90) by

$$\begin{aligned} s_{ij} &= \sigma_{ij} - \tfrac{1}{3}\sigma\delta_{ij} \\ e_{ij} &= \epsilon_{ij} - \tfrac{1}{3}\epsilon\delta_{ij} \\ \sigma &= \sigma_{11} + \sigma_{22} + \sigma_{33} \qquad \text{VC2-(4)} \\ \epsilon &= \epsilon_{11} + \epsilon_{22} + \epsilon_{33} \\ \epsilon_{ij} &= \tfrac{1}{2}[(\partial u_i/\partial x_j) + (\partial u_j/\partial x_i)] \end{aligned}$$

For the three-dimensional case, the relations between the elastic and viscoelastic theories have been studied by Radok (11). It is shown (11) that in a formal operator approach based on the Navier (15) equations, one has for the viscoelastic displacements,

$$(2PS + QR)(\partial\epsilon/\partial x_i) + 3QR\nabla^2 u_i = 0 \qquad \text{VC2-(5)}$$

The stresses must be obtained from the differential equations

$$PR\sigma_{ij} = \tfrac{1}{3}\delta_{ij}(PS - QR)\epsilon + QR\epsilon_{ij} \qquad \text{VC2-(6)}$$

and the basic characteristic times are determined by the operator P^pR^r of order $p + r$. In the case of a material of the Voigt type which behaves elastically in dilatation, one has $P = R = 1$, and the stresses σ_{ij} may be determined directly in terms of the displacements u_i. In all other cases one has to solve ordinary differential equations in order to obtain expressions for the stresses. The particular integrals of these equations, corresponding to the right-hand side of eq. (6), render the pseudo-characteristic times to be discussed later, while the complementary functions, i.e., the solutions of the homogeneous equations, render the basic characteristic times. Obviously, the initial conditions must be adequate to make the solutions of these ordinary differential equations unique, i.e., their number must be equal to the order $p + r$ of the operator $P^p R^r$.

In the two-dimensional case of plane strain (12), the general formulae in terms of functions of a complex variable $z = x_1 + ix_2$ are

$$(QR + PS)(\sigma_{11} + \sigma_{22}) = 2[\varphi'(z) + \overline{\varphi'(z)}]$$

$$(QR + PS)(\sigma_{22} - \sigma_{11} + 2i\sigma_{12}) = 2[\bar{z}\varphi''(z) + \psi'(z)] \quad \text{VC2-(7a)}$$

$$QS(QR + PS)(u_1 + iu_2) = (2QR + PS)\varphi(z) - PS[z\overline{\varphi'(z)} + \overline{\psi(z)}] \quad \text{VC2-(8a)}$$

where the stress–strain relations have been rewritten

$$P(\sigma_{22} - \sigma_{11} + 2i\sigma_{12}) = -2Q\frac{\partial}{\partial z}(u_1 - iu_2) \quad \text{VC2-(9a)}$$

$$R(\sigma_{11} + \sigma_{22}) = S\left[\frac{\partial}{\partial z}(u_1 + iu_2) + \frac{\partial}{\partial \bar{z}}(u_1 - iu_2)\right] \quad \text{VC2-(9b)}$$

Here the functions φ and ψ are the solutions of the corresponding elastic problem with the elastic constants replaced by the corresponding combinations of operators. The operators P, Q, R, and S in two dimensions are obviously not the same as those in three dimensions. There exist simple relationships between the operators in two and three dimensions which have no significance within the present study.

The formulae, eqs. (7a), render ordinary differential equations for the stresses σ_{11}, σ_{22}, σ_{12}. The operator $(Q^q R^r + P^p S^s)$ of order max $(q + r, p + s)$ determines the basic characteristic times of the material. The right-hand sides of eqs. (7a) are the elastic solutions in which the elastic constants must be replaced by their expressions in terms of the operators

$$2\mu = Q/P$$
$$2\mu/(1-2\nu) = S/R \quad \text{VC2-(10)}$$

When the material is incompressible, the preceding formulae reduce to

$$P(\sigma_{11} + \sigma_{22}) = 2[\varphi'(z) + \overline{\varphi'(z)}]$$

$$P(\sigma_{22} - \sigma_{11} + 2i\sigma_{12}) = 2[\bar{z}\varphi''(z) + \psi'(z)] \quad \text{VC2-(7b)}$$

$$Q(u_1 + iu_2) = \varphi(z) - z\varphi'(z) - \overline{\psi(z)} \quad \text{VC2-(8b)}$$

so that the basic characteristic times are determined by P. The single stress–strain relation is now eq. (9a). These formulae correspond to those given by Alfrey (1), except that the right-hand sides are expressions derived from the elastic solutions. When the results are presented in the form of eq. (7b), it is readily seen that viscoelastic and elastic stresses can only be identical if the material is of the Voigt type.

From the above, it follows that the essential nature of the basic characteristic times is that they also manifest themselves in the case of a homogeneous solid. If this were not the case, the general solution of the problem would not be sufficiently versatile to be adjustable to the number of initial conditions associated with the order of the operators of the material under consideration.

VC3. The Elastically Reinforced Tube. Influence of Geometry and Boundary Conditions

It has been shown in the last subsection that there occurs a natural division of the characteristic times of a viscoelastic body into those associated with the material and with the particular problem under consideration, respectively. In the discussion of the second group, a further subdivision of effects will be observed, namely, geometry and boundary conditions. The isolation of these phenomena demands a careful choice of problem.

Consider an infinite cylindrical tube of viscoelastic material reinforced externally by an elastic shell, the two materials being glued together over the common surface. Denote the internal and external radii of the viscoelastic tube by R_1 and R_2, respectively; their ratio by ρ; and the elastic constants of the reinforcing material by E_R and ν_R. The corresponding elastic problem is a generalization of Lamé's problem, while the condition on the external boundary has been discussed in detail by Radok (13). In viscoelastic stress analysis, this problem was first discussed by Radok and Woodward (14), and also by Lee, Radok, and Woodward (7). The advantage of this problem arises from the point that the elastic constants of the reinforcement and the ratio ρ make it possible to study independently the effects of geometry (size) and boundary conditions.

The elastic solution of this problem is (7):

$$\sigma_{rr} = p(t) \frac{[\alpha\nu(1+\nu) - E][(R_2/r)^2 - 1] - \alpha(1-\nu^2)[(R_2/r)^2 + 1]}{\alpha(1-\nu^2)(\rho^2+1) - [\alpha\nu(1+\nu) - E](\rho^2 - 1)}$$

$$\sigma_{\theta\theta} = p(t) \frac{\alpha(1-\nu^2)[(R_2/r)^2 - 1] - [\alpha\nu(1+\nu) - E][(R_2/r)^2 + 1]}{\alpha(1-\nu^2)(\rho^2+1) - [\alpha\nu(1+\nu) - E](\rho^2 - 1)} \qquad \text{VC3-(11)}$$

where

$$\alpha = E_R h/(1 - \nu_R{}^2)$$

h is the thickness of the casing, E and ν are the elastic constants of the material of the main tube, and $p(t)$ is the internal pressure as a function of the time. By use of the formal relationships between E and ν and the viscoelastic operators P, Q, R, and S of eq. (1)

$$E = \frac{3QS}{2PS + QR}$$

$$\nu = \frac{PS - QR}{2PS + QR} \qquad \text{VC3-(12)}$$

one can derive from the denominator of eq. (11) the operator which determines the pseudo-characteristic times of the reinforced viscoelastic tube. The general expression for this operator will not be shown here. Instead, consider the example of the Maxwell type material with elastic dilatational response for which $P = p_0 + p_1\partial/\partial t$, $Q = p$, $R = 1$, $S = s_0$. After some manipulations one finds the quadratic

$$[(1 - \rho^2 - \alpha p_1)(2p_1 s_0 + 1) - 3\alpha p_1 \rho^2]\tau^2 + [(1-\rho^2)2p_0 s_0 - \alpha p_0(3\rho^2 + 4p_1 s_0 + 1)]\tau - 2\alpha p_0{}^2 s_0 = 0 \qquad \text{VC3-(13)}$$

for the determination of the characteristic times $\tau_{1,2}$

In order to examine the influence of ρ and α on the roots of this equation, the coefficients p_0, p_1, and s_0 will be assigned the values, $p_0 = p_1 = (4/3) \times 10^{-5}$, $s_0 = 3 \times 10^5$ psi. In this case, the basic characteristic time of the material will be unity since the integrating factor corresponding to the operator $(\partial/\partial t + 1)$ is e^t. Further, letting

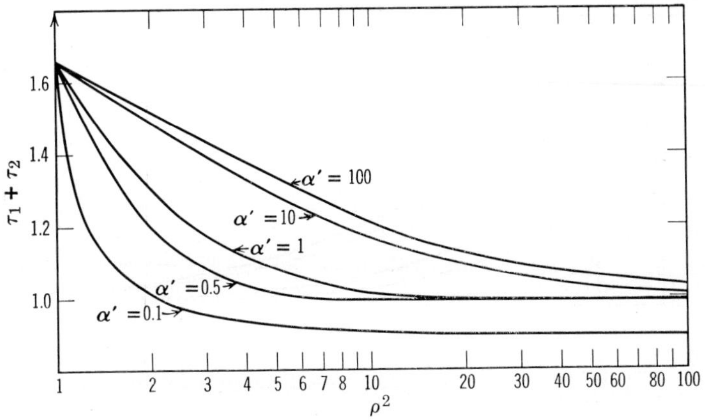

Fig. VC-1. Sum of characteristic times for reinforced tube.

$E_R = 3 \times 10^7$ psi, $\nu_R = 1/\sqrt{11}$ and $h/R_2 = 1/33$, one finds for steel casing $\alpha = 10^6$. Obviously, for $\alpha = 0$ and $\alpha = \infty$, the reinforcement vanishes and is rigid, respectively.

Substituting the above values for p_0, p_1, and s_0 into eq. (13), one finds

$$\tau^2 - A\tau + B = 0$$

where,

$$A = \frac{8(1 - \rho^2) - \alpha'(17 + 3\rho^2)}{3\alpha'\rho^2 - 9(1 - \rho^2 - \alpha')} = \tau_1 + \tau_2$$

$$B = \frac{8\alpha'}{3\alpha'\rho^2 - 9(1 - \rho^2 - \alpha')} = \tau_1\tau_2$$

$$\alpha' = (4 \times 10^{-5}/3)\alpha$$

Figures VC-1 and VC-2 show the sum and product of the characteristic times graphed against the ratio ρ^2 for various values of α'.

Now we consider the particular cases *a–d*.

(*a*) For $\alpha' = 0$, the viscoelastic tube is not strengthened elastically: $A = \tau_1 + \tau_2 = -8/9$, $B = 0$; i.e., there is only one pseudocharacteristic time which may be associated with the internal boundary.

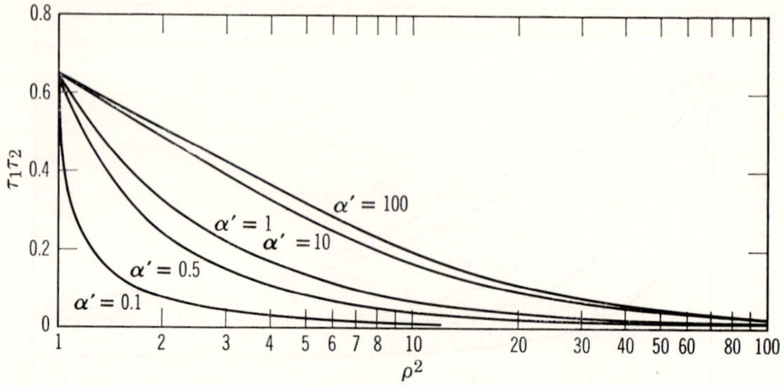

Fig. VC-2. Product of characteristic times for reinforced tube.

(*b*) For $\alpha' = \infty$, the viscoelastic tube is held in a rigid casing: $A = -(1 + B), B = 8/(3\rho^2 + 9)$; i.e., for any value of ρ^2, $\tau_1 = 1$, $\tau_2 = 8/(3\rho^2 + 9)$.

(*c*) For $\rho^2 = \infty$, A varies between $-8/9$ for $\alpha' = 0$ and -1 for $\alpha' = \infty$. For $\rho^2 = 0$, A varies from $-8/9$ for $\alpha' = 0$ to $-17/9$ for $\alpha' = \infty$. Since A can be shown to increase monotonically with increasing ρ^2 and α', the sums of the pseudo-characteristic times lie in the range: $8/9 < \tau_1 + \tau_2 < 17/9$.

(*d*) For $\rho^2 = \infty$, $B = 0$; for $\rho^2 = 1$, $B = 2/3$. Since B can be shown to decrease monotonically with increasing ρ^2 and to increase monotonically with α', the products of the pseudo-characteristic times lie in the range: $0 < \tau_1\tau_2 < 2/3$.

In any particular case, the pseudo-characteristic times can now be obtained from Figures VC-1 and VC-2. It is remarkable that these characteristic times lie so crowded into a narrow band. It would be interesting to confirm the existence of the pseudo-characteristic times by experiment and the most suitable configuration would have to involve characteristic times which are significantly different. For the value of $\alpha' = 40/3$, mentioned above, one finds (11) $\rho = 2$: $\tau_1 = 0.9849$; $\tau_2 = 0.3528$.

It is readily seen that the work of this subsection can be developed in an identical manner for the case of a spherical viscoelastic shell with an elastic casing. In this case (16)

$$\begin{aligned} \sigma_{rr} &= A - (2B/3r^3) \\ \sigma_{\theta\theta} &= A + (B/3r^3) \end{aligned} \qquad \text{VC3-(14)}$$

where, for the elastic case,

$$A = \frac{p(t)\,(2 + \beta)\rho^3}{2(1 - \rho^3) - \beta(2 + \rho^3)}$$

$$B = \frac{p(t)\,3R_2^3\,(1 - \beta)\rho^3}{2(1 - \rho^3) - \beta(2 + \rho^3)} \qquad \text{VC3-(15)}$$

$$\beta = (1 - \nu)/[\nu - \gamma(1 + \nu)\mu]$$

$$\gamma = R_2(1 - \nu_R)/E_R h$$

$\rho = R_2/R_1$ is the ratio of the external to the internal radii and all other quantities are the same as in the case of the tube. Assuming the viscoelastic material to be incompressible, i.e., setting $\nu = 1/2$, one finds from eq. (15) for the determination of the pseudo-characteristic times the operator

$$(1 - \rho^3)[2P - 3\gamma Q] - P(2 + \rho^3) = 0 \qquad \text{VC3-(16)}$$

As is to be expected, the characteristic times again depend on ρ and γ, and eq. (16) can be subjected to the same type of discussion as eq. (13).

An essential feature of the stress distributions in the reinforced tube and spherical shell is their high degree of symmetry, which eliminates the need to consider shear stresses. The case of inclusions, considered in the next subsection, extends the investigation to more complicated states of stress.

VC4. Effects of Single Inclusions

In the case of elastic materials, inclusion phenomena have found attention for many years (4,5,8). They have aroused renewed interest in recent years in connection with studies of crack formation (2). One of the difficulties in this area arises from the fact that ends of cracks represent stress raisers which may eventually turn out to be so small that the forces of molecular cohesion must be taken into account.

Cases of viscoelastic media with cylindrical inclusions in plane strain may be discussed on the basis of the elastic solutions mentioned above and the formulae of Subsection VC-2, in particular eqs. (7)–(9). Using functions of the complex variable $\varphi(z)$ and $\psi(z)$, Muskhelishvili (8) has shown that the states of stress and deformation in the bulk material surrounding circular inclusions may be described by

$$\varphi(z) = \frac{\Pi}{4}\left(z + \frac{\beta R^2}{z}\right)$$
$$\psi(z) = -\frac{\Pi}{2}\left(z + \gamma\frac{R^2}{z} + \frac{\delta R^4}{z^3}\right) \qquad \text{VC4-(17)}$$

where the values of the coefficients β, γ, and δ for various modes of attachment and types of inclusion are shown in Table VC-I.

For the viscoelastic case, the differential equations in time are more readily derived from these elastic solutions, since the process of satisfying the stress and displacement conditions at the interface involves only algebraic operations which are permissible for the linear operators P, Q, R, and S. As in the case of the tube, these operators combine to determine the pseudo-characteristic times, additional to the basic characteristic times discussed earlier.

Consider the two simplest models for incompressible viscoelastic materials; the Maxwell model with

$$P^1 = p_0 + p_1\frac{\partial}{\partial t}$$
$$Q^1 = \partial/\partial t \qquad \text{VC4-(18)}$$

TABLE VC-I[a]

$$\varphi(z) = \frac{\Pi}{4}\left(z + \frac{\beta R^2}{z}\right), \qquad \psi(z) = -\frac{\Pi}{2}\left(z + \frac{\gamma R^2}{z} + \frac{\delta R^4}{z^3}\right)$$

Material						Limits of solution for $r = R$, σ_{rr} = radial stress
Inclusion	Bulk	Mode of attachment	β	γ	δ	
Void	Elastic	Zero	2	1	-1	—
Rigid	Elastic	Attached	-2	0	1	—
Rigid	Elastic	Unattached	-1	0	0	$\sigma_{rr} < 0$
Elastic	Elastic	Attached	$\frac{2(\mu_1 - \mu_2)}{\mu_1 + \mu_2}$	0	$\frac{\mu_2 - \mu_1}{\mu_1 + \mu_2}$	—
Elastic	Elastic	Unattached	$\frac{2\mu_1 - \mu_2}{\mu_1 + \mu_2}$	0	$\frac{-\mu_1}{\mu_1 + \mu_2}$	$\sigma_{rr} < 0$

[a] μ_1, for bulk material; μ_2 for inclusion material.

and the Voigt model with

$$P^0 = 1$$
$$Q^1 = q_0 + q_1\, \partial/\partial t \qquad \text{VC4-(19)}$$

In these two cases, the characteristic times in the presence of an elastic inclusion are as follows.

For the Maxwell model:

$$\tau_1 = p_1/p_0$$
$$\tau_2 = [(\tfrac{1}{2}p_1 + \mu_2)/\mu_2]\tau_1 \qquad \text{VC4-(20a)}$$

For the Voigt model:

$$\tau_1 = q_1/q_0$$
$$\tau_2 = [q_0/(q_0 + 2\mu_2)]\tau_1 \qquad \text{VC4-(20b)}$$

In each case τ_1 corresponds to the relaxation and retardation times, respectively, of the homogeneous material. The quantities $\frac{1}{2}p_1$ and $\frac{1}{2}q_0$, respectively, play the role of the shear moduli of the materials, and hence will be assumed to be of the same order of magnitude as the shear modulus μ_2 of the elastic inclusion. It follows that the Maxwell material with an inclusion acquires a second, substantially larger relaxation time, while the Voigt material acquires a second, substantially smaller retardation time.

For example, assuming for both materials that

$$\tau_1 = 0.01$$

and

$$\mu_2 = \tfrac{1}{2}\mu_1,\ \mu_1,\ 2\mu_1$$

one finds for the two materials the corresponding pseudo-times;

Maxwell material:

$$3\tau_1,\ 2\tau_1,\ 3/2\tau_1$$

Voigt material:

$$\tfrac{2}{3}\tau_1,\ \tfrac{1}{2}\tau_1,\ \tfrac{1}{3}\tau_1$$

Introducing the two integrals

$$I_1(\tau_1) = \int_0^t e^{(s-t)/\tau_1}\, \Pi(s)\, ds$$
$$I_2(\tau_1,\tau_2) = \int_0^t e^{(\sigma-t)/\tau_1}\, I_1(\tau_2)\, d\sigma \qquad \text{VC4-(21)}$$

TABLE

$$\sigma_{11} + \sigma_{22} = E + F\frac{R^2}{r^2}\cos 2\theta,$$

Inclusion	Bulk	Mode of attachment	E	F
Void	Elastic	Zero	Π	-2Π
Rigid	Elastic	Attached	Π	2Π
Rigid	Elastic	Unattached	Π	Π
Elastic	Elastic	Attached	Π	$\dfrac{2\Pi(\mu_2 - u_1)}{\mu_1 + \mu_2}$
Elastic	Elastic	Unattached	Π	$\dfrac{-\Pi(2\mu_1 - \mu_2)}{\mu_1 + \mu_2}$
Void	Maxwell	Zero	$\dfrac{1}{p_1}\,\mathrm{I}_1(\tau_1)$	$\dfrac{-2}{p_1}\,\mathrm{I}_1(\tau_1)$
Rigid	Maxwell	Attached	$\dfrac{1}{p_1}\,\mathrm{I}_1(\tau_1)$	$\dfrac{2}{p_1}\,\mathrm{I}_1(\tau_1)$
Rigid	Maxwell	Unattached	$\dfrac{1}{p_1}\,\mathrm{I}_1(\tau_1)$	$\dfrac{1}{p_1}\,\mathrm{I}_1(\tau_1)$
Void	Voigt	Zero	Π	-2Π
Rigid	Voigt	Attached	Π	2Π
Rigid	Voigt	Unattached	Π	Π
Elastic	Maxwell	Attached	$\dfrac{1}{p_1}\,\mathrm{I}_1(\tau_1)$	$\dfrac{1}{p_1\tau_2}\left\{2(2\tau_1 - \tau_2)\mathrm{I}_1(\tau_1) + \dfrac{4}{r_2}(\tau_2 - \tau_1)\,\mathrm{I}_2(\tau_1, \tau_2)\right\}$
Elastic	Maxwell	Unattached	$\dfrac{1}{p_1}\,\mathrm{I}_1(\tau_1)$	$\dfrac{1}{p_1\tau_2}\left\{(3\tau_1 - 2\tau_2)\mathrm{I}_1(\tau_1) + \dfrac{3}{\tau_2}(\tau_2 - \tau_1)\mathrm{I}_2(\tau_1, \tau_2)\right\}$
Elastic	Voigt	Attached	Π	$-2\Pi + 4\left(\dfrac{1}{\tau_2} - \dfrac{1}{\tau_1}\right)\mathrm{I}_1(\tau_2)$
Elastic	Voigt	Unattached	Π	$-2\Pi + 3\left(\dfrac{1}{\tau_2} - \dfrac{1}{\tau_1}\right)\mathrm{I}_1(\tau_2)$

VC-II

$$\sigma_{22} - \sigma_{11} = \left(G\frac{R^2}{r^2} + H\frac{R^4}{r^4} + K\right)\cos 2\theta + L\frac{R^2}{r^2}$$

G	H	K	L
2Π	-3Π	$-\Pi$	Π
-2Π	3Π	$-\Pi$	0
$-\Pi$	0	$-\Pi$	0
$\frac{-2\Pi(\mu_2 - \mu_1)}{\mu_1 + \mu_2}$	$\frac{3\Pi(\mu_2 - \mu_1)}{\mu_1 + \mu_2}$	$-\Pi$	0
$\frac{\Pi(2\mu_1 - \mu_2)}{\mu_1 + \mu_2}$	$\frac{-3\Pi\mu_1}{\mu_1 + \mu_2}$	$-\Pi$	0
$\frac{2}{p_1} I_1(\tau_1)$	$\frac{-3}{p_1} I_1(\tau_1)$	$\frac{-1}{p_1} I_1(\tau_1)$	$\frac{1}{p_1} I_1(\tau_1)$
$\frac{-2}{p_1} I_1(\tau_1)$	$\frac{3}{p_1} I_1(\tau_1)$	$\frac{-1}{p_1} I_1(\tau_1)$	0
$\frac{-1}{p_1} I_1(\tau_1)$	0	$\frac{-1}{p_1} I_1(\tau_1)$	0
2Π	-3Π	$-\Pi$	Π
-2Π	3Π	$-\Pi$	0
$-\Pi$	0	$-\Pi$	0
$\frac{1}{p_1\tau_2}\left\{2(\tau_2 - \tau_1)I_1(\tau_1) - \frac{4}{\tau_2}(\tau_2 - \tau_1)I_2(\tau_1,\tau_2)\right\}$	$\frac{1}{p_1\tau_2}\left\{3(\tau_1 - \tau_2)I_1(\tau_1) + \frac{6}{\tau_2}(\tau_2 - \tau_1)I_2(\tau_1,\tau_2)\right\}$	$\frac{-1}{p_1}\left\{I_1(\tau_1) - \frac{2}{\tau_2} I_2(\tau_1,\tau_2)\right\}$	0
$\frac{1}{p_1\tau_2}\left\{(2\tau_2 - 3\tau_1)I_1(\tau_1) - \frac{1}{\tau_2}(3\tau_2 - 3\tau_1)I_2(\tau_1,\tau_2)\right\}$	$\frac{1}{p_1\tau_2}\left\{(3\tau_1 - 3\tau_2)I_1(\tau_1) + \frac{1}{\tau_2}(3\tau_2 - 3\tau_1)I_2(\tau_1,\tau_2)\right\}$	$\frac{-1}{p_1}\left\{I_1(\tau_1) + \frac{1}{\tau_2} I_2(\tau_1,\tau_2)\right\}$	0
$2\Pi + 4\left(\frac{1}{\tau_1} - \frac{1}{\tau_2}\right) I_1(\tau_2)$	$-3\Pi - 6\left(\frac{1}{\tau_1} - \frac{1}{\tau_2}\right) I_1(\tau_2)$	$-\Pi$	0
$2\Pi + 3\left(\frac{1}{\tau_1} - \frac{1}{\tau_2}\right) I_1(\tau_2)$	$-3\Pi - 3\left(\frac{1}{\tau_1} - \frac{1}{\tau_2}\right) I_1(\tau_2)$	$-\Pi$	0

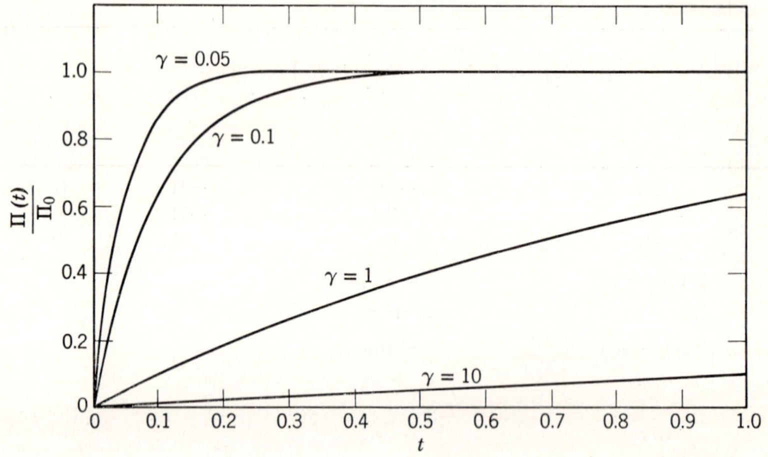

Fig. VC-3. Loading histories.

the general elastic and viscoelastic solutions, i.e., the expressions for the stresses and displacements in the bulk material for the cases of Table VC-I, can be presented in the form of Table VC-II with Π representing the loading varying with time.

Now let

$$\Pi(t) = \Pi_0(1 - e^{-t/\gamma}) \qquad \text{VC4-(22)}$$

with γ a parameter determining the initial rate of loading

$$(d\Pi/dt)_{t=0} = \Pi_0/\gamma \qquad \text{VC4-(23)}$$

Figure VC-3 shows that loading function for $\gamma = 0.05, 0.10, 1.0$, and 10.0 and demonstrates its usefulness for the present investigation.

The case of the elastic, attached inclusion will now be considered in detail for the Maxwell material with the relaxation time $\tau_1 = 0.01$ and an inclusion with $\mu_2 = \frac{1}{2}\mu_1$, so that $\tau_2 = 0.03$; and for the Voigt material with the retardation time $\tau_1 = 0.03$ and an inclusion with $\mu_2 = 2\mu_1$, so that $\tau_2 = 0.01$.

Figures VC-4 to 7 demonstrate the changes in the stress distributions around the inclusion in the two materials as functions of the time as well as of the distance from the inclusion. These curves illustrate well the typical behavior of the respective materials under two-dimensional conditions.

Figure VC-8 shows the time histories of the sum and difference of the direct stresses at a point on the interface between the same bulk materials and attached and unattached inclusions. It is important to note here that the curves for unattached inclusions have validity only while $\sigma_r < 0$ since, otherwise, detachment occurs and the imposed boundary conditions are no longer applicable. The presence of the second characteristic time is demonstrated by the greater complexities of these time histories which exhibit extreme values and points of inflection. Indeed, in some cases it can lead to a reversal of the stresses during the early stages of the loading process.

Clearly, in cases of several inclusions lying close together, the above effects will accumulate and lead to a finite number of characteristic times. As the number of inclusions increases, one will approach the need for a spectrum of such times. This is in agreement with the conclusions reached by Kolsky and Shi (6) that real materials cannot be represented over their entire loading range by low order operators. In fact, even if materials could be assumed to be pure and homogeneous, the finite extent and shape of specimen would influence the characteristic times. Further, the molecular structure of polymeric materials might be expected to lead to homogeneous media interspersed with stress raisers of the type discussed above.

These considerations now lead to the same kind of problems encountered in the study of the initiation and formation of cracks. As the size of stress raisers decreases, their effect must change. The existing theory does not allow for this possibility. In the case of polymeric substances, crosslinks between molecules may have a significance similar to that of stress raisers. (See Section IIIA.)

Extensions of this work to three-dimensional inclusions and compressible materials, while possible in principle, leads to much greater complexity of the analysis. The same observation applies to the case of viscoelastic inclusions and bulk materials with higher order operators. In all cases, the number of pseudo-characteristic times increases. For example, for attached and unattached inclusions in Maxwell and Voigt materials, the number of acquired relaxation and retardation times rises to four, and it is no longer possible to present simple formulae.

The study of three-dimensional spherical inclusions on the basis of the elastic results obtained by Goodier (4) is more complex than that for the plane case. Goodier's solutions are derived from Navier's

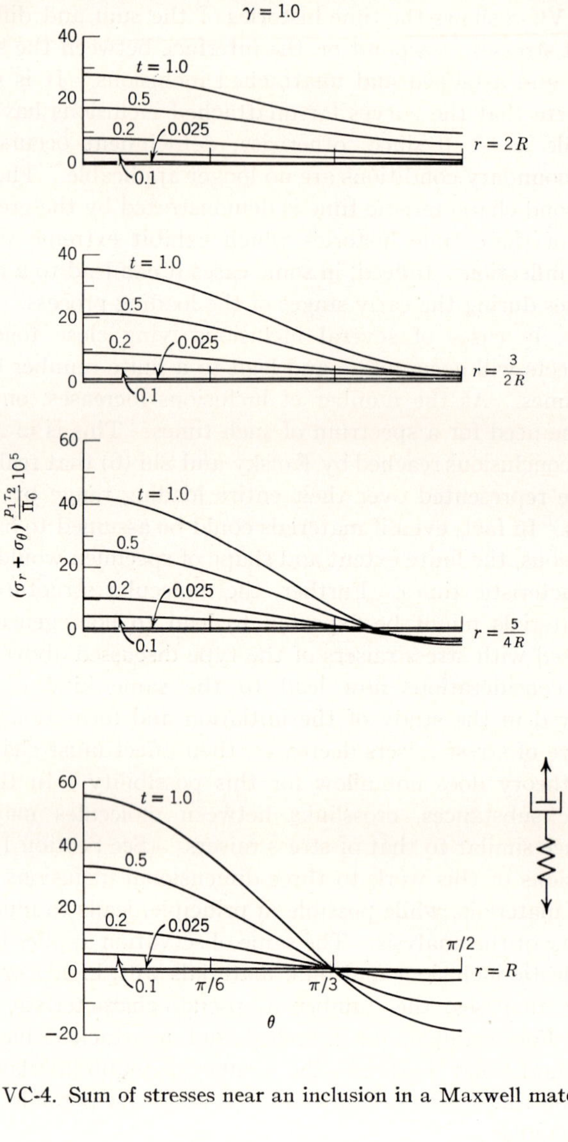

Fig. VC-4. Sum of stresses near an inclusion in a Maxwell material.

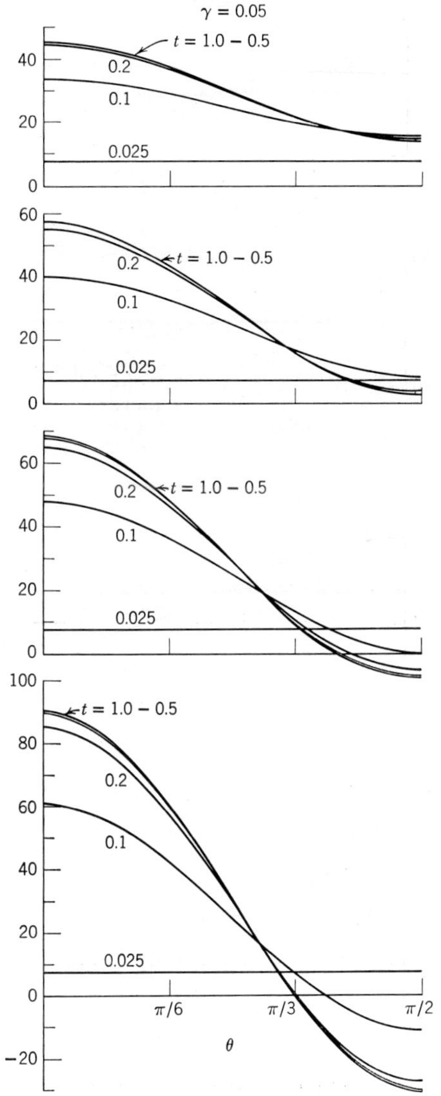

Fig. VC-4. *Continued.*

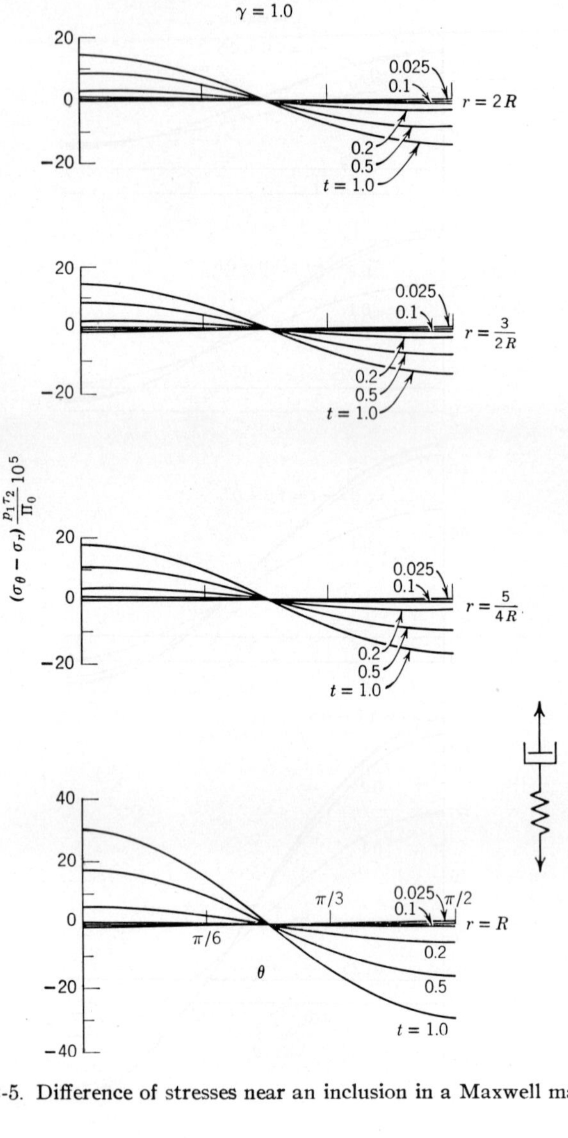

Fig. VC-5. Difference of stresses near an inclusion in a Maxwell material.

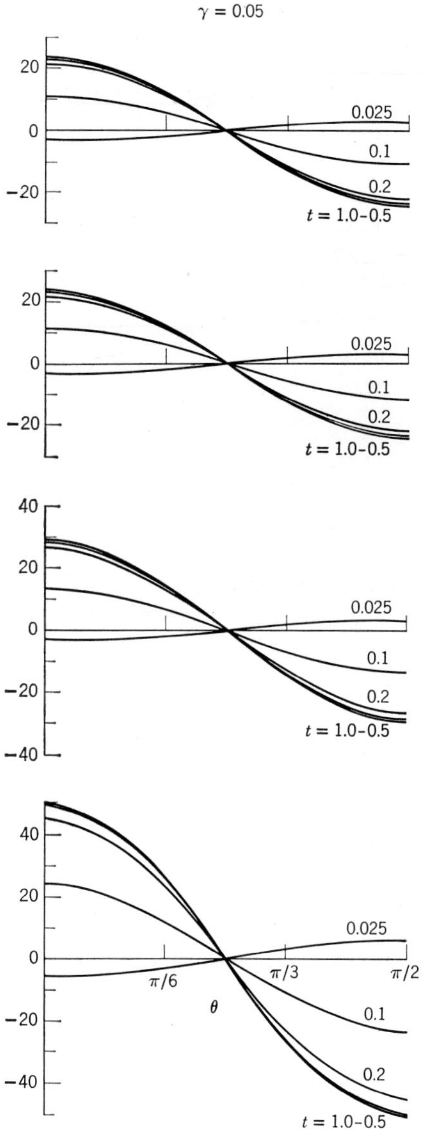

Fig. VC-5. *Continued.*

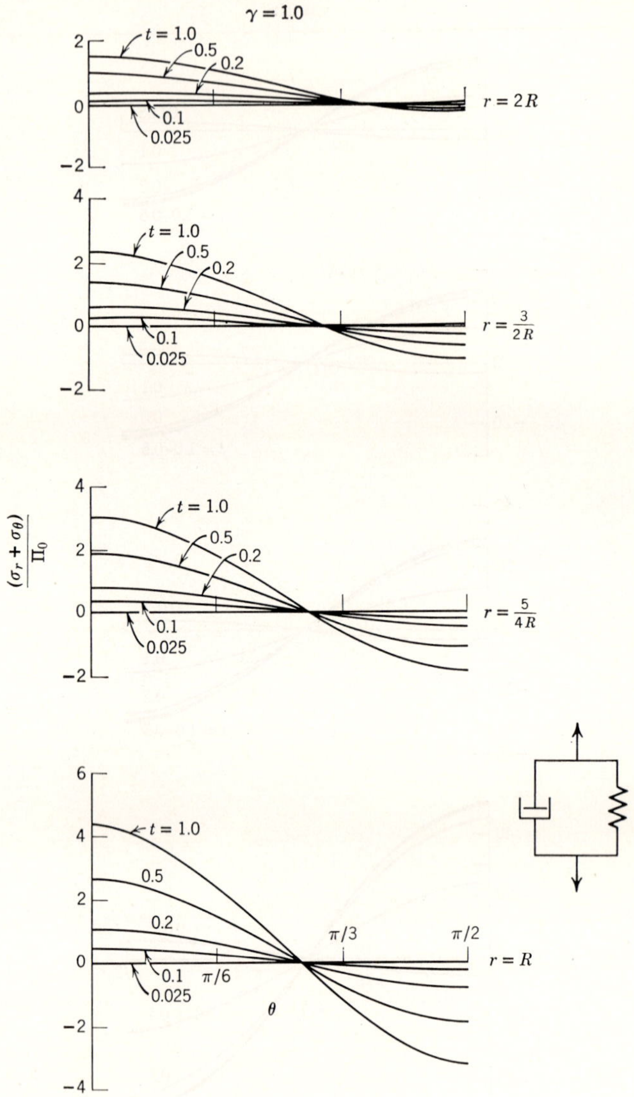

Fig. VC-6. Sum of stresses near an inclusion in a Voigt material.

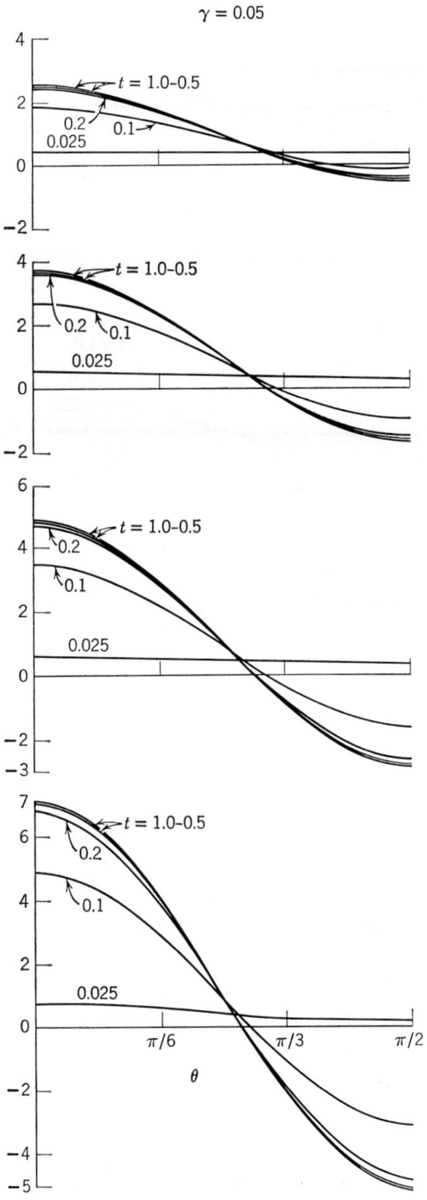

Fig. VC-6. *Continued.*

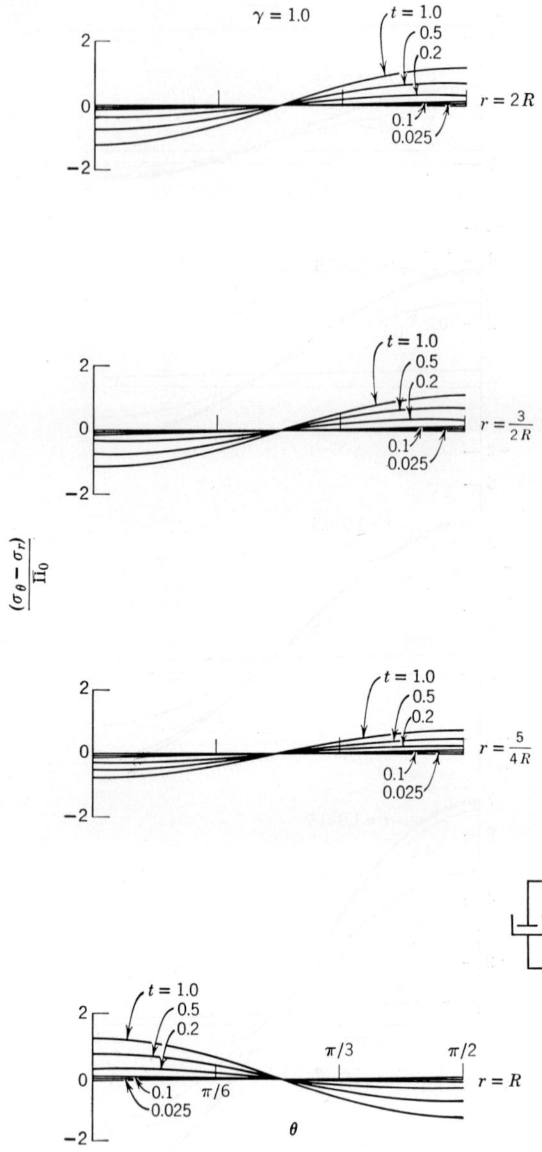

Fig. VC-7. Difference of stresses near an inclusion in a Voigt material.

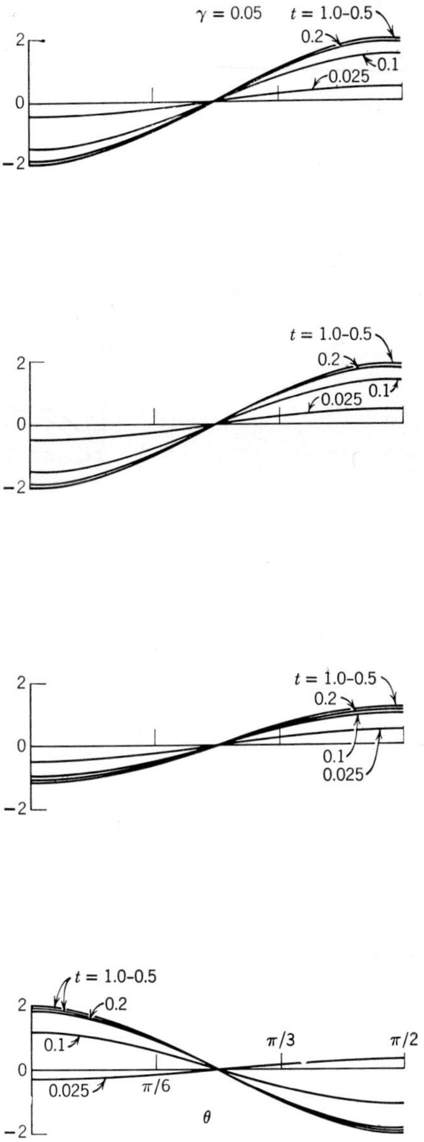

Fig. VC-7. *Continued.*

equations in terms of spherical harmonics and the evaluation of stress histories becomes very involved. On the other hand, the determination of the characteristic times does not differ basically from the preceding work.

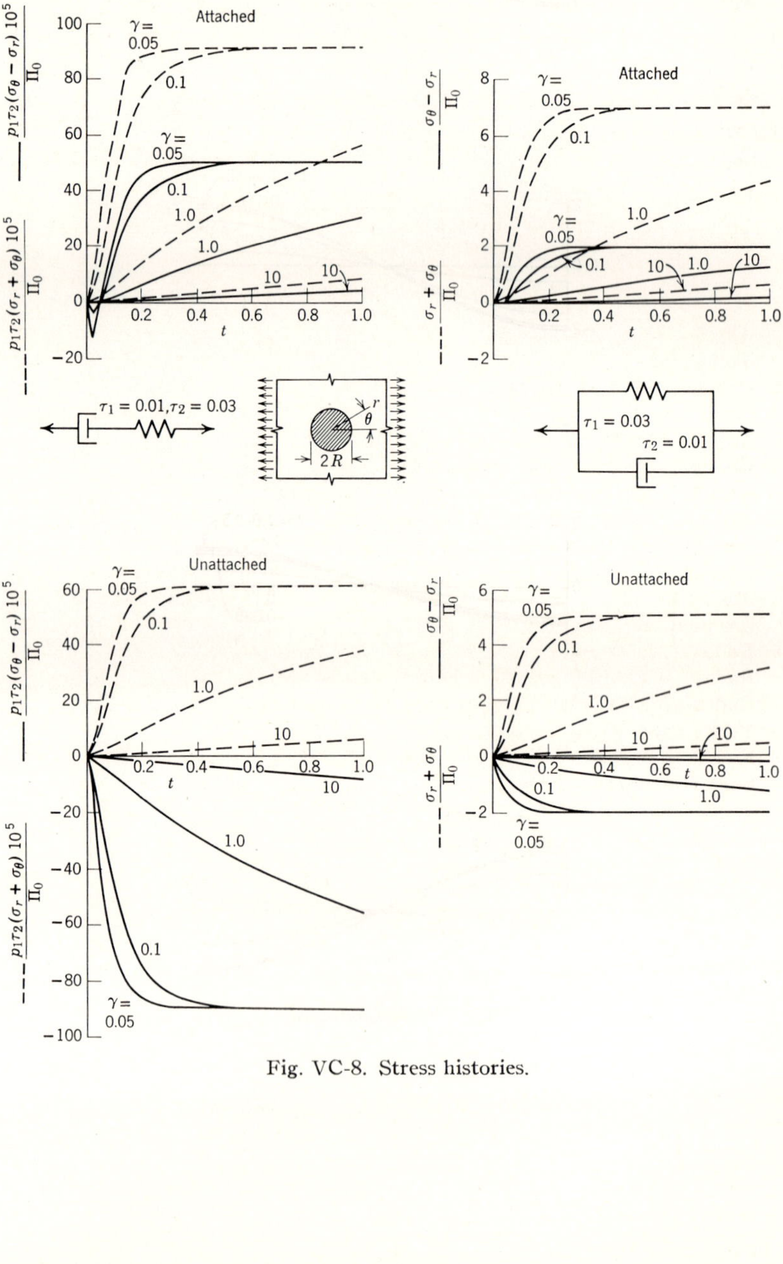

Fig. VC-8. Stress histories.

VC5. Conclusions

It is probably correct to state that the particular phenomena discussed in this section have been manifest in much of the work published previously under the heading of viscoelastic stress analysis. However, they seem to have been disregarded for the sake of quantitative stress analysis. In particular, the effects of boundaries and multi-connectivity on the spectrum of characteristic times of bodies made of viscoelastic materials should be more readily verified by experiment than the stress and displacement histories of idealized bodies subjected to idealized loading histories. It is the author's hope that such a confirmation may be attained by means of tests on tubes of the type discussed in Subsection VC3.

References

1. Alfrey, T., Jr., *Mechanical Behavior of High Polymers*, Interscience, New York, 1948.
2. Barenblat, G. I., *Prikl. Mat. Mekh.*, **20**, 4,475 (1959).
3. Corneliussen, A. H., E. F. Kainowitz, E. H. Lee, and J. R. M. Radok, Div. Appl. Math., Brown Univ., NOrd 18594/5, 1961.
4. Goodier, J. N., *J. Appl. Mech.*, **1**, 39 (1933).
5. Kirsch, G., *Z. Ver. Deut. Ing.*, **m42**, 29, 799 (1898).
6. Kolsky, H., and Y. Y. Shi, Brown University, Rept. 562(14)/5 (1958).
7. Lee, E. H., J. R. M. Radok, and W. B. Woodward, *Trans. Soc. Rheol.*, **3**, 41 (1959).
8. Muskhelishvili, N. I., *Some Basic Problems of the Mathematical Theory of Elasticity*, 2nd Engl. Ed., Noordhoff, Groningen, 1963.
9. Radok, J. R. M., and C. L. Tai, *J. Appl. Polymer Sci.*, **6**, 518 (1962).
10. Radok, J. R. M., *Quart. Appl. Math.*, **15**, 198 (1957).
11. Radok, J. R. M., Pibal Report 565, Polytechnic Institute of Brooklyn, New York, 1960; *Tekh* (*Yugoslavia*), **15**, 287 (1960).
12. Radok, J. R. M., *Plane Problems of Linear Viscoelasticity: Problems of Continuum Mechanics, 70th Birthday Volume for N. I. Muskhelishvili*, Philadeldelphia–Moscow, 1961, pp. 359–369.
13. Radok, J. R. M., *J. Appl. Mech.*, **22**, 249 (1955).
14. Radok, J. R. M., and W. B. Woodward, Div. of Appl. Math., Brown Univ., NOrd 11496, PA-TR/14, 1955.
15. Sokolnikoff, I. S., *Mathematical Theory of Elasticity*, McGraw-Hill, New York, 1946.
16. Williams, M. L., and R. J. Arenz, GALCIT SM59.1, California Institute of Technology, Pasadena, California, 1959.

VD. PLANE WAVES IN NONLINEAR VISCOELASTIC MATERIALS

Edward J. Mercado and Samuel Katz

The present section is intended as an introduction to the mathematical formalisms and techniques for handling the system of quasilinear partial differential equations which represents wave propagation in a nonlinear viscoelastic material. The results of several numerical calculations, based on the infinitesimal strain theory and performed at Gulf Research and Development Co. with an I.B.M. 7090 digital computer, are presented. These calculations utilize empirical, dynamic (time-dependent) constants deduced from published data for soils. This section of the book offers an *approach* to the specification of the formalisms of nonlinear viscoelasticity, even for the case of a finite strain, and pertains to polymeric bodies only by way of possible applications of that approach. See Subsection IC5a, p. 92.

VD1. Introduction

The system of partial differential equations that must be solved is: (*1*) a dynamic stress–strain law, also frequently referred to as the "constitutive equation," that relates stress and strain; (*2*) the equation expressing conservation of momentum, i.e., Newton's Law of Motion; (*3*) the strain-displacement law; and (*4*) the continuity equation. These equations must be solved subject to a prescribed set of initial and boundary conditions chosen to simulate actual testing procedures. When large deformations are involved, several known nonlinear effects can no longer be disregarded, and the corresponding terms in the above equations cannot simply be dropped as "negligible quantities of higher order." The *two* nonlinear effects considered here are: (*A*) under slow extension (or compression), deviation of the quasi-static, stress–strain law from Hooke's Law and, (*B*) introduction of the second-order terms in the strain-displacement law, i.e., a finite strain. The effect of these nonlinear terms on the analysis of the differential equations is disastrous since Fourier analysis (and other linear operator techniques) no longer yields useful results. Consequently, except in those cases where the quasi-linear, partial differential equations are homogeneous, explicit solutions are no longer obtainable and we must be content with numerical solutions to specific cases.

The equations of motion are best handled numerically by the Method of Characteristics. In essence, this technique transforms the set of partial differential equations with independent variables (x,t) to a set of ordinary differential equations along a set of curves in the (x,t) plane called the characteristics. Numerical integration along these curves also guarantees that errors will not propagate.

The first theoretical formulation combining nonlinear and time-rate effects was due to Malvern (6,7), whose primary interest was to explain some differences between observed plastic waves in metal bars and predictions from the plastic wave theory of von Kármán (19) and Taylor (18). These plastic waves were produced by longitudinal impact on one end of the bar and propagate under conditions of zero lateral stress. Guided by a physical theory of solids by Prandtl (11), Malvern deduced an empirical dynamic stress–strain law that may be written

$$E\epsilon_t - \sigma_t = k[\sigma - f(\epsilon)]$$

where σ is stress, $\epsilon = U_x$ is the gradient of particle displacement, or strain, E is Young's modulus determined statically, $\sigma = f(\epsilon)$ represents the quasi-static stress–strain function, and k is a dynamic constant of the material. In this theory, Malvern assumes that Young's modulus measured dynamically is equal to the statically determined modulus. This implies that an initial disturbance will propagate with elastic wave speed. The maximum strain level Malvern considered in his computations is $\cong 1\%$; consequently, the quadratic term in the strain-displacement law is negligible, and the infinitesimal strain approximation is used.

A power series approach may be used to obtain a family of dynamic laws describing the material and containing Malvern's (6) form as a special case. The power series approach suggests an extension of this which is valid for finite strain and further generalizes Malvern's theory by introducing an additional dynamic constant. The ratio of these two dynamic constants may be interpreted as a "dynamic" Young's modulus not necessarily equal to the classical Young's modulus. The present analysis is intended to apply to plane waves in which the strain is uniaxial. For this purpose E is interpreted as the ratio of longitudinal stress to the corresponding strain under conditions of zero lateral strain. The influence of lateral inertia has been investigated by Plass (11) and Tapley and Plass (17) under the simplifying assumption that the axial displacement is constant across any cross section of the bar of material and the radial displacement is assumed proportional to the distance from the axis. Numerical calculations based on this theory are included in reference 17.

VD2. Static Stress–Strain Laws

Numerous empirical formulas have been proposed in the literature to represent analytically the observed nonlinearity of static stress–strain relationships. Osgood (9) has reviewed many of the suggested nonlinear functions and finds that two formulas in particular should be more widely applicable than others.

$$\sigma/E = \alpha(1 + R\alpha)/(1 + S\alpha)$$

$$S > R > 0 \qquad \text{VD2-(1)}$$

$$\alpha = (\sigma/E) + R(\sigma/E)^S \qquad \text{VD2-(2)}$$

The latter is a special case of a formula proposed by Holmquist and Nadai (3). If eq. (1) provides an adequate fit, S is the intercept and R the negative slope of the straight line obtained by plotting $[(E/\sigma) - (1/\alpha)]$ versus $E\alpha/\sigma$. If eq. (2) provides a good fit of the observed static data, plotting log $[\alpha - (\sigma/E)]$ versus log (σ/E) will yield a straight line of slope S and intercept log R.

The above relations were chosen to illustrate the analysis because they have been used successfully to represent static stress-strain data and because they simplify the mathematical analysis, especially eq. (2). However a numerical relationship could be used if desired.

VD3. Dynamic Stress–Strain Laws

VD3a. Lowest Order Approximation: Infinitesimal Strain

We assume that, for a given set of experimental data and conditions, the stress can be written as some unknown function of the strain and strain rate, and that the strain may be expressed as a function of the stress and stress rate.* Hence we can write a general functional relationship, to be determined empirically, of the form

$$P(\epsilon, \epsilon_t) = Q(\sigma, \sigma_t) \qquad \text{VD3-(3)}$$

* The symbols ϵ_t and σ_t are represented by $\dot{\epsilon}$ and $\dot{\sigma}$, respectively, elsewhere in this volume when only the first derivative is implied. *Ed.*

We expand each side of eq. (3) in a Taylor's series about the point

$$\sigma_t = \epsilon_t = 0$$

obtaining

$$P(\epsilon,\epsilon_t) = P(\epsilon,0) + \frac{dP(\epsilon,0)}{d\epsilon_t}\epsilon_t + \ldots$$

$$Q(\sigma,\sigma_t) = Q(\sigma,0) + \frac{dQ(\sigma,0)}{d\sigma_t}\sigma_t + \ldots$$

The two leading terms of each expansion give, as a first approximation to the dynamic law,

$$Q(\sigma,0) + \frac{dQ(\sigma,0)}{d\sigma_t}\sigma_t = P(\epsilon,0) + \frac{dP(\epsilon,0)}{d\epsilon_t}\epsilon_t \qquad \text{VD3-(4)}$$

The coefficients in these Taylor expansions are only functions of σ and ϵ, respectively. Since these functional forms are unknown, we write for simplicity

$$\frac{1}{n!}\frac{d^nP(\epsilon,0)}{d\epsilon_t^n} = f_n(\epsilon)$$

$$\frac{1}{n!}\frac{d^nQ(\sigma,0)}{d\sigma_t^n} = g_n(\sigma)$$

$$n = 0, 1, 2, \ldots,$$

With this change of notation, eq. (4) becomes

$$g_0(\sigma) + g_1(\sigma)\,\sigma_t = f_0(\epsilon) + f_1(\epsilon)\,\epsilon_t \qquad \text{VD3-(5)}$$

For a static test, $\sigma_t = \epsilon_t = 0$, and eq. (5) reduces to

$$g_0(\sigma) = f_0(\epsilon) \qquad \text{VD3-(6)}$$

which defines the static stress–strain relation. We may consider eq. (6) as known, either from eqs. (1) and (2), with the empirical constants evaluated from the data or as a numerical relation. Since $g_1(\sigma)$ and $f_1(\epsilon)$ are still unknown, however, eq. (5) may be written in a more useful form by expanding these unknown coefficients in Taylor's series about the point $\sigma = \epsilon = 0$:

$$f_1(\epsilon) = f_1(0) + f_1'(0)\,\epsilon + \ldots \qquad \text{VD3-(7)}$$

$$g_1(\sigma) = g_1(0) + g_1'(0)\,\sigma + \ldots \qquad \text{VD3-(8)}$$

The lowest order approximation to the dynamic stress–strain law is then obtained by retaining only the first term in these expansions:

$$g_0(\sigma) + g_1(0)\ \sigma_t = f_0(\epsilon) + f_1(0)\ \epsilon_t \qquad \text{VD3-(9)}$$

With $g_1(0) = a$, $f_1(0) = b$, and dropping the zero subscripts on the other functions, eq. (9) becomes

$$g(\sigma) + a\sigma_t = f(\epsilon) + b\epsilon_t \qquad \text{VD3-(10)}$$

where a and b are the dynamic constants of the material. The ratio b/a may be interpreted as a dynamic Young's modulus E_d. The term E_d (or equivalently, b/a) is determined by measuring C, the velocity of propagation of the initial wavefront as $E_d = b/a = \rho C^2$. Hence, by comparing eq. (10) to Malvern's form for the dynamic stress-strain law, $1/a \sim k$, $b/a \sim E_d$.

Equation (10) is the lowest order approximation to the dynamic stress–strain law of a nonlinear viscoelastic material, since the only nonlinearity present is in the static properties of the material. If this nonlinearity were removed by expanding the static stress–strain curve in a Taylor's series about the point $\sigma = \epsilon = 0$ and only the linear terms retained, we would obtain the Maxwell and Voigt solids (4) by appropriate choice of the material constants.

For a Maxwell solid:

$$g(\sigma) = \sigma$$

$$g_1(0) = \tau = \frac{\text{Dashpot constant}}{\text{Spring constant}}$$

$$f(\epsilon) = 0$$

$$f_1(0) = \tau\epsilon$$

For a Voigt solid:

$$g(\sigma) = \sigma$$

$$g_1(0) = 0$$

$$f(\epsilon) = E\epsilon$$

$$f_1(0) = \tau E$$

Cf. Subsection IC4, pp. 75 and 81.

VD3b. Extension to Finite Strain

The previous analysis assumed the standard infinitesimal form of the strain-displacement law (15). If we had used the standard Lagrangian finite strain-displacement law,

$$\alpha = \epsilon + \tfrac{1}{2}\epsilon^2 = U_x + \tfrac{1}{2}U_x^2$$

the final result would be the same, except that α would replace ϵ. In terms of α, eq. (10) may be written

$$g(\sigma) + a\sigma_t = f(\alpha) + b\alpha_t \qquad \text{VD3-(11)}$$

In terms of the infinitesimal strain, eq. (11) becomes

$$g(\sigma) + a\sigma_t = h(\epsilon) + b(1 + \epsilon)\epsilon_t \qquad \text{VD3-(12)}$$

where

$$h(\epsilon) = f(\alpha) = f(\epsilon + \tfrac{1}{2}\epsilon^2)$$

VD3c. Higher Order Approximation

The lowest order dynamic law was obtained by retaining terms through the first order in σ_t and ϵ_t and the zeroth order in σ and ϵ in the series expansions. This suggests that the next higher order approximation to the dynamic stress–strain law would be obtained by retaining terms through the first order in all the series expansions. If the two leading terms of each series are retained, the resulting higher order dynamic law is

$$g_0(\sigma) + [g_1(0) + g_1'(0)\sigma]\sigma_t = f_0(\epsilon) + [f_1(0) + f_1'(0)\epsilon]\epsilon_t \qquad \text{VD3-(13)}$$

To simplify the notation, we set

$$f_0(\epsilon) = f(\epsilon)$$
$$f_1(0) = b_1$$
$$f_1'(0) = b_2$$
$$g_0(\sigma) = g(\sigma)$$
$$g_1(0) = a_1$$
$$g_1'(0) = a_2$$

Equation (13) may then be written

$$g(\sigma) + (a_1 + a_2\sigma)\sigma_t = f(\epsilon) + (b_1 + b_2\epsilon)\epsilon_t \qquad \text{VD3-(14)}$$

By comparing eqs. (12) and (14), we see that the higher order dynamic law reduces to the finite strain, dynamic law for $a_1 = a$, $a_2 = 0$, $b_1 = b_2 = b$, $f(\epsilon) = h(\epsilon)$.

VD3d. Plastic Strain Rate

With eq. (2) as the static stress–strain law, the basic constitutive equation defining the material [eq. (10)] takes the form,

$$\sigma_t - (b/a)\epsilon_t = (1/a)[\epsilon - g(\sigma)]$$

If this is interpreted as an elastic-plastic law as in Malvern's work, the elastic strain rate is

$$\epsilon_t' = (a/b)\sigma_t$$

while the simultaneously occurring plastic strain rate is

$$\epsilon_t'' = \epsilon_t - \epsilon_t' = (1/b)[g(\sigma) - \epsilon]$$

Thus, the plastic strain rate is assumed proportional to the difference between the instantaneous strain and the strain which would result if the instantaneous stress had been statically applied. In Malvern's theory, the plastic strain rate was a function of the difference between the instantaneous stress and the stress the material would support if the instantaneous strain had been produced statically.

To gain some insight into the difference between these alternative expressions for the plastic strain rate, calculations were performed with identical boundary conditions and physical constants—first with eq. (2) as the static law and, second, by solving eq. (2) for σ so that $\epsilon t''$ was proportional to $[\sigma - g^{-1}(\epsilon)]$.

VD4. Method of Characteristics

The dynamic behavior of a material is described by the simultaneous solution of the dynamic stress–strain law and the laws for the conservation of mass and momentum. Each of these equations is a first order quasi-linear partial differential equation of the form

$$\mathbf{L}_i = \mathbf{A}_{ij}\mathbf{U}_x{}^j + \mathbf{B}_{ij}\mathbf{U}_t{}^j + \mathbf{C}_i = 0$$

$$i, j = 0, 1, 2, \ldots, \qquad \text{VD4-(15)}$$

where we sum on repeated indices and $\mathbf{U}^j$ represents, in turn, stress, strain, and particle velocity. The coefficient matrices $\mathbf{A}_{ij}$, $\mathbf{B}_{ij}$, and

$\mathbf{C}_i$ may be functions of x, t, and $\mathbf{U}^j$. Following the procedure of Courant and Friedrichs (2), we seek a linear combination $\lambda_i \mathbf{L}_i = 0$ such that the partial derivatives of $\mathbf{U}^j$ with respect to x and t combine to total derivatives along some curve $x(l)$, $t(l)$ in the x–t plane. Such a curve is called a characteristic. The condition for $\lambda_i \mathbf{L}_i$ to be differentiated in some direction l is

$$\frac{\lambda_i \mathbf{A}_{ij}}{\lambda_k \mathbf{B}_{kj}} = \left(\frac{dx/dl}{dt/dl}\right) = \frac{dx}{dt} \qquad \text{VD4-(16)}$$

which may be written as

$$\lambda_i \mathbf{A}_{ij}\, dt - \lambda_i \mathbf{B}_{ij}\, dx = 0 \qquad \text{VD4-(17)}$$

Equations (17) constitute N homogeneous algebraic equations for the N unknowns λ_i. The necessary and sufficient condition for a nontrivial solution to exist is that the determinant of the coefficients vanish. The condition that

$$\left|\mathbf{A}_{ij}\, dt - \mathbf{B}_{ij}\, dx\right| = 0$$

yields an Nth degree algebraic equation in dx/dt. The roots of this polynomial constitute N ordinary differential equations for the characteristic curves. To find the transformed differential equations valid along the N directions, consider the linear combination $\lambda_i \mathbf{L}_i\, dx = 0$:

$$\lambda_i \mathbf{L}_i\, dx = \lambda_i \mathbf{A}_{ij} \mathbf{U}_x{}^j\, dx + \lambda_i \mathbf{B}_{ij} \mathbf{U}_t{}^j\, dx + \lambda_i \mathbf{C}_i\, dx = 0$$

By using eq. (16), this may be written

$$\lambda_i \mathbf{A}_{ij}[\mathbf{U}_x{}^j\, dx + \mathbf{U}_t{}^j\, dt] + \lambda_i \mathbf{C}_i\, dx = 0$$

which is equivalent to

$$\lambda_i \mathbf{A}_{ij}\, d\mathbf{U}^j + \lambda_i \mathbf{C}_i\, dx = 0 \qquad \text{VD4-(18)}$$

Another linearly independent equation can be obtained in a similar fashion from the combination $\lambda_i \mathbf{L}_i\, dt = 0$:

$$\lambda_i \mathbf{L}_i\, dt = \lambda_i \mathbf{A}_{ij} \mathbf{U}_x{}^j\, dt + \lambda_i \mathbf{B}_{ij} \mathbf{U}_t{}^j\, dt + \lambda_i \mathbf{C}_i\, dt = 0$$

Again, by using eq. (16), this may be written

$$\lambda_i \mathbf{B}_{ij}[\mathbf{U}_x{}^j\, dx + \mathbf{U}_t{}^j\, dt] + \lambda_i \mathbf{C}_i\, dt = 0$$

which is equivalent to

$$\lambda_i \mathbf{B}_{ij}\, d\mathbf{U}^j + \lambda_i \mathbf{C}_i\, dt = 0 \qquad \text{VD4-(19)}$$

We have obtained $N + 2$ equations for the N unknowns λ_i. Therefore, for a nontrivial solution to exist, all $N \times N$ subdeterminants of the $(N + 2) \times N$ coefficient matrix must vanish. The N ordinary differential equations so obtained along the N characteristic directions are the characteristic equations.

VD5. Infinitesimal Strain Wave Propagation: Loading Cycle

VD5a. Characteristic Equations

As a first approximation to the dynamic behavior of soils and polymers we use the lowest order dynamic stress–strain law and the equation for the conservation of momentum and continuity of particle displacement, valid for infinitesimal strains in a Lagrangian coordinate system:

$$g(\sigma) + a\sigma_t = f(\epsilon) + b\epsilon_t \qquad \text{VD3-(10)}$$

$$\epsilon_t = V_x \qquad \text{VD5-(20)}$$

$$\sigma_x = \rho V_t \qquad \text{VD5-(21)}$$

For this set of equations, comparison with eq. (15) yields

$$\mathbf{U}^j = \begin{bmatrix} \sigma \\ \epsilon \\ V \end{bmatrix}$$

$$\mathbf{A}_{ij} = \begin{bmatrix} 0 & 0 & 0 \\ 0 & 0 & -1 \\ 1 & 0 & 0 \end{bmatrix}$$

$$\mathbf{B}_{ij} = \begin{bmatrix} a & -b & 0 \\ 0 & 1 & 0 \\ 0 & 0 & -\rho \end{bmatrix}$$

$$\mathbf{C}_i = \begin{bmatrix} g(\sigma) - f(\epsilon) \\ 0 \\ 0 \end{bmatrix}$$

From eq. (17) the determinantal equation for the characteristic directions

$$\begin{vmatrix} -adx & bdx & 0 \\ 0 & -dx & -dt \\ dt & 0 & \rho dx \end{vmatrix} = 0$$

yields a cubic equation in dx/dt, the roots of which are three ordinary differential equations for the characteristic curves:

$$\left.\begin{aligned} dx/dt &= 0 \\ dx/dt &= \pm(b/a\rho)^{1/2} = \pm C \end{aligned}\right\} \qquad \text{VD5-(22)}$$

The characteristic equations are obtained by setting equal to zero all 3 × 3 subdeterminants of the 5 × 3 coefficient matrix made up of eqs. (17), (18), and (19). The ordinary differential equations obtained in this manner are as follows.

(*a*) Along $dx/dt = 0$:

$$d\sigma - (b/a)d\epsilon = -(1/a)[g(\sigma) - f(\epsilon)]dt \qquad \text{VD5-(23)}$$

(*b*) Along $dx/dt = +(b/a\rho)^{1/2}$:

$$d\sigma - \rho(b/a\rho)^{1/2}dV = -(1/a)[g(\sigma) - f(\epsilon)]dt \qquad \text{VD5-(24)}$$

(*c*) Along $dx/dt = -(b/a\rho)^{1/2}$:

$$d\sigma + \rho(b/a\rho)^{1/2}dV = -(1/a)[g(\sigma) - f(\epsilon)]dt \qquad \text{VD5-(25)}$$

VD5b. Propagation of Initial Elastic Rise

The strain rate associated with a step in stress (or strain, or particle velocity) applied at $x = 0$ is infinite. We *assume* this step discontinuity propagates into this material with velocity C_s controlled by a limiting form of the dynamic stress-strain law as $\epsilon_t \to \infty$. This step discontinuity does not necessarily follow a characteristic. (Discontinuities in all *derivatives* in σ, α, ϵ, and V must follow the characteristics, but not discontinuities, in the dependent variables.) From eq. (10),

$$\lim_{\epsilon_t \to \infty} \sigma_t/\epsilon_t = \lim_{\epsilon_t \to \infty} \{b/a - [f(\epsilon) - g(\sigma)]/\epsilon_t\} = b/a = \rho C^2$$

Therefore, in this case, the initial discontinuous elastic rise propagates with velocity C independently of the local state of stress or strain and does follow a characteristic.

Newton's Law of Motion for an element of mass containing a discontinuity in stress, strain or particle velocity propagating with velocity $C = (b/a\rho)^{1/2}$ is

$$d\sigma = -\rho C dV \qquad \text{VD5-(26)}$$

Continuity of displacement in this element yields

$$dV = -C d\epsilon \qquad \text{VD5-(27)}$$

By combining eqs. (26) and (27) we obtain

$$d\sigma = \rho C^2 d\epsilon \qquad \text{VD5-(28)}$$

Since the discontinuity is moving into an undisturbed region where $\sigma = \epsilon = V = 0$, eqs. (26), (27), and (28) can be immediately integrated to yield

$$\sigma = -\rho CV = \rho C^2 \epsilon \qquad \text{VD5-(29)}$$

These equations are valid across the discontinuous leading wave along which eq. (24) holds. Therefore we can use them to write eq. (24) entirely in terms of one variable at a time to obtain:

$$d\mathbf{U}^j/dt = -(1/2a)[g(\mathbf{U}^j) - f(\mathbf{U}^j/\rho C^2)] \qquad \text{VD5-(30)}$$

where

$$\mathbf{U}^j = \begin{bmatrix} \sigma \\ \rho C^2 \epsilon \\ -\rho CV \end{bmatrix}$$

This is an ordinary differential equation, in turn, for the stress, strain, and particle velocity along the wavefront.

The solutions for eq. (30), with eq. (1) as the static law, yield the following expressions for the jumps in stress, strain, and particle velocity along the wavefront:

$$\exp\{-t/2aE\} = \left(\frac{\sigma(x/C)}{\sigma_0}\right)^{1/A} \left(\frac{A + B\sigma(x/C)}{A + B\sigma_0}\right)^{\frac{AC - B}{AB}} \qquad \text{VD5-(31)}$$

where $A = 1 - E(a/b)$, $B = (Sa/b) - ER(a/b)^2$, and $C = Sa/b$;

$$\exp\{-t/2aE\} = \left(\frac{\epsilon(x/C)}{\epsilon_0}\right)^{1/A'} \left(\frac{A' + B'\epsilon(x/C)}{A' + B'\epsilon_0}\right)^{\frac{A'C' - B'}{A'B'}} \qquad \text{VD5-(32)}$$

where $A' = A$, $B' = (b/a)B$, and $C' = (b/a)C$;
and

$$\exp\{-t/2aE\} = \left(\frac{V(x/C)}{V_0}\right)^{1/A''} \left(\frac{A'' + B''V(x/C)}{A'' + B''V_0}\right)^{\frac{A''C'' - B''}{A''B''}} \qquad \text{VD5-(33)}$$

where $A'' = A$, $B'' = -\rho(b/a\rho)^{1/2}B$, $C'' = -\rho(b/a\rho)^{1/2}C$; and

$$\sigma_0 = \rho C^2 \epsilon_0 = -\rho C V_0$$

When eq. (2) is used as the static law, the resulting differential equations are again integrable and yield for the jumps in stress, strain, and particle velocity as functions of distance:

$$\sigma(x/C) = \frac{\sigma_0 A^{1/\delta} \exp\{-Ax/2aC\}}{[A + B\sigma_0^{\delta}(1 - \exp\{-Ax\delta/2aC\})]^{1/\delta}} \qquad \text{VD5-(34)}$$

$$\epsilon(x/C) = \frac{\epsilon_0 A^{1/\delta} \exp\{-Ax/2aC\}}{[A + B\sigma_0^{\delta}(1 - \exp\{-Ax\delta/2aC\})]^{1/\delta}} \qquad \text{VD5-(35)}$$

$$V(x/C) = \frac{V_0 A^{1/\delta} \exp\{-Ax/2aC\}}{[A + B\sigma_0^{\delta}(1 - \exp\{-Ax\delta/2aC\})]^{1/\delta}} \qquad \text{VD5-(36)}$$

where

$$A = (1/E) - (a/b)$$

$$B = R/E^S$$

$$\delta = S - 1$$

VD5c. Boundary Conditions

Equation (23) is independent of particle velocity. If the boundary condition at $x = 0$ is specified in terms of stress, we are left with an ordinary first order differential equation in ϵ; i.e., if $\sigma = h(t)$ at $x = 0$ is specified, $d\sigma = h'(t)\,dt$, and eq. (23) becomes

$$\frac{d\epsilon}{dt} + \frac{f(\epsilon)}{b} = \frac{1}{b}\{g[h(t)] + a\,h'(t)\} \qquad \text{VD5-(37)}$$

It is possible to choose $f(\epsilon)$, $g(\sigma)$, $h(t)$ so that an explicit solution to eq. (37) can be obtained in closed form.

While eq. (37) may be solved numerically for any static stress–strain law, the use of eqs. (1) and (2) permits an explicit solution for different boundary conditions. Comparison of eqs. (1) and (6) yields

$$g(\sigma) = \sigma/E$$

$$f(\epsilon) = \epsilon(1 + R\epsilon)/(1 + S\epsilon)$$

and eq. (37) then becomes

$$\frac{d\epsilon}{dt} + \frac{\epsilon(1 + R\epsilon)}{b(1 + S\epsilon)} = \frac{1}{b}\left[\frac{h(t)}{E} + a\,h'(t)\right] \qquad \text{VD5-(38)}$$

The corresponding relation for the stress–strain law given by eq. (2) is

$$\frac{d\epsilon}{dt} + \frac{\epsilon}{b} = \frac{1}{b}\left\{\frac{h(t)}{E} + R\left[\frac{h(t)}{E}\right]^S + a\,h'(t)\right\} \qquad \text{VD5-(39)}$$

The simplest boundary condition is $h(t) = \sigma_0$, a constant; thus $h'(t) = 0$. For this boundary condition eq. (38) becomes

$$(d\epsilon/dt) + [\epsilon(1 + R\epsilon)/b(1 + S\epsilon)] = \sigma_0/bE$$

Since this equation is separable, it may be integrated directly to give implicitly the time variation of the strain at $x = 0$, yielding

$$e^{t/bE} = \left[\frac{(\sigma_0/E) + F\epsilon_0 - R\epsilon_0^2}{(\sigma_0/E) + F\epsilon - R\epsilon^2}\right]^{S/2R}\left[\frac{-2R\epsilon + F - G}{-2R\epsilon + F + G}\right]^{\phi} \cdot \left[\frac{-2R\epsilon_0 + F + G}{-2R\epsilon_0 + F - G}\right]^{\phi} \qquad \text{VD5–(40)}$$

where

$$F = [(S\sigma_0/E) - 1]$$

$$G = [1 + (S\sigma_0/E)^2 + (2\sigma_0/E)(2R - S)]^{1/2}$$

$$\phi = (1/G)[1 + (SF/2R)]$$

$$\sigma_0 = \rho C^2\epsilon_0 = -\rho C V_0$$

While eq. (38) will in general not be separable when $h(t)$ is not a constant, it may still be solved by numerical methods.

When the stress–strain law, eq. (2), is used with the identical initial condition $h(t) = \sigma_0$, the corresponding relations are

$$(d\epsilon/dt) + (\epsilon/b) = (1/b)[(\sigma_0/E) + R(\sigma_0/E)^S] = g(\sigma_0)/b \qquad \text{VD5-(41)}$$

and

$$\epsilon(0,t) = [\epsilon_0 - g(\sigma_0)]e^{-t/b} + g(\sigma_0) \qquad \text{VD5-(42)}$$

Physically, eq. (42) may be interpreted as an instantaneous elastic deformation under the influence of the stress σ_0. After this elastic response, the material flows until the stress is balanced by the elastic restoring force $g^{-1}(\epsilon_\infty)$. Note that as $t \to \infty$, $\epsilon(0, t) \to g(\sigma_0)$, the equivalent plastic strain. This is shown in Figure VD-1.

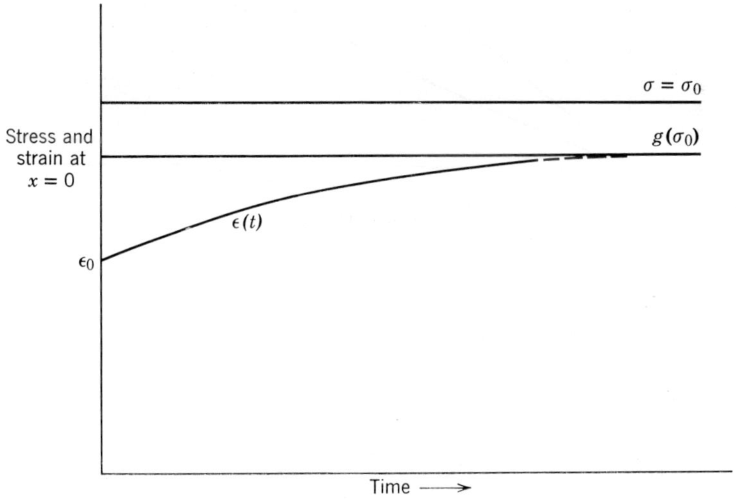

Figure VD-1.

Another relatively simple source function is a linear increase of stress with time, $h(t) = Bt$; and $h'(t) = B$. Equation (39) becomes

$$(d\epsilon/dt) + (\epsilon/b) = (1/b)[(Bt/E) + R(Bt/E)^s + aB] \qquad \text{VD5-(43)}$$

The solution is obtained with the method of variation of parameters. Applying the initial condition $\epsilon(0,0) = 0$, we obtain

$$\epsilon(0,t) = B\left(a - \frac{b}{E}\right)(1 - e^{-t/b}) + \frac{Bt}{E} + \frac{R}{b}\left(\frac{B}{E}\right)^S e^{-t/b}\cdot \sum_{n=0}^{\infty} \frac{t^{n+S+1}}{b^n(n+S+1)n!} \qquad \text{VD5-(44)}$$

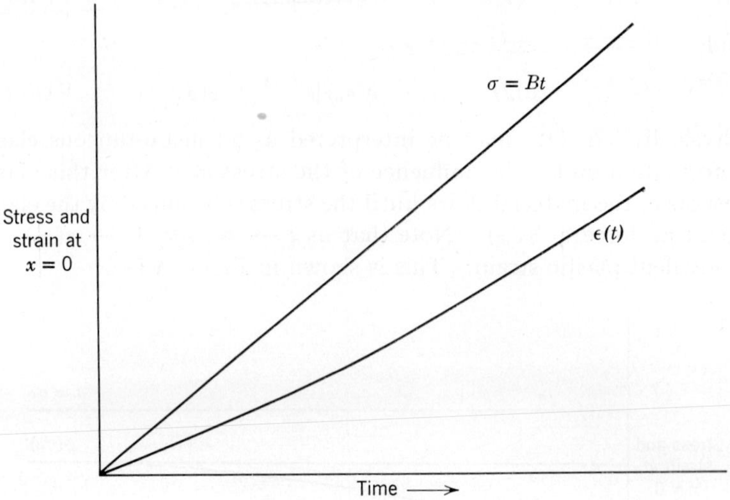

Figure VD-2.

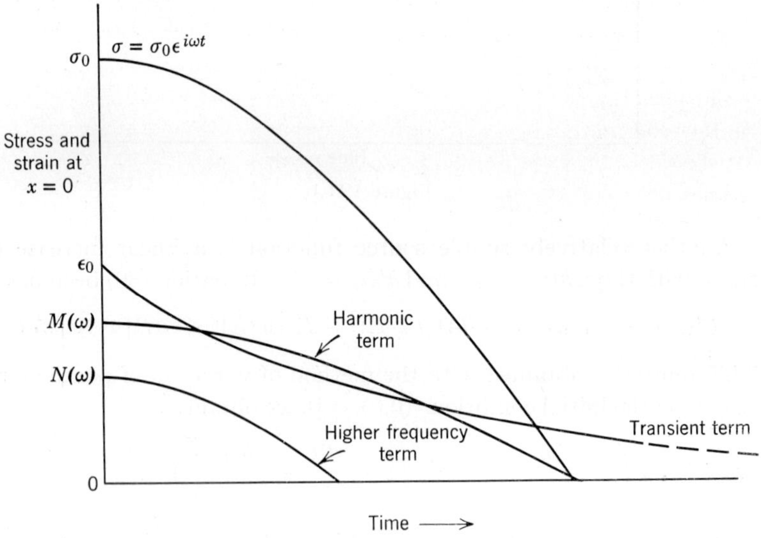

Figure VD-3.

A sketch of eq. (44) is shown in Figure VD-2. The strain rate is given by $B/\rho C^2$ at $t = 0$ and increases as $t \to \infty$.

For a simple harmonic source $h(t) = \sigma_0 e^{i\omega t}$ and $h'(t) = i\omega\, h(t)$. The solution is

$$\epsilon(0,t) = [\epsilon_0 - M(\omega) - N(\omega)]e^{-t/b} + M(\omega)\, e^{i\omega t} + N(\omega)\, e^{iS\omega t} \qquad \text{VD5-(45)}$$

where

$$M(\omega) = \frac{(\sigma_0/b)\,[(1/E) + ia\omega][(1/b) - i\omega]}{(1/b^2) + \omega^2}$$

$$\text{VD5-(46)}$$

$$N(\omega) = \frac{(R/b)(\sigma_0/E)^S[(1/b) - iS\omega]}{(1/b^2) + (S\omega)^2}$$

Equation (45) contains both real and imaginary terms as did the source function $h(t)$. We use the real part of eq. (45) if the real part of $h(t)$ is used and vice versa. The first term in eq. (45) is the transient solution of the differential equation and the two harmonic terms are the steady-state solution. If the static stress–strain curve is concave to the strain axis, $S \geq 1$, and a higher frequency term is generated. If it is concave to the stress axis, then $S < 1$ and a lower frequency term is generated. This is shown in Figure VD-3, with $S = 2$ for purposes of illustration.

The boundary condition at $x = 0$ could just as easily be specified in terms of strain and analogous equations developed for the variation of stress. Let $\epsilon(0,t) = k(t)$ be specified at $x = 0$; then eq. (23) becomes

$$\frac{d\sigma}{dt} + \frac{1}{a}\, g(\sigma) = \frac{b}{a}\left\{k'(t) + \frac{1}{b}\, f[k(t)]\right\} \qquad \text{VD5-(47)}$$

The equations analogous to eqs. (38) and (39) for static laws (*1*) and (*2*) are:

$$\frac{d\sigma}{dt} + \frac{\sigma}{Ea} = \frac{b}{a}\left\{k'(t) + \frac{k(t)\,[1 + Rk(t)]}{b[1 + Sk(t)]}\right\} \qquad \text{VD5-(48)}$$

and

$$\frac{d\sigma}{dt} + R\left(\frac{\sigma}{E}\right)^S + \frac{\sigma}{E} = \frac{b}{a}\left[k'(t) + \frac{k(t)}{b}\right] \qquad \text{VD5-(49)}$$

For the case $k(t) = \epsilon_0 =$ constant, the solution to eq. (48) is

$$\sigma/E = [(\sigma_0/E) - f(\epsilon_0)]e^{-t/a} + f(\epsilon_0) \qquad \text{VD5-(50)}$$

The solution to eq. (49) is left as a quadrature which in general will require numerical integration depending on the specific value for S.

$$\int_{\sigma_0}^{\sigma} d\sigma/[(\epsilon_0/a) - (\sigma/E) - R(\sigma/E)^S] = \int_0^t dt \qquad \text{VD5-(51)}$$

The stress response to either a linear increase of strain with time or a simple harmonic strain with time applied at $x = 0$ is obtained by solving eq. (48) or (49) with the appropriate definition of $k(t)$.

VD5d. Numerical Calculation: Infinitesimal Strain

To calculate the stress, strain, and particle velocity behind the wavefront we must extrapolate from the known values along the wavefront and from the initial value of stress, using numerical integration techniques. The simplest procedure for extending the solution out from the wavefront and the $x = 0$ boundary is to use the character-

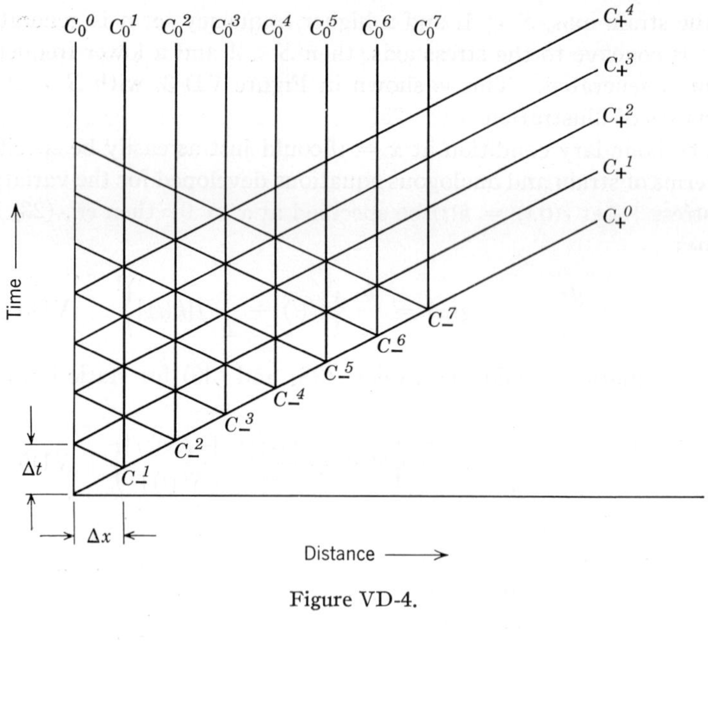

Figure VD-4.

istic curves as a coordinate system, making it possible to use the characteristic equations which are differentiated with respect to only one variable. For the infinitesimal strain case, the method simplifies even further because the differential equations defining the characteristic curves are independent of the final solution. Hence eq. (22) can be integrated immediately to yield

$$x = \text{constant}$$

$$x = Ct + \text{constant}$$

$$x = -Ct + \text{constant}$$

as the three families of characteristic curves. Following the nomenclature of Courant and Friedrichs these relations are, respectively, the C_0 characteristics, the C_+ characteristics, and the C_- characteristics. Figure VD-4 is a schematic diagram of the grid of characteristics and the numbering system whereby any grid point can be identified by the characteristics passing through it.

Define the value of the dependent variables at the point of intersection of $C_+{}^i$ and $C_0{}^j$ as σ_{ij}, ϵ_{ij}, V_{ij}. With this notation, the line separating the undisturbed and disturbed regions is the $C_+{}^0$ characteristic. The boundary condition is applied on $C_0{}^0$. For a discontinuous increase in stress at $t = 0$, the stress, strain, and particle velocity are known on $C_+{}^0$ from eqs. (31), (32), and (33) for stress–strain law (*1*) and eqs. (34), (35), and (36) for stress–strain law (*2*). For a continuous stress source, the initial values of stress, strain, and particle velocity along the wave front are all identically zero. The values of stress and strain are known on $x = 0$, that is, along $C_0{}^0$, by specifying the boundary condition and solving the resulting differential equation in strain.

VD6. Fort Peck Sand

VD6a. Material Constants

In order to illustrate the type of predictions that the present theory can make, several different materials with different constants, stress-strain laws, and viscoelastic response were considered. Among these is the Fort Peck Sand, having a static stress-strain response determined by Whitman et al. (21), as shown in Figure VD-5. Young's modulus, the initial slope of this curve, is 14.7×10^8 dynes/cm.^{2}, and the density is reported as 1.60 g./cm.^{3}. An analytical representation of the static stress-strain curve is obtained by fitting eq. (2) to the static data. The constants R and S in eq. (2) were determined as outlined in Subsection VD2 and are 6.86×10^3 and 2.22. With these values, eq. (2) adequately fits the data over the range of interest and is shown superimposed on the experimentally determined stress-strain curve in Figure VD-5.

There are no experimental data known to the authors from which the dynamic constants a and b may be directly determined. However, the experimentally observed fact that the initial wave front propagates with velocity greater than or equal to the elastic wave velocity requires $b/a \geqslant E$. On the basis of this relation, b/a is chosen

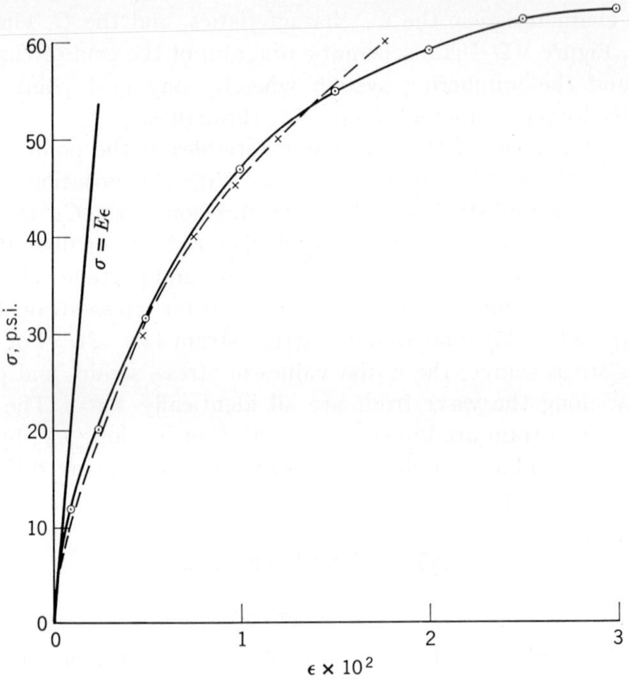

Fig. VD-5. Stress-strain curves for Fort Peck Sand. Experimental (——) and calculated (- -) values from the equation

$$\epsilon = \frac{\sigma}{21.4 \times 10^3} + 6.86 \times 10^3 \left(\frac{\sigma}{21.4 \times 10^3}\right)^{2.22}$$

as 14.9 × 10^8 dynes/cm.^{2}, which yields an initial wavefront velocity of 12,000 in./sec. One set of dynamic constants is found by assuming $1/a = 6.5 \times 10^3$ as suggested by Parkin (10) from an empirical study of Whitman's data. This yields $b = 3.32$. The calculation using these constants and eq. (2) as the static law is referred to as *Case I*. This value of b applies to a hypothetical material that approaches equilibrium very slowly, as is illustrated by the fact that at $x = 0$, almost 7 seconds are required for the material to reach 90% of the final strain. Since such slow flow is not observed experimentally in

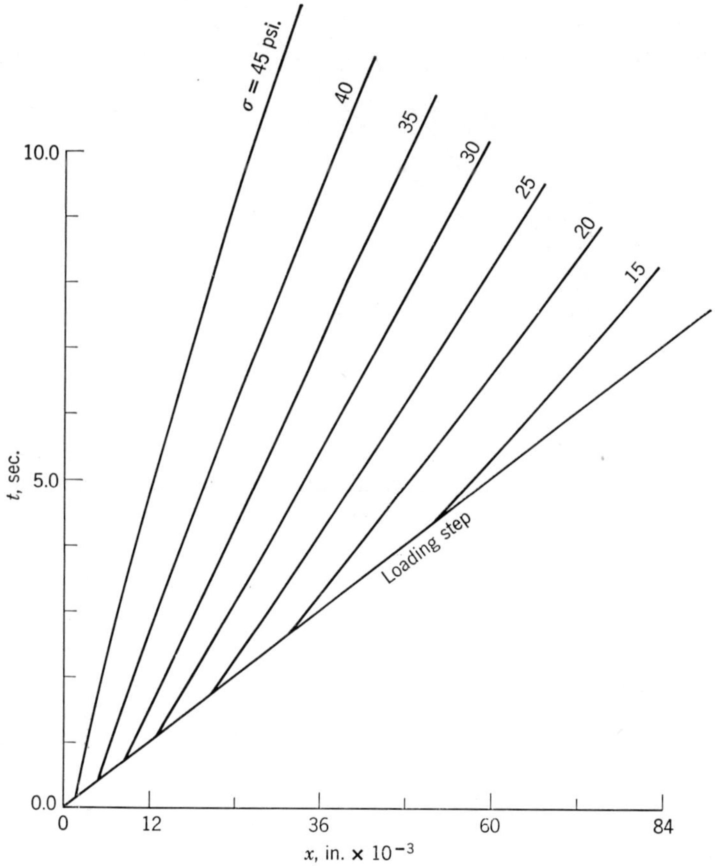

Fig. VD-6. Contours of constant stress, Fort Peck Sand, *Case I*.

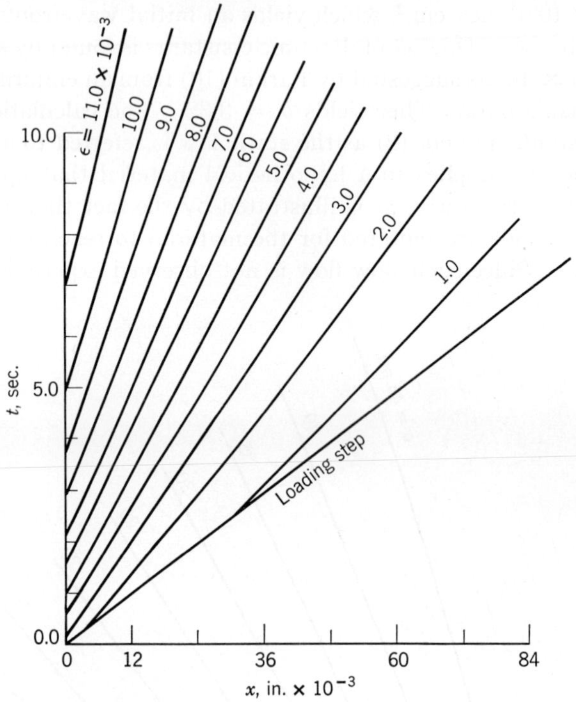

Fig. VD-7. Contours of constant strain, Fort Peck Sand, *Case I.*

soils and metals, another value for b may be obtained by requiring the strain to reach 90% of its final value in 0.005 sec. in the following manner. When eq. (2) represents the static stress-strain curve, the solution for $\epsilon(0,t)$ is eq. (42). By setting $\epsilon(0,0.005) = 0.9\, g(\sigma_0)$ and solving for b, one obtains $b = -0.005/\log \{0.1/[1 - \epsilon_0/g(\sigma_0)]\} = 2.39 \times 10^{-3}$. By requiring the initial wavefront velocity to be 12,000 in./sec., we obtain $a = b/\rho C^2 = 1.1 \times 10^{-7}$. The calculation using these more realistic values for a and b with eq. (2) as the static law is referred to as *Case II*. The calculation with the dynamic constants for *Case II* and the static law written as $\sigma = g^{-1}(\epsilon)$ is referred to as *Case III*.

VD6b. Step Response

The response of the Fort Peck Sand to a step in stress of 50 p.s.i. applied at $x = 0$ is numerically evaluated for the above two sets of

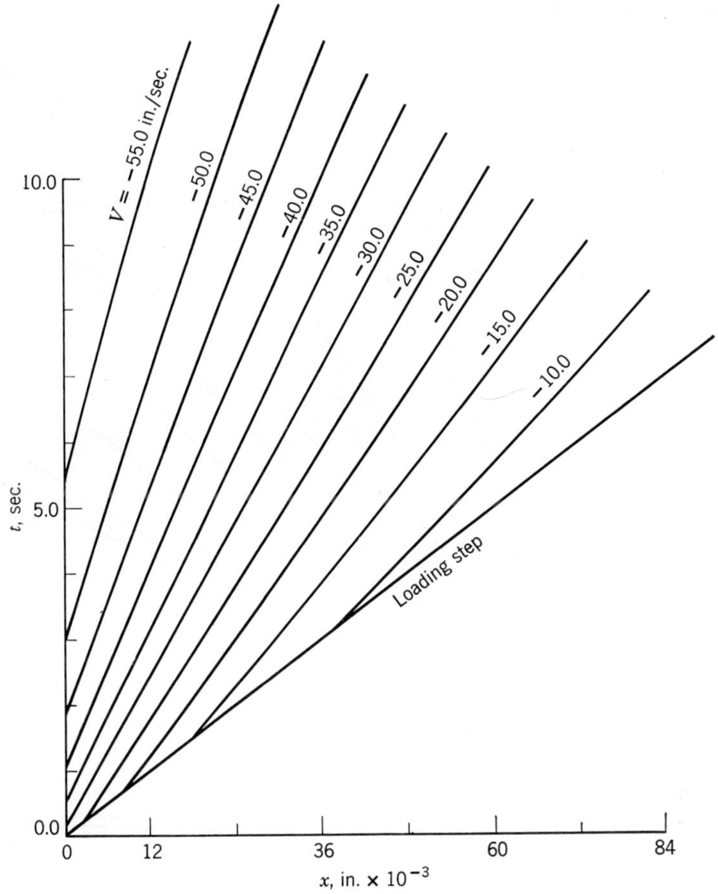

Fig. VD-8. Contours of constant particle velocity, Fort Peck Sand, *Case I*.

dynamic constants and two different forms for defining the plastic strain rate dependence. These solutions are compared to the centered simple wave solution obtained by neglecting time-rate effects.

Contours of constant stress, strain, and particle velocity using the set of dynamic constants of *Case I* (Figs. VD-6, 7, and 8) show the initial, slowly decaying, discontinuous elastic wavefront, followed by a region in which the stress, strain, and particle velocity increase toward their respective limits. In this region, lines of constant amplitude are not straight; however, their deviation from linearity is small over a considerable range of amplitude and strain rate. Because of this, the resulting contour patterns resemble a simple wave originating from some unknown origin.

The calculation is repeated using the second set of dynamic constants (*Case II*). These constants essentially contract the time scale

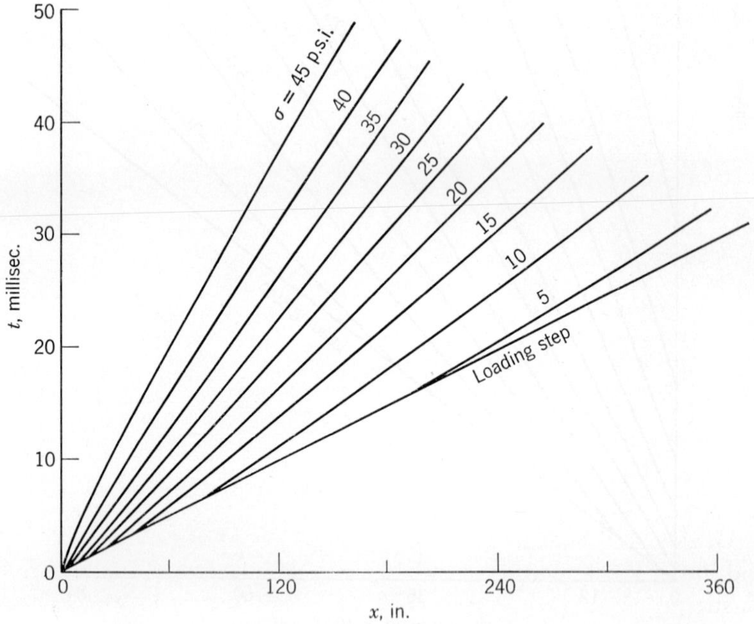

Fig. VD-9. Contours of constant stress, Fort Peck Sand, *Case II*.

compared to *Case I*, resulting in a much closer approximation to the centered simple wave solution. The resulting contours of stress, strain, and particle velocity (Figures VD-9, 10, and 11) show that the initial, discontinuous wavefront decays quite rapidly, subsiding to half its initial amplitude after approximately 14 inches. As in *Case I*, lines of constant amplitude behind the initial wavefront are not quite straight. These limited calculations suggest that, except for the infinite strain rates associated with discontinuities, phase velocities are relatively more sensitive to strain amplitude than to strain rate.

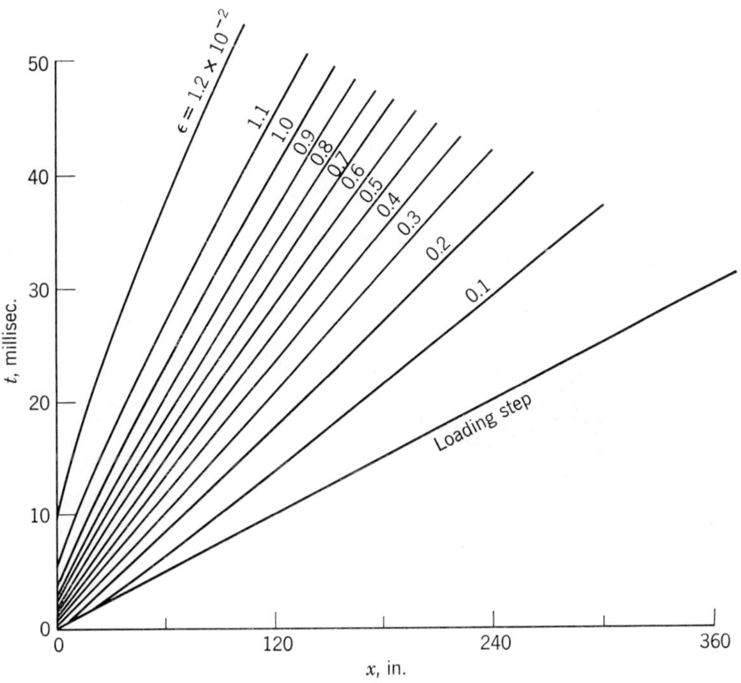

Fig. VD-10. Contours of constant strain, Fort Peck Sand, *Case II*.

As indicated above, the *Case III* calculation (Figures VD-12, 13, and 14) differs from the *Case II* calculation only in assuming the plastic strain rate proportional to the stress difference rather than the strain difference. Comparison of the contour patterns for these two cases shows that the material approaches equilibrium much more rapidly when the plastic strain rate is taken proportional to the strain difference.

For purposes of comparison, the centered simple wave solution based on the static stress-strain law [eq. (2)] is shown in Figure VD-15. Since it is based on the rate-independent plastic wave theory, it may be directly compared to the calculations for all three cases.

VD6c. Residual Strain Distributions

A commonly accepted procedure (6,7) for calculating the residual strain distribution is to assume a linear unloading stress-strain curve

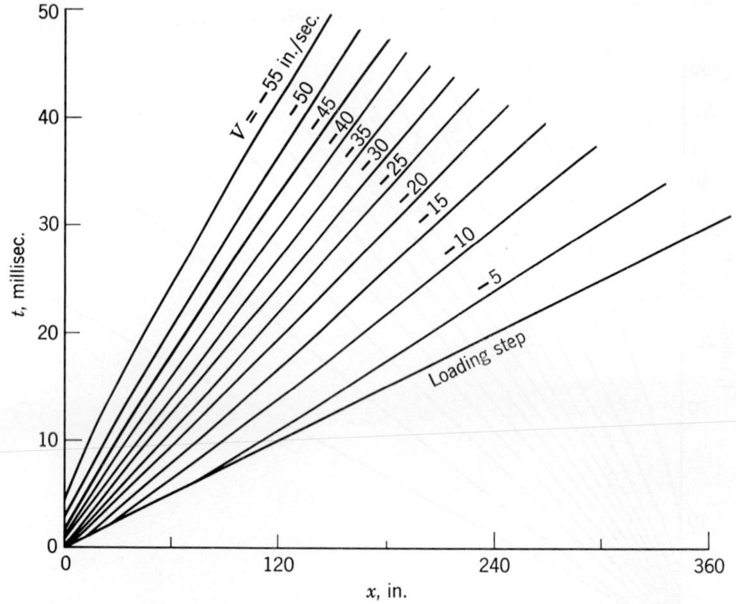

Fig. VD-11. Contours of constant particle velocity, Fort Peck Sand, *Case II*.

parallel to the initial slope of the loading cycle. Under this assumption, the residual strain distribution resulting from a rectangular stress pulse, t_0 seconds in duration, is

$$\epsilon_{res}(t_0,t_0 + x/C) = \epsilon(t_0,t_0 + x/C) - (a/b)\ \sigma(t_0,t_0 + x/C) \qquad \text{VD6-(52)}$$

With the loading cycle calculated from the rate-independent theory, the residual strain produced by a rectangular stress pulse of variable duration shows the familiar plateau of constant strain (4,15,16) near the impact boundary $x = 0$ (Fig. VD-16). Beyond this plateau, the residual strain rapidly approaches zero.

The residual strain distributions based on the Fort Peck Sand (*Case II*) are shown in Figure VD-17 for pulses of 10, 20, 30, and 40 milliseconds, respectively. A plateau of almost constant strain develops close to $x = 0$ for a pulse duration which is several times greater than the rise time. For this calculation, the rise time is arbitrarily chosen as the time required to attain 90% of the final strain: 0.005 sec. Thus the observed plateau of constant residual

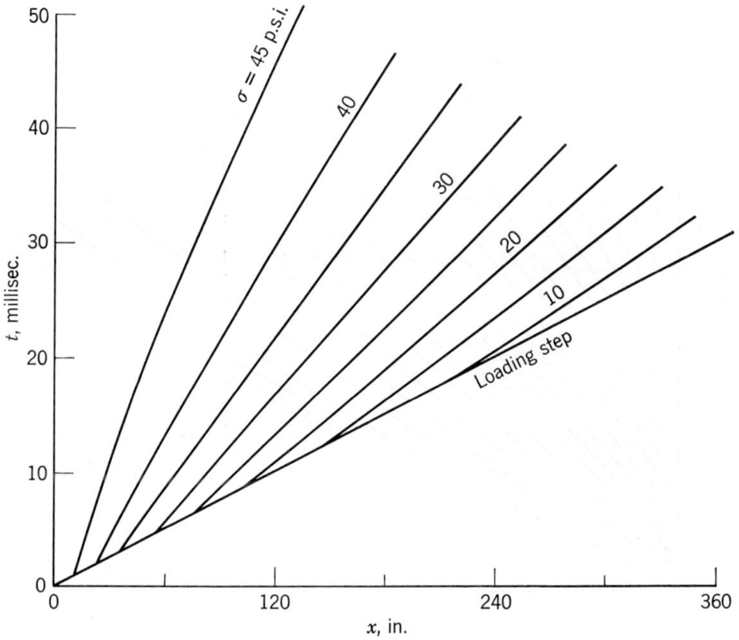

Fig. VD-12. Contours of constant stress, Fort Peck Sand, *Case III*.

strain is predicted in viscoelastic theory, if appropriate values of the dynamic constants are chosen and if the pulse is long enough for the material near the impact face to approach equilibrium.

Residual strain distributions resulting from impact tests on copper bars have been reported by Kolsky and Douch (4) (Fig. VD-18). The ratio of rise time to pulse duration for the experimentally recorded pulses which produced these residual strain distributions was of the order of four to five. The residual strain distributions calculated by applying eq. (52) to the *Case II* calculation are qualitatively similar in form when the corresponding ratio is of this order of magnitude ($t_0 = 30$ or 40 milliseconds).

VD6d. Stress-Strain Curves

Kolsky and Douch have determined the dynamic and static stress-strain curves for an aluminum alloy, pure aluminum, and copper (Figs. VD-19, 20, and 21). The tests on an aluminum alloy show

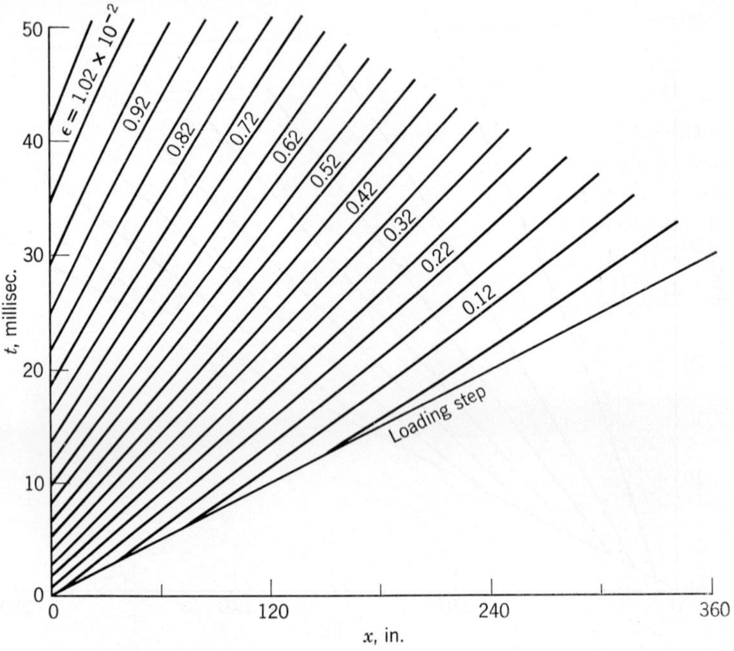

Fig. VD-13. Contours of constant strain, Fort Peck Sand, *Case III*.

that the values of stress and strain determined dynamically scatter evenly about the static curve. This alloy belongs to the class of strain rate-insensitive materials which are well described by the von Kármán-Taylor plastic wave theory.

Similar dynamic tests on pure aluminum show that it is strain rate sensitive (Fig. VD-20). However, there is no discernible difference between the dynamic and static Young's modulus. This may be interpreted in terms of the nonlinear viscoelastic theory as requiring only one dynamic viscoelastic constant. Thus it should be possible to describe the dynamic behavior of pure aluminum by Malvern's theory, utilizing only one dynamic constant.

On the other hand, the stress-strain curve for pure copper shows a small but significant increase in Young's modulus under dynamic test conditions (Fig. VD-21). This increase in Young's modulus suggests that at least two dynamic constants are required to describe adequately the dyanamic behavior of copper.

The data from which these dynamic stress-strain curves are constructed would be obtained from gages located at various distances from the impact face. Since, for strain rate-sensitive materials, there is no one-to-one correspondence between stress and strain, the stress-strain curve deduced from experimental data will be a function of recording position. To gain some insight into the importance of recording position, the stress-strain curves obtained from the *Case II*

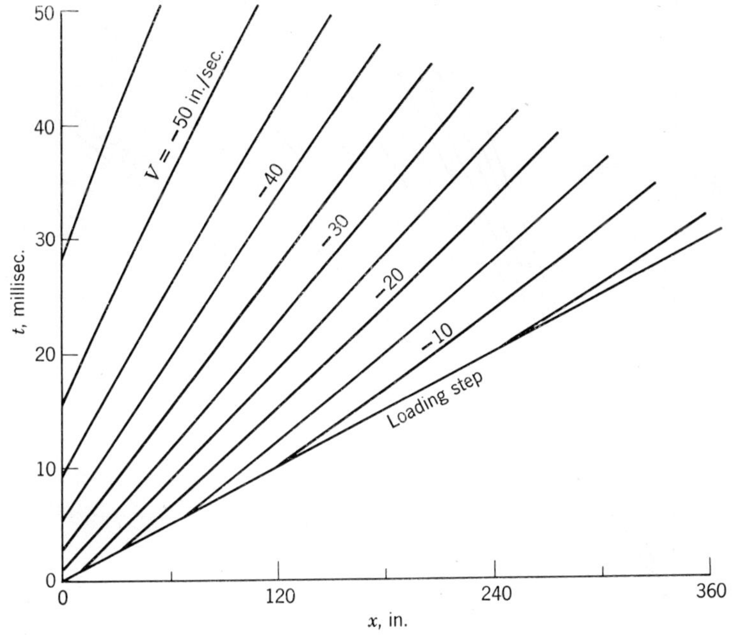

Fig. VD-14. Contours of constant particle velocity, Fort Peck Sand, *Case III*.

calculation for the Fort Peck Sand are shown in Fig. VD-22 for various distances from the impact face. There is an apparent increase in the yield point which decreases with distance from the origin. Above the apparent yield point, the dynamic curve roughly parallels the static curve giving the appearance of shifting the entire static curve upward. The calculation for *Case II* shows that for distances greater than 12 inches, the apparent stress-strain curve varies slowly with small changes in recording position.

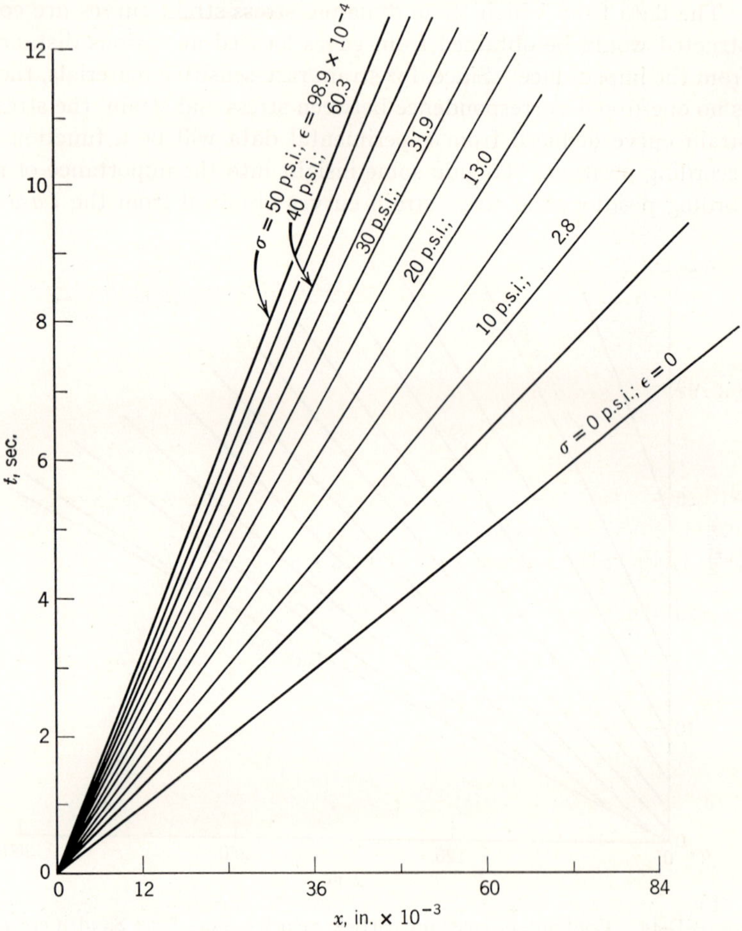

Fig. VD-15. Simple wave solution, Fort Peck Sand.

VD7. Finite Strain Wave Propagation: Loading Cycle

VD7a. Characteristic Equations

To analyze the dynamic behavior of soils and polymers under conditions of large deformations we use the higher order dynamic stress-strain law that corresponds to finite strain theory. With eq. (12) instead of eq. (10) as the dynamic stress–strain law, in conjunction

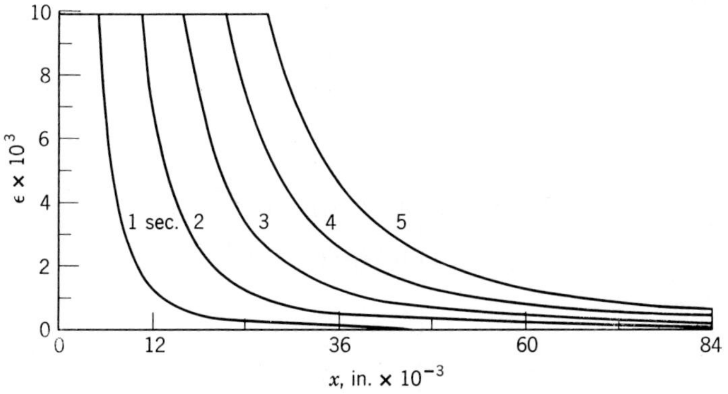

Fig. VD-16. Residual strain distribution for various pulse widths for the simple wave solution for the Fort Peck Sand.

with eqs. (20) and (21), new characteristic equations and directions are obtained by replacing b by $b(1 + \epsilon)$ and $f(\epsilon)$ by $f(\alpha) = f(\epsilon + \frac{1}{2}\epsilon^2)$, yielding the following relationships.

(a) Along $dx/dt = 0$:

$$d\sigma - (b/a)(1 + \epsilon)d\epsilon = -(1/a)[g(\sigma) - f(\alpha)]dt \quad \text{VD7-(53)}$$

(b) Along $dx/dt = +[b(1 + \epsilon)/a\rho]^{1/2}$:

$$d\sigma - \rho[b(1 + \epsilon)/a\rho]^{1/2}dV = -(1/a)[g(\sigma) - f(\alpha)]dt \quad \text{VD7-(54)}$$

(c) Along $dx/dt = -[b(1 + \epsilon)/a\rho]^{1/2}$:

$$d\sigma + \rho[b\,(1 + \epsilon)/a\rho]^{1/2}dV = -(1/a)[g(\sigma) - f(\alpha)]dt \quad \text{VD7-(55)}$$

This intuitive method for extending the infinitesimal strain case to the finite strain case may be justified as follows. We represent the dynamic behavior of finite-amplitude strain waves by the set of quasi-linear partial differential equations

$$g(\sigma) + a\sigma_t = f(\alpha) + b\alpha_t \quad \text{VD3-(11)}$$

$$\sigma_x = \rho V_t \quad \text{VD5-(21)}$$

$$\alpha_t - V_x = \epsilon\epsilon_t \quad \text{VD7-(56)}$$

$$\epsilon_t = V_x \quad \text{VD5-(20)}$$

The method of characteristics developed in a previous subsection is applicable to this system of equations by summing $N = 1$ up to and including 4. The coefficient matrices of the above system are

$$\mathbf{A}_{ij} = \begin{bmatrix} 0 & 0 & 0 & 0 \\ 1 & 0 & 0 & 0 \\ 0 & 0 & 0 & -1 \\ 0 & 0 & 0 & -1 \end{bmatrix}$$

$$\mathbf{B}_{ij} = \begin{bmatrix} a & -b & 0 & 0 \\ 0 & 0 & 0 & -\rho \\ 0 & 1 & -\epsilon & 0 \\ 0 & 0 & 1 & 0 \end{bmatrix}$$

$$\mathbf{C}_{i} = \begin{bmatrix} g(\sigma) - f(\alpha) \\ 0 \\ 0 \\ 0 \end{bmatrix}$$

where

$$\mathbf{U}^{j} = \begin{bmatrix} \sigma \\ \alpha \\ \epsilon \\ V \end{bmatrix}$$

The determinantal equation for the characteristic directions is

$$\begin{vmatrix} -adx & bdx & 0 & 0 \\ dt & 0 & 0 & \rho dx \\ 0 & -dx & \epsilon dx & -dt \\ 0 & 0 & -dx & -dt \end{vmatrix} = 0$$

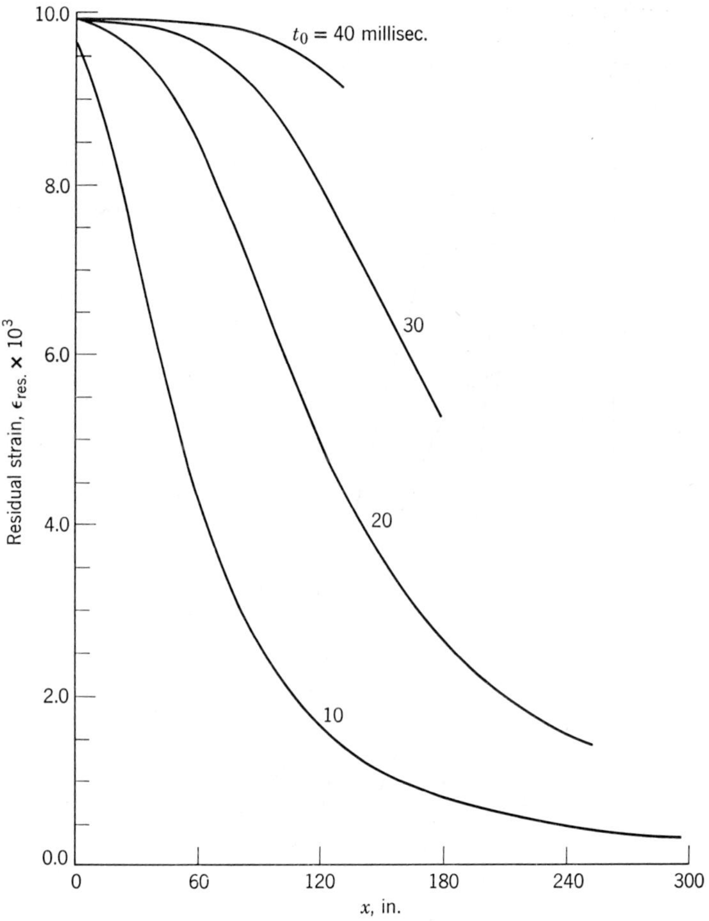

Fig. VD-17. Residual strain distributions for various pulse durations, *Case II.*

which, upon expanding and factoring, yields

$$(dx/dt)^2 = 0 \qquad \text{VD7-(57)}$$

$$(dx/dt) = \pm [b(1 + \epsilon)/a\rho]^{1/2}$$

The characteristic equations are found in the same manner as in the infinitesimal case, and the results are identical to eqs. (53), (54), and (55). Hence, the equations obtained with the higher order dynamic

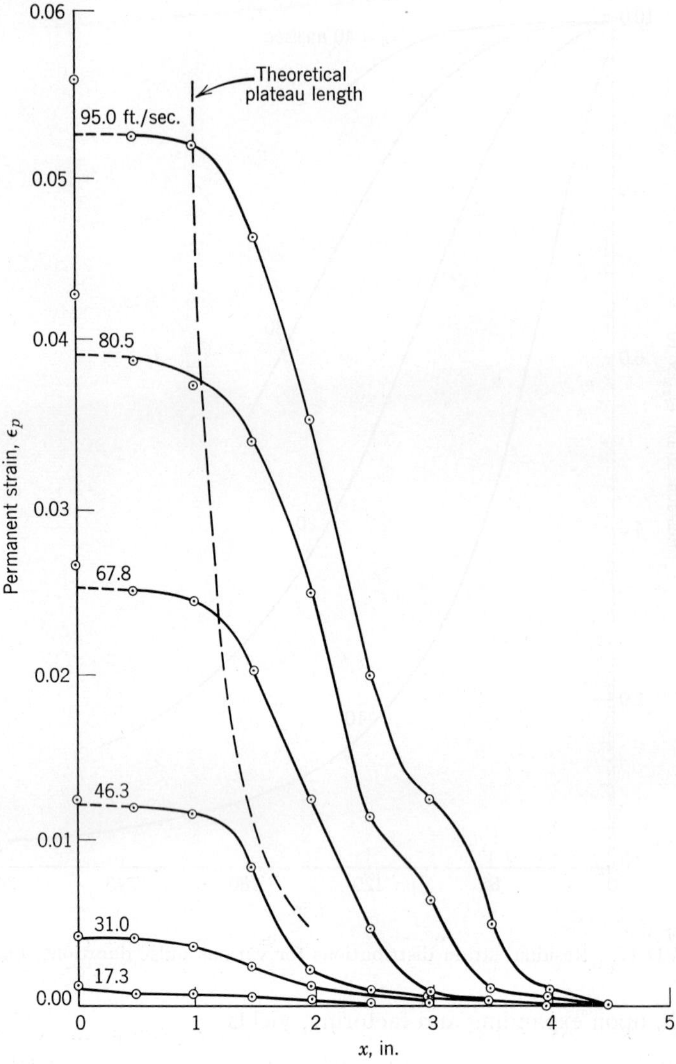

Fig. VD-18. Strain distribution in copper specimens. Data from Kolsky and Douch (4).

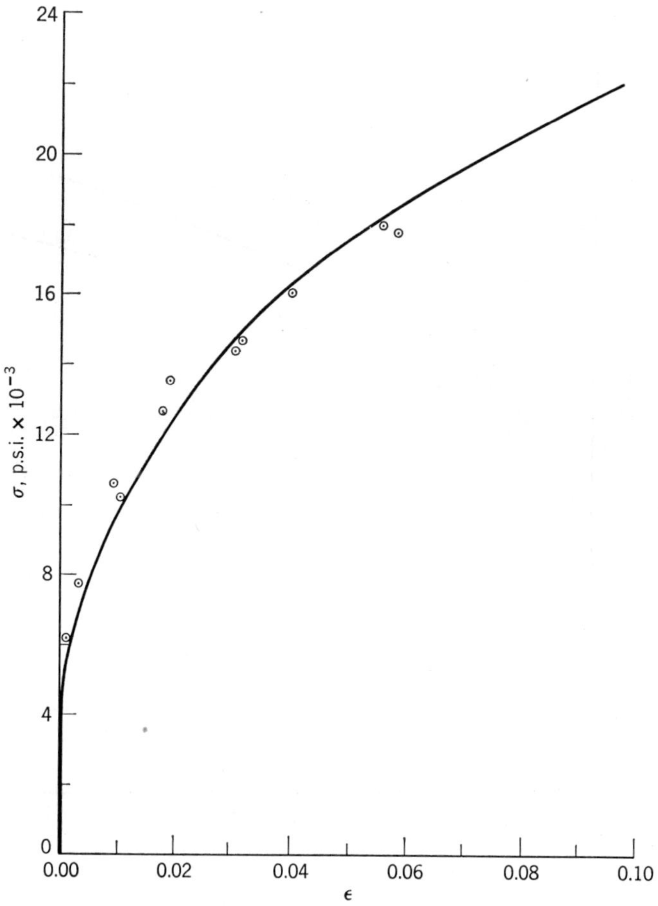

Fig. VD-19. Stress-strain curve for aluminum alloy. Static curve (——); experimental points (⊙). Data from Kolsky and Douch (4).

stress–strain law contain the finite amplitude strain theory as a special case.

VD7b. Propagation of Initial Elastic Rise

The case of a discontinuous step propagating into a region of either zero or constant strain is considerably more complicated than in the infinitesimal strain case because the characteristic net now depends

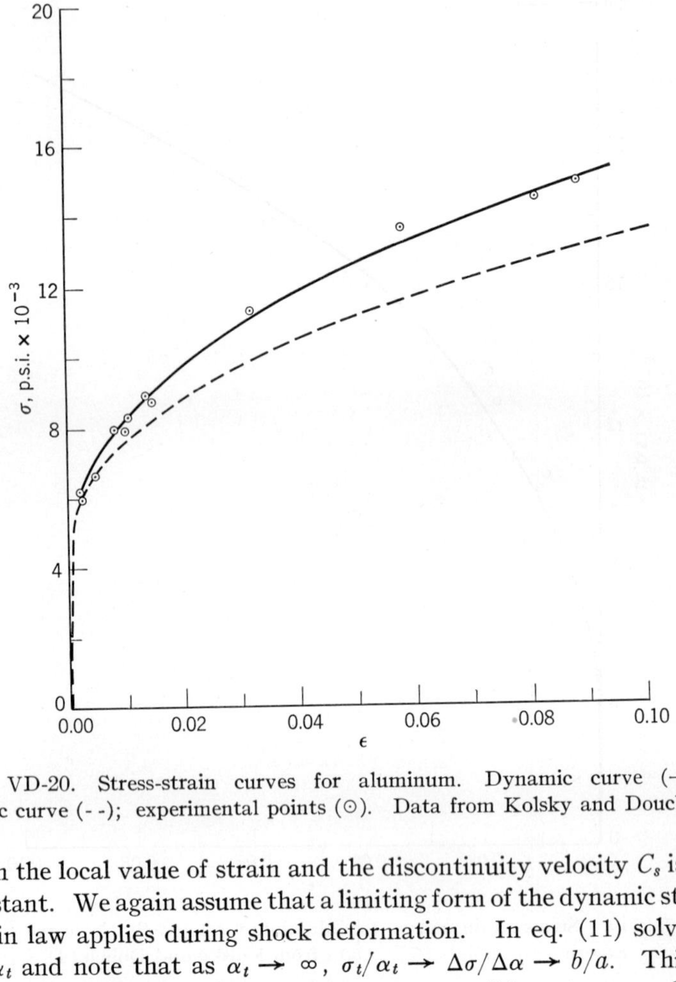

Fig. VD-20. Stress-strain curves for aluminum. Dynamic curve (——); Static curve (- -); experimental points (⊙). Data from Kolsky and Douch (4).

upon the local value of strain and the discontinuity velocity C_s is not constant. We again assume that a limiting form of the dynamic stress-strain law applies during shock deformation. In eq. (11) solve for σ_t/α_t and note that as $\alpha_t \to \infty$, $\sigma_t/\alpha_t \to \Delta\sigma/\Delta\alpha \to b/a$. This, in conjunction with the mechanical jump condition $\Delta\sigma = \rho C_s^2 \Delta\epsilon$, yields $C_s = (b/a\rho)^{1/2}\Delta\alpha/\Delta\epsilon = C\Delta\alpha/\Delta\epsilon$. However, $\Delta\alpha = \epsilon_2 - \epsilon_1 + \frac{1}{2}(\epsilon_2{}^2 - \epsilon_1{}^2) = \Delta\epsilon + \frac{1}{2}(\epsilon_2 + \epsilon_1)\Delta\epsilon = \Delta\epsilon[1 + \frac{1}{2}(\epsilon_2 + \epsilon_1)]$ from which we obtain $C_s = C[1 + \frac{1}{2}(\epsilon_2 - \epsilon_1)]$. For the special case of an initially undeformed material, $\epsilon_1 = 0$ and $C_s = C[1 + (\epsilon_2/2)]$. Thus, the initial elastic rise in finite strain theory will arrive before the initial elastic rise if the infinitesimal strain approximation is used. Since the initial elastic rise decays rapidly with distance traveled, the term

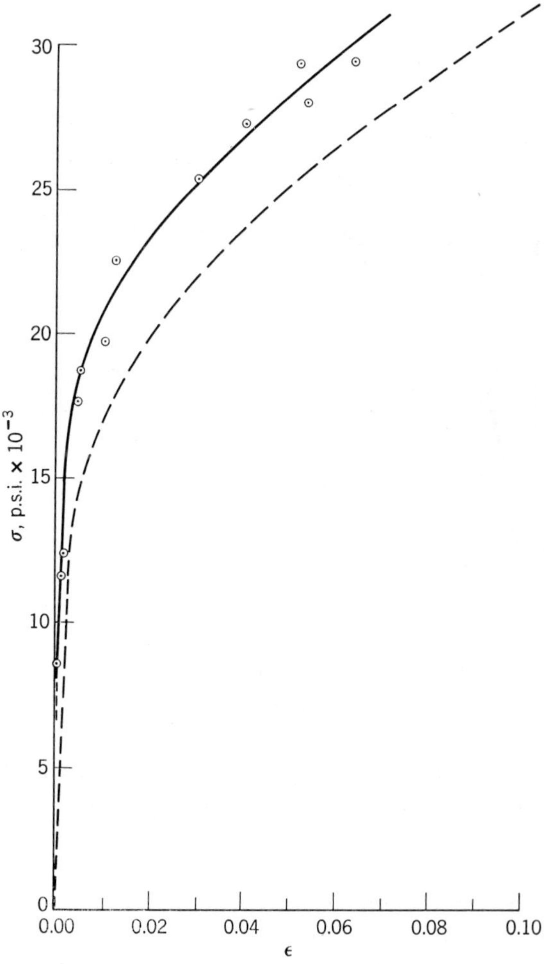

Fig. VD-21. Stress-strain curves for copper. Dynamic curve (——); static curve (- -); experimental points (⊙). Data from Kolsky and Douch (4).

ϵ_2 rapidly approaches zero, hence the velocity of the initial event measured between two stations some distance from the source will approximately yield C.

If the boundary condition at $x = 0$ is continuous, for example, a linear increase of stress with time, no discontinuity will be generated

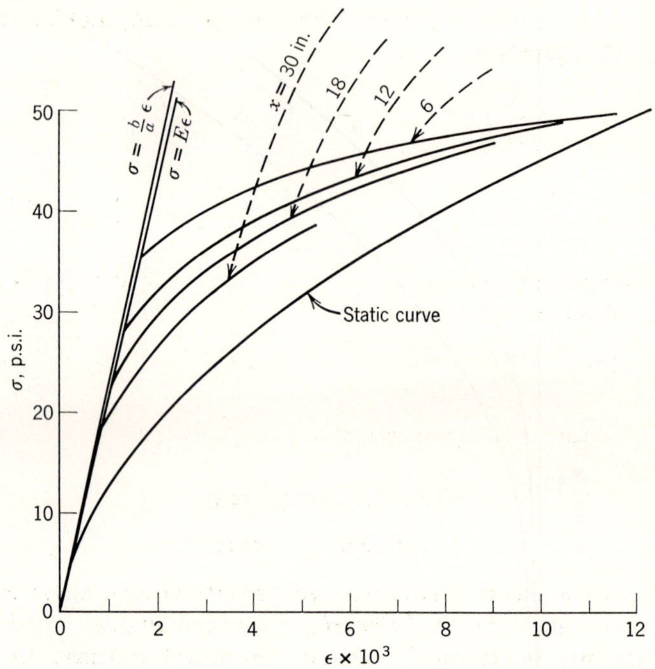

Fig. VD-22. Apparent stress-strain curves for various distances from the origin. Fort Peck Sand, *Case II.*

and the initial disturbance will propagate along the C_+^0 characteristic emanating from $x = t = 0$.

VD7c. Boundary Conditions

The strain at $x = 0$ can be determined once the boundary condition in stress is substituted into eq. (53). The resulting differential equation is identical in form to the equation obtained in the infinitesimal theory except that ϵ is replaced by α. Hence the solutions will be of the same form and we can carry over the results from the infinitesimal theory by replacing ϵ with α.

For the boundary condition $\sigma = \sigma_0$ for $t \geq 0$, the finite strain at $x = 0$ analogous to eq. (42) is:

$$\alpha(0,t) = [\alpha_0 - g(\sigma_0)]e^{-t/b} + g(\sigma_0) \qquad \text{VD7-(58)}$$

As $t \rightarrow \infty$, the material still approaches the strain $g(\sigma_0)$ as before.

For a linear increase of stress with time, $\sigma = Bt$, and the strain analogous to eq. (44) is:

$$\alpha(0,t) = B\left(a - \frac{b}{E}\right)(1 - e^{-t/b}) + \frac{Bt}{E} + \frac{R}{b}\left(\frac{B}{E}\right)^S e^{-t/b}.$$

$$\sum_{n=0}^{\infty} \frac{t^{n+S+1}}{b^n(n+S+1)n!} \qquad \text{VD7-(59)}$$

For a simple harmonic source $\sigma = \sigma_0 e^{i\omega t}$ the strain at $x = 0$ is given by the analog to eq. (45) and is:

$$\alpha(0,t) = [\alpha_0 - M(\omega) - N(\omega)]e^{-t/b} + M(\omega)\, e^{i\omega t} + N(\omega)\, e^{iS\omega t} \qquad \text{VD7-(60)}$$

where $M(\omega)$ and $N(\omega)$ are identical to those in eq. (46).

VD8. Unloading Cycle

VD8a. Infinitesimal Strain Case

The previous theory only considered discontinuities in stress, strain, and particle velocity at the beginning of the loading pulse which then propagate into undeformed material. In direct contrast, the unloading half of a complete stress cycle may involve discontinuities in the dependent variables that propagate into a region of nonconstant state.

Malvern has developed the theory for an unloading step generated at the boundary $x = 0$ assuming infinitesimal strain theory and that the material behaves elastically during unloading. We will parallel his development fairly closely, making the alterations required by the different assumption that the material remains viscoelastic during the unloading cycle. The changes required for finite strain theory will be treated separately.

The strain rate associated with an unloading step generated by instantaneously reducing the stress to zero on the boundary $x = 0$ at time t_0 is infinite. The resulting step discontinuity propagates with velocity $C = (b/a\rho)^{1/2}$ independently of the state of stress and strain into which it is moving, as was shown in Subsection VD5c. This is in agreement with results from the dynamic testing of metals where step discontinuities propagate into prestressed bars with elastic velocity (1).

For infinitesimal strain theory, the unloading step propagates along the C_+ characteristic emanating from the points $(0,t_0)$ where the stress is suddenly released, and it is possible to express the decay in amplitude of the unloading stress as an ordinary differential equation which must be solved numerically. For infinitesimal strain theory, the characteristic equations valid at the top and bottom of the unloading step are, respectively:

$$d\sigma_1 - \rho C dV_1 = -(1/a)[g(\sigma_1) - f(\epsilon_1)]dt \qquad \text{VD8-(61)}$$

$$d\sigma_2 - \rho C dV_2 = -(1/a)[g(\sigma_2) - f(\epsilon_2)]dt \qquad \text{VD8-(62)}$$

The mechanical jump condition may be written

$$\Delta\sigma = [(\sigma_1 - \rho C V_1)/2] + \text{constant}$$

or, in differential form, $d\sigma_1 - d\sigma_2 = \rho C[dV_2 - dV_1]$. On subtracting eq. (62) from eq. (61) and eliminating dV_2 by using the differential form for the mechanical jump condition, we obtain:

$$d\Delta\sigma/dt = -(1/2a)\{[g(\sigma_1) - g(\sigma_2)] - [f(\epsilon_1) - f(\epsilon_2)]\} \qquad \text{VD8-(63)}$$

Equation (63) is of the form $d\Delta\sigma/dt = F(\Delta\sigma,\sigma_1,\epsilon_1)$ since $\sigma_2 = \sigma_1 - \Delta\sigma$ and $\epsilon_2 = \epsilon_1 - [\Delta\sigma/\rho C^2]$, and may then be solved numerically as the necessary initial conditions are known at $x = 0$.

The variation of ϵ along $x = 0$ for $t \geq t_0$ may be determined from the solution of eq. (37) with $\sigma \equiv 0$, $d\sigma/dt \equiv 0$,

$$(d\epsilon/dt) + [f(\epsilon)/b] = 0 \qquad \text{VD8-(64)}$$

subject to the initial condition

$$\epsilon_2(t_0) = \epsilon_1(t_0) - [\sigma_1(t_0)/\rho C^2]$$

We now know σ, ϵ, and V along the C_+ characteristic passing through $(0,t_0)$ and the time variation of σ and ϵ along $x = 0$. This set of boundary values is identical to that considered for the loading cycle, hence the same numerical techniques for extending the solution away from these boundaries apply.

VD8b. Finite Strain Case

An unloading step in the finite strain theory case will not follow a characteristic, hence it is not possible to develop an expression relating the jumps in stress, strain, and particle velocity as in the infinitesi-

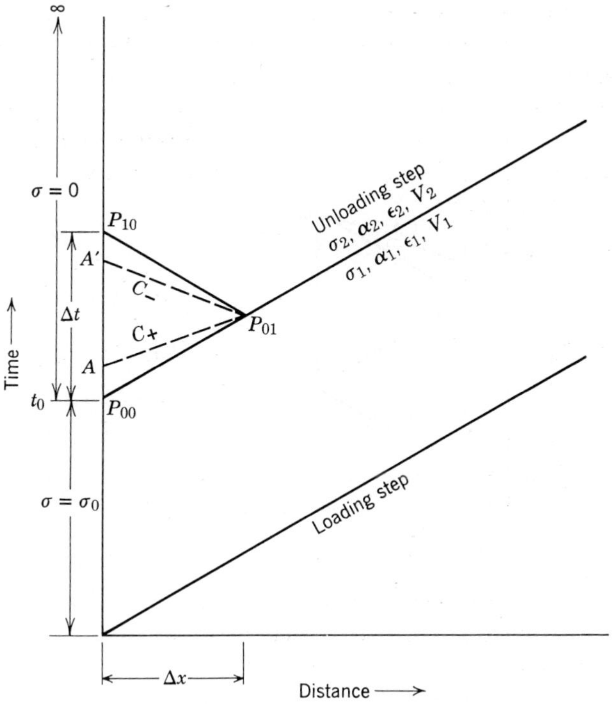

Figure VD-23.

mal strain theory case. A purely numerical technique such as used in work on non-uniform hydrodynamic shocks may be applicable, but this is a tedious, trial and error method. Perhaps a better way would be to calculate C_s at the boundary $x = 0$, and obtain another point P_{01} approximating the location of the step, as suggested in Figure VD-23. This approximation assumes C_s may be treated as constant along a length Δx. At P_{01}; σ_2, α_2, ϵ_2, and V_2 may be written in terms of σ_1, α_1, ϵ_1, V_1, and $\Delta\sigma$. Since we have considered α positive in our coordinate system, the C_+ and C_- characteristics extended backwards from P_{01} will intersect the boundary $x = 0$ at points A and A', respectively. As a first approximation, the values of σ, α, ϵ, and V at A may be assumed equal to those at P_{00} and the term $g(\sigma) - f(\alpha)$ assumed constant on the path $A - P_{01}$. Under these assumptions, a first approximation for $\Delta\sigma(P_{01})$ may be directly calculated

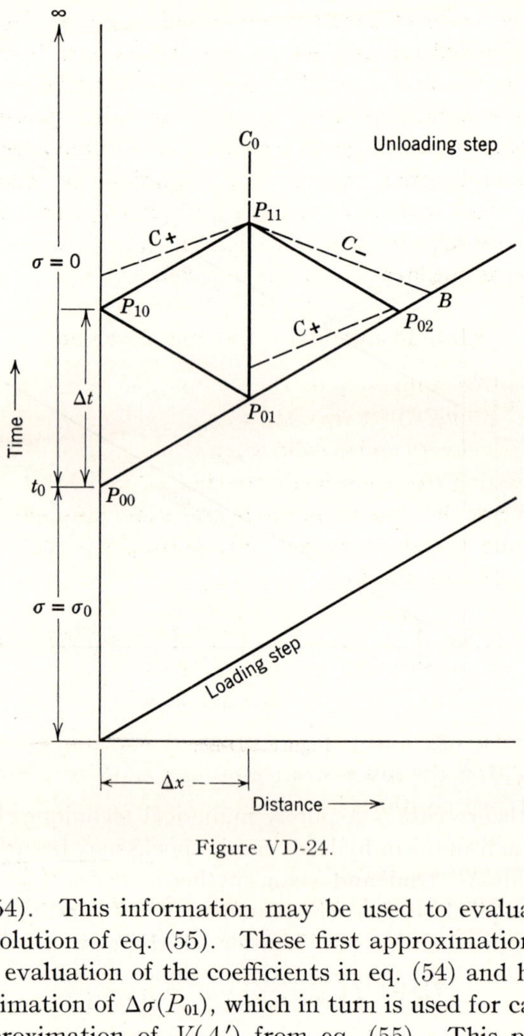

Figure VD-24.

from eq. (54). This information may be used to evaluate directly $V(A')$ by solution of eq. (55). These first approximations permit a more exact evaluation of the coefficients in eq. (54) and hence a second approximation of $\Delta\sigma(P_{01})$, which in turn is used for calculating a second approximation of $V(A')$ from eq. (55). This procedure is repeated until successive values for $\Delta\sigma(P_{01})$ and $V(A)$ agree within some prespecified accuracy.

The calculation at the next point P_{02} is illustrated in Figure VD-24. We again approximate the step path by calculating C_s at P_{01} and treat C_s constant in the next Δx interval. A first approximation for $\Delta\sigma(P_{02})$ is obtained by assuming σ, α, ϵ, and V at A equals σ, α, ϵ, and V at P_{01}

and g-f along A–P_{02} as constant and equal to g–f at P_{01}. Then eq. (54) may be solved for $\Delta\sigma(P_{02})$. These results may then be utilized to solve for σ, α, ϵ, and V at P_{11}. These results can be used in turn to evaluate more accurately $\Delta\sigma(P_{02})$, and the entire process repeated until successive iterations yield results of the desired accuracy.

In this step-by-step fashion, the variation in unloading step amplitude is evaluated along with the first strip of solution behind the step path. Subsequent strips of solution may be evaluated in the same manner as outlined for the loading cycle.

VD9. Evaluation of Dynamic Constants

An alternative approach to the evaluation of the dynamic constants is that provided by eq. (23) or (53), each of which is independent of particle velocity and is valid when x = constant. This suggests a way of evaluating the viscoelastic constants a and b if it is possible to measure directly the time variation of stress and strain at two or more positions along a test specimen. Integrating eq. (23) over a time interval x/C to t we obtain

$$\int_{\sigma(x/C)}^{\sigma(t)} d\sigma - (b/a) \int_{\epsilon(x/C)}^{\epsilon(t)} d\epsilon = -(1/a) \int_{x/C}^{t} \{g[\sigma(t)] - f[\epsilon(t)]\} dt \qquad \text{VD9-(65)}$$

Each of these integrals can be evaluated numerically from experimental data after the functional forms for g and f are chosen. Measurements at two positions in the tube would provide two equations for the two unknowns a and b.

With the empirical constants evaluated, it then becomes possible to utilize the above procedures to predict the stress, strain, and particle velocities produced by various types of loading.

VD10. Applications to Polymers

A theory of this nature might be expected to apply to those polymeric substances such as crosslinked rubbers and "highly crystalline" leathers which have an equilibrium stress-strain response.

Other types of polymeric materials that creep sparingly during the interval of several pulse-widths of the transient phenomena considered here may also be treated approximately by this theory. For example, those polymers which depend upon the oxygen-hydrogen bond for their viscoelastic and strength properties appear to obey a relation (8,16,20) between σ and ϵ of the type

$$\sigma = F\epsilon - k\epsilon^2$$

where F has values in the range 10^{10}–10^{11} dynes/cm.^{2} and k is approximately 10^{12} dynes/cm.^{2} This relation holds instantaneously upon the impulsive application of a stress, before any stress relaxation occurs, and provides the following relationship for the propagation velocity of an infinitesimal elastic step:

$$C = (F/\rho)^{1/2}$$

Comparison with the discussion under Subsection VD5b gives $F = b/a$.

The creep behavior of polymers has been extensively studied. The type of data needed for evaluating the empirical constants of eq. (16) is available from reports of J. C. Smith et al. (13) and from T. L. Smith (14).

VD11. Summary

Large amplitude stress pulses applied to many different materials generate both nonlinear and time-rate-dependent effects. A phenomenological theory developed by Malvern, incorporating both nonlinear and viscoelastic effects for application to metals, is modified to include finite strain theory and to introduce a dynamic (time-dependent) Young's modulus. Malvern's analysis of elastic unloading is modified by assuming the material remains viscoelastic during a complete stress cycle. The method of characteristics is used to obtain numerical solutions to the set of quasi-linear partial differential equations defining the dynamic response of the material.

Numerical solutions based on infinitesimal strain theory for the response to a step in stress applied at the surface of a half-space are presented in graphical form. The static (equilibrium) stress-strain curve is modeled after that of Fort Peck Sand. Computations are made for various empirical dynamic constants and forms of the nonlinear static stress-strain function. Residual strain distributions, calculated by assuming elastic unloading, show the experimentally observed plateau of constant residual strain produced by longitudinal impact at the impact end of a bar for a proper combination of dynamic constants and pulse durations. Comparison of features predicted theoretically with experimental testing of copper suggests that some metals may require two dynamic constants to describe their dynamic behavior.

The authors wish to express their gratitude to Prof. L. E. Malvern, Michigan State University, for his numerous helpful suggestions and criticisms during the course of this work and to Mrs. L. Lintner, of Gulf Research and Development Co. for programming the infinitesimal strain theory. We wish to thank the Ballistic Research Laboratory, Aberdeen, Maryland, for supporting this work at the Rensselaer Polytechnic Institute, Troy, New York.

VD12. List of Key Symbols

a, b	Dynamic (time-dependent) material constants
$\mathbf{A}_{ij}, \mathbf{B}_{ij}, \mathbf{C}_i$	Coefficient matrices
C	Wave velocity
C_s	Shock velocity
E	Young's modulus
$g(\sigma), f(\epsilon)$	Functions describing the static (equilibrium) response of a material
$g^{-1}(\epsilon)$	Inverse function of $g(\sigma)$
R, S	Constants defining the static (equilibrium) response of a material
t	Times
U	Particle displacements
$\mathbf{U}^j$	Matrix of dependent variables
V	Particle velocity $= dU/dt = U_t$
x	Lagrangian space variable
α	Finite strain $= U_x + \frac{1}{2}U_x^2 = \epsilon + \frac{1}{2}\epsilon^2$
ϵ	Infinitesimal strain $= dU/dx = U_x$
λ_i	Arbitrary constants
ρ	Density (g./cm.^{3})
σ	Stress = force/unit area
ω	Circular frequencies

References

1. Bell, J. F., Navy Contract NG-ONR-243 VIII, The Johns Hopkins Univ , Tech. Report No. 5, 1951.
2. Courant, R., and K. O. Friedrichs, *Supersonic Flow and Shock Waves*, Interscience, New York, 1948.

3. Holmquist, J. L., and A. Nadai, *Drilling Production Practice*, **20**, 392 (1939).
4. Kolsky, H., and L. L. Douch, *J. Mech. Phys. Solids*, **10**, 195 (1962).
5. Lister, M., in *Mathematical Methods for Digital Computers*, A. Ralston and H. S. Wilf, Eds., Wiley, New York, 1961.
6. Malvern, L. E., *Quart. Appl. Math.*, **8**, 405 (1950).
7. Malvern, L. E., *Trans. ASME, J. Appl. Mech.*, **18**, 203 (1951).
8. Nissan, A. H., *Trans. Faraday Soc.*, **53**, 700 (1957).
9. Osgood, W. R., *J. Aeronaut. Sci.*, **13**, 45 (1946).
10. Parkin, B. R., *J. Soil Mech. Found. Div., Am. Soc. Civil Engrs.*, **87**, No. SM3 (June 1961).
11. Plass, H. J., Jr., in *Plasticity*, E. H. Lee and P. S. Symonds, Eds., Pergamon, New York, 1961.
12. Prandtl, L., *Z. angew. Math. Mech*, **8**, 85 (1928).
13. Smith, J. C., J. M. Blandford, P. J. Shouse, and K. M. Towne, *Textile Res. J.*, **31**, 721 (1961); *ibid.*, **32**, 472 (1962).
14. Smith, T. L., STP 325, ASTM, 1962, p. 73.
15. Sokolnikoff, I. S., *Mathematical Theory of Elasticity*, McGraw-Hill, New York, 1956, pp. 29–32.
16. Sternstein, S. S., and A. H. Nissan, in *Formation and Structure of Paper*, British Paper and Board Makers' Association, London, England, 1961, p. 319.
17. Tapley, B. D., and H. J. Plass, Jr., in *Developments in Mechanics*, Vol. I, J. E. Lay and L. E. Malvern, Eds., Plenum Press, New York, 1961.
18. Taylor, G. I., British Official Report, R. C. 70, 1940.
19 von Kármán, Th., N.D.R.C. Report No. A-29, 1942.
20. Wellisch, E., L. Marker, and O. J. Sweeting, *J. Appl. Polymer Sci.*, **5**, 647 (1961).
21. Whitman, R. V., et al., M.I.T. Department of Civil Engineering Professional Paper P63-16, presented at the 32nd Symposium on Shock Vibration and Associated Environments, April 1963.

SUBJECT INDEX

D

E

N

Q

R

S